Soil Biology

Volume 48

Series Editor
Ajit Varma, Amity Institute of Microbial Technology,
Amity University Uttar Pradesh, Noida, UP, India

More information about this series at http://www.springer.com/series/5138

Mansour Ghorbanpour • Manika Khanuja •
Ajit Varma
Editors

Nanoscience and Plant–Soil Systems

Springer

Editors
Mansour Ghorbanpour
Department of Medicinal Plants,
Faculty of Agriculture and Natural Resources
Arak University
Arak, Iran

Manika Khanuja
Centre for Nanoscience and
Nanotechnology
Jamia Millia Islamia
New Delhi
India

Ajit Varma
Amity University Uttar Pradesh,
Amity Inst of Microbial Technology
Noida, Uttar Pradesh
India

ISSN 1613-3382 ISSN 2196-4831 (electronic)
Soil Biology
ISBN 978-3-319-46833-4 ISBN 978-3-319-46835-8 (eBook)
DOI 10.1007/978-3-319-46835-8

Library of Congress Control Number: 2017931978

Printed on acid-free paper

This Springer imprint is published by Springer Nature
The registered company is Springer International Publishing AG
The registered company address is: Gewerbestrasse 11, 6330 Cham, Switzerland

The updated original version of this book was revised. An erratum to the book can be found at
DOI 10.1007/978-3-319-46835-8_22

Foreword

Nanoscience coupled with nanotechnology is one of the most important emerging tools which can complement modern agriculture by providing new agrochemical agents and new delivery mechanisms to improve crop productivity. Nanoscience has found ways to control the release of nitrogen in agriculture fields, micromorphology of soil and characterization of soil minerals, rhizospheric nature, nutrient ion transport in soil–plant system, precision water farming, etc. In the era of climate change, it is a challenge to feed the rapidly increasing world population, with agricultural productivity facing various challenges. Thus, it is now very important to improve the crop productivity to cope with this upcoming problem of food security. Nanoscience (nanotechnology) promises to accelerate the development of biomass-to-fuel production technologies. Experts feel that the potential benefits of nanotechnology for agriculture, food, fisheries, and aquaculture need to be balanced against concerns for the soil, water, and environment and the occupational health of workers. Nanoparticles (size range from 1 to 100 nm) have unique physicochemical properties, i.e., high surface area, high reactivity, tunable pore size, and particle morphology; therefore, they have novel applications in diverse fields of science including medicine, physics, chemistry, materials science, and agriculture. The appropriate elucidation of physiological, biochemical, and molecular mechanism of nanoparticles in plant leads to better plant growth and development. Several countries have recognized the potential impression of nanotechnology could have on their economies and spending profoundly in research direction.

This book "Nanoscience and Plant–Soil Systems" edited by Drs Mansour Ghorbanpour, Manika Khanuja, and Ajit Varma is an enthusiastic celebration of a new paradigm "nanoscience in agricultural research." It is important to assemble the ever-improving methods based on nanotechnology and its role in plant soil system in a book under these new guidelines, i.e., practical aspects and immediate use in the laboratory and beyond. The chapters of this book are indeed an excellent and outstanding contribution. This book succeeds in presenting many concepts, methods, etc., which can broaden our understanding of the role of nanotechnology in plant–soil system. This comprehensiveness should make this book equally valuable to students, teachers, and researchers entering this field of nanoscience. I am sure readers in the fields of biotechnology, microbiology, agriculture, and nanotechnology would find this book very useful.

Overall, I am glad to see good coverage in this book. Congratulations and my best compliments to editors of the book who performed an outstanding work in getting valuable contributions from the team of global experts on the subject which has major implications not only for food security worldwide but also for the socioeconomic condition of communities affected by climate change at the basic grassroots level. The contributors are to be congratulated on their efforts, and readers are recommended to use this volume for a long time to come. The publisher also deserves for publishing this useful book.

Department of Science and Technology, Ashutosh Sharma
Government of India
New Delhi, India
June 23, 2016

Preface

There is general belief and admission that important, innovative, and novel ideas emerge over a cup of tea or a glass of beer and the weather must be congenial and most suitable for materializations of original ideas. The genesis of this book underlines the concept developed in 2015.

Technological advances and sociological changes are such that Science demands evolution. We believe that one reason for publishing of ideas is broadening one's view, through the examination of a text of wide and extensive coverage, nurtures one's capacity for learning and reflection. The study of microorganisms has become a valuable science in the last 100 years as it has provided the means to control a number of infectious diseases. In this direction, nanotechnology has emerged as a potential candidate. The ideas and concepts behind nanoscience and nanotechnology started with a talk entitled "There's Plenty of Room at the Bottom" by physicist Richard Feynman at an American Physical Society meeting at the California Institute of Technology (CalTech) on December 29, 1959, long before the term nanotechnology was used. Nanoscience and nanotechnology are the study and application of extremely small things and can be used across all the other science fields, such as chemistry, biology, physics, materials science, and engineering. Nanoparticles are gaining attention due to their low cost, simplicity, and eco-friendly nature.

In this volume entitled "Nanoscience and Plant–Soil Systems," the editors have accumulated various advanced approaches for studying the different soil microorganisms for the benefit of humankind. Currently, world agriculture scientists face a wide spectrum of challenges including climate change, urbanization, and environmental issues: accumulation of insecticides and pesticides, decay in soil organic matter, and sustainable use of natural resources. These challenges are going to be further intensified due to increase in food demand. Nanotechnology has significant benefits on food and agriculture system. Through nanotechnology, optimization of agriculture inputs (viz. nanopesticides, nanoherbicides) to enhance the effectiveness of the active ingredients including targeted delivery and release and less dosage per application and to reduce bi-products that otherwise degrade ecosystem can be achieved. This book is divided into three parts. In the first part which includes Chaps. 1–3, the authors give introduction to nanoscience and nanotechnology, how nanoparticles are being synthesized with their origin and activity, and also the application of these nanoparticles. The second part of the book which includes Chaps. 4–11 describes nanomaterials in soil environment with their applications as antimicrobial and bioremediating agents and also their effect on soil properties and soil microorganisms and how they act as a fertilizer. The last part of the book which includes Chaps. 12–19 describes the interaction of nanomaterials with plants and their effect on seed germination. Chapter 14 describes the role of nanoparticles on plant growth after interacting with a novel root endophyte *Piriformospora indica*. In Chaps. 15–21, the application of nanoparticles as a biofertilizer and pesticide and in plant disease control with the challenges faced and threats involved with the use of nanoscience plant soil system is elaborated.

We are grateful to the many people who helped to bring this volume to light. We wish to thank Jutta Lindenborn and Hanna Hensler-Fritton from Springer Heidelberg for generous assistance and patience in initializing the volume. Finally, specific thanks go to our families, immediate and extended, not forgetting those who have passed away, for their support or their incentives in putting everything together. Ajit Varma in particular is very thankful to Dr. Ashok K. Chauhan, Founder President of the Ritnand Balved Education Foundation (an umbrella organization of Amity Institutions), New Delhi, for the kind support and constant encouragement received. Special thanks are due to his esteemed friend and well-wisher Professor Dr. Sunil Saran, Director General, Amity Institute of Biotechnology, and Adviser to Founder President, Amity Universe; all faculty colleagues; and his Ph.D. students, research fellows (Uma Singhal and Manpreet Kaur Attri), and other technical staff.

This book will be useful to microbiologists, nanotechnologists, and ecologists if interpreted with caution. I am honored that the leading scientists who have extensive, in-depth experience and expertise in soil biology and nanotechnology took the time and effort to develop these excellent chapters. This select group of scientists is uniquely suited to write these chapters and have firsthand knowledge of the methods and techniques they have presented. This ensures that the methods presented are current, relevant, and readily applicable. I want to thank all contributing authors for their diligence and patience in bringing this book to fruition with such collegiality.

Arak, Iran — Mansour Ghorbanpour
New Delhi, India — Manika Khanuja
Noida, UP, India — Ajit Varma

Contents

List of Contributors

Nasibeh Amini Department of Natural Resources Engineering, Isfahan University of Technology, Isfahan, Iran

Rupesh Kumar Basniwal Amity Institute of Advanced Research and Studies (M&D), Amity University Uttar Pradesh, Noida, India

Cristina Buzea IIPB Medicine Corporation, Owen Sound, ON, Canada

Mansour Ghorbanpour Department of Medicinal Plants, Faculty of Agriculture and Natural Resources, Arak University, Arak, Iran

Marina Gladkova Soil Science Faculty, Lomonosov Moscow State University (MSU), Moscow, Russia

Pankaj Goyal Amity Institute of Microbial Technology, Amity University Uttar Pradesh, Noida, India

Mehrnaz Hatami Department of Medicinal Plants, Faculty of Agriculture and Natural Resources, Arak University, Arak, Iran

Sayed Mohsen Hosseini Department of Chemical Engineering, Faculty of Engineering, Arak University, Arak, Iran

Azamal Husen Department of Biology, College of Natural and Computational Sciences, University of Gondar, Gondar, Ethiopia

Jośko Izabela Institute of Plant Genetics, Breeding and Bionanotechnology, University of Life Sciences in Lublin, Lublin, Poland

Javad Hadian Medicinal Plants and Drug Research Institute, Shahid Beheshti University, G.C., Evin, Tehran, Iran

Ebrahim Karimi Department of Microbial Biotechnology, Agricultural Biotechnology Research Institute of Iran, Karaj, Iran

Mujeebur Rahman Khan Department of Plant Protection, Faculty of Agricultural Sciences, Aligarh Muslim University, Aligarh, India

Kamila Kydralieva Moscow Aviation Institute, Volokolamskoe shosse 4, Moscow, Russia

Sayed Siavash Madaeni Membrane Research Centre, Department of Chemical Engineering, Faculty of Engineering, Razi University, Kermanshah, Iran

Stefaniuk Magdalena Department of Environmental Chemistry, University of Marie Skłodowska-Curie, Lublin, Poland

Zaid Hameed Majeed Department of Civil Engineering, College of Engineering, University of Babylon, Babil, Iraq

Manika Khanuja Centre for Nanoscience and Nanotechnology, Jamia Millia Islamia, New Delhi, India

G. Ali Mansoori University of Illinois at Chicago, Chicago, IL, USA

Eugeny Milanovskiy Soil Science Faculty, Lomonosov Moscow State University (MSU), Moscow, Russia

Ehsan Mohseni Fard Department of Agronomy and Plant Breeding, Faculty of Agriculture, University of Zanjan, Zanjan, Iran

Ivan Pacheco Department of Pathology, Grey Bruce Health Services, Owen Sound, ON, Canada

Department of Pathology and Laboratory Medicine, Schulich School of Medicine and Dentistry, Western University, London, ON, Canada

Oleszczuk Patryk Department of Environmental Chemistry, University of Marie Skłodowska-Curie, Lublin, Poland

Tanveer Fatima Rizvi Department of Plant Protection, Faculty of Agricultural Sciences, Aligarh Muslim University, Aligarh, India

Mojtaba Salouti Biology Research Center, Zanjan Branch, Islamic Azad University, Zanjan, Iran

Nahid Sarlak Department of Chemistry, Lorestan University, khorram Abad, Iran

Mohsen Soleimani Department of Natural Resources, Isfahan University of Technology, Isfahan, Iran

Kizhaeral S. Subramanian Department of Nano Science & Technology, Tamil Nadu Agricultural University, Coimbatore, Tamilnadu, India

Mohd Raihan Taha Department of Civil and Structural Engineering, Faculty of Engineering and Built Environment, Universiti Kebangsaan Malaysia (UKM), Bangi, Selangor, Malaysia

Asghar Taherifar Department of Chemistry, Lorestan University, khorram Abad, Iran

Vera Terekhova Soil Science Faculty, Lomonosov Moscow State University (MSU), Moscow, Russia

Severtsov Institute of Ecology and Evolution (IPEE), Russian Academy of Sciences, Moscow, Russia

Pirogov Russian National Research Medical University (RNRMU), Moscow, Russia

M. Thirunavukkarasu Department of Nano Science & Technology, Tamil Nadu Agricultural University, Coimbatore, Tamilnadu, India

Uma Amity Institute of Microbial Technology, Amity University Uttar Pradesh, Noida, Uttar Pradesh, India

Ajit Varma Amity Institute of Microbial Technology, Amity University Uttar Pradesh, Noida, Uttar Pradesh, India

Amity Science, Technology & Innovation Foundation (ASTIF), Noida, Uttar Pradesh, India

Shweta Yadav Department of Zoology, School of Biological Science, Dr HS Gour Vishwavidyalaya (A Central University), Sagar, Madhya Pradesh, India

Neda Faghri Zonooz Department of Microbiology, Faculty of Sciences, Zanjan Branch, Islamic Azad University, Zanjan, Iran

Part 1
Nanoscience and Nanomaterials

Chapter 1
An Introduction to Nanoscience and Nanotechnology

G. Ali Mansoori

1.1 Introduction

Even though the scientific community is fascinated with the field of nanoscience, most of the ongoing discussions, definitions, and attentions are on nanotechnology. The shortest and most quoted definition of nanotechnology is the statement by the US National Science and Technology Council (NSTC 2000) which states: "*The essence of nanotechnology is the ability to work at the molecular level, atom by atom, to create large structures with fundamentally new molecular organization. The aim is to exploit these properties by gaining control of structures and devices at atomic, molecular, and supramolecular levels and to learn to efficiently manufacture and use these devices.*" In short nanotechnology is the ability to build micro and macro materials and products with atomic precision. Nanoscience is the study of properties and behavior of condensed materials in nanoscale and study of natural nanoscale phenomena such as the fascinating field of bio-systems, and involves investigating the peculiarities of nanosystems (Mansoori 2005).

The promise and essence of the nanoscale science and technology are based on the demonstrated fact that materials at the nanoscale have properties (i.e., mechanical, optical, chemical, and electrical) quite different than the bulk materials. For example, macromolecules and particles made of a limited number of molecules, in the size range of 1–50 nm, possess distinct chemical (i.e., reactivity, catalytic potential, etc.) and physical properties (i.e., magnetic, optical). Some of such properties are, somehow, intermediate between those of the smallest elements (atoms and molecules) from which they can be composed of and those of the macroscopic materials. Compared to bulk materials, it is demonstrated that

G.A. Mansoori (✉)
University of Illinois at Chicago, Room 218 Science and Engineering Offices,
851 S. Morgan St. (M/C 063), Chicago, IL 60607-7052, USA
e-mail: mansoori@uic.edu

M. Ghorbanpour et al. (eds.), *Nanoscience and Plant–Soil Systems*, Soil Biology 48,
DOI 10.1007/978-3-319-46835-8_1

nanoparticles possess enhanced performance properties when they are used in similar applications. An important application of nanoparticles is recognized to be the production of a new class of catalysts known as nanocatalysts. Significant advances are being made in this field contributing to the production and detailed understandings of the nature (composition, particle size, and structure) and role of nanoparticles as catalysts in improvement of chemical reaction performances. This is because a catalyst performance is a strong function of its particle sizes and size distributions. Surface morphology, surface-to-volume ratio, and electronic properties of materials could change appreciably due to particle size changes. There are many present and expected advances in nanoscience and nanotechnology with applications in agriculture, electronics, energy, medicine, etc., which are rapidly increasing (Ghorbanpour et al. 2015; Mansoori 2005, 2007; Mansoori et al. 2016).

1.2 The Importance of Nanoscale

The Greek word "nano" (meaning dwarf) refers to a reduction of size, or time, by 10^{-9}, which is 1000 times smaller than a micron. One nanometer (nm) is one billionth of a meter and it is also equivalent to ten Angstroms. As such a nanometer is 10^{-9} m and it is 10,000 times smaller than the diameter of a human hair. A human hair diameter is about 50 μm (i.e., 50×10^{-6} m) in size, meaning that a 50 nm object is about 1/1000th of the thickness of a hair. One cubic nanometer (nm^3) is roughly 20 times the volume of an individual atom. A nanosize particle compares to a basketball like a basketball to the size of the earth. Figure 1.1 shows size ranges for different nanoscale and microscale objects.

It is obvious that nanoscience and nanotechnology all deal with very-small-sized objects and systems. Officially, the United States National Science Foundation (Roco et al. 1999) defined nanoscience/nanotechnology as studies that deal with materials and systems having the following key properties: (1) Dimension: at least one dimension from 1 to 100 nm. (2) Process: designed with methodologies that show fundamental control over the physical and chemical attributes of molecular scale structures. (3) Building block property: they can be combined to form larger structures. Nanoscience, in a general sense, is quite natural in biological sciences considering that the sizes of many bio-entities we deal with (like DNA, RNA, proteins, enzymes, viruses) fall within the nanoscale range of 1–100 nm.

Nanoscale is regarded as a magical point on the dimensional scale: Structures in nanoscale (called *nanostructures*) are considered at the borderline of the smallest of human-made devices and the largest molecules of living systems. Our ability to control and manipulate nanostructures will make it possible to exploit new physical, biological, and chemical properties of systems that are intermediate in size, between single atoms, molecules, and bulk materials.

There are many specific reasons why nanoscale has become so important, some of which are the following (Mansoori 2005):

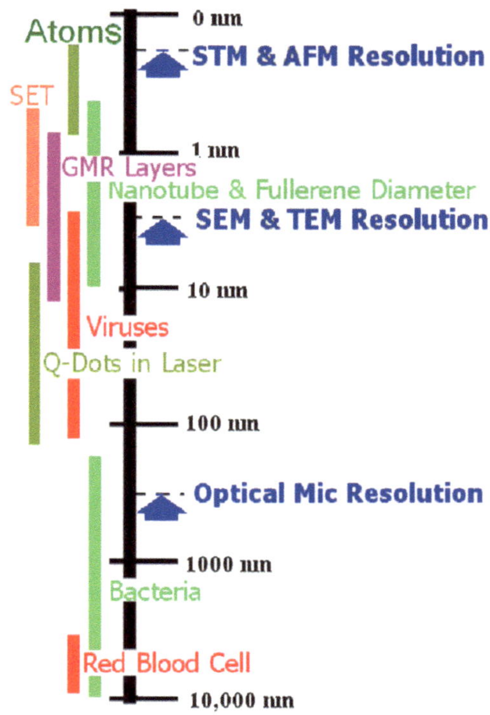

Fig. 1.1 Comparison of size ranges for several entities as compared to some nanotechnology devices and various microscope resolutions. *AFM* atomic force microscope, *GMR* giant magnetoresistive, *Q-DOTS* quantum dots, *SEM* scanning electron microscope, *SET* single-electron transistor, *STM* scanning tunneling microscope, *TEM* transmission electron microscope

1. Quantum mechanical (wavelike) properties of electrons inside matter are influenced by variations on the nanoscale. By nanoscale design of materials it is possible to vary their micro- and macroscopic properties, such as charge capacity, magnetization, and melting temperature, without changing their chemical composition.
2. A key feature of biological entities is the systematic organization of matter on the nanoscale. Developments in nanoscience and nanotechnology have allowed us to place man-made nanoscale things inside living cells (Ebrahimi and Mansoori 2014). It has also made it possible to study micro- and macrostructure of materials using molecular self-assembly (Xue and Mansoori 2010). This certainly is a powerful tool in materials science.
3. Nanoscale components have very high surface-to-volume ratio, making them ideal for use in composite materials, reacting systems, drug delivery, and energy storage.
4. Macroscopic systems made up of nanostructures can have much higher density than those made up of microstructures. They can also be better conductors of electricity. This can result in new electronic device concepts, smaller and faster circuits, more sophisticated functions, and greatly reduced power consumption simultaneously by controlling nanostructure interactions and complexity.

1.3 Historical Advances in Nanoscience and Nanotechnology

Although we have long been aware of many investigators who have been dealing with "nano"-sized entities, the historic birth of nanotechnology is commonly credited to Richard P. Feynman. Historically nanotechnology was for the first time formally recognized as a viable field of research with the landmark lecture delivered by Feynman, the famous Noble Laureate physicist, on December 29th 1959 at the annual meeting of the American Physical Society (Feynman 1960). His lecture was entitled "*There's Plenty of Room at the Bottom*—An invitation to enter a new field of physics." Feynman stated in his lecture that the entire encyclopedia of Britannica could be put on the tip of a needle and, in principle, there is no law preventing such an undertaking. Feynman described then the advances made in this field in the past and he envisioned the future for nanotechnology. His lecture was published in the February 1960 issue of Engineering and Science quarterly magazine of California Institute of Technology.

In his talk Feynman also described how the laws of nature do not limit our ability to work at the molecular level, atom by atom. Instead, he said, it was our lack of the appropriate equipment and techniques for doing so. Feynman in his lecture talked about "*How do we write small?*," "*Information on a small scale*," possibility to have "*Better electron microscopes*" that could take the image of an atom, doing things small scale through "*The marvelous biological system*," "*Miniaturizing the computer*," "*Miniaturization by evaporation*" example of which is thin-film formation by chemical vapor deposition, solving the "*Problems of lubrication*" through miniaturization of machinery and nanorobotics, "*Rearranging the atoms*" to build various nanostructures and nanodevices, and behavior of "*Atoms in a small world*" which included atomic scale fabrication as a bottom-up approach as opposed to the top-down approach that we are accustomed to. Bottom-up approach is self-assembly of machines from basic chemical building blocks which is considered to be an ideal through which nanotechnology will ultimately be implemented. Top-down approach is assembly by manipulating components with much larger devices which is more readily achievable using the current technology. It is important to mention that almost all of the ideas presented in Feynman's lecture and even more are now under intensive research by numerous nanotechnology investigators all around the world.

Feynman in 1983 talked about a scaleable manufacturing system that could be made which will manufacture a smaller scale replica of it (Feynman 1993). That in turn would replicate itself in smaller scale, and so on down to molecular scale. Feynman was subscribing to the "Theory of Self-Reproducing Automata" proposed by von Neumann, the 1940s' eminent mathematician and physicist who was interested in the question of whether a machine can self-replicate, that is, produce copies of itself (Von Neumann and Burks 1966). The study of man-made self-replicating systems has been taking place now for more than half a century. Much of this work is motivated by the desire to understand the fundamentals involved in self-replication and advance our knowledge of single-cell biological self-replications.

Some of the other important achievements about which Feynman mentioned in his 1959 lecture include the manipulation of single atoms on a silicon surface, positioning single atoms with a scanning tunneling microscope (STM), and the trapping of single, 3 nm in diameter, colloidal particles from solution using electrostatic methods.

In early 1960s there were other ongoing research on small systems but with a different emphasis. A good example is the publication of two books on "Thermodynamics of Small Systems" by T. L. Hill in early 1960s. Thermodynamics of small systems is now called "nanothermodynamics" (Hill 1964, 2001; Mansoori et al. 2005).

In 1960s when Feynman recognized and recommended the importance of nanotechnology the devices necessary for nanotechnology were not invented yet. At that time, the world was intrigued with space exploration, discoveries, and the desire and pledges for travel to the moon, partly due to political rivalries of the time and partly due to its bigger promise of new frontiers that man had also not captured yet. Research and developments in small (nano) systems did not sell very well at that time with the governmental research funding agencies and as a result little attention was put in it by the scientific community.

It is only appropriate to name the nanometer scale "the *Feynman* (ϕ nman) *scale*" after Feynman's great contribution and we suggested the notation "ϕ" for it like Å as used for Angstrom scale and μ as used for micron scale over 10 years ago (Mansoori 2005):

$$\text{One Feynman } (\phi) \equiv 1\,\text{nm} = 10\,\text{Å} = 10^{-3}\,\mu = 10^{-9}\,\text{m}$$

1.4 Some Key Inventions and Discoveries

1.4.1 Scanning Tunneling Microscope

Nanotechnology received its greatest momentum with the invention of STM in 1985 by G. K. Binnig and H. Rohrer, staff scientists at the IBM's Zürich Research Laboratory (Binnig and Rohrer 1985). That happened 41 years after Feynman's predictions. To make headway into a realm of molecule-sized devices, it would be necessary to survey the landscape at that tiny scale. Binning and Rohrer's STM offered a new way to do just that. STM allows imaging solid surfaces with atomic scale resolution. It operates based on tunneling current, which starts to flow when a sharp tip is mounted on a piezoelectric scanner approaches a conducting surface at a distance of about 1 nm (1 ϕ). This scanning is recorded and displayed as an image of the surface topography. Actually the individual atoms of a surface can be resolved and displayed using STM.

1.4.2 Atomic Force Microscope

After the invention of STM it was quickly followed by the development of a family of related techniques which, together with STM, may be classified in the general category of scanning probe microscopy (SPM) techniques. Of the latter technologies, the most important is undoubtedly the atomic force microscope (AFM) developed in 1986 (Binnig et al. 1986). Figure 1.2 shows schematic of two typical AFMs that we use in our laboratory at UIC. AFMs are a combination of the principle of STM and the stylus profilometer. It enables us to study nonconducting surfaces, because it scans van der Waals forces with its "atomic" tips. The main components of AFM are a thin cantilever with extremely sharp [1–10 nm (ϕ) in radius] probing tip, a 3D piezoelectric scanner, and optical system to measure deflection of the cantilever. When the tip is brought into contact with the surface or in its proximity, or is tapping the surface, it is being affected by a combination of the surface forces (attractive and repulsive). Those forces cause cantilever bending and torsion, which is continuously measured via the deflection of the reflected laser

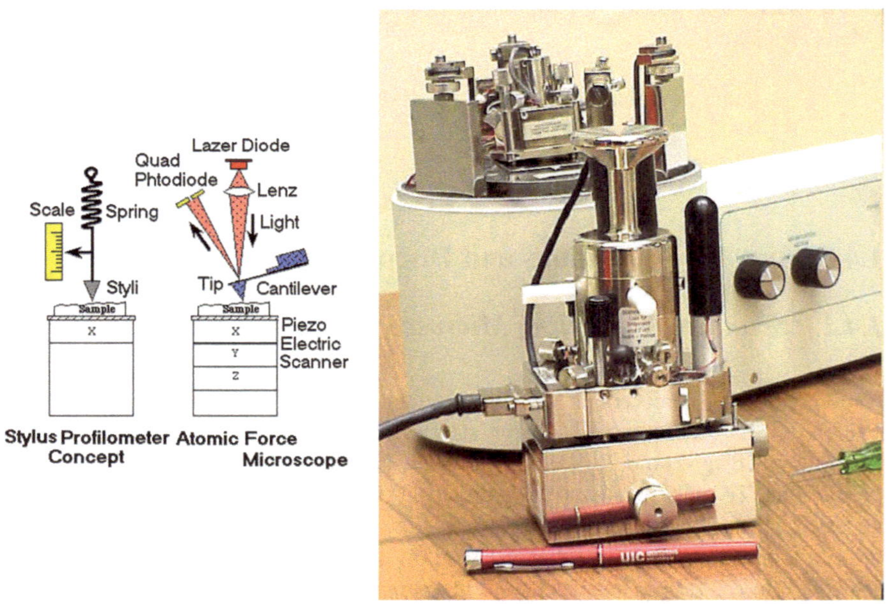

Fig. 1.2 Schematic of a typical AFM and its function as compared with a stylus profilometer. As it is shown an AFM has similarities to a conventional stylus profilometer, but with a much higher resolution in nano scale. In the *right-hand side* pictures of two AFMs are shown (Mansoori 2005)

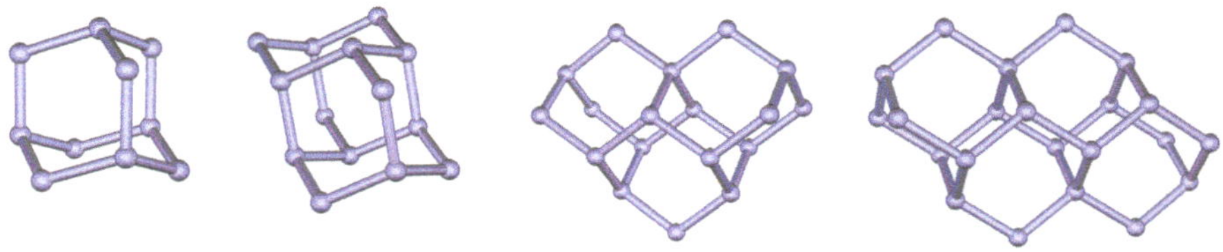

Fig. 1.3 Chemical structures of diamondoid molecules also known as nanodiamonds. These compounds have diamond-like fused-ring structures which can have many applications in nanotechnology. They have the same structure as the diamond lattice, i.e., highly symmetrical and strain free. The rigidity, strength, and assortment of their 3-D shapes make them valuable molecular building blocks (Mansoori et al. 2012)

beam. 3D scanner moves the sample or, in alternative designs, the cantilever, in three dimensions, thus scanning predetermined area of the surface. A vertical resolution of this tool is extremely high reaching 0.01 nm (ϕ) on the order of atomic radius.

1.4.3 Diamondoid Molecules (a.k.a. Nanodiamonds)

Diamondoid molecules are saturated hydrocarbons that have diamond-like fused-ring structures which can have applications in nanotechnology (Mansoori et al. 2012). They have the same structure as the diamond lattice, i.e., highly symmetrical and strain free. Diamondoids offer the possibility of producing a variety of nanostructural shapes. They have quite high strength, toughness, and stiffness compared to other known molecules. The smallest diamondoid molecule was first discovered and isolated from a Czechoslovakian petroleum in 1933. The isolated substance was named adamantane, from the Greek for diamond. This name was chosen because it has the same structure as the diamond lattice, highly symmetrical and strain free as shown in Fig. 1.3. It is generally accompanied by small amounts of alkylated adamantanes: 2-methyl-; 1-ethyl-; and probably 1-methyl-; 1,3-dimethyl; and others. From the nanobiotechnology point of view diamondoids are in the category of organic nanostructures (Mansoori et al. 2012).

The unique structure of adamantane is reflected in its highly unusual physical and chemical properties. The carbon skeleton of adamantane comprises a small cage structure. Because of this, adamantane, and diamondoid molecules in general, is commonly known as cage hydrocarbon. In a broader sense they may be described as saturated, polycyclic, cage-like hydrocarbons. The diamond-like term arises from the fact that their carbon atom structure can be superimposed upon a diamond lattice. The simplest of these polycyclic diamondoids is adamantane, followed by its homologues diamantane, tria-, tetra-, penta-, and hexa-mantane.

Diamondoid molecules are named as the building blocks for nanotechnology (Dahl et al. 2003; Mansoori et al. 2012). In Table 1.1 we report an alphabetical list

Table 1.1 Major applications of diamondoids and derivatives

Antiviral drug
Cages for drug delivery
Crystal engineering
Designing molecular capsules
Design of new antidotes
Diamondoid-DNA nanoarchitectures
Drug delivery (they can easily pass through blood-brain barrier due to their lipophilicity/hydrophobicity)
Drug targeting
Fighting infectious viral diseases (influenza, etc.)
Fighting infectious bacterial diseases (tuberculosis, etc.)
Fighting infectious parasitic diseases (malaria, etc.)
Fighting cancer (new antineoplastic agents)
Gene delivery
Hypoglycemic action (drugs for diabetes treatment, etc.)
In designing an artificial red blood cell, called respirocyte
In host-guest chemistry and combinatorial chemistry
MEMS
Molecular machines
Molecular probe
Nanodevices
Nanorobotics
Nanofabrication
Nanocomposites
Nanomodule
NEMS
Neuroprotective effect (for Alzheimer's disease, etc.)
Organic MBBs in formation of nanostructures
Pharmacophore-based drug design
Polymer, copolymers
Positional assembly
Power cells overcharge protecting compounds
Preparation of fluorescent molecular probes
Prodrugs
Rational design of multifunctional drug systems and drug carriers
Self-assembly: DNA-directed self-assembly, host-guest chemistry
Shape-targeted nanostructures
Synthesis of supramolecules with manipulated architecture
The only semiconductors which show a negative electron affinity
Microelectronics (thermally conductive films in integrated circuit packaging, low-k dielectric layers in integrated circuit multilevel interconnects, thermally conductive adhesive films, thermally conductive films in thermoelectric cooling devices, passivation films for integrated circuit devices (ICs), and field emission cathodes)

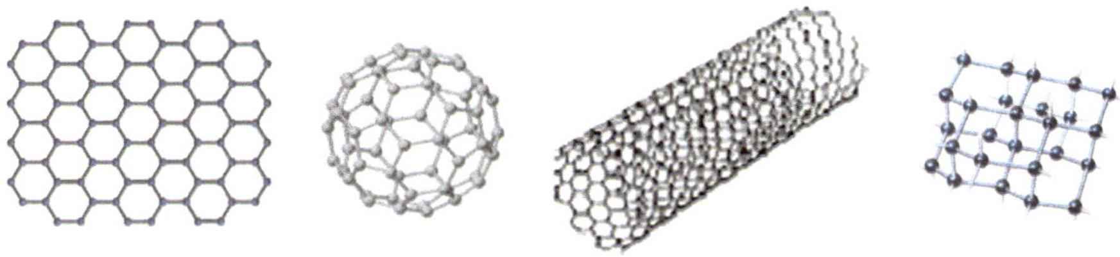

Fig. 1.4 Nano allotropes of carbon

of applications of diamondoids in biomedicine, materials science, and nanotechnology (Mansoori et al. 2012; Mansoori 2013; Ramezani and Mansoori 2007).

1.5 Nano Allotropes of Carbon

1.5.1 Fullerene (a.k.a. Buckyballs)

Buckminsterfullerene (or fullerene), C60, as is shown in Fig. 1.4 is a nano allotrope of carbon, which was discovered in 1985 by Kroto and collaborators (Kroto et al. 1985). These investigators used laser evaporation of graphite and they found Cn clusters (with $n > 20$ and even numbers) of which the most common were found to be C60 and C70. For this discovery, Curl, Kroto, and Smalley were awarded the 1996 Nobel Prize in Chemistry. Later fullerenes with larger number of carbon atoms (C_{76}, C_{80}, C_{240}, etc.) were also synthesized. Since the time of discovery of fullerenes about three decades ago, a great deal of investigation has gone into these nanostructures. Several more efficient and less expensive methods to produce fullerenes have been developed (Eliassi et al. 2002; Valand and Patel 2015). Availability of low-cost fullerene will pave the way for further research into practical applications of fullerene and its role in nanotechnology.

1.5.2 Carbon Nanotubes

Carbon nanotube was discovered in 1991 (Iijima 1991) using an electron microscope while studying cathodic material deposition through vaporizing carbon graphite in an electric arc-evaporation reactor under an inert atmosphere during the synthesis of fullerenes (Iijima and Ichihashi 1993). The nanotubes produced by Iijima and coworkers appeared to be made up of a perfect network of hexagonal graphite, Fig. 1.4, rolled up to form a hollow tube. The nanotube diameter range is from one to several nanometers which is much smaller than its length range which is from one to a few micrometers. A variety of manufacturing techniques have since been developed to synthesize and purify carbon nanotubes with tailored characteristics and functionalities. Controlled production of single-walled carbon nanotubes is one of the favorite forms of carbon nanotube which has many present and future applications in nanoscience and nanotechnology. Laser ablation chemical vapor deposition joined with metal-catalyzed disproportionation of suitable carbonaceous feedstock are often used to produce carbon nanotubes (Morris and Iniewski 2013).

1.5.3 Graphene

Graphene is a two-dimensional nano allotrope of carbon, one atom thick, composed of hexagonal lattice, Fig. 1.5. One atom forms each of its vertex, very strong, flexible, transparent, and conductive of electricity. Graphene may be considered as an indefinitely large aromatic molecule, the limiting case of the family of flat polycyclic aromatic hydrocarbons. Graphene was first studied at a theoretical level in 1940s, but it wasn't practically pursued until the 1970s (Morris and Iniewski 2013). Graphite is made of millions of layers of graphene. In 2003 Geim and Novoselov extracted graphene from graphite using Scotch tape. They published their research the following year (Novoselov et al. 2004). They received the 2010 Nobel Prize for Physics for their investigations on graphene. Graphene is presently considered as a wonder material with many superlatives to its name. Since it is one atom thick, it is the thinnest material in use and quite strong and stiff. Graphene electrical conductivity is reported to be six orders higher than copper and its thermal conductivity is also quite high. It is impermeable to gases and reconciles such conflicting qualities as brittleness and ductility. Electron transport in graphene is described by a Dirac-like equation, which allows the investigation of relativistic quantum phenomena in a benchtop experiment. Presently there are many projects worldwide investigating the potential of graphene for a variety of applications. Graphene research is expanding quite fast and several publications appear in the literature about graphene every day which makes it a real struggle to keep up with the developments. Research on graphene's electronic properties is now quite matured. It is expected that graphene will continue to stand out as a truly unique molecular building block (MBB) in the field of nanotechnology. Research on various properties of graphene is in progress, and it may bring up new phenomena (Morris and Iniewski 2013).

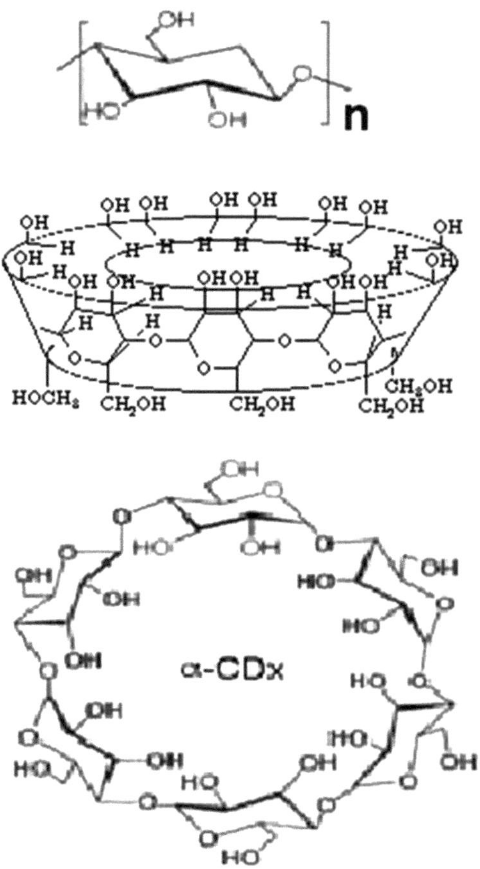

Fig. 1.5 Chemical formula and structure of cyclodextrins—For $n = 6$ it is called α-CDx, $n = 7$ is called β-CDx, $n = 8$ is called γ-CDx. Cyclodextrins are cyclic oligosaccharides. Their shape is like a truncated cone and they have a relatively hydrophobic interiors. They have the ability to form inclusion complexes with a wide range of substrates in aqueous solution. This property has led to their application for encapsulation of drugs in drug delivery (Mansoori 2005)

1.6 Metallic and Oxide Molecular Building Blocks of Nanotechnology

Michael Faraday is credited for the first person who recognized the existence of metallic nanoparticles (MNPs), especially gold nanoparticles, in solution in 1857. Later in 1908, Gustav Mie gave a quantitative explanation of MNP color. MNPs, which are in size between molecular and metallic states, possess specific electronic structure which includes local density of states; plasmon excitation; quantum confinement; short-range ordering; increased number of kinks; a large number of low-coordination sites (such as corners and edges), having a large number of "dangling bonds"; and consequently specific and chemical properties and the ability to store excess electrons. Among all MNPs gold nanoparticles and silver nanoparticles have found more extensive and interesting applications in the fields of agriculture, energy, and medicine (Hatami and Ghorbanpour 2014; Ghorbanpour and Hatami 2015; Prasad et al. 2014, 2016; Mansoori et al. 2010; Vahabi et al. 2011).

Considering the existence of a variety of metals, we are able to form a large diversity of oxide nanoparticles (ONPs) from such oxides as Al_2O_3, MgO, ZrO_2, CeO_2, TiO_2, ZnO, Fe_2O_3, Fe_3O_4, and SnO, just to name a few. Oxides generally possess a vast number of structural geometries with varied electronic structure that can exhibit metallic, semiconductor, or insulator character. ONPs can exhibit unique physical and chemical properties due to their limited size and a high density of corner or edge surface sites. For example, during the past two decades, research and development in the area of synthesis and applications of different nanostructured titanium dioxide have become tremendous (Aghdam et al. 2015; Ghorbanpour 2015; Ghorbanpour et al. 2015; Hatami et al. 2014; Khataee and Mansoori 2011). TiO_2 nanomaterials can assume the forms of nanoparticles, nanorods, nanowires, nanosheets, nanofibers, and nanotubes. Many applications of TiO_2 nanomaterials are closely related to their optical properties. Examples of applications of nanostructured titanium dioxide include in dye-sensitized solar cells, hydrogen production and storage, sensors, rechargeable batteries, electrocatalysis, self-cleaning and antibacterial surfaces, cancer treatment, photocatalytic removal of various pollutants, and TiO_2-based nanoclays.

1.7 Biological Molecular Building Blocks of Nanotechnology

At the same time that physical scientists and engineers have been experimenting with organic and inorganic nanostructures as mentioned above, bio- and medical scientists and engineers have been making their own advances with other nanoscale structures known as biological molecular building blocks (MBBs) like cyclodextrins, liposomes, and monoclonal antibodies (Mansoori 2005; Mansoori et al. 2007). These nanostructures have specific applications in bio-systems/drug delivery and targeting. Cyclodextrins, as shown in Fig. 1.5, are cyclic oligosaccharides. Their shape is like a truncated cone and they have a relatively hydrophobic interior. They have the ability to form inclusion complexes with a wide range of substrates in aqueous solution. This property has led to their application for encapsulation of drugs in drug delivery. Liposomes, which are also in nanoscale size range as shown in Fig. 1.6, self-assemble based on hydrophilic and hydrophobic properties and they encapsulate drugs inside. Many commercially available anticancer drugs are cyclodextrins or liposome loaded with 100 or less nanometer in diameter. A monoclonal (derived from a single clone) antibody molecule consists of four protein chains, two heavy and two light, which are folded to form a Y-shaped structure, Fig. 1.7. It is about ten nanometers in diameter. This small size is important, for example, to ensure that intravenously administered these particles can penetrate small capillaries and reach cells in tissues where they are needed for treatment.

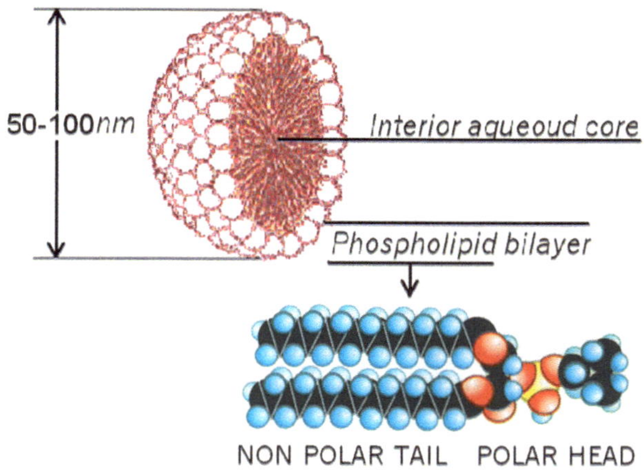

Fig. 1.6 Cross section of a liposome composed of the phospholipid distearoyl phosphatidylcholine (DSPC) and cholesterol—a spherical bilayer which is quite similar to a micelle. It is made of a bilayer with an internal aqueous compartment. Liposome vesicles can be used as drug carriers and loaded with a great variety of molecules, such as small drug molecules, proteins, nucleotides, and even plasmids (Mansoori 2005)

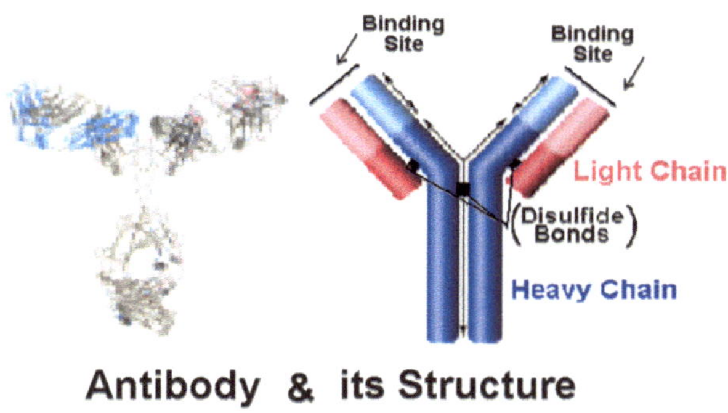

Fig. 1.7 Antibody and its structure (Mansoori 2005)

1.8 Conclusions and General Discussion

The atomic scale and cutting-edge fields of nanoscience and nanotechnology, which are considered to lead us to the next industrial revolution, are likely to have a revolutionary impact on the way things will be done, designed, and manufactured in the future.

Results of research and developments in these fields are entering into all aspects of our lives including, but not limited to, aerospace, agriculture, defense, energy, environment, materials, manufacturing, and medicine. It is truly an atomic and molecular approach for building biologically, chemically, and physically stable structures one atom or one molecule at a time. Presently some of the active nanoscience and nanotechnology research areas include nanolithography, nanodevices, nanorobotics, nanocomputers, nanopowders, nanostructured catalysts and nanoporous materials, molecular manufacturing, nanolayers, molecular nanotechnology, medicine such as Alzheimer's disease (Nazem and Mansoori 2008, 2014) and cancer (Ebrahimi and Mansoori 2014; Mansoori et al. 2010; Mansoori

2007) prediction, prevention and treatment through nanotechnology, nanobiology, and organic nanostructures to name a few.

We have known for many years that several existing technologies depend crucially on processes that take place on the nanoscale. Adsorption, lithography, ion exchange, catalysis, drug design, plastics, and composites are some examples of such technologies. The "nano" aspect of these technologies was not known and, for most part, they were initiated accidentally by mere luck. They were further developed using tedious trial-and-error laboratory techniques due to the limited ability of the times to probe and control matter on nanoscale. Investigations at nanoscale were left behind as compared to micro and macro length scales because significant developments of the nanoscale investigative tools have been made only recently.

The above-mentioned technologies, and more, stand to be improved vastly as the methods of nanoscience and nanotechnology develop. Such methods include the possibility to control the arrangement of atoms inside a particular molecule and, as a result, the ability to organize and control matter simultaneously on several length scales. The developing concepts of nanoscience and nanotechnology seem pervasive and broad. It is expected to influence every area of science and technology, in ways that are clearly unpredictable.

Advances in nanoscience and nanotechnology will also help solve other technology and science problems. For example, we have realized the benefits that nanostructuring can bring to (Mansoori 2005):

(a) Wear-resistant tires made by combining nanoscale particles of inorganic clays with polymers as well as other nanoparticle-reinforced materials.
(b) Greatly improved printing brought about by nanoscale particles that have the best properties of both dyes and pigments as well as advanced ink jet systems.
(c) Vastly improved new generation of lasers, magnetic disk heads, nanolayers with selective optical barriers, and systems on a chip made by controlling layer thickness to better than a nanometer.
(d) Design of advanced chemical and biosensors.
(e) Nanoparticles to be used in medicine with vastly advanced drug delivery and drug-targeting capabilities.
(f) Chemical-mechanical polishing with nanoparticle slurries, hard coatings, and high-hardness cutting tools.
(g) Methods of nanotechnology could provide a new dimension to the control and improvement of living organisms.
(h) Photolithographic patterning of matter on the microscale has led to the revolution in microelectronics over the past few decades. With nanotechnology, it is becoming possible to control matter on every important length scale, enabling tremendous new power in material design.
(i) Biotechnology is being influenced by research in nanoscience and nanotechnology greatly. It is anticipated that, for example, this will revolutionize healthcare to produce ingestible systems that will be harmlessly flushed from the body if the patient is healthy but will notify a physician of the type and location of diseased cells and organs if there are problems.

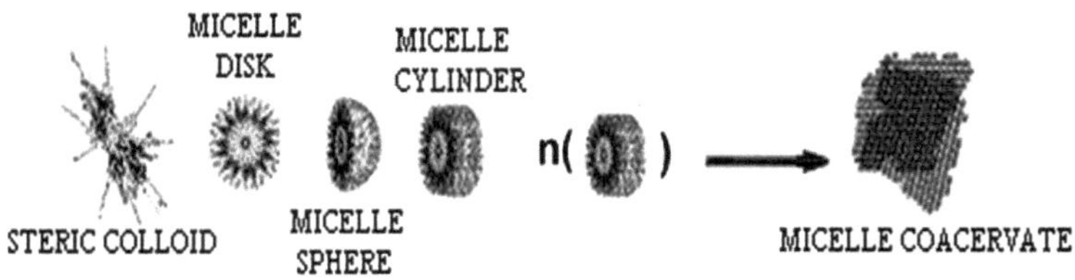

Fig. 1.8 Organic nanostructure self-assemblies of various shapes (Mansoori 2002; Priyanto et al. 2001)

(j) Micro and macro systems constructed of nanoscale components are expected to have entirely new properties that have never before been identified in nature. As a result, by altering the design of the structure of materials in the nanoscale range we would be able to systematically and appreciably modify or change selected properties of matter at macro and micro scales. This would include, for example, production of polymers or composites with most desirable properties which nature and existing technologies are incapable of producing.

(k) Robotic spacecraft that weigh only a few pounds are being flown out for various exploratory missions.

(l) Nanoscale traps will be constructed that will be able to remove pollutants from the environment and deactivate chemical warfare agents. Computers with the capabilities of current workstations will be the size of a grain of sand and will be able to operate for decades with the equivalent of a single wristwatch battery.

(m) There are many more observations in the areas of agricultural applications (Ghorbanpour et al. 2015; Ghorbanpour and Hadian 2015; Ghorbanpour and Hatami 2015), inks and dyes, protective coatings, dispersions with optoelectronic properties, nanostructured catalysts, high-reactivity reagents, medicine, electronics, structural materials, and energy conversion, conservation, storage, and usage (Mansoori et al. 2016) which are also worth mentioning.

(n) Many large organic molecules are known to form organic nanostructures of various shapes as shown in Figs. 1.3 and 1.8, the driving force of which is the intermolecular interaction energies between such macromolecules (Mansoori 2002; Priyanto et al. 2001; Rafii-Tabar and Mansoori 2004). There has been an appreciable progress in research during the past few years on organic nanostructures, such as thin-film nanostructures, which have excellent potential for use in areas that are not accessible to more conventional, inorganic nanostructures. The primary attraction of organic nanostructures is their potential for molding and coating, and the extreme flexibility that they have in being tailored to meet the needs of a particular application. The organic nanostructure materials are easily integrated with conventional inorganic nanostructures (like semiconductor devices), thereby providing additional functionality to existing photonic circuits and components. Some progress has been made in understanding the formation and behavior of organic nanostructures that might be formed to serve as elements of nanomaterials and also on synthetic strategies for creating such structures (Mansoori 2002; Priyanto et al. 2001; Rafii-Tabar and Mansoori 2004). The ultimate goal is to achieve a better understanding of the fundamental molecular processes and properties of these nanostructures which are dominated by grain boundaries and interfaces. In understanding the behavior and the properties of these nanostructures the potential for technological applications will be considered.

Many other unpredictable advances resulting from nanotechnology are inevitable. Thus, the future prospects for nanotechnology actually represent a revolutionary super-cutting-edge field that is expected to eventually become the foundation for such currently disparate areas as, and many others that we cannot even foresee at this time. It is then no wonder that it is considered to lead the humanity to the next industrial revolution.

References

Aghdam MTB, Mohammadi H, Ghorbanpour M (2015) Effects of nanoparticulate anatase titanium dioxide on physiological and biochemical performance of *Linum usitatissimum* (Linaceae) under well-watered and drought. Braz J Bot 23:1–8

Binnig G, Rohrer H (1985) The scanning tunneling microscope. Sci Am 253:50–56

Binnig G, Quate CF, Gerber C (1986) Atomic force microscope. Phys Rev Lett 56:930

Dahl JE, Liu SG, Carlson RMK (2003) Isolation and structure of higher diamondoids, nanometer-sized diamond molecules. Science 299:96–99

Ebrahimi N, Mansoori GA (2014) Reliability for drug targeting in cancer treatment through nanotechnology. Int J Med Nano Res 1:2378–3664

Eliassi A, Eikani MH, Mansoori GA (2002) Production of single-walled carbon nanotubes. In: Proceedings of the 1st conference on nanotechnology – the next industrial revolution, vol 2, p 160, March 2002

Feynman RP (1960) There's plenty of room at the bottom – an invitation to enter a new field of physics. Eng Sci Mag Cal Inst Technol 23:22

Feynman RP (1993) Infinitesimal machinery. J Microelectromech Syst 2:1–4

Ghorbanpour M (2015) Major essential oil constituents, total phenolics and flavonoids content and antioxidant activity of Salvia officinalis plant in response to nano-titanium dioxide. Indian J Plant Physiol 20:249–256

Ghorbanpour M, Hadian J (2015) Multi-walled carbon nanotubes stimulate callus induction, secondary metabolites biosynthesis and antioxidant capacity in medicinal plant. Carbon 94:749–759

Ghorbanpour M, Hatami M (2015) Changes in growth, antioxidant defense system and major essential oils constituents of *Pelargonium graveolens* plant exposed to nano-scale silver and thidiazuron. Indian J Plant Physiol 20:116–123

Ghorbanpour M, Hatami M, Hatami M (2015) Activating antioxidant enzymes, hyoscyamine and scopolamine biosynthesis of *Hyoscyamus niger* L. plants with nano-sized titanium dioxide and bulk application. Acta Agric Slov 105:23–32

Hatami M, Ghorbanpour M (2014) Defense enzyme activities and biochemical variations of *Pelargonium zonale* in response to nanosilver application and dark storage. Turk J Biol 38:130–139

Hatami M, Ghorbanpour M, Salehiarjomand H (2014) Nano-anatase TiO_2 modulates the germination behavior and seedling vigority of some commercially important medicinal and aromatic plants. J Biol Env Sci 8:53–59
Hill TL (1964) Thermodynamics of small systems, vol 1. W.A. Benjamin, New York, pp 103–106
Hill TL (2001) A different approach to nanothermodynamics. Nano Lett 1:273–275
Iijima S (1991) Helical microtubules of graphitic carbon. Nature 345:56–58
Iijima S, Ichihashi T (1993) Single-shell carbon nanotubes of 1-nm diameter. Nature 363:603–605
Khataee AR, Mansoori GA (2011) Nanostructured titanium dioxide materials (Properties, preparation and applications), vol 4. World Scientific, Hackensack, pp 200–230
Kroto HW, Heath JR, O'Brien SC, Curl RF, Smalley RE (1985) C_{60}: Buckminsterfullerene. Nature 318:162–163
Mansoori GA (2002) Organic nanostructures and their phase transitions. In: Proceedings of the 1st conference on nanotechnology – the next industrial revolution, vol 2, p 345
Mansoori GA (2005) Principles of nanotechnology, vol 2. World Scientific, Hackensack, pp 300–349
Mansoori GA (2007) Nanotechnology in cancer prevention, detection and treatment: bright future lies ahead. World Rev Sci Technol Sustain Dev 4:226–257
Mansoori GA (2013) The molecular lego of biomedicine, materials science and nanotechnology. J Bioanal Biomed 5:1–3
Mansoori GA, Vakili-Nezhaad GR, Ashrafi AR (2005) Some mathematical concepts applicable in nanothermodynamics. Int J Pure Appl Math Sci 2:58–61
Mansoori GA, George TF, Assoufid L, Zhang G (eds) (2007) Molecular building blocks for nanotechnology: from diamondoids to nanoscale materials and applications. Topics in Applied Physics #109, Springer
Mansoori GA, Brandenburg KS, Shakeri-Zadeh A (2010) A comparative study of two folate-conjugated gold nanoparticles for cancer nanotechnology applications. Cancers 2:1911–1928
Mansoori GA, de Araujo PLB, de Araujo ES (2012) Diamondoid molecules with applications in biomedicine, materials science, nanotechnology and petroleum science. World Scientific, Hackensack
Mansoori GA, Enayati N, Agyarko L (2016) Energy: sources, utilization, legislation, sustainability, Illinois as the model state. World Scientific, Hackensack
Morris JE, Iniewski K (eds) (2013) Graphene, carbon nanotubes, and nanostructures: techniques and applications (devices, circuits, and systems). CRC Press, Boca Raton
Nazem A, Mansoori GA (2008) Nanotechnology solutions for Alzheimer's disease: (Advances in research tools, diagnostic methods and therapeutic agents). J Alzheimers Dis 13:199–223
Nazem A, Mansoori GA (2014) Nanotechnology building blocks for intervention with Alzheimer's disease pathology: implications in disease modifying strategies. J Bioanal Biomed 6:009–014
Novoselov KS, Geim AK et al (2004) Electric field effect in atomically thin carbon films. Science 306:666–669
NSTC (2000) National nanotechnology initiative: leading to the next industrial revolution. A report by the interagency working group on nanoscience, engineering and technology committee on technology. National Science and Technology Council, Washington, DC
Prasad R, Kumar V, Prasad KS (2014) Nanotechnology in sustainable agriculture: present concerns and future aspects. Afr J Biotechnol 13:705–713
Prasad R, Pandey R, Barman I (2016) Engineering tailored nanoparticles with microbes: quo vadis. WIREs Nanomed Nanobiotechnol 8:316–330
Priyanto S, Mansoori GA, Suwono A (2001) Measurement of property relationships of nanostructure micelles and coacervates of asphaltene in a pure solvent. Chem Eng Sci 56:6933–6939
Rafii-Tabar H, Mansoori GA (2004) Interatomic potential models for nanostructures. Encycl Nanosci Nanotechnol 4:231–248

Ramezani H, Mansoori GA (2007) Diamondoids as molecular building blocks for biotechnology (Wet nanotechnology, drug targeting and gene delivery). Top Appl Phys 109:44–71

Roco MC, Williams S, Alivisatos P (eds) (1999) Nanotechnology research directions: IWGN workshop report – vision for nanotechnology R&D in the next decade. WTEC, Loyola College in Maryland

Vahabi K, Mansoori GA, Karimi S (2011) Biosynthesis of silver nanoparticles by fungus *Trichoderma reesei* (A route for large-scale production of AgNPs). Nanotechnol/Insci J 1:65–79

Valand NN, Patel MB (2015) Fullerenes chemistry and its applications. Scholars' Press, Cambridge

Von Neumann J, Burks AW (1966) Theory of self-reproducing automata. University of Illinois Press, Champaign, IL

Xue Y, Mansoori GA (2010) Self-assembly of diamondoid molecules and derivatives (MD simulations and DFT calculations). Int J Mol Sci 11:288–303. doi:10.3390/ijms11010288

Chapter 2
Biosynthesis of Metal and Semiconductor Nanoparticles, Scale-Up, and Their Applications

Mojtaba Salouti and Neda Faghri Zonooz

2.1 Introduction

Nanotechnology is a general term that refers to the techniques and methods for studying, designing, and fabricating devices at the level of atoms and molecules. Nanostructures are at the leading edge of rapidly developing field of nanotechnology and recently there is a great emphasis on nanomaterial synthesis methods. A number of chemical and physical approaches are available for synthesis of nanomaterials. Unfortunately, many of these methods involve low material conversions, high energy requirements, use of hazardous chemicals and toxic solvents, and generation of hazardous by-products which become hard to apply in some fields (Hulkoti and Taranath 2014). For example, gold nanoparticles produced by chemical methods are toxic due to hazardous precursor chemicals and have difficulties to apply medically. Hence, there is an increasing need to develop high-yield, low-cost, nontoxic, biocompatible, and environmentally benign processes for synthesis of nanostructures where the biological approaches for synthesis of nanomaterials gain importance.

It is well known that many organisms can provide inorganic metabolites either intra- or extracellularly. For example, unicellular organisms such as surface-layer bacteria produce gypsum and calcium carbonate layers that synthesize siliceous materials. Even though many biotechnological applications such as the remediation of toxic metals employ microorganisms such as bacteria and fungi, such

M. Salouti (✉)
Biology Research Center, Zanjan Branch, Islamic Azad University, Zanjan, Iran
e-mail: saloutim@iauz.ac.ir; saloutim@yahoo.com

N. Faghri Zonooz
Department of Microbiology, Faculty of Sciences, Zanjan Branch, Islamic Azad University, Zanjan, Iran
e-mail: nfaghri@yahoo.com

M. Ghorbanpour et al. (eds.), *Nanoscience and Plant–Soil Systems*, Soil Biology 48,
DOI 10.1007/978-3-319-46835-8_2

microorganisms are recently found as possible eco-friendly manufactories. The processes devised by nature for the synthesis of inorganic materials on nanolength scale have contributed to the development of a relatively new and largely unexplored area of research based on the use of microbes in the biosynthesis of nanomaterials (Mandal et al. 2006; Kalishwaralal et al. 2010). Biosynthesis refers to the phenomenon that takes place by means of biological processes or enzymatic reactions. Biosynthesis methods employing either biological microorganisms or plant extracts have emerged as a simple and viable alternative to chemical and physical synthesis procedures (Pantidos and Horsfall 2014; Salunke et al. 2016). As the changing of conditions of nanomaterial synthesis by chemical and physical procedures leads to high production rate and controls the size, shape, dispersity, and chemical composition of nanomaterials, the same can happen in biological methods by optimizing the culture conditions such as pH, incubation temperature and time, concentration of metal ions, and the amount of biological material (Shakouri et al. 2016). There is also the possibility of producing genetically engineered microbes that overexpress specific reducing agents and thereby control the size, shape, dispersity, and chemical composition of biological nanostructures.

However, there are many secrets about the biochemical and molecular mechanisms of biological processes that should be revealed (Keat et al. 2015). It appears that strategies such as enzymatic oxidation or reduction, sorption on the cell wall, and in some cases subsequent chelating with extracellular peptides or polysaccharides have been developed and used by microorganisms. Hence, there is a growing need to understand the basics of biological synthesis methods to facilitate the application of new methodology to laboratory and industrial needs (Park et al. 2016). In this chapter, we present a detailed outlook about different biological resources available in nature for synthesis of metal and semiconductor nanostructures, the processes of nanostructured material biosynthesis, and methods of synthesis optimization.

2.2 Biological Synthesis of Nanomaterials

Biological synthesis process, referred to as green technology, can be used to obtain stable nanostructures from biological resources (Gericke and Pinches 2006a). Different organisms (prokaryotes and eukaryotes) can be used to synthesize nanomaterials. Regarding the kind of organism, various cultures for isolation would be used. The antifungal antibiotics (such as Nystatin) are used for the isolation of bacteria, and antibacterial antibiotics (such as chloramphenicol) are used for the isolation of fungi. Then, the biomass is separated by centrifuging. This biomass (or probably cell-free supernatant) can be used as a resource for nanomaterial synthesis. These extracts are the resources for nanomaterial synthesis. However, despite the stability, biological nanomaterials are not monodispersed and the rate of synthesis is slow. To overcome these problems, several factors such as microbial cultivation methods and the extraction techniques have to be optimized and the combinatorial approaches such as photobiological methods may be used (Narayanan and Sakthivel 2010). Overall, biosynthesis of nanomaterials is generally considered better than chemical and physical synthesis methods, because:

1. Biological nanomaterial synthesis would have greater commercial viability and large savings in reductant and energy costs and high production rate in comparison with conventional methods (Mukherjee et al. 2002).
2. Large-scale production by chemical and physical methods usually results in producing nanoparticles larger than several micrometers while the biological synthesis can be successfully used for production of small nanoparticles in large-scale operations (Joerger et al. 2000).
3. It is a clean, nontoxic, and eco-friendly method (Senapati et al. 2005).
4. Physical methods need high temperatures and chemical methods need high pressures which are hard situations to provide (Bansal 2004).

2.2.1 Nanomaterial Biosynthesis by Bacteria

Bacteria are prokaryotic microorganisms that have simple structure and lack cell nucleus. Among the microorganisms, bacteria have received the most attention in the area of biosynthesis of nanomaterials. One major advantage of having bacteria as nanomaterial synthesizers, apart from the ease of handling, is that they can be easily modified using genetic engineering techniques for overexpression of specific enzymes due to lack of nucleus (Vaidyanathan et al. 2010). At the present, there is much more attention in the development of protocols for the synthesis of different nanomaterials using bacteria which is discussed.

2.2.2 Gold Nanoparticles

The use of gold compounds and gold nanoparticles, with respect to their potential therapeutic applications such as antiangiogenesis agent, antimalarial agent, and antiarthritic agent, has driven various breakthrough in the field of nanotechnology (Mukherjee et al. 2005; Navarro et al. 1997; Tsai et al. 2007). The studies revealed that *Bacillus subtilis* 168 was able to reduce Au^{3+} ions to produce octahedral gold particles of nanoscale dimensions within bacterial cells by incubation of the cells with gold chloride (Beveridge and Murray 1980). Kashefi et al. (2001) showed that in Fe(III)-reducing bacterium, *Geobacter ferrireducens*, gold was precipitated intracellularly in periplasmic space. In addition, they showed that the bacteria like *Stenotrophomonas maltophilia*, *Plectonema boryanum*, *Pyrobaculum islandicum*, *Thermotoga maritima*, *G. sulfurreducens*, and *Pyrococcus furiosus* are capable of precipitating gold by reducing Au(III) to Au(0) with hydrogen as the electron donor. Bioreduction of chloroauric acid ($HAuCl_4$) to Au nanoparticles by *Escherichia coli* DH5α was reported in 2007 (Du et al. 2007). The accumulated particles on the cell surface were mostly spherical with little other morphologies of triangle and quasi-hexagon. Husseiny et al. (2007) demonstrated the extracellular synthesizing of gold nanoparticles by *Pseudomonas aeruginosa* during incubation with gold solution.

He et al. (2008) showed that *Rhodopseudomonas capsulata* could produce gold nanostructures in different dimensions and morphologies extracellularly. Cai et al. (2011) showed that *Magnetospirillum gryphiswaldense* MSR-1 was successfully used to reduce gold ions to a zero-valent metal in a water environment, and more importantly could accumulate them into spherical nanoparticles on the cell surface through biosorption. *Shewanella oneidensis* is a gram-negative bacterium that was known to have considerable potential for the bioremediation of environmental contaminants with metal-reducing capability (Suresh et al. 2011). The γ-proteobacterium *S. oneidensis* was reported to be able to reduce tetrachloroaurate (III) ions to produce discrete extracellular spherical gold nanocrystallines.

Actinomycetes, though have been classified as prokaryotes, as they share important characteristics of fungi, are popularly known as ray fungi. A novel extremophilic actinomycete, *Thermomonospora* sp. was found to synthesize extracellular spherical gold nanoparticles (Ahmad et al. 2003b). The *Streptomycetes* are also members of the bacterial order *Actinomycetales* that can synthesize gold nanoparticles. For example, Faghri Zonooz et al. (2012) showed that *Streptomyces* sp. ERI-3 was able to produce gold nanoparticles extracellularly. These nanoparticles had cylindrical and spherical morphology. In a similar study, Khadivi Derakhshan et al. (2012) reported that *Streptomyces griseus* could synthesize spherical gold nanoparticles extracellularly (Fig 2.1).

Therefore, it can be considered that most of the bacteria are able to synthesize gold nanoparticles in different sizes and morphologies.

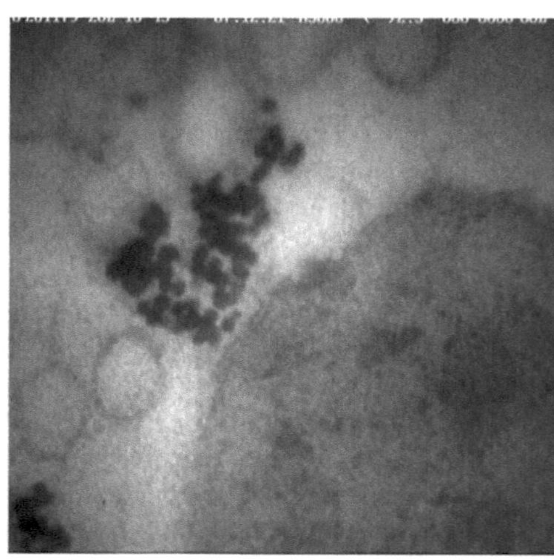

Fig. 2.1 TEM micrograph shows the synthesis of spherical gold nanoparticles by *Streptomyces griseus* extracellular

2.2.3 Silver Nanoparticles

Silver nanoparticles have several important applications in the field of bio-labeling, sensors, antimicrobial agents, and filters. The silver nanoparticles are capable of purifying drinking water, degrading pesticides, and killing human pathogenic bacteria (Srikar et al. 2016). Several bacterial strains were reported as silver resistant and may even accumulate silver at the cell wall to as much as 25 % of the dry weight biomass, thus suggesting their use for the industrial recovery of silver from ore materials. For example, the silver-resistant bacteria strain *Pseudomonas stutzeri* AG259 was found to produce a small number of monoclinic crystalline α-form silver sulfide acanthite (Ag_2S) with the composition of silver and sulfur in the ratio 2:1 (Klaus-Joerger et al. 1999). Fu et al. (2000) reported biosorption and bioreduction of Ag(I) on cell surface in *Lactobacillus sp*. Similarly, Zhang et al. (2005) reported that dried cells of *Corynebacterium* sp. SH09 could produce silver nanoparticles on the cell wall with diamine silver complex $(Ag(NH_3)_2)^+$. Silver nanoparticles were found to be produced by dried cells of *Aeromonas sp*. SH10, which reduced $(Ag(NH_3)_2)^+$ to Ag^0 by a fast process (Mouxing et al. 2006). These particles were monodispersed and uniform in size and remained stable for more than 6 months without aggregation and precipitation. Shahverdi et al. (2007) showed that the culture supernatant of Enterobacteria (*Klebsiella pneumonia*, *E. coli*, and *Enterobacter cloacae*) synthesized silver nanoparticles rapidly by reducing Ag^+ to Ag^0. Kalishwaralal et al. (2008) reported that *B. licheniformis* was able to form extracellular Ag nanomaterials in various sizes and morphologies. Barud et al. (2008) demonstrated the formation of homogenous silver-containing bacterial cellulose (BC) membranes obtained from BC-hydrated membranes of *Acetobacter xylinum* cultures soaked in silver ion with triethanolamine (Ag^+-TAE) solution. *Morganella* sp., a silver-resistant bacterium isolated from the insect gut belonging to Enterobacteriaceae, was reported to produce spherical silver nanoparticles (Parekh et al. 2008). Pugazhenthiran et al. (2009) showed that an airborne *Bacillus* sp. isolated from the atmosphere was able to reduce Ag^+ ions to Ag^0. This bacterium accumulated metallic silver in the periplasmic space of cells.

Biological reduction of aqueous silver ions by extracellular components of *Streptomyces hygroscopicus* (Sadhasivam et al. 2010), *Plectonema boryanum* UTEX 485, a filamentous cyanobacterium, was led to biosynthesis of silver nanoparticles (Lengke et al. 2007). Sivalingama et al. (2012) showed that when aqueous silver nitrate ($AgNO_3$) solution was treated with cell-free supernatant of a novel *Streptomyces* sp. BDUKAS10, an isolate of mangrove sediment could synthesize spherical silver and gold nanoparticles extracellularly. Furthermore, cell-free supernatant of *Streptomyces* sp. ERI-3 was reported to produce silver nanoparticles extracellularly (Faghri Zonooz and Salouti 2011). These nanoparticles were spherical and monodispersed (Fig 2.2).

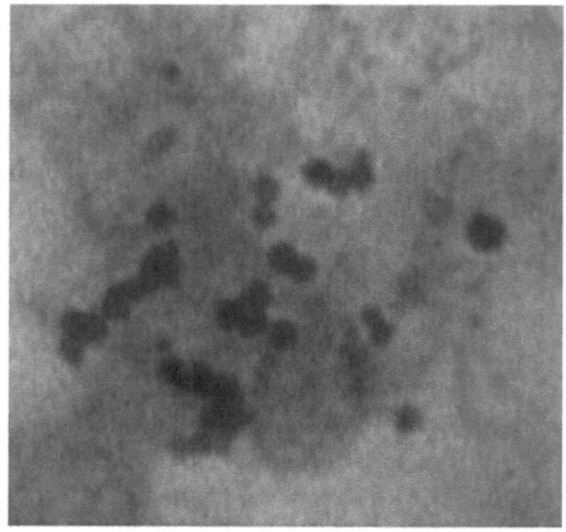

Fig. 2.2 TEM micrograph shows the synthesis of spherical silver nanoparticles by cell-free supernatant of *Streptomyces* sp. *ERI-3*

2.2.4 Magnetic Nanoparticles

Magnetite is a common product of bacterial iron reduction and can be a potential physical indicator of biological activity in geological settings (Faivre and Schuler 2008). The mineralization processes are highly controlled by the magnetotactic bacteria, leading to the formation of uniform, species-specific magnetic nanoparticles (Miot et al. 2009). Single-domain tiny magnetic particles, which exhibit octahedral shapes, are formed exclusively outside of the bacterial cells by a thermophilic fermentive bacterial strain *Thermoanaerobacter ethanolicus* (TOR-39) (Narayanan and Sakthivel 2010). Transition metals such as Co, Cr, and Ni may be substituted for magnetic crystals biosynthesized in the TOR-39 by the way of the electrochemical process. Mann et al. (1984) studied about *Aquaspirillum magnetotacticum*, a microaerophilic bacterium isolated from sediments that produced crystals of ordered single-domain magnetite (Fe_3O_4) particles. Magnetic Fe sulfide nanoparticles were also synthesized using sulfate-reducing bacteria, where particles having the size of a few nanometers formed on the surface (Watson et al. 2001). Furthermore, Bharde et al. (2005) reported the production of magnetite crystals *by Actinobacter* spp., a non-magnetotactic aerobic bacterium, extracellularly. Figure 2.3 shows the SEM images of Fe_3O_4 nanoparticles synthesized by magnetotactic bacteria.

A multicellular magnetotactic bacterium, *Candidatus Magnetoglobus multicellularis*, was found to interact with geomagnetic field with the use of biomineralized magnetic nanocrystals (Perantoni et al. 2009). *Geobacter metallireducens* GS-15 was able to produce magnetite nanocrystals attached to the cell wall (Narayanan and Sakthivel 2010).

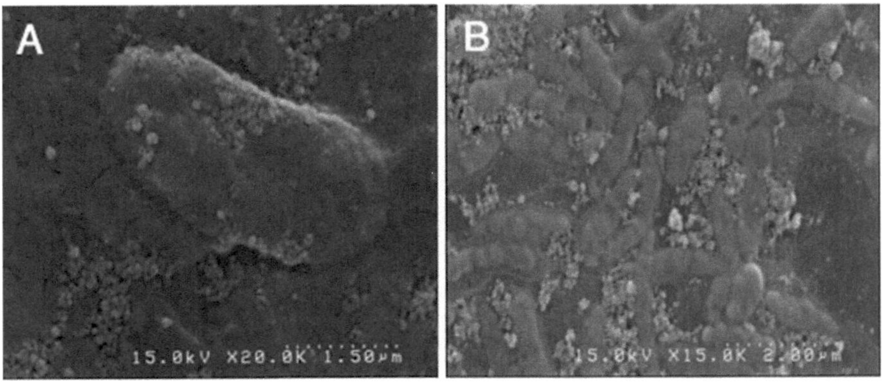

Fig. 2.3 SEM images show the production of Fe_3O_4 nanoparticles by magnetotactic bacteria: attached to the cell membrane (**a**) and extracellular (**b**)

2.2.5 *Other Nanomaterials*

Bacteria not normally exposed to large concentrations of metal ions may also be used to grow various nanomaterials (Lezcano et al. 2010). For example, Konishi et al. (2007a) showed that when growth-decoupled, resting cells of metalion-reducing bacteria, *S. algae*, was incubated anaerobically in aqueous solution of H_2PtCl_6, it reduced $PtCl_6^{-2}$ ions in the presence of lactate as the electron donor to metallic platinum. Biocomposites of nanocrystalline such as silver and the bacteria may be thermally treated to yield a carbonaceous (cermet) nanomaterial with interesting optical properties for potential application in functional thin-film coating. This type of carbonaceous material is composed primarily of graphite carbon and up to 5 % by weight (of the dry biomass) of silver (Klaus-Joerger et al. 2001). A platinum group metal nanoparticle was produced by sulfate-reducing bacterium, *Desulfovibrio desulfuricans* NCIMB 8307 (Yong et al. 2002). Suzuki et al. (2002) showed that *Desulfosporosinus* sp., a Gram-positive sulfate-reducing microbe isolated from sediments, when incubated with mobile hexavalent uranium U(VI) reduced to tetravalent uranium U(IV) which precipitated uraninite. De Windt et al. (2005) showed that another iron-reducing bacterium, *Shewanella oneidensis* in the presence of formate as the electron donor, reduced Pd(II) to Pd(0) nanoparticles on the cell wall and inside the periplasmic space. Table 2.1 shows the list of bacteria reported for synthesis of various nanomaterials.

Table 2.1 List of bacteria reported for synthesizing nanomaterials

Name of bacterium	Nanoparticles	Localization	References
Pseudomonas stutzeri	Se	Periplasmic space	Lortie et al. (1992)
Clostridium thermoaceticum	CdS	Cell surface	Cunningham and Lundie (1993)
Rhodospirillum rubrum	Se	Cell outside and inside	Kessi et al. (1995)
Wolinella succinogenes	Se	Cell outside and inside	Tomei et al. (1995)
Enterobacter cloacae SLD1a-1	Se	Cell outside and inside	Losi and Frankenberger (1997)
Klebsiella pneumonia	CdS	Cell surface	Smith et al. (1998)
Pseudomonas stutzeri AG259	Ag, Ag_2S	Periplasmic space	Joerger et al. (2000)
Bacillus licheniformis	Ag	Periplasmic space	Joerger et al. (2000)
Lactobacillus sp.	Ag	Contour	Fu et al. (2000)
Desulfobacteraceae	ZnS	Cell outside	Labrenz et al. (2000)
Plectonema boryanum UTEX485	Au	Membrane vesicles	Kashefi et al. (2001)
Stenotrophomonas maltophilia			
Pyrobaculum islandicum			
Thermotoga maritima			
G. sulfurreducens			
Pyrococcus furiosus			
Geobacter ferrireducens			
Escherichia coli	CdS	Cell inside	Wang et al. (2001)
Sulfate-reducing bacteria	Magnetite	Cell outside	Watson et al. (2001)
Desulfovibrio desulfuricans	Pd	Cell outside	Yong et al. (2002)
Lactobacillus strains	Ag, Au, Ag–Au	Cell outside	Nair and Pradeep (2002)
Thermomonospora sp.	Au	Cell outside	Ahmad et al. (2003a, 2003b)
Escherichia coli	CdS	Cell inside	Sweeney et al. (2004)
Corynebacterium sp. *SH09*	Ag	Cell wall	Zhang et al. (2005)
Stenotrophomonas sp.	Se	Cell outside	Gregorio et al. (2005)
Stenotrophomonas maltophilia SELTE02	Se	Cell outside and inside	Gregorio et al. (2005)
Aeromonas sp. *SH10*	Ag	Cell inside	Mouxing et al. (2006)
Escherichia coli DH5α	Au	Cell surface	Du et al. (2007)
Rhodobacter capsulatus	Au	Cell outside	Feng et al. (2007)
Shewanella algae	Pt	Cell envelope	Konishi et al. (2007a)
Pseudomonas aeruginosa	Au	Cell outside	Husseiny et al. (2007)
Rhodopseudomonas capsulate	Au	Cell outside	He et al. (2008)
P. aeruginosa SNT1	Se	Contour	Yadav et al. (2008)

(continued)

Table 2.1 (continued)

Name of bacterium	Nanoparticles	Localization	References
Acetobacter xylinum	Ag	Cellulose fiber	Barud et al. (2008)
Morganella sp.	Ag	Cell outside	Parekh et al. (2008)
Bacillus licheniformis	Ag	Cell outside	Kalishwaralal et al. (2008)
Rhodobacter sp. *strain SW2*	nano-goethite	Cell outside	Miot et al. (2009)
Escherichia coli	Ag	Cell outside	Gurunathan et al. (2009)
Streptomyces hygroscopicus	Ag	Cell outside	Sadhasivam et al. (2010)
Streptomyces sp. *ERI-3*	Ag	Cell outside	Faghri Zonooz and Salouti (2011)
Magnetospirillum gryphiswaldense	Au	Membrane-enclosed	Cai et al. (2011)
Streptomyces sp. *BDUKAS10*	Ag	Cell outside	Sivalingama et al. (2012)
Streptomyces ERI-3	Au	Cell outside	Faghri Zonooz et al. (2012)
Aeromonas hydrophila	ZnO	Cell outside	Jayaseelan et al. (2013)
Escherichia coli K12	Au	Cell outside	Sirvastava et al. (2013)
Bacillus megaterium	Ag	Cell outside	Shivai Karkaj et al. (2013)
Bacillus spp.	Ag	Cell outside	Elbeshehy et al. (2015)
Streptomyces fulvissimus	Au	Cell outside	Soltani Nejad et al. (2015)
Escherichia coli	Au	Cell outside	Gholami-Shabani et al. (2015)

2.3 Nanomaterial Biosynthesis by Fungi

Fungi are eukaryotic, spore-producing, achlorophyllous organisms that are usually filamentous, branched somatic structures with hyphae surrounded by cell walls. The mycosynthesis of nanomaterials, or myconanotechnology, is at the interface between mycology and nanotechnology and includes an exciting new applied interdisciplinary science with considerable potential due to the wide range and diversity of fungi (Moghaddam et al. 2015). The use of fungi is potentially exciting since they secrete large amounts of enzymes and are simpler to deal with in the laboratory (for this reason, *Streptomycetes* which have common features between fungi and bacteria have high potential to produce nanoparticles). Fungal systems or myconanofactories have already been exploited for the synthesis of nanostructures of silver, gold, zirconium, silica, titanium, iron (magnetite), and platinum (Sastry et al. 2003).

2.3.1 Gold Nanoparticles

The use of fungi in the synthesis of nanoparticles is relatively a new addition to the list of microorganisms capable of nanoparticle biosynthesis (Yadav et al. 2015).

Shankar et al. (2003) showed that an endophytic fungus (*Colletotrichum* sp.) growing in the geranium leaves, when exposed to aqueous chloroaurate ions, produced gold nanoparticles. *Trichothecium* sp. was found to accumulate gold nanoparticles intracellularly by Ahmad et al. (2005). When the fungus *Trichothecium* sp. was cultured in static condition, it reduced Au^{+3} to form gold nanoparticles.

The plant pathogenic fungal strain *Fusarium oxysporum* behaved differently (Anil Kumar et al. 2007). The reduction of the metal ions occurred extracellularly, resulting in the rapid formation of highly stable gold and silver nanoparticles. Apart from individual metal nanoparticles, it has been shown that when the biomass of *F. oxysporum* is exposed to equimolar solutions of $HAuCl_4$ and $AgNO_3$, highly stable Au–Ag alloy nanoparticles of varying mole fractions can be achieved. The extract of saprophytic straw mushroom fungus, *Volvariella volvacea* was also used to produce silver, gold, and gold–silver nanoparticles formed by co-reduction of both metal ions (Philip 2009). Intra- and extracellular biosynthesis of gold nanoparticles by *Epicoccum nigrum* (Sheikhloo et al. 2011) and *Rhizopus oryza* (Sheikhloo et al. 2012) was reported too (Fig. 2.4).

The intracellular synthesis of gold nanoparticles was achieved with the fungi *Penicillium chrysogenum* (Sheikhloo and Salouti 2011) and *Phoma macrostoma* (Sheikhloo and Salouti 2012) when the fungal biomass was exposed to aqueous $HAuCl_4$ solution (Fig. 2.5).

2.3.2 Silver Nanoparticles

Fungi such as *Fusarium semitectum* (Basavaraja et al. 2008), *Fusarium solani*, *Penicillium fellutanum*, *Cladosporium cladosporioides*, *Trichoderma viride*, *Coriolus versicolor*, and *Phanerochaete chrysosporium* (Gade et al. 2010) were

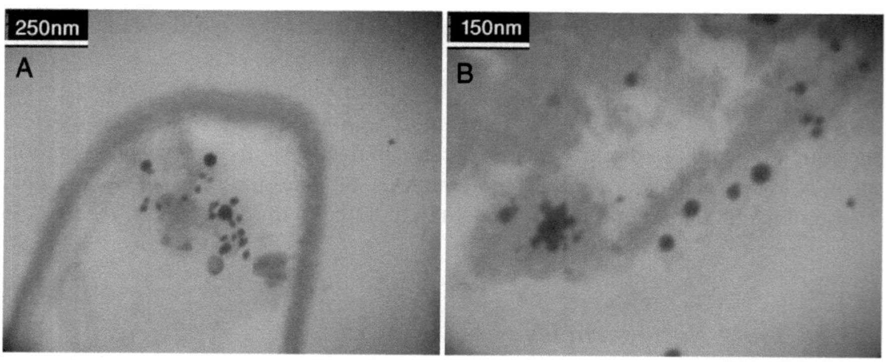

Fig. 2.4 TEM images of the thin sections of (**a**) *Rhizopus oryza*, and (**b**) *Eicoccum nigrum* cells after reaction with chloroaurate ions at different magnifications. Gold nanoparticles were synthesized both intra- and extracellularly

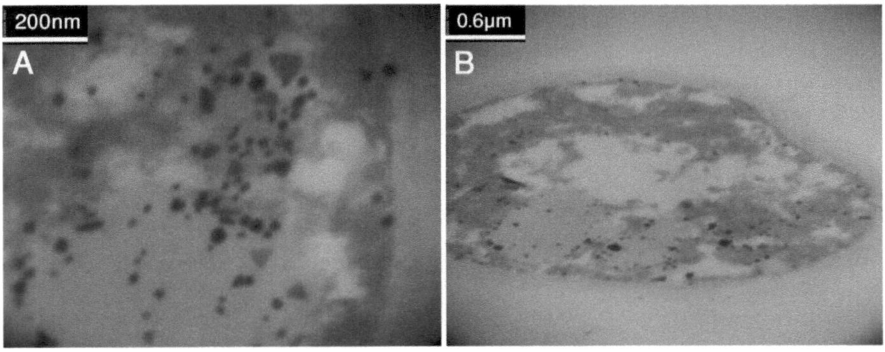

Fig. 2.5 TEM micrographs show the intracellular biosynthesis of gold nanoparticles by (**a**) *Penicillium chrysogenum* and (**b**) *Phoma macrostoma*

reported to be able to synthesize silver nanoparticles extracellularly when incubated with the aqueous solution of silver salt. The soil-born *Aspergillus fumigates* was reported to produce silver nanoparticles extracellularly (Kuber and D'Souza 2006; Ranjbar Navazi et al. 2010). Mukherjee et al. (2008) showed that the fungal filtrate of *Trichoderma asperellum* could synthesize spherical silver nanoparticles when challenged with silver nitrate. Similarly, *Aspergillus niger*, isolated from soil, was reported to produce silver nanoparticles (Gade et al. 2008). Extracellular mycosynthesis of silver nanoparticles by *Fusarium acuminatum* (USM-3793), isolated from infected ginger, was reported by Ingle et al. (2008). Also, Shaligram et al. (2009) demonstrated that when culture filtrate of *Penicillium brevicompactum* WA2315 was treated with silver ions, the ions were reduced to silver nanoparticles with spherical morphology and high stability for many weeks.

Mycelia-free water extract of a fungal strain, *Amylomyces rouxii* strain KSU-09 isolated from the roots of date palm (*Phoenix dactylifera*), suspended in water for 72 h facilitated the production of stable and spherical silver nanoparticles (Musarrat et al. 2010). Verma et al. (2010) reported the extracellular production of silver nanoparticles, polydispersed spherical or hexagonal particles ranging from 10 to 25 nm in size, by endophytic fungus *Aspergillus clavatus*, isolated from surface-sterilized stem tissues of *Azadirachta indica*, when incubated with an aqueous solution of 1 mM $AgNO_3$.

Extracellular biosynthesis of silver nanoparticle using mycelial mats of *Tricholoma crassum* (Ray et al. 2011) was resulted in production of monodispersed nanoparticles with high concentration. The nanoparticles were mostly spherical with a few of hexagonal shape. Guangquan et al. (2012) showed that silver nanoparticles were synthesized by reducing of aqueous Ag^+ ion with the culture supernatants of *Aspergillus terreus*. Extracellular synthesis of silver nanoparticles was reported by fungi *Penicillium chrysogenum* and *Penicillium expansum* by Mohammadi and Salouti (2013), when challenged with silver nitrate. The details of some fungi reported for the synthesis of nanomaterials are given in Table 2.2.

Table 2.2 List of fungi reported for synthesizing nanomaterials

Name of fungus	Nanoparticles	Localization	References
Fusarium oxysporum	CdS	Cell wall	Ahmad et al. (2002)
Colletotrichum sp.	Au	Cell inside	Shankar et al. (2003)
Verticillium sp.	Ag	Cell outside	Sastry et al. (2003)
Fusarium oxysporum	Zr	Cell outside	Bansal (2004)
Trichothecium sp.	Au	Cell inside and outside	Ahmad et al. (2005)
Aspergillus fumigates	Ag	Cell outside	Kuber and D'Souza (2006)
Fusarium oxysporum	Ag	Cell outside	Duran et al. (2007)
Capsicum annum	Ag	Cell outside	Li et al. (2007)
Fusarium oxysporum	Ag, Au, Ag–Au	Cell outside	Anil Kumar et al. (2007)
Fusarium oxysporum	Ag	Cell outside	Mohammadian et al. (2007)
Fusarium semitectum	Ag	Cell outside	Basavaraja et al. (2008)
T. asperellum	Ag	Cell outside	Mukherjee et al. (2008)
Fusarium semitectum	Ag	Cell outside	Basavaraja et al. (2008)
Penicillium brevicompactum WA 2315	Ag	Cell outside	Shaligram et al. (2009)
Volvariella volvacea	Au, Ag, Au–Ag	Cell outside	Philip (2009)
Amylomyces rouxii strain KSU-09	Ag	Cell outside	Musarrat et al. (2010)
Fusarium acuminatum	Ag	Cell outside	Gade et al. (2010)
Aspergillus clavatus	Ag	Cell outside	Verma et al. (2010)
Alternaria alternate	Se	Cell outside	Sarkar et al. (2011)
Epicoccum nigrum	Au	Cell inside	Sheikhloo et al. (2011)
Phoma macrostoma	Au	Cell inside	Sheikhloo and Salouti (2012)
Aspergillus flavus	TiO_2	Cell outside	Rajakumar et al. (2012)
Rhizopus oryza	Au	Cell outside	Sheikhloo et al. (2012)
Phoma macrostoma	Au	Cell outside	Sheikhloo and Salouti (2012)
Saccharomyces cerevisiae MTCC 2918	ZnS	Cell inside	Sandana Mala and Rose (2014)
Aspergillus terreus IF0	Au	Cell outside	Priyadarshini et al. (2014)
Penicillum chrysogenum	Ag	Cell outside	Mohammadi and Salouti (2015)
Penicillium expansum	Ag	Cell outside	Mohammadi and Salouti (2015)
Candida albicans	Ag	Cell outside	Saminathan (2015)
Curvularia lunata	Ag	Cell outside	Ramalingmam et al. (2015)
Aspergillus flavus	Au	Cell outside	Shakouri et al. (2016)

(continued)

Table 2.2 (continued)

Name of fungus	Nanoparticles	Localization	References
Trichoderma viride	Ag	Cell outside	Elgorban et al. (2016)
Rhizopus stolonifer	Ag	Cell outside	Abdel Rahim et al. (2016)

2.4 Nanomaterial Biosynthesis by Yeasts

Yeasts are fungi (eukaryotic microorganisms) that grow as single cells, although some species with yeast forms may become multicellular. They produce daughter cells either by budding (the budding yeasts) or by binary fission (the fission yeasts) and differ from most fungi, which grow as threadlike hyphae. But this distinction is not a fundamental one, because some fungi can alternate between a yeast phase and a hyphal phase, depending on environmental conditions (Moghaddam et al. 2015). The studies showed that yeasts are able to synthesize metal nanoparticles. For example, baker's yeast, *Saccharomyces cerevisiae*, was reported to biosorb and reduce Au^{+3} to elemental gold in the peptidoglycan layer of the cell wall in situ by the aldehyde group present in reducing sugars (Lin et al. 2005). Similarly, a number of different yeasts were reported to form gold nanoparticles of spherical, triangular, and hexagonal morphologies throughout the cell mainly in the cytoplasm (Gericke and Pinches 2006b). Agnihotri et al. (2009) reported that the tropical marine yeast, *Yarrowia lipolytica* NCIM 3589, was able to synthesize gold nanoparticles associated with the cell wall.

Though yeasts have been used to synthesize intracellular nanoparticles for several years, recently silver nanoparticles have been synthesized extracellularly by a silver-tolerant yeast strain, MKY3, when challenged with Ag^{+} ions in the log phase of growth. Extracellular biosynthesis of silver nanoparticles was obtained by cell-free extract (CFE) of *Candida viswanathii* by Kaler et al. (2011). The investigation of Mourato et al. (2011) showed that biosynthesis of Ag and Au nanoparticles was obtained using extremophilic yeasts isolated from acid mine drainage in Portugal. Table 2.3 shows the list of some yeasts reported for synthesizing various nanomaterials.

2.5 Nanomaterial Biosynthesis by Algae

Algae are a very large and diverse group of simple, typically autotrophic organisms, ranging from unicellular to multicellular forms and they are photosynthetic like plants. There are very few reports regarding alga-mediated synthesis of nanomaterials to date (Sicard et al. 2010). Torres de Araujo et al. (1986) reported the natural synthesis of magnetite nanoparticles by a magnetotactic alga of the genus *Anisonema* (Euglenophyceae) isolated from a coastal mangrove swamp in northeastern Brazil. Davis et al. (2003) reported production of nanoparticles by

Table 2.3 List of yeasts reported for synthesizing nanomaterials

Name of yeast	Nanoparticles	Localization	References
Schizosaccharomyces pombe Candida glabrata	CdS	Cell inside	Dameron et al. (1989)
Torulopsis sp.	PbS	Cell inside and outside	Kowshik et al. (2002)
Saccharomyces cerevisiae	Au	Cell wall	Lin et al. (2005)
Yarrowia lipolytica NCIM 3589	Au	Cell outside	Agnihotri et al. (2009)
Saccharomyces cerevisiae	Sb_2O_3	Cell outside	Jha et al. (2009)
Candida viswanathii	*Ag*	Cell outside	Kaler et al. (2011)
Extremophilic yeast strain	Ag, Au	Cell outside	Mourato et al. (2011)
Rhodosporidium diobovatum	PbS	Cell wall	Seshadri et al. (2011)
Candida diversa strain JA1	Ag, ZnO	Cell outside	Chauhan et al. (2014)
Saccharomyces cerevisiae	Ag	Cell outside	Roy et al. (2015)
Saccharomyces cerevisiae	Ag	Cell outside	Niknejad et al. (2015)

Chlorella vulgaris. Suspension of dried cells of alga *C. vulgaris* in $HAuCl_4$ solution accumulated elemental gold in the cells. A marine alga, *Sargassum wightii Greville*, was used by Singaravelu et al. (2007) to produce highly stable gold nanoparticles within short period in comparison with other biological methods. These extracellular nanoparticles were predominantly monodispersed and spherical. The presence of extracellular polysaccharides in seaweed *S. wightii* is 35 % which may facilitate the stabilization of nanoparticles. Common Anabaena and *Calothrix cyanobacteria* and *Klebsormidium* green microalgae were used to produce akaganeite β-FeOOH nanorods of well-controlled size intracellularly (Brayner et al. 2009). Rajasulochana et al. (2010) showed that the reaction of gold ions with the *Kappaphycus alvarezii* biomass under stationary conditions resulted in the rapid extracellular formation of gold nanoparticles. Barwal et al. (2011) showed that *Chlamydomonas reinhardtii* cell-free extract mediated synthesis of silver nanoparticles. *Turbinaria conoides* is a very common brown alga found throughout the Pacific and Indian Ocean and is known for its rigidity. Vijayaraghavan et al. (2011) showed that this alga was able to do biosorption and subsequent bioreduction of Au(III) to Au(0). Dahoumane et al. (2012) showed that a *Klebsormidium flaccidum* microalga was able to tolerate large metal concentrations to synthesize gold nanoparticles. Brayner et al. (2012) showed that *Euglena gracilis microalga was* able to biosynthesize two-line ferrihydrite nanoparticles intracellularly. The photosynthetic activity increase of *E. gracilis* after addition of Fe(II)/Fe (III) ions showed that the culture growth remains constant in the presence of iron ions. Table 2.4 shows the synthesis of different nanostructures by some algae.

Table 2.4 List of algae reported for synthesizing nanomaterials

Name of alga	Nanoparticles	Localization	References
Chlorella vulgaris	Au	Cell inside	Hosea et al. (1986)
Chlorella pyrenoidosa	Ag	Cell inside	Aziz et al. (2015)
Sargassum wightii Greville	Ag	Cell outside	Singaravelu et al. (2007)
Common *Anabaena Calothrix* cyanobacteria *Klebsormidium*	Akaganeite β-FeOOH	Cell outside	Brayner et al. (2009)
Klebsormidium flaccidum	Au	Cell inside	Sicard et al. (2010)
Kappaphycus alvarezii	Au	Cell outside	Rajasulochana et al. (2010)
Turbinaria conoides	Au	Cell outside	Vijayaraghavan et al. (2011)
Sargassum muticum	ZnO	Cell outside	Azizi et al. (2013)
Ecklonia cava	Au	Cell outside	Venkatesan et al. (2014)
Spirulina platensis	Ag	Cell outside	Sharma et al. (2015)
Turbinaria conoides Sargassum tenerrimum	Au	Cell outside	Ramakrishna et al. (2016)

2.6 Nanomaterial Biosynthesis by Plants

Plants, also called green plants, are living organisms of the kingdom Planate including multicellular groups. The rate of reduction of metal ions using plants has been found to be much faster as compared with other biological systems and the stable formation of metal nanoparticles has been reported. Generally, using plants for nanoparticle synthesis can be advantageous over other biological systems such as microbial route, because it eliminates the elaborate process of maintaining cell cultures and can also be suitably scaled up for large-scale synthesis of nanoparticles (Ahmed and Ikram 2015; Ahmed et al. 2016a). The various nanostructures synthesized by plants are mentioned below.

2.6.1 Gold Nanoparticles

The first report of plants synthesizing gold nanoparticles appeared when alfalfa seedling was shown to uptake gold from metal-enriched nutrient media (Gardea-Torresdey et al. 2002). This study demonstrated that Au(III) ions were reduced in the solid media to Au(0) by live alfalfa plant, and then the metal atoms were absorbed into the plant, where growth of nanoparticles took place. This method can also be very efficient in decontaminating soil polluted with heavy metal ions. Thus, plants now play an important role in the remediation of toxic metals through the reduction of metal ions. Singh et al. (2010) showed Au and Ag nanoparticle biosynthesis by reducing the aqueous solution of $AuCl_4$ and $AgNO_3$ with clove extract, called "Lavang" in India. One interesting aspect here is that reduction time is quite small (few minutes instead of hours as compared with other natural precursors). Sun-dried zero-calorie sweetener herb (*Stevia rebaudiana*) leaves were used by Mishra et al. (2010) for gold nanoparticle synthesis. Most of the previous researches on synthesis of nanostructures using plant extracts employed a broth resulting from boiling fresh plant leaves. It has been found that the sun-dried biomass has some advantages over the broth. The approach using the broth suffers from two main drawbacks. Firstly, most of the leaves are seasonal so that fresh leaves would not be readily available for the bioreduction all the time. Secondly, it is fairly difficult to control some parameters accurately such as the optimum boiling time when the broth is attained. Smitha et al. (2009) showed that *Cinnamomum zeylanicum* leaf broth was able to form gold nanocrystals. Philip (2010) showed that *Mangifera indica* leaf extract was able to form gold nanocrystals in spherical shape in relatively short time.

The reduction of gold ions was reported by Narayanan and Sakthivel (2010) using coriander leaf extract which resulted in the formation of stable gold nanoparticles extracellularly. The rate of reaction for the synthesis of nanoparticles by this method (12 h) was more rapid than the microbe-mediated synthesis (24–120 h) that had been reported earlier. Rapid green synthesis of gold nanoparticles was achieved using *Rosa hybrid* petal extract in 5 min at room temperature by Noruzi et al. (2011). This reaction is one of the most rapid reactions which have ever been reported in green synthesis of gold nanoparticles at room temperature.

2.6.2 Silver Nanoparticles

Various reports are available for the synthesis of AgNPs using different parts of the plant including leaves, bark, roots, flower, and fruits. It has been demonstrated that the extract of *Rosmarinus officinalis* L., commonly referred to as rosemary (belongs to mint family), is capable of producing silver nanoparticles and these particles are quite stable in solution (Sulaiman et al. 2013). *Rosmarinus officinalis* L. has a long list of claims pertaining to its medicinal usage including antibacterial and antioxidant properties (Minnunni et al. 1992; Karamanoli et al. 2000; Ozcan 2003; Faixov and Faix 2008).

The studies showed that leaf extract of *Chenopodium album* was able to perform one-step synthesis of silver nanoparticles (Amarendra and Krishna 2010). Kaviya et al. (2011) showed that peel extract of *Citrus sinensis* was able to synthesize silver nanoparticles. As the modern method using agriculture waste, silver nanoparticles were synthesized by employing an aqueous peel extract of *Annona squamosa* in $AgNO_3$. *Annona squamosa* (Annonaceae) is well known for its edible tropical fruits, custard apple, and mostly distributed in America and Asia (Kumar et al. 2012). Bioreduction of silver ions by *Coleus amboinicus Lour* leaf extract resulted in the synthesis of nanoparticles, extracellularly (Vadivel and Suja 2012).

C. amboinicus Lour, commonly known as Indian borage, is a tender fleshy perennial medicinal plant that contains many phytochemicals such as carvacrol (monoterpenoid), caryophyllene (bicyclic sesquiterpene), patchoulane, and flavanoids (quercetin, apigenin, luteolin, salvigenin, and genkwanin) which have enormous medicinal applications.

Silver nanoparticles were successfully synthesized from aqueous $AgNO_3$ solution through a simple green route using the leaf extract of *C. grandis* by Arunachalam et al. (2012) as a reducing as well as capping agent. *Coccinia grandis L.* (Cucurbitaceae) is a climbing perennial herb distributed almost all over the world.

Although the biosynthesis of nanomaterials by plants such as *Coccinia grandis* (Arunachalam et al. 2012), *Azadirachta indica* (Neem) (Shankar et al. 2004a), *Emblica officinalis*, *Tamarind*, *pear* fruit extract (Ghodake et al. 2010), *Pinus densiflora*, *Diospyros kaki*, *Ginkgo biloba*, *Magnolia kobus*, *Platanus orientalis* and *parthenium hysterophorus* L. (Thombre et al. 2013), *Terminalia Catappa* (TC), *Barbated Skullcup* (BS) herb extract, bark extract, *Crossandra infundibuliformis* (Kaviya et al. 2012), *Acalypha indica* (Krishnaraj et al. 2010), *Podophyllum hexandrum* (Jeyaraj et al. 2013), *Mentha piperita* (Rai et al. 2009), Geranium (*Pelargonium graveolens*) (Shankar et al. 2003), *Trachyspermum ammi* and *Papaver somniferum* (Vijayaraghavan et al. 2012), *Cassia fistula* (Liqin et al. 2010), *Banana* peel extract (Ashok et al. 2010), *Phyllanthus amarus* (Kasthuri et al. 2009), *Lantana camara* (Sivakumar et al. 2012), *Moringa oleifera* (Prasad and Elumalai 2011), *Catharanthus roseus* (Panneerselvam et al. 2012), *Eucalyptus hybrida* (Dubay et al. 2009), *Cassia auriculata* (Udayasoorian et al. 2011) and three categories of plants, (a) xerophyte (*Bryophyllum* sp.), (b) mesophyte (*Cyprus* sp.), and (c) hydrophyte (*Hydrilla* sp.), have been reported, possibilities in plant-mediated biological synthesis of nanomaterials have to be fully explored. Table 2.5 shows the synthesis of different nanostructures by some plants.

2.7 Semiconductor Biological Synthesis

A quantum dot (the so-called semiconductor) is a portion of matter whose excitons are confined in all three spatial dimensions. Consequently, such materials have electronic properties intermediate between those of bulk semiconductors and those of discrete molecules. Metal semiconductor nanocrystals have wide applications in various fields of research particularly in biomedical fields (Walling et al. 2009; Michalet et al. 2005; Kairdolf et al. 2013). CdS nanoparticles have been successfully synthesized using a wide range of organisms like algae (Scarano and Morelli 2003), fungi (Dameron et al. 1989; Ahmad et al. 2002), yeast (Roy et al. 2015), and bacteria (Prasad et al. 2007; Bai et al. 2009; Mubarak Ali et al. 2012) and plant (Liu et al. 2009).

Table 2.5 List of plants reported for synthesizing nanomaterials

Name of plant	Produced nanoparticles	Localization	References
Alfalfa	Au	Cell outside	Gardea-Torresdey et al. (2002)
Geranium leaf	Au	Cell outside	Shankar et al. (2003)
Neem (*Azadirachta indica*) leaf	Au, Ag, Au–Ag	Cell outside	Shankar et al. (2004a)
Lemongrass	Au	Cell outside	Shankar et al. (2004c)
Phyllanthin	Au, Ag	Cell outside	Kasthuri et al. (2009)
Clove	Au, Ag	Cell outside	Singh et al. (2010)
Stevia rebaudiana	Au	Cell outside	Mishra et al. (2010)
Mangifera indica leaf	Au	Cell outside	Philip (2010)
Pear fruit	Au	Cell outside	Ghodake et al. (2010)
Chenopodium album leaf	Ag	Cell outside	Amarendra and Krishna (2010)
Cassia fistula leaf	Ag	Cell outside	Liqin et al. (2010)
Mangosteen leaf	Ag	Cell outside	Veerasamy et al. (2011)
Mangosteen leaf	Ag	Cell outside	Veerasamy et al. (2011)
Citrus sinensis peel	Ag	Cell outside	Kaviya et al. (2011)
Annona squamosa peel	Ag	Cell outside	Mohana Roopan et al. (2012)
Coccinia grandis leaf	Ag	Cell outside	Arunachalam et al. (2012)
Crossandra infundibuliformis leaf	Au	Cell outside	Kaviya et al. (2012)
Rosmarinus officinalis	Ag	Cell outside	Sulaiman et al. (2013)
Parthenium hysterophorus L.	Ag	Cell outside	Thombre et al. (2013)
Podophyllum hexandrum	Ag	Cell outside	Jeyaraj et al. (2013)
Psidium guajava	TiO_2	Cell outside	Rajakumar et al. (2014)
Acorus calamus rhizome	Ag	Cell outside	Sudhakar et al. (2015)
Pelargonium Mentha	Au	Cell outside	Jafarizad et al. (2015)
Apple	Ag	Cell outside	Ali et al. (2016)
Azadirachta indica aqueous leaf	Ag	Cell outside	Ahmed et al. (2016b)
Momordica cochinchinensis Rhizome	Au	Cell outside	Lakshmanan et al. (2016)
Croton bonplandianum Baill. leaf	Ag	Cell outside	Khanra et al. (2016)

2.7.1 Semiconductor Biosynthesis by Bacteria

Quantum dots such as CdS and ZnS can be synthesized by bacteria. For example, Cunningham and Lundie (1993) demonstrated that *Clostridium thermoaceticum* in late-exponential to early-stationary phase precipitated bright yellow CdS crystals on the surface of cells as well as in the medium within 24–48 h after the addition of $CdCl_2$ and 0.05 % cysteine hydrochloride. A combination of geochemical and microbial processes led to ZnS biomineralization in a complex natural system (Labrenz et al. 2000). Synthesis of cadmium sulfide super lattices using self-assembled bacterial S-layers was reported by Shenton et al. (1997). Cadmium removal by a new strain of *Pseudomonas aeruginosa* in aerobic culture (Wang et al. 1997) and aerobic sulfide production and cadmium precipitation by *Escherichia coli* (Wang et al. 2001) were reported too.

Smith et al. (1998) produced "bio-semiconductor" CdS nanoparticles on the cell surface of *Klebsiella pneumoniae*. When *Klebsiella pneumoniae* was exposed to Cd^{2+} ions in the growth medium, CdS was formed on the cell surface. Some years later, Sweeney et al. (2004) showed that *E. coli* can intracellularly accumulate semiconductor nanocrystals by incubation with $CdCl_2$ and sodium sulfide. The production of nanocrystal was 20-fold higher when *E. coli* cells were grown to stationary phase compared with late logarithmic phase.

Another metalloid semiconductor, tellurium belongs to chalcogen family that has been reduced from tellurite to elemental tellurium by two anaerobic bacteria: *Bacillus selenitireducens* and *Sulfurospirillum barnesii* (Narayanan and Sakthivel 2010). In case of *B. selenitireducens*, initially formed nanorods were clustered together to form larger rosettes, but with *S. barnesii* small irregularly shaped extracellular nanospheres were formed.

Selenium, a nonmetal chemical element, with photo-optical and semiconducting properties has applications in photocopiers and microelectronic circuit devices. Few selenite- and selenate-respiring bacteria such as *Sulfurospirillum barnesii*, *B. selenitireducens*, and *Selenihalanaerobacter shriftii* were reported to synthesize extracellular stable uniform nanospheres of elemental selenium with monoclinic crystalline structures (Oremland et al. 2004). These bacteria have also produced small amounts of Se intracellularly. Yadav et al. (2008) showed that *Pseudomonas aeruginosa* SNT1, isolated from rhizospheric seleniferous soil, biosynthesized nanostructured selenium by biotransforming selenium oxyanions to spherical amorphous, red selenium both intra- and extracellularly. In addition, a facultative anaerobic bacterium, *Enterobacter cloacae* SLD1a-1 (Losi and Frankenberger 1997); a purple non-sulfur bacterium, *Rhodospirillum rubrum* in oxic and anoxic conditions (Kessi et al. 1995); and *Wolinella succinogenes* (Tomei et al. 1995) were found to bioreduce selenite to selenium both inside and outside the cell with various morphologies like spherical, fibrillar, and granular structure. *Escherichia coli* deposited elemental selenium both in periplasmic space and cytoplasm (Gerrard et al. 1974) and *Pseudomonas stutzeri* aerobically reduced selenite to elemental selenium (Lortie et al. 1992). *Stenotrophomonas maltophilia* SELTE02, a strain isolated from the rhizospheric soil of Astragalus bisulcatus, selenium hyper accumulator legume, showed promising transformation of selenite (SeO_3^{-2}) to elemental selenium (Se°) in the cytoplasm or in the extracellular space (Gregorio et al. 2005). *Rhodopseudomonas palustris*, *Gluconacetobacter xylinus*, and *Rhodobacter sphaeroides* were reported to synthesize CdS, ZnS, and PbS quantum dots (Narayanan and Sakthivel 2010; Shenton et al. 1997). Prasad et al. (2007) showed that titanium nanoparticles were produced extracellularly using the culture filtrate of *Lactobacillus* strains.

Hunter and Manter (2008) reported a bacterial strain *Tetrathiobacter kashmirensis*, which bioreduced selenite to elemental red selenium under aerobic conditions. A 90 kDa protein present in the cell-free extract was believed to be responsible for this bioreduction. Jayaseelan et al. (2013) found that *Aeromonas hydrophila* was able to synthesize ZnO nanoparticles with different morphologies.

2.7.2 Semiconductor Biosynthesis by Fungi

Besides metal nanoparticles, fungi can produce different metal sulfides and metal oxide nanoparticles. Even more exciting is that the exposure of *Fusarium oxysporum* to the aqueous $CdSO_4$ solution yields CdS quantum dots extracellularly (Ahmad et al. 2002). Another significant application of this fungus is in the synthesis of zirconia nanoparticles (Bansal 2004). The culture filtrate of the fungus, *Alternaria alternate*, was found to be able to reduce sodium selenate to produce selenium nanoparticles (Sarkar et al. 2011). Rajakumar et al. (2012) showed biosynthesis of TiO_2 nanoparticles by a novel procedure using *Aspergillus flavus* as a reducing and capping agent.

2.7.3 Semiconductor Biosynthesis by Yeasts

It has been recognized that among the eukaryotes, yeasts are explored mostly in the biosynthesis of the semiconductor nanoparticles. Dameron et al. (1989) showed that CdS quantum dots were synthesized intracellularly in *Schizosaccharomyces pombe* yeast cells that exhibited ideal diode characteristics. Biogenic CdS nanoparticles were used in the fabrication of a heterojunction with poly (ρ-phenylenevinylene). The synthesis of lead sulfide (PbS) nanocrystallites was found in *Torulopsis* species (Kowshik et al. 2002). The intracellular synthesis of stable and monodisperse PbS nanoparticles was obtained by a marine yeast, *Rhodosporidium diobovatum* (Seshadri et al. 2011).

2.7.4 Semiconductor Biosynthesis by Algae

There are very few investigations regarding semiconductor biosynthesis by algae. For example, phytochelation (PC)-coated CdS nanocrystallites were formed in a marine phytoplanktonic alga *Phaeodactylum tricornutum* in response to Cd (Mandal et al. 2006). Azizi et al. (2013) showed that brown marine macro alga

Sargassum muticum aqueous extract was able to synthesize zinc oxide nanoparticles.

2.8 Instruments for Nanomaterial Characterization

To understand the special properties of nanoparticle systems, it has become increasingly important to develop techniques for characterizing such materials at the nanometer level. The first question asked about nanomaterials (nanoparticles and quantum dots) is concerned with dispersity, aggregation state, size, and morphology. Among the techniques commonly used, transmission electron microscopy (TEM) is necessary for nanostructure studies. The particle growth can be directly seen by in situ observations. When energy-dispersive X-ray microanalysis (EDX) is used in conjunction with TEM, localized elemental information can be obtained (Gallezot and Leclereq 1994; Harada et al. 1994). High-resolution TEM (HRTEM) can provide information not only on the particle size and shape (similar to TEM), but also on the crystallography of the nanomaterials.

Gold, silver, and copper nanoparticles all have characteristic colors related with their particle size. Thus for these metals, observation of the UV-vis spectrum can be suitable. The UV-vis spectrum changes during the reduction can provide quite important information (Mulvaney et al. 1993; Yonezawa and Toshima 1995). The analysis of extended X-ray absorption fine structure (EXAFS) allows determination of local structural parameters, such as interatomic distance and coordination number, which are difficult to measure by any other method.

Infrared (IR) spectroscopy has been widely applied to the investigation of the surface chemistry of adsorbed small molecules (Ghosh et al. 2004). X-ray diffraction (XRD) gives structural information of nanoparticles, including qualitative elemental information (Faghri Zonooz et al. 2012). For a rationalization of catalytic properties, the surface composition and structure is indispensable information and quantitative X-ray photoelectron spectroscopy (XPS) is a powerful tool in the elucidation of the surface composition. One of the most revealing analytical methods for the study of composition of nanostructures is energy-dispersive X-ray spectroscopy (EDX), which is usually coupled with a transmission electron microscope with high resolution. EDX is a kind of electron probe microanalysis (EPMA) or X-ray microanalysis (XMA) method, which has higher sensitivity than the usual EPMA or XMA techniques. This method provides analytical data that cannot be obtained by the other three methods mentioned above.

Nuclear magnetic resonance (NMR) spectroscopy of metal isotopes is a powerful technique for understanding the electronic environment of metal atoms in nanomaterials by virtue of the NMR shifts caused by free electrons (knight shifts) (Bucherand and van der Link 1988).

Over the past decade, atomic force microscopy (AFM) and near-field optical scanning microscopy (NSOM) have evolved into new frontiers of science with significant impact on various areas of research. Several articles applying these techniques have been extensively reported in the literature (Lehenkari et al. 2000; Fotiadis et al. 2002). AFM can be used to obtain two- or three-dimensional images for a wide range of nanostructures. In addition, the relative heights of the structural features on the surfaces of objects obtained by AFM enable quantitative study of surface modifications (Dammer et al. 1995; Shi et al. 2001; Rotsch and Radmacher 2000). Near-field scanning optical microscope (NSOM), on the other hand, is an imaging technique that combines the high-resolution scanning probe microscopy (SPM) and fluorescence microscopy (Pohl et al. 1984; Betzig and Chichester 1993).

2.9 Nanomaterial Applications

Nanomaterials have many applications at different fields as mentioned below:

2.9.1 Applications of Metal Nanoparticles

The following are just some applications of most important metal nanoparticles (Halawani 2016; Taton 2002; Mandal et al. 2006):

2.9.1.1 Gold Nanoparticles

Electronics
Gold nanoparticles are designed for using as conductors from printable inks to electronic chips. Nanoscale gold nanoparticles are being used to connect resistors, conductors, and other elements of an electronic chip.

Photodynamic Therapy
Near-IR-absorbing gold nanoparticles (including gold nanoshells and gold nanorods) produce heat when excited by light at wavelengths from 700 to 800 nm. When light is applied to a tumor-containing gold nanoparticles, the particles rapidly heat up, killing tumor cells in a treatment also known as hyperthermia therapy.

Therapeutic Agent Delivery
The large surface area-to-volume ratio of gold nanoparticles enables their surface to be coated with hundreds of molecules (including therapeutics, targeting agents, and antifouling polymers).

Diagnostics
Gold nanoparticles are also used to detect biomarkers in the diagnosis of heart diseases, cancers, and infectious agents. They are also common in lateral flow immunoassays; a common household example is the home pregnancy test.

Catalysis
Gold nanoparticles are used as catalysts in a number of chemical reactions. The surface of a gold nanoparticle can be used for selective oxidation or in certain cases the surface can reduce a reaction.

2.9.1.2 Silver Nanoparticles

Diagnostic Applications
Silver nanoparticles are used in biosensors and numerous assays where they can be used as biological tags for quantitative detection.

Antibacterial Applications
Silver nanoparticles are incorporated in apparel, footwear, paints, wound dressings, appliances, cosmetics, and plastics for their antibacterial properties.

Optical Applications
Silver nanoparticles are used to efficiently harvest light and for enhanced optical spectroscopies including metal-enhanced fluorescence (MEF) and surface-enhanced Raman scattering (SERS).

Copper nanoparticle application research is ongoing to discover their potential dielectric, magnetic, electrical, optical, imaging, catalytic, biomedical, and bioscience properties (Gurunathan et al. 2013).

2.9.1.3 Magnetic Nanoparticles

The key applications of magnetite nanoparticles (iron oxide nanoparticles) are as follows:

- In magnetic resonance imaging (MRI) to provide enhanced contrast at very low concentrations in the nanomolar range for studying tumors
- As a targeted delivery vehicle and as a drug delivery coating for nanoscale anticancer drugs
- For magnetic data storage
- In coatings, plastics, nanowires, nanofibers, and textiles and in specific alloy and catalyst applications

2.10 Applications of Semiconductors

Semiconductors, tiny light-emitting particles on the nanometer scale, are rapidly emerging as a new class of fluorescent probes for biomolecular and cellular imaging (Chan et al. 2002; Kairdolf et al. 2013). In comparison with organic dyes and fluorescent proteins, quantum dots have unique optical and electronic properties such as size-tunable light emission, improved signal brightness, resistance against photobleaching, and simultaneous excitation of multiple fluorescence colors (Bruchez et al. 1998; Chan and Nie 1998). Extensive research has been directed towards developing QDs for using in biodetection and bioimaging. In particular, using bioconjugated QDs as fluorescent probes, a research has achieved real-time imaging of single-cell surface receptors and noninvasive detection of small tumors in live animal models (Dahan et al. 2003; Gao et al. 2004). A further advantage is that multicolor QD probes can be used to image and track multiple molecular targets simultaneously. This is a very important feature because most complex human diseases such as cancer and atherosclerosis involve a large number of genes and proteins. QDs have applications in drug delivery to living cells.

2.11 The Synthesis Location of Biogenic Nanomaterials

Intracellular or extracellular microbial synthesis of nanomaterials depends on the localization of the reductive components of the cell. In this respect, nanomaterial biosynthesis is classified as either intracellular or extracellular.

2.11.1 Intracellular Biosynthesis

Intracellular biosynthesis requires in vivo synthesis in the cells, which is a time-limiting factor. Detoxification process of hazardous materials which are mediated by some enzymatic reactions may be involved in bioreduction of metals and their deposition inside of the cells. In this mode, nanostructures should be separated from the cells after getting synthesized by a designed method (Mata et al. 2009). The sophisticated instruments would be required to isolate the nanostructures from the biomass into the cell-free filtrate. It is possible to release the intracellular nanostructures via ultrasound treatment of the biomass-nanostructure composite or via reaction with suitable detergents into the cell-free filtrate. Some traditional methods can be used for final recovery of nanomaterials from the cell-free filtrate including precipitation and filtration, electrochemical treatments, reverse osmosis, ion-exchange resins, and evaporations (Sathishkumar et al. 2009).

The biosynthetic nanomaterials that are found attached to the cell surface (cell membrane) are also classified as intracellular synthesis and therefore can be recovered through sonication or using detergents (Yashhiro 2006). Magnetotactic bacteria have the ability to navigate along the Earth's magnetic field (Blakemore 1975), a distinct feature, which facilitates the separation of nanostructures from first bacteria and later aqueous solution. Although in case of intracellular production the accumulated particles are of particular dimension and with less polydispersity, extracellular production of nanomaterials has more commercial applications in various biomedical fields than intracellular synthesis (Vadivel and Suja 2012).

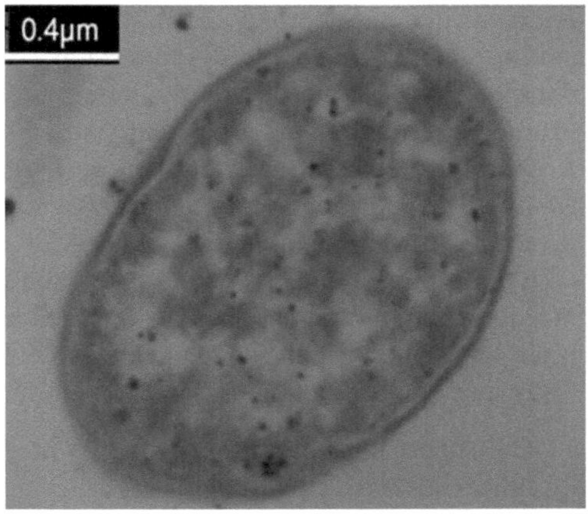

Fig. 2.6 TEM micrograph shows the production of silver nanoparticles by *Bacillus* sp. intra- and extracellularly

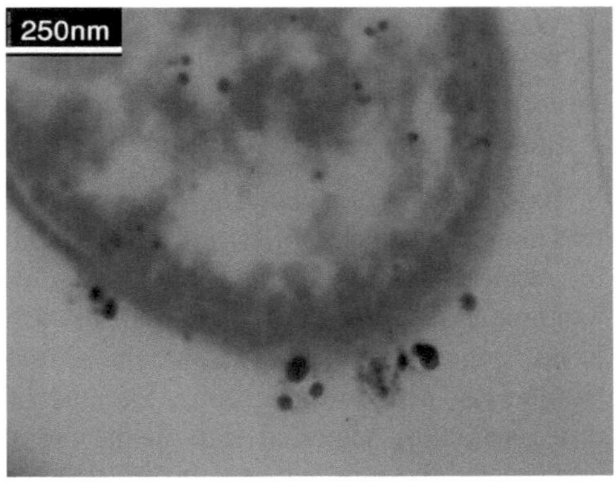

Fig. 2.7 TEM micrograph shows the formation of silver nanoparticles by *Bacillus* sp. attached to the cell surface

The location of nanomaterial deposition can be controlled by optimization of biosynthesis process conditions. For example, in synthesizing gold nanoparticles by mesophilic bacterium *Shewanella algae* at pH 7, gold nanoparticles were synthesized in the periplasmic space of *S. algae* cells. When the solution pH was decreased to 1, gold nanoparticles were precipitated outside of cells (Konishi et al. 2007b). Figure 2.6 shows intracellular biosynthesis of silver nanomaterials and Fig. 2.7 shows the formation of silver nanoparticles attached to the cell surface by *Bacillus* sp.

2.11.2 *Extracellular Biosynthesis*

When the cell wall-reductive enzymes or soluble secreted components are involved in the reductive process, it is obvious to find the nanomaterials extracellularly

(Narayanan and Sakthivel 2010). In extracellular biosynthesis, two different preparation statuses are observed: rapid synthesis and slow synthesis. The former can be done in a few minutes, while the latter requires several hours or even days. As an example, formation of silver nanoparticles in 5 min by culture supernatant of *Klebsiella pneumonia* is classified as a rapid synthesis, while its formation in 24 h by mycelia mat of *Phaenerochaete chrysosporiom* is classified as a slow synthesis. In extracellular biosynthesis of nanomaterials, the synthesis does not take place in biological growth phases and does not require in vivo conditions. In this approach, biological reagents required for the biosynthesis are presented in the bioliquids (Mollazadeh Moghaddam 2010). In extracellular biosynthesis, any of the following options can be used instead of the culture as a matrix solution:

1. The supernatant of culture which is prepared from centrifuging the biological culture after its growth.
2. Sterile supernatant which comes from sterilizing the supernatant by filtering that makes it completely free from microbes.
3. Water containing cell biomass.
4. Water which has kept biomass for 1 day: In this mode the biological materials are released from the biomass into the water, and this water can be used as a biosynthetic factor for creating the nanostructures.

Extracellular biosynthesis, in comparison with intracellular biosynthesis, has two main advantages. First, since the nanomaterials are formed inside the biomass by intracellular mode, there is an additional step of processing to release the nanomaterials from the biomass. In extracellular biosynthesis however, this step is no longer necessary. Second, the extracellular biosynthesis is a cheaper and simpler downstream processing. Because of these advantages, and because sometimes achieving nanomaterials in the biomass is not feasible in the intracellular method, much focus has been given to the development of extracellular process for biosynthesis of nanostructures. Figure 2.8 shows extracellular biosynthesis of gold nanoparticles by *Streptomyces sp. ERI-3* and Fig. 2.9 shows extracellular biosynthesis of gold nanoparticles by *Streptomyces griseus*.

2.12 Mechanisms and Involved Agents in Nanomaterial Biosynthesis

The exact reaction mechanism leading to the formation of nanomaterials by biological systems is yet to be elucidated. The ability of these systems to nanostructure synthesis might result from specific mechanisms of resistance. The mechanisms include efflux systems, alteration of solubility and toxicity via reduction or oxidation, bioabsorption, bioaccumulation, extracellular complexation or precipitation of metals, and lack of specific metal transport systems. For example, one mechanism that is used to create metal nanoparticles by microorganisms is bioreduction.

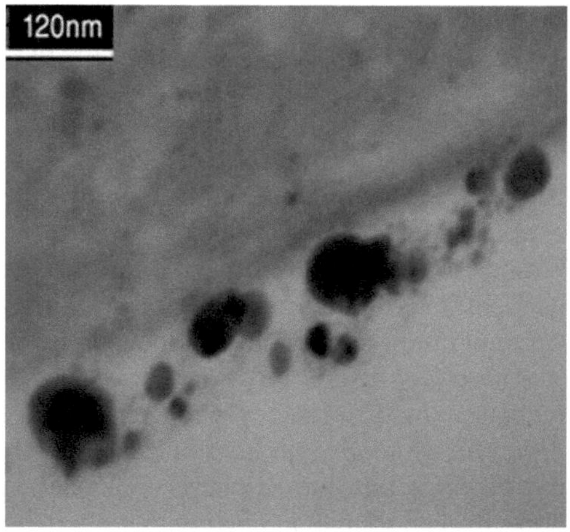

Fig. 2.8 TEM micrograph shows the production of gold nanoparticles by *Streptomyces sp. ERI-3* extracellularly

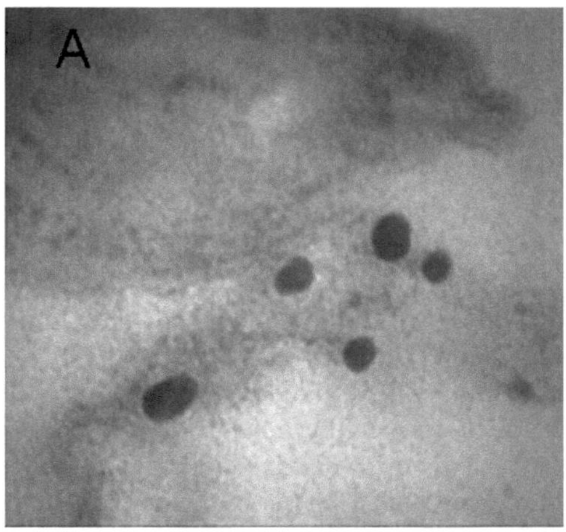

Fig. 2.9 TEM micrograph shows the formation of gold nanoparticles by *Streptomyces griseus* extracellularly

Although it is known that biological systems play an important role in remediation of toxic metals through reduction of the metal ions, investigation of biosynthesis mechanism was considered interesting recently. Agents involved in biosynthesis process of nanomaterials are classified into different kinds, such as enzymes, proteins, and carbohydrates (Mollazadeh Moghaddam 2010).

2.12.1 Enzymes

For the probable reduction mechanisms during nanomaterial biosynthesis, there are different understandings. Ahmad et al. (2003a) postulated an enzymatic process involving certain NADH-dependent reductase for Ag nanoparticle synthesis. Similarly, Kumar et al. (2007) suggested the likely role of α-NADPH-dependent sulfite reductase in the reduction of $AuCl_4^-$ to zero-valent gold. Based on the same experiments, it was found that the gold nanoparticles synthesized by *Shewanella oneidensis* are likely fabricated by the aid of reducing enzymes present in the bacterial cell membrane and capped by a detachable protein/peptide coat (Suresh et al. 2011). As these particles were negatively charged, which could be one of the reasons for their long-term stability, the electrostatic repulsive forces between the nanoparticles might protect them from getting closer, thereby preventing them from agglomeration or clumping in aqueous suspension.

The effect of the nitrogen source on the cellular activity of ferredoxin-nitrate reductase in different cyanobacteria was examined. In the unicellular species *Anacystis nidulans*, nitrate reductase was repressed in the presence of ammonium but de novo enzyme synthesis took place in media containing either nitrate or non-nitrogen source, indicating that nitrate was not required as an obligate inducer. Nitrate reductase in *A. nidulans* was freed from ammonium repression by L-methionine-D, L-sulfoximine, an irreversible inhibitor of glutamine synthetase. Ammonium-promoted repression appears therefore to be indirect; ammonium has to be metabolized through glutamine synthetase to be effective in the repression of nitrate reductase. Unlike the situation in *A. nidulans*, nitrate appeared to play an active role in nitrate reductase synthesis in the filamentous nitrogen-fixing strains *Anabaena* sp. strain 7119 and Nostoc sp. strain 6719, with ammonium acting as an antagonist with regard to nitrate (Herrero et al. 1981). Another report suggested that nitrate reductase is the enzyme responsible for the synthesis of AgNPs (Konohana et al. 1993).

Among hyperthermophilic and mesophilic dissimilatory Fe(III)-reducing bacteria and archaea like *Pyrobaculum islandicum*, *Thermotoga maritima*, *G. sulfurreducens*, and *Pyrococcus furiosus*, gold was precipitated by reducing gold(III) to metallic gold in the presence of hydrogen (Kashefi et al. 2001). In *Aeromonas hydrophila*, some organic groups, such as amide and ionized carboxyl in the cell wall, play an important role in the process of ZnO biosorption and the pH-sensitive oxidoreductase enzymes act as reducing agents (Jayaseelan et al. 2013).

Fungi can also accumulate metals by physicochemical and biological mechanisms including extracellular binding by metabolites and polymers, binding to specific polypeptides, and metabolism-dependent accumulation. Fungal species can readily synthesize metal nanoparticles extracellularly using high levels of nitrate reductase. This reductase gains electrons from NADH and oxidizes it to NAD^+. The enzyme is then oxidized by the simultaneous reduction of metal ions. The nitrate reductase was apparently essential for ferric iron reduction. Many fungi that exhibit these characteristic properties, in general, are capable of reducing Au (III) or Ag(I). The enzymatic route of synthesis of silver hydrosol using α-NADPH-dependent nitrate reductase from *Fusarium oxysporum* with capping peptide, phytochelatin, was demonstrated recently. Duran et al. (2007) reported that apart from enzymes, quinine derivatives of naphthoquinones and anthraquinones also act as redox centers in the formation of silver nanoparticles. A similar finding was also reported in the reduction of gold(III) chloride to metallic gold by α-NADPH-dependent sulfite reductase and phytochelatin (Anil Kumar et al. 2007). Gholami-Shabani et al. (2015) developed a cell-free viable approach for synthesis of gold nanoparticles using α-NADPH-dependent sulfite reductase purified from *Escherichia coli*. A dimeric hydrogenase enzyme of *F. oxysporum* that showed optimum activity at pH 7.5 and 38 °C and passively reduced H_2PtCl_6 to platinum nanoparticles was also reported (Govender et al. 2010). In conclusion, the studies of enzyme structure and the genes which code these enzymes may help to improve our understanding of how nanomaterial biosynthesis is carried out.

2.12.2 Proteins

Protein assays show that the proteins are as primary biomolecules in the biogenic entities involved in providing function of Au(III) reduction (Li et al. 2007). The results with *Capsicum annuum* L. extract indicated that the proteins which have amine groups play a reducing and controlling role in the formation of AgNPs in the solution, and that the secondary structure of the proteins changes after reaction with Ag^+ ions. Similarly the phenomenon of a change in secondary structure of proteins was also reported during rapid synthesis of metallic nanoparticles of silver by reduction of aqueous Ag^+ ions using the culture supernatant of *Klebsiella pneumonia*, *Escherichia coli*, and *Enterobacter cloacae* (*Enterobacteriaceae*) (Shahverdi et al. 2007).

In another work by Xie et al. (2007), single-crystalline gold nanoplates were produced by treating an aqueous solution of chloroauric acid with the extract of the unicellular green alga *Chlorella vulgaris* at room temperature. The results suggest proteins as the primary biomolecules involved in providing the dual function of Au (III) reduction and the size- and shape-controlled synthesis of the nanogold crystals. A 29 kDa "gold shape-directing protein (GSP)" present in the extract of green algae *Chlorella vulgaris* was also used in the bioreduction and in the synthesis of shape- and size-controlled distinctive triangular and hexagonal gold nanoparticles (Thakkar et al. 2009). A controlled and up-scalable route for the biosynthesis of silver nanoparticles mediated by fungal proteins of *Coriolus versicolor* has been undertaken for the first time by Sanghi and Verma (2009).

All magnetotactic bacteria contain magnetosomes, which are intracellular structures comprising magnetic iron mineral crystals. The biomineralization of magnetosome particles is achieved by a complex mechanism that involves the uptake and accumulation of iron and the deposition of the mineral particle with a specific size and morphology. The *mam* genes appear to be conserved in a large gene cluster within several magnetotactic bacteria (*Magnetospirillum* species and strain MC-1) and may be involved in magnetite biosynthesis (Schuler 1999). At magnetotactic microorganisms, it appears that the intracellular formation of ferrihydrite within ferritin protein is well established as the main strategy for iron storage by bacteria, algae, higher plants, and animals (Grunberg et al. 2001; Briat et al. 2010). A 90 kDa protein present in the cell-free extract of *Tetrathiobacter kashmirensis* was believed to be responsible for bioreduction to synthesize elemental selenium (Narayanan and Sakthivel 2010). Marshall et al. (2006) found that c-type cytochrome (*MtrC*) on the outer membrane of dissimilatory metal-reducing bacterium, *Shewanella oneidensis* MR-1, was involved in the reduction of U(VI). Kathiresan et al. (2009) isolated a rhizospheric fungus, *Penicillium fellutanum*, from mangrove root-soil of *Rhizophora annamalayana*, that was able to produce silver nanoparticles and it was found that the fungal protein of 70 kDa was involved in nanoparticle formation. Similarly, by employing the fungal isolate *P. brevicompactum* WA2315, the biosynthesis of silver nanoparticles was reported from compactin (Shaligram et al. 2009). In addition, Balaji et al. (2009) showed the production of extracellular silver nanoparticles by fungus *Cladosporium cladosporioides* by the same mode.

Cellular oxidoreductive proteins of *Chlamydomonas reinhardtii* control the biosynthesis of silver nanoparticles. In vivo-biosynthesized AgNPs by *C. reinhardtii* were localized in the peripheral cytoplasm and at one side of flagella root the site of pathway of ATP transport and its synthesis-related enzymes. This provides an evidence for the involvement of oxidoreductive proteins in biosynthesis and stabilization of AgNPs. Alterations in size distribution and decrease of synthesis rate of AgNPs in protein-depleted fractions confirmed the involvement of cellular proteins in AgNP biosynthesis. Spectroscopic and SDS-PAGE analyses indicated the association of various proteins of *Chlamydomonas reinhardtii* mediated in vivo and in vitro-biosynthesized AgNPs (Barwal et al. 2011). The studies showed that the peptides and/or proteins carried out the dual function of effective Au(III) reduction and successful capping of the GNPs by extracellular biological synthesis of nanoparticles using the aqueous extract of the brown algae *Laminaria japonica* (Ghodake and Lee 2011). The carotenoids of *Shewanella oneidensis* embedded in plasma membrane were found to be involved in the biosorption of Au $^{+3}$ to Au^0 on the plasma membrane and extracellular (Suresh et al. 2011). In addition, the algal pigments of *Turbinaria conoides* such as fucoxanthins, a kind of carotenoid rich in hydroxyl groups, could also have participated in the gold reduction (Vijayaraghavan et al. 2011). Therefore, according to the findings, proteins play a vital role in nanomaterial biosynthesis in some organisms.

2.12.3 Carbohydrates

Greene et al. (1986) reported that two steps were involved in Au(III) removal by bacterial biomass, namely, the redox reaction of Au(III) to Au(I) that follows by reducing Au(I) to Au(0). Shankar et al. (2004b) believed that the reduction of tetrachloroaurate ions was caused by the reducing sugar present in the geranium plant extract, while the growth of gold nanotriangle crystal was the result of an interaction between gold nanoparticles and aldehydes/ketones present in the extract.

2.12.4 Other Biological Agents

The studies showed that some other agents apart from the enzymes and proteins involve in biosynthesis process of nanoparticles (Wei et al. 2004; Rodriguez et al. 2008a). It was believed that terpenoids with Neem leaf broth are surface-active molecules that are responsible for stabilizing the nanoparticles, and that reaction of the metal ions is possibly facilitated by reducing sugars and/or terpenoids present in the broth (Shankar et al. 2004a).

It has been suggested that thymol gets adsorbed on the metal nanoparticle surface serving as a capping agent due to the π electrons (Vijayaraghavan et al. 2012). The main constituent in *Trachyspermum ammi* is thymol. Thymol is the predominant constituent in the carom seed which possesses aromatic ring in its structure. The corroborating reports suggested that terpenoids present in *Cinnamomum camphora* and *Cinnamomum burmannii* leaves were found to be involved in the biosynthesis of Ag nanoparticles (Huang et al. 2007; Shan et al. 2007). The crude extract leaves of *Diospyros peregrine*, *Coccinia grandis*, and *Swietenia macrophylla* contain triterpenoids, alkaloids, and tannins that act as reducing agents for the synthesis of Ag nanoparticles (Dewanjee et al. 2007). At bioreduction of silver ions by *Coleus amboinicus* leaf extract, the involvement of aromatic amines, amide (II) groups, and secondary alcohols was confirmed in capping and reduction of silver nanoparticles (Vadivel and Suja 2012). The reductive groups such as hydroly and aldehyde on the *Magnetospirillum gryphiswaldense* MSR-1 may also attribute to the reduction of Au(III) to Au(0) (Cai et al. 2011).

Apart from the synthesis of nanoparticles, the stabilization of synthesized particles is also important. Several stabilizing agents such as starch (Rodriguez et al. 2008a, b), chitosan (Mubarak Ali et al. 2012), 3-mercaptopropionic acid (MPA), mercaptosuccinic acid (MSA), and glutathione (GSH) have been used to stabilize CdS nanoparticles. These stabilizing agents provide the thiol group that would bind Cd through the S/H group. Similarly r-phycoerythrin has also this property (Brekhovskikh and Bekasova 2005). *Phormidium tenue*, a marine cyanobacterium, is a rich source of phycoerythrin, the C-phycoerythrin (Mubarak Ali et al. 2012).

Brown algae *Turbinaria conoides* mainly consist of alginic acid, which constitutes 10–40 % of the dry weight of algae (Davis et al. 2003). The alginic acids are linear carboxylated copolymers constituted by different proportions of 1,4-linked β-D-mannuronic acid (M-block) and α-L-guluronic acid (G-block). Among the different functional groups, carboxyl groups are abundant. Several investigators reported that these functional groups play an important role in metal biosorption at different pH conditions (Davis et al. 2000). Other functional groups present in the brown seaweed cell wall are amino, sulfonate, phosphonate, and hydroxyl groups (Vijayaraghavan et al. 2005).

The micrograph of virgin *T. conoides* by scanning electron microscopy revealed important information on the surface morphology (Vijayaraghavan et al. 2011). Surface protuberance and microstructures were observed, which is thought to be due to calcium and other salt crystalloid deposition. After Au binding, the surface of *T. conoides* appeared flattened in comparison with the raw sample. This supports our earlier explanation that when virgin *T. conoides* is exposed to Au solution at acidic pH, H^+ ions may replace some of the alkali and alkaline earth metals naturally present in the cell wall through ion-exchange mechanism. The second stage, which was started 1 h after *T. conoides* contacted with $HAuCl_4$ solution, involved reduction of Au(III) to Au(0) on the surface of *T. conoides*. The mechanism of Au(III) reduction most likely involved oxidation of biomass groups such as hydroxyl. Hydroxyl groups are very abundant in polysaccharides of the brown algal cell wall. At the end, besides the primary explanation about probable mechanisms of nanomaterial biosynthesis, many things about the biochemical and molecular mechanism of these processes remain unknown that should be revealed.

2.13 Optimization of Nanomaterial Biosynthesis

The control of size, shape, dispersity, and composition of nanomaterials during synthesis is an important criterion in the area of nanoparticle biosynthesis (Gericke and Pinches 2006a, b; Shakouri et al. 2016). Depending on the size of nanoparticles, their applications branch out, because nanoparticles present a higher surface-to-volume ratio with decreasing size. Particular emphasis has been placed on the control of shape, because in many cases it allows properties to be fine-tuned with a greater versatility that gives the particles a unique nature (Gurunathan et al. 2009).

Since the polydispersity is a major concern, it is important to optimize the conditions for monodispersity in a biological process. Regarding the biological synthesis approaches, the insights gained from strain selection, optimizing the conditions such as pH, incubation temperature and time, concentration of metal ions (substrate), and the amount of biological material has come up to give hope in implementation of these approaches in large scale and high quality and for commercial applications. There are also the possibilities of producing genetically engineered microbes that overexpress specific reducing agents and thereby enhance formation rate of biological nanoparticles. One of the options to achieve desirable production of biogenic nanomaterials is to change the different parameters of biosynthetic reaction (Narayanan and Sakthivel 2010; Kalishwaralal et al. 2010). Effect of different parameters on the synthesis of nanomaterials can be described as below:

2.13.1 Concentration of Substrate (Initial Metal Solution)

The possibility of controlling the reaction rate and particle size was investigated by changing the composition of the reaction mixture. In order to confirm whether the concentration of $AgNO_3$ plays an important role in the synthesis rate and size of nanoparticles by the culture supernatant of *Escherichia coli* or not, the different concentrations of $AgNO_3$ were used. The maximum synthesis of AgNPs was performed with respect to Ag^+ ion concentration in the range of 1–10 mM. This was reflected with an increase in the nanoparticle production up to a concentration of 5 mM; however, the production rate was found to decrease at higher concentrations of Ag^+ ions. The control experiments involving different concentrations of $AgNO_3$ showed no synthesis of nanoparticles. The results clearly indicated that 5 mM concentration of Ag^+ ions was most appropriate for the maximum synthesis of AgNPs from the culture supernatant of *E. coli* (Gurunathan et al. 2009). TEM images showed that the size of AgNPs decreased with increasing concentrations of $AgNO_3$. However, when the concentration of $AgNO_3$ was more than 5 mM, the size of AgNPs was altered. The increase in concentration of $AgNO_3$ up to 5 mM resulted in size-controlled synthesis with the particle size being around 15 nm as revealed by the particle size analysis. This indicated that the size of AgNPs can be modulated by the concentration of $AgNO_3$.

In the recent study, the culture medium of *Streptomyces* sp. ERI-3 was furnished with optimal concentration of $HAuCl_4$ that resulted in a high yield of gold nanoparticles. The effect of different concentrations (1.5–4 mM) showed that 3 mM $HAuCl_4$ was optimum concentration for maximum synthesis of gold nanoparticles (Faghri Zonooz et al. 2012). The effect of the initial metal concentration on the gold nanoparticle formation by *Magnetospirillum gryphiswaldense* was studied within a range of 20–600 mg/l. The results of TEM showed that the initial concentration had a clear effect on the size of gold nanoparticles. As the initial concentration was increased, the average diameter of gold nanoparticles was increased from 12 to 50 nm (Cai et al. 2011).

In optimization of conditions for the synthesis of an isotrophic gold nanostructures using cell-free extract of *Rhodopseudomonas capsulata*, different concentrations of gold ions were used. At lower concentration, spherical gold nanoparticles were produced in the size of 10–20 nm. But highly networked structures of gold nanowires with 50–60 nm were synthesized with higher concentrations of gold ions (He et al. 2008). Thus, according to the mentioned examples, concentration of substrate is an important factor to control size, shape, and synthesis rate of nanomaterials.

2.13.2 Reaction Temperature

The studies showed that the optimum temperature for cell growth and nanomaterial accumulation was different. For example, at the $AgNO_3$ concentration of 5 mM, it was evident that increasing the temperature of the reaction solution, up to 60 °C, results in an increase in the synthesis rate of AgNPs by *E. coli*. The enhanced rate of synthesis of AgNPs might be the direct result of the effect of temperature on a key enzyme present in the culture supernatant of *E. coli*. The formation rate of AgNPs was related to the incubation temperature of the reaction mixture and the particle growth increase at a higher rate with increasing the temperature levels (Gurunathan et al. 2009). Faghri Zonooz et al. (2012) reported that different temperatures affect the rate of GNP synthesis by the culture supernatant of *Streptomyces* sp. ERI-3. The results of this study showed that 30 °C was the best temperature for maximum synthesis of GNPs at the shortest time. According to above examples, a decrease in the yield of nanoparticles was observed when the incubation temperature was higher or lower than the optimum incubation temperature. Therefore, reaction temperature plays an important role in enhancing the biosynthesis rate of nanoparticles.

2.13.3 pH of the Reaction Mixture

In general, the reduction of metallic ions is sensitive to the pH of the reaction solution as it may affect the morphology of the product via the formation of certain nanoparticles. In formation of AgNPs by the culture supernatant of *E. coli*, the concentration of $AgNO_3$ was maintained at 5 mM and the reaction temperature at 60 °C; when pH was increased from 8 to 12, the maximum synthesis was observed at pH 10, the time of synthesis greatly reduced, and the yield of AgNPs was significantly enhanced. Therefore, this study showed that the optimum pH for synthesis of AgNPs was 10. This was in agreement with the earlier reports saying that the addition of an alkaline ion is necessary to carry out the reductive reaction of metal ions (Gurunathan et al. 2009). In the absence of hydroxide ion, the time taken for reduction of Ag^+ ions was longer, indicating the requirement of OH^- ions for the reduction reaction. The effect of pH on nanoparticle synthesis has been explained previously in the case of bioreduction of trivalent aurum and the synthesis of platinum nanoparticles. Similarly in these experiments, when hydroxide ion was added, there was a rapid increase in silver conversion and the time taken was less than 30 min. In addition, the mean diameter of AgNPs increased when the pH of the reaction mixture was increased from 10 to 12.

The effect of pH on the formation of gold nanoparticles by *Magnetospirillum gryphiswaldense* was investigated within different pH ranges (1.5–4). The TEM characterization at invariable initial metal concentration of 80 mg/l showed that the size of gold nanoparticles on the surface of cells at different pH values was clearly different. As the pH increased, the gold particles exhibited larger particle diameters. Different pH conditions would regulate the proton concentration resulting in the control of gold nanoparticle size. In fact, the effect of pH can be explained by the surface charge of the bacteria. It was reported that the chemical groups, including amino, sulfhydryl, and carboxylic groups, on the cell walls might play an important role in reducing the Au(III) ions and the Au(III) ions could bind to biomass through these functional groups (Cai et al. 2011).

In synthesizing gold nanoparticles by mesophilic bacterium *Shewanella algae* at pH 7, gold nanoparticles of 10–20 nm were synthesized in the periplasmic space of *S. algae* cells. When pH of solution was decreased to 2, gold nanoparticles of about 20 nm were precipitated outside of the cells. So here, by controlling and optimization of pH, the size of biogenic gold particles and the location of its deposition can be controlled (Konishi et al. 2007b).

In another study, the influence of pH (4–8) on the production of gold nanoparticles by *Streptomyces* ERI-3 was studied (Faghri Zonooz et al. 2012). The results showed that the maximum production of nanoparticles was at pH 6. Moreover, by changing the pH of the reaction medium from 4 to 8, a variety of aqueous solution colors including greenish-blue, red, dark purple, light pink, and orange were resulted that were likely due to the gold nanoparticle formation with different morphologies and sizes. Figure 2.10 shows color changes of the gold nanoparticle solution synthesized by *Streptomyces* ERI-3 at various pH. According to the above explanations, the metabolic activities of microorganisms are very sensitive to changes in pH.

2.13.4 Incubation Time of the Reaction Mixture

The time interval lapsed since the beginning of the bioreductive reaction was found to be an important factor in controlling the morphology and size of biogenic nanomaterials. For example, it was reported that in production of nanoparticles by *Shewanella algae*, after 1 h, there was a large population of well-dispersed, spherical gold nanoparticles with the mean size of 9.6 nm. Gold nanoplates with an edge length of 100 nm were appeared after 6 h, and 60 % of the total nanoparticle population was due to gold nanoplates with an edge length of 100–200 nm after 24 h (Ogi et al. 2010). It was shown that reaction time was an important parameter in controlling the morphology and size of nanostructures synthesized by *Bacillus megaterium*. The studies showed that after 30 min of reaction time, the particle size and shape of the gold nanoparticles have not been obviously changed, and well-separated nanoparticles with occasional aggregation (<2.5 nm) can clearly be seen in the solution. After 1 h of reaction, there were small amounts of larger spherical gold particles (5–8 nm) in the solution. After 3 h of reaction, the produced gold particles were mostly spherical and irregular in shape. After 6 h of reaction, the gold nanoparticles were formed on the cell wall and in the solution as well as in the cytoplasmic membrane, and the number of gold nanoparticles was clearly higher in

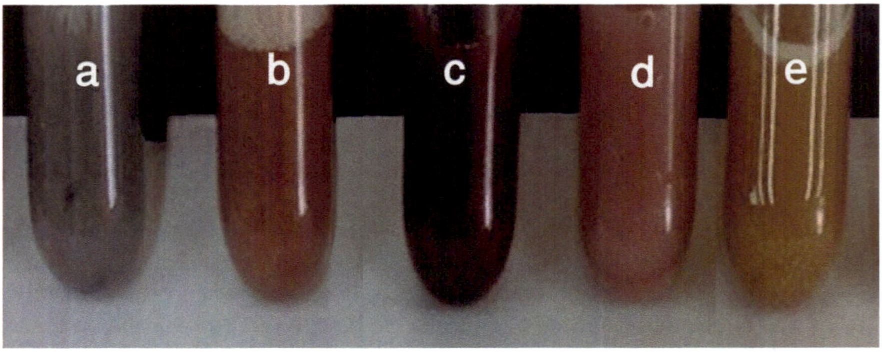

Fig. 2.10 Gold nanoparticle biosynthesis at different pH. The color change of the reaction solutions can be seen at various pH due to the formation of gold nanoparticles with different morphologies and sizes [(**a**) pH = 4, (**b**) pH = 5, (**c**) pH = 6, (**d**) pH = 7, (**e**) pH = 8]

the cell wall than in the cytoplasmic membrane. Some larger hexagonal, triangular, spherical, and anisotropic gold particles were observed in addition to some smaller spherical particles. In consequence, monodispersed spherical gold nanoparticles (<2.5 nm) via the modulation of reaction time could be formed by using *Bacillus megaterium* (Wen et al. 2009).

In another study, the influence of incubation time on the production of gold nanoparticles by *Streptomyces* ERI-3 showed that the reduction of the $Aucl_4^-$ ions with 3 mM concentration and pH 6 at 30 °C was nearly completed after 12 h (in comparison with 96 h before concentration optimization) (Faghri Zonooz et al. 2012). It is necessary to note that besides the effect of optimization processes on nanoparticle production, the shape and size of biogenic nanomaterials depend on the biological species involved. For example, geranium leaf broth reacted with aqueous chloroaurate ions induced a variety of gold nanoparticle shapes including rods, flat sheets, and triangles, while its endophytic fungus (*Colletotrichum* sp.) produced essentially spherical nanoparticles under the same conditions (Shankar et al. 2003). Yet another plant, lemon grass (leaf broth) induced the formation of a high percentage of single-crystalline gold nanotriangles under similar conditions. Quite surprisingly, any parts of the plant affect the nature of nanoparticle synthesis (Shankar et al. 2005). However, there is still much scope for improvement in bio-based methods for nanomaterial synthesis, particularly in relation to improving the monodispersity of the nanomaterials and modulating their size and shape, as well as in reducing the time required for nanomaterial synthesis.

Besides the optimization of nanomaterial biosynthesis conditions for shape, size, dispersity, and chemical composition using the mentioned findings, it is appropriate to obtain the highest biomass at the shortest time. The obvious example for this purpose is investigation of *Streptomyces* sp. ERI-3 growth rate in different culture media including Luria Bertani (LB), nutrient broth, YPD, YPG, M17, YT(2X), and terrific broth under various temperatures and pH (Faghri Zonooz et al. 2012). The measurement of the biomass growth of *Streptomyces* sp. ERI-3 spectrophotometrically by taking optical density reading at 650 nm after appropriate dilution of samples showed that bacterium had optimum growth in M17 medium, incubation temperature 28 °C, and pH 7. Thus, this medium was considered as main medium to obtain biomass of *Streptomyces* sp. ERI-3 for nanomaterial biosynthesis.

2.14 Industrial Production of Nanomaterials

One of the important challenges in nanomaterial production is scaling up laboratory processes to the industrial scale. Large-scale biological synthesis of nanoparticles has always been a big challenge, because the use of biomass for the large-scale synthesis requires various power-consuming processes such as harvesting the biomass. In addition, the purification of the nanoparticles is a difficult process.

Although currently there are no reports on very large quantity of synthesis of nanomaterials, some prototypes have been designed in continuous-flow reactors similar to chemical methods (Huang et al. 2008). In this regard, continuous-flow reactors are generally favored over batch reactors. Decomposition of organometallic precursor's inorganic solvents is one of the most commonly used approaches to nanoparticles because the narrow size distribution can be achieved in such reaction systems. A flow reactor, on the other hand, can generate products on a continuous basis once the reaction reaches steady state and is more appropriate for a large-scale production than the batch reactor. Generally, industrial production of nanoparticles needs further investigations.

2.15 Future Perspectives

Today, nanostructures have drawn the attention of scientists because of their extensive applications in new technologies. Actually, almost all kinds of biological systems can be used for nanoparticle and semiconductor production. It is apparently helpful using fungi for nanostructure biosynthesis in industrial and semi-industrial scales, because they form branching filaments of cells which become a network of strands called mycelium and secrete large amounts of enzymes (synthesizer agents of nanostructures). The bacteria can also find potential in nanostructure production similar to fungi by genetic modifications. Synthesis of nanomaterials can be much easier and faster using the plants, because there is no need to prepare cell culture and only different parts of plants should be prepared. However, cellular, biochemical, and molecular mechanisms that mediate the synthesis of biological nanomaterials should be studied in details to increase the rate of synthesis and improve properties of nanostructures. The future focus should assess the exact conditions for the industrial production of nanoparticles and semiconductors with biological methods. Multimodality imaging probes could be created by integrating semiconductors with paramagnetic or superparamagentic agents. By correlating the deep imaging capabilities of MRI with ultrasensitive optical imaging, a surgeon could visually identify tiny tumors or other small lesions during an operation and remove the diseased cells and tissue completely. Another desired multifunctional device would be the combination of a semiconductor imaging agent with a therapeutic agent. Not only would this allow tracking of pharmacokinetics, but also the diseased tissue could be treated and monitored simultaneously and in real time. These combinations are only a few possible achievements for the future.

References

Abdel Rahim K, Younis Mahmoud S, Mohamed Ali A (2016) Extracellular biosynthesis of silver nanoparticles using *Rhizopus stolonifer*. Saudi J Biol Sci (in press)

Agnihotri M, Joshi S, Kumar AR, Zinjarde S, Kulkarni S (2009) Biosynthesis of gold nanoparticles by the tropical marine yeast *Yarrowia lipolytica* NCIM 3589. J Mater Lett 63:1231–1234

Ahmad A, Mukherjee P, Mandal D, Senapati S, Khan MI, Kumar R, Sastry M (2002) Enzyme mediated extracellular synthesis of CdS nanoparticles by the fungus, *Fusarium oxysporum*. J Am Chem Soc 124:12108–12109

Ahmad A, Mukherjee P, Senapati S, Mandal D, Khan MI, Kumar R, Sastry M (2003a) Extracellular biosynthesis of silver nanoparticles using the fungus *Fusarium oxysporum*. Colloids Surf B Biointerfaces 28:313–318

Ahmad A, Senapati S, Khan MI, Kumar R, Sastry M (2003b) Extracellular biosynthesis of monodisperse gold nanoparticles by a novel extremophilic actinomycete, *Thermomonospora* sp. J Langmuir 19:3550–3553

Ahmad A, Senapati S, Khan MI, Kumar R, Sastry M (2005) Extra-intracellular biosynthesis of gold nanoparticles by an alkalotolerant fungus, *Trichothecium* sp. J Biomed Nanotechnol 1:47–53

Ahmed S, Ikram S (2015) Synthesis of gold nanoparticles using plant extract: an overview. Nano Res Appl 1:1–6

Ahmed S, Ahmad M, Swami BL, Ikram S (2016a) A review on plants extract mediated synthesis of silver nanoparticles for antimicrobial applications: a green expertise. J Adv Res 7:17–28

Ahmed S, Saifullah AM, Swami BL, Ikram S (2016b) Green synthesis of silver nanoparticles using *Azadirachta indica* aqueous leaf extract. J Rad Res Appl Sci 9:1–7

Ali Z, Yahya R, Sekaran S, Puteh R (2016) Green synthesis of silver nanoparticles using apple extract and its antibacterial properties. Adv Mater Sci Eng 2016:1–6

Amarendra DD, Krishna G (2010) Biosynthesis of silver and gold nanoparticles using *Chenopodium album* leaf extract. Colloids Surf A Physicochem Eng Asp 369:27–33

Anil Kumar S, Kazemian Abyaneh M, Gosavi SW, Kulkarni SK, Pasricha R, Ahmad A, Khan MI (2007) Nitrate reductase-mediated synthesis of silver nanoparticles from $AgNO_3$. Biotechnol Lett 29:439–445

Arunachalam R, Dhanasingh S, Kalimuthu B, Uthirappan M, Rose C, Mandal AB (2012) Phytosynthesis of silver nanoparticles using *Coccinia grandis* leaf extract and its application in the photocatalytic degradation. Colloids Surf B Biointerfaces 94:226–230

Ashok B, Bhagyashree J, Ameeta R, Smita Z (2010) Banana peel extract mediated novel route for the synthesis of silver nanoparticles. Colloids Surf A Physicochem Eng Asp 368:58–63

Aziz N, Faraz M, Pandey R, Shakir M, Fatma T, Varma A, Barman I, Prasad R (2015) Facile algae-derived route to biogenic silver nanoparticles: synthesis, antibacterial, and photocatalytic properties. J Langmuir 31:11605–12

Azizi S, Ahmad MB, Namvar F, Mohamad R (2013) Green biosynthesis and characterization of zinc oxide nanoparticles using brown marine macroalga *Sargassum muticum* aqueous extract. Mater Lett 116:275–277

Bai HJ, Zhang ZM, Guo Y, Yang GE (2009) Biosynthesis of cadmium sulfide nanoparticles by photosynthetic bacteria *Rhodopseudomonas palustris*. Colloids Surf B Biointerfaces 1:142–146

Balaji DS, Basavaraja S, Deshpande R, Mahesh BD, Prabhakar BK, Venkataraman A (2009) Extracellular biosynthesis of functionalized silver nanoparticles by strains of *Cladosporium cladosporioides* fungus. Colloids Surf B Biointerfaces 68:88–92

Bansal V (2004) Biosynthesis of zirconia nanoparticles using the fungus *Fusarium oxysporum*. J Mater Chem 14:3303–3305

Barud HS, Barrios C, Regiani T, Marques RFC, Verelst M, Dexpert-Ghys J (2008) Self-supported silver nanoparticles containing bacterial cellulose membranes. J Mater Sci Eng C 28:515–518

Barwal L, Ranjan P, Kateriya S, Yadav SC (2011) Cellular oxido-reductive proteins of *Chlamydomonas reinhardtii* control the biosynthesis of silver nanoparticles. J Nanobiotechnol 9:56–61

Basavaraja S, Balaji SD, Legashetty A, Rasab AH, Venkatraman A (2008) Extracellular biosynthesis of silver nanoparticles using the fungus *Fusarium semitectum*. Mater Res Bull 43:1164–1170

Betzig E, Chichester RJ (1993) Single molecules observed by near-field scanning optical microscopy. Science 262:1422–1425

Beveridge TJ, Murray RGE (1980) Sites of metal deposition in the cell wall of *Bacillus subtilis*. J Bacteriol 141:876–887

Bharde A, Wani A, Shouche Y, Joy PA, Prasad BLV, Sastry M (2005) Bacterial aerobic synthesis of nanocrystalline magnetite. J Am Chem Soc 127:9326–9327

Blakemore R (1975) Magnetotactic bacteria. Science 190:377–379

Brayner R, Yepremian C, Djediat C, Coradin T, Herbst F, Livage J, Fievet F, Coute A (2009) Photosynthetic microorganisms-mediated synthesis of akaganeite (b-FeOOH). Langmuir 25:10062–10067

Brayner R, Coradin T, Beaunier P, Greneche JM, Djediat C, Yepremian C, Coute A, Fievet F (2012) Intracellular biosynthesis of superparamagnetic 2-lines ferri-hydrite nanoparticles using *Euglena gracilis* microalgae. Colloids Surf B Biointerfaces 93:20–23

Brekhovskikh AA, Bekasova OD (2005) CdS nanoparticles in R-phycoerythrin, a protein matrix. Inorg Mater 41:331–337

Briat JF, Duc C, Ravet K, Gaymard F (2010) Ferritins and iron storage in plants. Biochim Biophys Acta Gen Subj 1800:806–814

Bruchez M, Moronne M, Gin P, Weiss S, Alivisatos AP (1998) Semiconductor nanocrystals as fluorescent biological labels. Science 281:2013–2016

Bucherand JP, Van der Link JJ (1988) Electronic properties of small supported Pt particles: NMR study of Pt hyperfine parameters. Phys Rev B 38:11038

Cai F, Li J, Sun J, Ji Y (2011) Biosynthesis of gold nanoparticles by biosorption using *Magnetospirillum gryphiswaldense* MSR-1. J Chem Eng 175:70–75

Chan WC, Nie S (1998) Quantum dot bioconjugates for ultrasensitive nonisotopic detection. Science 281:2016–2018

Chan WC, Maxwell DJ, Gao X, Bailey RE, Han M, Nie S (2002) Luminescent quantum dots for multiplexed biological detection and imaging. Curr Opin Biotechnol 13:40–46

Chauhan R, Reddy A, Abraham J (2014) Biosynthesis and antimicrobial potential of silver and zinc oxide nanoparticles using *Candida diversa* strain JA1. Pharm Chem 6:39–47

Cunningham DP, Lundie LL (1993) Precipitation of cadmium by *Clostridium thermoaceticum*. J Appl Environ Microbiol 59:7–14

Dahan M, Levi S, Luccardini C, Rostaing P, Riveau B, Triller A (2003) Diffusion dynamics of glycine receptors revealed by single-quantum dot tracking. Science 302:442–445

Dahoumane SA, Djediat C, Yepremian C, Coute A, Fievet F, Coradin T, Brayner R (2012) Recycling and adaptation of *Klebsormidium flaccidum* microalgae for the sustained production of gold nanoparticles. Biotechnol Bioeng 109:284–288

Dameron CT, Reese RN, Mehra RK, Kortan AR, Carroll PJ, Steigerwald ML, Brus LE, Winge DR (1989) Biosynthesis of cadmium sulphide quantum semiconductor crystallites. Nature 338:596–598

Dammer U, Popescu O, Wagner P, Anselmetti D, Guntherodt HJ, Misevic GN (1995) Binding strength between cell adhesion proteoglycans measured by atomic force microscopy. Science 267:1173–1175

Davis TA, Volesky B, Vieira R (2000) Sargassum seaweed as biosorbent for heavy metals. Water Res 34:4270–4278

Davis TA, Volesky B, Mucci A (2003) A review of the biochemistry of heavy metal biosorption by brown algae. Water Res 37:4311–4330

De Windt W, Aelterman P, Verstraete W (2005) Bioreductive deposition of palladium (0) nanoparticles on *Shewanella oneidensis* with catalytic activity towards reductive dechlorination of polychlorinated biphenyls. J Environ Microbiol 7:314–325

Dewanjee S, Kundu M, Maiti A, Majumdar R, Majumdar A, Mandal SC (2007) In vitro evaluation of antimicrobial activity of crude extract from plants *Diospyros peregrine*, *Coccinia grandis* and *Swietenia macrophylla*. Trop J Pharm Res 6:773–778

Du L, Jiang H, Liu X, Wang E (2007) Biosynthesis of gold nanoparticles assisted by *Escherichia coli* DH5α and its application on direct electrochemistry of haemoglobin. J Electrochem Commun 9:1165–1170

Dubay M, Bhadauria S, Kushwah BS (2009) Green synthesis of nanosilver particles from extract of *Eucalyptus hybrida* (Safeda) Leaf. Dig J Nanomater Biostruct 4:537–543

Duran N, Marcato PD, Alves OL, De Souza GIH, Esposito E (2007) Antibacterial effect of silver nanoparticles produced by fungal process on textile fabrics and their effluent treatment. J Biomed Nanotechnol 3(203):208

Elbeshehy EKF, Elazzazy AM, Aggelis G (2015) Silver nanoparticles synthesis mediated by new isolates of *Bacillus* spp., nanoparticle characterization and their activity against Bean Yellow Mosaic virus and human pathogens. Front Microbiol 6:453

Elgorban AM, Al-Rahmah AN, Rushdy Sayed S, Hirad A, Abdel-Fattah Mostafa A, Bahkali AH (2016) Antimicrobial activity and green synthesis of silver nanoparticles using *Trichoderma viride*. Biotechnol Biotechnol Equip 30:299–304

Faghri Zonooz N, Salouti M (2011) Extracellular biosynthesis of silver nanoparticles using cell filtrate of *Streptomyces* sp. ERI-3. J Sci Iran F 18:1631–1635

Faghri Zonooz N, Salouti M, Shapouri R, Nasseryan J (2012) Biosynthesis of gold nanoparticles by *Streptomyces sp. ERI-3* supernatant and process optimization for enhanced production. J Cluster Sci 23:375–382

Faivre D, Schuler D (2008) Magnetotactic Bacteria and Magnetosomes. J Chem Rev 108:4875–4898

Faixov ZR, Faix S (2008) Biological effects of Rosemary (*Rusmarinus officinalis)* essential oil (a review). Folia Vet 3:135–139

Feng Y, Yu Y, Wang Y, Lin X (2007) Biosorption and bioreduction of trivalent aurum by photosynthetic bacteria *Rhodobacter capsulatus*. Curr Microbiol 55:402–408

Fotiadis D, Scheuring S, Muller SA, Engel A, Muller DJ (2002) Imaging and manipulation of biological structures with the AFM. Micron 33:385–397

Fu JK, Liu YY, Gu PY, Tang D, Lin Z, Yao B (2000) Spectroscopic characterization on the biosorption and bioreduction of Ag(I) by Lactobacillus sp. A09. J Acta Phys Chim Sin 16:779–782

Gade AK, Bonde PP, Ingle AP, Marcato PD, Duran N, Rai MK (2008) Exploitation of *Aspergillus niger* for synthesis of silver nanoparticles. J Biobased Mater Bioenergy 2:243–247

Gade A, Ingle A, Whiteley C, Rai M (2010) Mycogenic metal nanoparticles: progress and applications. Biotechnol Lett 32:593–600

Gallezot P, Leclereq C (1994) Characterization of catalysts by conventional and analytical electron microscopy. In: Imelik B, Vedrine JC (eds) Catalyst characterization. Plenum, New York, p 509

Gao X, Cui Y, Levenson RM, Chung LW, Nie S (2004) *In vivo* cancer targeting and imaging with semiconductor quantum dots. Nat Biotechnol 22:969–976

Gardea-Torresdey JL, Parsons JG, Gomez E, Peralta-Videa J, Troiani HE, Santiago P, Jose Yacaman M (2002) Formation and growth of Au nanoparticles inside live alfalfa plants. Nano Lett 401:397–401

Gericke M, Pinches A (2006a) Microbial production of gold nanoparticles. J Gold Bull 39:22–28

Gericke M, Pinches A (2006b) Biological synthesis of metal nanoparticles. Hydrometallurgy 83:132–140

Gerrard TL, Telford JN, Williams HH (1974) Detection of selenium deposits in *Escherichia coli* by electron microscopy. J Bacteriol 119:1057–1060

Ghodake GS, Lee DS (2011) Biological synthesis of gold nanoparticles using the aqueous extract of the brown algae *Laminaria japonica*. J Nanoelectro Optoelectro 6:268–271

Ghodake GS, Deshpande NG, Lee YP, Jin ES (2010) Pear fruit extract-assisted room-temperature biosynthesis of gold nanoplates. Colloids Surf B Biointerfaces 75:584–589

Gholami-Shabani M, Shams-Ghahfarokhi M, Gholami-Shabanid Z, Akbarzadeh A, Riazi G, Ajdari S, Amani A, Razzaghi-Abyaneh M (2015) Enzymatic synthesis of gold nanoparticles using sulfite reductase purified from *Escherichia coli*: a green eco-friendly approach. Process Biochem 50:1076–1085

Ghosh SK, Mandal M, Kundu S, Nath S, Pa T (2004) Bimetallic Pt–Ni nanoparticles can catalyze reduction of aromatic nitro compounds by sodium borohydride in aqueous solution. J Appl Catal A 286:61–66

Govender Y, Riddin TL, Gericke M, Whiteley CG (2010) On the enzymatic formation of platinum nanoparticles. J Nanopart Res 12:261–271

Greene B, Hosea M, McPherson R, Henzl M, Alexander MD, Darnall DW (1986) Interaction of gold (I) to gold (III) complexes with algal biomass. Environ Sci Technol 20:627–632

Gregorio SD, Lampis S, Vallini G (2005) Selenite precipitation by a rhizospheric strain of *Stenotrophomonas* sp. isolated from the root system of *Astragalus bisulcatus*: a biotechnological perspective. J Environ Int 31:233–241

Grunberg K, Wawer C, Tebo BM, Schuler D (2001) A large gene cluster encoding several magnetosome proteins is conserved in different species of magnetotactic bacteria. Appl Environ Microbiol 67:4573–4582

Guangquan L, Dan H, Yongqing Q, Buyuan Guan G, Song G, Yan C, Koji Yokoyama Y, Li Wang W (2012) Fungus-mediated green synthesis of silver nanoparticles using *Aspergillus terreus*. Int J Mol Sci 13:466–476

Gurunathan S, Kalishwaralal K, Vaidyanathan R, Venkataraman D, Pandian SRK, Muniyandi J, Hariharan N, Eom SH (2009) Biosynthesis, purification and characterization of silver nanoparticles using *Escherichia coli*. Colloids Surf B Biointerfaces 74:328–335

Gurunathan S, Han JW, Eppakayala V, Jeyaraj M, Kim JH (2013) Cytotoxicity of biologically synthesized silver nanoparticles in MDA-MB-231 human breast cancer cells. Biomed Res Int 2013:535796–535801

Halawani EM (2016) Nanomedicine opened new horizons for metal nanoparticles to treat multi-drug resistant organisms. Int J Curr Microbiol App Sci 5:397–414

Harada M, Asakura K, Toshima N (1994) Structural analysis of polymer-protected platinum/rhodium bimetallic clusters using extended X-ray absorption fine structure spectroscopy. Importance of microclusters for the formation of bimetallic clusters. J Phys Chem 98:2653–2662

He SY, Zhang Y, Guo ZR, Gu N (2008) Biological synthesis of gold nanowires using extract of *Rhodopseudomonas capsulata*. Biotechnol Prog 24:476–480

Herrero A, Flores E, Guerrero MG (1981) Regulation of nitrate reductase levels in the cyanobacteria Anacystis nidulans, *Anabaena* sp. strain 7119, and *Nostoc* sp. strain 6719. J Bacteriol 145:175–180

Hosea M, Greene B, Mcpherson R, Henzl M, Darnall DW (1986) Accumulation of elemental gold on the alga Chlorella vulgaris. Inorg Chim Acta 123:161–165

Huang J, Li Q, Sun D, Lu Y, Su Y, Yang X, Wang H, Wang Y, Shao W, He N, Hong J, Chen C (2007) Biosynthesis of silver and gold nanoparticles by novel sun dried *Cinnamomum camphora* leaf. Nanotechnology 18:105104

Huang J, Lin L, Li Q, Sun D, Wang Y, Li Y, He N, Yang K, Yang X, Wang H, Wang W, Lin W (2008) Continuous flow biosynthesis of silver nanoparticles by lixivium of Sun dried *Cinnamomum camphora* leaf in tubular micro reactors. Ind Eng Chem Res 47:6081–6090

Hulkoti NI, Taranath T (2014) Biosynthesis of nanoparticles using microbes: a review. Colloids Surf B Biointerfaces 121:474–483

Hunter WJ, Manter D (2008) Bio-reduction of selenite to elemental red selenium by *Tetrathiobacter kashmirensis*. J Curr Microbiol 57:83–88

Husseiny MI, Abd El-Aziz M, Badr Y, Mahmoud MA (2007) Biosynthesis of gold nanoparticles using *Pseudomonas aeroginosoa*. Spectrochim Acta A 67:1003–1006

Ingle A, Gade A, Pierrat S, Sonnichsen C, Rai M (2008) Mycosynthesis of silver nanoparticles using the fungus *Fusarium acuminatum* and its activity against some human pathogenic bacteria. J Curr Nanosci 4:141–144

Jafarizad A, Safaee K, Gharibian S, Omidi Y, Ekinci D (2015) Biosynthesis and *In vitro* study of gold nanoparticles using mentha and pelargonium extracts. Proc Mater Sci 11:224–230

Jayaseelan C, Abdul Rahumana A, Vishnu Kirthi A, Marimuthu S, Santhoshkumar T, Bagavan A, Gaurav K, Karthik L, Bhaskara Rao KV (2013) Novel microbial route to synthesize ZnO nanoparticles using *Aeromonas hydrophila* and their activity against pathogenic bacteria and fungi. Spectrochim Acta A 90:78–84

Jeyaraj M, Rajesh M, Arun R, Mubarak Ali D, Sathishkumar G, Sivanandhan G, Kapil Dev G, Manickavasagam M, Premkumar K, Thajuddin N, Ganaphati A (2013) An investigation on the cytotoxicity and caspase-mediated apoptotic effect of biologically synthesized silver nanoparticles using *Podophyllum hexandrum* on human cervical carcinoma cells. Colloids Surf B Biointerfaces 102:708–717

Jha AK, Prasad K, Prasad K (2009) A green low-cost biosynthesis of Sb_2O_3 nanoparticles. Biochem Eng J 43:303–306

Joerger R, Klaus-Joerger T, Granqvist CG (2000) Biologically produced silver-carbon composite materials for optically functional thin-film coatings. J Adv Mater 12:407–409

Kairdolf BA, Smith AM, Stokes TH, Wang MD, Young AN, Nie S (2013) Semiconductor quantum dots for bioimaging and biodiagnostic applications. Annu Rev Anal Chem (Palo Alto, Calif) 6:143

Kaler A, Nankar R, Bhattacharyya MS, Banerjee UC (2011) Extracellular biosynthesis of silver nanoparticles using aqueous extract of *Candida viswanathii*. J Bionanosci 5:53–58

Kalishwaralal K, Deepak V, Ram Kumar Pandian SB, Mohd B, Gurunathan S (2008) Extracellular biosynthesis of silver nanoparticles by the culture supernatant of *Bacillus licheniformis*. Mater Lett 62:4411–4413

Kalishwaralal K, GopalramS VR, Deepak V, Ram Kumar Pandian S, Gurunathan S (2010) Optimization of alpha-amylase production for the green synthesis of gold nanoparticles. Colloids Surf B Biointerfaces 77:174–180

Karamanoli K, Vokou D, Menkissoglu U, Constantinidou IH (2000) Bacterial colonization of the phyllosphere of Mediterranean aromatic plants. J Chem Ecol 26:2035–2048

Kashefi K, Tor JM, Nevin KP, Lovley DR (2001) Reductive precipitation of gold by dissimilatory Fe(III)-reducing bacteria and archaea. J Appl Environ Microbiol 67:3275–3279

Kasthuri J, Kathiravan K, Rajendiran N (2009) Phyllanthin-assisted synthesis of silver and gold nanoparticles. J Nanopart Res 11:1075–1085

Kathiresan K, Manivanan S, Nabeel MA, Dhivya B (2009) Studies on silver nanoparticles synthesized by a marine fungus, *Penicillium fellutanum* isolated from coastal mangrove sediment. Colloids Surf B Biointerfaces 71:133–137
Kaviya S, Santhanalakshmi J, Viswanathan B, Muthumary J, Srinivasan K (2011) Biosynthesis of silver nanoparticles using citrus sinensis peel extract and its antibacterial activity. Spectrochim Acta A 79:594–598
Kaviya S, Santhanalakshmi J, Viswanathan B (2012) Biosynthesis of silver nano-flakes by *Crossandra infundibuliformis* leaf extract. Mater Lett 67:64–66
Keat CL, Aziz A, Ahmad M, Eid AM, Elmarzugi NA (2015) Biosynthesis of nanoparticles and silver nanoparticles. Bioresour Bioprocess 2:47
Kessi J, Ramuz M, Wehrli E, Spycher M, Bachofen R (1995) Reduction of selenite and detoxification of elemental selenium by the phototrophic bacterium *Rhodospirillum rubrum*. J Appl Environ Microbiol 65:4734–7440
Khadivi Derakhshan F, Dehnad A, Saluoti M (2012) Extracellular biosynthesis of gold nanoparticles by metal resistance bacteria: *Streptomyces griseus*. Synth React Inorg Met 42:868–871
Khanra K, Panja S, Choudhuri I, Chakraborty A, Bhattacharyya N (2016) Antimicrobial and cytotoxicity effect of silver nanoparticle synthesized by *Croton bonplandianum* Baill. leaves. Nanomed J 3:15–22
Klaus-Joerger T, Joerger R, Olsson E, Granqvist CG (1999) Silver-based crystalline nanoparticles, microbially fabricated. Proc Natl Acad Sci USA 96(24):13611–13614
Klaus-Joerger T, Joerger R, Olsson E, Granqvist CG (2001) Bacteria as workers in the living factory: metal-accumulating bacteria and their potential for materials science. Trends Biotechnol 19:15–20
Konishi Y, Ohno K, Saitoh N, Nomura T, Nagamine S, Hishida H (2007a) Bioreductive deposition of platinum nanoparticles on the bacterium *Shewanella algae*. J Biotechnol 128:648–653
Konishi Y, Tsukiyama T, Tachimi T, Saitoh N, Nomura T, Nagamine S (2007b) Microbial deposition of gold nanoparticles by the metal-reducing bacterium *Shewanella algae*. Electrochim Acta 53:186–192
Konohana T, Murakami S, Nanmori T, Aoki K, Shinke R (1993) Increase in nitrate reductase activity with ammonium chloride in *Bacillus licheniformis* by shaking culture. Biosci Biotechnol Biochem 57:2170–2171
Kowshik CM, Vogel W, Urban J, Kulkari SK, Paknikar KM (2002) Microbial synthesis of semiconductor PbS nanocrystallites. Adv Mater 14:815–818
Krishnaraj C, Jagan EG, Rajasekar S, Selvakumar P, Kalaichelvan PT, Mohan N (2010) Synthesis of silver nanoparticles using *Acalypha indica* leaf extracts and its antibacterial activity against water borne pathogens. Colloids Surf B Biointerfaces 76:50–56
Kuber CB, D'Souza SF (2006) Extracellular biosynthesis of silver nanoparticles using the fungus *Aspergillus fumigates*. Colloids Surf B Biointerfaces 47:160–164
Kumar SA, Kazemian Abyaneh M, Gosavi SW, Kulkarni SK, Ahmad A, Khan MI (2007) Sulfite reductase-mediated synthesis of gold nanoparticles capped with phytochelatin. Appl Biochem 47:191–195
Kumar R, Roopan SM, Prabhakarn A, Khanna VG, Chakroborty S (2012) Agricultural waste *Annona squamosa* peel extract: biosynthesis of silver nanoparticles. Spectrochim Acta A Mol Biomol Spectrosc 90:173–176
Labrenz M, Druschel GK, Thomsen-Ebert T, Gilbert B, Welch SA, Kemner KM (2000) Formation of sphalerite (ZnS) deposits in natural biofilms of sulfate-reducing bacteria. J Sci 290:1744–1747
Lakshmanan A, Umamaheswari C, Nagarajan NS (2016) A facile phyto-mediated synthesis of gold nanoparticles using aqueous extract of *Momordica cochinchinensis* rhizome and their biological activities. J Nanosci Technol 2:76–80
Lehenkari PP, Charras GT, Nykanen A, Horton MA (2000) Adapting atomic force microscopy for cell biology. Ultramicroscopy 82:289–295

Lengke MF, Fleet ME, Southam G (2007) Biosynthesis of silver nanoparticles by filamentous cyanobacteria from a Silver(I) Nitrate complex. Langmuir 23:2694–2699

Lezcano M, Gonzalez F, Ballester A, Blazquez ML, Munoz JA, Garcia-Balboa C (2010) Biosorption of Cd (II), Cu (II), Ni (II), Pb (II) and Zn (II) using different residual biomass. Chem Ecol 26:1–17

Li S, Shen Y, Xie A, Yu X, Qiu L, Zhang L, Zhang Q (2007) Green synthesis of silver nanoparticles using *Capsicum annum*. Green Chem 9:852–858

Lin Z, Wu J, Xue R, Yang Y (2005) Spectroscopic characterization of Au3+ biosorption by waste biomass of *Saccharomyces cerevisiae*. Spectrochim Acta A 61:761–765

Liqin L, Wenta W, Jiale H, Qingbiao L, Daohua S, Xin Y, Huixuan W, Ning H, Yuanpeng W (2010) Nature factory of nanowires: plant-mediated synthesis using broth of *Cassia fistula* leaf. Chem Eng J 162:852–858

Liu JK, Luo CX, Yang XH, Zhang XY (2009) One-step, low-temperature fabrication of CdS quantum dots by watermelon rind: a green approach. J Mater Lett 63:124–126

Lortie L, Gould WD, Rajan S, McCready RGL, Cheng KJ (1992) Reduction of selenate and selenite to elemental selenium by a *Pseudomonas stutzeri* isolate. J Appl Environ Microbiol 58:4042–4044

Losi ME, Frankenberger WT (1997) Reduction of selenium oxyanions by *Enterobacter cloacae* SLD1a-1: isolation and growth of the bacterium and its expulsion of selenium particles. J Appl Environ Microbiol 63:3079–3084

Mandal D, Bolander ME, Mukhopadhyay D, Sarkar G, Mukherjee P (2006) The use of microorganisms for the formation of metal nanoparticles and their application. Appl Microbiol Biotechnol 69:485–492

Mann S, Frankel RB, Blakemore RP (1984) Structure, morphology, and crystal growth of bacterial magnetite. J Nat 310:405–407

Marshall MJ, Beliaev AS, Dohnalkova AC, Kennedy DW, Shi L, Wang Z (2006) c-type cytochrome-dependent formation of U(IV) nanoparticles by *Shewanella oneidensis*. PLoS Biol 4:1324–1333

Mata YN, Torres E, Blazquez ML, Ballester A, Gonzalez F, Munoz JA (2009) Gold (III) biosorption and bioreduction with the brown alga Fucus vesiculosus. J Hazard Mater 166:612–618

Michalet X, Pinaud FF, Bentolila LA, Tsay JM, Doose S, Li JJ, Sundaresan G, Wu AM, Gambhir SS, Weiss S (2005) Quantum dots for live cells, in vivo imaging, and diagnostics. Science 307:538–544

Minnunni M, Wolleb U, Mueller O, Pfeifer A, Aeschbacher HU (1992) Natural antioxidants as inhibitors of oxygen species induced mutagenicity. Mutat Res 269:193–200

Miot J, Benzerara K, Obst M, Kappler A, Hegler F, Shadler S, Bouchez C, Guyot F, Morin G (2009) Extracellular iron biomineralization by photoautotrophic iron-oxidizing bacteria. J Appl Environ Microbiol 75:5586–5591

Mishra AN, Bhadauria S, Gaur MS, Pasricha R, Kushwah BS (2010) Synthesis of gold nanoparticles by leaves of zero-calorie sweetener herb (*Stevia rebaudiana*) and their nanoscopic characterization by spectroscopy and microscopy. Inter J Green Nanotechnol 1:118–124

Moghaddam AB, Namvar F, Moniri M, Tahir PM, Azizi S, Mohamad R (2015) Nanoparticles biosynthesized by fungi and yeast: a review of their preparation, properties, and medical applications. Molecules 20:16540–16565

Mohammadi B, Salouti M (2013) Extracellular bioynthesis of silver nanoparticles by *Penicillium chrysogenum* and *Penicillium expansum*. Synth React Inorg Met 45:844–847

Mohammadi B, Salouti M (2015) Extracellular bioynthesis of silver nanoparticles by *penicillium chrysogenum* and *penicillium expansum*. Synth React Inorg Met 45:844–847

Mohammadian A, Shojaosadati SA, Rezaee MH (2007) *Fusarium oxysporum* mediates photogeneration of silver nanoparticles. Sci Iran 14:323–326

Mohana Roopan S, Kumar R, Prabhakarn A, Gopiesh Khanna V, Chakroborty S (2012) Agricultural waste annona squamosa peel extract: biosynthesisof silver nanoparticles. Spectrochim Acta A 90:173–176

Mollazadeh Moghaddam K (2010) An introduction to microbial metal nanoparticle preparation method. J Young Investig 19:1–7

Mourato A, Gadanho M, Lino AR, Tenreiro R (2011) Biosynthesis of crystalline silver and gold nanoparticles by extremophilic yeasts. J Bioinorg Chem App 2011:1–8

Mouxing FU, Qingbiao LI, Daohua SUN, Yinghua LU, Ning HE, Xu D (2006) Rapid preparation process of silver nanoparticles by bioreduction and their characterizations. Chin J Chem Eng 14:114–117

Mubarak Ali D, Gopinath V, Rameshbabu N, Thajuddin N (2012) Synthesis and characterization of CdS nanoparticles using c-phycoerythrin extracted from marine cyanobacteria. J Mater Lett 74:8–11

Mukherjee P, Senapati S, Mandal D, Ahmad A, Khan MI, Kumar R, Sastry M (2002) Extracellular synthesis of gold nanoparticles by the fungus *Fusarium oxysporum*. Chembiochem 3:461–463

Mukherjee P, Bhattacharya R, Wang P, Wang L, Basu S, Nagy JA, Atala A, Mukhopadhyay D, Soker S (2005) Antiangiogenic properties of gold nanoparticles. Clin Cancer Res 11:3530–3534

Mukherjee P, Roy M, Mandal BP, Dey GK, Mukherjee PK, Ghatak J (2008) Green synthesis of highly stabilized nanocrystalline silver particles by a non-pathogenic and agriculturally important fungus *T. asperellum*. Nanotechnology 19:075103

Mulvaney P, Giersig M, Henglein A (1993) Electrochemistry of multilayer colloids: preparation and absorption spectrum of gold-coated silver particles. J Phys Chem 97:7061–7064

Musarrat J, Dwivedi S, Raj Singh B, Al-Khedhairy AA, Azam A, Naqvi A (2010) Production of antimicrobial silver nanoparticles in water extracts of the fungus *Amylomyces rouxii* strain KSU-09. J Bioresour Technol 101:8772–8776

Nair B, Pradeep T (2002) Coalescence of nanoclusters and formation of submicron crystallites assisted by *Lactobacillus* strains. Cryst Growth Des 2:293–298

Narayanan KB, Sakthivel N (2010) Biological synthesis of metal nanoparticles by microbes. Adv Colloid Interface Sci 156:1–13

Navarro M, Perez H, Sanchez-Delgado RA (1997) Toward a novel metal-based chemotherapy against tropical diseases. synthesis and antimalarial activity *in vitro* and *in vivo* of the new gold-chloroquine complex [Au(PPh3)(CQ)] PF6. J Med Chem 40:1937–1939

Niknejad F, Nabili M, Daie Ghazvini R, Moazeni M (2015) Green synthesis of silver nanoparticles: advantages of the yeast *Saccharomyces cerevisiae* model. Curr Med Mycol 1:17–24

Noruzi M, Zare D, Khoshnevisan K, Davoodi D (2011) Rapid green synthesis of gold nanoparticles by *Rosa hybrida* petal extract at room temperature. Spectrochim Acta A 79:1461–1465

Ogi T, Saitoh N, Nomura T, Konishi Y (2010) Room-temperature synthesis of gold nanoparticles and nanoplates using *Shewanella algae* cell extract. J Nanopart Res 12:2531–2539

Oremland RS, Herbel MJ, Blum JS, Langley S, Beveridge TJ, Ajayan PM (2004) Structural and spectral features of selenium nanospheres produced by Se-respiring bacteria. J Appl Environ Microbiol 70:52–60

Ozcan K (2003) Antioxidant activities of rosemary, sage, and sumac extracts and their combinations on stability of natural peanut oil. J Med Food 6:267–270

Panneerselvam C, Ponarulselvam S, Murugan K, Kalimuthu K, Thangamani S (2012) Synthesis of silver nanoparticles using leaves of *Catharanthus roseus* Linn. G. Don and their antiplasmodial activities. Asian Pac J Trop Biomed 3:574–580

Pantidos N, Horsfall LE (2014) Biological synthesis of metallic nanoparticles by bacteria, fungi and plants. J Nanomed Nanotechnol 2014:1–10

Parekh RY, Singh S, Prasad BLV, Patole MS, Sastry M, Shouche YS (2008) Extracellular synthesis of crystalline silver nanoparticles and molecular evidence of silver resistance from

Morganella sp.: towards understanding biochemical synthesis mechanism. J Chem Bio Chem 9:1415–1422
Park TJ, Lee KG, Lee SY (2016) Advances in microbial biosynthesis of metal nanoparticles. Appl Microbiol Biotechnol 100:521–534
Perantoni M, Esquivel DMS, Wajnberg E, Acosta-Avalos D, Cernicchiaro G, Lins de Barros H (2009) Magnetic properties of the microorganism Candidatus Magnetoglobus multicellularis. J Naturwiss 96:685–690
Philip D (2009) Biosynthesis of Au, Ag and Au-Ag nanoparticles using edible mushroom extract. Spectrochim Acta A 73:374–381
Philip D (2010) Rapid green synthesis of spherical gold nanoparticles using *Mangifera indica* leaf. Spectrochim Acta A 77:807–810
Pohl DW, Denk W, Lanz M (1984) Optical stethoscopy-image recording with resolution lambda/20. Appl Phys Lett 44:651–653
Prasad T, Elumalai EK (2011) Biofabrication of Ag nanoparticles using *Moringa oleifera* leaf extract and their antimicrobial activity. Asian Pac J Trop Biomed 1:439–442
Prasad K, Jha AK, Kulkarni AR (2007) *Lactobacillus* assisted synthesis of titanium nanoparticles. J Nanoscale Res Lett 2:248–250
Priyadarshini E, Pradhan N, Sukla LB, Panda PK (2014) Controlled synthesis of gold nanoparticles using *Aspergillus terreus* IF0 and its antibacterial potential against gram negative pathogenic bacteria. J Nanotechnol 2014:1–9
Pugazhenthiran N, Anandan S, Kathiravan G, Prakash NKU, Crawford S, Ashokkumar M (2009) Microbial synthesis of silver nanoparticles by *Bacillus* sp. J Nanopart Res 11:1811–1815
Rai M, Yadava A, Gadea A (2009) Silver nanoparticles as a new generation of antimicrobials. Biotechnol Adv 27:76–83
Rajakumar G, Abdul Rahuman A, Roopan SM, Gopiesh Khanna V, Elango G, Kamaraj C, Abduz Zahir A, Velayutham K (2012) Fungus-mediated biosynthesis and characterization of TiO_2 nanoparticles and their activity against pathogenic bacteria. Spectrochim Acta A 91:23–29
Rajakumar G, Marimuthu S, Vishnu Kirthi A, Velayutham K, Thomas J, Venkatesan J, Kim SK (2014) Green synthesis of titanium dioxide nanoparticles using *Psidium guajava* extract and its antibacterial and antioxidant properties. Asian Pac J Trop Med 2014:968–976
Rajasulochana P, Dhamotharan R, Murugakoothan P, Murugesan S, Krishnamoorthy P (2010) Biosynthesis and characterization of gold nanoparticles using the alga *Kappaphycus alvarezii*. J Int J Nanosci 9:511–515
Ramakrishna M, Rajesh Babu D, Gengan RM, Chandra S, Rao GN (2016) Green synthesis of gold nanoparticles using marine algae and evaluation of their catalytic activity. J Nanostruct Chem 6:1–13
Ramalingmam P, Muthukrishnan S, Thangaraj P (2015) Biosynthesis of silver nanoparticles using an endophytic fungus, *Curvularia lunata* and its antimicrobial potential. J Nanosci Nanoeng 1:241–247
Ranjbar Navazi Z, Pazouki M, Halek FS (2010) Investigation of culture condition for biosynthesis of silver nanoparticles using *Aspergillus fumigatus*. Iran J Biotechnol 8:56
Ray S, Sarkar S, Kund S (2011) Extracellular biosynthesis of silver nanoparticles using the mycorrhizal mushroom *Tricholoma crassum* (berk) its antimicrobial activity against pathogenic bacteria and fungus, including multidrug resistant plant and human bacteria. Dig J Nanomater Biostruct SACC 6:1289–1299
Rodriguez P, Munoz-Aguirr N, San-Martin Martinez E, De la Cruz GG, Tomas SA, Zelaya Angel O (2008a) Synthesis and spectral properties of starch capped CdS nanoparticles in aqueous solution. J Cryst Growth 310:160–164
Rodriguez P, Munoz-Aguirre N, San-Martin Martinez E, Gonzalez OZ, Mendoza J (2008b) Formation of CdS nanoparticles using starch as capping agent. Appl Surf Sci 225:740–742
Rotsch C, Radmacher M (2000) Drug-induced changes of cytoskeletal structure and mechanics in fibroblasts: an atomic force microscopy study. Biophys J 78:520–535

Roy K, Sarkar CK, Ghosh CK (2015) Photocatalytic activity of biogenic silver nanoparticles synthesized using yeast (*Saccharomyces cerevisiae*) extract. Appl Nanosci 5:953–959

Sadhasivam S, Shanmugam P, Yun KS (2010) Biosynthesis of silver nanoparticles by *Streptomyces hygroscopicus* and antimicrobial activity against medically important pathogenic microorganisms. J Colloids Surf B Biointerfaces 81:358–362

Salunke BK, Sawant SS, Lee S, Kim BS (2016) Microorganisms as efficient biosystem for the synthesis of metal nanoparticles: current scenario and future possibilities. World J Microbiol Biotechnol 32:88

Saminathan K (2015) Biosynthesis of silver nanoparticles from dental caries causing fungi *Candida albicans*. Int J Curr Microbiol App Sci 4:1084–1091

Sandana Mala JG, Rose C (2014) Facile production of ZnS quantum dot nanoparticles by *Saccharomyces cerevisiae* MTCC 2918. J Biotechnol 170:73–78

Sanghi R, Verma P (2009) Biomimetic synthesis and characterisation of protein capped silver nanoparticles. Bioresour Technol 100:501–504

Sarkar J, Dey P, Saha S, Acharya K (2011) Mycosynthesis of selenium nanoparticles. Micro Nano Lett 6:599–602

Sastry M, Ahmad A, Khan MI, Kumar R (2003) Biosynthesis of metal nanoparticles using fungi and actinomycete. Curr Sci 85:162–170

Sathishkumar M, Sneha K, Won SW, Cho CW, Kim S, Yun YS (2009) *Cinnamon zeylanicum* bark extract and powder mediated green synthesis of nano-crystalline silver particles and its bactericidal activity. Colloids Surf B Biointerface 73:332–338

Scarano G, Morelli E (2003) Properties of phytochelatin-coated CdS nanochrystallites formed in a marine phytoplanktonic alga (*Phaeodactylum tricornutum*, Bohlin) in response to Cd. Plant Sci 165:803–810

Schuler D (1999) Formation of magnetosomes in magnetotactic bacteria. J Mol Microbiol Biotechnol 1:79–86

Senapati S, Ahmad A, Mandal D, Khan MI, Kumar R, Sastry M (2005) Extracellular biosynthesis of bimetallic Au–Ag alloy nanoparticles. Small 1:517–520

Seshadri S, Saranya K, Kowshik M (2011) Green synthesis of lead sulfide nanoparticles by the lead resistant marine yeast, *Rhodosporidium diobovatum*. J Biotechnol Prog 27:1464–1469

Shahverdi A, Minaeian S, Shahverdi SR, Jamalifar H, Nohi AA (2007) Rapid synthesis of silver nanoparticles using culture supernatants of Enterobacteria: a novel biological approach. Proc Biochem 42:919–923

Shakouri V, Salouti M, Mohammadi B, Faghri Zonooz N (2016) Procedure optimization for increasing biosynthesis rate of gold nanoparticles by *Aspergillus flavus* Supernatant. Synth React Inorg Met (in press)

Shaligram SN, Bule M, Bhambure R, Singhal SR, Singh KS, Szakacs G, Pandey A (2009) Biosynthesis of silver nanoparticles using aqueous extract from the compactin producing fungi. Process Biochem 44:939–943

Shan S, Cai YZ, Brooks JD, Corke H (2007) Antibacterial properties and major bioactive components of cinnamon stick (*Cinnamomum burmannii*): activity against foodborne pathogenic bacteria. Agric Food Chem 55:5484–5490

Shankar SS, Ahmad A, Pasrichaa R, Sastry M (2003) Bioreduction of chloroaurate ions by Geranium leaves and its endophytic fungus yields gold nanoparticles of different shapes. J Mater Chem 13:1822–1826

Shankar SS, Rai A, Ahmad A, Sastry M (2004a) Rapid synthesis of Au, Ag, and bimetallic Au core-Ag shell nanoparticles using Neem (*Azadirachta indica*) leaf broth. J Colloid Interface Sci 275:496–502

Shankar SS, Rai A, Ahmad A, Sastry M (2004b) Biosynthesis of silver and gold nanoparticles from extracts of different parts of the geranium plant. Appl Nano Sci 1:69–77

Shankar SS, Rai A, Ankamwar B, Singh A, Absar Ahmad A, Sastry M (2004c) Biological synthesis of triangular gold nanoprisms. Nat Mater 3:482–488

Shankar SS, Rai A, Ahmad A, Sastry M (2005) Controlling the optical properties of lemongrass extract synthesized gold nanotriangles and potential application in infrared-absorbing optical coatings. Chem Mater 17:566–572

Sharma G, Jasuja ND, Kumar M, Ali MI (2015) Biological synthesis of silver nanoparticles by cell-free extract of *Spirulina platensis*. J Nanotechnol 2015:1–6

Sheikhloo Z, Salouti M (2011) Intracellular biosynthesis of gold nanoparticles by the fungus *Penicillium Chrysogenum*. Int J Nanosci Nanotechnol 7:102–105

Sheikhloo Z, Salouti M (2012) Intracellular biosynthesis of gold nanoparticles by fungus *Phoma macrostoma*. Synth React Inorg Met 42:65–67

Sheikhloo Z, Salouti M, Katiraee F (2011) Biological synthesis of gold nanoparticles by fungus *Epicoccum nigrum*. J Cluster Sci 22:661–665

Sheikhloo Z, Salouti M, Farahmandkia Z, Mahmazi S, Einlou A (2012) Intra-extracellular biosynthesis of gold nanoparticles by fungus *Rhizopus oryza*. J Zanjan Univ Med Sci Health Serv 20:47–56

Shenton W, Pum D, Sleytr UB, Mann S (1997) Synthesis of cadmium sulphide superlattices using self-assembled bacterial s-layers. Nature 389:585–587

Shi D, Somlyo AV, Somlyo AP, Shao Z (2001) Visualizing filamentous actin on lipid bilayers by atomic force microscopy in solution. J Microsc 201:377–382

Shivai Karkaj O, Salouti M, Sorouri Zanjani R, Khadivi Derakhshan F (2013) Extracellular deposition of silver nanoparticles by *Bacillus megaterium*. Synth React Inorg Met 43:903–906

Sicard C, Brayner R, Margueritat J, Hemadi M, Coute A, Yepremian DC, Coradin T (2010) Nano-gold biosynthesis by silica-encapsulated micro-algae: a "'living" bio-hybrid material. J Mater Chem 20:9342–9347

Singaravelu G, Arockiamary JS, Ganesh Kumar V, Govindaraju K (2007) A novel extracellular synthesis of monodisperse gold nanoparticles using marine alga, *Sargassum wightii Greville*. J Colloids Surf B Biointerfaces 57:97–101

Singh AK, Talat M, Singh DP, Srivastava ON (2010) Biosynthesis of gold and silver nanoparticles by natural precursor clove and their functionalization with amine group. J Nanopart Res 12:1667–1675

Sirvastava SK, Yamada R, Ogino C, Kondo A (2013) Biogenic synthesis and characterization of gold nanoparticles by Escherichia coli K12 and its heterogeneous catalysis in degradation of 4-nitrophenol. Nano Scale Res 8:70

Sivakumar P, Nethra Devi C, Renganathan S (2012) Synthesis of silver nanoparticles using *Lantana camara* fruit extract and its effect on pathogens. Asian J Pharm Clin Res 5:97–101

Sivalingama P, Joe Antonyb J, Sivab D, Achiramanb S, Anbarasua K (2012) Mangrove *Streptomyces sp. BDUKAS10* as nanofactory for fabrication of bactericidal silver nanoparticles. Colloids Surf B Biointerfaces 98:12–17

Smith PR, Holmes JD, Richardson DJ, Russell DA, Sodeau JR (1998) Photophysical and photochemical characterisation of bacterial semiconductor cadmium sulfide particles. J Chem Soc Faraday Trans 94:1235–1241

Smitha SL, Philip D, Gopchandran KG (2009) Green synthesis of gold nanoparticles using *Cinnamomum zeylanicum* leaf broth. Spectrochim Acta A 74:735–739

Soltani Nejad M, Shahidi Bonjar G, Khaleghi N (2015) Biosynthesis of gold nanoparticles using *Streptomyces fulvissimus* isolate. Nanomed J 2:153–159

Srikar SK, Giri DD, Pal DB, Mishra PK, Upadhyay SN (2016) Green synthesis of silver nanoparticles: a review. Green Sustain Chem 6:34–56

Sudhakar C, Selvam K, Govarthanan M, Senthilkumar B, Sengottaiyan A, Stalin M, Selvankumar T (2015) Acorus calamus rhizome extract mediated biosynthesis of silver nanoparticles and their bactericidal activity against human pathogens. J Genet Eng Biotechnol 13:93–99

Sulaiman GM, Mohammad AAW, Abdul-Wahed HE, Ismail MM (2013) Biosynthesis, antimicrobial and cytotoxic effects of silver nanoparticles using *Rusmarinus officinalis* extract. Dig J Nanomater Bios 8:273–280

Suresh AK, Pelletier DA, Wang W, Broich ML, Moon JW, Gu B, Allison DP, Joy DC, Phelps TJ, Doktycz MJ (2011) Biofabrication of discrete spherical gold nanoparticles using the metal-reducing bacterium *Shewanella oneidensis*. J Acta Biomater 7:2148–2152

Suzuki Y, Kelly SD, Kemner KM, Banfield JF (2002) Radionuclide contamination: nanometre-size products of uranium bioreduction. J Nat 419:134

Sweeney RY, Mao C, Gao X, Burt JL, Belcher AM, Georgiou G (2004) Bacterial biosynthesis of cadmium sulfide nanocrystals. J Chem Biol 11:1553–1559

Taton TA (2002) Nanostructures as tailored biological probes. Trends Biotechnol 20:277–279

Thakkar KN, Mhatre SS, Parikh RY (2009) Biological synthesis of metallic nanoparticles. Nanomed Nano 6:257–262

Thombre R, Leksminarayanan P, Hegde R, Parekh F, Francis G, Mehta PN, Zunjarrao R (2013) A facile method for green synthesis of stabilized silver nanoparticles and its *in vitro* antagonistic applications. J Nat Prod Plant Resour 3:36–40

Tomei FA, Barton LL, Lemanski CL, Zocco TG, Fink NH, Sillerud LO (1995) Transformation of selenate and selenite to elemental selenium by *Wolinella succinogenes*. J Ind Microbiol 14:329–336

Torres de Araujo FF, Pires MA, Frankel RB, Bicudo CEM (1986) Magnetite and magnetotaxis in algae. J Biophys 50:375–378

Tsai CY, Shiau AL, Chen SY, Chen H, Cheng PC, Chang MY, Chen DH, Chou CH, Wang CR, Wu CL (2007) Amelioration of collagen-induced arthritis in rats by nanogold. Arthritis Rheum 56:544–554

Udayasoorian C, Kumar RV, Jayabalakrishnan M (2011) Extracellular synthesis of silver nanoparticles using leaf extract of *Cassia auriculata*. Dig J Nanomater Biostruct 6:537–543

Vadivel S, Suja S (2012) Green synthesis of silver nanoparticles using *Coleus amboinicus* lour, antioxitant activity and *in vitro* cytotoxicity against Ehrlich's Ascite carcinoma. J Pharm Res 5:1268–1272

Vaidyanathan R, Gopalram S, Kalishwaralal K, Deepak V, Ram Kumar Pandian S, Gurunathan S (2010) Enhanced silver nanoparticle synthesis by optimization of nitrate reductase activity. Colloids Surf B Biointerfaces 75:335–341

Veerasamy R, Zi Xin T, Gunasagaran S, Foo Wei Xiang T, Fang Chou Yang E, Jeyakumar N, Arumugam Dhanaraj S (2011) Biosynthesis of silver nanoparticles using *mangosteen* leaf extract and evaluation of their antimicrobial activities. J Saudi Chem Soc 15:113–120

Venkatesan J, Manivasagan P, Kim SK, Vishnu Kirthi A, Marimuthu S, Abdul Rahuman A (2014) Marine algae-mediated synthesis of gold nanoparticles using a novel *Ecklonia cava*. Bioprocess Biosyst Eng 37:1591–1597

Verma VC, Kharwa RN, Gange AC (2010) Biosynthesis of antimicrobial silver nanoparticles by the endophytic fungus *Aspergillus clavatus*. J Nanomed 5:33–40

Vijayaraghavan K, Jegan J, Palanivelu K, Velan M (2005) Batch and column removal of copper from aqueous solution using a brown marine alga *Turbinaria ornate*. Chem Eng J 106:177–184

Vijayaraghavan K, Mahadevan A, Sathishkumar M, Pavagadhi S, Balasubramanian R (2011) Biosynthesis of Au(0) from Au(III) via biosorption and bioreduction using brown marine alga *Turbinaria conoides*. Chem Eng J 167:223–227

Vijayaraghavan K, Kamala Nalini SP, Udaya Prakash N, Madhankumar D (2012) One step green synthesis of silver nano/microparticles using extracts of *Trachyspermum ammi* and *Papaver somniferum*. Colloids Surf B Biointerfaces 94:114–117

Walling AM, Novak AJ, Shepard RE (2009) Quantum dots for live cell and *in vivo* imaging. Int J Mol Sci 10:441–491

Wang CL, Michels PC, Dawson SC, Kitisakkul S, Baross JA, Keasling JD, Clark DS (1997) Cadmium removal by a new strain of *Pseudomonas aeruginosa* in aerobic culture. Appl Environ Microbiol 63:4075–4078

Wang CL, Lum AM, Ozuna SC, Clark DS, Keasling JD (2001) Aerobic sulfide production and cadmium precipitation by *Escherichia coli* expressing the *Treponema denticola* cysteine desulfhydrase gene. Appl Microbiol Biotechnol 56:425–430

Watson JHP, Croudace IW, Warwick PE, James PAB, Charnock JM, Ellwood DC (2001) Adsorption of radioactive metals by strongly magnetic iron sulfide nanoparticles produced by sulfate-reducing bacteria. Sep Sci Technol 36:2571–2607

Wei Q, Kang SZ, Mu J (2004) Green synthesis of starch capped CdS nanoparticles. J Physiochem Eng A 247:125–127

Wen L, Lin Z, Gu P, Zhou J, Yao B, Chen G, Fu J (2009) Extracellular biosynthesis of monodispersed gold nanoparticles by a SAM capping route. J Nanopart Res 11:279–288

Xie JP, Lee JY, Wang DIC, Ting YP (2007) Identification of active biomolecules in the high-yield synthesis of single-crystalline gold nanoplates in algal solutions. Small 3:672–682

Yadav V, Sharma N, Prakash R, Raina KK, Bharadwaj LM, Prakash NT (2008) Generation of selenium containing nano-structures by soil bacterium, *Pseudomonas aeruginosa*. J Biotechnol 7:299–304

Yadav A, Kon K, Kratosova G, Duran N, Ingle AP, Rai R (2015) Fungi as an efficient mycosystem for the synthesis of metal nanoparticles: progress and key aspects of research. Biotechnol Lett 37:2099–2120

Yashhiro Y (2006) Microbial synthesis of noble metal nanoparticles using metal-reducing bacteria. J Soc Powder Technol 43:515–521

Yonezawa T, Toshima N (1995) Mechanistic consideration of formation of polymer-protected nanoscopic bimetallic clusters. J Chem Soc Faraday Trans 91:4111–4119

Yong P, Rowsen NA, Farr JPG, Harris IR, Macaskie LE (2002) Bioreduction and biocrystallization of palladium by *Desulfovibrio desulfuricans* NCIMB 8307. J Biotechnol Bioeng 80:369–379

Zhang H, Li Q, Lu Y, Sun D, Lin X, Deng X (2005) Biosorption and bioreduction of diamine silver complex by Corynebacterium. J Chem Technol Biotechnol 80:285–290

Chapter 3
Nanomaterial and Nanoparticle: Origin and Activity

Cristina Buzea and Ivan Pacheco

3.1 Introduction

Nanomaterials are defined as materials with external dimensions in the nanoscale or with internal structure or surface structure in the nanoscale. This would qualify most of the materials as nanomaterials, as their internal structure is modulated at the nanoscale.

A nanoparticle can be defined as a particle with at least one external dimension in the nanoscale. The term nanoscale can be defined as a size range from about 1 nm to less than 1 μm (Buzea et al. 2007). Particles larger than nanoparticles are called microparticles, while the ones smaller than nanoparticles are atoms and molecules. One must emphasize that the properties of a material in nanoform can be quite different that those of its bulk (heat of fusion, melting temperature, band energy gap, electronic structure and catalytic activity, magnetic properties, fluorescence, tensile strength, etc.) (Auffan et al. 2009; Buzea et al. 2007). The size at which nanoparticles physical properties begin to deviate from its bulk counterpart is about 100 nm (Bakshi et al. 2015), however it is not always a rule (Buzea et al. 2007). Hence, some authors are trying to limit the size of nanoparticles to sizes for which the properties of particles differ from those of their bulk counterparts.

C. Buzea (✉)
IIPB Medicine Corporation, 2656 8th Ave East, Owen Sound, ON, Canada, N4K 6S5
e-mail: cristinabuzea@mdcorporation.ca

I. Pacheco
Department of Pathology, Grey Bruce Health Services, 1800 8th St East, Owen Sound, ON, Canada, N4K 6M9

Department of Pathology and Laboratory Medicine, Schulich School of Medicine and Dentistry, Western University, London, ON, Canada, N6A 5C1
e-mail: ipacheco@gbhs.on.ca; iblandin@uwo.ca

M. Ghorbanpour et al. (eds.), *Nanoscience and Plant–Soil Systems*, Soil Biology 48,
DOI 10.1007/978-3-319-46835-8_3

3.2 Nanoparticle Classification

From the point of view of their dimensionality, nanomaterials encompass broadly three types: with one, two, and three dimensions in the nanoscale. Those with one dimension in the nanoscale are very thin films or coatings attached on a substrate. Those with two dimensions in the nanoscale can be porous films with nanoscale pores, long aspect ratio fibers, wires, or tubes. The third type nanomaterials are membranes with nanopores and nanoparticles. This chapter focuses only on nanoparticle and long aspect ratio particles.

From the point of view of their provenance, inorganic nanoparticles qualify as natural and anthropogenic, as seen in Fig. 3.1. Natural nanoparticles can have geologic sources, incomplete combustion, biochemical cycling, and abiotic chemical precipitation. Geological sources of natural nanoparticles encompass wind erosion, dust storms, glaciations, earthquakes, and volcanic activity. Natural nanoparticles are generated via incomplete combustion in forest fires and earthquakes. An example of abiotic chemical precipitation is sea spray. Biochemical cycling of nanoparticles can be mediated by bacteria, plants, or fungi. Anthropogenic sources of nanoparticles can be unintentional and intentional. The unintentional can be due to incomplete combustion from heating and cooking, automobiles, industry, wear, corrosion (such as tires or construction materials), and garbage disposal of products containing nanoparticles. The intentional nanoparticles are released as pesticides, fertilizers, water or soil remediation.

Sometimes is hard to predict the origin of environmental nanoparticle, however several markers elements can be used to determine the most probable source of inorganic nanoparticles, as shown in Table 3.1 (Calvo et al. 2013). For example, anthropogenic nanoparticles from the steel industry will have Cr, Ni and Mo markers, while the ones from the petrochemical industry those of Ni and V. Incomplete combustion from vehicle tailpipe have platinum group elements tracers, as well as Ce, Mo, and Zn.

Origins, composition, and morphology of various types of natural and anthropogenic nanoparticles are discussed in this chapter.

3.3 Natural Nanoparticles

The term "nanoparticle" is relatively a new one. One should keep in mind that when searching the earlier literature for publications on natural nanoparticle keywords other than nanoparticles should be used, as this term is a relatively new one. For instance to increase the search yield, particle, particulate matter, dust or the specific composition of mineral particles should be used.

When are think of nanoparticle we associate them mainly with nanotechnology and manufactured nanoparticles. However, naturally formed nanoparticles are ubiquitous in nature and can be found in 10,000 years old glacial ice cores or in

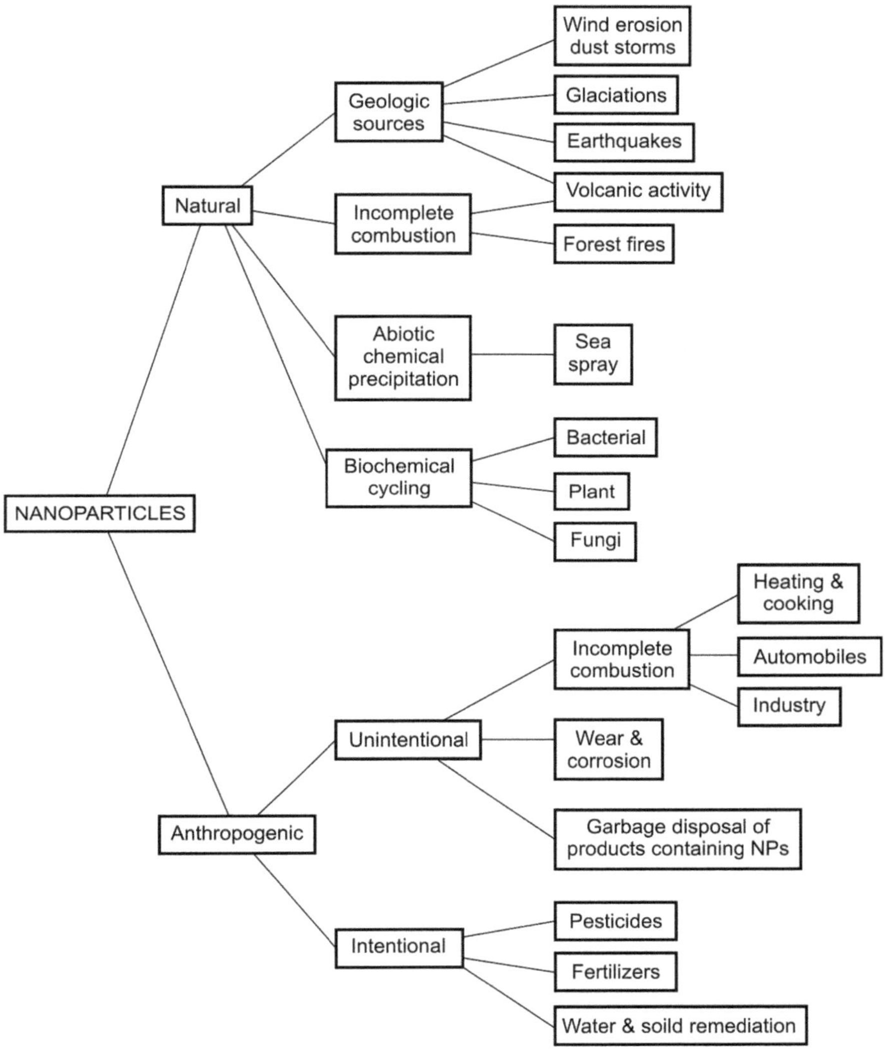

Fig. 3.1 Sources of natural and anthropogenic nanoparticles

the Cretaceous–Tertiary boundary layer (Murr et al. 2004; Verma et al. 2002; Handy et al. 2008a).

For example, studies of samples from the Cretaceous–Tertiary (K-T) boundary layer from Italy shows nanoparticles made of magnetically ordered iron materials, hematite, and a paramagnetic silicate, with a size of about 16–27 nm (Verma et al. 2002). It is believed that these nanoparticles formed due to the impact between Earth and a meteorite, and remained as aerosols for a long time before settling.

Table 3.1 Inorganic marker elements associated with emission sources or processes that generate nanoparticles

Tracers, secondary aerosols		Elements
Crustal or geological		Elements associated with feldspars, quartz, micas, and their weathering products (mostly clay minerals), i.e., Si, Al, K, Na, Ca, Fe and associated trace elements such as Ba, Sr, Rb, and Li. In addition, accessory silicates and carbonates, sulfates, oxides, hydroxides, and phosphates
Anthropogenic	Steel industry	Cr, Ni and Mo
	Copper metallurgy	Cu and As
	Ceramic industries	Ce, Zr, and Pb
	Heavy industry (refinery, coal mine, power stations)	Ti, V, Cr, Co, Ni, Zn, As, and Sb
	Petrochemical industry	Ni and V
	Oil burning	V, Ni, Mn, Fe, Cr, As, S, and SO_4^{2-}
	Coal burning	Al, Sc, Se, Co, As, Ti, Th, S, Pb, and Sb
	Iron and steel industries	Mn, Cr, Fe, Zn, W, and Rb
	Non-ferrous metal industries	Zn, Cu, As, Sb, Pb, and Al
	Cement industry	Ca
	Refuse incineration	K, Zn, Pb, and Sb
	Biomass burning	K and Br
	Firework combustion	K, Pb, Ba, Sb, and Sr
	Vehicle tailpipe	Platinum group elements, Ce, Mo, and Zn
	Automobile gasoline	Ce, La, Pt, SO_4^{2-}, and NO_3
	Automobile diesel	S, SO_4^{2-}, and NO^3
	Mechanical abrasion of tires	Zn
	Mechanical abrasion of brakes	Ba, Cu, and Sb

Adapted from Calvo A. I. et al., Research on aerosol sources and chemical composition: Past, current and emerging issues. Atmospheric Research, 2013, 120–121, 1–28, Copyright (2013), with permission from Elsevier (Calvo et al. 2013)

Natural nanoparticles can be organic and inorganic. The focus of this chapter is on inorganic nanoparticles. Natural nanoparticles sources can be classified as geologic, incomplete combustion, abiotic precipitation, and biochemical cycling. These sources of natural nanoparticles can involve bottom-up processes, which begin with molecular or ionic species, or top-down processes which begin with a larger precursor. Examples of bottom-up approaches are metallic nanoparticles formation via biomineralization or halide and hydrous sulfate nanoparticles from evaporation of sea spray (Sharma et al. 2015; Hochella et al. 2008). Examples of top-down approaches are nanoparticles generated by wind erosion in deserts, nanoparticles from incomplete combustion of matter.

Nowadays, the mineral dust generated by blowing wind from arid and agricultural lands constitutes probably the largest source of natural nanoparticles (Hochella et al. 2008). Other considerable source of natural nanoparticles are those generated by volcanoes and forest fires as ash, and biological debris (Bernhardt et al. 2010).

The composition of naturally generated nanoparticles is varied, ranging from: metals, metal oxides, and their compounds, alloys, carbon allotropes and silicates (Sharma et al. 2015; Guo and Barnard 2013; Hough et al. 2011). Noble metal nanoparticles are found in various environments. Figure 3.2 shows schematics, scanning electron microscopy (SEM) images and transmission electron microscopy (TEM) images of several types of natural nanoparticles: (a), (b) imogolite, a naturally occurring aluminosilicate nanotube in volcanic soils (Wilson et al. 2008; Yamamoto et al. 2005), (c) natural hydrous iron oxide aggregates on a mixture of organic fibrils found between 6.5 and 7.5 m in the water column (Taillefert et al. 2000), (d) weathered quartz fracture with Au nanoparticles (Hough et al. 2011), (e) bacteriomorphic gold (Brugger et al. 2013). Sometimes mineral nanoparticles can form in aqueous environments via nanobiomineralization, as aggregates with natural organic matter. Mineral weathering can generate nanoparticles of Fe and Mn oxides or ferrihydrite (Bakshi et al. 2015). Weathering reactions result in the formation of nanoparticle aggregates and coatings. Hence, nanoparticles in the environment have a dynamical chemical composition, size, and shape.

3.3.1 Nanoparticles from Wind Erosion and Dust Storms

Natural processes, such as weathering, can result in the formation of nanoparticles in soil, atmosphere, and aquatic environments. Weathering is a process of physical abrasion or chemical dissolution of rock material that results in the formation of nanoparticles and microparticles at the solid–air or solid–liquid interface. Weathering existed on Earth since 4 billion years from the time Earth was cold enough to have an atmosphere (Handy et al. 2008a). Today, the mineral dust generated by blowing wind from arid and agricultural lands constitutes probably the largest source of natural nanoparticles (Hochella et al. 2008).

Wind erosion is a top-down process of generating nanoparticles from larger precursors. Nanoparticle generation occurs via aeolian erosion in the desert, and on lands that is not covered by vegetation. Dust deposition occurs annually and increases with extended drought and land disturbance in source regions (Painter et al. 2007). Nanoparticles generated from dust storms are the largest amount of natural nanoparticles in the atmosphere.

Dust storms originate in drylands, the main sources being the Sahara and Kalahari (Africa), Arabian desert (Middle East), Gobi desert (Asia), Patagonian desert (South America), Western US deserts (North America) and Great Victoria desert (Australia). Desert dust has an effect on human health and ecosystems

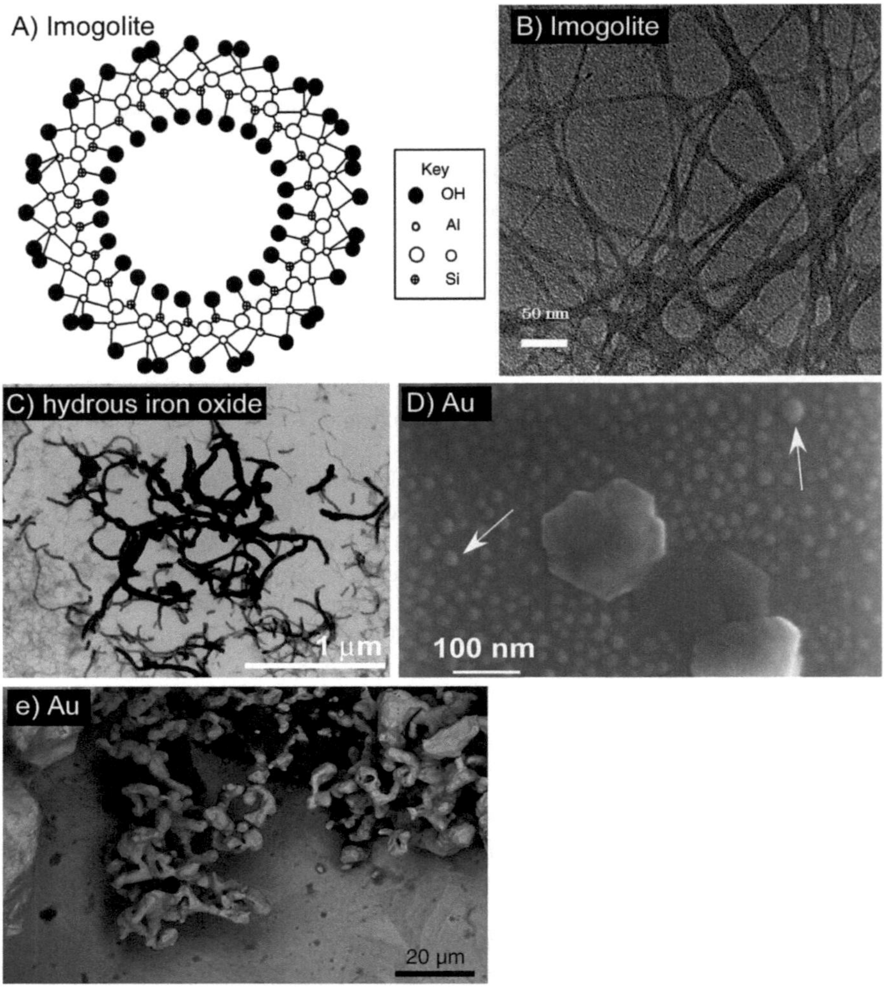

Fig. 3.2 (**a**) Schematics of imogolite structure. Reprinted from Wilson M. A. et al., Nanomaterials in soils. Geoderma, 146, 2008, 291–302. Copyright (2008), with permission from Elsevier (Wilson et al. 2008). (**b**) TEM image of imogolite, a naturally occurring aluminosilicate nanotube. Reprinted from Yamamoto K. et al., Preparation and properties of poly(methyl methacrylate)/imogolite hybrid via surface modification using phosphoric acid ester. Polymer, 2005, 46, 12386–12392. Copyright (2005), with permission from Elsevier (Yamamoto et al. 2005). (**c**) Natural hydrous iron oxide aggregates on a mixture of organic fibrils found between 6.5 and 7.5 m in the water column. Reprinted from Taillefert M. et al., Speciation, reactivity, and cycling of Fe and Pb in a meromictic lake. Geochimica Et Cosmochimica Acta, 2000, 64, 169–183, Copyright (2000), with permission from Elsevier (Taillefert et al. 2000). (**d**) Image of the surface of a weathered quartz fracture which shows 20 nm nanospheres (*white arrows*) in the background next to coarser Au nanoparticles. Reprinted from Hough R. M. et al., Natural gold nanoparticles, Ore Geology Reviews, 2011, 42, 55–61, Copyright (2011), with permission from Elsevier (Hough et al. 2011). (**e**) SEM image of bacteriomorphic gold on a gold grain from soil at the Platina Lead at Fifield. Reprinted from Brugger J. et al., Can biological toxicity drive the contrasting behavior of platinum and gold in surface environments? Chemical Geology 2013, 343, 99–110, Copyright (2013), with permission from Elsevier (Brugger et al. 2013)

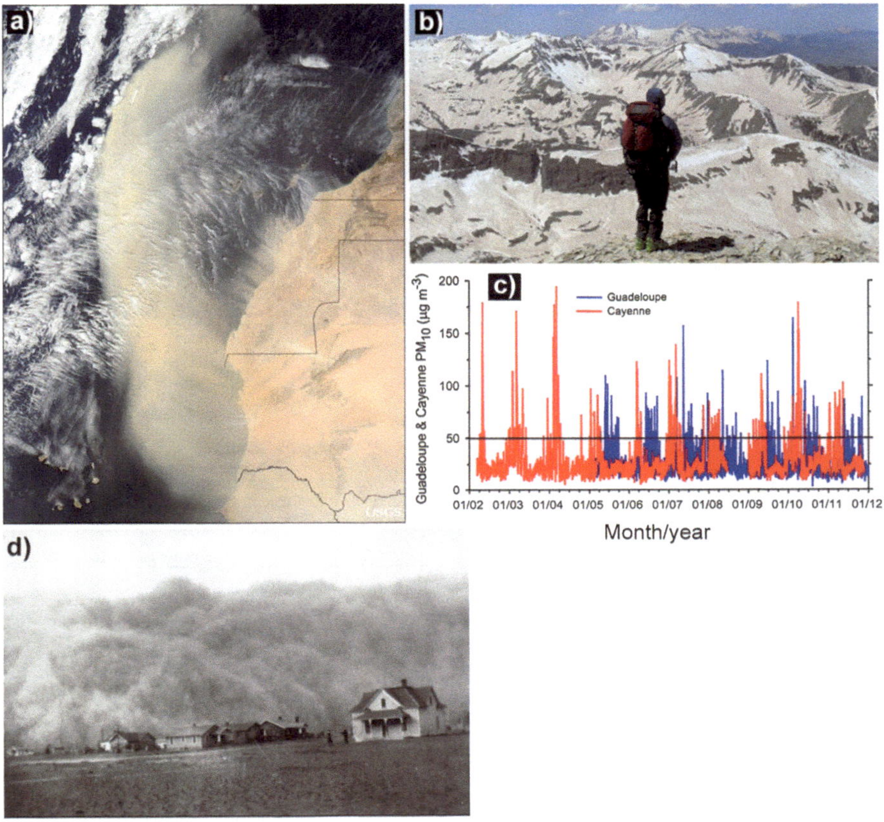

Fig. 3.3 (**a**) A large Saharan dust storm leaves the West African coast to move over the Atlantic Ocean. Courtesy Moderate Resolution Imaging Spectroradiometer (MODIS) aboard NASA's Terra satellite on March 2004. (**b**) Dust deposits on snow covered surface in the San Juan Mountains, Colorado, 12 March 2009. Courtesy of U.S. Geological Survey. (**c**) Plot of the daily average PM10 concentrations (μg/m^3) at Cayenne and Guadeloupe. The horizontal line is the World Health Organization value Air Quality Guideline for 24 h. Reprinted from Prospero J. M. et al., Characterizing the annual cycle of African dust transport to the Caribbean Basin and South America and its impact on the environment and air quality. Global Biogeochemical Cycles, 2014, 28, 757–773, Copyright (2014) with permission from John Wiley and Sons (Prospero et al. 2014). (**d**) Dust storm near Stratford, Texas, 1935. NOAA George E. Marsh Album. Reprinted from Morman S. A. and Plumee G. S., The role of airborne mineral dusts in human disease. Aeolian Research, 9, 203–212, Copyright (2013), with permission from Elsevier (Morman and Plumlee 2013)

specially those situate downwind (Goudie 2014). Dust storms occur many times throughout the year in many regions. Dust particles from dust storms are able to transport pollutants and allergens over immense distances across continents.

Naturally occurring nanoparticles from dust storms have not only a local or regional environmental effect, but also global consequences, as seen in Fig. 3.3.

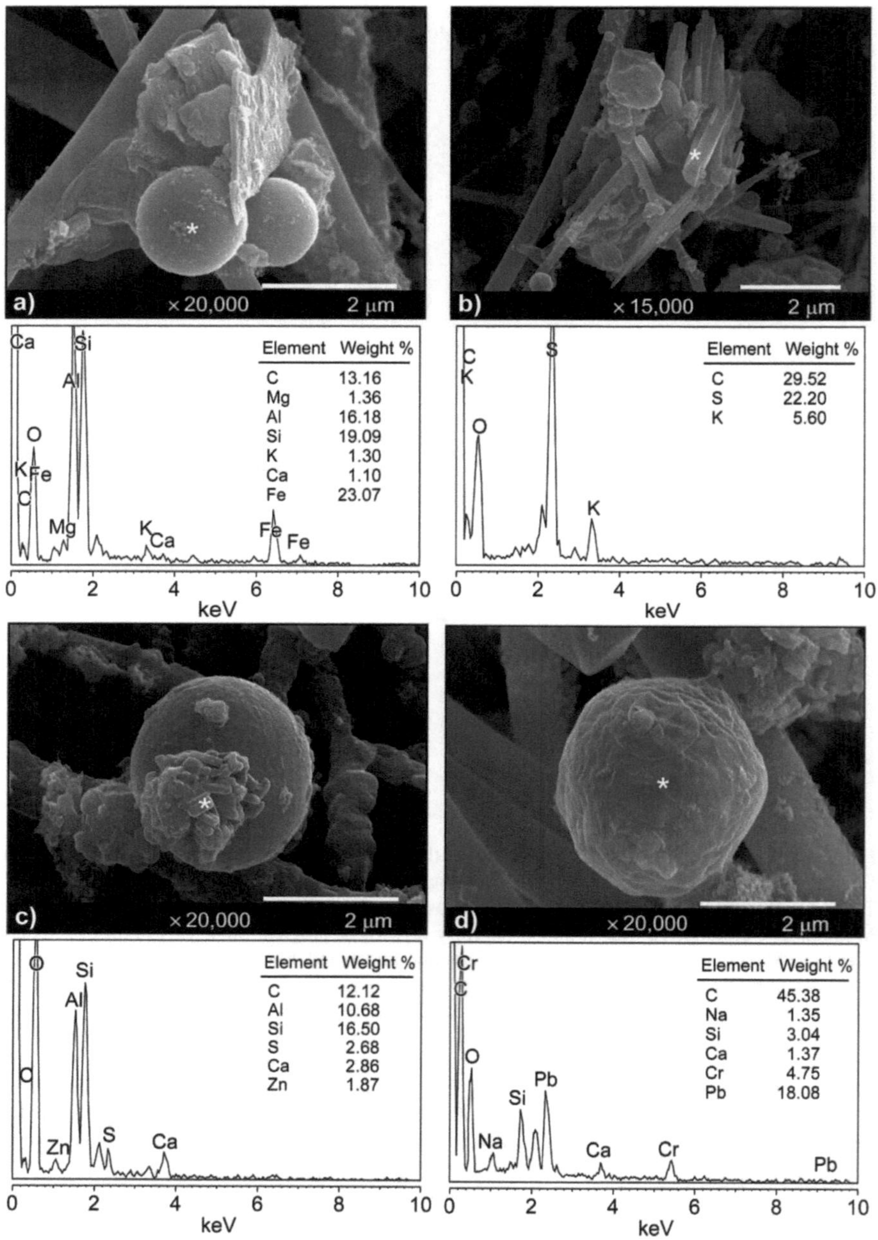

Fig. 3.4 SEM images and composition determined by energy dispersive X-ray spectroscopy (EDS) of different particles from Asian dust storm samples. (**a**) Spherical particle of carbon enriched iron oxides of combustion origin. (**b**) Elongated particles containing sulfur. (**c**) Spherical particle with sulfur-containing nanoparticles, an anthropogenic signature, suggesting the possible adsorption of sulfates on the surface of wind-blown dust particles. (**d**) some heavy metals (e.g., Cr

Dust from dessert storms in Africa are transported across the oceans and seas and are able to affect ecosystems worldwide, as illustrated in Fig. 3.3a, c (Prospero and Mayol-Bracero 2013; Prospero et al. 2014). An example of dust deposition on mountain snow cover is shown in Fig. 3.3b. African mineral dust is transported to the Caribbean basin, showing a strong seasonal cycle, as seen in Fig. 3.3c (Prospero et al. 2014). The measurement of particulate matter at two different locations, at Cayenne (French Guiana) and Guadeloupe, show that the high levels of particles are almost entirely due to African dust.

Figure 3.4 shows examples of scanning electron microscope images and composition of particles from Asian dust storm samples (Kim et al. 2012). Carbon and crustal elements magnesium, aluminum, silicon and calcium are major components in most particles.

3.3.2 Nanoparticles from Volcanic Activity

A class of natural nanoparticle are those generated by volcanic activity. During a volcano eruption gases and large amounts of ash, composed of nanoparticles and microparticles, are ejected in high in the atmosphere, up to a height of 18,000 m (Buzea et al. 2007). During a single volcanic eruption hundreds of tons of ash can be ejected. Due to the air circulation in the upper troposphere and stratosphere, ash can distribute worldwide, blocking and scattering sun radiation. Figure 3.5 shows examples of scanning electron microscope images of volcanic ash and the 2011 Eruption of Shinmoedake volcano in Japan.

Figure 3.6 shows transmission electron microscope images of nanoparticles trapped in ice core drilled from Greenland and estimated to be about 10,000 years old. Some of these core samples include particles related to volcanic eruptions and dust-fall (Murr et al. 2004). Interesting to note that the arctic airborne particulate matter as well as carbon dioxide and methane concentrations are highly correlated. The ice cores of glacier ice are able to shed some light on their concentrations and antiquity (Murr et al. 2004). Various techniques can allow the identification of the time of the ice cores, ranging from 100 years to several millions years old ice identified in Antarctica. The study of particulate matter trapped in ice cores drilled from Greenland from a depth of 506 m, estimated to be about 10,000 years old, showed the composition of nanoparticles to be of carbon, including carbon nanotubes, silica, alumina, and iron-based compound, among others (Murr et al. 2004). Some clusters had a much more complex chemistry, including C, O, Na, S,

Fig. 3.4 (continued) and Pb) on the surface of spherical nanoparticles. The *asterisk* (*) in the photographs indicates the position where EDS analysis was carried out. Reprinted from Kim W. et al., Asian dust storm as conveyance media of anthropogenic pollutants. Atmospheric Environment, 2012, 49, 41–50. Copyright (2012), with permission from Elsevier (Kim et al. 2012)

Fig. 3.5 (**a**) Scanning electron microscope image of volcanic ash from the first volcanic eruption of Mount St. Helen, Washington state, USA in 1980. Courtesy Louisa Howard, http://remf.dartmouth.edu/imagesindex.html. (**b**) SEM image of ash microparticle with nanoparticles on its surface. The ash particle was erupted by Augustine volcano in 2006 and collected during the ashfall in Homer, Alaska by John Paskievitch, AVO. The image was acquired by PavelIzbekov using ISI-40 Scanning Electron Microscope at the Advanced Instrumentation Laboratory, University of Alaska Fairbanks. (**c**) 2011 Eruption of Shinmoedake, Japan. Courtesy U. S. Geological Survey

Cl, K, Ca, Cu, and Zn. The carbon nanotubes are indicative of incomplete combustion processes, such as burning natural gas, while alumina is indicative of weathered alumino silicates (Murr et al. 2004).

3.3.3 Nanoparticle Weathering

Research on aerosol sources and their chemical composition shows that aerosol particles may be removed from the atmosphere by dry and/or wet deposition (Calvo et al. 2013). They can enter soil, interact with the substances from soil, release ions and be available for uptake by plants (Coutris et al. 2012). The size of aerosol

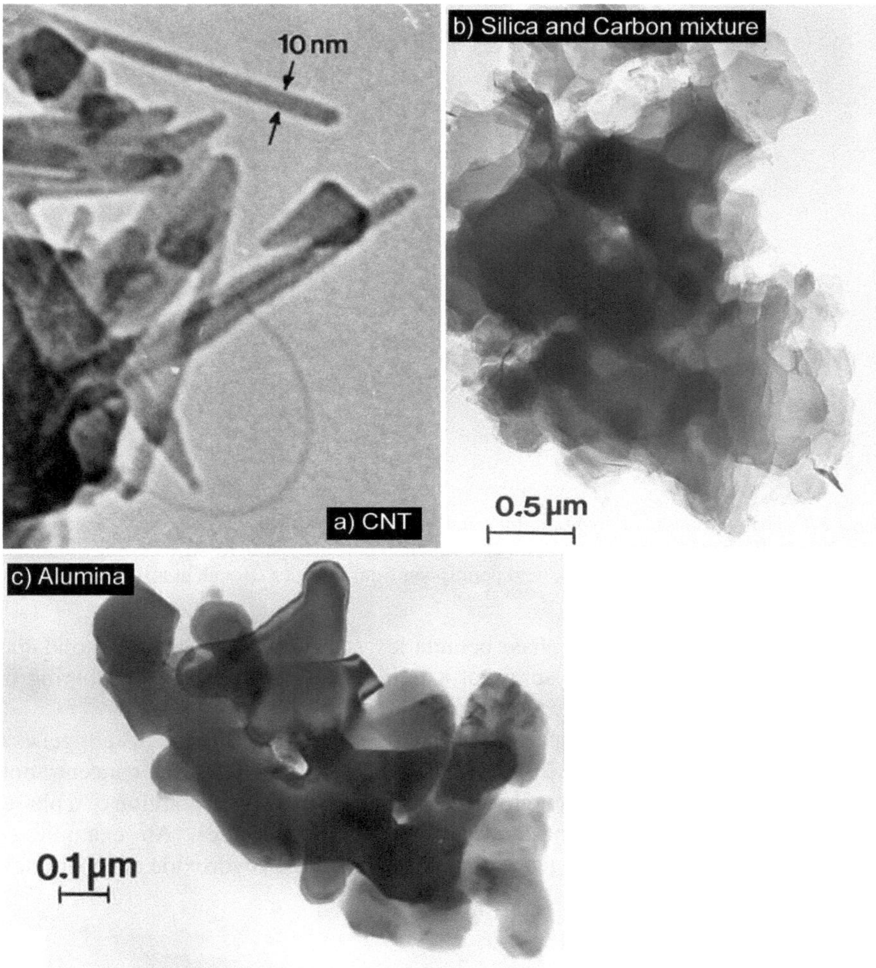

Fig. 3.6 TEM images of nanoparticles trapped in ice core drilled from Greenland and estimated to be about 10,000 years old. (**a**) carbon nanotubes. (**b**) silica SiO_2 and carbon C mixture, and (**c**) Alumina Al_2O_3 nanoparticles. Reprinted from Murr L. E. et al., Chemistry and nanoparticulate compositions of a 10,000 year-old ice core melt water. Water Research, 2004, 38, 4282–4296, Copyright (2004), with permission from Elsevier (Murr et al. 2004)

particle varies from a few nanometers to several tens of microns (Calvo et al. 2013). Size is an important parameter that dictates its environmental behavior.

In many cases, nanoparticles occur via a combination of processes. An example would be weathering, which is a combination of mechanical processes, such as erosion, combined with dissolution and precipitation (an example is shown in Fig. 3.7), or volcanic activity that involves fast cooling of fumes and explosions.

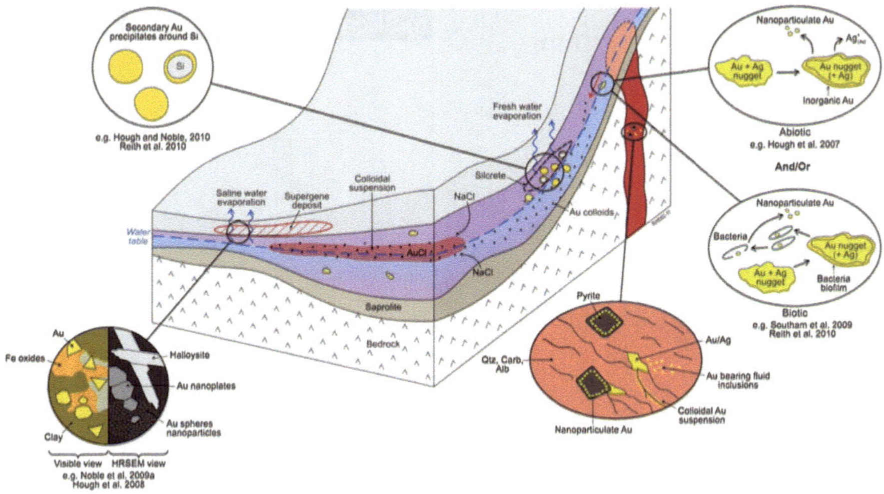

Fig. 3.7 The formation of hydrothermal and supergene gold deposits via gold nanoparticle weathering. Reprinted from Hough R. M. et al., Natural gold nanoparticles, Ore Geology Reviews, 2011, 42, 55–61, Copyright (2011), with permission from Elsevier (Hough et al. 2011)

Usually nanoparticles form at phase boundaries, such as wind erosion at a solid–gas interface, evaporation of sea spray, at a liquid–gas interface, and weathering of rocks at a solid–liquid interface (Sharma et al. 2015).

The formation of new nanoparticles from dissolved species is called authigenesis or neoformation (Handy et al. 2008a). This process occurs when the concentration of dissolved species is high enough and exceed the saturation in solution of a phase, and results in the nucleation and growth of nanoparticles. An example of nanoparticles formed by authigenesis are clay and iron oxyhydroxide nanoparticles.

3.3.4 *Biogenic Nanoparticles: From Bacteria*

Organisms, especially microorganisms, generate a large amount of inorganic nanoparticles in the environment, such as iron, silicon and calcium based minerals (Sharma et al. 2015). These biogenic nanoparticles are ubiquitous. The morphology of these particles varies from wires, see for example Fig. 3.8 (Leung et al. 2013), to spherical or cuboidal shapes, as seen in Fig. 3.9 (Fairbrother et al. 2013; Reith et al. 2014; Lengke et al. 2006; Alphandéry 2014).

Microorganisms that produce nanoparticles are bacteria, fungi, and yeast (Prasad et al. 2016). Table 3.2 gives some examples of microorganisms able to synthesize nanoparticles and the composition of nanoparticles. The composition of nanoparticles produced by microorganism is varied, ranging from precious metals to magnetic materials, such as Au, Ag, Ba, Hg, Se, Cd, Te, CdS, Cu, U, Tc, Cr, Co,

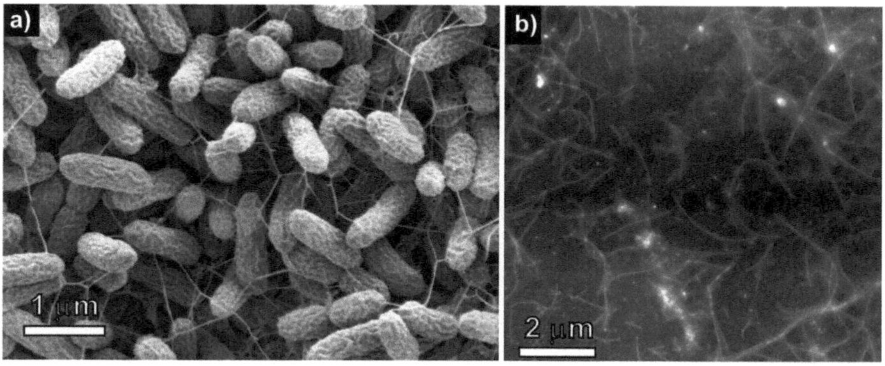

Fig. 3.8 Nanowires generated by *Shewanella oneidensis* MR-1 strain. (**a**) SEM image of *S. oneidensis* exhibiting large amounts of nanowires. (**b**) atomic force microscopy image of nanowires onto a flat surface. Reprinted with permission from Leung K. M. et al., Shewanella oneidensis MR-1 Bacterial Nanowires Exhibit p-Type, Tunable Electronic Behavior. Nano Letters, 2013, 13, 2407–2411. Copyright (2013) American Chemical Society (Leung et al. 2013)

Mn, Pd, Pt, Zr, Fe_3O_4, Fe_3S_4, Zn, TiO_2, among other (Nath and Banerjee 2013; Quester et al. 2013).

Some types of bacteria, as a response to environmental stresses (such as toxicity of metal ions), have developed a defense mechanism that allows them to survive and grow in environments with high metal ion concentrations (Quester et al. 2013; Prasad et al. 2016). As a result, many bacterial species are currently used to produce metallic nanoparticles as an alternate route to simple chemical synthesis. This method of nanoparticle biosynthesis can use intact cells and cell extracts. In Table 3.2 are presented types of bacteria used to produce a large variety of metallic nanoparticles. The synthesis of nanoparticles by bacteria can be an intracellular or extracellular process, depending on the type of bacteria and material.

Biologically induced mineralization (BIM) is one of the processes involved in nanoparticle formation by organisms (Sharma et al. 2015). During this process nanoparticles are formed as a result of metabolic processes, the nanoparticles having no particular function in the functioning of the organisms, being only in a substrate attached to bacteria, or interacting with bacterial cell wall.

Biologically controlled mineralization (BCM) is another process when, in contrast to the previously discussed one, the growth of the nanoparticles is controlled by the organisms, being usually formed inside the cells (Sharma et al. 2015). Nanoparticles formed during this process are usually crystalline with a narrow particle-size distribution and serve different functions within the organisms, such as iron storage, tissue hardening, electron transfer mediators to promote oxidation and reduction under anaerobic conditions. For example, magnetotactic bacteria generates magnetite nanoparticles that are used for navigation, or in *Shewanella oneidensis* hematite nanoparticles facilitate respiration (Sharma et al. 2015).

Some bacteria developed a strategy for efficient electron transfer and energy distribution by generating electrically conductive appendages having the shape of

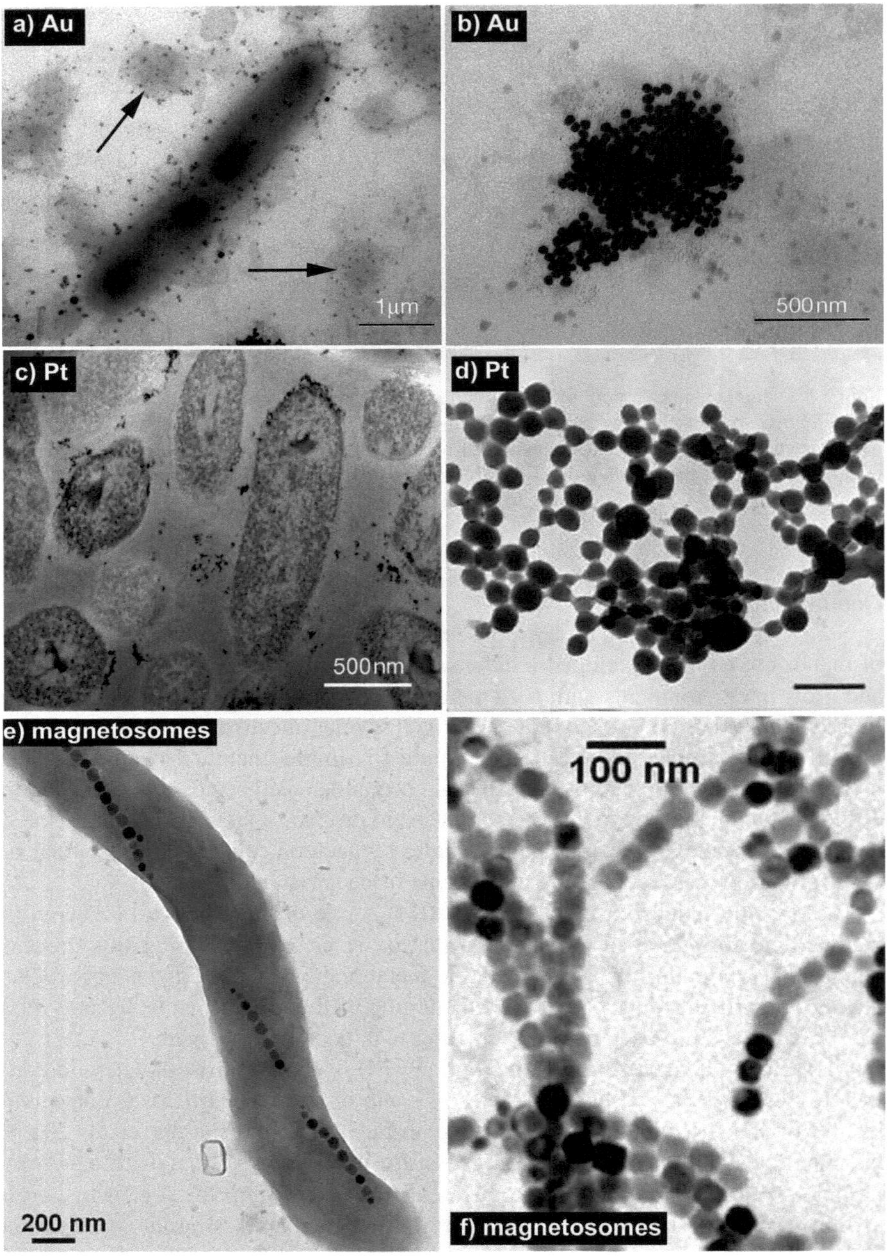

Fig. 3.9 (**a**) TEM of *Cupriavidus metallidurans* bacterium capable of precipitating gold nanoparticles from aqueous gold complexes. (**b**) agglomerate of Au nanoparticles synthesized by *Cupriavidus metallidurans* bacterium. Images (**a**) and (**b**) are reprinted with permission from Fairbrother L. et al., Biomineralization of Gold in Biofilms of *Cupriavidus metallidurans*. Environmental Science & Technology, 2013, 47, 2628–2635. Copyright (2013) American Chemical Society (Fairbrother et al. 2013). (**c**) TEM image of *C. metallidurans* exposed to Pt(II)-chloride

nanowires (Gorby et al. 2006). This is the case for the mesophilic bacterium *Shewanella oneidensis*, cyanobacterium Synechocystis PCC6803, and the thermophilic bacterium Pelotomaculum (Gorby et al. 2006; Leung et al. 2013).

Magnetosomes are nanoparticles synthesized intracellularly by magnetotactic bacteria, as seen in Fig. 3.9e, f (Alphandéry 2014). Magnetotactic bacteria use the magnetosomes as tiny compasses to navigate using the geomagnetic field. These bacteria are ubiquitous in sediments of fresh water, marine and hypersaline habitats, being found on all continents (Lefevre and Bazylinski 2013). The structure of a magnetosome is composed of a magnetic nanoparticle surrounded by lipid bilayer membrane. The magnetic nanoparticles are morphologically uniform, with a size between 35 and 120 nm, with a narrow size distribution, for the most studied species of magnetotactic bacteria. The composition of the core nanoparticle is high purity and crystallinity magnetite (Fe_3O_4) or greigite (Fe_3S_4) (Lefevre and Bazylinski 2013). The magnetosome has usually a single domain magnetic nanoparticle, as opposed to chemically synthesized iron oxide nanoparticles which are usually multidomain. As shown in Fig. 3.9e, the magnetosomes are ordered in chains inside the magnetotactic bacteria. The magnetic nanoparticles are coated by lipids and proteins, which gives it a total negative charge and good water dispersability.

Via their interactions with some metals, microorganisms are involved in metal cycling, determining their mobility in surface environments (Brugger et al. 2013). For example Au has a high mobility in many near-surface environments due to their interaction with metal-resistant bacteria.

3.3.5 Biogenic Nanoparticles: From Plants

Plants have different behavior towards excess metals present in the growth media. Metal-tolerant ones limit the uptake of nanoparticles into the photosynthetic tissue by restricting the transport of metals across the root endodermis and storing them on the root cortex (Manceau et al. 2008). Hyper-accumulating plants uptake and translocate excess amounts of metals in harvestable parts (Manceau et al. 2008).

Fig. 3.9 (continued) showing Pt nanoparticles in the cytoplasm and periplasm. Reprinted from Reith F. et al., Platinum in Earth surface environments. Earth-Science Reviews, 2014, 131, 1–21, Copyright (2014), with permission from Elsevier (Reith et al. 2014). (**d**) TEM image of Pt nanoparticles forming long beadlike chains in cyanobacteria-PtCl4° systems. Scale bar 0.25 μm. Reprinted with permission from Lengke M. F. et al., Synthesis of Platinum Nanoparticles by Reaction of Filamentous Cyanobacteria with Platinum(IV)–Chloride Complex. Langmuir, 2006, 22, 7318–7323. Copyright (2006) American Chemical Society (Lengke et al. 2006). (**e**) Magnetosomes inside a magnetotactic bacteria. (**f**) magnetosomes extracted from the magnetotactic bacteria. Images (**e**) and (**f**) are reprinted from Alphandery E., 2014, Applications of Magnetosomes Synthesized by Magnetotactic Bacteria in Medicine, Frontiers in Bioengineering and Biotechnology, 2, 5 (Alphandéry 2014)

Table 3.2 Microorganisms able to synthesize nanoparticles and the composition of nanoparticles (NPC)

NPC	Bacteria and actinomycetes	Fungi	Yeast
Au	*Brevibacterium casei*, *Bacillus licheniformis*, *Bacillus subtilis*, *Brevibacterium casei*, *Candida utilis*, *Cupriavidus metallidurans*, *Escherichia coli*, *Desulfovibrio desulfuricans*, *Neurospora crassa*, *Plectonema boryanum*, *Pseudomonas stutzeri*, *Pseudomonas aeruginosa*, *Rhodococcus* sp., *Rhodopseudomonas capsulate*, *Sargassum wightii*, *Shewanella algae*, *Shewanella oneidensis*, *Streptomyces hygroscopicus*, *Thermomonospora* sp., *Yarrowia lipolytica*	*Alternaria alternata*, *Aspergillus niger*, *Colletotrichum* sp., *Coriolis versicolor*, *Fusarium oxysporum*, *Neurospora crassa*, *Rhizopus oryzae*, *Trichoderma viride*, *Verticillium luteoalbum*, *Volvariella volvaceae*	*Pichia jadinii* (*Candida utilis*), *Yarrowia lipolytica*
Ag	Aeromonas spp., *Aspergillus flavus*, *Aspergillus fumigatus*, *Brevibacterium casei*, *Bacillus cereus*, *Bacillus subtilis*, *Bacillus licheniformis*, *Bacillus thuringiensis*, *Brevibacterium casei*, *Corynebacterium glutamicum*, *Escherichia coli*, *Fusarium oxysporum*, *Geobacter sulfurreducens*, *Lactobacillus* sp. from yoghurt, Morganella sp., *Morganella psychrotolerans*, *Neurospora crassa*, *Shewanella oneidensis*, *Phaenerochaete chrysosporium*, *Plectonema boryanum*, *Proteus mirabilis*, *Staphylococcus aureus*, *Streptomyces albidoflavus*, *Streptomyces hygroscopicus*, *Trichoderma viride*, *Ureibacillus thermosphaericus*	*Amylomyces rouxii* KSU-09, *Aspergillus clavitus*, *Aspergillus flavus*, *Aspergillus fumigatus*, *Aspergillus terreus*, *Aspergillus niger*, *Cladosporium cladosporioides*, *Coriolis versicolor*, *Fusarium oxysporum*, *Neurospora crassa*, *Penicillium fellutanum*, Penicillium strain J3, Phanerochaete, *Phoma glomerata*, *Pleurotus sajor caju*, *Rhizopus stolonifera*, *Streptomyces* sp., *Chrysosporium*, *Trichoderma asperellum*, *Trichoderma viride*, *Verticillium*, *Volvariella volvaceae*	Yeast strain MKY3
Ba		*Fusarium oxysporum*	
Hg	Enterobacter sp.	*Aspergillus versicolor*	
Se	Shewanella sp.		
Cd	*Escherichia coli*	*Fusarium oxysporum*	

(continued)

Table 3.2 (continued)

NPC	Bacteria and actinomycetes	Fungi	Yeast
			Candida glabrata, *Schizosaccharomyces pombe*
Te	*Escherichia coli*		
CdS	*Escherichia coli*, *Lactobacillus* sp., *Rhodopseudomonas palustris*	*Fusarium oxysporum*	*Candida glabrata*, *Schizosaccharomyces pombe*
Cu	*Mycobacterium smegmatis*		
U	*Pyrobaculum islandicum*		
Tc	*Pyrobaculum islandicum*		
Cr	*Pyrobaculum islandicum*		
Co	*Pyrobaculum islandicum*		
Co_3O_4	*Brevibacterium casei*		
Mn	*Pyrobaculum islandicum*		
Pd	*Bacillus sphaericus*, *Cupriavidus metallidurans*, *Cupriavidus necator*, *Desulfovibrio desulfuricans*, *Desulfovibrio fructosovorans*, *Shewanella oneidensis*, *Paracoccus denitrificans*, *Plectonema boryanum*, *Pseudomonas putida*		
Pt	*Shewanella algae*, *Plectonema boryanum*	*Fusarium oxysporum*	
Zr			Yeast
Fe_3O_4	Actinobacter spp., Shewanella sp., *Thermoanaerobacter ethanolicus*, Thermoanaerobacter sp., various magnetotactic bacteria		
Fe_3S_4	Actinobacter spp., various magnetotactic bacteria		
ZnS	*Rhodobacter sphaeroides*		
Zn		Fusarium spp.	
ZnO	*Bacillus cereus*		
TiO_2	*Bacillus subtilis*, Lactobacillus sp., Lactobacillus sp. from yoghurt	*Aspergillus flavus*	
UO_2	*Anaeromyxobacter dehalogenans*, *Shewanella oneidensis*		

(continued)

Table 3.2 (continued)

NPC	Bacteria and actinomycetes	Fungi	Yeast
TcO_2	*Anaeromyxobacter dehalogenans*		

Data adapted with permission from Nath and Banerjee (2013) and Quester et al. (2013)

Hydraulic lift of nanoparticles by deep-rooted vegetation is a mechanism of transport of nanoparticles from deeper sediments to the earth surface (Lintern et al. 2013). It has been shown that Eucalyptus trees are able to translocate Au from mineral deposits at more than 30 m depth (Lintern et al. 2013). Gold nanoparticles were found in the foliage of the studied Eucalyptus tree.

Rice (*Oryza sativa*) uptakes silica from the soil as silicic acid and accumulates around cellulose micro-compartments (Mohammadinejad et al. 2016). As the second largest farmed crop in the world, rice has the potential to become a significant source of silica nanoparticles (Mohammadinejad et al. 2016). Rice husks are considered an important agricultural residue, 1 t of husks for every 5 t of rice. Some authors suggest the use of rice husks as a source of silica nanoparticles (Mohammadinejad et al. 2016). Silica nanoparticles derived from burning the husks is about 20 % of the dry weight of their weight.

Common wetlands plants can transform cationic copper into nanoparticles in and near their roots with the help of fungi when grown in contaminated soils (Manceau et al. 2008). Table 3.3 shows examples of plants that are able to synthesize nanoparticles together with the composition of nanoparticles (Nath and Banerjee 2013; Quester et al. 2013).

3.4 Anthropogenic Nanoparticles

3.4.1 Examples of Engineered Nanoparticles

Engineered nanoparticles can have different compositions: carbon based, metals and their oxides, polymers, etc. Carbon based nanomaterials encompass carbon different allotropes: carbon black, an amorphous form of carbon, graphite, nanodiamonds, carbon nanotubes (CNTs), and fullerenes, such as C60, C70 (Bhatt and Tripathi 2011; Nowack and Bucheli 2007). Examples of carbon allotropes that can be found in nanoform are shown in Fig. 3.10 (Balasubramanian and Burghard 2005). The size of these nanomaterials can range from smaller than 1 nm, 0.7 nm for fullerenes, while carbon nanotubes can have diameter of about 1 nm and varying lengths. Carbon nanotubes can be single walled carbon nanotubes (SWCNTs), or multi-walled carbon nanotubes (MWCNTs) (Balasubramanian and Burghard 2005). SWCNTs are formed of rolled up single-layered graphene sheets while MWCNTs can have several concentric layers with varying diameter, as shown in Fig. 3.10e, g, respectively.

Table 3.3 Plants that synthesize nanoparticles and the composition of nanoparticles

Nanoparticle composition	Plants	Algae
Au	Alfalfa plant (*Medicago sativa*), *Aloe vera*, *Avena sativa*, *Azadirachta indica*, black tea, *Camelia sinensis*, *Camelia sinensis*, *Cassia fistula*, *Cinnamomum camphora*, *Citrus paradisi*, *Coriandrum sativum*, *Cymbopogon flexuosus*, *Diopyros kaki*, *Emblica officinalis*, *Hibiscus rosasinensis*, *Magnolia kobus*, *Pelargonium graveolens*, *Psidium guajava*, *Sesbania drummondii*, *Terminalia chebula*	*Chlorella vulgaris*, *Fucus vesiculosus*, *Sargassum myriocystum*, *Sargassum wightii*
Ag	Alfalfa plant (*Medicago sativa*), *Aloe vera*, *Annona squamosa*, *Azadirachta indica*, Black tea, *Brassica juncea*, *Camellia sinensis*, *Camellia sinensis*, *Capsicum annum*, *Cinnamomum camphora*, *Cinnamomum zeylanicum*, *Citrullus colocynthis*, *Citrus sinensis*, *Coccinia grandis*, *Dioscorea bulbifera*, *Emblica officinalis*, *Ficus racemose*, *Helianthus annus*, *Hibiscus rosasinensis*, *Jatropha curcas* (latex), *Lippia citriodora*, *Mimosa pudica*, *Murraya koenigii*, *Musa paradisiaca*, *Musa paradisiaca*, *Nelumbo nucifera*, *Origanum vulgare*, *Pandanus odorifer*, *Pelargonium graveolens*, *Punica granatum*, Purified Apiin (from henna leaves), *Rhizophora mucronata*, *Solanum torvum*, *Syzygium cumini*, *Terminalia chebula*, *Tribulus terrestris*, *Vinca rosea*	*Chlorella vulgaris*, *Sargassum wightii*, Urosporasp.
Ni	Alfalfa plant (*Medicago sativa*), *Brassica juncea*, *Helianthus annus*	
Zn	Alfalfa plant (*Medicago sativa*), *Brassica juncea*, *Helianthus annus*	
Cu	Alfalfa plant (*Medicago sativa*), *Brassica juncea*, *Helianthus annus*, *Calotropis procera*	
Co	Alfalfa plant (*Medicago sativa*), *Brassica juncea*, *Helianthus annus*	
Pd	*Gardenia jasminoides* Ellis	
Ti	Alfalfa plant (*Medicago sativa*)	
TiO_2	*Annona squamosa*	
Sm	Alfalfa plant (*Medicago sativa*)	
Se	Lemon plant	
FeO	Alfalfa plant (*Medicago sativa*)	

Data adapted with permission from Nath and Banerjee (2013) and Quester et al. (2013)

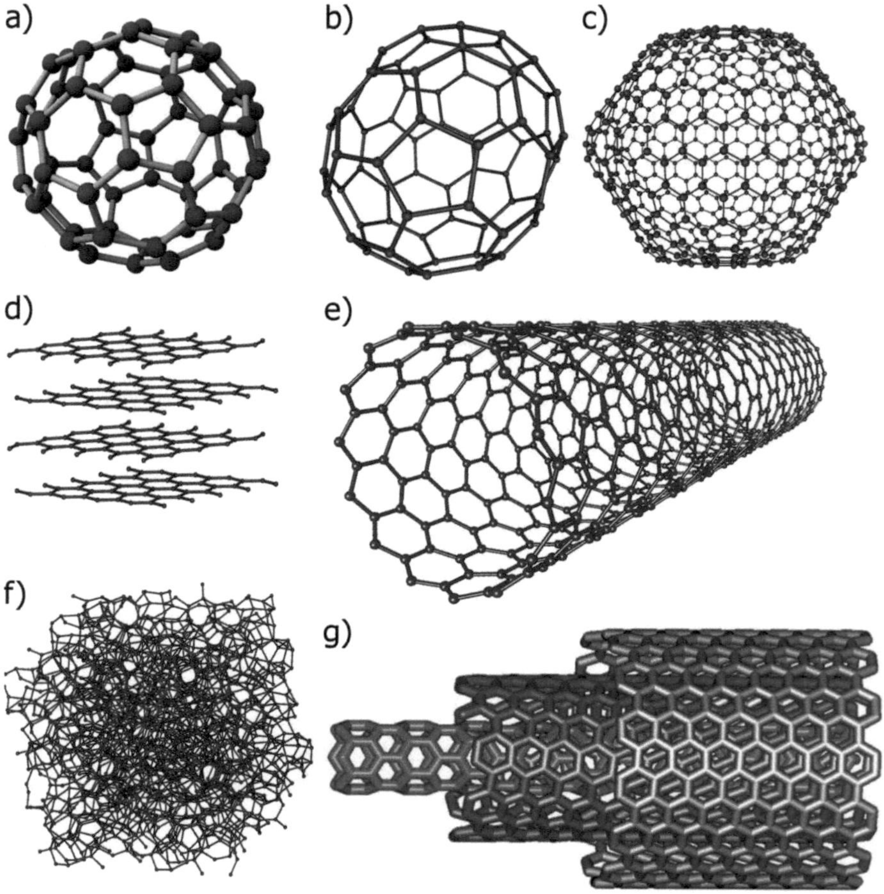

Fig. 3.10 Structure of various carbon allotropes in nanoform. (**a**) fullerene C60, (**b**) C70, (**c**) C540, (**d**) graphite, (**e**) single-walled carbon nanotube SWCNT, (**f**) amorphous carbon, (**g**) multi-walled carbon nanotubes. Image (**a**) courtesy of James Hedberg, images (**b**), (**c**), (**d**),(**e**), and (f) courtesy of Michael Ströck, (**g**) reproduced with permission from Balasubramanian and Burghard (2005), John Wiley and Sons

Various types of nanoparticles are currently synthesized for various applications. They range from metals, metal oxides (such as TiO_2 and ZnO), non-metals, such as carbon and silica. Figure 3.11 shows transmission electron microscope images of various anthropogenic nanoparticles: rattle-type Au/CdS nanostructures (Xia and Tang 2012), MWCNTs (Larue et al. 2012), silica nanoparticles (Nair et al. 2011), cobalt oxide nanoparticles, zinc oxide nanoparticles (Ghodake et al. 2011), ceria nanoparticles (Zhang et al. 2011), and rare earth oxides nanoparticles La_2O_3, Gd_2O_3, Yb_2O_3 (Ma et al. 2010).

Magnetic nanoparticles are being used extensively in various fields, ranging from medicine to water remediation. Examples of magnetic nanoparticles are Fe,

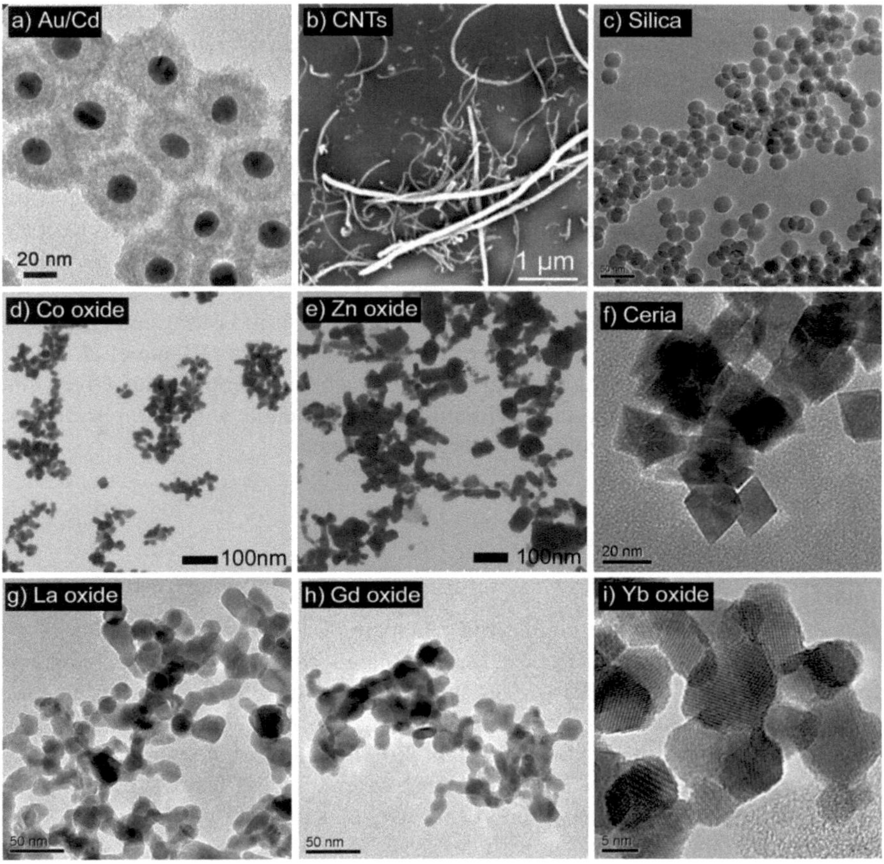

Fig. 3.11 TEM images of various anthropogenic nanoparticles. (**a**) Rattle-type Au/CdS nanostructures. Reprinted from reference (Xia and Tang 2012) with permission from John Wiley and Sons. (**b**) MWCNTs. Reprinted from Larue C. et al., Quantitative evaluation of multi-walled carbon nanotube uptake in wheat and rapeseed. Journal of Hazardous Materials, 227, 155–163, Copyright (2012), with permission from Elsevier (Larue et al. 2012). (**c**) Silica nanoparticles. Reprinted from Nair R. et al., Uptake of FITC Labeled Silica Nanoparticles and Quantum Dots by Rice Seedlings: Effects on Seed Germination and Their Potential as Biolabels for Plants. Journal of Fluorescence, 21, 2057–2068. Copyright (2011). With permission from Springer (Nair et al. 2011). (**d**) Cobalt oxide nanoparticles. (**e**) Zinc oxide nanoparticles. Images (**d**) and (**e**) are reprinted from Ghodake G. et al., Hazardous phytotoxic nature of cobalt and zinc oxide nanoparticles assessed using Allium cepa. Journal of Hazardous Materials, 186, 2011, 952–955, Copyright (2011), with permission from Elsevier (Ghodake et al. 2011). (**f**) Ceria nanoparticles. Reproduced from Zhang Z. Y. et al., Uptake and distribution of ceria nanoparticles in cucumber plants. Metallomics, 2011, 3, 816–822 with permission of The Royal Society of Chemistry (Zhang et al. 2011). (**g**) La_2O_3, (**h**) Gd_2O_3, (**i**) Yb_2O_3. Images (**g**), (**h**), and (**i**) are reprinted from Ma Y. et al., Effects of rare earth oxide nanoparticles on root elongation of plants. Chemosphere, 78, 273–279, Copyright (2010), with permission from Elsevier (Ma et al. 2010)

Co, Ni, Fe_3O_4, $CoFe_2O_4$, $HoMnO_3$, $MnFe_2O_4$, FePt (Kolhatkar et al. 2013; Chekli et al. 2016). Interesting to note that nanoparticles made of materials that are not magnetic in bulk form may acquire magnetic properties when in nanoform (Buzea et al. 2007). This is the case for nanoparticles with size of few nanometers made of Au, Pd (Hori et al. 1999), Pt (Sakamoto et al. 2011).

Quantum dots are a class of nanoparticles made of semiconductor nanocrystals with special optical properties, usually covered in a shell (silica). The role of the shell is to protect the core from oxidation and to mitigate the quantum dots toxicity. The core can be made of cadmium selenide CdSe, cadmium telluride CdTe, indium phosphide InP, or zinc selenide ZnSe (Farre et al. 2011).

Nowadays engineered nanoparticles are fabricated in industrial amounts. Each class of materials, such as SWCNTs exceed 1000 t per year (Klaine et al. 2008). Nanotechnology is one of the most rapidly growing fields of application. Given this, the entry of nanoparticles in the environment is unavoidable and its implication on the ecosystem health is becoming an increasing concern.

3.4.2 Pesticides and Fertilizers

A major way to introduce engineered nanoparticles in agricultural fields is the intentional use of nanoparticles as agrichemicals for crop protection and soil remediation (Deng et al. 2014). Agrichemicals encompass nanosystems that act as delivery devices for specific target tissues, pesticides additives that enhance solubility of active ingredients (Deng et al. 2014). Nanotechnology in the pesticide field comprise either very small particles of pesticides, nanocages, or small nanoparticles with pesticidal properties when in nanoform (Khot et al. 2012). Pesticides containing silver nanoparticles are used for their properties of supressing harmful organisms.

Nanoparticles with compositions of TiO_2, ZnO, Ag, Zn, Au are being developed for pesticide remediation due to their photocatalytic activity (Aragay et al. 2012; Gogos et al. 2012; Kah and Hofmann 2014). Other nanomaterials, such as Fe_3O_4, are being used for pollutant removal due to their magnetic properties (Aragay et al. 2012). Some research papers study the effects of carbon nanotubes and fullerenes on some pesticide uptake by agricultural plants (De La Torre-Roche et al. 2013).

The idea that nanomaterials can be applied in agriculture started to evolve about a decade or so ago, after the year 2000 (Gogos et al. 2012). There is an increasing number of research papers and patents involving nanomaterials for plant protection or fertilizers, as shown in Table 3.4 (Gogos et al. 2012; Aragay et al. 2012; Kah and Hofmann 2014; Servin et al. 2015). The composition of these nanomaterials is carbon-based (carbon nanotubes, liposomes, organic polymers), titanium dioxide, silver, silica, and alumina (Gogos et al. 2012). They serve as additives (for controlled release) and as active constituents.

Gogos et al. analyzed publications and patents of nanomaterials used for plant protection and fertilization (Gogos et al. 2012). They found that nanomaterials can be in the form of solid nanoparticles or nonsolid structures (lipid or polymer). As

Table 3.4 Nano-enabled products/patents in agriculture

Name of product/patent	Product type	Nanoparticle composition	Patent number	Inventors
Active nano-grade organic fine humic fertilizer and its production	Active organic Fertilizer	Nano-fermented active organic fertilizer	CN1472176-A	Wu (2002)
Application of hydroxide nano rare earth to produce fertilizer products	Fertilizer	Hydroxide of nano rare earth	CN1686955-A	Wang et al. (2007)
Application of oxide nano rare earth in fertilizer	Fertilizer	Nano rare earth	CN1686957-A	Wang et al. (2005)
Environment-friendly carbon-nano synergistic complex fertilizers	Fertilizer	Carbon nanomaterials	US 0174032-A1	Jian and Zhiming (2011)
Nano-composite superabsorbent containing fertilizer nutrients used in agriculture	Fertilizer	Nano-composite carbohydrate graft copolymer	US 0139347-A1	Barati (2010)
Nano-diatomite and Zeolite ceramic crystal powder	Fertilizer	Nano diatomite and zeolite	US 0115469 -A1	Yu (2005)
Nano-leucite for slow-releasing nitrogen fertilizer and Green environment	Fertilizer	Potassium aluminum silicate (Leucite) NPs occluded by calcium ammonium nitrates	US 0190226-A1	Farrukh and Naseem (2014)
Nano long-acting selenium fertilizer	Fertilizer	Nano-selenium	US 0326153-A1	Ying et al. (2012)
Nano-micron foam plastic mixed polymer fertilizer adhesive coating agent preparation method	Fertilizer	Nano-micron-foamed plastic organic compound mixed polymer	CN1631952-A	Zhang (2006a)
New method for preparation of controlled release special fertilizer comprises mixing and granulating Ximaxi clay minerals, coating with various fertilizers, trace elements, and additives	Fertilizer	Nano-clay	CN1349958-A	Li (2002)
Nonmetallic nano/microparticles coated with metal, process and applications	Fertilizer	Core of the nonmetallic nano/microparticles is selected from inorganic material such as silica and barium sulfate The metal coating is selected from Ag or	US 0047546-A1	Malshe and Malshe (2010)

(continued)

Table 3.4 (continued)

Name of product/patent	Product type	Nanoparticle composition	Patent number	Inventors
		transition/noble metals Cu, Ni, Ag, Pd, Os, Ru, Rh		
Process compromises combining soil repairing technique and nanobiological fertilizer to promote growth of microbes, improve soil, and remove residual herbicides	Biological Fertilizer	Nano-class biological fertilizer	CN1413963-A	Min (2003)
Production of novel precision customized control release fertilizers	Controlled release fertilizers	Transition metal silicates	US 8375629 B2	Prasad (2013)
Production technology of nano-clay–polyester mixed polymer fertilizer coating cementing agent	Controlled releasing fertilizer; soil improver	Nano-clay polyester mixed polymer	CN1414033-A	Zhang (2004b)
Production technology of coating cement for nanosulfonate lignin mixture fertilizer	Coating cement for Controlled release Fertilizer	Nano-sulfonated lignin mixture water solution	CN1417173-A	Zhang (2004a)
Preparation of nanometer-scale olefin/starch mixed polymer fertilizer covering agent	Slow release Fertilizer	Nano-level nonhomogeneous phase mixed polymer of hydroxyethylmethacrylate	CN1546543-A	Zhang (2006b)
Silicon nanocarrier for delivery of drug, pesticides, and herbicides, and for wastewater treatment	Pesticide	Nano-silicon carrier	US 0225412-A1	Sardari et al. (2013)
HeiQ AGS-20	Pesticide	Silver–silica composite material	US 0294919-A1 *Product available in the market	Jaynes et al. (2012)
Nano-Argentum 10	Fertilizer/antifungal/bug repellent	Silver	*Product available in the market	(Nano-Argentum)

Reproduced from Journal of Nanoparticle Research, A review of the use of engineered nanomaterials to suppress plant disease and enhance crop yield, Servin A. et al., 17, 2015. Copyright (2015). With permission of Springer (Servin et al. 2015)

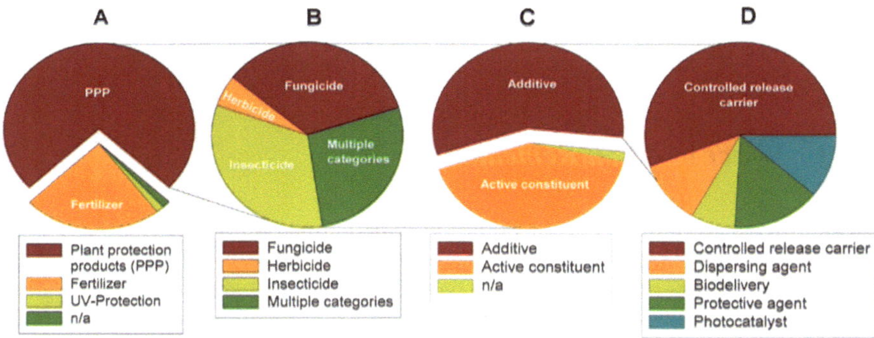

Fig. 3.12 Nanomaterials applications in agriculture (**a**), types of plant protection products—PPP containing nanoparticles, (**b**), general functions of nanomaterials in PPP (**c**), and tasks of nanomaterials additives in PPP (**d**). Reprinted with permission from Gogos A. et al., Nanomaterials in Plant Protection and Fertilization: Current State, Foreseen Applications, and Research Priorities. Journal of Agricultural and Food Chemistry, 2012, 60, 9781–9792. Copyright (2012) American Chemical Society (Gogos et al. 2012)

seen in (Fig. 3.12a), plant protection products were the topic of 74 % of all papers, with the rest on fertilizers and UV protection. 35 % of the plant protection products were used as fungicides, 33 % as insecticides, and 27 % had multiple function (Fig. 3.12b). Nanomaterials were the active component in 41 % of the plant protection products, and additives in 57 %, acting as controlled release carriers (Fig. 3.12c, d).

Another suggested application of nanoparticles in agriculture that is still in the developmental stages is for genetically modified crops with nanoparticles acting as vectors for DNA (Kah et al. 2013). The increase interest for the utilization of nanoparticles in agriculture should be addressed by regulatory bodies in order to control their use as nanopesticides or fertilizers (Kookana et al. 2014). A large number of patents involving nanoparticles have been already filed in agriculture and food industry and many research papers explore the use of nanoparticles for improved seed germination and crop yield. For example, carbon nanotubes are researched for improving the germination of some seeds (Alexandru et al. 2012). However, note should be made that multiple studies have proved that carbon nanotubes are quite toxic to humans and animals. Thus, the true benefit of carbon nanotubes in agriculture seems quite controversial and perhaps dangerous.

In the future, it is reasonable to speculate that the unregulated use of engineered nanoparticles in agriculture and food industry may pose a threat to food safety and health of unawares consumers. The accumulation of nanoparticles in soils and groundwater may potentially lead to accumulation of nanoparticles in plants. The long term effects of nanoparticle accumulation in plants is still largely unknown, but as research suggest, they may undergo morphological, physiological, genetic, and epigenetic changes that may alter plant growth and nutritional status. The introduction of engineered nanoparticles in the food chain from plants grown in contaminated soils may have serious repercussion upon the health of animals and humans.

3.4.3 Water and Soil Remediation

Nanoparticles are prospected for pollution remediation, water treatment and purification due to their high surface area, high adsorption capacity and unique structure and electronic properties (Gupta and Saleh 2013). They are able to remove organic and inorganic contaminants. Several types of materials are used for their adsorption of pollutants: metal-containing particles, carbon nanotubes, fullerenes, and zeolites (Gupta and Saleh 2013).

The magnetism of some nanoparticles makes them unique agents for water treatment (Xu et al. 2012). Adsorption processes together with magnetic separation makes some magnetic nanoparticles extensively used in water treatment and environmental cleanup. Most commonly used in soil remediation is nano-zero-valent iron due to its high reactivity (high surface area) and its ability to detoxify a wide range of contaminants, such as chlorinated organic solvents, organochlorine pesticides, and polychlorinated biphenyls (Deng et al. 2014).

The composition of nanoparticles used as heavy metals pollutant adsorbents include Fe_2O_3, Fe_3O_4, SiO_2, and Al_2O_3 (Bakshi et al. 2015). Nanoparticles of Fe and Al oxides or oxyhydroxides can be useful in the removal of As from water (Bakshi et al. 2015).

Recently, carbon nanomaterials, like carbon nanotubes and fullerene, are being broadly used for advanced treatment of wastewaters and are being researched as promising adsorbents for water treatment (Gupta and Saleh 2013). There is ongoing research regarding CNTs adsorbent properties against organic compounds such as pesticides, polycyclic aromatic hydrocarbons (PAHs), antibiotics, herbicides. They are also able to adsorb metal ions (Cu, Ni, Co, V, Ag, Cd and rare earth elements).

3.4.4 Incomplete Combustion

Accidental release of anthropogenic nanoparticles near urban areas is a result of traffic (such as exhaust emission due to incomplete combustion, road surface abrasion, brake and tire wear), industrial activities (incomplete combustion emissions from power plants, oil refineries, dust from mining), building (dust from excavations and demolitions) and from housing (incomplete combustion particles from heating and cooking) (Buzea et al. 2007). The anthropogenic particles in rural areas are mainly from biomass burning and various farming activities.

Nanoparticles generated via incomplete combustion of vehicle fuel are a large source of pollutants. In addition to the inherent release of nanoparticles due to incomplete burning of the fuel, nanocerium nanoparticles are being used as fuel additives to improve fuel efficiency and to reduce emissions of fine particulate matter, carbon monoxide, nitrogen oxides, and hydrocarbon species (Erdakos et al. 2014). The concentrations of cerium oxide is estimated to be between 0.32 and 1.12 μg/g at a distance of 26 m from the edge of highway, and 0.28–0.98 μg/g for a distance of 96 m from the highway (Park et al. 2008).

Nanoparticles synthesized by incomplete combustion are also used in different products (Heiligtag and Niederberger 2013). For example, carbon black, produced by the incomplete combustion of natural gas, is produced worldwide in huge amounts (10 million metric tons) and it is used as rubber reinforcement in car tires (Heiligtag and Niederberger 2013).

3.4.5 *Particles in the Environment and their Long Range Transport*

In environmental sciences particles are labeled as a function of their size. For example, particulate matter with components having diameters less than 0.1 μm (100 nm) is labeled $PM_{0.1}$ (often called ultra-fine particles) and results mainly from combustion processes. PM_{10} is particulate matter having a diameter smaller than 10 μm while $PM_{2.5}$ has size smaller than 2.5 μm. PM_{10} and $PM_{2.5}$ are mainly aggregates of ultra-fine particles and larger particles mostly mechanically generated (Buzea et al. 2007).

The particulate air matter may change its properties and chemical composition during its transport and aging. Its lifetime in the atmosphere can range from hours to weeks (Sielicki et al. 2011). The particulate matter in the atmosphere varies in amount and composition depending on its geographical location (Sielicki et al. 2011). The anthropogenic contribution to particulate matter is considerable, especially in highly populated areas of the globe. For example, in Netherlands between 70 and 95 % of particulate matter, with agriculture, transportation, and industry being the major sources of pollution (Hendriks et al. 2013).

Anthropogenic nanoparticles generation has an important effect not only locally, but also globally. They can be transported over large distances. For example, measurements of the dustfall in the high snowfields of the Caucasus is correlated to the European and Soviet industrialization, with little variation from 1790 to 1930, followed by an enormous 19-fold increase from 1930 to 1963 and a leveling off during World War II (Murr et al. 2004).

Due to their adverse effect on human health, particulate matter aerosols are monitored around the world. In order to protect human health, maximum thresholds standards are set by governments around the world for the fraction of particles with aerodynamic diameter smaller than 2.5 μm (Calvo et al. 2013). For example, in Europe a maximum threshold value of 20 μg/m^3 must be attained by 2020.

3.4.6 Nano-Waste

Many human activities generate nano-waste, such as research and development or various industries (Yadav et al. 2014). Nano-waste can also enter the environment from the disposal of consumer products that contain nanomaterials discarded via municipal or industrial waste treatment plants. Nanoparticles from waste water are not efficiently removed in water treatment plants, therefore they can enter the environment (Yadav et al. 2014; Aziz et al. 2015; Brar et al. 2010). The sludge from the waste water treatment plants is an important exposure source of engineered nanomaterials (Cornelis et al. 2014).

A study that model the flows of nanoparticles in the recycling system in Switzerland showed that occupational exposure, release of nanoparticles in landfills and incineration plants are the most important factors to be considered in the risk assessment (Caballero-Guzman et al. 2015).

Atmospheric nanoparticles can enter the water reservoir and the soil. Drinking water treatment does not remove entirely nanoparticles from the water. A study reported that only 2–20 %, 3–8 %, and 48–99 % of Ag, TiO_2, and ZnO nanoparticles are removed by a conventional drinking water treatment (Chalew et al. 2013). Therefore, a large amount of nanoparticles are still detectable in finished water.

3.4.7 Nanoparticles in the Construction Industry

Nanoparticles are also used in the construction industry (Pacheco-Blandino et al. 2012). Nanoparticles can be incorporated as nanocomposites in concrete. Silica, alumina, and zinc oxide nanoparticles nanocomposites in concrete confer it increased strength, water penetration and accelerated hydration. Titanium dioxide nanoparticles is used in concrete for their self-cleaning, antimicrobial, and pollution remediation properties. The incorporation of carbon nanotubes in concrete, polymers, or glass may result in a nanocomposite with improved properties, such as self-sensing, electromagnetic field shielding, and increased strength. Iron oxide nanoparticles incorporated in construction materials can offer self-sensing capabilities as well. Coating with different nanoparticles can greatly improve the properties of the covered material (Pacheco-Blandino et al. 2012). For example, photocatalytic TiO_2 and ZnO nanoparticles are used for pollution remediation, for their antifog, anticorrosion, and antimicrobial properties. Cu and Zn nanoparticles in coatings are anticorrosive, while Ag, Cu nanoparticles are antimicrobial. Silica nanoparticle coatings are scratch resistant, fireproof, and antireflective, while W_2O_3 can be used in coatings to adjust light transmittance (Pacheco-Blandino et al. 2012).

3.4.8 Nanoparticles in the Food Industry

Nanotechnology is becoming widely used in the food industry (Dasgupta et al. 2015). The use of nanotechnology in foods and dietary supplements encompasses food additives, food contact materials, and encapsulation (Prasad et al. 2014).

Nanoparticles of Ag, ZnO, TiO_2 are used as antimicrobial coating for refrigerators, storage containers, equipment for food processing (Rossi et al. 2014; Chellaram et al. 2014).

Smart packaging incorporates sensors that inform on the existence of specific chemicals and mitigate their effects, and can give information on the product shelf life. The smart packaging industry is about 80 % composed of oxygen, pathogen scavengers, moisture absorbers and barrier packing product (Chellaram et al. 2014).

Active food packaging serves as a passive barrier and also as active components via a desirable direct interaction with the food (antimicrobial or antioxidative agent) (Dasgupta et al. 2015). Food packaging with nano-coatings improve barrier properties. Food packaging uses mostly nanoparticles of clay or silver. Fewer products use silicon dioxide and titanium dioxide (Bouwmeester et al. 2014). Some nanoparticles are added to the matrix to confer antimicrobial activity or preservation via oxygen absorption. Nanoclays are used in polymers to improve barrier protection and preserve food by maintaining it in a low oxygen environment in order to inhibiting microbial growth (Chellaram et al. 2014).

Antibacterial nanoparticles are incorporated in food packaging which may extend food shelf life (Chellaram et al. 2014). For example the incorporation of silver or titania nanoparticles into food packaging serves as antimicrobial. The sunlight initiates photocatalysis of titania nanoparticle which promotes peroxidation of cell membrane phospholipids of bacteria (Dasgupta et al. 2015). Titania nanoparticles incorporated into food packaging can also act as O_2 scavengers, reducing food browning and rancid flavors. Silver is one of the most used nanomaterial in food industry and packaging. Silver can be used in refrigerators and food storage boxes due to its antibacterial properties. Nanoparticles can migrate from food packaging coatings into food, such as the case of Ag nanoparticles into food samples that were stored in contact with nano-silver coated surfaces (Metak et al. 2015).

Nanocomposites containing polymer incorporated nanoparticles are used to improve the strength of materials, durability, and biodegradation.

Some nanoparticles used as food additives have been available in products on the market for a long period of time, such as silicon dioxide SiO_2—E551, titanium dioxide TiO_2—E171, and Ag—E174, and are permitted as food additive without a definition of the particle size (Rossi et al. 2014; Bouwmeester et al. 2014; Lim et al. 2015; Wang et al. 2013). Silver nanoparticles can be used as food colorant (Wang et al. 2013). The anatase crystalline phase of titanium dioxide is mostly used as whitening agent and for texture (Wang et al. 2013). Silica is utilized in the

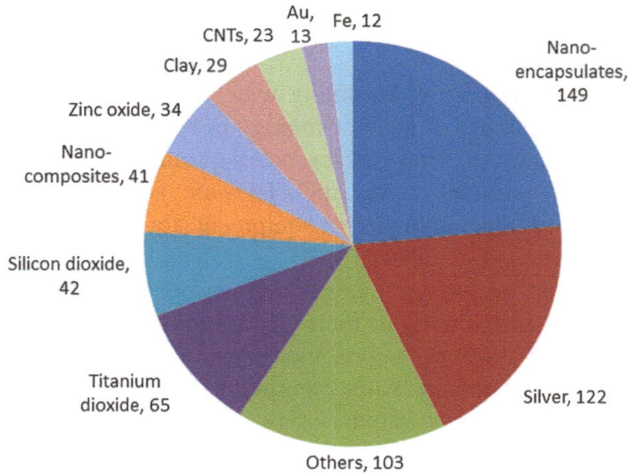

Fig. 3.13 Overview of the inventory of currently used and foreseen applications of nanomaterials in agriculture and food/feed production, including the number of records. Adapted from Aschberger et al. (2015). doi:10.1088/1742-6596/617/1/012032

production process of beer and wine, as a clearing, anticaking, and anti-clumping agent (Wang et al. 2013; Dekkers et al. 2011; Lim et al. 2015).

Other metals in nano-form marketed as food or health supplements are selenium, calcium, iron, copper, gold, platinum, silver, molybdenum, palladium, titanium, and zinc (Bouwmeester et al. 2014).

European Food Safety Authority (EFSA) commissioned an inventory of currently used and reasonably foreseen applications of NM in agriculture and food/feed production, published in 2015 (Aschberger et al. 2015). The majority of applications of nanotechnology are used in food (about 90 %), mostly as food additives and food contact materials. About 9 % are used in agriculture and 3 % in feed. The report found 55 types of inorganic and organic nanomaterials, of which the most used are nano-encapsulates, silver and titanium dioxide. This information is depicted in Fig. 3.13 (Aschberger et al. 2015).

Table 3.5 shows examples of nanomaterials and nanoparticles used in food, food additives and food packaging (Blasco and Pico 2011; Pico and Blasco 2012). Studies had shown that nanoparticles are in general toxic to humans and they are associated with a multitude of diseases (Buzea et al. 2007), therefore food contamination with nanoparticles or the environmental pollution via inadvertent release of nanoparticles from food packaging is highly undesirable.

3.4.9 Cosmetics

Nanoparticles have been used in cosmetics since ancient times, as powders or hair-dyes. Currently several types of nanoparticles are incorporated in cosmetics due to

Table 3.5 Nanomaterials and nanoparticles used in food, food additives, and food packaging

Product name	Manufacturer	Nanomaterial	Claim
Food packaging materials			
Durethan KU 2-2601	Bayer	Silica in a polymer-based nanocomposite	Nanoparticles of silica in the plastic prevent the penetration of oxygen and gas of the wrapping, extending the product shelf life
Hite brewery beers: three-layer, 1.6 L beer bottle	Honeywell	Honeywells Aegis OX nylon-based nanocomposite	Oxygen and carbon dioxide barrier, Clarity, Recyclability, Ease of Perform, Processability, Flavor/odor/aroma barrier, Structural integrity, Delamination resistance
Millar beers: Lite Genuine draft Ice house	Nanocor	Imperm nylon/nanocomposite barrier technology produced by Nanocor	Imperm is a plastic imbedded with clay nanoparticles that makes bottles less likely to shatter and increases shelf life to up to 6 months
Nano plastic wrap	Songsing Nanotechnology	Nano zinc light catalyst	Biodegradable after use, Compostable to European standards EN13432, Made from renewable and sustainable resources (non-GM corn starch), Water dispersible, will not pollute local groundwater systems or waterways, In use since 2002
Constantia multifilm N-COAT	Constantia Multifilm	Nanocomposite polymer	A clear laminate with outstanding gas-barrier proprieties developed primarily for the nuts, dry food, and snack markets
DuPont light stabilizer 210	Du Pont	Nano TiO_2	UV-protected plastic food packaging
Adhesive form MacDonalds burger containers	Ecosynthetix	50–150 nm starch nanospheres	The adhesive requires less water as well as less time and energy to dry
Food additives AdNano	Evonik (Degussa)	Nano ZnO (food grade)	
Aerosil, Sipernat	Evonik (Degussa)	Silica (food grade)	Free flow add for powdered ingredients in the food industry
AquaNovaNovaSol	AquaNova	Product micelle (capsule) of lipophilic, water-insoluble substances	An optimum carrier system of hydrophobic substances for a higher and faster intestinal and dermal

(continued)

Table 3.5 (continued)

Product name	Manufacturer	Nanomaterial	Claim
			resorption and penetration of active ingredients
Bioral Omega-3 nanocochleates	BioDelivery Sciences International	Nano-cochleates as small as 50 nm	Effective means for the addition of omega-3 fatty acids for use in cake, muffins, pasta, noodles, soup, cookies, cereals, chips, and candy bars
NanoCoQ10	Pharmanex	Nano coQ10	Nanotechnology to deliver highly bioavailable coenzyme Q10, making it up to 10 times more bioavailable than other forms of CoQ10
Nano self-assembled, structured liquids	Nutralease	Nanomicelles for encapsulation of nutraceuticals	Improved bioavailability means nutraceuticals are released into membrane between the digestive system and the blood
Solu E 200 BASF	BASF	Vitamin E nano-solution using NovaSOl	Solubilization of fat-soluble vitamins
Synthetic Lycopene	BASF	LycoVit 10 % (<200 nm synthetic lycopene)	
Food and beverages			
Nano tea	Shenzen Become Industry & Trading Co	Nanoparticles (160 nm)	Patent No.: 0100033.3—Three-step preparation method and its application for nanotea. Patent No.:02100314.9/00244295.7—Multi-layer, swinging nano-ball milling procedures
Nano slim	Nano Slim	Nano-diffuse Technology	Orosolic acid (derived from the *Lagerstroemia speciosa* plant)
Nanoceuticals slim shake chocolate and vanilla	RBC Lifescience	Nanodusters	
Fortified fruit juice	High Vive.com	300 nm iron (SunActive Fe)	
Daily Vitamin Boost	Jamba Juice Hawaii	300 nm iron (SunActive Fe)	22 essential vitamins and minerals and 100 %, or more of your daily needs of 18 of them!
Oat chocolate nutritional drink mix and	Toddler Health	300 nm iron (SunActive Fe)	Toddler health is an all-natural balanced nutritional drink for children

(continued)

Table 3.5 (continued)

Product name	Manufacturer	Nanomaterial	Claim
oat vanilla nutritional drink mix			from 13 months to 5 years. One serving of Toddler Health helps little ones meet their daily requirements for vitamins, minerals, and protein
Canola active oil	Shemen	Nano-sized self-assembled structured liquid micelles	
Maternal water	La posta del Aguila	Nanoparticle colloidal silver ions	Especially for the baby and the mother in pregnancy period, the mineral water is without chemical treatments, by using technology nanoparticle colloidal silver ions

Reprinted from Blasco C and Pico Y, 2011, Determining nanomaterials in food, Trends in Analytical Chemistry, 30, 84–99, Copyright (2011), with permission from Elsevier (Blasco and Pico 2011). The data from the last row in the table is taken from Pico and Blasco (2012)

their special properties (Borowska and Brzoska 2015). For example, Ag and Au nanoparticles are used in face and body care products, while Cu-Silica nanoparticles are used in deodorants due to their antibacterial properties. Alumina nanoparticles are incorporated in mineral foundations and concealers due to their properties of diffusing light. Many sunscreens contain titanium dioxide and zinc oxide due to their properties of UV filtering.

3.4.10 *Engineered Nanoparticles Applications and Environmental Accumulation*

While there is an increasing number of products containing nanoparticles on the market, the assessment of risk posed by nanomaterials and related legislations that should limit the use of toxic nanoparticles in products are in lagging behind. Many consumer products contain nanoparticles that will most likely enter the environment after improper recycling and disposal of consumer products. The buildup of engineered nanoparticles in the environment may affect soil-based food crop quality and yield. The effect of nanoparticle on edible plants is very important because of their uptake by plants. Once nanoparticles entered the food chain, they may potentially affect the health of humans and animals in ways that we cannot predict empirically.

Table 3.6 Engineered nanoparticles, their common applications, and possible environmental accumulation

Type of nanoparticle	Applications	Environmental accumulation			
		Soil	Water	Waste	Air
Carbon black	Additives to tyres, rubber, paints	+	+	+	+
CNTs	Electronics, sensors, conductive coatings, supercapacitors, catalysts, battery, fuel cell electrodes, additives to tyres, lubricants, plastics, water purification systems, orthopedic implants, adhesives and composites, sporting goods, components in aircrafts, aerospace, automotive industry, adsorption of contaminants	+	+	+	+
Fullerenes	Cosmetics, sorption of organic compounds (e.g., naphthalene), for the removal of organometallic compounds, in medicine	+	+	+	–
Fe and its oxides	Cosmetics and personal care products, environmental remediation of water, sediments, and soils, bioremediation for some pesticides and herbicides, concrete additives, biochemical assays, bio-manipulation	+	+	+	+
Ag	Antimicrobial agent in various products: toothpastes, baby-products, wound dressings, socks, and other textiles, in air filters, air conditioning, vacuum cleaners, and washing machines, paints, coatings, food packaging, plastics, varnish	+	+	+	+
Au	In tumor therapy, catalyst, flexible conducting inks or films	+	+	+	–
TiO_2	In cosmetics, skin care products, sunscreen lotions, solar cells, paints and coatings, environmental remediation, for the removal of various organics	+	+	+	–
ZnO	In cosmetics and skin care products, sunscreens, in bottle coatings, paints	+	+	+	–
CeO_2	Fuel catalyst, gas sensors, solar cells, oxygen pumps In lubricants and fuel additives In metallurgical and glass/ceramic applications	+	+	+	+
Pt, Pd	Catalysts, lubricants and fuel additives, automotive exhaust converters	+	+	+	+
Sn	Paints	+	+	+	–
Al and Al_2O_3	Metallic coatings and plating, batteries, grinding, fire protection, metal and biosorbent, flame retardant paints	+	+	+	+
Cu	Microelectronics, environmental remediation	–	–	+	–
Se	Nutraceuticals, health supplements	+	+	+	–
Ca	Nutraceuticals, health supplements	+	+	+	–
Mg	Nutraceuticals, health supplements	+	+	+	–

(continued)

Table 3.6 (continued)

Type of nanoparticle	Applications	Environmental accumulation			
		Soil	Water	Waste	Air
MoS_3	Catalyst, lubricants and fuel additives	+	+	+	+
QD	Medical imaging and targeted therapeutics, solar cells, photovoltaic cells, security inks, photonics, telecommunications	+	+	+	–
SiO_2	In porous form as a carrier (agrochemicals) Fireproof glass, UV-protection, flame retardant paints, coatings, varnish, ceramics, electronics, pharmaceutical products, dentistry, polishing	+	+	+	+

Data adapted from references: (1) Bhatt I and Tripathi B. N., 2011, Interaction of engineered nanoparticles with various components of the environment and possible strategies for their risk assessment. Chemosphere, 82, 308–317, Copyright (2011), with permission from Elsevier (Bhatt and Tripathi 2011), (2) Boxall et al. (2007), (3) Brar et al. (2010), (4) Gladkova and Terekhova (2013), (5) Gottschalk et al. (2015), (6) Hincapie et al. (2015), (7) Khin et al. (2012), (8) Tang and Lo (2013), (9) Bystrzejewska-Piotrowska et al. (2009)

Table 3.6 shows some of the most used engineered nanoparticles, their applications and possible environmental accumulation in soil, water, waste, and air (Bhatt and Tripathi 2011; Boxall et al. 2007; Brar et al. 2010; Gladkova and Terekhova 2013; Gottschalk et al. 2015; Hincapie et al. 2015; Khin et al. 2012; Tang and Lo 2013; Bystrzejewska-Piotrowska et al. 2009). The interaction, transport, and eco-toxicology of nanoparticles are dynamically altered by abiotic factors (pH, salinity, water hardness, temperature) (Handy et al. 2008b).

3.5 Nanoparticle Relative Toxicity Index on Humans and Animals

There are already various studies on the effects of nanoparticles on plants, with most of them reporting detrimental effects, and a minority showing some beneficial effects. While some plants exhibit benefits from a type of nanoparticle, other plants are negatively impacted by the same nanoparticles. Moreover, many nanoparticles are found to be extremely toxic to humans and animals and even though they might show some minor benefits for the growth of some plants, they may lead to detrimental effects if these plants containing nanoparticles were to be ingested by humans or animals. For example, Ag nanoparticles that are starting to be used increasingly in agriculture have a toxicity index higher than those of asbestos in animal cells, as shown in Table 3.7 (Soto et al. 2005). Also, carbon nanotubes, which have similar morphologies and sizes with asbestos, as shown in Fig. 3.14

Table 3.7 The relative toxicity index (RCI) on murine macrophage cells of nanoparticles varies with composition, their particle and mean aggregate size

Material	Mean aggregate size (μm)	Mean particle size (nm)	RCI (at 5 μg/ml)	RCI (at 10 μg/ml)
Ag	1	30	1.5	0.8
Ag	0.4	30	1.8	0.1
Al_2O_3	0.7	50	0.7	0.4
Fe_2O_3	0.7	50	0.9	0.1
ZrO_2	0.7	20	0.7	0.6
TiO_2 rutile	1	Short fibers 5–15 nm diam	0.3	0.05
TiO_2 anatase	2.5	20	0.4	0.1
Si_3N_4	1	60	0.4	0.06
Asbestos Chrysotile	7	Fibers 20 nm diam. Up to 500 aspect ratio	1	1
Carbon black	0.5	20	0.8	0.6
SWCNT	10	100 nm diam.	1.1	0.9
MWCNT	2	15 nm diam.	0.9	0.8

Adapted with permission from Soto K. F. et al., 2005, Comparative in vitro cytotoxicity assessment of some manufactured nanoparticulate materials characterized by transmission electron microscopy, Journal of Nanoparticle Research, 7, 145–169. Copyright (2005) Springer (Soto et al. 2005)

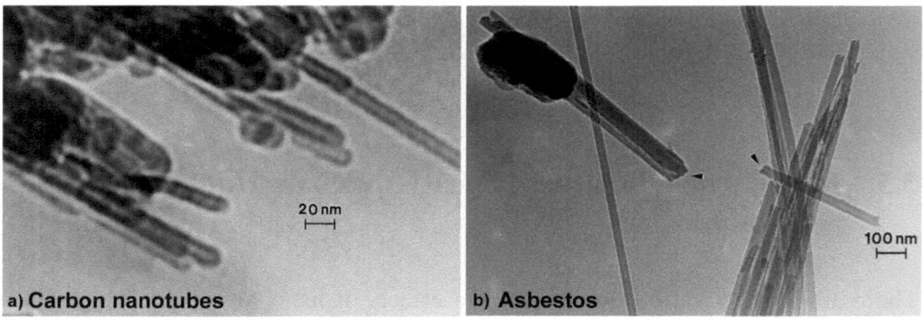

Fig. 3.14 TEM bright-field images of (**a**) carbon nanotube aggregates extracted from a natural gas flame exhaust, (**b**) chrysotile asbestos fibers. Reproduced from Murr L. E. & Soto K. F., TEM comparison of chrysotile (asbestos) nanotubes and carbon nanotubes. Journal of Materials Science, 2004, 39, 4941–4947. Copyright (2004). With permission from Springer (Murr and Soto 2004)

(Murr and Soto 2004), have similar toxicities in animal cells (Soto et al. 2005). For a comprehensive review of nanoparticles toxicity on humans see reference Buzea et al. (2007).

References

Alexandru SB, Mariya VK, Biris AS, Khodakovskaya MV (2012) Method of using carbon nanotubes to affect seed germination and plant growth. US Patent Application Ep20100830332
Alphandéry E (2014) Applications of magnetosomes synthesized by magnetotactic bacteria in medicine. Front Bioeng Biotechnol 2:5
Aragay G, Pino F, Merkoci A (2012) Nanomaterials for sensing and destroying pesticides. Chem Rev 112:5317–5338
Aschberger K, Gottardo S, Amenta V, Arena M, Moniz FB, Bouwmeester H, Brandhoff P, Mech A, Pesudo LQ, Rauscher H, Schoonjans R, Vettori MV, Peters R, Iop (2015) Nanomaterials in food – current and future applications and regulatory aspects. 4th International conference on safe production and use of nanomaterials. Iop, Bristol
Auffan M, Rose J, Bottero JY, Lowry GV, Jolivet JP, Wiesner MR (2009) Towards a definition of inorganic nanoparticles from an environmental, health and safety perspective. Nat Nanotechnol 4:634–641
Aziz N, Faraz M, Pandey R, Sakir M, Fatma T, Varma A, Barman I, Prasad R (2015) Facile algae-derived route to biogenic silver nanoparticles: synthesis, antibacterial and photocatalytic properties. Langmuir 31:11605–11612
Bakshi S, He ZLL, Harris WG (2015) Natural nanoparticles: implications for environment and human health. Crit Rev Environ Sci Technol 45:861–904
Balasubramanian K, Burghard M (2005) Chemically functionalized carbon nanotubes. Small 1:180–192
Barati A (2010) Nano-composite superabsorbent containing fertilizer nutrients used in agriculture. US 20100139347 A1. US Patent Application US 12/701,613
Bernhardt ES, Colman BP, Hochella MF, Cardinale BJ, Nisbet RM, Richardson CJ, Yin LY (2010) An ecological perspective on nanomaterial impacts in the environment. J Environ Qual 39:1954–1965
Bhatt I, Tripathi BN (2011) Interaction of engineered nanoparticles with various components of the environment and possible strategies for their risk assessment. Chemosphere 82:308–317
Blasco C, Pico Y (2011) Determining nanomaterials in food. Trac-Trends Anal Chem 30:84–99
Borowska S, Brzoska MM (2015) Metals in cosmetics: implications for human health. J Appl Toxicol 35:551–572
Bouwmeester H, Brandhoff P, Marvin HJP, Weigel S, Peters RJB (2014) State of the safety assessment and current use of nanomaterials in food and food production. Trends Food Sci Technol 40:200–210
Boxall AB, Tiede K, Chaudhry Q (2007) Engineered nanomaterials in soils and water: how do they behave and could they pose a risk to human health? Nanomedicine 2:919–927
Brar SK, Verma M, Tyagi RD, Surampalli RY (2010) Engineered nanoparticles in wastewater and wastewater sludge – evidence and impacts. Waste Manag 30:504–520
Brugger J, Etschmann B, Grosse C, Plumridge C, Kaminski J, Paterson D, Shar SS, Ta C, Howard DL, De Jonge MD, Ball AS, Reith F (2013) Can biological toxicity drive the contrasting behavior of platinum and gold in surface environments? Chem Geol 343:99–110
Buzea C, Pacheco I, Robbie K (2007) Nanomaterials and nanoparticles: sources and toxicity. Biointerphases 2:Mr17–Mr71
Bystrzejewska-Piotrowska G, Golimowski J, Urban PL (2009) Nanoparticles: their potential toxicity, waste and environmental management. Waste Manag 29:2587–2595
Caballero-Guzman A, Sun T, Nowack B (2015) Flows of engineered nanomaterials through the recycling process in Switzerland. Waste Manag 36:33–43
Calvo AI, Alves C, Castro A, Pont V, Vicente AM, Fraile R (2013) Research on aerosol sources and chemical composition: past, current and emerging issues. Atmos Res 120–121:1–28
Chalew TEA, Ajmani GS, Huang HO, Schwab KJ (2013) Evaluating nanoparticle breakthrough during drinking water treatment. Environ Health Perspect 121:1161–1166

Chekli L, Bayatsarmadi B, Sekine R, Sarkar B, Shen AM, Scheckel KG, Skinner W, Naidu R, Shon HK, Lombi E, Donner E (2016) Analytical characterisation of nanoscale zero-valent iron: a methodological review. Anal Chim Acta 903:13–35
Chellaram C, Murugaboopathi G, John AA, Sivakumar R, Ganesan S, Krithika S, Priya G (2014) Significance of nanotechnology in food industry. APCBEE Proc 8:109–113
Cornelis G, Hund-Rinke K, Kuhlbusch T, Van Den Brink N, Nickel C (2014) Fate and bioavailability of engineered nanoparticles in soils: a review. Crit Rev Environ Sci Technol 44:2720–2764
Coutris C, Joner EJ, Oughton DH (2012) Aging and soil organic matter content affect the fate of silver nanoparticles in soil. Sci Total Environ 420:327–333
Dasgupta N, Ranjan S, Mundekkad D, Ramalingam C, Shanker R, Kumar A (2015) Nanotechnology in agro-food: from field to plate. Food Res Int 69:381–400
De La Torre-Roche R, Hawthorne J, Deng YQ, Xing BS, Cai WJ, Newman LA, Wang Q, Ma XM, Hamdi H, White JC (2013) Multiwalled carbon nanotubes and C-60 fullerenes differentially impact the accumulation of weathered pesticides in four agricultural plants. Environ Sci Technol 47:12539–12547
Dekkers S, Krystek P, Peters RJB, Lankveld DPK, Bokkers BGH, Van Hoeven-Arentzen PH, Bouwmeester H, Oomen AG (2011) Presence and risks of nanosilica in food products. Nanotoxicology 5:393–405
Deng YQ, White JC, Xing BS (2014) Interactions between engineered nanomaterials and agricultural crops: implications for food safety. J Zhejiang Univ Sci A 15:552–572
Erdakos GB, Bhave PV, Pouliot GA, Simon H, Mathur R (2014) Predicting the effects of nanoscale cerium additives in diesel fuel on regional-scale air quality. Environ Sci Technol 48:12775–12782
Fairbrother L, Etschmann B, Brugger J, Shapter J, Southam G, Reith F (2013) Biomineralization of gold in biofilms of *Cupriavidus metallidurans*. Environ Sci Technol 47:2628–2635
Farre M, Sanchis J, Barcelo D (2011) Analysis and assessment of the occurrence, the fate and the behavior of nanomaterials in the environment. Trac-Trends Anal Chem 30:517–527
Farrukh MA, Naseem F (2014) Nano-Leucite for slow release nitrogen fertilizer and green environment. US 20140190226 A1. US Patent Application US 13/738,727
Ghodake G, Seo YD, Lee DS (2011) Hazardous phytotoxic nature of cobalt and zinc oxide nanoparticles assessed using *Allium cepa*. J Hazard Mater 186:952–955
Gladkova MM, Terekhova VA (2013) Engineered nanomaterials in soil: sources of entry and migration pathways. Moscow Univ Soil Sci Bull 68:129–134
Gogos A, Knauer K, Bucheli TD (2012) Nanomaterials in plant protection and fertilization: current state, foreseen applications, and research priorities. J Agric Food Chem 60:9781–9792
Gorby YA, Yanina S, Mclean JS, Rosso KM, Moyles D, Dohnalkova A, Beveridge TJ, Chang IS, Kim BH, Kim KS, Culley DE, Reed SB, Romine MF, Saffarini DA, Hill EA, Shi L, Elias DA, Kennedy DW, Pinchuk G, Watanabe K, Ishii S, Logan B, Nealson KH, Fredrickson JK (2006) Electrically conductive bacterial nanowires produced by *Shewanella oneidensis* strain Mr-1 and other microorganisms. Proc Natl Acad Sci USA 103:11358–11363
Gottschalk F, Lassen C, Kjoelholt J, Christensen F, Nowack B (2015) Modeling flows and concentrations of nine engineered nanomaterials in the Danish environment. Int J Environ Res Public Health 12:5581–5602
Goudie AS (2014) Desert dust and human health disorders. Environ Int 63:101–113
Guo HB, Barnard AS (2013) Naturally occurring iron oxide nanoparticles: morphology, surface chemistry and environmental stability. J Mater Chem A 1:27–42
Gupta VK, Saleh TA (2013) Sorption of pollutants by porous carbon, carbon nanotubes and fullerene- an overview. Environ Sci Pollut Res 20:2828–2843
Handy RD, Owen R, Valsami-Jones E (2008a) The ecotoxicology of nanoparticles and nanomaterials: current status, knowledge gaps, challenges, and future needs. Ecotoxicology 17:315–325

Handy RD, Von Der Kammer F, Lead JR, Hassellov M, Owen R, Crane M (2008b) The ecotoxicology and chemistry of manufactured nanoparticles. Ecotoxicology 17:287–314

Heiligtag FJ, Niederberger M (2013) The fascinating world of nanoparticle research. Mater Today 16:262–271

Hendriks C, Kranenburg R, Kuenen J, Van Gijlswijk R, Kruit RW, Segers A, Van Der Gon HD, Schaap M (2013) The origin of ambient particulate matter concentrations in the Netherlands. Atmos Environ 69:289–303

Hincapie I, Kunniger T, Hischier R, Cervellati D, Nowack B, Som C (2015) Nanoparticles in facade coatings: a survey of industrial experts on functional and environmental benefits and challenges. J Nanopart Res 17:287

Hochella MF, Lower SK, Maurice PA, Penn RL, Sahai N, Sparks DL, Twining BS (2008) Nanominerals, mineral nanoparticles, and earth systems. Science 319:1631–1635

Hori H, Teranishi T, Nakae Y, Seino Y, Miyake M, Yamada S (1999) Anomalous magnetic polarization effect of Pd and Au nano-particles. Phys Lett A 263:406–410

Hough RM, Noble RRP, Reich M (2011) Natural gold nanoparticles. Ore Geol Rev 42:55–61

Jaynes BS, Sgande ME, Fenton RJ, Stadler UL, Mamak M, Choi S (2012) Antimicrobial silver silica composite. US 20120294919 A1. US 13/470,881

Jian L, Zhiming Z (2011) Environment-friendly carbon-nano synergistic complex fertilizers. US 20110174032 A1. US Patent Application US 12/672,951

Kah M, Hofmann T (2014) Nanopesticide research: current trends and future priorities. Environ Int 63:224–235

Kah M, Beulke S, Tiede K, Hofmann T (2013) Nanopesticides: state of knowledge, environmental fate, and exposure modeling. Crit Rev Environ Sci Technol 43:1823–1867

Khin MM, Nair AS, Babu VJ, Murugan R, Ramakrishna S (2012) A review on nanomaterials for environmental remediation. Energ Environ Sci 5:8075–8109

Khot LR, Sankaran S, Maja JM, Ehsani R, Schuster EW (2012) Applications of nanomaterials in agricultural production and crop protection: a review. Crop Prot 35:64–70

Kim W, Doh S-J, Yu Y (2012) Asian dust storm as conveyance media of anthropogenic pollutants. Atmos Environ 49:41–50

Klaine SJ, Alvarez PJJ, Batley GE, Fernandes TF, Handy RD, Lyon DY, Mahendra S, Mclaughlin MJ, Lead JR (2008) Nanomaterials in the environment: behavior, fate, bioavailability, and effects. Environ Toxicol Chem 27:1825–1851

Kolhatkar AG, Jamison AC, Litvinov D, Willson RC, Lee TR (2013) Tuning the magnetic properties of nanoparticles. Int J Mol Sci 14:15977–16009

Kookana RS, Boxall ABA, Reeves PT, Ashauer R, Beulke S, Chaudhry Q, Cornelis G, Fernandes TF, Gan J, Kah M, Lynch I, Ranville J, Sinclair C, Spurgeon D, Tiede K, Van Den Brink PJ (2014) Nanopesticides: guiding principles for regulatory evaluation of environmental risks. J Agric Food Chem 62:4227–4240

Larue C, Pinault M, Czarny B, Georgin D, Jaillard D, Bendiab N, Mayne-L'hermite M, Taran F, Dive V, Carriere M (2012) Quantitative evaluation of multi-walled carbon nanotube uptake in wheat and rapeseed. J Hazard Mater 227:155–163

Lefevre CT, Bazylinski DA (2013) Ecology, diversity, and evolution of magnetotactic bacteria. Microbiol Mol Biol Rev 77:497–526

Lengke MF, Fleet ME, Southam G (2006) Synthesis of platinum nanoparticles by reaction of filamentous cyanobacteria with Platinum(Iv)–Chloride complex. Langmuir 22:7318–7323

Leung KM, Wanger G, El-Naggar MY, Gorby Y, Southam G, Lau WM, Yang J (2013) *Shewanella oneidensis* Mr-1 bacterial nanowires exhibit P-type, tunable electronic behavior. Nano Lett 13:2407–2411

Li Y (2002) Ximaxi controlled release special fertilizer and its preparation. CN 1349958 A. China Patent Application CN 01106895

Lim JH, Sisco P, Mudalige TK, Sanchez-Pomales G, Howard PC, Linder SW (2015) Detection and characterization of Sio_2 and Tio_2 nanostructures in dietary supplements. J Agric Food Chem 63:3144–3152

Lintern M, Anand R, Ryan C, Paterson D (2013) Natural gold particles in eucalyptus leaves and their relevance to exploration for buried gold deposits. Nat Commun 4:2274
Ma YH, Kuang LL, He X, Bai W, Ding YY, Zhang ZY, Zhao YL, Chai ZF (2010) Effects of rare earth oxide nanoparticles on root elongation of plants. Chemosphere 78:273–279
Malshe VC, Malshe AP (2010) Non-metallic nano/micro particles coated with metal, process and applications. US 20100047546 A1. US 12/227,716
Manceau A, Nagy KL, Marcus MA, Lanson M, Geoffroy N, Jacquet T, Kirpichtchikova T (2008) Formation of metallic copper nanoparticles at the soil-root interface. Environ Sci Technol 42:1766–1772
Metak AM, Nabhani F, Connolly SN (2015) Migration of engineered nanoparticles from packaging into food products. Lwt - Food Sci Technol 64:781–787
Min K (2003) Compound application technology for nano-biological fertilizer and soil conditioner. CN 1413963 A. China Patent Application
Mohammadinejad R, Karimi S, Iravani S, Varma RS (2016) Plant-derived nanostructures: types and applications. Green Chem 18:20–52
Morman SA, Plumlee GS (2013) The role of airborne mineral dusts in human disease. Aeol Res 9:203–212
Murr LE, Soto KF (2004) TEM comparison of chrysotile (asbestos) nanotubes and carbon nanotubes. J Mater Sci 39:4941–4947
Murr LE, Esquivel EV, Bang JJ, De La Rosa G, Gardea-Torresdey JL (2004) Chemistry and nanoparticulate compositions of a 10,000 year-old ice core melt water. Water Res 38:4282–4296
Nair R, Poulose AC, Nagaoka Y, Yoshida Y, Maekawa T, Kumar DS (2011) Uptake of FITC labeled silica nanoparticles and quantum dots by rice seedlings: effects on seed germination and their potential as biolabels for plants. J Fluoresc 21:2057–2068
Nano-Argentum NG. Available: Http://Www.Nanosys.Ch/Global/Kolloide/Nano_Argentum_Flakon_E.Html#Reload
Nath D, Banerjee P (2013) Green nanotechnology – a new hope for medical biology. Environ Toxicol Pharmacol 36:997–1014
Nowack B, Bucheli TD (2007) Occurrence, behavior and effects of nanoparticles in the environment. Environ Pollut 150:5–22
Pacheco-Blandino I, Vanner R, Buzea C (2012) Toxicity of nanoparticles. In: Pacheco-Torgal F, Jalali S, Fucic A (eds) Toxicity of building materials. Woodhead, Cambridge
Painter TH, Barrett AP, Landry CC, Neff JC, Cassidy MP, Lawrence CR, Mcbride KE, Farmer GL (2007) Impact of disturbed desert soils on duration of mountain snow cover. Geophys Res Lett 34:L12502
Park B, Donaldson K, Duffin R, Tran L, Kelly F, Mudway I, Morin JP, Guest R, Jenkinson P, Samaras Z, Giannouli M, Kouridis H, Martin P (2008) Hazard and risk assessment of a nanoparticulate cerium oxide-based diesel fuel additive – a case study. Inhal Toxicol 20:547–566
Pico Y, Blasco C (2012) Nanomaterials in food, which way forward? In: Farre M, Barcelo D (eds) Analysis and risk of nanomaterials in environmental and food samples. Elsevier, Oxford
Prasad DY (2013) Production of novel precision customized control release fertilizers. US 8375629 B2. US 12/312,328
Prasad R, Kumar V, Prasad KS (2014) Nanotechnology in sustainable agriculture: present concerns and future aspects. Afr J Biotechnol 13:705–713
Prasad R, Pandey R, Barman I (2016) Engineering tailored nanoparticles with microbes: quo vadis. Wires Nanomed Nanobiotechnol 8:316–330
Prospero JM, Mayol-Bracero OL (2013) Understanding the transport and impact of African dust on the Caribbean Basin. Bull Am Meteorol Soc 94:1329–1337
Prospero JM, Collard FX, Molinie J, Jeannot A (2014) Characterizing the annual cycle of African dust transport to the Caribbean Basin and South America and its impact on the environment and air quality. Global Biogeochem Cycles 28:757–773

Quester K, Avalos-Borja M, Castro-Longoria E (2013) Biosynthesis and microscopic study of metallic nanoparticles. Micron 54–55:1–27

Reith F, Campbell SG, Ball AS, Pring A, Southam G (2014) Platinum in earth surface environments. Earth-Sci Rev 131:1–21

Rossi M, Cubadda F, Dini L, Terranova ML, Aureli F, Sorbo A, Passeri D (2014) Scientific basis of nanotechnology, implications for the food sector and future trends. Trends Food Sci Technol 40:127–148

Sakamoto Y, Oba Y, Maki H, Suda M, Einaga Y, Sato T, Mizumaki M, Kawamura N, Suzuki M (2011) Ferromagnetism of Pt nanoparticles induced by surface chemisorption. Phys Rev B 83:4420

Sardari S, Soltani S, Mirzazadeh R (2013) Silicon nanocarrier for delivery of drug, pesticides and herbicides, and for waste water treatment. US 20130225412 A1. US 13/406,538

Servin A, Elmer W, Mukherjee A, De La Torre-Roche R, Hamdi H, White JC, Bindraban P, Dimkpa C (2015) A review of the use of engineered nanomaterials to suppress plant disease and enhance crop yield. J Nanopart Res 17:1–21

Sharma VK, Filip J, Zboril R, Varma RS (2015) Natural inorganic nanoparticles – formation, fate, and toxicity in the environment. Chem Soc Rev 44:8410–8423

Sielicki P, Janik H, Guzman A, Namiesnik J (2011) Particulate matters to be used as a standard monitoring method for air dust pollution. Crit Rev Anal Chem 41:314–334

Soto KF, Carrasco A, Powell TG, Garza KM, Murr LE (2005) Comparative in vitro cytotoxicity assessment of some manufactured nanoparticulate materials characterized by transmission electron microscopy. J Nanopart Res 7:145–169

Taillefert M, Lienemann CP, Gaillard JF, Perret D (2000) Speciation, reactivity, and cycling of Fe and Pb in a Meromictic lake. Geochim Cosmochim Acta 64:169–183

Tang SCN, Lo IMC (2013) Magnetic nanoparticles: essential factors for sustainable environmental applications. Water Res 47:2613–2632

Verma HC, Upadhyay C, Tripathi A, Tripathi RP, Bhandari N (2002) Thermal decomposition pattern and particle size estimation of iron minerals associated with the Cretaceous-Tertiary boundary at Gubbio. Meteorit Planet Sci 37:901–909

Wang J, Yang LX, Fan Y, Wu Y, Zheng W, Zhao F (2005) Application of oxide nano rare earth. CN 1686957 A. China Patent Application CN 200510066394

Wang J, Yang LXS, Fan Y, Wu Y, Zheng W, Zhao F, Han X (2007) Application of hydroxide of nano rare earth. CN 1314629 C. CN 200510066392

Wang HF, Du LJ, Song ZM, Chen XX (2013) Progress in the characterization and safety evaluation of engineered inorganic nanomaterials in food. Nanomedicine 8:2007–2025

Wilson MA, Tran NH, Milev AS, Kannangara GSK, Volk H, Lu GQM (2008) Nanomaterials in soils. Geoderma 146:291–302

Wu X (2002) Active nano grade organic fine humic fertilizer and its production. CN 1472176 A. China Patent Application CN 02139321

Xia YS, Tang ZY (2012) Monodisperse hollow supraparticles via selective oxidation. Adv Funct Mater 22:2585–2593

Xu PA, Zeng GM, Huang DL, Feng CL, Hu S, Zhao MH, Lai C, Wei Z, Huang C, Xie GX, Liu ZF (2012) Use of iron oxide nanomaterials in wastewater treatment: a review. Sci Total Environ 424:1–10

Yadav T, Mungray AA, Mungray AK (2014) Fabricated nanoparticles: current status and potential phytotoxic threats. In: Whitacre DM (ed) Reviews of environmental contamination and toxicology. Springer, New York

Yamamoto K, Otsuka H, Wada SI, Sohn D, Takahara A (2005) Preparation and properties of poly (methyl methacrylate)/imogolite hybrid via surface modification using phosphoric acid ester. Polymer 46:12386–12392

Ying X, Liu Y, Tian W (2012) Preparation of a nano long-acting selenium fertilizer. US 8246713 B2. US 12/879,813

Yu E (2005) Nano diatomite and zeolite ceramic crystal powder. US 20050115469 A1. US Patent Application US 10/351,518
Zhang FT (2004a) Production technology of coating cement for nanometer sulfonated lignin mixture fertilizer. CN 1164531 C. China Patent Application CN 02149247
Zhang FT (2004b) Production technology of nano-clay-polyester mixed polymer fertilizer coating cementing agent. CN 1171948 C. China Patent Application CN 02126009
Zhang F (2006a) Nano-micron foam plastic mixed polymer fertilizer adhesive coating agent preparation method. CN 1279095 C. China Patent Application CN 200410088477
Zhang FT (2006b) Nanometer-scale olefin-starch mixed polymer fertilizer envelope cementing agent production method. CN 1279074 C. CN 200310116857
Zhang ZY, He X, Zhang HF, Ma YH, Zhang P, Ding YY, Zhao YL (2011) Uptake and distribution of ceria nanoparticles in cucumber plants. Metallomics 3:816–822

Part II
Nanomaterials in Soil Environment

Chapter 4
Engineered Nanomaterials' Effects on Soil Properties: Problems and Advances in Investigation

Vera Terekhova, Marina Gladkova, Eugeny Milanovskiy, and Kamila Kydralieva

4.1 Sources of Entry and Migration Pathways of Engineered Nanomaterials in Soil

4.1.1 Introduction

The latest achievements in the field of nanotechnologies and the corresponding growth in the use of nanomaterials (NMs) in many industries as well as in the production of consumer goods inevitably have led to their dispersion into the environment. It is evident that such a wide introduction of nanoparticles (NPs) in our life and their expansion and accumulation in natural habitats gives grounds to consider them as a particular type of pollutants. Specialists have come to the conclusion that the processes of nanoparticle transfer with air and water flows

V. Terekhova (✉)
Soil Science Faculty, Lomonosov Moscow State University (MSU), 1-16, Leninskiye Gory, 119234 Moscow, Russia

Severtsov Institute of Ecology and Evolution (IPEE), Russian Academy of Sciences, Moscow, Russia

Pirogov Russian National Research Medical University (RNRMU), Moscow, Russia
e-mail: vterekhova@gmail.com

M. Gladkova • E. Milanovskiy
Soil Science Faculty, Lomonosov Moscow State University (MSU), 1-16, Leninskiye Gory, 119234 Moscow, Russia
e-mail: marika230489@gmail.com; milanovskiy@gmail.com

K. Kydralieva
Moscow Aviation Institute, Volokolamskoe shosse 4, 125993 Moscow, Russia
e-mail: k_kamila@mail.ru

M. Ghorbanpour et al. (eds.), *Nanoscience and Plant–Soil Systems*, Soil Biology 48,
DOI 10.1007/978-3-319-46835-8_4

and their accumulation in soil, water, and bottom sediment differ significantly from the behavior of larger particles. The active development of works on analyzing engineered nanomaterials in natural habitats helps in clarifying the following questions: What are their distribution pathways? Do artificial NMs retain their properties (size, original structure, and reactivity) in water, air, soil, and sedimentation objects? What are the consequences of nanoparticle expansion in a liquid medium? What are the distinctions in the effects of NPs and molecular and atomic forms of the same material on biota under water and soil conditions? Several definitions of nanomaterials and different variants of their classification occur in the publications (Klaine et al. 2008). The generally accepted sign that characterizes the relation of objects to nanomaterials is the dimension of constituent particles at an interval of 1–100 nm, in at least one dimension. This definition is right for natural colloids (superdispersed particles in the air, biological objects such as viruses, etc.), like water and soil colloids, as well as materials constructed using nanomaterials.

Soil colloids have been studied for many decades. These are primary nanoparticles in the structure of clay and organic substances, iron oxides, and other minerals, which play an important role in biogeochemical processes. Particular attention has been devoted to analyzing their effect on soil formation and change in the structure of soils (Fedotov and Shalaev 2012).

4.1.2 Sources of Nanoparticle Entry into the Environment

The sources of nanoparticle entry into the environment can be divided into natural and anthropogenic. Natural processes have been sources of nanomaterial entry into the environment for thousands of years. They include forest fires, sandstorms, dust, muddled waters, formation of aerosols, clustering in gases, volcanic bursts, and salt evaporation, as well as biological objects (viruses, products of vital functions, films, colloids, etc.) (Krichevskiy 2010).

Many anthropogenic objects and processes are sources of the so-called unintentional nanoparticle entry into the environment. They include the following: the combustion of waste and fuel with combustion catalysts in transport vehicle engines and at power plants, domestic waste, and mining operations, open pits and mines, industrial production and emissions, construction, welding, soldering, preparation of food, etc. Environmental objects are contaminated during the production, transportation, and use of different hygiene products and household chemical goods (sun shielding instruments, detergents), motor tire resin, typographic dyes, textile products, etc.

The development of some industries and nanotechnologies has led to abrupt growth in the quantity of engineered nanoparticles in the environment. Today, nanotechnologies have become a source for the intentional expansion of a significant amount of nanomaterials in different natural habitats. They include purification and processing using NMs from polluted groundwater, land reclamation, and application for agricultural needs. Nanoparticles may get into soils due to using NMs in soil and water purification systems for agricultural needs (as nanofertilizers, pesticides, seed treatment preparations, materials for agro-films, preparation of hydroponic solutions, etc.). Such materials include fullerenes, nanotubes, inorganic nanocrystals, quantum points, nanofilms, micelles, colloids, specific-action drugs, etc.

Many natural sources of nanoparticles can create local ecological problems, but they are included in the evolutional respect among the factors that affect the environment only periodically, without breaking the general laws of the development of cyclical successions in natural systems. Engineered materials are becoming constantly acting factors, which can give birth to global problems in the environment.

4.1.3 Pathways of Nanoparticle Entry into Soil

The pathways of nanoparticle entry into soil characterize their entry, accumulation (content), and migration. NPs can enter soil with atmospheric precipitation, sedimentation in the form of dust and aerosols, direct soil absorption of gaseous compounds, abscission of leaves or as a result of anthropogenic activity, etc. After NMs get into a water system through sewage or industrial emissions, nanoparticles can accumulate in plant organisms (e.g., in algae), as well as in organisms of invertebrates (plankton, benthos, crustaceous) that are the primary links of a food chain, and then they can pass into organisms of water vertebrates taking part in the human food chain. In a land ecosystem, NPs can accumulate in soil, surface water, sewage, and groundwater. The results of studies in this field are generalized in Table 4.1.

Coming from different sources, pollutants ultimately get on the surface of soil, and their further fate depends on its chemical and physical properties. Pollutant components stay in soils much longer than in other biospheric objects. Many examples of direct anthropogenic and man-caused environmental effects on soil are known; therefore, soil is as sort of an object depositing anthropogenic pollutants and a source of the secondary pollution of water and other environmental objects. The pollution of soils with nanomaterials presents a serious risk of getting into the human organism and tissues of land plants and animals. The entry of NPs into any biocenosis component can lead to their introduction into other objects of this system and transfer through the food chain. With allowance for the ecological functions of soil and its role in substance turnover, the following migration pathways of nanoparticle entry into this object (Fig. 4.1) are marked out: (1) the translocation pathway that characterizes the transition of a substance from land plants and NP waste; (2) the water migration pathway that characterizes the capability of a substance to migrate from groundwater, sewage, and water sources; and (3) the air migration pathway that characterizes the transition of a substance from the atmospheric air (Venitsianov et al. 2003).

Table 4.1 Environmental objects in which nanoparticles of different types can accumulate

Branch, application	Type of a nanomaterial	Environmental object					
Cosmetic products, personal hygiene products	TiO_2, ZnO, fullerene C60, Fe_2O_3, Ag	Soil	Surface water	Groundwater	Sewage	Waste	Air
Catalytic agents, lubricants, combustion catalysts	CeO_2, Pt, MoS_3	–	+	–	+	–	–
Dyes and covers	TiO_2, SiO_2, Ag, quantum points	–	+	–	+	–	+
Water treatment, recovery of the environment	Fe, Fe–Pd	–	+	–	+	–	+
Agrochemical preparations	SiO_2 (porous) as a carrier	+	+	+	+	–	–
Food package	Ag, nanoclay, TiO_2	+	+	–	–	–	+
Pharmaceutical preparations	Nanopreparations, filling agents	+	–	–	–	+	–

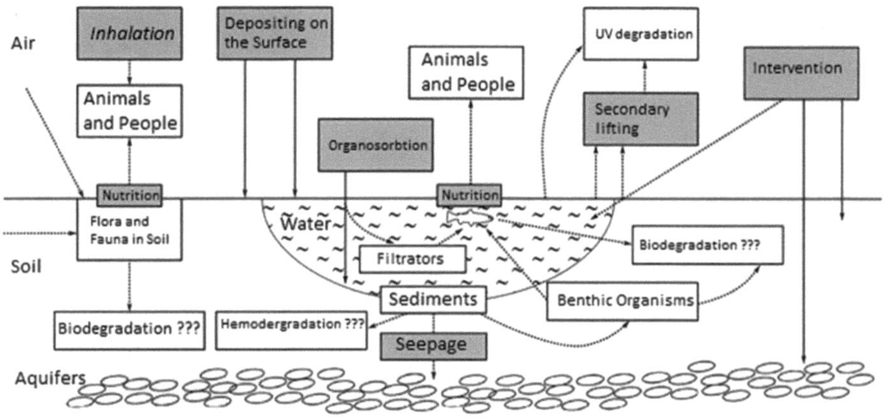

Fig. 4.1 Possible entry and migration pathways of nanoparticles proved experimentally (*a solid line*) and supposed pathways (*a dotted line*) (Krichevskiy 2010)

At present, there are few studies on revealing the nanoparticle migration processes in soil. Nevertheless, pollutants absorbed or included in colloids are known to be transported, when favorable conditions are formed for this, for example, if natural soil colloids are carriers of metals through soil profiles (Cornelis et al. 2010; Darlington et al. 2009; Fang et al. 2009; Klaine et al. 2008). A model has been developed for the migration of metal-containing nanoparticles in the NP waste–soil system (Fedotov and Shalaev 2012). Distribution along a soil profile has been ascertained to take place as follows: at the initial stages of NP introduction, the humus layer accounts for their maximum concentration; subsequently, the maximum concentration moves to the underlying layers, usually being affected by acid filtration water. The capability of metal-containing NPs to migrate in a solution has been shown to represent a high danger. This fact agrees with the experimental evidences of their having a much greater bioavailability in liquid media in comparison with natural soil (Handy et al. 2012). The depth to which waste components migrate along a soil profile and the level of their content in a filtrate are indicators of the water migration danger.

Study of the biological activity of nanomaterials in connection with different conditions of their entry into soil is very significant to understand the effect of nanoparticles on the environment. Engineered NPs can get into soil in the dissolved state (*A*), which can entail the following: (a) bioaccumulation by roots of land plants; (b) accumulation by invertebrates and, correspondingly, toxicity; and (c) microbial toxicity. The direct absorption of solid particles (*B*) can cause the toxicity of the following: (a) roots of land plants, (b) invertebrates, and (c) microbes. The direct entry of NPs (*C*) into soil can result in their sorption/aggregation or in migration along the profile. Nanoparticles of engineered nanomaterials differ in bioavailability depending on their aggregative state. They have a greater bioavailability in the dissolved form and consequently are more toxic, since they are not included in the structure of soil. One of the suggested

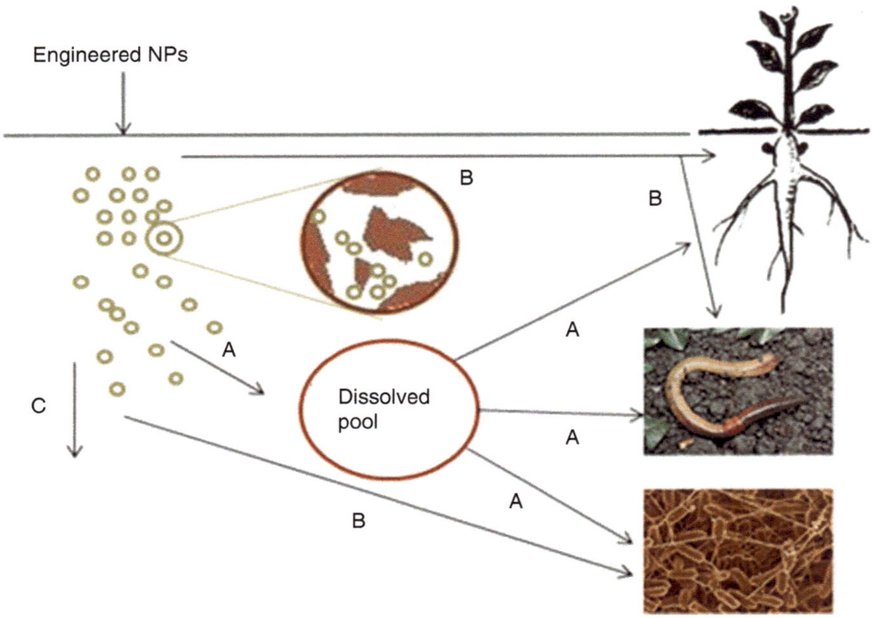

Fig. 4.2 Bioavailability of nanoparticles under different conditions of their entry into the soil environment: *A* is the dissolved pool of NPs; *B* is the direct absorption of solid NPs; *C* is the direct entry of NPs, migration along the profile

methods for detecting NPs according to accumulation and solution in soil can be used when estimating their bioavailability for plants. However, in such experiments root exudates can affect the distribution of NPs in the rhizosphere, thus changing their relative bioavailability. Nonetheless, test methods clearly should take into account the diversity of natural soils such as their pH, clay content, cation exchange capacity, texture, amount/type of organic matter, and mineralogy, as well as include a standard soil in the test (Handy et al. 2012) (Fig. 4.2).

Our research has shown that humic substances, which are known to be used to detoxify different pollutants of organic and inorganic origin (Perminova 2008), can affect the biological activity of nanomaterials. The toxic effect of some concentrations of detonation nanodiamonds (DNDs-U) and titanium nanodioxide (nano-TiO_2) has been ascertained to be removed in the presence of an industrial humic preparation potassium humate from leonardite of the POW HUMUS mark (Le-PhK) produced by the Humintech company (Germany). The detoxification properties of Le-PhK (5 mg/L) manifest themselves in biological tests with higher plants (*Brassica juncea*) in case of its being applied jointly with DNDs-U and nano-TiO_2 at the same concentration; at other concentrations of nanoparticles (50 and 100 mg/L), the effect is ambiguous (Gladkova and Terekhova 2013). The discovered difference in the direction of the action speaks for the complex interactions between humic substances and nanomaterials. At this stage, it is evidently necessary to carry out further research and to accumulate significant information on the factors affecting the detoxification capability of both natural and industrial humic substances.

There is relatively little information on the entry and expansion pathways of engineered nanomaterials and their subsequent fate in water and land ecosystems currently. The difficulty in detecting engineered nanoparticles in the presence of natural colloids is one of the obstacles in studying NMs. It is rather difficult to detect the behavior of engineered nanoparticles. Second, the interaction of nanoparticles and natural colloids will affect the behavior of NPs depending on their concentration (Boxall et al. 2007). In soil, natural colloids have much lower concentrations, but NP concentrations would likely be much lower also because of increased aggregation and sedimentation at higher ionic strengths (Klaine et al. 2008). With allowance for the difficulties in detecting NPs in soil, studies on soil organisms have become an alternative method for proving this interaction, which analytic methods have shown that they accumulate here. Thus, the experiments using ^{14}C-labeled single-layer nanotubes (SLNTs) have fixed their absorption by *Eisenia fetida* worms and then the purification of animal organisms. An attempt has been made to investigate the absorption of silver nanoparticles by nematode tissues (Meyer et al. 2010) and gold and copper nanoparticles by rainworms (Unrine et al. 2008, 2010).

However, complex methods, which will hardly be applied in normative testing, were used for this to isolate and characterize natural NPs in soil (the so-called soil colloids) (Gimbert et al. 2006, 2007). Reliable detection requires that particles be isolated from the solid soil phase (desorption) and dispersed in a water suspension, which is a significant analytical problem. At present, works aimed at its solution are very few in number (Klaine et al. 2008). Thus, the opinion has been advanced that the major theses of colloid chemistry can help in studying the behavior of nanoparticles. Relying upon the research on natural water colloids with a dimension of several nanometers, foreign specialists have drawn the conclusion that they can behave analogously to industrial NPs (Gustafsson and Gschwemd 1997; Lead and Wilkinson 2007; Madden et al. 2006). This comparison is possible first because colloids are given to aggregation and ultimately aggregated into particles > 1 mkm, which are sufficiently large, and sedimentation is dominant in their transfer (Honeyman and Santschi 1992).

The results of analyzing the literature indicate that the problem of the expansion of nanomaterials in the environment is becoming ever more acute. Soil seems to be the most useful for the development of reliable methods for analyzing and revealing the content of nanoparticles. Being a very specific part of the biosphere, it not only geochemically accumulates pollution components but also plays the role of a natural buffer that controls the transfer of chemical elements and compounds into the atmosphere, hydrosphere, and living matter. Coming from different sources, pollutants ultimately get on the surface of soil, and their subsequent fate depends on its chemical and physical properties.

The problems of detecting nanoparticles in soil are related not so much to the well-known technical difficulties, which require expensive special equipment and a high level of technical qualification of specialists, as to the fact that soil, which is a complex multiphase system, contains a great number of mineral and organic macro- and micro-components as well as natural nanoparticles. The modern methods do not permit the bioavailability of engineered nanoparticles to be investigated in detail. However, judging from indirect indicators (the change in standard test functions in biotests), the bioavailability of nanoparticles in natural habitats can be asserted to depend primarily on their sizes and degree of aggregation, and, when they are found in soil, the diversity of natural soil properties (acidity, presence of cations, organic substances, etc.) acquires paramount significance.

4.2 Problem of Bioassay Engineered Nanomaterials in Soil

Because of the recent development and rapid advent of nanotechnologies, great attention is paid to the effect of engineered nanomaterials (NMs) on living organisms. Both Russian and foreign researchers pay emphasis to the search for potential methods of assessing the effect of synthetic products of nanotechnologies on natural complexes and on the functioning of the main links of the trophic chain and separate organisms. The complexity of the soil organo-mineral composition and the unpredictable dynamics of soil properties in time and space create problems in the structural and functional analysis of the biotic complex of soil under the impact of conventional pollutants, whose chemical transformations are well understood. Available data indicate that, to assess the impact of nanoparticles on soil components, the existing methods should be adapted and new methods have to be developed. This work is devoted to the analysis of the behavior of engineered NMs in soil and the description of the methods for their ecotoxicological assessment. It is known that the behavior of nanoparticles in natural media differs from that of coarser particles of the same material. As a rule, NMs more easily enter into chemical reactions with other environmental components compared to coarser objects of the same composition; they are capable of forming complexes with previously unknown properties. An important factor for assessing NMs' impact on living organisms is the effect of the NMs' interactions. The inverted dose–response ratio, or the U-shaped curve describing this relationship in ecotoxicological studies of dispersed systems, is largely due to the formation of aggregates at high concentrations and the increase in the content of free nanoparticles under dilution. The biotic and abiotic transformations of any chemical compounds, including NMs, during bioassays can give different results: (1) the formation of more toxic products, including those with delayed action or new properties; (2) the formation of products with higher hazard indices compared to the original substance; (3) the formation of products whose toxicity is similar to that of the original substance; and (4) the formation of less toxic products. Therefore, the revelation and investigation of the adaptation of living organisms and their resistance to the action of engineered nanoparticles in the soil, which is a depositing medium for pollutants from different sources, are of special importance under the increasing technogenic impact.

4.2.1 *Problems and Advances in the Studies of the Impact of Nanoparticles on the Environment*

One of the problems in the study of nanoparticles is related to the revelation of engineered nanoparticles in the soil in the presence of natural colloids. Natural nanoparticles in the soil (referred to as soil colloids) are difficult to separate and characterize (Noack et al. 2000; Gimbert et al. 2006, 2007). Their reliable detection includes the separation of particles from the soil solid phase (desorption) and their dispersion in a water suspension, which represents a serious analytical problem. Works aimed at its solution are extremely scarce now (Klaine et al. 2008). An alternative method for confirming the presence of nanoparticles in soil is the study of the responses of soil organisms. In some cases, analytical methods proved the accumulation of nanoparticles in tissues of living organisms. Thus, the uptake of ^{14} C-labeled monolayer carbon nanotubes by *Eisenia fetida* earthworms was experimentally confirmed (Petersen et al. 2008). Gold and copper nanoparticles were found in the earthworm tissues (Unrine et al. 2008; Yang and Watts 2005). Attempts were made to analyze the adsorption of silver nanoparticles by nematode tissues (Meyer et al. 2010). Complicated analytical methods and sophisticated equipment were used for this purpose, which are hardly suitable for the wide distribution of these methods in conservation practice. In the resolution of the 2005 seminar of the European Centre for Ecotoxicology and Toxicology of Chemicals (ECOTOC) devoted to the biosafety of NMs, it was specially emphasized that the nature, surface area (including the state of aggregates and agglomerates), and shape of nanoparticles should be taken into consideration in the study of the toxic effect of NMs. It is inadvisable to use a single dimension (e.g., mass, surface area, or particle size) to characterize nanoparticles (Masycheva et al. 2008). The conventional assessment of the toxicity is more informative than the physicochemical methods of analysis. Therefore, along with analytical methods, bioassay procedures find increasing use in the assessment of the environmental effect of nanoparticles. Bioassays provide advanced information about problems before the appearance of obvious changes in natural ecosystems It is advisable to include representatives of the main trophic levels in the system of bioassays (Terekhova 2011).

4.2.1.1 Effects of Nanoparticles on the Environments

There are opposite opinions about the safety of nanoparticles for living objects: some authors declare the complete harmlessness of NMs; other authors, on the contrary, express extreme concern over the distribution of products of new technologies and are at an alarm. This again emphasizes the poor knowledge and complexity of the identification of NMs and their effects not only in soil but also in aerial and aqueous environments and organisms.

Under real conditions almost all nanoparticles in water environments form conglomerates, which undergo sedimentation and elimination from active processes. One relates these processes to the peculiar self-purification of water environments, which would erase the problem of assessing the effect of high concentrations of nanoparticles. However, the hazard of such concentrated precipitates for soil-inhabiting and benthic organisms cannot be excluded. At this stage, there is neither a clear idea of the hazard of NMs nor any general concepts of the possible mechanisms or theory explaining the effect of NMs on living cells. It is essential that, although authors hold significantly different views on the environmental hazard of nanoparticles, most of them recognize the existence of this hazard. Daniel Watts wrote "as far as the use of nanoparticles increases, their risk also increases." At the same time, "we should attentively examine this domain and probably undertake some serious measures." Let us consider some studies on the toxicity of NMs widely used in the production area.

4.2.1.2 Carbon Nanoparticles

Fullerenes and/or carbon nanotubes are most widely used in different areas. For example, the annual production of single-wall carbon nanotubes carbon nanotubes stronger than steel by 460 times reached 1000t in 2011. Studies of the toxic effects of fullerenes give contradictory results. Some authors express concern that these particles can damage microorganisms, "whose disappearance can cause a real ecological catastrophe" (Klaine et al. 2008). However, evidence of the harmlessness of fullerenes for soil is more abundant. A recent study performed at Purdue University showed that fullerenes are harmless for microorganisms and are adsorbed by soil without damaging it. No effect on soil organisms was recorded in the analysis of fullerenes (nC_{60}) by Tong et al., who supposed that the interaction of nC_{60} with soil organic matter ensures the neutralization of the potential toxic effect of fullerenes (Unrine et al. 2008). It was shown that fullerenes modified by amino groups have toxic effects. In particular, the inhibition of test functions in the assays with *Escherichia coli* reached 60 % compared to the control (coliform bacteria untreated with NMs). Toxic and other unfavorable effects of NMs can be manifested in different forms. Carbon nanotubes show varying degrees of toxicity depending upon their arrival into animal organisms (Allsopp et al. 2007; Donaldson et al. 2006).

However, the data obtained on their harmfulness for living organisms are also contradictory. Some authors showed that ultra dispersed diamonds from detonation synthesis have no carcinogenic or mutagenic properties. Due to their high adsorption capacity and other specific properties, hyperactive sorbents act as immobilizers of biologically active substances (Schrand et al. 2007). Our studies showed that, in spite of the carbon nature of nanodiamonds, which imparts them a specific affinity for organic elements of the environment, they have a toxic effect when present in specific concentrations. This was in particular revealed in the study of the effect of synthetic nanodiamonds (PL-D-G02, PlasmaChem GmbH) on the growth and fluorescence of a standardized algal culture of *Chlorella vulgaris*. Studies performed according to the standardized procedure recommended for ecological control, where the assessment is based on the changes in two different test functions (an increase in the cell population of the microalgae and fluorescence), showed that the revelation of nanodiamond toxicity largely depends on the selected method of the toxic effect's registration, as well as on some other factors. The direct counting of cells under a microscope, rather than fluorescence indices, is obviously a more reliable method of studying the biosafety of nanodiamonds, which inevitably form aggregates of different sizes in suspensions. It is interesting that a lower concentration of synthetic nanodiamonds (0.0005 %) had a higher damaging effect than a higher concentration of 0.005 %. This conclusion is based on the analysis of changes in the cell number and the fluorescence study of algae. This perfectly agrees with observations of other authors and our data obtained for other test species. Finer particles and aggregates of nanodiamonds were also found to be more toxic for other test organisms, including *Paramecium caudatum*. This can be related to the higher aggregation of nanodiamond particles in the incubation medium and, hence, the lower ability of coarse aggregates to penetrate into the cells of living organisms (Karateeva et al. 2009).We have performed thorough studies of the biological activity of Russian industrial detonation nanodiamonds (DNDs-U produced by OOO SKN, Snezhinsk, Chelyabinsk Oblast) differing in the size of the free particles in water suspensions. In experiments on three nanodiamond samples with mean particle sizes of 15, 30, and 100 nm (at 5, 50, and 500 mg/L), we studied the responses of standardized test organisms of the main trophic levels: (1) producers, higher plants (the leaf mustard *Brassica juncea*); (2) consumers, infusoria (the slipper animalcule *Paramecium caudatum*); and (3) reducers, bacteria (a luminescent strain of *Escherichia coli*). A relationship between the toxicity of the DNDs-U samples and the size of the particles was revealed. An increase in toxicity with the decrease in particle size (100–30–15 nm) was observed, as in some other known cases (Gladkova 2011).

4.2.1.3 Metal-Containing Nanoparticles

The safety of metal-containing NMs attracts no less attention than that of carbon NMs. Nanotitanium dioxide is a material widely distributed in consumption products and the nanoindustry. The studies of the uptake and accumulation of titanium dioxide nanoparticles in test cultures (chlorella and daphnia) showed their high accumulation rate and concentration in phyto- and zooplanktons. Thus, the content of titanium in algal cells exceeded that in the environment by more than 200 times. In daphnia organisms, the content of titanium was half as high as in chlorella but 100 times higher than in the environment. The study of the effect of titanium dioxide nanoparticles (nano-TiO_2) with a mean particle size <75 nm in a water suspension (a mixture of two crystalline TiO_2 modifications, anatase and rutile, Sigma-Aldrich, USA) in three bioassay systems revealed uncertain effects, although toxicity was more frequently detected at the studied concentrations of

5, 50, and 500 mg/L. For example, a phytotest with mustard seedlings showed that the impact of nano-TiO_2 suspension (5, 50, and 500 mg/L) inhibited the development of leaf mustard roots. In an experiment with paramecia, it was found that nano-TiO_2 at concentration of 5 mg/L had a low stimulating effect on the development of a protozoan culture, while an acute toxic effect was manifested data concentration of 50 mg/L. In a bioassay with luminescent bacteria, a low concentration of nano-TiO_2 (5 mg/L) suppressed the fluorescence of bacteria (evidence of a toxic effect), but a stimulation of fluorescence was observed at concentrations of 50 and 500 mg/L with the higher stimulation being observed at 50 mg/L (Gladkova 2011). Thus, no linear dose–effect relationship was revealed in most of our experiments on the analysis of biotic responses to the concentrations of nanoparticles in water suspensions, as was observed by many authors for other NMs. This can be due to the peculiar interaction mechanism of nanoparticles and the relationship between the aggregation of nanoparticles and the degree of dilution, i.e., the minimum distance between nanoparticles in the dispersed system. This relationship can also be explained by different impacts of NMs on living organisms. In a series of experiments on studying the responses of soil microorganisms to the antibacterial properties of silver nanoparticles (Benjamin Colman, Duke University), the almost complete suppression of the development of nitrogen-fixing bacteria was observed a month after the treatment. This group of microorganisms was more susceptible to silver nanoparticles by a million times compared to other microorganisms. Another experiment showed that the activity of bacterial enzymes degrading organic substances in soil treated with silver nanoparticles decreased by 34 % compared to the untreated soil. These data indicate that a profound study of the interaction between nanoparticles and soil components is necessary.

4.2.1.4 Effect of Soil Properties on the Manifestation of NMs Toxicity

Data on the behavior of nanoparticles in different soil environments are gradually being accumulated in the literature. It was experimentally shown that soil is a reliable filter for the migration of nanoparticles if it contains an increased amount of clay or has a high ionic strength. In a series of experiments performed at the Georgia Institute of Technology, water-containing fullerenes was passed through vessels filled with sand, sediment, glass microgranules, and other materials; it was revealed that even sand retains up to 80 % of the nanoparticles. It was also shown that the filtration of nanoparticles depends on the water's composition. It is interesting that the presence of humic acids or surfactants allowed nanoparticles to freely pass through sand. Under hydroponic conditions, toxic effects of engineered NMs on higher plants were frequently observed (Meyer et al. 2010). In soil, the phytotoxicity of NMs for the grown test plants is minimum if any (Baun et al. 2008; Fernandes et al. 2007).

4.2.2 Possible Mechanisms of the Impact of Nanoparticles on Living Organisms

Most mechanisms of the toxic action of NMs are unclear; however, relatively well-defined concepts are reported in the literature for some of them.

Studies of the quantitative uptake and accumulation of NMs by whole organisms showed that nanoparticles mainly arrive into multicellular animals by ingestion and absorption through intestinal walls (Donaldson et al. 2006; Fountain and Hopkin 2001; Stampoulis et al. 2009). Most works on assessing the possible migration of nanoparticles in animal tissues were performed with model test organisms. The first works dealt with well-studied species widely used in ecotoxicology, in which species these processes could be observed in an optical microscope. The absorption of fluorescent carboxylated nanoparticles by daphnia (*Daphnia magna*) and their translocation from the intestinal tract to fat deposits were demonstrated. The mechanism of this absorption remains in the focus of the attention of researchers (Lin et al. 2007).Varied mechanisms for the development of the toxic effect of nanoparticles are determined by their specific physicochemical properties, which depend not only on their size but also on the adhesive, catalytic, optical, electrical, and quantum-mechanical properties, as well as their geometry, size distribution, and organization in the nano object.

Many NMs are capable of inducing active oxygen species due to their physical nature (Roberts et al. 2007; Lyon et al. 2005; Klaine et al. 2008). The mechanism of the impact of nano objects on living structures is related to both the formation of free radicals in their presence and the appearance of complexes with nucleic acids. The induction of active oxygen is considered as the main mechanism of the toxic effect of TiO_2 nanoparticles; the reactivity depends not only on the size of the nanoparticles but also on the structure of the TiO_2 (a crystalline or amorphous one) (Kai et al. 2003). Some NMs are capable of penetrating through tissue barriers into cells and interact with intracellular components (Kapustka et al. 2006). Some types of NMs (dendrimers of different degrees of generation) can disturb membrane structures and make them permeable.

It has been shown that nanoparticles can penetrate into cells in different ways. Some authors observed simple diffusion through the cell membrane (Reevesa et al. 2008), and other authors reported endocytosis (Klaine et al. 2008) or adhesion (Terekhova and Gladkova 2013; Lin et al. 2007). Nanoparticles arriving into an organism can act as catalysts for the formation of toxic compounds, even if they themselves are harmless. Similar phenomena are typical for TiO_2 and ZnO nanoparticles catalyzing photooxidation and oxide nanoparticles of iron and some other metals causing metal (most frequently, zinc) fever. As for higher plants, it is believed that the sensitivity of plants to NMs is based on the capacity to filter and accumulate nanoparticles. The revealed toxicity mechanisms of nanotechnological products are difficult to classify, because they differ even within a class of materials. For example, fullerol (hydroxylated fullerene $C_{60}(O)_x$) generates single oxygen and can behave as a powerful oxidant in biological systems, but it reveals no cytotoxicity (Reevesa et al. 2008). The covering of fullerene with polyvinyl pyrrolidone is accompanied by the formation of nanoparticles also generating singlet oxygen, which can cause the peroxidation of lipids and the damage of cell DNA (Kang 2008). Other studies of water suspensions of fullerene showed their antibacterial activity in the absence of light and oxygen and thus denied the exceptional effect of singlet oxygen at the manifestation of toxicity. Different mechanisms are reported for explaining the toxicity of silver nanoparticles. Some of them relate the toxicity of these particles to changes in the penetrability of cell covers, because the adhesion of nanoparticles to the surface of cells affects the properties of membranes. It is not excluded that the silver nanoparticles penetrating within bacteria damage their DNA and can release toxic Ag^{+} ions during the interaction with the cell. Some authors disagree even concerning the interpretation of the toxicity mechanisms for the same nanoparticles. Several authors relate the suppression of growth of five different plant species (cabbage, carrot, corn, cucumber, and soybean) by aluminum nanoparticles (Al, 13 nm, 2 mg/ml) to the presence of free hydroxyl groups on the surface of particles, while other authors suppose that the phytotoxicity is due only to the increased solubility of aluminum nanoparticles (Noack et al. 2000). A special problem is related to the assessment of NM toxicity in soil and the effect of soil properties on the biological activity of nanoparticles. This involves the complicated development of methodological approaches and the formation of a system for estimating the ecological toxicity of nanoparticles in terrestrial cenoses. The aging and changes of nanoparticles during long-term experiments with soil organisms significantly hamper the studies of their toxicity. The studied material can be transformed in the soil within several weeks or months. It is known that this problem is also typical for conventional chemical pollutants. However, the interaction of nanoparticles with the soil also involves specific features of NMs. For example, unstable nanoparticles can be completely eliminated during an experiment on the revelation of acute and chronic toxicity with the use of test plants. This was observed in experiments with silver nanoparticles. Assays with a short exposure of test organisms are necessary to minimize the effect of aging. Nematode bioassays (e.g., ASTME2172, ISO/DIS10872) are promising (Asli and Neumann 2009; Jiang et al. 2008). The determination of test functions susceptible to nanoparticles in soils was repeatedly discussed in ecotoxicological works. It is considered difficult to reveal the biological activity of nanoparticles from the test parameters used for detecting the effect of conventional chemical pollutants (the survival and propagation of pedobionts). Some authors are sure that such common test functions as seed germination and seedling root growth have a limited sensitivity to NMs In separate cases, the behavior of soil-inhabiting animals can be considered as a sensitive test function. However, the correct interpretation of behavioral changes is very important in soil bioassays. For example, earthworms can cease to feed and move in the contaminated soil. This protecting mechanism prevents the negative effect. A conclusion about the absence of acute toxicity can be drawn in this case, which will be a false negative result. In this context, it is recommended to select more sensitive test species and not focus efforts on searching for more sensitive test parameters in the standard test organisms. For example, springtails, which showed good results in the study of metal toxicities in soils (Geiser et al. 2005), can be also very sensitive to metal-containing NMs (Hong et al. 2004). Some authors focus attention on the sensitivity to NMs of such plant species as the adzuki bean (*Phaseolus radiatus*), tomatoes (*Lycopersicon esculentum*), and *Arabidopsis thaliana*. Biochemical or metabolic measurements are recommended in this case, e.g., for the content of chlorophyll (Meyer et al. 2010), the respiration rate, or the nitrogen fixation by legumes.

The diversity of the developed engineered NMs, the absence of common priorities for assessing their safety, and the unsuitability of the conventional toxicological (hygienic) characteristics for nanosized structures result in the necessity for searching for and using new approaches in econanotoxicology. The preparation of natural samples and the composition of the incubation medium for standard test cultures require special attention. The range of bioassay procedures designed for the ecotoxicological assessment of soils should obviously be based on the responses of soil-inhabiting organisms (pedobionts). Contact methods, rather than eluate methods, are more reliable for determining the effects of NMs in soil, including from the responses of microorganisms. However, authors rarely set themselves the task to develop procedures suitable for legitimate decision making and practical use. The natural diversity of soils, the pH variations, the clay content, the cation exchange capacity, the texture, the mineralogy, and the organic matter should obviously be taken into consideration in the creation of standardized assays for the determination of the effects of nanoparticles in soils. The effect of the organic matrix on the toxicity of NMs was repeatedly manifested, including in our works with nanodiamonds from detonation synthesis and nanotitanium dioxide (Gladkova and Terekhova 2014; Gimbert et al. 2006, 2007). Animated discussions still accompany proposals for the creation and use of model soil samples (Hong et al. 2004) for comparing the toxicity of different preparations and concentration effects of NMs in different countries.

The analysis of the literature data showed that the assessment of the implications of the NM distribution in the environment remains an open problem. This is largely due to the insufficient methodological supply of their identification in natural environments, especially in soils. There is no universally accepted theory explaining the mechanism of the effects of any nanosized structures with consideration for the structural features of their surface and reactivity. There are no reasons for hampering the development of nanotechnologies and the propagation of NMs in soils taking into account the imperfection of the methodological approaches to the analysis of their toxicity. To overcome nanophobia and extreme views on the problem considered, we should extend ecotoxicological studies to all produced NMs, accumulate experimental data, and gradually select the sets of test systems the most adequate for the analysis of the biological safety of NMs in soils.

4.3 The Biological Activity Modulation of Engineered Nanomaterials in Soils Under the Humic Substances' Influence

In our experiments test responses of three trophic level organisms (producers, consumers, and reducers) on nanomaterials of different natures: carbon containing (nanodiamonds) and metal containing (nanodioxide titanium and nanomagnetite) adding humate in water were analyzed. Water extracts from natural and artificial soils were used durring experiments. The objective of this research is to study engineering carbon- and metal-containing nanomaterials' toxicity change under humic substances' influence.

Widespread engineered nanomaterials and their accumulation in environments give grounds to consider them as a special kind of pollutants. Currently the most effective areas of humic substances' (HS) application are known. Their use as detoxicants of organic and inorganic pollutants is one of the most important (Kaniskin et al. 2011; Tan 2003). Nanomaterials' biological activity in soils and HS influence on nanomaterials remain poorly understood despite of considerable attention given to the nanomaterials' study in environments.

In our work following materials were investigated: (1) humate "POW HUMUS" (Le-PhK, K-humate originated from leonardite, "Humintech", Germany); (2) carbon-containing nanomaterials—nanodiamonds produced by industrial detonation synthesis of high explosives (DNDs, different size free particles in aqueous suspensions up to 15, 30, and 100 nm, "SNK", Snezhinsk, Chelyabinsk region, Russia); (3) metal-containing NMs—nanodioxide titanium (nano-TiO2, <25 nm, "Sigma-Aldrich", US); (4) metal-containing NMs—nanomagnetite (nano-Fe3O4, 30 nm, MAI, Russia). Nanomaterials' concentration varied in range of 5–500 mg/L; humate concentration was 5 mg/L in water.

The research is based on standard environmental soil control methods recommended for industrial and state issues. The bioassay of standardized test cultures represented by different trophic levels such as producers (higher plants *Brassica juncea* L.), consumers (infusorium *Paramecium caudatum* Ehrenberg), and reducers (bacterial biosensor—genetically modified strain of *Escherichia coli*) was carried out.

In one set of experiments, the test responses of organisms on nanomaterials in water (0.5–500 mg/L) and humic preparation's response reactions to them were analyzed. In another set of experiments, the nanoparticles' toxicity and humate's response reactions to it in water extracts of podzolic soil (Chashnikovo, Moscow region, A horizon) and artificial soil, model soil prepared in accordance with ISO 11268-1, were investigated.

Bioassay showed that soil contaminated with nanomaterials exhibit inhibiting and stimulating biological activity. Biotic response level fluctuations in nanoparticles' presence in water and in soil sample extracts were noticed. Depending on the type of medium and nanomaterials, humate's detoxication effect on test cultures varies.

In addition, nanomaterials' bioassay in water on test cultures of different trophic levels with and without HS was performed.

Nanodioxide titanium inhibited producers' test functions (higher plants–root length) in all range of concentrations (0.5–500 mg/L). At the same time, humate in all concentrations, except 50 mg/L, relieves inhibition, stimulating root growth and seed germination. Nano-TiO_2 has a stimulating effect on infusorium and bacteria test cultures, and the HS presence increased twice more stimulating effect compared to higher plants.

Nanomagnetite except of nanodioxide titanium stimulated the development of higher plants at all concentrations, except 50 and 100 mg/L, which showed an inhibitory effect. HS effect on the nano-Fe_3O_4 bioactivity at different test cultures appeared ambiguous: at high concentrations (100 and 500 mg/L), inhibition of higher plants' roots and bacterial luminescence stimulation was observed, and at low concentrations, on the contrary, inhibition of bacterial luminescence and stimulation of the plant roots and infusorium's survival.

Adding humate to nanodiamonds (particle size 15–100 nm) mitigated toxic effects. These effects are more evident in concentration of 500 mg/L in nanodiamond water suspension.

Thus, research has shown that the toxic effect of nanomaterials in water was nearly removed in the presence of humate Le-PhK (5 mg/L). In some cases, HS combined with nanomaterials increases toxic effects in concentrations 50 and 500 mg/L.

Bioassays of nanomaterials with and without HS on test cultures of different trophic levels in soil extracts from natural and artificial soils revealed the followings.

Nanodioxide titanium in the extract of podzolic soil is almost neutral for higher plants. Humate addition caused stimulating effect at all concentrations (0.5–500 mg/L) from 8 to 35 %. Phytotesting on model soil extract showed stimulatory effect in all concentrations. Humate stimulates further growth of *Brassica juncea* roots for 3–14 %. Bioassay on infusoria showed that nano-TiO_2 has mostly inhibitory effect except of 0.5 mg/L, in which stimulation was showed. Adding HS eliminates this inhibition with the exception of 0.5 mg/L, in which it certainly inhibited survival of infusoria not only in the extract of podzolic soil but also in artificial soil. Nano-TiO_2 significantly increased the bacteria luminescence in both media, and the addition of humate further enhanced this effect.

Nanomagnetite phytotesting mainly shows little stimulatory effect in podzolic soil extract, excepting of 500 mg/L. Adding humate doesn't affect nanomagnetite's nature of impact. Nano-Fe_3O_4 impact is neutral in the artificial soil medium. Adding humate at low concentrations depresses higher plant root development. Natural soil extract with 100 mg/L nanomagnetite concentration showed acute toxic effect on infusorium's survival, but humate completely eliminated this effect. However, in the range of lower concentrations (5–10 mg/L), nanomagnetite toxicity increases. Nano-Fe_3O_4 reduces infusorium's survival in the range 100–500 mg/L in the model soil medium. HS further enhances toxicity at 500 mg/L. Bacteria luminescence inhibited in the whole range of concentrations (0.5–100 mg/L) in podzolic soil media and in the range of 100–500 mg/L in artificial soil medium. Humate exhibits inhibitory effect in both mediums.

Nanomagnetite's exposure stepwise nature has been established. Equal inhibitory activity is typical for 0.5 mg/L concentration and ten times bigger concentrations (e.g., 5 mg/L). Equal stimulating activity is typical for 1 mg/L concentrations and ten times bigger concentrations (e.g.. 10 mg/L). Such dependence is difficult to explain by the basic of different soil matrixes. This can be attributed to different mechanisms of impact in each concentration range (Gladkova and Terekhova 2014).

Determined by a number of peculiarities, concentrations of nanomaterials differed by an order or two have a similar effect; the bioactivity sign changes from concentration to the concentration "stimulation-inhibition." Average zone concentration effect in some cases is lower than in small concentration; it was also noticed in other researches (Terekhova and Gladkova 2013).

We may also conclude that the bioactivity of engineering nanomaterials entering environments (water or soil, enriched with natural organic matter) can be modified by the presence of humic substances. Generalizing humate (5 mg/L) impact data positive effect is clearly evident in conjunction (1) with nanomagnetite on infusoria in water (10 mg/L), bacteria (500 mg/L), extraction from podzolic soil on infusoria (0.5 and 100 mg/L), and from artificial soil (100 mg/L) and (2) with nanodioxide titanium on infusorium (1 mg/L), bacteria (10 mg/l), extraction from podzolic soil on higher plants (10 mg/L), and infusorium (1 and 50 mg/L). This confirms the universality of detoxication properties of humates (Yakimenko and Terekhova 2011) and expressed in neutralizing nanoparticles toxic effect.

The obtained bioassay data of the three nanomaterial types (nano-TiO_2, nano-Fe_3O_4, and DNDs-U) showed that toxicity depends on the physical nature of the nanoparticles (metal or carbon containing), size, and ability to form aggregates.

4.4 Influence of Nanomaterials on Soil Structure and Mechanical Properties: Effect with and Without Humic Substances

Research in the field of environmental behavior of nanomaterials has been increasing over the past decade due to their unique physical and chemical properties and to an expected rise in their production in the future.

The question of their fate and impact on soils has become a major concern since poorly understood interactions of nanomaterials with the soil particles. Impact of nanomaterials on the fate of other pollutants in soil remains controversial. There is almost no data on the effects of nanomaterials on soil structure and physical and chemical properties with different humus status.

In experimental research (Gladkova et al. 2015) we applied metal-containing nanomaterials, nanomagnetite (nano-Fe_3O_4), which are characterized 30 nm in size

(MAI, Russia). Concentration of nano-Fe_3O_4 was 500 mg/kg in soil. Among HS we have chosen "POW HUMUS" (Le-PhK) (K-humate, originated from leonardite) manufactured by German company "Humintech." Concentration of humate Le-PhK was 100 mg/kg in soil.

This experimental study aimed to reveal the rheological properties of structural bonds between gray-humus soil (Botanical garden, MSU, Moscow, Russia) particles in samples treatment by nanomagnetite with addition of humate potassium and without it.

Determination of rheological parameters was carried out by amplitude sweep test on a modular rheometer of MCR-302 (Anton Paar, Austria) (Markgraf et al. 2006; Khaydapova and Milanovskiy 2013; Khaydapova et al. 2013). The following parameters were determined: elastic modulus, viscosity modulus, and point of destruction of structure at which the elastic modulus becomes equal to the viscosity modulus ($G' = G''$-crossover).

The results of rheological studies using a MCR 302 modular research rheometer of soil samples are shown in Fig. 4.3. It was found out that the soil with nanomagnetite has more elastic properties ($G' - 3.95 \times 10^5$ Pa) than the original (control) samples ($G' - 1.48 \times 10^5$ Pa).

Adding humate to the soil with nanomagnetite enhances the strength of the structure.

The destruction of the structure (the point of equality models $G' = G''$) for the original soil deformation occurs at 13.7 % and with nano-Fe_3O_4 and humate is much less (1.88 %).

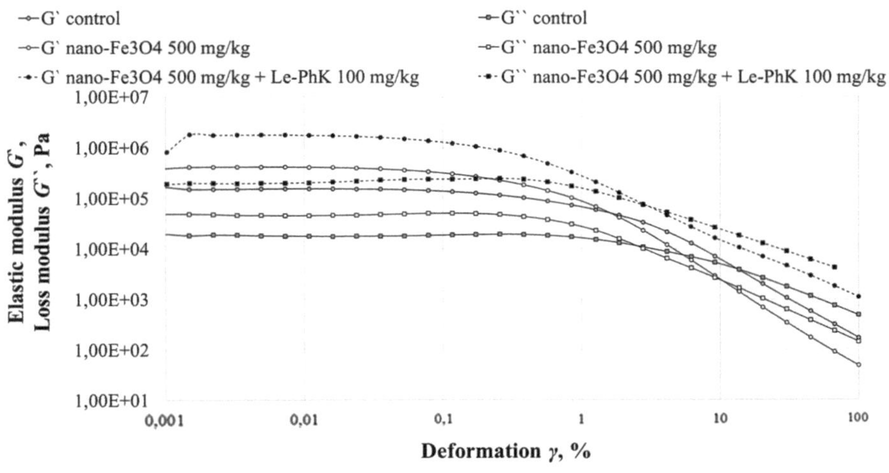

Fig. 4.3 Depending on the loss modulus and elastic modulus from the deformation

4.5 Conclusion

It can be concluded that the previously identified differences in toxicity effect nanomagnetite in soils by adding humates associated not only with the expected change in the specific surface of the particles (our preliminary results) but also with the physical and chemical characteristics of the rheological interaction between soil particles and engineered nanoparticles in the presence of humate.

Acknowledgment Authors thank Dr. Dolgar Khaydapova for contribution in rheological investigation of soil samples. The bioassay experiments in this work were supported by Russian Science Foundation (grant 14-50-00029).

References

Allsopp M, Walters A, Santino D (2007) Nanotechnologies and nanomaterials in electrical and electronic goods: a review of uses and health concerns. Green peace, Res Lab. December, 22p

Asli S, Neumann PM (2009) Colloidal suspensions of clay or titanium dioxide nanoparticles can inhibit leaf growth and transpiration via physical effects on root water transport. Plant Cell Environ 32:577–584

Baun A, Sorensen SN, Rasmussen RF et al (2008) Toxicity and bioaccumulation of xenobiotic organic compounds in the presence of aqueous suspensions of aggregates of nano-C60. Aquat Toxicol 86:379–387

Boxall AB, Chaudhry Q, Sinclair C et al (2007) Current and future predicted environmental exposure to engineered nanoparticles. Rep. Central Science Lab., Dep. Environment and Rural Affairs, York, UK

Cornelis G, Ryan B, Mc Laughlin MJ et al (2010) A method for determining the partitioning of manufactured silver and cerium oxide nanoparticles in soil environments. Environ Chem 7:298–308

Darlington TK, Neigh AM, Spencer MT et al (2009) Nanoparticle characteristics affecting environmental fate and transport through soil. Envrion Toxicol Chem 28:1191–1199

Donaldson K, Aitken R, Tran L et al (2006) Carbon nanotubes: review of their properties in relation to pulmonary toxicology and workplace safety. Toxicol Sci 92:5–22

Fang J, Shan X-q, Wen B et al (2009) Stability of titania nanoparticles in soil suspensions and transport in saturated homogeneous soil columns. Environ Pollut 157:1101–1109. doi:10.1016/j.envpol.2008.11.006

Fedotov GN, Shalaev VS (2012) Nanostructured organization of soils: academician Dobrovol'skii GV (ed) Student's book. Moscow, 520 p

Fernandes TF, Christofi N, Stone V (2007) The environmental implications of nanomaterials. In: Monteiro-Riviere NA, Tran CL (eds) Nanotoxicology: characterization, dosing and health effects. CRC, Boca Raton, FL

Fountain MT, Hopkin SP (2001) Continuous monitoring of *Folsomia candida* (Insecta: Collembola) in a metal exposure test. Ecotoxicol Environ Saf 48:275–286

Geiser M, Rothen-Rutishauser B, Kapp N et al (2005) Ultrafine particles cross cellular membranes by non phagocytic mechanisms in lungs and in cultured cells. Environ Health Perspect 113:1555–1560

Gimbert LJ, Haygarth PM, Beckett R et al (2006) The influence of sample preparation on observed particle size distributions for contrasting soil suspensions using flow field-flow fractionation. Environ Chem 3:184–191

Gimbert LJ, Hamon RE, Casey PS, Worsfold PJ (2007) Partitioning and stability of engineered nanoparticles in soil suspensions using field-flow fractionation. Environ Chem 4:8–10
Gladkova MM (2011) Effects of carbon- and metal-containing nanomaterials on test-organisms of main trophical level. In: Proceedings of the international conference on man and environment: enemies or friends, June 22–24, Pushchino, Moscow, pp 256–259
Gladkova MM, Terekhova VA (2013) Engineered nanomaterials in soil: sources of entry and migration pathways. Moscow Univ Soil Sci Bull 68:129–134
Gladkova MM, Terekhova VA (2014) The biological activity modulation of engineered nanomaterials in soils under the humic substances influence. In: Book of abstracts: Natural organic matter: structure-dynamic innovative applications, 17th Meeting of the International Humic Substances Society, Ioanina, Greece, 1–5 September, pp 212–213
Gladkova MM, Milanovskiy EYu, Khaydapova DD, Terehova VA (2015) Influence of nanomaterials on soil structure and mechanical properties: of effect with and without addition of humate substances. In: International soil science congress on "soil science in international year of soils 2015", Sochi, Russia, 19–23 October
Gustafsson O, Gschwemd G (1997) Aquatic colloids: concepts, definitions and current challenges. Limnol Oceanogr 42:517–528
Handy RD, Cornelis G, Fernandes T et al (2012) Ecotoxicity test methods for engineered nanomaterials: practical experiences and recommendations from the bench. Environ Toxicol Chem 31:15–31. doi:10.1002/etc.706
Honeyman BD, Santschi PH (1992) The role of particles and colloids in the transport of radionuclides and trace metals in the ocean. In: Buffle J, van Leeuwen HP (eds) Environmental particles, vol 1. Levis, Boca Raton, FL, pp 379–423
Hong S, Bielinska AU, Mecke A et al (2004) Interaction of poly (amidoamine) dendrimers with supported lipid bilayers and cells: hole formation and the relation to transport. Bioconjugate Chem 15:774–782
Jiang J, Oberdrster G, Elder A et al (2008) Does nanoparticle activity depend upon size and crystal phase? Nanotoxicology 2:33–42
Kai Y, Komazawa A, Miyajima N et al (2003) Fullerene as a novel photoinduced antibiotic. Fullerenes Nanotubes Carbon Nanostruct 11:79–87
Kang SJ (2008) Titanium dioxide nanoparticles trigger P53-mediated damage response in peripheral blood lymphocytes. Environ Mol Mutagen 49:399–405
Kaniskin MA, Izosimov AA, Terekhova VA et al (2011) Influence of humic substances on bioactivity of soils and phosphor gypsum. Theor Appl Ecol 1:87–95
Kapustka L, Eskew D, Yocm JM (2006) Plant toxicity testing to derive ecological soil screening levels for cobalt and nickel. Environ Toxicol Chem 25:865–874
Karateeva AV, Terekhova VA, Matorin DN et al (2009) Changes in the growth parameters and fluorescence of the culture of green protococcialga *Chlorellavulgaris* Beijer under the impact of synthetic nanodiamonds. Byul Mosk Obshch Ispytat Prirody Otdel Biol 114:68–73
Khaydapova DD, Milanovskiy EYu (2013) Influence of organic matter on rheological properties of chernozem. In: Book of abstracts of the IV international conference on colloid chemistry and physicochemical mechanics, 30 June–05 July 2013, pp 531–532
Khaydapova DD, Milanovskiy EYu, Shein EV (2013) Impact of anthropogenic load on rheological properties of typical chernozems (Kursk region, Russia). In: Soil degradation, advances in geoecology 42. Catena, Reiskirchen, pp 62–71
Klaine SJ, Alvarez PJ, Batley GE et al (2008) Nanomaterials in the environment: behavior, fate, bioavailability, and effects. Environ Toxicol Chem 27:1825–1851
Krichevskiy GE (2010) Nanotechnologies: dangers and risks. Inspecting principles for nanotechnologies and nanomaterials. Nanotekhnol Okhrana Zdorov'ya 2(3):4
Lead JR, Wilkinson KJ (2007) Environmental colloids and particles: current knowledge and future developments. In: Wilkinson KJ, Lead UR (eds) Environmental colloids and particles: behavior, structure and characterization, vol 10. Wiley, Chichester, UK, pp 1–16

Lin S, Keskar D, Wu Y et al (2007) Detection of phospholipid-carbon nanotube translocation using fluorescence energy transfer. App Phys Lett 89:143111–143118
Lyon DY, Fortner JD, Sayes CM et al (2005) Bacterial cell association and antimicrobial activity of a C-60 water suspension. Environ Toxicol Chem 24:2757–2762
Madden AS, Hochella MF, Luxton TP (2006) Insights for size-dependent reactivity of hematite nanomineral surfaces through Cu^{2+} sorption. Geochim Cosmochim Acta 70:4095–4104
Markgraf W, Horn R, Peth S (2006) An approach to rheometry in soil mechanics-structural changes in bentonite, clayey and silty soils. Soil Tillage Res 91:1–14
Masycheva VI, Danilenko ED, Belkina AO et al (2008) Nanomaterials. Problems of regulation. Remedium 9:12–16
Meyer JN, Lord CA, Yang XY et al (2010) Intracellular uptake and associated toxicity of silver nanoparticles in *Caenorhabditis elegans*. Aquat Toxicol 100:140–150
Noack AG, Grant CD, Chittleborough DJ (2000) Colloid movement through stable soils of low cation-exchange capacity. Environ Sci Technol 34:2490–2497
Perminova IV (2008) Humic matters is the challenge to chemists of 21st century. Chem Life 1:50–55
Petersen EJ, Huang Q, Weber WJ (2008) Bioaccumulation of radio-labeled carbon nanotubes by *Eisenia fetida*. Environ Sci Technol 42:3090–3095
Reevesa JF, Daviesa SJ, Dodda NJF, Jha ND (2008) Hydroxyl radicals (OH) are associated with titanium dioxide (TiO_2) nanoparticle-induced cytotoxicity and oxidative DNA damage in fish cells. Mutat Res 640:113–122
Roberts AP, Mount AS, Seda B et al (2007) In vivo biomodification of lipid-coated carbon nanotubes by *Daphnia magna*. Environ Sci Technol 41:3025–3029
Schrand AM, Huang H, Carlson C (2007) Are diamond nanoparticles cytotoxic? J Phys Chem Toxicol Lett 111(1):2–7. doi:10.1021/jp066387v
Stampoulis D, Sinha SK, White JC (2009) Assay-dependent phytotoxicity of nanoparticles to plants. Environ Sci Technol 43:9473–9479
Tan KH (2003) Humic matter in soil and the environment: principles and controversies. CRC Press, New York, NY, p. 386
Terekhova VA (2011) Soil bioassay: problems and approaches. Eur Soil Sci 44(2):173–179. doi:10.1134/S1064229311020141
Terekhova VA, Gladkova M (2013) Engineered nanomaterials in soil: problems in assessing their effect on living organisms. Eurasian Soil Sci 46:1203–1210
Unrine J, Bertsch P, Hunyadi S (2008) Bioavailability, trophic transfer, and toxicity of manufactured metal and metal oxide nanoparticles in terrestrial environments. In: Grassian VH (ed) Nanoscience and nanotechnology: environmental and health impacts. Wiley, Hoboken, NJ, pp 345–366
Unrine JM, Tsyusko OV, Hunyadi S (2010) Effects of particle size on chemical speciation and bioavailability of Cu to earthworms (*Eisenia fetida*) exposed to Cu nanoparticles. J Environ Qual 39:1942–1953
Venitsianov EV, Vinnichenko VN, Guseva TV (2003) Ecological monitoring: step by step, Zaik EA (ed). RCHTU, Moscow, 252p
Yakimenko OS, Terekhova VA (2011) Humic preparations and the assessment of their biological activity for certification purposes. Eurasian Soil Sci 44:1222–1230. doi:10.1134/S1064229311090183
Yang L, Watts DJ (2005) Particle surface characteristics may play an important role in phytotoxicity of alumina nanoparticles. Toxicol Lett 158:122–132

Chapter 5
Nanomaterial Effects on Soil Microorganisms

Ebrahim Karimi and Ehsan Mohseni Fard

5.1 Introduction

The benefits of nanotechnology are evident in all fields of science and technology. The range of nanotechnology products is now extensive and can be broken down into a number of different compound classes, including carbonaceous nanomaterials; metal oxides; semiconductor materials, including quantum dots; zerovalent metals such as iron, silver, and gold; and nanopolymers, such as dendrimers. A variety of products are now being generated, including nanoparticles (NPs) as well as nanofibers, nanowires, and nanosheets, and the range and types of nanomaterials (NMs) are continually expanding. In recent estimates till October 2013, the nanotechnology-based consumer products inventory grows to 1628 products or product lines (Thul and Sarangi 2015). The scale of application of NMs is very broad, including healthcare, agriculture, transport, energy, materials, and information and communication technologies (Prasad et al. 2014).Given the increasing production and application of NMs of all types, the potential for their release in the environment and subsequent effects on ecosystem is becoming an increasing concern that needs to be addressed, especially by regulatory agencies. For this, we need to collect scientific data around the fate and effects of manufactured NMs in the environment. These information will guide the setting

E. Karimi (✉)
Department of Microbial Biotechnology, Agricultural Biotechnology Research Institute of Iran, Agricultural Research, Education and Extension Organization (AREEO), Karaj, Iran
e-mail: ekarimi@abrii.ac.ir

E. Mohseni Fard
Department of Agronomy and Plant Breeding, Faculty of Agriculture, University of Zanjan, Zanjan, Iran
e-mail: ehsan_mfai@yahoo.com

M. Ghorbanpour et al. (eds.), *Nanoscience and Plant–Soil Systems*, Soil Biology 48,
DOI 10.1007/978-3-319-46835-8_5

of regulatory guidelines that will provide adequate protection to ecosystems while permitting the advantages that nanotechnology offers to be developed.

NMs can enter the environment through different pathways like agricultural amendments of sewage sludge, atmospheric deposition, landfills, or accidental spills during industrial production. The leakage of NMs into the environment, especially soil, is one of the most serious threats to microbial communities in ecosystems. Microorganisms in soil are crucial to the maintenance of soil function in both natural and agricultural soils due to their involvement in such key processes as soil structure formation, decomposition of organic matter, toxin removal, and the cycling of carbon, nitrogen, phosphorus, and sulfur. In addition, they play key roles in suppressing soilborne plant diseases, in promoting plant growth, and in changes in vegetation (Garbeva et al. 2004).Changes in the composition and structure of soil microflora can be critical for the functional integrity of soil. For example, in the nitrification process, ammonium nitrogen is converted to nitrite and then to nitrate by ammonia-oxidizing and nitrite-oxidizing bacteria, respectively. The deletion of these bacteria from the environment leads to decreased nitrogen removal and interferes with plant growth. Therefore, protection of the environment and beneficial microorganisms from NMs is very important, and the scientific community should pay attention to the adverse effects of the NMs on microorganisms, in spite of their beneficial commercial use (Hajipour et al. 2012).

Recent years have seen an enormous increase in the number of documents, indicating an exponential increase over the past decade in NM-related research, in terms of manufacturing, applications, exposure, and hazard (Fig. 5.1). These scientific data have focused on different aspects of NMs, such as their novel applications as adsorbents, ion exchangers and disinfectants in water and air for removing ions, organic compounds, and pathogens, as well as assessing risks associated with them to human health, ecology, and environment (Kumar et al. 2014). NM effects toward microorganisms have been demonstrated in numerous in vitro studies; but the assessment of their environmental impact is still in its early stages. It should be noted that the effects of many NPs have not been studied yet or only in a single study (Al_2O_3, CeO_2, quantum dots, SiO_2, SnO_2) (Fig. 5.2), whereas a significant amount of these NPs is susceptible to be released to soils. Some NPs have been more studied than others (Ag, TiO_2, ZnO) (Fig. 5.2), and paradoxically, these are not necessarily the most produced and used NPs. The overall number of publications on each class of NP remains still limited to date (≤ 6) (Fig. 5.2), and thus, it is still difficult to generalize the results. More research is needed, especially through assays using more environmentally realistic concentrations of NMs based on the predicted concentrations in modelization researches and using more realistic exposure conditions (Simonin and Richaume 2015). It has been demonstrated that the toxic effects on soil microbial community are highly dependent on both the NMs considered and the soil properties. The soil properties seem to play an important role for the bioavailability of NMs, especially the clay and organic matter content. The identification of soil parameters controlling the bioavailability of NMs is fundamental for a better environmental risk assessment. Therefore, in the following sections, we will present an exhaustive literature review of the effects of

Fig. 5.1 Trend of published scientific articles over time in various nanomaterial-related areas (data obtained from Scopus on July 4, 2104, using keywords mentioned in legends of this figure) (Kumar et al. 2014)

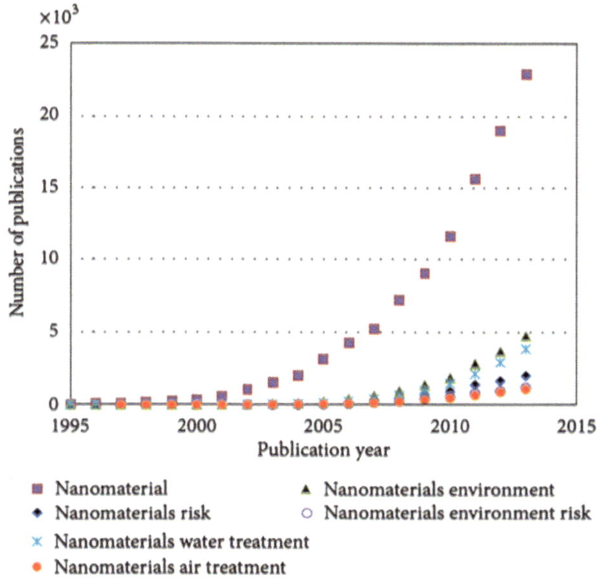

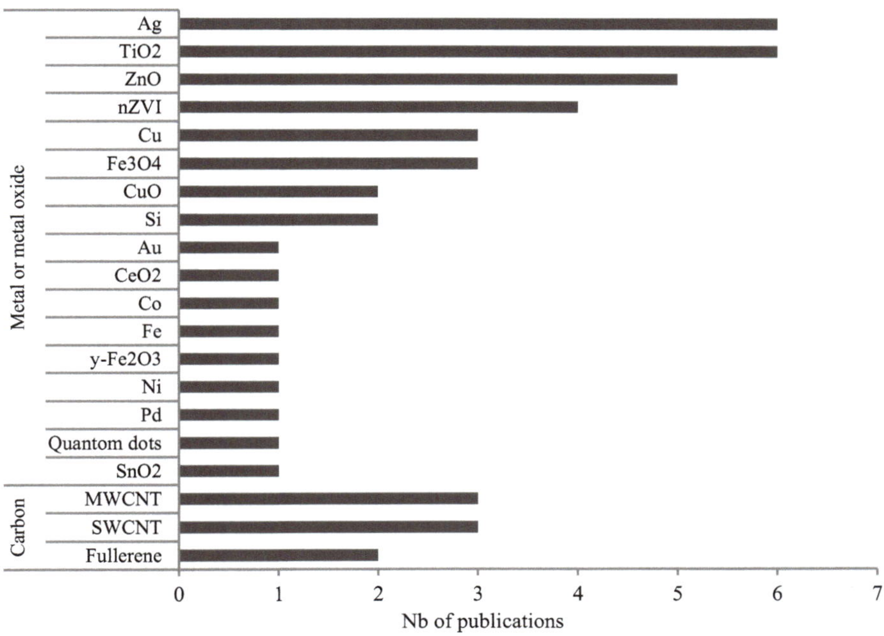

Fig. 5.2 Number of publications studying the impact of metal/metal oxide or carbon nanoparticles on soil microbial communities. 31 publications were available in July 2014 (Simonin and Richaume 2015)

NMs on soil microorganisms to get a better understanding of the ecological ramifications of nanomaterials.

5.2 Fate and Bioavailability of Nanomaterials in Soil

Soil is one of the important environments that may be a final sink for NMs. New and current studies have shown that they can be toxic to soil organisms, but most of these investigations occurred using pristine NM at higher concentrations than environmentally relevant and in standard tests, often inartificial soils or in hydroponic settings. The actual risk of a NM exposure in a natural soil depends on the bioavailability, which is the actual NM concentration to which organisms are exposed to and which may lead to effects. If there is no exposure, there is no risk. The bioavailable concentration of NM is lower than the total concentration in most realistic environments, particularly in soils, where there is a high concentration of reactive surfaces with which the NM can interact and a range of chemical environments in which the speciation of NM (their form of occurrence) can be transformed before they reach an organism (Soni et al. 2015). Fate and bioavailability of NMs as a main subject in nanotoxicology will be discussed in the next paragraphs.

Principally, NMs may enter the environment via many pathways due to the diversity of their applications. The sources of NMs to the environment are complex, consisting of both point and diffuse releases. During industrial production and transportation, accidental spills may happen. Emissions to the atmosphere may result in deposition to soils and waters from different sources (e.g., waste incineration). Further deliberate additions to the environment may occur via the use of NMs in soil and water remediation technologies and agriculture (e.g., as pesticides or fertilizers).The major source of NMs deposition onto land is currently through the disposal of wastewater treatment plant (WWTP) sludge, where NMs that are released from consumer products into wastewater may partition into sewage sludge during the wastewater treatment process (Mueller and Nowack 2008). NMs in WWTP sludge are mostly aggregated with bacteria, but they can also be associated with iron oxides or other inorganic particles and/or given the presence of humic acids in WWTP sludge, probably also coated with dissolved organic matter (DOM). Exposure via WWTP can also have implications for the composition of NMs, because silver or zinc oxide NMs, for example, may react in WWTP with ubiquitous sulfide or phosphate to form sparingly soluble Ag_2S, ZnS, or $Zn_3(PO_4)_2$NPs (Cornelis et al. 2014).

NMs possess a number of key properties that are believed to exert important controls on their environmental behavior, fate, and ecotoxicity. These include physical properties, particularly size and shape, and chemical properties such as the acid–base character of the surface and the aqueous solubility of the metal. These properties in turn will determine the extent to which NMs undergo transformations that will control their fate, behavior, and ecotoxicity in the environment. Such processes will include aggregation/agglomeration, sorption to surfaces, and dissolution to the ionic metal. Furthermore, NMs are frequently manufactured with surface coatings, which may modify their intrinsic behavior. The existence of these multiple characteristics implies that characterization of NP form and presentation in the environment will be important to understanding their behavior, fate, and ecotoxicity. Soil represents a relatively complex medium for the understanding of the physicochemical behavior of manufactured NPs. In comparison with the dissolved phase, in which behavior can be understood largely in terms of particle stability against aggregation, soils present a solid matrix with which NPs may interact, as well as an aqueous phase, which may contain appreciable amounts of natural colloidal/particulate material. In the context of ecotoxicity, a key issue is the understanding of how specific organisms are exposed to NPs present in different phases (soil, soil water) and how the presentation of the NPs within these phases further influences exposure. The effective exposure level of organisms may not be assessable by elemental mass concentrations alone, and additional information on the presentation of the NPs is likely to be essential for understanding and predicting their effects (Tourinho et al. 2012). Because much of the existing work on the behavior of NPs has been done in aqueous environments, we include examples of such work in this section. Such work is itself directly relevant to soils because NP behavior in the aqueous phase of soils may be of considerable importance for their transport and bioavailability. Furthermore, patterns of NP behavior in the aqueous phase may provide useful information regarding their presentation in the solid matrix itself.

The assessment of the form and presentation of NPs in environmental matrices, and soils especially, is currently hampered by the relative lack of appropriate procedures for their characterization, and the development of suitable techniques for better characterization poses a great challenge. For relatively simple analyses, such as the aggregation state of NPs in aqueous solution, it is possible to use dynamic light scattering or microscopy-based techniques such as scanning and transmission electron microscopy and atomic force microscopy (AFM). However, quantifying the state of NPs in soils, compared with solutions, is hampered by the need for knowledge of the NP presentation within the soil matrix, rather than in aqueous extracts of the soil. Currently most characterization techniques are limited by being applicable only in the aqueous phase (Tiede et al. 2009).Stone et al. (2010) have proposed that the main NP properties that should be evaluated in ecotoxicological studies to describe exposure are dispersibility, agglomeration/aggregation, dissolution rate, size, surface area and charge, and surface chemistry. These properties are likely to be key in controlling NP stability and consequently on their transport through the environment and availability to organisms. Therefore, quantification of these properties, and of how they are modified by NP interactions with soils, will enable more accurate assessment of which NP properties influence bioavailability and toxicity across soil types, concentration ranges, and time frames.

5.2.1 Aggregation and Agglomeration

The role of aggregation and agglomeration for the fate and behavior of NMs in the environment has been highlighted in numerous studies. Aggregates are defined as clusters of particles held together by strong chemical bonds or electrostatic interactions. Agglomerated particles are held together by weaker forces (caused by Van der Waals forces) and can be a reversible process (Fig. 5.3). However, in papers on both water and soil exposure studies with NPs, the terminology for aggregates and aggregation is often used in cases where possibly only agglomeration has occurred (i.e., where no actual permanent sintering of the particles takes place). In the remainder of this section, we have followed the terminology used by the authors of the original papers, but attention should be given to the correct use of terminology in future work, because this will be important for evaluating size-specific effects in terms of toxicity (Hartmann et al. 2014).

In the environment, physical forces (e.g., Brownian motion, gravity, and fluid motion) and NP properties (e.g., surface properties and particle size) will affect NP agglomeration and aggregation. The particles are constantly colliding with each other because of Brownian motion, and agglomeration will occur when the energy of either motion or attraction exceeds the energy of repulsion. To aggregate, the cores of particles must make contact; thus, the aggregation rate is believed to be proportional to the probability of collision between two particles. Aggregation may result in the formation of particle flocks that are of sufficient size to sediment out of solution by gravity (Tourinho et al. 2012). The NP aggregate size in solution depends on properties such as initial particle size and concentration (Fig. 5.4). Phenrat et al. (2006), using iron NPs, found that higher concentrations (60 mg/L) resulted in higher aggregation rates and stability of aggregate size in comparison

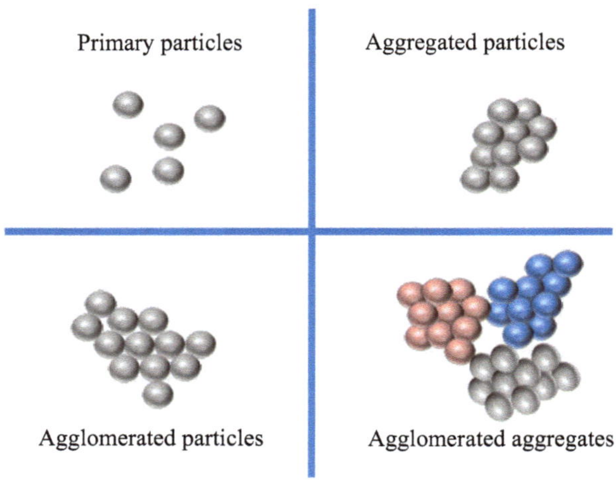

Fig. 5.3 Difference between primary particles, aggregated particles, agglomerated particles, and agglomerated aggregates of particles (Hartmann et al. 2014)

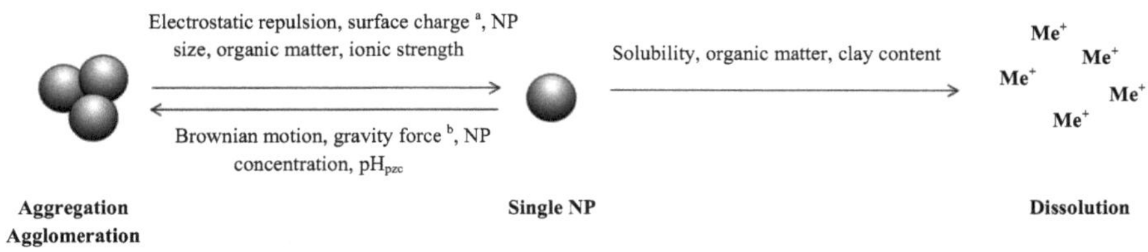

Fig. 5.4 Factors affecting the processes of aggregation/agglomeration and dissolution of single nanoparticles in the environment. (**a**) Considering similar surface charge, (**b**) acting only on larger particles (Tourinho et al. 2012)

with lower concentrations (2 mg/L). Zinc oxide NPs dispersed in aqueous solution aggregated in a wide range of sizes, resulting in aggregates almost tenfold larger than the primary NPs (Pipan-Tkalec et al. 2010). Nevertheless, not all particles were incorporated into aggregates, and individual NPs were also detected in suspensions (Wang et al. 2009; Lin and Xing. 2008). Thus, the size distribution of aggregates may vary among particle types. For example, titanium dioxide NPs showed a uniform distribution and agglomeration (Jemec et al. 2008), but zinc oxide NPs showed a wide distribution size and aggregation (Pipan-Tkalec et al. 2010).

5.2.2 *Surface Coating*

The chemistry of the medium will influence the electrostatic surface charge of the particles, thereby affecting agglomeration/aggregation rates and particle stability (Fig. 5.4). In the absence of a surface coating, metal-based NPs have charged surfaces resulting from the presence of hydroxyl (–OH) groups that can take up and release protons and can take up dissolved chemical species such as metal ions and ligands. The sign and magnitude of the surface charge will be determined by the intrinsic chemistry of the surface groups and the chemical composition of the solution, in particular the pH and the concentrations of binding species. Surface charging results in the formation of an electrical double layer, comprising the charged surface sites and a diffuse layer containing ions attracted from the solution to the particle surface in response to the charge. The electrical potential at the interface of the diffuse layer and the bulk solution (the zeta potential) can be measured, and its variation with solution chemistry can effectively be used as a surrogate for the variation in particle surface charge with solution chemistry. The zeta potential of uncoated metal or metal oxide particles typically decreases from positive values at low pH to negative values at high pH. As the pH approaches the isoelectric point, or the "point of zero charge (pzc)," where particle charge/zeta potential approaches zero, the aggregation rate will increase (Jiang et al. 2009; Guzman et al. 2006) because of the lowering of the electrostatic repulsive forces between particles. A suspension of homogenously charged particles will be stable when the magnitude of the zeta potential is greater than 30 mV. Jiang et al. (2009) found that the hydrodynamic diameter of TiO_2NP aggregates varied with the solution pH, being greatest at the isoelectric point (pH 6.0). Fabrega et al. (2009) reported that silver NPs stabilized by citrate coating showed no variation in aggregate size over the pH range 6–9, consistent with the findings of El Badawy et al. (2010) for similarly stabilized silver NPs. Here, the invariant aggregate size may be explained by the relatively small variation in zeta potential of the particles in this pH range (El Badawy et al. 2010). Small aggregates of silver NPs adsorbed humic acid, and this adsorption resulted in disaggregation (or disagglomeration) of the NPs (Fabrega et al. 2009).

The presence of a surface coating on NPs may significantly modify their surface chemistry, compared with the uncoated equivalent. For example, El Badawy et al. (2010) measured the contrasting surface-charging behavior of different types of silver NPs, one uncoated and the remainder coated, with substances imparting contrasting surface chemistries. Under environmental conditions, the stability of such coatings over time will be important in determining how long the particles maintain the manufactured surface properties. Properties influencing coating stability will include the reversibility of the coating process and the biodegradability of the coating over time. Classical colloid theory can be applied to metal and metal oxide NPs to help explain their stability. The DLVO (Derjaguin, Landau, Verwey, and Overbeek) theory considers stability as a function of the repulsive (i.e., electrostatic) and attractive forces (i.e., Van der Waals) to which a particle is subjected. Although DLVO theory has generally proved unsatisfactory for the quantitative prediction of colloid behavior in complex natural environments, the conceptual framework that it provides for considering such behavior can be useful in explaining trends in observed behavior in such systems. Some applications of the theory have been made to explain the stability of NPs in aqueous environments (Tourinho et al. 2012).

5.2.3 Dissolution and Transformation

Most fate investigations on NMs used pristine materials, but recent studies show that the chemical nature of these materials changes in the environment (e.g., after being released from nano-enabled products). Dissolution is the most widely investigated transformation reaction, possibly because the solubility of particles theoretically increases as their size decreases, which has been experimentally verified for Ag and ZnO NPs. However, the effect of size on dissolution rate in a realistic environment, even for relatively soluble NMs such as silver, zinc oxide, or quantum dots, is probably confounded by simultaneous deposition, aggregation, coating by organic matter, and/or transformation to sparingly soluble materials. Ions dissolved from NM are often adsorbed by soil particles or complex with chelating agents, which may accelerate NM dissolution and/or transformation. The dissolution rate of less soluble NM such as CeO_2 is most likely not significantly enhanced in natural soils, and these NMs therefore likely accumulate if not leached out of the soil. Relatively soluble NMs are often transformed to thermodynamically stable compounds in a range of environments. As described previously, silver transforms into Ag_2S in WWTP forming Ag_2S NPs of a similar size than the original Ag NM, and this process occurs independently of the original Ag NM coating. A similar process occurs also in soils. Other NMs such as ZnO may lose their NP character during transformation. It is found that, for instance, during aging of ZnO or ZnS containing sewage sludge, most Zn became associated with iron oxides, a process that may also occur in soils. Biodegradation of carbon-based NM in soils is possible, but scarcely studied, and half-lives are likely to vary depending on locally available enzymes in soil pores. Not only the core material of NM is changed over time in soils, but most likely also the coating. Small, electrostatically adsorbed organic acids such as citrate are easily degraded in a range of environments, but even covalently bound large coatings can be degraded. Such degradation possibly occurs abiotically by oxidation combined with dissolution, but it has been proposed that reactive oxygen species (ROS) that are naturally produced in soils (e.g., fungi) can also mediate coating degradation. The decomposition of coatings likely results in reduced NM mobility in soils unless the coating is replaced by naturally occurring natural organic matter (NOM) (Cornelis et al. 2014).

5.2.4 Soil Properties

NM behavior within soil systems will be further complicated by the presence of the solid phase. Soil components such as humic molecules or clay particles will themselves have charged surfaces, which will influence the association of NMs with the solid phase. Such soil components may also form colloids in the aqueous phase, which will interact with NMs. For example, humic molecules desorbing into the aqueous phase may sorb to NM surfaces and so influence NM stability. All of these processes are themselves strongly influenced by overriding properties of the soil system, particularly the pH and the ionic strength of the aqueous phase. Currently, work on the behavior of metal-based NPs in soils has focused on soil suspensions rather than intact soils. For example, Gimbert et al. (2007) studied the particle size distribution of ZnO NPs in smaller than 1-μm size suspensions extracted from a high pH soil 0, 7, and 14 days after spiking. The NPs were found to quickly equilibrate between the aqueous and solid phases, and concentrations in the smaller than 1-μm fraction were stable during the experimental period.

The aggregation rate of TiO_2 NPs in soil suspensions has been found to be negatively correlated to soil properties, such as dissolved organic matter and clay contents, and positively correlated to the ionic strength, zeta potential, and pH. NPs sorbed less strongly to soils of low ionic strength and high dissolved organic matter content, suggesting that these factors may affect the bioavailability of metal-based NPs in soils, assuming that bioavailability is related to the particle fraction suspended in the pore water and not the fraction associated with the soil matrix.

Other studies showed that aggregation of titanium dioxide NPs occurred at an ionic strength higher than 4.5 mM and that increasing ionic strength from 1 to 100 mM caused a 50-fold increase in TiO_2NP diameter (Tourinho et al. 2012).

The ionic strength of the medium affects the stability of the diffuse layer in the electrical double layer. An increase in the ionic strength will lead to a decrease in the electrical double layer thickness, which favors particle association and leads to increasing agglomeration. The effects of ionic strength on the aggregation of AgNPs (uncoated, citrate, and sodium borohydride-coated particles) were observed for suspensions with pH higher than 7. This demonstrates that here the particles were stabilized by electrostatic repulsion, caused by the predominance of negative forms (i.e., anions) in the medium and the negative charge of the particles. In contrast, the agglomeration of AgNPs coated with polyvinylpyrrolidone was not influenced by increasing ionic strength. This reflects that here the particles were stabilized because of steric repulsions caused by the uncharged coating material, the effectiveness of which was not influenced by ionic strength.

Overall, the stability and sorptive behavior of NMs in soils are likely to be of importance for transport, fate, and toxicity; yet current researches have focused on behavior at relatively short timescales, and there has been a focus on investigations at soil/solution ratios that are low compared with both ecotoxicity testing and field conditions and at relatively short timescales. Whether the outcomes of such investigations can be robustly applied to higher soil/solution ratios and longer, more environmentally relevant timescales is unclear. More research is needed on NM behavior in intact soils and over timescales of months to years (Tourinho et al. 2012).

In soils, dissolved or particulate organic matter can sorb to NP surfaces. This sorption may influence particle properties in a number of ways. Humic substances are negatively charged at environmental pHs, and thus their sorption will make the overall particle–humic conglomerate negatively charged. This may increase particle stability in solution, reducing aggregation and settling. The alteration of the surface charge may also decrease particle affinity for cell membranes and thus reduce their bioavailability and uptake. Steric hindrance effects may also contribute to the enhanced stability of humic-coated NPs. Conversely, Ghosh et al. (2008) showed that at low pH, humic acid caused aggregation of Al_2O_3 NPs. Here, the charge of the humic acid appeared sufficiently low to allow its aggregation because of hydrophobic interactions; thus, humic acid-coated particles also became susceptible to aggregation. Kool et al. (2011) presented transmission electron microscopy images showing zinc oxide NPs bound to solid-phase organic matter in a soil with a pH of 5.5, suggesting that under suitable conditions organic matter may destabilize particle dispersions. The overall effect of humic substance sorption on particle stability and bioavailability appears to be a complex function of factors, particularly the soil pH and the intrinsic hydrophobicity of the humic substances.

Because of the limitations of analyzing NPs in the soil matrix, most studies characterize the particles in their pristine form before addition to the soil or in the aqueous solution used to contaminate the soil. In these studies, the particles are typically found to be agglomerated/aggregated. Manzo et al. (2010) analyzed a ZnO

NP-contaminated soil by the Brunauer–Emmett–Teller method and found that the particles were not aggregated. The author attributed this result to the spiking procedure, which consisted of mixing a dry powder of the NPs with dry soil. However, the solution extracted from soil samples wetted after spiking showed larger particles, in which a bimodal peak could be observed in a dynamic light scattering analysis ranging from 103 to 470 nm. Therefore, some attention should be given to this issue when designing experiments and choosing the spiking procedures. Although the review by Handy et al. (2012) on practical experiences and recommendations does mention soil, only little attention has been given to soil experiments. However, prior studies with NPs in soils have presented different contamination methodologies, including mixing the NPs powder directly in with the soil, adding a stock dispersion made in distilled water to soil (Unrine et al. 2010), or preparing a stock dispersion in soil elutriate that is then mixed in with the soil (Kool et al. 2011). Developing standard procedures for soil dosing with NPs is needed to improve the comparability of multiple test results. Regarding soil variability, Handy et al. (2012) recommended the use of an artificial soil as a benchmark for exposure evaluation of NPs. This is a useful recommendation to compare the behavior and toxicity of different types of NPs in a consistent soil medium. For a better understanding of NP behavior as a function of soil properties, however, studies on multiple soils are additionally needed (Tourinho et al. 2012).

5.3 Effect of Nanomaterials on Soil Microorganisms

Despite the potential application of NMs in many fields like medicine, agriculture, etc., to fully exploit the promised benefits of nanotechnology requires that we improve our understanding of the environmental ramifications (especially soil) of NMs regarding effects that could influence the performance of nontarget and beneficial soil microbes. Therefore, the following section synthesizes available studies on NMs interactions with soil microbes.

5.3.1 Effect on Soil Microbial Diversity

Microbial characteristics of soils are being evaluated increasingly as sensitive indicators of soil health because of the clear relationships between microbial diversity, soil and plant quality, and ecosystem sustainability (Hill et al. 2000). Soil hosts an immense diversity of microorganisms (individual taxa commonly described as operational taxonomic units (OTUs)) of bacteria, fungi, and archaea. Microbial diversity encompasses genetic variability within taxa (species) and the number (richness) and the relative abundance (evenness) of taxa and functional groups (guilds) in communities (Torsvik and Øvreas 2002). That is why investigating the impact of NMs on microbial diversity is crucial to provide information on how and why soil ecosystem functioning is affected. A panel of techniques, such as fluorescent in situ hybridization (FISH), phospholipid fatty acid (PLFA) profiles, denaturing gradient gel electrophoresis (DGGE), terminal restriction fragment length polymorphism (T-RFLP) analysis, or next-generation sequencing (NGS), have been used to evaluate the impact of NMs on microbial diversity.

5.3.1.1 Metal and Metal Oxide Nanoparticles

Metal or metal oxide NPs may be responsible for a biodiversity loss and a modification of soil microbial community composition. Although many researches focused on the effect of these NPs on bacterial communities, more work is still required to assess the impact of NPs on fungal and archaeal communities (Simonin and Richaume 2015; Prasad et al. 2016).

Silver NPs could alter bacterial community structure, after short-term exposure of sewage sludge containing 0.14 mg/kg of silver NP (Colman et al. 2013). Kumar et al. (2011) also observed that silver NPs modified bacterial community in an arctic soil but with a higher concentration (660 mg/kg). Plants generally depend on soil bacteria and fungi to help mine nutrients from the soil. A study finds that silver NPs negatively impact on the growth of plants and kill the soil microbes that sustain them (Zeliadt 2010). Ge et al. (2012) studied the impact of TiO_2 and ZnO NPs on soil bacteria. The alteration of bacterial communities was reported to be in a dose-dependent manner, with some taxa increasing as a proportion of the community, whereas more taxa decreasing that resulted in reduced diversity. A field study also reported that zerovalent copper and zinc oxide NPs had no effect on PLFA profiles and on microbial community composition as determined by pyrosequencing (Collins et al. 2012).Changes in soil bacterial community composition due to the presence of Fe_3O_4NPs were observed after a short incubation (24 or 48 h) as well as after 15 and 30 days (Ben-Moshe et al. 2013; He et al. 2011). Cloning sequencing of DGGE bands indicated a stimulation of specific groups of *Actinobacteria*, *Duganella*, *Streptomycetaceae*, or *Nocardioides* (He et al. 2011). These groups facilitate the decomposition of organic matter, which could explain the concomitant soil invertase and urease increases measured during this experiment. Different concentrations of TiO_2 and ZnO NPs decreased soil bacterial diversity after 60 days of incubation (Ge et al. 2011, 2012). The pyrosequencing data indicated that some of the declining taxa are known to be associated to nitrogen fixation (*Rhizobiales*, *Bradyrhizobiaceae*, and *Bradyrhizobium*) and methane oxidation (*Methylobacteriaceae*), while some positively impacted taxa are known to be associated with the decomposition of recalcitrant organic pollutants (*Sphingomonadaceae*) and biopolymers including protein (*Streptomycetaceae* and *Streptomyces*). The role of these taxa for soil functioning suggests potential consequences on ecosystem-scale processes. Nogueira et al. (2012) assessed the effect of five inorganic nanomaterials (TiO_2, $TiSiO_4$, CdSe/ZnS quantum dots, gold nanorods, and Fe/Co magnetic fluid) on soil bacterial community structure using DGGE. After 30 days of soil exposure, TiO_2 and gold nanorods induced the highest changes in the structural diversity of bacterial community. The limited effect of $TiSiO_4$, CdSe/ZnS quantum dots, and Fe/Co magnetic fluid NPs on DGGE profiles was attributed to their zeta potential values reflecting an unstable state. Hence, once added to the soil, they may have interacted with soil components, becoming unavailable to exert toxic effects.

5.3.1.2 Carbon Nanoparticles (Fullerene and Carbon Nanotubes)

Fullerene NPs had no effect on microbial diversity (Tong et al. 2007) or induced only slight modification of *Eubacteria* and *Kinetoplastida* (protozoans) community structure on DGGE profiles (20–30 % of dissimilarity, Johansen et al. 2008). Pyrosequencing data indicated that MWCNT (10 g/kg soil) induced an enrichment of potential degraders of recalcitrant contaminants (PAH) *Rhodococcus*, *Cellulomonas*, *Nocardioides*, and *Pseudomonas*, while some bacterial genera like *Derxia*, *Holophaga*, *Opitutus*, and *Waddlia* were decreased (Shrestha et al. 2013). Using a comparative metagenomic analysis of bacterial communities, Khodakovskaya et al. (2013) found that the diversity and richness of bacterial communities were not affected by MWCNTs, while a significant modification of the bacterial composition was observed. SWCNTs induced a modification of microbial community composition resulting in a decrease of Gram-positive and Gram-negative bacterial biomass and fungal biomass as well (Jin et al. 2014). Rodrigues et al. (2013) reported also a modification of the fungal community structure after 14 days of soil exposure to SWCNT (250 and 500 mg/kg). Consistent with activity and abundance measurements, carbon NP scan alters soil microbial community structure but only in the presence of high concentrations (>250 mg/kg).

5.3.1.3 Nanoscale Zerovalent Iron

The impact of nanoscale zerovalent iron (nZVI) on microbial diversity was investigated using FISH, DGGE, and PLFA analysis. Fajardo et al. (2012) did not report a broad bactericidal effect of nZVI but observed significant shifts in the structure and phylogenetic composition of the soil microbial community after 72 h of incubation. The FISH assays provided evidence that nZVI exerts a selective pressure on the microbial community, promoting the dominance of some microbial groups (*Archaea*, *α-Proteobacteria*, and low G + C Gram-positive bacteria) or the decrease of other ones (β- and *γ-Proteobacteria* and subclasses). DGGE profiles also indicated a significant modification of bacterial community composition after 28 days in the presence of 10 g/kg of nZVI (Tilston et al. 2013). Pawlett et al. (2013) observed that nZVI caused a modification of PLFA profiles in all soil textures tested, but that these effects were modulated by the organic matter content of the soil. These studies suggest that nZVI could induce a significant modification of soil microbial community structure, affecting bacteria, archaea, and fungi populations on the short term (<4 months).

5.3.2 Effect on Soil Beneficial Microbes

The number of studies examining the impact of NMs on beneficial soil microbes, such as nitrifying bacteria, nitrogen-fixing bacteria, arbuscular mycorrhizal fungi (AMFs), or plant growth-promoting rhizobacteria (PGPRs), is limited. Considering the critical ecosystem services that beneficial microbes deliver and the possibility that they may be more sensitive to soil contaminants than higher organisms (Giller et al. 2009), a need exists to further examine how the accumulation of NMs in soil might affect these organisms (Judy and McNear 2015).

5.3.2.1 Arbuscular Mycorrhizal Fungi

AMFs are ubiquitous soil microorganisms that are symbiotic with the roots of over 90 % of land plants. These fungi are beneficial for plant growth because they improve plant nutrient acquisition by supplying mineral nutrients, especially phosphorus (P). Additionally, AMFs confer heavy metal resistance to plants, improve soil structure, protect the plants from pathogens, and suppress aggressive agricultural weeds. Very little research to date has examined the effects of NMs on mycorrhizal colonization of plant roots. One such study examined how mycorrhizal colonization of sunflower (*Helianthus annuus*) responded to the presence of pristine silver NPs (Dubchak et al. 2010). In this study, the authors reported that silver NPs inhibited mycorrhizal colonization of *Helianthus annuus* at a soil concentration of approximately 150 mg/kg. Another more recent study by Feng et al. (2013), which examined how pristine iron oxide (FeO) and silver NPs affected AMF colonization of clover (*Trifolium repens*) roots in perlite(s) and mix, reported significant biomass reduction as a result of exposure to 3.2 mg/kg FeO NPs. Iron oxide NPs lowered the glomalin content and acquisition of P by AMF, resulting in reduced clover biomass. Glomalins are glycoproteins produced on the hyphae and spores of AMF. As components of soil organic matter, glomalins play a role in connecting mineral particles together, thereby improving soil quality (Gillespie et al. 2011). Similarly, effects of AgNPs on clover AM were found: reduced glomalin and biomass and lowered ability of AM transfer of P to plant at very low NP concentrations. Curiously, these deleterious effects were diminished at higher AgNP concentrations; instead the NPs enhanced AMF ability to alleviate metal stress by stimulating AMF growth, leading to decrease in plant Ag content and the activities of plant antioxidant enzymes (Feng et al. 2013). Such behavior of NPs could be attributed to their enhanced agglomeration at high concentration (Keller et al. 2010), which not only diminish their ion-release potential but also reduce the concentration of NPs small enough to penetrate cells.

Judy et al. (2015) investigated the effects of Ag_2S NPs, polyvinylpyrrolidone-coated AgNPs (PVP-Ag), and Ag^+ on AMF, their colonization of tomato (*Solanum lycopersicum*), and overall microbial community structure in biosolid-amended soil. Concentration-dependent uptake was measured in all their treatments. Plants exposed to 100 mg/kg PVP-AgNPs and Ag^+ exhibited reduced biomass and greatly reduced mycorrhizal colonization. Bacteria, actinomycetes, and fungi were inhibited by all treatment classes, with the largest reductions measured in 100 mg/kg PVP-AgNPs and Ag^+. Overall, Ag_2S NPs were less toxic to plants, less disruptive to plant–mycorrhizal symbiosis, and less inhibitory to the soil microbial community than PVP-AgNPs or Ag^+. However, significant effects were observed at 1 mg/kg Ag_2S NPs, suggesting that the potential exists for microbial communities and the ecosystem services they provide to be disrupted by environmentally relevant concentrations of Ag_2S NPs.

5.3.2.2 Plant Growth-Promoting Rhizobacteria

Rhizospheric bacteria (Rhizobacteria) exert beneficial effect on plant growth are called as PGPRs. They are free-living, soilborne bacteria, isolated from the rhizosphere zone and when applied to seeds or crops enhance the growth of the plant. They are known to participate in many important processes, such as the biological control of plant pathogens, nutrient cycling, and/or seedling growth. These bacteria help plant growth by a combination of physiological attributes such as asymbiotic N_2 fixation; phytohormone production, namely, indole-3-acetic acid (IAA), cytokinin, gibberellins, and solubilizing insoluble mineral phosphate; and siderophore production (Hayat et al. 2010; Hinsinger et al. 2009). Although many NMs have already been reported to have antimicrobial properties, studies on the ecotoxicological behavior of NMs on PGPRs are scanty. There is thus an urgent need for research on interactions between NMs and PGPRs not only under laboratory conditions but also under natural conditions (Mishra and Kumar 2009).

Studies by Karunakaran et al. (2013) showed TiO_2 but not ZrO_2 (zirconia) NPs inhibited the growth of studied PGPRs. Also, *Bacillus subtilis* and *Pseudomonas fluorescens* were susceptible to the toxicity of Al_2O_3, TiO_2, ZnO, and SiO_2NPs; ZnO caused complete mortality in the strains, while Al_2O_3, TiO_2, and SiO_2 NPs resulted in mortality rates of between 40 and 70 %, depending on the strain. CuO, ZnO, NiO_2, and Sb_2O_3, in that order, all inhibited the growth of *B. subtilis* (Baek and An. 2011). Karunakaran et al. (2014) focused on the ecotoxicological behavior of nanosilica and bulk silica (SiO_2) and alumina (Al_2O_3) particles on PGPR (*B. megaterium*, *B. brevis*, *P. fluorescens*, and *Azotobacter vinelandii*) and soil nutrient contents. Their findings showed that nano and bulk SiO_2 particles were nontoxic toward studied PGPRs up to 1000 mg/L concentration. In addition, bulk Al_2O_3 particles were less toxic, whereas Al_2O_3 NPs were highly toxic in the order of *A. vinelandii* < *P. fluorescens* < *B. megaterium* < *B. brevis*. Their results revealed that the size of the particles plays a key role in bacterial toxicity. Moreover, nano-Al_2O_3 particles led to a decrease in microbial population of the soil, leading to decrease in available forms of nutrients.

Culturability of an engineered biosensor strain of *P. putida* KT2440 was compromised by Ag, CuO, and, to a lesser extent, ZnO NPs, with concomitant reduction in light production, indicating the effect of the NPs on primary metabolism (Gajjar et al. 2009). Similarly, culturability or growth of *P. chlororaphis* O6 (*Pc*O6) was diminished by these NPs, with differential outcomes on the production of secondary metabolites. ZnO NPs were found to decrease indole-3-acetic acid (IAA) but increase siderophore production, while an opposite effect was seen with CuO NPs (Dimkpa et al. 2015; Dimkpa 2014). Fang et al. (2013) also reported similar alterations in microbial growth, production of IAA, siderophores, and the antifungal compound, phenazine, by ZnO NPs.

There are some antibacterial reports about graphene oxide (GO) NPs such as *E. coli* and *P. aeruginosa*, but none has reported on soil bacteria especially PGPRs. Gurunathan (2015) demonstrated the negative effect of GO NPs on isolated rhizobacteria from the soil (five *Bacillus* species: *B. megaterium*, *B. cereus*, *B. subtilis*, *B. mycoides*, and *B. marisflavi*). In this study, GO NPs showed antibacterial activity against all test strains. The dose- and time-dependent effect showed that *B. megaterium* was the most sensitive toward GO, whereas *B. marisflavi* was the least sensitive strain. The biochemical information showed significant agreement with cell viability and indicated that GO has a significant toxicological effect on rhizobacteria. They therefore suggested that GO materials should be managed with care, and their disposal in the environment should be prevented.

5.3.2.3 Nitrogen-Fixing Bacteria

One of the most important beneficial roles of soil microbes in soil is in plant nutrition for which N_2 fixation is significant. It is required for the biosynthesis of basic molecules that are the building blocks for essential macromolecules such as nucleic acids and proteins (Dimkpa 2014). However the N_2 fixation could be impacted by NPs. Uptake of manufactured CeO_2 NPs into roots and root nodules could eliminate N_2 fixation potentials and impaired soybean growth (Priester et al. 2012). Similarly, WO_3 NPs (tungsten), but not TiO_2 NPs, were detrimental to the growth of the N_2-fixing bacterium, *Azotobacter vinelandii*, especially under Mo-limiting conditions. WO_3 NPs not only induced the production of catechol-type siderophores but also resulted in enhanced siderophore-mediated uptake of the WO_3 ions released from the NPs, with decrease in cellular Mo levels (Allard et al. 2013).The effect of this process on N_2 cycle is apparent from the standpoints of (1) Mo being a cofactor in the nitrogenase responsible for converting N_2 to NH^{4+}, and (2) WO_3 being capable of displacing Mo in the enzyme active site, to generate an inactive enzyme ([Allard et al. (2013)] and references therein). In contrast to *A. vinelandii*, growth, N_2 fixation, and storage by the cyanobacterium, *Anabaena variabilis*, were affected by TiO_2 NPs (Cherchi and Gu 2010). Fan et al. (2014) observed the impact of TiO_2 NPs on *Rhizobium*–legume symbiosis using garden peas and *Rhizobium leguminosarum* bv. *viciae* 3841 and found that TiO_2 NPs exert morphological changes in bacterial cells. Further, it was noticed that the interaction between these two organisms was disrupted in the form of root nodule development and the subsequent delay in onset of nitrogen fixation. TiO_2 NPs caused structural distortions on the cell surface of *R. leguminosarum*, slowing nodule development, and onset of N_2 fixation in the symbiotic interaction with garden pea. These effects were accompanied by alteration in the polysaccharide composition of cell walls of infected nodules, as well as decrease in lateral root proliferation, but not on plant germination nor root elongation (Fan et al. 2014). Both CeO_2 and ZnO NPs were toxic to *Sinorhizobium meliloti*, inhibiting growth and viable cell counts. Changes in the protein and polysaccharide structures of extracellular polymeric substances (EPSs) in the cell walls were detected with both NPs. The two NPs however exhibited different toxicity mechanisms to a related strain, *S. melba*, whereby CeO_2 NPs were bacteriostatic, while ZnO NPs were bactericidal (Bandyopadhyay and Peralta-Videa 2012). Judy et al. (2015) studied the impacts of amending soil with biosolids (WWTP) containing a mixture of NMs on the growth of *Medicago truncatula*, its symbiosis with *S. meliloti*, and on soil microbial community structure. Large reductions in nodulation frequency, plant growth, and significant shifts in soil microbial community composition were found for the NM treatment compared to the bulk/dissolved metal treatment.

5.3.2.4 Nitrifying Bacteria

Nitrifying bacteria, a group of ubiquitous and important autotrophic microorganisms, convert ammonia to nitrate in two steps called nitrification (Roh et al. 2009). The first step of nitrification, from ammonia to nitrite, is carried out by ammonia-oxidizing bacteria (AOB), and the second step, from nitrite to nitrate, is carried out by nitrite-oxidizing bacteria (NOB). Some studies have shown that they are more sensitive to silver NPs than heterotrophic bacteria and low concentrations ($\leq$1 mg/L) of silver NPs can effectively inhibit their activity (Choi and Hu 2008; Liang et al. 2010). Choi and Hu (2008) and Yang et al. (2014) demonstrated the inhibition of growth and abundance of nitrifying bacteria by AgNPs. Yuan et al. (2013) linked the inhibition of *Nitrosomonas europaea* growth to alterations in expression of genes involved in energy production and the nitrification process. Similarly, decreased NH^3 oxidation as well as destabilization of the outer membrane of this bacterium by AgNPs also has been reported (Radniecki et al. 2011). Arnaout and Gunsch (2012) have investigated the toxicity of silver NPs with different surface coatings to the nitrifier bacteria, *N. europaea*, and an increase in heavy metal stress response was observed. The shock loading of 1 mg/L AgNPs to activated sludge inhibited 41 % of nitrification and led to shift or loss of nitrifying bacteria in the reactor (Liang et al. 2010).

The toxicity of NMs to nitrifying bacteria has been studied under pure culture media, and this toxicological information is often difficult to extrapolate to the ecosystem scale. In this field, Masrahi et al. (2014) investigated the impact of Ag chemical speciation (Ag^+ and AgNPs [50-nm uncoated and 15-nm polyvinylpyrrolidone (PVP)-coated AgNPs]) to soil nitrification kinetics using a batch soil slurry nitrification method along with sorption isotherm and dissolution experiments. Results showed Ag-based compounds had some inhibitory effect to the soil nitrification process and also the concentration and the chemical species was critical in the observed toxicity to the nitrification process. At 1 mg/L of $[Ag]_{total}$, nanoparticles were far more toxic than Ag^+. This difference in toxicity is caused by Ag^+-complexation processes with inorganic and organic soil components. In particular, soft basic ligands in soils (e.g., thiol functional groups of organic matter) likely chelated with 1 mg/L of Ag^+, thus effectively decreasing the toxicity of Ag^+ in soils, whereas Ag(0) NPs do not as readily complex with soft ligands in soils. They revealed that within NPs, PVP-coated 15-nm was far more toxic than uncoated 50-nm AgNPs. PVP-capped AgNPs were highly dispersed and released more Ag^+ than uncapped NPs as was evident in the dissolution experiments. The reactivity of PVP-capped AgNPs induced greater toxicity to nitrifying bacteria. They also interestingly found that at 10 mg/L of $[Ag]_{total}$, PVP-coated NPs were most effective in suppressing the nitrification process than Ag^+ in the aerobic system. The oxidative dissolution is the likely cause of AgNP toxicity. However there is no straightforward explanation for the decreased toxicity in Ag^+ under the same concentration. Although the exact mode of PVP-coated AgNP toxicity mechanism is not known, the interactions of AgNPs with soils might hold the key to understand the pronounced NP toxicity to bacteria in the adsorbent system. Masrahi et al. (2014) compared the dose–response relationship of Ag^+/AgNPs to bacteria with the literature values and revealed that the toxicity of Ag^+ and nano-Ag in their study was much lower than that observed in laboratory-pure culture media. According to these studies (Radniecki et al. 2011; Choi and Hu 2008, 2009), 0.08 mg/L of Ag^+ and 1 mg/L of PVA-coated AgNP decreased the nitrification process by 50 and 86 %, respectively. Compared with the Masrahi et al. (2014) soil nitrification study, the toxicity of 1 mg/L of Ag^+ was not statistically different from that of the control. This difference in the dose–response relationship is likely caused by the partitioning processes of Ag^+ and AgNPs in soils.

Although considerable progress has been made in elucidating factors affecting AgNP toxicity, there is less information regarding the toxicity mechanism of AgNPs to test microorganisms, especially nitrifying bacteria. In this field, Yuan et al. (2013) used three different silver NP suspensions (7 ± 3 nm, 7 ± 3 nm, and 40 ± 14 nm, coated with adenosine triphosphate disodium (Na_2-ATP), polyvinyl alcohol (PVA), and Na_2ATP, respectively) to study their toxicity to pure nitrifying bacteria (*N. europaea* ATCC 19718). For all different AgNPs, large aggregates were gradually formed after addition of AgNPs into the media containing *N. europaea*. The scanning electron microscopy and energy dispersive X-ray (EDX) spectroscopy of the microstructures suggested that bacterial cells and electrolytes had significant effects on AgNP aggregation (Fig. 5.5). Size- and coating-dependent inhibition of ammonia oxidation by AgNPs was observed, and their analysis suggested that the inhibition was not only due to the released dissolved silver but also the dispersity of AgNPs in the culture media. Electron microscopy images showed AgNPs could cause the damage of cell wall of *N. europaea* and make the nucleoids disintegrated and condensed next to cell membrane (Fig. 5.6). TEM–EDX analysis suggested AgNPs entered the cell interior (Fig. 5.6e, arrow, and Fig. 5.6f). Similar data were also reported, which found

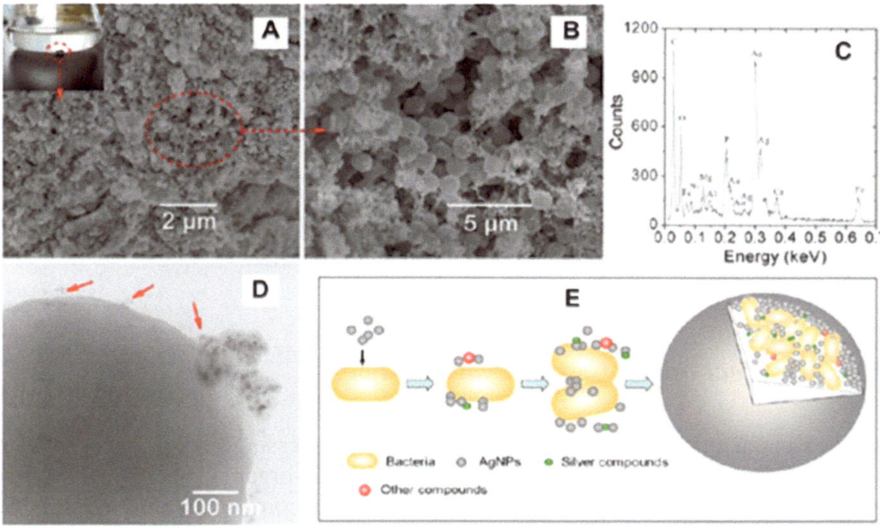

Fig. 5.5 (**a**, **b**) SEM images and (**c**) EDX analysis of the visible aggregates formed in toxicity tests. (**d**) TEM image showed adsorption of PVA onto bacteria cells in the aggregates. (**e**) Schematic diagram of the probable formation process of the aggregates in the culture medium with *N. europaea* (Yuan et al. 2013)

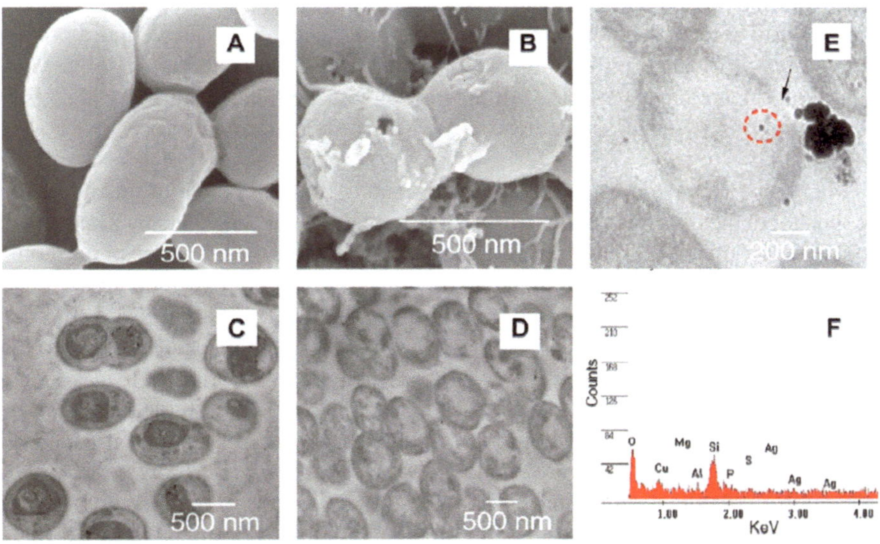

Fig. 5.6 Morphological changes of *N. europaea* before and after PVA exposure. SEM images of (**a**) untreated and (**b**) treated *N. europaea*. TEM images of (**c**) untreated and (**d**, **e**) treated *N. europaea*. (**f**) EDX analysis of the *dark spot* in the *circle* of (**e**) detected silver (Yuan et al. 2013)

AgNPs could lead *E. coli* to the formation of "pits" in the cell walls and could enter the periplasm through the pits and destroy the cell membrane and then could cause condensing of DNA and the leakage of the cytoplasmic component (Chen et al. 2011). Surface-enhanced Raman scattering (SERS) signals also implied the damage of cell membrane caused by AgNPs. Further protein expression analysis revealed that AgNPs would inhibit important protein functions, including biosynthesis, gene expression, energy production, and nitrification to further cause toxicity to *N. europaea*. Their data explain the susceptibility of *N. europaea* to inhibition by silver NPs and the possible interaction between each other.

5.3.3 Effect on Soil Exoenzyme Activity

One essential microbial function in soils is the processing and recovery of key nutrients from detrital inputs and accumulated soil organic matter. This often requires the activity of enzymes to process complex compounds into assimilable subunits (Caldwell 2005). Enzymes found in soils are either intracellular (i.e., found within live organisms) or extracellular (i.e., are released by organisms). Extracellular enzymes (exoenzymes) can be found both in soil solution and bound to soil components. Microbial diversity can be expressed through a variety of soil enzymes with functional significance over a range of biogeochemical processes and metabolic pathways. Soil enzyme assays have demonstrated potential for the early detection of anthropogenic or natural disturbances as well as a proven sensitivity for evaluating the impacts of trace metals in contaminated soils. Measurements of enzyme activities within key biogeochemical cycles can therefore improve our knowledge of the effects of NMs on soil microbial processes (Peyrot et al. 2014).

Many studies have investigated the microbial toxicity of NMs but much less are known about the exoenzymes in soils treated with them. Experiments by Zheng et al. (2011) report on the inhibition by TiO_2 NPs (50 mg/L) of the activities of ammonia monooxygenase and nitrite oxidoreductase, enzymes potentially involved in the N_2 cycling process, following a 70-day exposure. However, no significant effect of TiO_2 NPs on the activities of exopolyphosphatase and polyphosphate kinase, nor on the transformation of intracellular polyhydroxyalkanoates and glycogen, was found. The authors related these findings to the impacts of TiO_2 NPs on biological N and P removal, as well as the depletion of NH_3-oxidizing bacterial population by the NPs. Zinc oxide NP doses that were nonlethal to *Bacillus subtilis* and *Pseudomonas aeruginosa* completely inhibited or lowered the activities of enzymes involved in starch degradation, denitrification, and urea degradation (Santimano and Kowshik 2013). Du et al. (2011) noted the inhibition of soil protease, catalase, peroxidase, but not urease, activity in the presence of TiO_2 and ZnO NPs. The inhibition by NPs of enzymes involved in the alleviation of oxidative stress could decimate soil microbial populations under abiotic stress conditions. Tong et al. (2007) reported negligible impacts of fullerene (C_{60} 1000 μg/g soil) on the activities of acid phosphatase, β-glucosidase, urease, and dehydrogenase. Cullen et al. (2011) found increased dehydrogenase activity, but a minimal effect of fluorescein diacetate hydrolase in soil treated with nano-zerovalent iron (nZVI, 10,000 μg/g soil). Kim et al. (2013) observed a disturbing effect of CuO NPs on dehydrogenase, phosphatase, and β-glucosidase activities in soil planted with cucumber and maize. Concomitantly, the CuO as well as ZnO NPs reduced plant biomass. It is instructive that similar levels of the effects were not observed with micron-size CuO, highlighting the greater reactivity of nanoscale materials. Chung et al. (2011) reported inhibited activities of phosphatase, β-*N*-acetylglucosaminidase, β-glucosidase, cellobiohydrolase, and xylosidase with 5000 μg/g of multiwalled carbon nanotubes (MWCNTs).

Due to the wide application of silver NPs, in this paragraph, we will discuss and review its impact on soil enzymes. Silver NP contamination of soil could directly affect numerous soil microorganisms but could also have indirect effects through its actions on soil enzymes. There is very little information about the effect of silver NPs on soil exoenzyme activities, which reflect the potential of soil to support biochemical processes. Some studies reported no inhibition for soil enzyme activities during their treatments, but the others described negative effects. Hansch and Emmerling (2010) evaluated the effects of AgNPs on the activities of leucine aminopeptidase, β-cellobiohydrolase, acid phosphatase, β-glucosidase, chitinase, and xylosidase. They reported no effect of AgNPs for all the enzymes tested, with the exception of slight decrease of the leucine aminopeptidase activity. In this study, they used up to 0.32 μg/g, and such a low concentration induced no effect. It would appear that the concentration range of AgNPs used in their research was too low to influence the enzyme activities. Also, there is a possibility that their fluorometric enzyme assay, which followed the method of Marx et al. (2001), may have suffered from interference due to the intrinsic fluorescence of AgNPs (Maali et al. 2003). Shin et al. (2012) provided evidence of the inhibitory effects of AgNPs on the activities of soil exoenzymes. In their study, six exoenzymes related to nutrient cycles (urease, acid phosphatase, arylsulfatase, β-glucosidase) and the overall microbial activity (dehydrogenase, fluorescein diacetate hydrolase) were tested in soils treated with AgNPs (1, 10, 100, and 1000 μg/g) and silver ion (0.035, 0.175, 0.525, 1, and 1.5 μg/g). The urease activity was observed to be the most sensitive to AgNPs, which was inhibited as low as 1 μg AgNP/g. Its activity was reduced to approximately 90 % at 1000 μg/g as soon as the AgNPs were applied to the soil and did not recover after 7 days. This means that AgNPs clearly inhibit the nitrogen cycling in the soil ecosystem. Also, AgNPs negatively affected the electron transfer by suppressing the dehydrogenase activity. Particularly, more than 50 % of the dehydrogenase and urease activities were inhibited at the highest exposure concentration, with 7d-IC50 (inhibition concentration) values estimated to be 107.98 (62.82–185.61) and 14.20 (8.78–22.97) μg/g, respectively. The other 7d-IC50 values for FDA hydrolase, acid phosphatase, arylsulfatase, and β-glucosidase were not calculable because their inhibition rates were less than 50 % within the exposure concentration ranges of the Shin et al. study, as shown for the enzyme activities in other studies (An and Kim 2009). Overall, the activities of all the soil enzymes tested were negatively affected by AgNP concentrations of 100 and 1000 μg/g. AgNPs affected the S cycle (arylsulfatase), C cycle (β-glucosidase), and P cycle (acid phosphatase), as well as the microbial decomposer activity, as reflected by the FDA hydrolase in the soil environment. Silver ions (Ag^+) dissolved from AgNPs in the soil were found to be 0.16, 0.13, and 1.31 μg Ag^+/g soil for 0, 1, and 1000 μg AgNPs/g soil, respectively; therefore, Ag^+ exposure concentrations of 0.035, 0.175, 0.525, 1, and 1.5 μg/g were prepared to cover the concentration range of dissolved silver ions. There was no significant change in any of the enzyme activities in the soil treated up to 1.5 μg Ag^+/g. This indicated that dissolved silver ions from AgNPs did not influence the enzyme activities tested in this study, but negative effects were observed due to the AgNPs themselves.

Peyrot et al. (2014) investigated the effects of AgNPs in soil ecosystems by measuring hydrolase activities (phosphomonoesterase, arylsulfatase, β-D-glucosidase, and leucine aminopeptidase) based on a sensitive fluorometric technique. Enzyme assays were applied to agricultural soil samples that were amended or not with compost made from deciduous tree leaves to specifically investigate the role of soil organic matter. Soil samples were treated with AgNPs and Ag^+ (as acetate) at equivalent total soil Ag concentrations. In general, an inhibition of the enzyme activities was observed as a function of Ag concentrations in the soil, with less inhibition being observed for the organic matter-amended soil: *Arylsulfatase* showed the greatest sensitivity to the Ag treatments. In the soil treatment without organic matter, inhibition was observed in all of the AgNP and Ag^+ (as acetate) treatments. In contrast, in the organic matter-amended soils, the enzyme activity was significantly inhibited only for the highest Ag treatment (31.25 mg Ag/kg for both Ag^+, as acetate, and AgNPs). Phosphomonoesterase appeared to be the least affected enzyme. No significant effect was observed as a result of the Ag^+ (as acetate) treatments in either soil. For an addition of 6.25 mg/kg of AgNPs, its activity decreased to 32.9 % of the control values in the unamended soil and to 49.2 % of the control values in the organic matter-amended soil. Similarly, no significant effect as a result of Ag^+ (as acetate) was observed for *β-D-glucosidase* activity in either soil. In the unamended soil, a decrease in its activity to 43 % of the control was observed for the highest concentration of added AgNPs, whereas no effect was observed in the organic matter-amended soil. *Leucine aminopeptidase* activity was not significantly affected by the addition of Ag^+ (as acetate) in either soil. In contrast, for an addition of 1.25 mg/kg of AgNPs to the unamended soil, its activity decreased significantly to 53.8 % of the control value. The Ag chemical speciation measurements suggested that AgNPs caused greater toxic effects to soil enzymes at the low Ag concentrations. For the larger concentrations of total soil Ag, causes of the negative effects on enzyme activities are less obvious, but it is possible that colloidal forms of Ag play a role. The addition of organic matter to the soils seemed to enhance enzyme activities, but its protective role could not necessarily be attributed to the complexation of free Ag because concentrations of dissolved Ag were similar in both amended and unamended soils.

5.3.4 Potential Mechanisms of Toxicity for Soil Microbes

While toxicity mechanisms have not yet been completely elucidated for most NMs, possible mechanisms include formation of reactive oxygen species (ROS), disruption of membranes or membrane potential, oxidation of proteins, genotoxicity, interruption of energy transduction, and release of toxic constituents (Klaine et al. 2008) (Fig. 5.7). Furthermore, morphological and physicochemical characteristics of the NMs have been proven to exert an effect on their antimicrobial activities. For example, it is known that the small NPs have the strongest bactericidal effect and the positive surface charge of the metal NPs facilitates their binding to the negatively charged surface of the bacteria which may result in an enhancement of the bactericidal effect (Table 5.1) (Maleki Dizaj et al. 2014).

5.3.4.1 Cell Damage Through Reactive Oxygen Species

The generation of ROS is an important toxicity mechanism of NMs. ROS include oxygen radicals that have one or more unpaired electrons, such as superoxide anion ($O_2^{-\bullet}$), peroxide ($O_2^{=\bullet}$), hydroxyl radical ($^\bullet OH$), and singlet oxygen ($1O_2$). They are, for example, formed in mitochondria as oxygen is reduced along the electron transport chain. But despite their beneficial activities, radicals possess an unpaired electron, which makes them highly reactive and can clearly be toxic to cells and then be able to damage, cell membranes, cellular organelles, and all macromolecules including lipids, proteins, and nucleic acids contained in DNA and RNA (Maleki Dizaj et al. 2014; Gohargani and Ghasemi 2013; Prasad and Swamy 2013; Aziz et al. 2015).

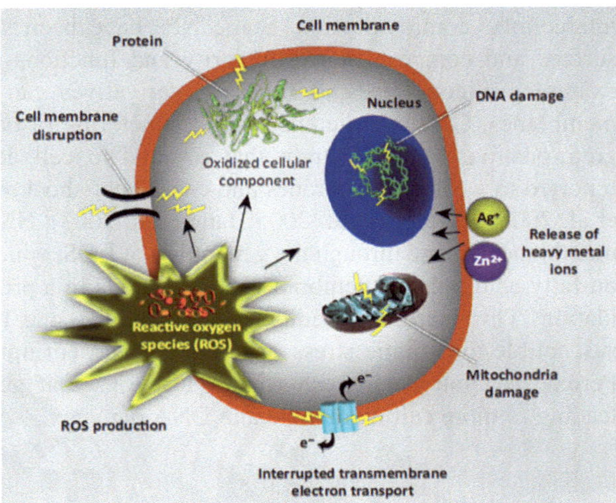

Fig. 5.7 Mechanisms of toxicity of nanoparticles against bacteria. NPs and their ions (e.g., silver and zinc) can produce free radicals, resulting in induction of oxidative stress (i.e., reactive oxygen species, ROS). The produced ROS can irreversibly damage bacteria (e.g., their membrane, DNA, and mitochondria), resulting in bacterial death (Hajipour et al. 2012)

Table 5.1 Proposed mechanism of antimicrobial action for some nanoparticles (Maleki Dizaj et al. 2014)

NMs	Proposed mechanism	The factors that influence antimicrobial activity
AgNPs	Ion release: induction of pits and gaps in the bacterial membrane; interact with disulfide or sulfhydryl groups of enzymes that lead to disruption of metabolic processes. DNA loses its replication ability and the cell cycle halts at the G2/M phase owing to the DNA damage (in the case of Ag_2O)	Particle size and shape of particles
ZnO NPs	ROS generation on the surface of the particles; zinc ion release, membrane dysfunction; and NP internalization into cell	Particle size and concentration
TiO_2 NPs	Oxidative stress via the generation of ROS; lipid peroxidation that causes to enhance membrane fluidity and disrupts the cell integrity	Crystal structure, shape, and size
Au NPs	Attachment of these NPs to membrane which change the membrane potential and then cause the decrease the ATP level and inhibition of tRNA binding to the ribosome	Roughness and particle size
Si NPs	Influencing the cell functions such as cell differentiation, adhesion, and spreading	Particle size and shape
CuO NPs	Crossing of NPs from the cell membrane and then damaging the vital enzymes of bacteria	Particle size and concentration
MgO and CaO NPs	Damaging the cell membrane and then causing the leakage of intracellular contents and death of the bacterial cells	Particle size, pH, and concentration

5.3.4.2 Damage to Membrane Integrity

The cell membrane is a semipermeable barrier that serves important cellular functions, such as regulation of material transport, energy transduction, and intercellular communication. Some NPs have been shown to attach to the cell surface and compromise the integrity and functions of the cell membrane. For example, silicon NPs and fullerene derivatives can embed themselves in the membranes. Carboxyfullerene caused puncturing of bacterial cell membrane in a gram-positive bacterial strain that resulted in cell death. Gold NPs have been reported to weaken membranes and cause heat-shock responses in *E. coli* (Hwang et al. 2008; Kloepfer et al. 2005; Jang et al. 2003). NMs can also indirectly cause membrane damage through the generation of ROS, which can oxidize double bonds on fatty acid tails of membrane phospholipids in a process known as lipid peroxidation. This increases membrane permeability and fluidity, making cells more susceptible to osmotic stress or hindering nutrient uptake (Cabiscol et al. 2000). Peroxidized fatty acids can trigger reactions that generate other free radicals, leading to more cell membrane and DNA damage.

5.3.4.3 Protein Destabilization and Oxidation

NM–protein interactions have been optimized for a variety of biomedical applications. For example, quantum dots are used to target and fluorescently label proteins for imaging. Multiwalled carbon nanotubes were used in biosensing applications to immobilize and optimize lactate dehydrogenase and alcohol dehydrogenase. The toxicological interactions between NMs and proteins are related to either the NM physically interacting with proteins or the NM producing ROS or other damaging radicals. The structure and activity of glucose oxidase was altered using electrodes containing gold NPs or SWCNTs. Nanomaterials generate ROS can damage iron–sulfur clusters that behave as cofactors in many enzymes, leading to Fenton chemistry that catalyzes the production of more ROS generation. Reactive oxygen species can also lead to the formation of disulfide bonds between sulfur-containing amino acids, thus disturbing the structure and function of the protein (Klaine et al. 2008).

5.3.4.4 Impact on DNA and Gene Expression

Interactions of NMs with nucleic acids have applications in DNA labeling or DNA cleavage. Nucleotides can be tagged with NPs, such as quantum dots, which act as labeling agents for bioimaging applications. As with NMs that are made to traverse the cell membrane, iron oxide NPs can be modified into nonviral NP gene transfection vectors to carry genetic information into the cell. Photosensitive metallic and metal oxides that generate ROS as well as fullerenes are used for photodynamic therapy, targeting cells and DNA. In contrast to the beneficial applications of NM–DNA conjugation, fullerenes have been reported to bind DNA and cause deformation of the strand, adversely impacting the stability and function of the molecule. Quantum dots can also nick supercoiled DNA. Some NMs indirectly damage DNA because of ROS production, which can lead to DNA strand breaks, cross-linking, and adducts of the bases or sugars. Titanium dioxide NPs, such as those used in sunscreen, generate oxygen radicals that can nick supercoiled DNA. Photosensitive fullerenes can cleave double-stranded DNA on exposure to light, although this is highly dependent on the type of the fullerene derivative. Despite these results, only a few studies on the genotoxicity of NMs using Ames tests or any other established protocol have been published, and little is known about the potential mutagenic effect of NMs (Gohargani and Ghasemi 2013; Klaine et al. 2008). [The Ames test was developed in the 1970s by Bruce Ames as a fast and sensitive assay of the ability of a chemical compound or mixture to induce mutations in DNA. Bruce Ames published his work in a series of papers, including "Identifying Environmental Chemicals Causing Mutations and Cancer" in the journal *Science* (Vol. 204, No. 4393, pp. 587–593, 1979)].

In the previous paragraph, we discussed around DNA damages, but there is another impact on DNA revealed by scientists: "NPs effects on gene expression."

Nanoparticles or their ions potentially can impact DNA replication and even gene expression. In *E. coli*, Yang et al. (2009) reported the binding of DNA by silver NPs in the cytoplasm, with impairment of DNA replication. Silver NPs at sublethal levels did not significantly affect the expression in *Pseudomonas stutzeri*, *Azotobacter vinelandii*, or *Nitrosomonas europaea* of a suite of N_2-fixing (*nifD*, *nifH*, *vnfD*, *anfD*) and denitrifying (*narG*, *napB*, *nirH*, and *norB*) genes, whereas other genes involved in nitrification, namely, *amoA1* and *amoC2*, were upregulated in *N. europaea* (Yang et al. 2013). The stimulation of nitrification genes without the concomitant stimulation of those for denitrification (conversion of NO_3^- to N_2) could have implications for both NO_3^- buildup and N_2 availability for subsequent fixation. Microarray-based researches with *E. coli* and silver NPs indicate that NPs could have transcriptome-wide ramifications in bacteria. Among other molecular perturbations, the stimulation of expression of stress-related genes and genes for Fe, S, and Cu balance was reported. The effect of silver NPs on genes involved in the regulation of other metals suggested an effect of silver NPs on the cellular metal homeostasis. Similarly, proteomic studies on the exposure of *Bacillus thuringiensis* to silver NPs reported the accumulation of envelope protein precursors, suggesting that the proton motive force was being affected. Other proteins with modified production in the presence of silver NPs included those involved in oxidative stress tolerance, metal detoxification, transcription and elongation processes, protein degradation, cytoskeleton remodeling, and cell division. In *Pseudomonas chlororaphis* O6, the expression of genes involved in the periplasmic maturation and secretion of fluorescent pyoverdine siderophores was repressed by copper oxide NPs, correlating with the reduced production of siderophores observed in the presence of copper oxide NPs (Dimkpa 2014).

5.3.4.5 Interruption of Energy Transduction

Electron transport phosphorylation and energy transduction processes may be disrupted if membrane integrity is compromised or if a redox-sensitive NM contacts membrane-bound electron carriers and withdraws electrons from the transport chain. Fullerene derivatives have been reported to inhibit *E. coli* respiration of glucose. Cerium dioxide NPs may themselves be transformed after contact with living cells, oxidize membrane components involved in the electron transport chain, and cause cytotoxicity (Klaine et al. 2008).

5.3.4.6 Release of Toxic Components

Certain NMs cause toxicity to bacterial cells by releasing harmful components, such as heavy metals or ions. Quantum dots (QD) are semiconductor nanocrystals that contain noble or transition metals, such as CdSe, CdTe, CdSeTe, ZnSe, InAs, or PbSe, in their core; CdS or ZnS in their shell; and an organic coating. Uptake of QD by *E. coli* and *Bacillus subtilis* has been reported (Kloepfer et al. 2005). Although no acute cytotoxic effects were observed in that study, in the absence of an efficient efflux system, QD uptake will lead to accumulation of potentially toxic metals in the cells, where they have long residence times and cause toxic effects. In addition to the metals of the core/shell, some organic coatings may also be toxic (Hoshino et al. 2004). Release of silver ions has been implicated in toxicity of silver NPs. It is believed that silver ions interact with thiol groups of proteins, resulting in inactivation of vital enzymes. Silver ions have also been shown to prevent DNA replication and affect the structure and permeability of the cell membrane (Klaine et al. 2008).

5.3.5 Factors Influencing Microbial Toxicity

Researchers have begun to characterize interactions between a wide variety of engineered NMs and microbes with the goal of finding NM toxicity and its connection to the material's physical and chemical properties. Many physical properties of NMs are interrelated, and subtle changes in size, shape, and surface coating can significantly alter interactions with biological systems. Similarly, other factors such as the method of NM synthesis, dose, the presence or absence of additives, the solubility of the material, and the intrinsic properties of microorganisms can also influence the biological impact of the NM.

5.3.5.1 Influence of Parent Material

While the nature of the parent material can be a major determinant of potential microbial toxicity, direct comparisons of relative toxicity of one material to another are not straightforward as size, shape, surface coating, and synthesis methodology can affect toxicity. Strict control over physical characteristics is difficult, and synthesis methodology can lead to differences in surface coating and unintended toxic materials. For example, different manufacturing processes may incorporate additives, detergents, and solvent chemicals that are not fully eliminated from the final product. Studies on the bactericidal efficacy of engineered AgNPs with diverse sizes, surface coatings, and synthesis using different methods suggested that the solvent remnants used in the synthesis can lead to false toxicity interpretations. Even though Ag-resistant *E. coli* was used, the remnants of formaldehyde led to killing of the Ag-resistant bacterium. Thus, the apparent biological properties of NMs may depend in part on other constituents present in the formulation or on differences in chemical coatings on the NM (Suresh et al. 2013).

Living organisms require trace amounts of metal ions, but higher doses are known to be toxic. While most metals have low solubility in aqueous environments, "dissolution" of metal NPs into ions can be toxic to microorganisms (Li et al. 2011; Sotiriou and Pratsinis 2010). Although the molecular mechanism of toxicity can differ for different ions and species, the dissolution of NPs into ions is often a primary step and a common cause for NP toxicity. The microbial toxicity of metal ions such as silver, copper, nickel, and zinc has long been recognized. Not surprisingly, NPs formed from these metals are often found to be toxic. The material commonly used for its microbial toxicity is silver, and ample literature on the application of silver as an antimicrobial agent, either solely or in combinatorial forms, is available (Marambio-Jones and Hoek 2010). Increasing evidence supporting silver NP "dissolution" as the origin of its toxicity is unfolding. The correlation between NP toxicity and that of its dissolved ion is observed with other materials as well. NPs comprised of materials such as iron, gold, palladium, silver sulfide, and platinum are frequently observed to be relatively nontoxic. These materials are poorly soluble, and their metal ions are frequently observed to be nontoxic or inert. Based on this issue, it is clear that NP dissolution into ions can lead to toxicity. However, the dissolution of NPs into ions is not the only means of toxicity. The ability to generate ROS underlies another recognized mechanism of NP-mediated microbial toxicity. Engineered NMs (e.g., CdSe, CdTe, TiO_2, 1 ZnO, Ag, CuO, ZnS, SiO_2, etc.) can contain physically or chemically redox-active surfaces that can react with molecular oxygen to generate ROS that are implicated in the toxic response of a number of biological systems. While the ability of metal or metal oxide NP to liberate dissolved ions, the toxicity of these ions, and/or the ability of the NP to generate ROS are key determinants of NP toxicity, the contribution of the NP itself to toxicity remains unclear and an area for further investigation (Suresh et al. 2013).

5.3.5.2 Influence of Size and Shape

These properties are important factors influencing toxicity of NMs. As particle size decreases, the ratio of surface area to mass increases, and changes to the physical–chemical properties of the NP occur (Grassian 2008). In general, a trend of increased toxicity with decreased particle size has been observed with toxic NPs. This is consistent with the increased reactivity of smaller particles and is well illustrated in the case of silver and zinc oxide NPs. For example, evaluations on the interaction of nanosilver upon immobilization on nanostructured silica particles against *E. coli* showed size-dependent release of Ag^+ ions. Released Ag^+ was prominent when fine particles of dimensions less than 10 nm were used (Sotiriou and Pratsinis 2010). Shape is another important characteristic that affects the properties of engineered NPs. Particles with uneven and rough surfaces, or with irregular shapes, can have corners and edges that are biologically and chemically reactive. Atoms at these locations have a lower bonding coordination (weaker bonds) than bulk atoms, and therefore their reactivity and interactions with microorganisms are affected. Some commonly used shapes of engineered NPs are spheres, rods, squares, hexagons, triangles, and pentagons prepared using various synthesis methodologies. The increased reactivity of edge containing NPs and their correlation to increased toxicity are supported by several studies. For example, truncated triangular silver nanoplates displayed the strongest biocidal activity when compared to that of similar spherical and rod-shaped particles against *E. coli* (Pal et al. 2007).

5.3.5.3 Influence of Surface Coating

Engineered NPs are invariably surrounded by a shell or capping material that acts as a stabilizing, biocompatibility, and/or reactivity agent. This is an important parameter in determining the environmental and biological fate of a NM as the surface is the primary mode of contact. Surface coating can affect the charge on the NP that in turn can affect the affinity of the material to the cell surface. Recently, surface charge has been suggested as the most important factor governing the toxicity of silver NPs. In investigations of four different surface coatings, that ranged from highly positive to highly negative, the toxicity decreased as particle charge decreased (El Badawy et al. 2010). Additionally, the surface coating can influence the dissolution or release of ions from the NPs. These surface properties are also strongly influenced, or altered, by environmental conditions (e.g., pH, ionic strength, presence of organics).

5.3.5.4 Particular Properties of Microorganisms

It is necessary to know NMs effects on microorganisms are not equal and similar. Numerous studies have compared and contrasted the response of different microbial species to NMs. Complicating these studies are the distinct physiological characteristics of different microorganisms that affect their growth and tolerance to NP-induced stress. What is clear is that different microbial species can show contrasting susceptibilities to potentially toxic NMs. For example, investigations on the comparative toxicity of engineered Ag nanocrystallites on Gram-negative and Gram-positive bacteria and Gram-negative *E. coli* and *S. oneidensis* were found to be more resistant than Gram-positive *B. subtilis*. Numerous other studies have also noted the susceptibility differences of Gram-positive and Gram-negative organisms to diverse forms of NPs (Suresh et al. 2013). Typically, the Gram-positive organisms have been found to be more sensitive to potentially toxic NMs, and this increased sensitivity is likely attributed to differences in the bacterial cell membrane and cell wall structures. The lipopolysaccharides of the outer membrane of Gram-negative bacteria may provide resistance against NPs (Yoon et al. 2007; Brayner et al. 2006; Qi et al. 2004). In an investigation on the comparative toxicity of CdTe QDs on Gram-negative (*E. coli* and *P. aeruginosa*) and Gram-positive (*S. aureus* and *B. subtilis*) bacterial strains, the higher sensitivity of Gram-positive organisms was noted. Even though the release of heavy metal ions (Cd^{2+}) was observed, the main cause of toxicity was attributed to the generation of hydroxyl radicals (Cooper et al. 2010). However, the increased sensitivity to QDs of Gram-positive over Gram-negative bacteria is debatable. Differential toxicity of microorganisms to other NMs has been noted. TiO_2 and Ag-TiO_2 NPs have been observed to be more toxic to *B. subtilis* when compared to *P. putida* due to the lack of lipopolysaccharide membrane (Li et al. 2011).

The rate of bacterial growth can influence the tolerance of bacteria against NPs. Fast-growing bacteria are more susceptible than slow-growing bacteria to antibiotics and NPs (Mah and O'Toole 2001). It is possible that the tolerance property of slow-growing bacteria is related to the expression of stress-response genes (Lu 2009). *B. subtilis* and *Pseudomonas putida* can physically adapt to nC_{60} (Fang et al. 2007). *P. putida* increases cyclopropane fatty acids and decreases unsaturated fatty acid levels, but *B. subtilis* increases the transition temperature and membrane fluidity in the presence of nC_{60}. These physiological adaptation responses of bacteria help to protect the bacterial membrane against oxidative stress. *Shewanella oneidensis* MR-1 has excellent resistance against several concentrations of Cu^{2+} and Cu-doped TiO_2 NPs (Wu 2010) because of the production of extracellular polymeric substances (EPSs) under NP stress. This bacterium is able to absorb NPs on the cell surface and to decrease the amount of ionic Cu in the culture medium. Therefore, this bacterium can be regarded as a promising candidate for cleaning of metal oxide NPs from the environment. TiO_2 and Al_2O_3 NPs are able to be internalized by *E. coli* and *Cupriavidus metallidurans* CH34, but these NPs are toxic only against *E. coli* (Simon-Deckers et al. 2009). The resistance mechanism of *C. metallidurans* CH34 is not yet understood completely. The tolerance mechanism of this bacterium may be related to physical properties of their peptidoglycan layer and/or products of genes that are located in the plasmids and are able to stabilize the plasma membrane or efflux of NPs. Many bacteria are able to tolerate NO NPs using various mechanisms. For example, *P. aeruginosa*, *E. coli*, and *Salmonella typhimurium* induce the expression of genes that are responsible for repairing of DNA and altering the metal homeostasis in the presence of NO NPs. In this condition, *K. pneumoniae* produces the enzyme flavohemoglobin, which neutralizes nitrosative stress. Some microbes can develop special structures like biofilm to protect them from negative conditions. Biofilms are a complex microbial community that form by adhesion to a solid surface and by secretion of a matrix (proteins, DNA, and extrapolysaccharide), which cover and protect the bacterial cell community. Exposure to AgNPs may prevent colonization of new bacteria onto the biofilm and decrease the development and succession of the biofilm (Hajipour et al. 2012).

5.4 Research and Risk Assessment of Nanomaterials by Advanced Tools

Many factors influence microbial community, and they can be classified into two groups, i.e., abiotic factors and biotic factors. Abiotic factors include both physical and chemical factors such as water availability, salinity, oxic/anoxic conditions, temperature, pH, pressure, chemical pollution, heavy metals, pesticides, antibiotics, etc. In general, all environmental variations affect in different ways and to different degrees, resulting in a shift in the diversity profile (Fakruddin and Mannan 2013). Evaluating soil microbial diversity can prepare much valuable information about soil variations, and it will help us to get true data for managing them. For this, many culture-dependent and culture-independent methods have developed. There are other tools to collecting exact data for NM–microbe interactions. Using these tools, a microbiologist can monitor live cells in real time. In the following paragraphs, we will introduce these techniques and will describe their especial advantages and disadvantages.

5.4.1 Atomic Force Microscopy

5.4.1.1 AFM Imaging in Microbiology

Most microorganisms have a well-defined cell wall that has presumably evolved during the course of evolution by selection in response to environmental and ecological pressures. As microbial surfaces form a boundary between the external environment and the cell, they have several important functions, including determining cell shape, growth, and division, enabling cells to resist turgor pressure, acting as molecular sieves, and mediating molecular recognition and cellular interactions. Therefore, studying the structure, properties, and functions of microbial surfaces is an exciting and continuously expanding field of microbiology (Dufrene 2004). In order to characterize microbial interactions with various surfaces and interfaces, we need to know their morphology and physicochemical properties. Because of the small size of microorganisms, the physical properties of their surfaces have been difficult to study. Quantitative and qualitative information on physical properties can be obtained by electron microscopy techniques, X-ray photoelectron spectroscopy, infrared spectroscopy, contact angle, and electrophoretic mobility measurements. Most of these methods, however, require cell manipulation prior to examination (staining and drying), which may seriously compromise the validity of the analysis. Furthermore, the information is generally obtained from large numbers of cells and not at the level of individual cells. Thus, there is clearly a need for new, nondestructive tools capable of probing single-cell surfaces at high resolution. As a result, the need for more sophisticated techniques was evident, and researchers started to get their questions successfully answered with the discovery of the AFM (Dufrene 2002).

At first, AFM has emerged as an imaging technique that provided 3D topographic views and structural details of microbial surfaces in air. A few years later, it became possible to image biological samples in their native environment and to measure interaction forces in these systems. The AFM has been improved with increasingly sophisticated sensing and actuating features to optimize its performance and allow new techniques, such as single-molecule force spectroscopy

(SMFS), single-cell force spectroscopy (SCFS), and chemical force microscopy (CFM) to emerge (Dorobantu et al. 2012).

The AFM consists of four major components: an AFM tip, a piezoelectric scanner, an optical deflection system (laser diode and photodetector), and an electrical feedback system. The heart of the AFM is the probing tip mounted at the end of a flexible cantilever, which is usually made of silicon or silicon nitride. Depending on the AFM system used, either the cantilever or the sample is mounted on a piezoelectric scanner that moves precisely in three dimensions, allowing high-resolution 3D positioning. The basic idea of the AFM technique is to raster scan a sharp tip over the surface of interest while sensing the interaction between the tip and the sample (Fig. 5.8). The cantilever deflection is usually detected by the optical deflection system. A laser beam is reflected off the backside of the cantilever and collected into a four quadrant photodetector, which records the position of the reflected beam. The magnitude of the deflected beam changes in response to the interaction force between the tip and the sample. The deflection signal is digitally processed to reconstruct a topographic image or an interaction force profile of the sample (Liu and Wang 2010).

AFM imaging has emerged in the past years as an important additional tool to standard light and electron microscopy techniques for its high-resolution imaging and simple sample preparation of microbiological systems in both air and fluid environments. Different AFM imaging modes are available (contact mode, noncontact mode, and intermittent mode), which differ mainly in the way the tip moves over the sample (Fig. 5.8). The most widely used mode is contact mode in which the AFM tip remains in direct contact with the sample when translated over its surface. In this mode, two types of images can be obtained, namely, height and deflection. The height image is generated by recording the cantilever deflection as the tip is scanned over the surface of the sample. It provides calibrated height information and allows accurate measurements of surface roughness. The deflection image is generated by adjusting the sample height using a feedback loop that keeps the deflection of the cantilever constant while scanning the AFM tip over the

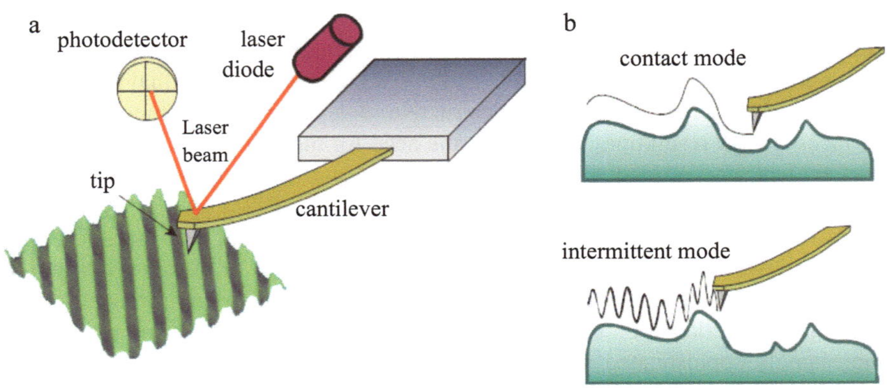

Fig. 5.8 (**a**) Basic setup of AFM (**b**) contact and intermittent modes

sample surface. In many cases, small cantilever deflections do occur because the feedback loop is not perfect, and then, the resulting error signal is used to produce the deflection image. This image is more sensitive to fine surface details. Contact mode of operation is associated with high lateral forces, which can damage the soft biological samples or remove them from the substrate. Consequently, minimizing the force generated by the AFM tip is necessary to prevent biological sample damage. This can be achieved by using intermittent mode or tapping mode. In this mode, the AFM probe is excited externally near its resonance frequency, and the amplitude and phase of the cantilever are monitored as the tip is scanned over the sample surface. In tapping mode, the tip interacts with the surface intermittently at the end of its downward movement which helps minimizing the biological sample damage. Three types of images can be recorded in tapping mode, namely, height, phase, and amplitude. The height image provides topographic information similar to contact mode. The amplitude image is equivalent to the deflection image in contact mode and highlights edges and fine surface structure at the expense of height information. The phase image records the phase lag of the cantilever oscillation relative to the driving signal and emphasizes variations in sample composition and properties such as elasticity and adhesion (Dorobantu et al. 2012).

Exciting new opportunities are offered by AFM force spectroscopy, which can probe the ultrastructure and local properties of biological samples under physiological conditions being accompanied by high force sensitivity and positional precision. In this mode of operation, the cantilever deflection is recorded as a function of the vertical displacement of the sample, while the AFM tip is pushed toward the sample and retracted from it. During the vertical movement of the AFM tip, a force versus separation distance curve is generated that emphasizes the force experienced by the AFM tip as a function of the tip-sample separation distance. Such a force curve can be exploited to gain insights into a variety of surface properties such as cell wall elasticity, turgor pressure, cell surface hydrophobicity, and cell surface charges. Notably, the force–distance curves recorded at multiple locations of the (x, y) plane generate spatially resolved maps of microbial physicochemical properties with nanometer-scale lateral resolution (Schaer-Zammaretti and Ubbink 2003). The data can then be represented either in adhesions maps (Radmacher et al. 1994) or in elasticity maps (Schaer-Zammaretti and Ubbink 2003).

The force spectroscopy technique has evolved over the years to provide three different force modes, namely, chemical force microscopy (CFM), single-cell force spectroscopy (SCFS), and single-molecule force spectroscopy (SMFS). In these techniques, the AFM tip is functionalized with chemical groups, biological molecules, or viruses or even replaced by a living cell or particle. The characteristic unbinding force observed during tip retraction is the key parameter that provides information on specific receptor–ligand interactions (SMFS), on the spatial distribution of chemical groups (CFM), or about the forces that govern cell–cell and cell–substrate interactions (SCFS). CFM is a force spectroscopy modality which offers a means to resolve local chemical properties and interactions of living microorganisms with nanoscale resolution, and this method has, for instance, been used to probe local surface hydrophobicity and charges of individual cells (Dorobantu et al. 2009; Noy 2006). Chemical functionalization of the AFM tips with specific functional groups provides new avenues for understanding the structure–function relationship of cell surfaces and has made it possible to study the interactions between bacterial cell surfaces and antimicrobial drugs (Dague et al. 2007). A crucial challenge in cell biology is to understand how cell surface-bound molecules are organized and how they interact with their environment (Dupres et al. 2009). These questions can be answered using SMFS which analyzes individual molecules in complex systems by simultaneously localizing and manipulating single molecules on live cells and then revealing events and properties that would otherwise be inaccessible. In SMFS, the molecule of interest is tethered to the AFM tip via distensible cross-linkers and the complimentary pair to the substrate (Francius et al. 2009). On retraction of the tip from the surface, the forces involved in the unbinding can be monitored in real time (Horejs et al. 2011). The force curve recorded in SMFS often shows a nonlinear elongation force, reflecting stretching of flexible molecules on the sample (and on the tip), until the adhesion "pull-off" force is observed. The elongation force may be described using models from statistical mechanics, i.e., the worm-like chain (WLC) and the freely jointed chain (FJC) models. The WLC model describes the polymer as an irregular curved filament, which is linear on the scale of the persistence length, a parameter which represents the stiffness of the molecule. In the FJC model, the polymer is considered as a series of rigid, orientationally independent statistical (Kuhn) segments, connected through flexible joints. Proteins that function like continuous deformable rods are generally well described by the WLC model, while polysaccharides that behave more as series of loosely connected segments can be described by the FJC model (Dorobantu et al. 2012). In SCFS, a microbial cell is immobilized on the AFM cantilever replacing the AFM tip, and the force between the obtained cell probe and other cells or substrates is measured (Helenius et al. 2008).

5.4.1.2 Cell Immobilization

The application of AFM in microbial research is challenging because of the difficulties encountered in immobilizing microbes to flat surfaces without changing their surface properties or viability. Firm immobilization of live microorganisms on solid substrates is a crucial prerequisite for successful AFM imaging and force measurements. The method selected must immobilize the cells strong enough to enable them to withstand the lateral friction forces exerted by the tip during scanning without denaturing the cell surface (Dague 2011). One major problem with microbial samples is that they are soft and flexible causing an additional hurdle when probed under liquid. In many cases, substrate surface modification is required to ensure firm attachment of the biological samples. Therefore, immobilizing the cells is one of the primary challenges encountered when analyzing microorganisms. In the early studies, AFM imaging was conducted with dried microbes, as sample preparation in this case was easy. However, drying the cells can generate misleading information. The past two decades have seen tremendous advance in developing new immobilization protocols for microbial cells and improving the existent ones. The most exciting applications of AFM in microbiology deal with imaging of living cells in liquid environments, where interactions between microbes and their surroundings can be investigated (Fukuma et al. 2005).

The most commonly used immobilization methods that do not require drying are physical entrapment in a porous membrane (Dufrene 2008); aluminum oxide filters (Yao et al. 2002); physical confinement of cells in microwells (Kailas et al. 2009); immobilization by electrostatic interactions to surfaces that have been coated with positively charged substances such as polyethylenimine (PEI) (Velegol and Logan 2002), poly-L-lysine (Schaer-Zammaretti and Ubbink 2003), or gelatin (Beckmann and Venkataraman 2006); covalent binding to amine or carboxyl-terminated surfaces (Dorobantu et al. 2008); and adsorption to surfaces coated with highly adhesive polyphenolic proteins (Meyer et al. 2010). Living cells could be immobilized with all of these approaches, but many cells detached when immobilized by electrostatic interactions (Meyer et al. 2010) and the reactive groups used in covalent binding and the reagents used for cross-linking are known to affect cell viability. For immobilizing viable cells, entrapment and adsorption techniques are found to be more effective. However, physical entrapment in membranes is only suitable for coccoid cells.

One of the most successful methods used for immobilization of living microbes under physiological conditions is based on highly adhesive polyphenolic proteins, which facilitate strong immobilization of the cells without affecting cell viability (Meyer et al. 2010). Dague et al. used a generic method for the assembly of living cells within the patterns of microstructured, functionalized polydimethylsiloxane (PDMS) stamps using convective/capillary deposition. Their results reveal the immense potential of microorganism assembly on functionalized, microstructured PDMS stamps by convective/capillary deposition for performing accurate AFM experiments on living cells (Dague 2011).

5.4.1.3 Example Studies by AFM

AFM has been extensively applied to evaluate morphological effects of treatments on microbes (exposure to antimicrobial substances, stress conditions, etc.). Da Silva and Teschke (2003) compared the effect of PGLa, an antimicrobial peptide isolated from hemocytes of frog skin, on *E. coli* with those of ethylene diamine tetraacetic acid. Alsteenns et al. (2008) described the effect of antimycobacterial antibiotics on the cell envelope of *Mycobacterium bovis* by using AFM in three complementary modes: topographic imaging, chemical force microscopy, and immunogold detection.

Liu et al. (2009) evaluated the effects of ferricyanide on *E. coli* DH5, at cell growth and viable rate. Cell collapse and rough cell surface were observed with AFM after the bacteria were exposed in 50 and 100 mM ferricyanide, demonstrating the inhibition effect of ferricyanide on microorganisms.

To elucidate the antimicrobial activities of chitosan with different molecular weights, Eaton and Fernandes (2008) and Fernandes et al. (2009) investigated the morphology changes of two Gram-positive (*Staphylococcus aureus* and *Bacillus cereus*) and one Gram-negative (*Escherichia coli*) microorganisms before and after chitosan treatments. Chitosans with high molecular weight surrounded microbe cells and formed a polymer layer during the treatments. The polymer layer prevented their uptake of nutrients and eventually led to their death, which was confirmed by cell wall collapse observed with AFM, but this effect did not significantly influence the *B. cereus* spores as they can survive for extended periods without nutrients. On the other hand, chitosans with low molecular weight (chit oligosaccharides) probably affected microbes by penetrating the cells, which provoked more visible damage. Rougher cell surface and lysed cells were observed with AFM after the treatments, but the use of chit oligosaccharides by itself on *B. cereus* spores was not enough for the destruction of a large number of cells. The observation also revealed the response strategies used by the bacteria. The microbe clusters increased both in amount and size during the treatments to resist the effects of chitosans.

La Storia et al. (2011) studied the antimicrobial activity of carvacrol against some Gram-positive and Gram-negative bacterial strains via AFM. In this investigation, AFM images of cells of all strains treated with carvacrol exhibited appreciable modifications (changes in cell surface structure and significant length and diameter reduction of the microorganisms). For example, carvacrol-treated cells of *Brochothrix thermosphacta* 1R2 showed a loss of the original shape and had a marked bumpy surface, *Carnobacterium maltaromaticum* 9P frequently showed division septa, and the surface of *E. coli* and *Pseudomonas fragi* cells appeared wrinkled, jagged, and with numerous granules which were not visible in control cells (Fig. 5.9).

Dubrovin et al. (2008) developed protocols to investigate bacteriophage infection of bacterial cells using AFM and studied different phases of this process for three types of bacterial hosts: Gram-negative *E. coli* 057, *Salmonella enteritidis* 89, and Gram-positive *B. thuringiensis* 393. The whole lytic cycles of the three hosts, from phage adsorption on the cells and flagella to complete cell lysis accompanied by the release of a large number of newly formed phages, was observed in AFM images of infected cells. For example, the AFM images of *E. coli* cells during the lytic cycles are shown in Fig. 5.10. These experiments have demonstrated AFM is an effective technique for investigation of phage infection of bacteria, demonstrating its ability to resolve phage particles in detail, characterize and compare bacterial surfaces in different phases of infection, and easily distinguish intact cells from infected ones.

AFM is also widely used in observation of microorganism behaviors. Yuan and Pehkonen (2009) used AFM to study biofilm colonization dynamics of *Pseudomonas* NCIMB 2021 and *Desulfovibrio desulfuricans* on 304 stainless steel coupons. The results showed that the biofilm formed on the stainless steel coupons by the two strains of bacteria increased in coverage, heterogeneity, and thickness with exposure time. The corrosion pits formed by *D. desulfuricans* were deeper than that

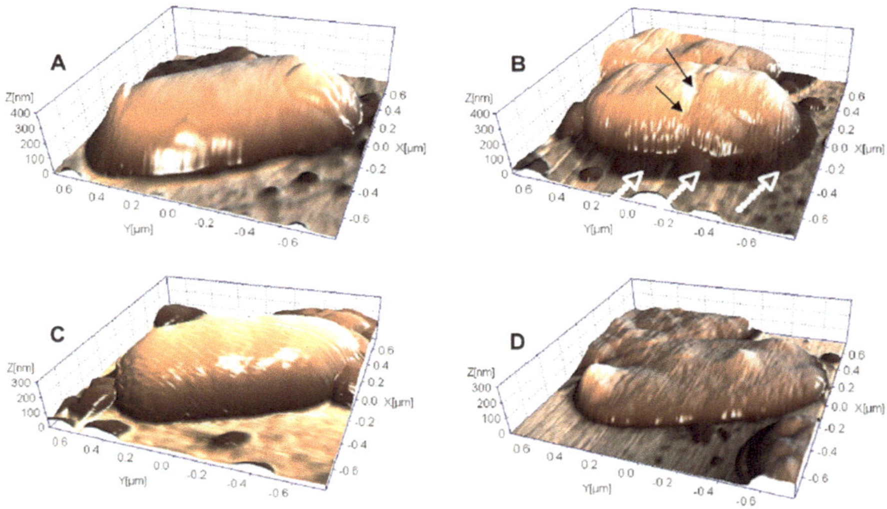

Fig. 5.9 Three-dimensional rendering of a representative Gram-positive and Gram-negative bacteria; *C. maltaromaticum* 9P (**a**, control cell; **b**, carvacrol-treated cell) and *P. fragi* 6P2 (**c**, control cell; **d**, carvacrol-treated cell) (La Storia et al. 2011)

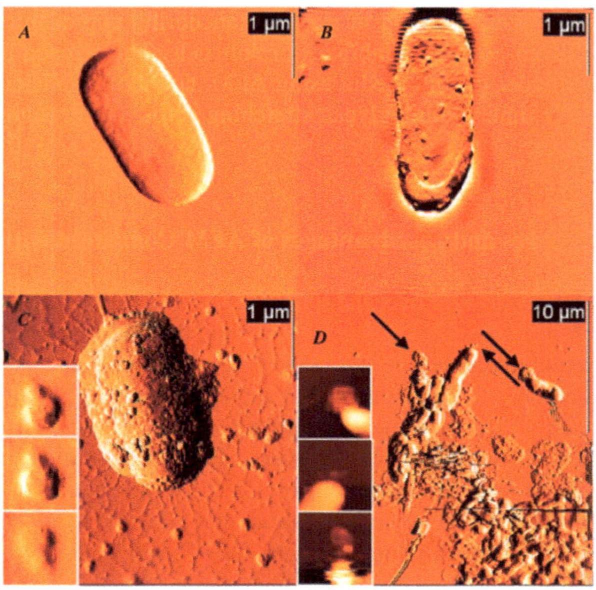

Fig. 5.10 AFM images of *E. coli* cells during its whole lytic cycles caused by phage infection. (**a**) Before mixing with phages, (**b–d**) incubated with bacteriophages A157 at 37 °C (**b**) for 5 min, (**c**) for 30 min (*insets* demonstrate zoomed bacteriophages), and (**d**) for 60 min. *Arrows* show the regions of phage release. *Insets* in panel **d** present zoomed regions of phage release, indicated by the *arrows*. *Insets* in panel **d** are height images (height range is 300 nm); others are deflection images (Dubrovin et al. 2008)

induced by *Pseudomonas* NCIMB 2021, which was mainly attributed to the enhanced corrosion by biogenic sulfide ions. Zhao et al. (2009) investigated the morphology of a strain of marine bacteria, *Shewanella* sp., on different self-assembled monolayers. Interesting fingerprints were observed on different surfaces. Evidence of morphological changes and recessed spots surrounding the bacteria were observed on hexadecanethiol, a hydrophobic compound monolayer, whereas these phenomena did not occur on more hydrophilic monolayers.

One of the most attractive advantages of AFM over other nanoscale microscopy is its capability of monitoring live cells in real time. Typically, a liquid cell is employed to keep live microbes in buffer solutions. This provides the opportunity for direct in vitro investigation of microbial processes. Hao et al. (2009) used AFM to carry out an in situ observation of *Acidiphilium acidophilum* on a pyrite surface in solution during an investigation of phospholipid effects on pyrite oxidation. A syringe pump was integrated with an AFM liquid cell to flow solutions across samples during measurements. Displacement of microcolonies was observed after phospholipid addition, whereas individually bound bacteria showed little displacement. The results revealed that the majority of the bacteria that are displaced from the pyrite surface were initially bound in microcolonies. To image dynamic cellular processes, Kailas et al. (2009) mechanically trapped coccoid cells of the bacterial species *S. aureus* in a lithographically patterned silicon wafer which consists of a square lattice of holes of ~450 nm depth and diameter ranging from 1 to 1.8 mm. Confinement effects are reduced in this new anchoring approach compared with trapping the cells in porous membranes or soft gels. The trapped *S. aureus* cells were monitored under growth media using AFM, and a cell division was successfully imaged over time. Peptidoglycan stretching across the septum was observed during the division.

5.4.1.4 Advantages and Disadvantages of AFM Compared with SEM

AFM has several advantages over the scanning electron microscope (SEM). Unlike the SEM which provides a 2D image of a sample, the AFM provides a 3D surface profile. Additionally, samples viewed by AFM do not require any special treatments (such as metal/carbon coatings) that would damage the sample. While a SEM needs an expensive vacuum environment for proper operation, most AFM modes can work perfectly well in ambient air or even a liquid environment. This property makes it possible to study biological macromolecules and even living organisms. In principle, AFM can provide higher resolution than SEM. It has been shown to give true atomic resolution in ultrahigh vacuum (UHV) and, more recently, in liquid environments.

A disadvantage of AFM compared with the scanning electron microscope (SEM) is the single scan image size. In one pass, the SEM can image an area on the order of square millimeters with a depth of field on the order of millimeters. Whereas the AFM can only image a maximum height on the order of 10–20 μm and a maximum scanning area of about 150 × 150 μm. One method of improving the scanned area size for AFM is by using parallel probes in a fashion similar to that of millipede data storage. The scanning speed of an AFM is also a limitation. Traditionally, an AFM cannot scan images as fast as SEM, requiring several minutes for a typical scan, while a SEM is capable of scanning at near real time, although at relatively low quality. The relatively slow rate of scanning during AFM imaging often leads to thermal drift in the image making the AFM microscope less suited for measuring accurate distances between topographical features on the image. However, several fast-acting designs were suggested to increase microscope scanning productivity including what is being termed videoAFM (reasonable quality images are being obtained with videoAFM at video rate: faster than the average SEM). To eliminate image distortions induced by thermal drift, several methods have been introduced.

AFM images can also be affected by hysteresis of the piezoelectric material and cross talk between the x, y, and z axes that may require software enhancement and filtering. Such filtering could "flatten" out real topographical features. However, newer AFMs utilize closed-loop scanners which practically eliminate these problems. Some AFMs also use separated orthogonal scanners (as opposed to a single tube) which also serve to eliminate part of the cross-talk problems. As with any other imaging technique, there is the possibility of image artifacts, which could be induced by an unsuitable tip, a poor operating environment, or even by the sample itself. These image artifacts are unavoidable; however, their occurrence and effect on the results can be reduced through various methods. Due to the nature of AFM probes, they cannot normally measure steep walls or overhangs. Specially made cantilevers and AFMs can be used to modulate the probe sideways as well as up and down (as with dynamic contact and noncontact modes) to measure sidewalls, at the cost of more expensive cantilevers, lower lateral resolution and additional artifacts.

5.4.2 Analyzing Soil Microbial Communities

Microorganisms are invisible to the naked eye, and their diversity (taxonomic and functional) in soil is stunning. To finding negative environmental impacts and better soil management, there is a need for more detailed and predictive understanding of the microbial communities. However, all researches of soil microorganisms are hampered by the heterogeneity of the soil composition, the vast numbers (c. 10^9 individual cells) and diversity ($>10^6$ distinct taxa) of microorganisms present in each gram of soil, and the paucity of knowledge concerning the majority of the microbiota (Schmidt and Waldron 2015; Hirsch et al. 2010). The last 25 years have brought many new techniques in community profiling and cultivation-independent methods to studying soil microflora. Many investigators argue that it is now possible to explore the "black box" of soil microorganisms. Traditional soil microbiological methods involve selectively cultivating organisms using specific media and conditions in the laboratory. Isolated microbes can then be tested for physiological properties and identified and classified using morphological, physiological, biochemical, and genetic assessments. These approaches, however, have allowed only a glimpse of microbial diversity because only an estimated 1 % of soil bacterial populations are cultivable using current methods. There is uncertainty about the number of soil fungi that exist, but it is estimated that 17 % of known fungal species can be cultivated. Moreover, it appears that these cultivable fractions are perhaps not representative of the total community. Knowledge of the diversity of microorganisms is thus biased toward the small percentage of organisms that can be grown in the laboratory. Currently developed molecular tools are now being used to estimate the diversity of soil microflora, compare the composition of communities, and begin to detect differences in the functioning of communities. This new knowledge can apply to increase our understanding of disturbances on soil communities and consequently ecosystem processes (Hirsch et al. 2010; Leckie 2005). Briefly, methods to measure microbial diversity in soil can be categorized into two groups: (a) conventional- and biochemical-based techniques and (b) molecular-based techniques (Fakruddin and Mannan 2013; Kirk et al. 2004).

5.4.2.1 Conventional and Biochemical Methods

Most methods described below can be used for either bacteria or fungi, although some are specific to one or the other (Table 5.2).

Table 5.2 Advantages and disadvantages of conventional and biochemical-based methods to study soil microbial diversity (Fakruddin and Mannan 2013; Kirk et al. 2004)

Method	Advantages	Disadvantages
Plate counts	Fast Inexpensive	Unculturable microorganisms not detected Bias toward fast-growing individuals Bias toward fungal species that produce large quantities of spores
CLPP/SCSU	Fast Highly reproducible Relatively inexpensive Differentiate between microbial communities Generates large amount of data Option of using bacterial, fungal plates or site-specific carbon sources (BIOLOG)	Only represents culturable fraction of community Favors fast-growing organisms Only represents those organisms capable of utilizing available carbon sources Potential metabolic diversity, not in situ diversity Sensitive to inoculum density
PLFA/FAME	No culturing of microorganisms, direct extraction from soil Follow specific organisms or communities	If using fungal spores, a lot of material is needed Can be influenced by external factors Possibility results can be confounded by other microorganisms

Plate Counts

Standard culture techniques to characterize microbial ecology involve isolation and characterization of microorganisms using commercial growth media such as nutrient agar, potato dextrose agar, Luria–Bertani medium, and tryptic soy agar. These plate count methods are fast and inexpensive and can directly provide information on the active, heterotrophic component of the population. Several improved cultivation procedures and culture media have been devised that mimic natural environments in terms of nutrients (composition and concentration), oxygen gradient, pH, etc. to maximize the cultivable fraction of microbial communities. For example, a technique has been devised for the cultivation of uncultured microorganisms from different environments including seawater and soil that involved encapsulation of cells in gel microdroplets for large-scale microbial cultivation under low nutrient flux conditions. Nonetheless, not all "uncultured" organisms are cultivable, and many of them remain "unculturable." These organisms, although viable in their natural environments, do not grow under laboratory conditions and remain in a "viable but nonculturable" (VBNC) stage (Oliver 2005; Kirk et al. 2004).

CLPP/SCSU

One of the more widely used culture-dependent methods for analyzing soil microbial communities has been that of community-level physiological profiles (CLPP) [also known as sole-carbon-source utilization (SCSU)]. CLPP method is a means of studying the physiological diversity present in soils. These profiles reflect how the microbial communities could potentially utilize a range of carbon substrates. Differences in utilization patterns are interpreted as differences in the major active members of the microbial community. For example, in the BIOLOG® system, 95 different carbon sources are used to produce a metabolic profile of microorganisms. As it is simple, using an automated measuring apparatus and yielding a great deal of information about important functional attributes of microbial communities, the technique has become popular; nonetheless, the analysis and interpretation of such data are often complicated. There are also other drawbacks. The BIOLOG systems only assess the metabolic diversity of "culturable" bacteria, and soil fungi and slow-growing microorganisms have minimal influence on the microbial metabolic profile. Additionally, the BIOLOG sole C-source test plates contain high concentrations of carbon sources and triphenyltetrazolium chloride (TTC). Moreover, the plates are buffered at nearly neutral pH, which is quite different from the pH of some acid or alkaline soils, leading to some limitations for partial microorganisms that have adapted well to acidic or alkaline soils. Many of these factors have presented disadvantages when determining soil microbial community structure, and therefore the usefulness of the information can be questioned (Bing-Ru et al. 2006).

PLFA/FAME Analysis

Phospholipid fatty acid (PLFA) analysis has been used as a culture-independent method of assessing the structure of soil microbial communities and determining gross changes that accompany soil disturbances such as cropping practices, pollution, fumigation, and changes in soil quality. Phospholipid fatty acids are potentially useful signature molecules due to their presence in all living cells. In microorganisms, phospholipids are found exclusively in cell membranes and not in other parts of the cell as storage products. This is important because cell membranes are rapidly degraded and the components of phospholipid fatty acids are rapidly metabolized following cell death. Consequently, phospholipids can serve as important indicators of active microbial biomass as opposed to nonliving microbial biomass. An essential consideration in the use of these molecules to describe microbial communities is that unique fatty acids are indicative of specific groups of organisms. Our knowledge of such signature molecules comes from the use of fatty acid analysis for bacterial taxonomy, in which specific fatty acid methyl esters (FAMEs) have been used as an accepted taxonomic discriminator for species identification. Furthermore, phospholipid fatty acids are easily extracted from microbial cells in soil allowing access to a greater proportion of the microbial community resident in soil than would otherwise be accessed during culture-dependent methods of analysis. The presence and abundance of these signature fatty acids in soil reveals the presence and abundance of particular organisms or groups of organisms in which those signatures can be found. Despite the usefulness of this method, there are some important limitations. First, appropriate signature molecules are not known for all organisms in a soil sample, and, in a number of cases, a specific fatty acid present in a soil sample cannot be linked with a specific microorganisms or group of microorganisms. In general, the method cannot be used to characterize microorganisms to species. Second, since the method relies heavily on signature fatty acids to determine gross community structure, any variation in these signatures would give rise to false community estimates created by artifacts in the methods. Third, bacteria and fungi produce widely different amounts of PLFA, and the types of fatty acids vary with growth conditions and environmental stresses. Although signature PLFAs can be correlated with the presence of some groups of organisms, they may not necessarily be unique to only those groups under all conditions. Consequently, this could give rise to false community signatures (Fakruddin and Mannan 2013; Hill et al. 2000).

5.4.2.2 Molecular-Based Techniques

Due to the nonculturability of the major fraction of natural microbial communities, the overall structure of the community has been difficult to interpret. Therefore, the primary source of information for these uncultured but viable organisms is molecules such as nucleic acids. DNA is one of the most important molecules that many researchers used to analyze microbial diversity.

Over the last few decades, a wide variety of molecular techniques have been developed for describing diversity of microorganisms (Table 5.3). Some of these approaches provide only partial community analysis. These techniques generally include polymerase chain reaction (PCR)-based methods where total DNA/RNA extracted from an environmental sample is used as a template for the characterization of microorganisms. In principle, the PCR product thus generated reflects a mixture of microbial gene signatures from all organisms present in a sample, including the VBNC fraction. PCR amplification of conserved genes such as 16S rRNA from an environmental sample has been used extensively in microbial ecology primarily because these genes (1) are ubiquitous, i.e., present in all prokaryotes, (2) are structurally and functionally conserved, and (3) contain variable and highly conserved regions. Other conserved genes such as RNA polymerase beta subunit (*rpo*B), gyrase beta subunit (*gyr*B), recombinase A (*rec*A), and heat-shock protein (*hsp*60) have also been used in microbial investigations and to differentiate some bacterial species. In these approaches, the PCR products amplified from environmental DNA are analyzed primarily by genetic fingerprinting (DGGE/TGGE, SSCP, ARDRA, T-RFLP, and RISA/ARISA), Q-PCR, FISH, DNA microarrays, and other or by a combination of these techniques. There are other techniques providing whole community analysis of microorganisms. These strategies offer a more comprehensive view of genetic diversity compared to PCR-based molecular approaches that target only a single or few genes. These techniques attempt to analyze all the genetic data present in total DNA extracted from an environmental sample. One of the famous approaches in this field is metagenomics (Thomas et al. 2012; Rastogi and Sani 2011).

Denaturing Gradient Gel Electrophoresis/Temperature Gradient Gel Electrophoresis

Temperature gradient gel electrophoresis (TGGE) are two similar methods for studying microbial diversity. These techniques were originally developed to detect point mutations in DNA sequences. Muyzer et al. (1993) expanded the use of DGGE to study microbial genetic diversity. DNA is extracted from soil samples and amplified using PCR with universal primers targeting part of the 16S or 18S rRNA sequences. The 5′ end of the forward primer contains a 35–40 base-pair GC clamp to ensure that at least part of the DNA remains double stranded. This is necessary so that the separation on a polyacrylamide gel with a gradient of increasing concentration of denaturants (formamide and urea) will occur based on melting behavior of the double-stranded DNA. Theoretically, DGGE can separate DNA with one base-pair difference. TGGE uses the same principle as DGGE except the gradient is temperature rather than chemical denaturants. It is estimated that DGGE can only detect 1–2 % of the microbial population representing dominant species present in an environmental sample. In addition, DNA fragments of different sequences may have similar mobility characteristics in the polyacrylamide gel. Therefore, one band may not necessarily represent one species, and one

Table 5.3 Advantages and disadvantages of some molecular-based methods to study soil microbial diversity (Fakruddin and Mannan 2013; Kirk et al. 2004)

Method	Advantages	Disadvantages
Guanine plus cytosine (G+C)	Not influenced by PCR biases Includes all DNA extracted Quantitative Includes rare members of community	Requires large quantities of DNA Dependent on lysing and extraction efficiency Coarse level of resolution
Nucleic acid reassociation and hybridization	Total DNA extracted Not influenced by PCR biases Can study DNA or RNA Can be studied in situ	Lack of sensitivity Sequences need to be in high copy number for detection Dependent on lysing and extraction efficiency
DNA microarrays and DNA hybridization	Same as nucleic acid hybridization Thousands of genes can be analyzed If using genes or DNA fragments, increased specificity	Only detect the most abundant species Need to culture organisms Only accurate in low-diversity systems
Denaturing and temperature gradient gel electrophoresis (DGGE and TGGE)	Large number of samples can be analyzed simultaneously Reliable, reproducible, and rapid	PCR biases Dependent on lysing and extraction efficiency Way of sample handling can influence community, i.e., the community can change if stored too long before extraction One band can represent more than one species (co-migration) Only detects dominant species
Single-strand conformation polymorphism (SSCP)	Same as DGGE/TGGE No GC clamp No gradient	PCR biases Some ssDNA can form more than one stable conformation
Amplified ribosomal DNA restriction analysis (ARDRA) or restriction fragment length polymorphism (RFLP)	Detect structural changes in microbial community	PCR biases Banding patterns often too complex
Terminal restriction fragment length polymorphism (T-RFLP)	Simpler banding patterns than RFLP Can be automated large number of samples Highly reproducible Ability to compare differences between microbial communities	PCR biases Dependent on extraction and lysing efficiency Type of *Taq* can increase variability Choice of universal primers Choice of restriction enzymes will influence community fingerprint

(continued)

Table 5.3 (continued)

Method	Advantages	Disadvantages
Ribosomal intergenic spacer analysis (RISA)/automated ribosomal intergenic spacer analysis (ARISA)	Highly reproducible community profiles	PCR biases Requires large quantities of DNA (for RISA)

bacterial species may also give rise to multiple bands because of multiple 16S rRNA genes with slightly different sequences. Specific DGGE/TGGE bands can also be excised from gels, reamplified and sequenced or transferred to membranes, and hybridized with specific primers to provide more structural or functional diversity information. By sequencing bands, one can obtain information about the specific microorganisms in the community (Leckie 2005; Kirk et al. 2004).

Single-Strand Conformation Polymorphism

Single-strand conformation polymorphism (SSCP) also relies on electrophoretic separation based on differences in DNA sequences and allows differentiation of DNA molecules having the same length but different nucleotide sequences. This technique was originally developed to detect known or novel polymorphisms or point mutations in DNA. In this method, single-stranded DNA separation on polyacrylamide gel was based on differences in mobility resulting from their folded secondary structure (heteroduplex). As formation of folded secondary structure or heteroduplex and hence mobility is dependent on the DNA sequences, this method reproduces an insight of the genetic diversity in a microbial community. All the limitations of DGGE are also equally applicable for SSCP. Again, some single-stranded DNA can exist in more than one stable conformation. As a result, same DNA sequence can produce multiple bands on the gel. However, it does not require a GC clamp or the construction of gradient gels; therefore, it is a more simple and straightforward technique than DGGE. Similar to DGGE, the DNA bands can be excised from the gel, reamplified, and sequenced. It has been used to study bacterial or fungal community diversity (Fakruddin and Mannan 2013).

Amplified Ribosomal DNA Restriction Analysis

This method is based on DNA sequence variations present in PCR-amplified 16S rRNA genes. The PCR product amplified from environmental DNA is generally digested with four-base-pair-cutting restriction endonucleases, and different fragment lengths are detected using agarose or non-denaturing polyacrylamide gel electrophoresis in the case of community analysis. Although amplified ribosomal DNA restriction analysis (ARDRA) provides little or no information about the type of microorganisms present in the sample, the method is still useful for rapid monitoring of microbial communities over time or to compare microbial diversity in response to changing environmental conditions. One of the major limitations of ARDRA is that restriction profiles generated from complex microbial communities are sometimes too difficult to resolve by agarose/PAGE. The ARDRA technique was applied for assessing the effect of copper contamination on the microbial communities in soil. Whole community ARDRA profiles showed a lower diversity in copper-contaminated soil compared with control soil with no contamination (Rastogi and Sani 2011).

Terminal Restriction Fragment Length Polymorphism

T-RFLP is a technique that addresses some of the limitations of RFLP (Thies 2007). This technique is an extension of the RFLP/ARDRA analysis and provides an alternate method for rapid analysis of microbial community diversity in various environments. It follows the same principle as RFLP except that one PCR primer is labeled (5′ end) with a fluorescent dye, such as TET (4,7,2′,7′-tetrachloro-6-carboxyfluorescein) or 6-FAM (phosphoramidite fluorochrome 5-carboxyfluorescein). PCR is performed on sample DNA using universal rDNA primers, one of which is fluorescently labeled. Fluorescently labeled terminal restriction fragment length polymorphism (FLT-RFLP) patterns can then be created by digestion of labeled amplicons using restriction enzymes. Fragments are then separated by gel electrophoresis using an automated sequence analyzer. Each unique fragment length can be counted as an operational taxonomic unit (OTU), and the frequency of each OTU can be calculated. The banding pattern can be used to measure species richness and evenness as well as similarities between samples. T-RFLP, like any PCR-based method, may underestimate true diversity because only numerically dominant species are detected due to the large quantity of available template DNA. Incomplete digestion by restriction enzymes could also lead to an overestimation of diversity. Despite these limitations, some researchers are of the opinion that once standardized T-RFLP can be a useful tool to study microbial diversity in the environment, while others feel that it is inadequate (Fakruddin and Mannan 2013). T-RFLP is limited not only by DNA extraction and PCR biases but also by the choice of universal primers. None of the presently available universal primers can amplify all sequences from eukaryote, bacterial, and archaeal domains. Additionally, these primers are based on existing 16S rRNA, 18S rRNA, or internal transcribed spacer (ITS) databases, which until recently contained mainly sequences from "culturable" microorganisms and therefore may not be representative of the true microbial diversity in a sample (Rudi et al. 2007). In addition, different enzymes will produce different community fingerprints (Dunbar et al. 2000). Tiedje et al. (1999) reported five times greater success at detecting and tracking specific ribotypes using T-RFLP than DGGE.

Ribosomal Intergenic Spacer Analysis/Automated Ribosomal Intergenic Spacer Analysis

In ribosomal intergenic spacer analysis (RISA) and automated ribosomal intergenic spacer analysis (ARISA), the intergenic spacer (IGS) region [intergenic spacer region (ISR)] between the small and large ribosomal subunits is amplified by PCR, denatured and separated on a polyacrylamide gel under denaturing conditions. This region may encode tRNAs and is useful for differentiating between bacterial strains and closely related species because of heterogeneity of the IGS length and sequence. Sequence polymorphisms are detected by silver staining in RISA. In ARISA, fluorescently labeled forward primer is detected automatically. Both methods can deduce highly reproducible microbial community profiles. Limitations of RISA include requirement of large quantities of DNA, relatively longer time requirement, insensitivity of silver staining in some cases, and low resolution. ARISA has increased sensitivity than RISA and is less time-consuming, but traditional limitation of PCR also applies for ARISA (Bing-Ru et al. 2006; Fisher and Triplett 1999).

Quantitative PCR

Q-PCR or real-time PCR has been used in microbial investigations to measure the abundance and expression of taxonomic and functional genemarkers (Smith and Osborn 2009). Unlike traditional PCR, which relies on end-point detection of amplified genes, Q-PCR uses either intercalating fluorescent dyes such as SYBR Green or fluorescent probes (TaqMan) to measure the accumulation of amplicons in real time during each cycle of the PCR. Software records the increase in amplicon concentration during the early exponential phase of amplification which enables the quantification of genes (or transcripts) when they are proportional to the starting template concentration. When Q-PCR is coupled with a preceding reverse transcription (RT) reaction, it can be used to quantify gene expression (RT-Q-PCR). Q-PCR is highly sensitive to starting template concentration and measures template abundance in a large dynamic range of around six orders of magnitude. Several sets of 16S and 5.8S rRNA gene primers have been designed for rapid Q-PCR-based quantification of soil bacterial and fungal microbial communities (Fierer et al. 2005). Q-PCR has also been successfully used in environmental samples for quantitative detection of important physiological groups of bacteria such as ammonia oxidizers, methane oxidizers, and sulfate reducers by targeting *amoA*, *pmoA*, and *dsrA* genes, respectively (Foti et al. 2007).

Fluorescent In Situ Hybridization

FISH allows the direct identification and quantification of specific and/or general taxonomic groups of microorganisms within their natural microhabitat by whole cell hybridization with oligonucleotide probes. After that, the labeled cells are viewed by scanning confocal laser microscopy (SCLM). SCLM surpasses epifluorescence microscopy in sensitivity and has the ability to view the distribution of several taxonomic groups simultaneously as a three-dimensional image. Because whole cells are hybridized, artifacts arising from biases in DNA extraction, PCR amplification, and cloning are avoided. FISH provides a more accurate quantification of cells as compared to the rough estimates obtained from dot blot assays in which microbial DNA is blotted onto a membrane than probed with the fluorescent oligonucleotide probe (Hill et al. 2000). A large number of molecular probes targeting rRNA genes have been reported at various taxonomic levels (Amann et al. 1995). The FISH probes are generally 18–30 nucleotides long and contain a fluorescent dye at the 5′ end that allows detection of probe bound to cellular rRNA. In addition, the intensity of fluorescent signals is correlated to cellular rRNA contents and growth rates, which provide insight into the metabolic state of the cells. FISH can be combined with flow cytometry for a high-resolution automated analysis of mixed microbial populations. FISH can be used to visualize soil microorganisms that have not yet been cultured and is useful in studying the ecological distribution of microorganisms throughout diverse habitats. When using FISH to examine all members within a given taxon, one must keep in mind that the probe being used is only as good as the representative members that were used to generate it. Other, noncultured organisms may not be detected with this probe, or cross hybridization to related organisms may occur (Hill et al. 2000). However, with respect to sensitivity, some limitations to the standard FISH method that prevents detection of cells with low ribosome content have been noted. Low physiological activity was often correlated with low ribosome content per cell; therefore, slow-growing or starving cells may not be detected (Amann et al. 1995). To overcome this limitation, FISH has adopted a tyramine signal amplification technique, which allowed the analysis of slow-growing microorganisms. Also, a disadvantage of nucleic acid hybridization or FISH is the lack of sensitivity unless sequences are present in high copy number (Bing-Ru et al. 2006).

DNA Microarrays

Microarray technology has been used to identify thousands of microbial species in natural environments using specific probes. This technique is used to provide a high-throughput and comprehensive view of microbial communities in environmental samples. The PCR products amplified from total environmental DNA are directly hybridized to known molecular probes, which are attached on the microarrays (Gentry et al. 2006). After the fluorescently labeled PCR amplicons are hybridized to the probes, positive signals are scored by the use of confocal laser scanning microscopy. The microarray technique allows samples to be rapidly evaluated with replication, which is a significant advantage in microbial community analyses. In general, the hybridization signal intensity on microarrays is directly proportional to the abundance of the target organism. Cross hybridization is a major limitation of microarray technology, particularly when dealing with environmental samples. In addition, the microarray is not useful in identifying and detecting novel prokaryotic taxa. DNA microarrays used in microbial ecology could be classified into two major categories depending on the probes: (a) 16S rRNA gene microarrays and (b) functional gene arrays (FGA).

Metagenomics

Each organism in an environment has a unique set of genes in its genome; the combined genomes of all the community members make up the "metagenome." Metagenomics is defined as the direct genetic analysis of genomes contained with an environmental sample. It is also known by other names such as population genomics, environmental genomics, community genomics, and microbial ecogenomics. Arguably, one of the most remarkable events in the field of microbial ecology in the past decade has been the advent and development of metagenomics. The field initially started with the cloning of environmental DNA, followed by "functional expression screening," and was then quickly complemented by direct random shotgun "sequencing" of environmental DNA (Thomas et al. 2012; Riesenfeld et al. 2004).

Over the past decade, metagenomic shotgun sequencing has gradually shifted from classical Sanger sequencing technology to next-generation sequencing (NGS). Sanger sequencing, however, is still considered the gold standard for sequencing, because of its low error rate, long read length (>700 bp), and large insert sizes (e.g., >30 Kb for fosmids or bacterial artificial chromosomes (BACs)). All of these aspects will improve assembly outcomes for shotgun data, and hence Sanger sequencing might still be applicable if generating close-to-complete genomes in low-diversity environments is the objective. A drawback of Sanger sequencing is the labor-intensive cloning process in its associated bias against genes toxic for the cloning host and the overall cost per gigabase (approximately, US$400,000). The rapid and substantial cost reduction in next-generation sequencing has dramatically accelerated the development of sequence-based metagenomics. In fact, the number of metagenome shotgun sequence datasets has exploded in the past few years. In the future, metagenomics will be used in the same manner as 16S rRNA gene fingerprinting methods to describe microbial community profiles. It will therefore become a standard tool for many laboratories and scientists working in the field of microbial ecology (Thomas et al. 2012).

Extraction of Nucleic Acids from Soil for Metagenomics

Soil metagenomic analyses are usually initiated by the isolation of environmental DNAs. Obtaining large quantities of pure DNA from the environment is a prerequisite for metagenomics. Microbial nucleic acid extraction from soil is complicated by the presence of humic substances, organic matter, and clay particles that are able to bind to nucleic acids and inhibit their purification. Moreover, humic and fulvic acids are also capable of inhibiting post-extraction downstream enzymatic analyses; thus an important step is the removal of organic contaminants that co-extract with nucleic acids. An additional consideration for RNA extractions is the ubiquitous RNases present in soil which must be deactivated rapidly to prevent degradation of the sample. Usually, RNA extractions involve an additional step to inactivate RNase before extraction of nucleic acids. Despite these difficulties, numerous techniques have been developed for DNA and RNA extraction from soils, many of which are now available commercially. Protocols can be divided into indirect and direct nucleic acid extractions. The first step of an indirect extraction is to extract the microbial cells from the soil matrix before cell lysis. This type of extraction has been criticized for showing bias, for example, methane-oxidizing and ammonia-oxidizing bacteria have been shown to be more difficult to dislodge from soil particles compared to the majority of other soil bacteria, and actinomycete spores may be underrepresented. On the other hand, indirect extraction enabled isolation of high molecular weight DNA for metagenomic library construction and revealed a different community structure, in comparison to direct extraction based on in situ lysis of microbes in soil (Hirsch et al. 2010).

The method of microbial cell lysis is an important step for either approach, and this choice depends on the type of biological structure being primarily extracted from spores or vegetative material, the size of DNA fragments required for downstream applications (metagenomic library construction requires larger fragments than, e.g., 16S ribotyping), and the properties of the soil. Soils with high organic matter can pose particular problems due to their high humic acid content. Similarly, soils with a high clay content present more problems when extracting nucleic acids than sandy soils due to their higher organic matter content and their hygroscopic propensity to bind water. Aggressive methods such as bead beating of soil samples provide effective disruption of cell walls and membranes in situ, and increasing vigor is needed as the soil humic acid and clay content increases notwithstanding a consequential decrease in the average DNA fragment size. Other methods of cell lysis include sonication, grinding–freezing–thawing, solubilization of cell walls and membranes by detergents, or degradation of these structures by boiling and/or enzymatic means. It is important to recognize that biological structures differ in their susceptibility to aggressive disruption methods such as bead beating, and so the protocol adopted for a given sample may depend on the target community. For example, extraction of DNA from spores may require vigorous disruption, but this will result in highly fragmented DNA from easily lysed cells. A protocol optimized to extract genomic DNA from the majority microbial community will be biased against both tougher and more fragile propagules: in order to maximize the diversity represented by genomic DNA extracted from soil, a variety of methods may be necessary. Once cells have been lysed, a number of methods may be used to extract and purify the nucleic acids, several using commercial kits (Thakuria et al. 2009; Saleh-Lakha et al. 2005; Robe et al. 2003; Roose-Amsaleg et al. 2001; Cullen and Hirsch 1998).

DNA extracted from soil represents the total metagenome, including components that are no longer viable, whereas RNA is synthesized only by actively growing cells, and it degrades relatively rapidly once produced: it arises from, and can thus identify, the functioning members of soil microbial communities. In prokaryotes, messenger RNA (mRNA) is usually very short-lived and indicates which genes are active at the time of extraction, but ribosomal RNA (rRNA) is more stable as it possesses secondary structure and is associated with ribosomal proteins so in theory; it could survive for months in moribund or dead cells in soil. However, in cells that are, or have recently been active, there are many thousands of molecules of rRNA. Thus, analysis of rRNA abundance and diversity has been used to indicate the dominant active population in soil, despite recognition that the number of ribosome varies between groups. More precise information relevant to particular functions can be obtained from mRNA, but it presents more technical difficulties during extraction from soil; nevertheless it opens exciting possibilities for future investigation. Both mRNA and rRNA can be converted to DNA using the enzyme reverse transcriptase (RT), but can also be hybridized to microarrays directly if sufficient material is obtained (Hirsch et al. 2010; McGrath et al. 2008; Janssen 2006).

DNA Preparation Techniques for Metagenomics

Different sequencing methods have various protocols, but the initial DNA sample preparation generally includes three similar steps: (a) the DNA molecules are fragmented to produce pieces that are small enough to sequence, (b) the fragments are given blunt ends to aid further processing, and (c) adaptors are ligated to the fragments. Fragmentation can be mechanical or enzymatic, and the former can be further classified into nebulization, hydrodynamic shearing, and ultrasonication. Mechanical fragmentation techniques yield more random and readily controlled overlapping DNA fragments compared to enzymatic approaches. In this way, fragmented DNA strands need repair and end polishing before adaptors can be ligated. Enzymatic fragmentation, a more currently developed method, can retrieve random fragments of the desired length with less DNA input. For example, the "dsDNA Fragmentase" from New England Biolabs (NEB) creates random double-stranded breaks in DNA with its mutant *Vibrio vulnificus* nuclease and mutant T7 endonuclease and acts at a constant rate, allowing the size of the fragments to be controlled in a time-dependent manner (Di Bella et al. 2013). After fragmentation step, adaptor sequences are ligated on the ends of the fragments, allowing them to be attached to a solid surface for sequencing (e.g., a tagged glass slide or bead). These adaptors are specific to the sequencing platform and allow amplification by acting as primers. In a new method, "tagmentation" (NextEra™) employs in vitro transposition with a specialized "Transposome™" enzyme that fragments DNA strands, repairs ends, and attaches sequence tags in the same step. The adaptors added can be platform-specific, making it an appropriate approach for preparing samples for both "Roche 454" and "Illumina" platforms (Caruccio 2011).

Next-Generation Sequencing

The first-generation DNA sequencing technology was developed by Sanger et al. (1977) based on the selective incorporation of chain-terminating dideoxynucleotides, and the first automatic sequencing machine (AB370) was produced by Applied Biosystems in 1987 (Liu et al. 2012). The Sanger sequencing technique completed the first bacterial genome sequencing in 1995 (Fleischmann et al. 1995) and constituted the main part of the Human Genome Project in 2001 (Collins et al. 2003), which in turn promoted further development of sequencing technology. With the launch of the Genome Sequencer 20 system by 454 Life Sciences in 2005, second-generation sequencing techniques came into recognition with massively parallel analysis, high-throughput, and reduced cost. This significant advance greatly reduced the difficulty of sequencing (Metzker 2005) and made it possible to analyze bacterial genomes in hours or days rather than months or years (Loman et al. 2012). While most second-generation sequencing techniques rely on a sequencing-by-synthesis design, the newly emerged third-generation sequencing techniques are performed on a single-molecule basis with no initial DNA amplification step. In this part we will discuss the major systems of second- and third-generation sequencing. The features of each next-generation sequencing device are summarized in Table 5.4. While high-throughput sequencing is increasingly

Table 5.4 Comparison of next-generation sequencing techniques (Di Bella et al. 2013)

Sequencing devices	Chemistry	Read length (bp)	Run time	Throughput per run	Reads per run
High-end instruments					
454 GS FLX +(Roche)	Pyrosequencing	700	23 h	700 Mb	~1,000,000 shotgun ~700,000 amplicon
HiSeq 2000/ 2500 (Illumina)	Reversible terminator	2 × 150	High output: ~11 days Rapid run: ~27 h	High output: 600 Gb Rapid run: 120 Gb	High output: 3 billion × 2 Rapid run: 600 million × 2
5500xl W SOLiD (Life Technologies)	Ligation	1 × 75 Frag 2 × 50 MP	8 days	~320 Gb	1.4 billion × 2
Bench-top devices					
454 GS Junior (Roche)	Pyrosequencing	400	8 h	35 Mb	100,000 shotgun 70,000 amplicon
Ion PGM (Life Technologies)	Proton detection	100 or 200	3 h	100 Mb (314 chip) 1 Gb (316 chip) 2 Gb (318 chip)	400,000–550,000 (314 chip) 2–3 million (316 chip) 4–5.5 million (318 chip)
MiSeq (Illumina)	Reversible terminator	2 × 250	27 h	8.35 Gb	6.8 million (LRGC routinely getting N15 M)

popular, fast, reliable, and cheap, for some experiments, easier and even less expensive methods such as microarrays may be appropriate. However, the depth and relatively unbiased method of this technique is, in many cases, replacing microarray technologies. There are many different platforms, each with advantages and disadvantages. The appropriate selection depends on the aims of the experiment. In some cases, a combination of techniques could provide more complete coverage of the genome. In order to achieve efficiencies in time and cost, sequencing of microbial samples usually employs barcoding and multiplexing of samples. Understanding the complexity of samples is crucial to determine whether multiplexing is needed and how many samples can be tested in a single run. If using a multiplexing method, the concentration of each sample should be similar so that equal amounts of data are achieved for each multiplexed sample.

The Roche 454 Sequencing Technique

454 Life Sciences developed the GS 20 system in 2005, which introduced the sequencing-by-synthesis approach to DNA sequencing. In 2007, the company released the GS FLX system and its 2008 upgrade, GS FLX Titanium, produced a fivefold increase in throughput. The Roche 454 platform employs pyrosequencing (Liu et al. 2012; Nowrousian 2010). In this method, DNA libraries are fragmented to a size between 400 and 800 base pairs, ligated to adaptors, and denatured into single strands. The single strands are captured by amplification beads and amplified by emulsion PCR (Berka et al. 2010). The beads are then transferred to a picotiter plate, where each dNTP is washed over the plate one at a time, and the release of pyrophosphate when one is incorporated drives a reaction, turning luciferin into oxyluciferin and generating visible light (Froehlich et al. 2010). The light signal recorded when the base is washed over the plate allows the system to determine the exact base being added. The Roche 454 system has a long read length and relatively high speed. The newest GS FLX Titanium XL+ sequencer can reach a length up to 1000 bp, with a throughput of 700 Mb. One major shortcoming is that the cost per base on this platform is about ten times more expensive than Illumina's HiSeq 2000. It also has quite low throughput and automation and relatively high error rates. The use of 454 sequencing is especially problematic in homopolymeric tracts, that is, regions where one nucleotide is repeated many times, because of the way it detects incorporation of nucleotides. In homopolymeric tracts, many bases are incorporated at once, and the intensity corresponding signal is in theory proportional to the number of nucleotides added. However, it is not very precise, so the number of nucleotides detected is not always the number added, resulting in insertion and deletion (indel) mutations at homopolymeric tracts. This quality makes it a less preferable option for 16S rDNA amplicon sequencing (Kunin et al. 2010). However, this technique's long read length is useful for sequencing repetitive or palindromic sequences, as well as scaffolding for metagenomics, since long read lengths are easier to assemble.

The Illumina HiSeq 2000 Platform

Like Roche 454, HiSeq 2000 also uses sequencing by synthesis following amplification of the input DNA. Briefly, the sample DNA is fragmented, and adapters are ligated onto the ends of the DNA fragments. The DNA is selected for size, denatured into single strands, and attached to a flow cell, where bridge PCR occurs to form clusters of identical DNA fragments on the cell. The amplified DNA is made single-stranded (Mardis 2008) and is then sequenced. Nucleotides tagged with fluorescent dyes and cleavable blocking group are washed over the flow cell; if they are incorporated, the cluster fluoresces with the wavelength associated with that particular nucleotide, and the dye and blocking group are cleaved to allow incorporation of the next nucleotide. The Illumina technique has the greatest output and lowest reagent cost: the HiSeq 2000 can reach a throughput of 600 G bases per run, and the cost per million bases is only $0.02 (Liu et al. 2012). It is also more accurate than the Roche 454 sequencing platform, since it is not prone to indels at homopolymeric sites, and the read count is steadily improving (Luo et al. 2012). The highly automated library preparation and concentration measurement also greatly reduces hands-on time. The major disadvantages of this method are its relatively short read length and long run time.

Life Technologies/Applied Biosystems SOLiD System

Sequencing by Oligonucleotide Ligation Detection (SOLiD), acquired by Applied Biosystems in 2006, utilizes a system, whereby DNA libraries are prepared, fragmented, and ligated to a P1 adaptor with known starting sequences. The fragments are then attached to the magnetic beads and go through emulsion PCR. The resulting PCR product-containing beads are then covalently bound to a glass slide, whereby sequencing is performed by ligating di-base probes, which are fluorescently labeled. SOLiD produces short reads, but due to the di-base ligation method, it has a very high accuracy after filtering. The SOLiD 5500xl system, released in 2010, produces 85 bp reads with 99.99 % accuracy (Liu et al. 2012). The newest double-flowchip 5500xl W Genetic Analyzer could reach a throughput as high as 320 Gb per run, with the wildfire upgrade in 2012 further improving its workflow and throughput. The major disadvantages of SOLiD are its short read length, its long run time, and the color space mapping of the resulting DNA sequence.

Bench-Top Sequencing Devices

Although high-end sequencing machines can deliver a high-throughput and long read lengths, they are also bulky, expensive, and usually can only be afforded by large centers. On the other hand, there are also modestly priced, bench-top instruments with throughputs and read lengths decent enough for microbial applications, appearing on the market, and these could be useful for some basic research and clinical applications (Table 5.4).

Third-Generation Sequencing Techniques

While next-generation sequencing techniques are still rapidly evolving, third-generation sequencing techniques have brought interesting innovation to the field. In contrast to most second-generation sequencing techniques, which depend on the production of libraries of clonally amplified templates, third-generation sequencing does not amplify the DNA. These techniques have the potential to be less costly, to be less time-consuming, and to have fewer biases from the amplification step, while also capturing their data in real time (Liu et al. 2012):

The Helicos Heliscope Sequencer System The first single-molecule DNA sequencer, the Heliscope Sequencer System, was launched by Helicos Biosciences in 2008. As with Illumina sequencing, the template DNA is immobilized, and synthesis occurs by reversible chain-terminating nucleotides, which are labeled. This system has previously been used to sequence DNA and RNA. Disadvantages of this platform are its high error rates (>5 %) and short read length (about 32 bases), making its sequences difficult to analyze (Di Bella et al. 2013).

Single-molecule real-time sequencing Single-molecule real-time (SMRT) sequencing developed by Pacific Bioscience uses fluorescent dye-modified nucleotides and a zero-mode waveguide (ZMW). The SMRT cell consists of millions of ZMWs, each containing one DNA template. DNA polymerase is fixed at the bottom of ZMWs with a biotin–streptavidin linkage and forms a complementary strand as in normal DNA replication during the reaction, cleaving off the fluorescent dye previously linked to the terminal phosphate of the nucleotide. This process emits light signals, which is captured by a built-in camera as videos on a real-time basis. This is useful since both the color and intensity are measured, which can give information not only on the sequence but the structure; this technique has thus been used for studying epigenetic base modifications. The utilization of DNA polymerase's natural abilities enables fast cycle time and very long reads. It also has simple sample preparation and low reagent costs. SMRT has had other applications to examine biochemical properties of transcription and translation. The major drawbacks of SMRT include high raw error rates (>10 %), low throughput, and expensive and difficult setup (Di Bella et al. 2013). Recently, Koren et al. (2012) reported a combined Illumina/SMRT algorithm to address these limitations, with a read accuracy over 99.9 %.

Oxford nanopore sequencing Nanopore sequencing enables direct reading of unlabeled DNA by threading it through a nanoscale-sized pore (nanopore). Biologically, nanopores are usually present as protein channels for ion exchange. Any substances that move through the channel can cause changes of the current across the channel due to their different conductivity, which are monitored and recorded. The identity of each deoxyribonucleoside monophosphate (dNMP) can be determined by their size difference. The advantage of nanopore sequencing is that it can reach a read length more than 5 kbp and a speed of 1 bp/ns. Since there is no need for fluorescent modification of the bases, it also reduces cost and potential biases. The mechanism of electrophysiological detection also reduces bias related to enzymatic activities. Like other third-generation techniques, the single molecular sequencing nature of nanopore sequencing greatly decreases the hands-on preparation time including cloning and amplification. Oxford Nanopore Technologies have launched GridION, a commercial sequencing device on an electronic-based platform, and MinION, a disposable portable device for electronic single-molecule sensing. The major drawbacks of nanopore technologies include its relatively low throughput, high translocation velocity, and the lack of nucleotide specificity (Di Bella et al. 2013).

References

Allard P, Darnajoux R, Phalyvong K, Bellenger JP (2013) Effects of tungsten and titanium oxide nanoparticles on the diazotrophic growth and metals acquisition by *Azotobacter vinelandii* under molybdenum limiting condition. Environ Sci Technol 47:2061–2068

Alsteenns D, Verbelen C, Dague E, Raze D, Baulard AR, Dufrene YF (2008) Organization of the mycobacterial cell wall: a nanoscale view. Pflugers Arch Eur J Physiol 456:117–125

Amann RI, Ludwig W, Schleifer KH (1995) Phylogenetic identification and in situ detection of individual microbial cells without cultivation. Microbiol Rev 59:143–169

An YJ, Kim M (2009) Effect of antimony on the microbial growth and the activities of soil enzymes. Chemosphere 74:654–659

Arnaout CL, Gunsch CK (2012) Impacts of silver nanoparticle coating on the nitrification potential of *Nitrosomonas europaea*. Environ Sci Technol 46:5387–5395

Aziz N, Faraz M, Pandey R, Sakir M, Fatma T, Varma A, Barman I, Prasad R (2015) Facile algae-derived route to biogenic silver nanoparticles: synthesis, antibacterial and photocatalytic properties. Langmuir 31:11605–11612

Baek YW, An YJ (2011) Microbial toxicity of metal oxide nanoparticles (CuO, NiO_2, ZnO, and Sb_2O_3) to *Escherichia coli*, *Bacillus subtilis*, and *Streptococcus aureus*. Sci Total Environ 409:1603–1608

Bandyopadhyay S, Peralta-Videa JR (2012) Comparative toxicity assessment of CeO_2 and ZnO nanoparticles towards *Sinorhizobium meliloti*, a symbiotic alfalfa associated bacterium: use of advanced microscopic and spectroscopic techniques. J Hazard Mater 241:379–386

Beckmann MA, Venkataraman S (2006) Measuring cell surface elasticity on enteroaggregative *Escherichia coli* wild type and dispersin mutant by AFM. Ultramicroscopy 106:695–702

Ben-Moshe T, Frenk S, Dror I, Minz D, Berkowitz B (2013) Effects of metal oxide nanoparticles on soil properties. Chemosphere 90:640–646

Berka J, Chen YJ, Leamon JH, Lefkowitz S et al (2010) Bead emulsion nucleic acid amplification. USA Patent 7842457B2

Bing-Ru L, Guo-Mei J, Jian C, Gang W (2006) A review of methods for studying microbial diversity in soils. Pedosphere 16:18–24

Brayner R, Ferrari-Iliou R, Brivois N et al (2006) Toxicological impact studies based on *Escherichia coli* bacteria in ultrafine ZnO nanoparticles colloidal medium. Nano Lett 6:866–870

Cabiscol E, Tamarit J, Ros J (2000) Oxidative stress in bacteria and protein damage by reactive oxygen species. Int Microbiol 3:3–8

Caldwell BA (2005) Enzyme activities as a component of soil biodiversity: a review. Pedobiologia 49:637–644

Caruccio N (2011) Preparation of next-generation sequencing libraries using Nextera technology: simultaneous DNA fragmentation and adaptor tagging by *in vitro* transposition. Methods Mol Biol 733:241–255

Chen MW, Yang ZW, Wu HM, Pan X, Xie XB, Wu CB (2011) Antimicrobial activity and the mechanism of silver nanoparticle thermosensitive gel. Int J Nanomed 6:2873–2877
Cherchi C, Gu A (2010) Impact of titanium dioxide nanomaterials on nitrogen fixation rate and intracellular nitrogen storage in *Anabaena variabilis*. Environ Sci Technol 44:8302–8307
Choi O, Hu Z (2008) Size dependent and reactive oxygen species related nanosilver toxicity to nitrifying bacteria. Environ Sci Technol 42:4583–4588
Choi O, Hu Z (2009) Nitrification inhibition by silver nanoparticles. Water Sci Technol 59:1699–1702
Chung H, Son Y, Yoon TK, Kim S, Kim W (2011) The effect of multi-walled carbon nanotubes on soil microbial activity. Ecotoxicol Environ Saf 74:569–575
Collins FS, Morgan M, Patrinos A (2003) The human genome project: lessons from large-scale biology. Science 300:286–290
Collins D, Luxton T, Kumar N, Shah S, Walker VK, Shah V (2012) Assessing the impact of copper and zinc oxide nanoparticles on soil: a field study. PLoS One 7:e42663
Colman BP, Arnaout CL, Anciaux S et al (2013) Low concentrations of silver nanoparticles in sewage sludge cause adverse ecosystem responses under realistic field scenario. PLoS One 8: e57189
Cooper D, Dumas E, Carlini L, Gao C et al (2010) Interfacial charge transfer between CdTe quantum dots and Gram negative vs. Gram positive bacteria. Environ Sci Technol 44:1464–1470
Cornelis G, Hund-Rinke K, Kuhlbusch T, Brink NVD, Nickle C (2014) Fate and bioavailability of engineered nanoparticles in soils: a review. Crit Rev Environ Sci Technol 44:2720–2764
Cullen DW, Hirsch PR (1998) Simple and rapid method for direct extraction of microbial DNA from soil for PCR. Soil Biol Biochem 30:983–993
Cullen LG, Tilston EL, Mitchell GR, Collins CD, Shaw LJ (2011) Assessing the impact of nano- and micro-scale zerovalent iron particles on soil microbial activities: particle reactivity interferes with assay conditions and interpretation of genuine microbial effects. Chemosphere 82:1675–1682
Dague E (2011) Assembly of live microorganisms on microstructured PDMS stamps by convective/capillary deposition for AFM bioexperiments. Nanotechnology 22:395102
Dague E, Alsteens D, Latge J, Verbelen C, Raze D, Baulard AR, Dufrene YF (2007) Chemical force microscopy of single live cells. Nano Lett 7:3026–3030
Da Silva A, Teschke O (2003) Effect of the antimicrobial peptide PGLa on live *Escherichia coli*. Biochem Biophys Acta Mol Cell Res 1643:95–103
Di Bella JM, Bao Y, Gloor GB, Burton JP, Reid G (2013) High throughput sequencing methods and analysis for microbiome research. J Microbiol Methods 95:401–414
Dimkpa CO (2014) Can nanotechnology deliver the promised benefits without negatively impacting soil microbial life? J Basic Microbiol 54:889–904
Dimkpa CO, Hansen T, Stewart J, McLean JE et al (2015) ZnO nanoparticles and root colonization by a beneficial pseudomonad influence essential metal responses in bean (*Phaseolus vulgaris*). Nanotoxicology 9:271–278
Dorobantu LS, Bhattacharjee S, Foght JM, Gray MR (2008) Atomic force microscopy measurement of heterogeneity in bacterial surface hydrophobicity. Langmuir 24:4944–4951
Dorobantu LS, Bhattacharjee S, Foght JM, Gray MR (2009) Analysis of force interactions between AFM tips and hydrophobic bacteria using DLVO theory. Langmuir 25:6968–6976
Dorobantu LS, Goss GG, Burrell RE (2012) Atomic force microscopy: a nanoscopic view of microbial cell surfaces. Micron 43:1312–1322
Du W, Sun Y, Ji R et al (2011) TiO_2 and ZnO nanoparticles negatively affect wheat growth and soil enzyme activities in agricultural soil. J Environ Monit 13:822–828
Dubchak S, Ogar A, Mietelski JW, Turnau K (2010) Influence of silver and titanium nanoparticles on arbuscular mycorrhiza colonization and accumulation of radiocaesium in *Helianthus annuus*. Span J Agric Res 8:S103–S108

Dubrovin EV, Voloshin AG, Kraevsky SV, Ignatyuk TE, Abramchuk SS et al (2008) Atomic force microscopy investigation of phage infection of bacteria. Langmuir 24:13068–13074

Dufrene YF (2002) Atomic force microscopy, a powerful tool in microbiology. J Bacteriol 184:5205–5213

Dufrene YF (2004) Using nanotechniques to explore microbial surfaces. Nat Rev Microbiol 2:451–460

Dufrene YF (2008) Atomic force microscopy and chemical force microscopy of microbial cells. Nat Protoc 3:1132–1138

Dunbar J et al (2000) Assessment of microbial diversity in four southwestern United States soils by 16S rRNA gene terminal restriction fragment analysis. Appl Environ Microbiol 66:2943–2950

Dupres V, Alsteens D, Andre G, Verbelen C, Dufrene YF (2009) Fishing single molecules on live cells. Nanotoday 4:262–268

Eaton P, Fernandes JC (2008) Atomic force microscopy study of the antibacterial effects of chitosans on *Escherichia coli* and *Staphylococcus aureus*. Ultramicroscopy 108:1128–1134

El Badawy AM, Luxton TP, Silva RG, Scheckel KG, Suidan MT, Tolaymat TM (2010) Impact of environmental conditions (pH, ionic strength, and electrolyte type) on the surface charge and aggregation of silver nanoparticles suspensions. Environ Sci Technol 44:1260–1266

Fabrega J, Fawcett SR, Renshaw JC, Lead JR (2009) Silver nanoparticle impact on bacterial growth: effect of pH, concentration, and organic matter. Environ Sci Technol 43:7285–7290

Fajardo C, Ortíz LT, Rodríguez-Membibre ML, Nande M, Lobo MC, Martin M (2012) Assessing the impact of zero-valent iron (ZVI) nanotechnology on soil microbial structure and functionality: a molecular approach. Chemosphere 86:802–808

Fakruddin MD, Mannan KSB (2013) Methods for analyzing diversity of microbial communities in natural environments. Ceylon J Sci (Biol Sci) 42:19–33

Fan R, Huang YC, Grusak MA, Huang CP, Sherrier DJ (2014) Effects of nano-TiO_2 on the agronomically-relevant Rhizobium-legume symbiosis. Sci Total Environ 466–467:503–512

Fang J, Lyon DY, Wiesner MR, Dong J, Alvarez PJJ (2007) Effect of a fullerene water suspension on bacterial phospholipids and membrane phase behavior. Environ Sci Technol 41:2636–2642

Fang T, Watson JL, Goodman J, Dimkpa CO et al (2013) Does doping with aluminum alter the effects of ZnO nanoparticles on the metabolism of soil pseudomonads? Microbiol Res 168:91–98

Feng Y, Cui X, He S, Dong G, Chen M, Wang J, Lin X (2013) The role of metal nanoparticles in influencing arbuscular mycorrhizal fungi effects on plant growth. Environ Sci Technol 47:9496–9504

Fernandes JC, Eaton P, Gomes AM, Pintado ME, Malcata FX (2009) Study of the antibacterial effect of chitosan on *Bacillus cereus* (and its spores) by atomic force microscopy imaging and nanoindentation. Ultramicroscopy 109:854–860

Fierer N, Jackson JA, Vilgalys R, Jackson RB (2005) Assessment of soil microbial community structure by use of taxon-specific quantitative PCR assays. Appl Environ Microbiol 71:4117–4120

Fisher MM, Triplett EW (1999) Automated approach for ribosomal intergenic spacer analysis of microbial diversity and its application to freshwater bacterial communities. Appl Environ Microbiol 65:4630–4636

Fleischmann RD, Adams MD, White O, Clayton RA, Kirkness EF et al (1995) Whole-genome random sequencing and assembly of *Haemophilus influenzae* Rd. Science 269:496–512

Foti M, Sorokin DY, Lomans B, Mussman M et al (2007) Diversity, activity, and abundance of sulfate-reducing bacteria in saline and hypersaline soda lakes. Appl Environ Microbiol 73:2093–3000

Francius G, Alsteens D, Dupres V, Lebeer S, De Keersmaecker S et al (2009) Stretching polysaccharides on live cells using single molecule force spectroscopy. Nat Protoc 4:939–946

Froehlich T, Heindl D, Roesler A (2010) Miniaturized high-throughput nucleic acid analysis. European Patent 2224014A1

Fukuma T, Kobayashi K, Matsushige K, Yamada H (2005) True atomic resolution in liquid by frequency-modulation atomic force microscopy. Appl Phys Lett 87:66–70

Gajjar P, Pettee B, Britt DW, Huang W et al (2009) Antimicrobial activities of commercial nanoparticles against an environmental soil microbe, *Pseudomonas putida* KT2440. J Biol Eng 3:9

Garbeva P, Van Veen JA, Vvan Elsas JD (2004) Microbial diversity in soil: selection of microbial populations by plant and soil type and implications for disease suppressiveness. Annu Rev Phytopathol 42:243–270

Ge YG, Schimel JP, Holden PA (2011) Evidence for negative effects of TiO_2 and ZnO nanoparticles on soil bacterial communities. Environ Sci Technol 45:1659–1664

Ge Y, Schimel JP, Holden PA (2012) Identification of soil bacteria susceptible to TiO_2 and ZnO nanoparticles. Appl Environ Microbiol 8:6749–6758

Gentry TJ, Wickham GS, Schadt CW, He Z, Zhou J (2006) Microarray applications in microbial ecology research. Microb Ecol 52:159–175

Ghosh S, Mashayekhi H, Pan B, Bhowmik P, Xing B (2008) Colloidal behavior of aluminum oxide nanoparticles As affected by pH and natural organic matter. Langmuir 24:12385–12391

Giller KE, Witter E, McGrath SP (2009) Heavy metals and soil microbes. Soil Biol Biochem 41:2031–2037

Gillespie AW, Farrell RE, Walley FL et al (2011) Glomalin-related soil protein contains nonmycorrhizal-related heat-stable proteins, lipids and humic materials. Soil Biol Biochem 43:766–777

Gimbert LJ, Hamon RE, Casey PS, Worsfold PJ (2007) Partitioning and stability of engineered ZnO nanoparticles in soil suspensions using flow field-flow fractionation. Environ Chem 4:8–10

Gohargani J, Ghasemi H (2013) Ecotoxicity of nanomaterials in soil. Ann Biol Res 4:86–92

Grassian VH (2008) When size really matters: size-dependent properties and surface chemistry of metal and metal oxide nanoparticles in gas and liquid phase environments. J Phys Chem C 112:18303–18313

Gurunathan S (2015) Cytotoxicity of graphene oxide nanoparticles on plant growth promoting rhizobacteria. J Ind Eng Chem 32:282–291

Guzman KA, Finnegan MP, Banfield JF (2006) Influence of surface potential on aggregation and transport of titania nanoparticles. Environ Sci Technol 40:7688–7693

Hajipour MJ, Fromm KM, Ashkarran AA, De Aberasturi D, De Larramendi IR, Rojo T, Serpooshan V, Parak WJ, Mahmoudi M (2012) Antibacterial properties of nanoparticles. Trends Biotechnol 30:499–511

Handy RD, Cornelis G, Fernandes T et al (2012) Ecotoxicity test methods for engineered nanomaterials: practical experiences and recommendations from the bench. Environ Toxicol Chem 31:15–31

Hansch M, Emmerling C (2010) Effects of silver nanoparticles on the microbiota and enzyme activity in soil. J Plant Nutr Soil Sci 173:554–558

Hao J, Murphy R, Lim E, Schoonen MAA, Strongin DR (2009) Effects of phospholipid on pyrite oxidation in the presence of autotrophic and heterotrophic bacteria. Geochim Cosmochim Acta 73:4111–4123

Hartmann NB, Skjolding LM, Hansen SF, Kjølholt J, Gottschalck F, Baun A (2014) Environmental fate and behaviour of nanomaterials. The Danish Environmental Protection Agency, Denmark, p 114

Hayat R, Ali S, Amara U, Khalid R, Ahmed I (2010) Soil beneficial bacteria and their role in plant growth promotion: a review. Ann Microbiol 60:579–598

He S, Feng Y, Ren H, Zhang Y, Gu N, Lin X (2011) The impact of iron oxide magnetic nanoparticles on the soil bacterial community. J Soils Sediments 11:1408–1417

Helenius J, Heisenberg C, Gaub HE, Muller DJ (2008) Single-cell force spectroscopy. J Cell Sci 121:1785–1791

Hill GT, Mitkowski NA, Aldrich-Wolfe L et al (2000) Methods for assessing the composition and diversity of soil microbial communities. Appl Soil Ecol 15:25–36

Hinsinger P, Glyn Bengough A, Vetterlein D, Young IM (2009) Rhizosphere: biophysics, biogeochemistry and ecological relevance. Plant Soil 321:117–152

Hirsch PR, Mauchline TH, Clark IM (2010) Culture-independent molecular techniques for soil microbial ecology. Soil Biol Biochem 42:878–887

Horejs C, Ristl R, Tscheliessnig R, Sleytr UB, Pum D (2011) Single-molecule force spectroscopy reveals the individual mechanical unfolding pathways of a surface layer protein. J Biol Chem 286:27416–27424

Hoshino A, Fujioka K, Oku T, Suga M et al (2004) Physicochemical properties and cellular toxicity of nanocrystal quantum dots depend on their surface modification. Nano Lett 4:2163–2169

Hwang ET, Lee JH, Chae YJ, Kim BC, Sang BI, Gu MB (2008) Analysis of the toxic mode of action of silver nanoparticles using stress-specific bioluminescent bacteria. Small 4:746–750

Jang H, Pell LE, Korgel BA, English DS (2003) Photoluminescence quenching of silicon nanoparticles in phospholipid vesicle bilayers. J Photochem Photobiol A Chem 158:111–117

Janssen PH (2006) Identifying the dominant soil bacterial taxa in libraries of 16S rRNA, and 16S rRNA genes. Appl Environ Microbiol 72:1719–1728

Jemec A, Drobne D, Remskar M, Sepcic K, Tisler T (2008) Effects of ingested nano-sized titanium dioxide on terrestrial isopods (*Porcellio scaber*). Environ Toxicol Chem 27:1904–1914

Jiang J, Oberdorster G, Biswas P (2009) Characterization of size, surface charge, and agglomeration state of nanoparticle dispersions for toxicological studies. J Nanoparticle Res 11:77–89

Jin L, Son Y, DeForest JL, Kang YJ, Kim W, Chung H (2014) Single-walled carbon nanotubes alter soil microbial community composition. Sci Total Environ 446:533–538

Johansen A, Pedersen AL, Jensen KA, Karlson U et al (2008) Effects of C_{60} fullerene nanoparticles on soil bacteria and protozoans. Environ Toxicol Chem 27:1895–1903

Judy JD, McNear DH (2015) Nanomaterials in biosolids inhibit nodulation, shift microbial community composition, and result in increased metal uptake relative to bulk/dissolved metals. Environ Sci Technol. doi:10.1021/acs.est.5b01208

Judy JD, Kirby JK, Creamer C, McLaughlin MJ, Fiebiger C, Wright C, Cavagnaro TR, Bertsch PM (2015) Effects of silver sulfide nanomaterials on mycorrhizal colonization of tomato plants and soil microbial communities in biosolid-amended soil. Environ Pollut 206:256–263

Kailas L, Ratcliffe EC, Hayhurst EJ, Walker MG, Foster SJ, Hobbs JK (2009) Immobilizing live bacteria for AFM imaging of cellular processes. Ultramicroscopy 109:775–780

Karunakaran G, Suriyaprabha R, Manivasakan P, Yuvakkumar R et al (2013) Impact of nano and bulk ZrO_2, TiO_2 particles on soil nutrient contents and PGPR. J Nanosci Nanotechnol 13:678–685

Karunakaran G, Suriyaprabha R, Manivasakan P, Rajendran V, Kannan N (2014) Influence of nano and bulk SiO_2 and Al_2O_3 particles on PGPR and soil nutrient contents. Curr Nanosci 10:604–612

Keller AA, Wang H, Zhou D, Lenihan HS et al (2010) Stability and aggregation of metal oxide nanoparticles in natural aqueous matrices. Environ Sci Technol 44:1962–1967

Khodakovskaya MV, Kim BS, Kim JN et al (2013) Carbon nanotubes as plant growth regulators: effects on tomato growth, reproductive system, and soil microbial community. Small 9:115–123

Kim S, Sin H, Lee S, Lee I (2013) Influence of metal oxide particles on soil enzyme activity and bioaccumulation of two plants. J Microbiol Biotechnol 23:1279–1286

Kirk JL, Beaudette LA, Hart M, Moutoglis P, Klironomos JN, Lee H, Trevors JT (2004) Methods of studying soil microbial diversity. J Microbiol Meth 58:169–188

Klaine SJ, Alvarez PJJ, Batley GE et al (2008) Nanomaterials in the environment: behavior, fate, bioavailability, and effects. Environ Toxicol Chem 27:1825–1851

Kloepfer JA, Mielke RE, Nadeau JL (2005) Uptake of CdSe and CdSe/ZnS quantum dots into bacteria via purine-dependent mechanisms. Appl Environ Microbiol 71:2548–2557

Kool PL, Diez Ortiz M, van Gestel CAM (2011) Chronic toxicity of ZnO nanoparticles, non-nano ZnO and $ZnCl_2$ to *Folsomia candida* (Collembola) in relation to bioavailability in soil. Environ Pollut 159:2713–2719

Koren S, Schatz MC, Walenz BP, Martin J, Howard JT, Ganapathy G, Wang Z, Rasko DA, McCombie WR, Jarvis ED, Adam MP (2012) Hybrid error correction and de novo assembly of single-molecule sequencing reads. Nat Biotechnol 30:693–700

Kumar N, Shah V, Walker VK (2011) Perturbation of an arctic soil microbial community by metal nanoparticles. J Hazard Mater 190:816–822

Kumar P, Kumar A, Fernandes T, Ayoko GA (2014) Nanomaterials and the environment. J Nanomater. Article ID 528606, pp 4

Kunin V, Engelbrektson A, Ochman H, Hugenholtz P (2010) Wrinkles in the rare biosphere: pyrosequencing errors can lead to artificial inflation of diversity estimates. Environ Microbiol 12:118–123

La Storia A, Ercolini D, Marinello F et al (2011) Atomic force microscopy analysis shows surface structure changes in carvacrol-treated bacterial cells. Res Microbiol 162:164–172

Leckie SE (2005) Methods of microbial community profiling and their application to forest soils. Forest Ecol Manag 220:88–106

Li M, Zhu LZ, Lin DH (2011) Toxicity of ZnO nanoparticles to *Escherichia coli*: mechanism and the influence of medium components. Environ Sci Technol 45:1977–1983

Liang ZH, Das A, Hu ZQ (2010) Bacterial response to a shock load of nanosilver in an activated sludge treatment system. Water Res 44:5432–5438

Lin D, Xing B (2008) Root uptake and phytotoxicity of ZnO nanoparticles. Environ Sci Technol 42:5580–5585

Liu S, Wang Y (2010) Application of AFM in microbiology: a review. Scanning 32:61–73

Liu C, Sun T, Zhai YM, Dong SJ (2009) Evaluation of ferricyanide effects on microorganisms with multi-methods. Talanta 78:613–617

Liu L, Li Y, Li S, Hu N, He Y, Pong R, Lin D, Lu L, Law M (2012) Comparison of next-generation sequencing systems. J Biomed Biotechnol. doi:10.1155/2012/251364

Loman NJ, Misra RV, Dallman TJ, Constantinidou C, Gharbia SE, Wain J, Pallen MJ (2012) Performance comparison of benchtop high-throughput sequencing platforms. Nat Biotechnol 30:434–439

Lu C (2009) Slow growth induces heat-shock resistance in normal and respiratory-deficient yeast. Mol Biol Cell 20:891–903

Luo C, Tsementzi D, Kyrpides N, Read T, Konstantinidis KT (2012) Direct comparisons of Illumina vs. Roche 454 sequencing technologies on the same microbial community DNA sample. PLoS One 7:e30087

Maali A, Cardinal T, Treguer-Delapierre M (2003) Intrinsic fluorescence from individual silver nanoparticles. Physica E 17:559–560

Mah TF, O'Toole GA (2001) Mechanisms of biofilm resistance to antimicrobial agents. Trends Microbiol 9:34–39

Maleki Dizaj S, Lotfipour F, Barzegar-Jalali M, Zarrintan MH, Adibkia K (2014) Antimicrobial activity of the metals and metal oxides nanoparticles. Mater Sci Eng C 44:278–284

Manzo S, Rocco A, Carotenuto R et al (2010) Investigation of ZnO nanoparticles ecotoxicological effects towards different soil organisms. Environ Sci Pollut Res 18:756–763

Marambio-Jones C, Hoek EMV (2010) A review of the antibacterial effects of silver nanomaterials and potential implications for human health and the environment. J Nanopart Res 12:1531–1551

Mardis ER (2008) The impact of next-generation sequencing technology on genetics. Trends Genet 24:133–141

Marx MC, Wood M, Jarvis SC (2001) A microplate fluorimetric assay for the study of enzyme diversity in soils. Soil Biol Biochem 33:1633–1640

Masrahi A, VandeVoort AR, Arai Y (2014) Effects of silver nanoparticle on soil-nitrification processes. Arch Environ Contam Toxicol 66:504–513

McGrath KC, Thomas-Hall SR, Cheng CT, Leo L, Alexa A, Schmidt S, Schenk PM (2008) Isolation and analysis of mRNA from environmental microbial communities. J Microbiol Meth 75:172–176

Metzker ML (2005) Emerging technologies in DNA sequencing. Genome Res 15:1767–1776

Meyer RL, Zhou X, Tang L et al (2010) Immobilisation of living bacteria for AFM imaging under physiological conditions. Ultramicroscopy 110:1349–1357

Mishra VK, Kumar A (2009) Impact of metal nanoparticles on the plant growth promoting rhizobacteria. Dig J Nanomater Biostruct 4:587–592

Mueller NC, Nowack B (2008) Exposure modeling of engineered nanoparticles in the environment. Environ Sci Technol 42:4447–4453

Muyzer G, Waal ECD, Uitterlinden AG (1993) Profiling of complex microbial populations by denaturing gradient gel electrophoresis analysis of polymerase chain reaction-amplified genes coding for 16S rRNA. Appl Environ Microbiol 59:695–700

Nogueira V, Lopes I, Rocha-Santos T, Santos AL et al (2012) Impact of organic and inorganic nanomaterials in the soil microbial community structure. Sci Total Environ 424:344–350

Nowrousian M (2010) Next-generation sequencing techniques for eukaryotic microorganisms: sequencing-based solutions to biological problems. Eukaryot Cell 9:1300–1310

Noy A (2006) Chemical force microscopy of chemical and biological interactions. Surf Interface Anal 38:1429–1441

Oliver JD (2005) The viable but nonculturable state in bacteria. J Microbiol 43:93–100

Pal S, Tak YK, Song JM (2007) Does the antibacterial activity of silver nanoparticles depend on the shape of the nanoparticle? A study of the gram-negative bacterium *Escherichia coli*. Appl Environ Microbiol 73:1712–1720

Pawlett M, Ritz K, Dorey RA et al (2013) The impact of zero-valent iron nanoparticles upon soil microbial communities is context dependent. Environ Sci Pollut Res 20:1041–1049

Peyrot C, Wilkinson KJ, Desrosiers M, Sauve S (2014) Effects of silver nanoparticles on soil enzyme activities with and without added organic matter. Environ Toxicol Chem 33:115–125

Phenrat T, Saleh N, Sirk K, Tilton RD, Lowry GV (2006) Aggregation and sedimentation of aqueous nanoscale zerovalent iron dispersions. Environ Sci Technol 41:284–290

Pipan-Tkalec Z, Drobne D et al (2010) Zinc bioaccumulation in a terrestrial invertebrate fed a diet treated with particulate ZnO or $ZnCl_2$ solution. Toxicology 269:198–203

Prasad R, Swamy VS (2013) Antibacterial activity of silver nanoparticles synthesized by bark extract of Syzygium cumini. J Nanoparticles. doi:10.1155/2013/431218

Prasad R, Kumar V, Prasad KS (2014) Nanotechnology in sustainable agriculture: present concerns and future aspects. Afr J Biotechnol 13:705–713

Prasad R, Pandey R, Barman I (2016) Engineering tailored nanoparticles with microbes: quo vadis. WIREs Nanomed Nanobiotechnol 8:316–330

Priester JH, Ge Y, Mielke RE et al (2012) Soybean susceptibility to manufactured nanomaterials with evidence for food quality and soil fertility interruption. Proc Natl Acad Sci U S A 109: E2451–E2456

Qi L, Xu Z, Jiang X, Hu C, Zou X (2004) Preparation and antibacterial activity of chitosan nanoparticles. Carbohydr Res 339:2693–2700

Radmacher M, Cleveland J, Fritz M, Hansma H, Hansma P (1994) Mapping interaction forces with the atomic-force microscope. Biophys J 66:2159–2165

Radniecki TS, Stankus DP, Neigh A, Nason JA, Semprini L (2011) Influence of liberated silver from silver nanoparticles on nitrification inhibition of *Nitrosomonas europaea*. Chemosphere 85:43–49

Rastogi G, Sani RK (2011) Molecular techniques to assess microbial community structure, function, and dynamics in the environment. In: Ahmad I et al (eds) Microbes and microbial technology: agricultural and environmental applications. Springer, New York, pp 29–57

Riesenfeld CS, Schloss PD, Handelsman J (2004) Metagenomics: genomic analysis of microbial communities. Annu Rev Genet 38:525–552

Robe P, Nalin R, Capellano C, Vogel TM, Simonet P (2003) Extraction of DNA from soil. Eur J Soil Biol 39:183–190

Rodrigues DF, Jaisi DP, Elimelech M (2013) Toxicity of functionalized single-walled carbon nanotubes on soil microbial communities: implications for nutrient cycling in soil. Environ Sci Technol 47:625–633

Roh H, Subramanya N, Zhao FM, Yu CP, Sandt J, Chu KH (2009) Biodegradation potential of wastewater micropollutants by ammonia-oxidizing bacteria. Chemosphere 77:1084–1089

Roose-Amsaleg CL, Garnier-Sillam E, Harry M (2001) Extraction and purification of microbial DNA from soil and sediment samples. Appl Soil Ecol 18:47–60

Rudi K, Zimonja M, Trosvik P, Naes T (2007) Use of multivariate statistics for 16S rRNA gene analysis of microbial communities. Int J Food Microbiol 120:95–99

Saleh-Lakha S, Miller M, Campbell RG, Schneider K, Elahimanesh P, Hart MM, Trevors JT (2005) Microbial gene expression in soil: methods, applications and challenges. J Microbiol Meth 63:1–19

Sanger F, Air GM, Barrell BG, Brown NL, Coulson AR, Fiddes CA, Hutchison CA, Slocombe PM, Smith M (1977) Nucleotide sequence of bacteriophage phi X174 DNA. Nature 265:687–695

Santimano MC, Kowshik M (2013) Altered growth and enzyme expression profile of ZnO nanoparticles exposed non-target environmentally beneficial bacteria. Environ Monit Assess 185:7205–7214

Schaer-Zammaretti P, Ubbink J (2003) Imaging of lactic acid bacteria with AFM—elasticity and adhesion maps and their relationship to biological and structural data. Ultramicroscopy 97:199–208

Schmidt TM, Waldron C (2015) Microbial diversity in soils of agricultural landscapes and its relation to ecosystem function. In: Hamilton SK, Doll JE, Robertson GP (eds) The ecology of agricultural landscapes: long-term research on the path to sustainability. Oxford University Press, New York, pp 135–157

Shin YJ, Kwak JI, An YJ (2012) Evidence for the inhibitory effects of silver nanoparticles on the activities of soil exoenzymes. Chemosphere 88:524–529

Shrestha B, Acosta-Martinez V, Cox SB, Green MJ, Li S, Canas-Carrell JE (2013) An evaluation of the impact of multiwalled carbon nanotubes on soil microbial community structure and functioning. J Hazard Mater 261:188–197

Simon-Deckers A et al (2009) Size-, composition- and shape-dependent toxicological impact of metal oxide nanoparticles and carbon nanotubes toward bacteria. Environ Sci Technol 43:8423–8429

Simonin M, Richaume A (2015) Impact of engineered nanoparticles on the activity, abundance, and diversity of soil microbial communities: a review. Environ Sci Pollut Res 22:13710–13723

Smith CJ, Osborn AM (2009) Advantages and limitations of quantitative PCR (Q-PCR)-based approaches in microbial ecology. FEMS Microbiol Ecol 67:6–20

Soni D, Naoghare PK, Saravanadevi S, Pandey RA (2015) Release, transport and toxicity of engineered nanoparticles. In: Whitacre DM (ed) Reviews of environmental contamination and toxicology. Springer, Cham, pp 1–48

Sotiriou GA, Pratsinis SE (2010) Antibacterial activity of nanosilver ions and particles. Environ Sci Technol 44:5649–5654

Stone V, Nowack B, Baun A, van den Brink N, von der Kammer F et al (2010) Nanomaterials for environmental studies: classification, reference material issues, and strategies for physico-chemical characterisation. Sci Total Environ 408:1745–1754

Suresh AK, Pelletier DA, Doktycz MJ (2013) Relating nanomaterial properties and microbial toxicity. Nanoscale 5:463–474

Thakuria D, Schmidt O, Mac Siurtain M, Egan D, Doohan FM (2009) Importance of DNA quality in comparative soil microbial community structure analyses. Soil Biol Biochem 40:1390–1403

Thies JE (2007) Soil microbial community analysis using terminal restriction fragment length polymorphisms. Soil Sci Soc Am J 71:579–591

Thomas T, Gilbert J, Meyer F (2012) Metagenomics—a guide from sampling to data analysis. Microb Inform Exp 2:1–12
Thul ST, Sarangi BK (2015) Implications of nanotechnology on plant productivity and its rhizospheric environment. In: Siddiqui MH, Al-Whaibi MH, Mohammad F (eds) Nanotechnology and plant sciences. Springer, Cham, pp 37–54
Tiede K, Hassellov M, Breitbarth E, Chaudhry Q, Boxall ABA (2009) Considerations for environmental fate and ecotoxicity testing to support environmental risk assessments for engineered nanoparticles. J Chromatogr A 1216:503–509
Tiedje JM, Asuming-Brempong S, Nusslein K, Marsh TL, Flynn SJ (1999) Opening the black box of soil microbial diversity. Appl Soil Ecol 13:09–122
Tilston EL, Collins CD, Mitchell GR, Princivalle J, Shaw LJ (2013) Nanoscale zerovalent iron alters soil bacterial community structure and inhibits chloroaromatic biodegradation potential in Aroclor 1242-contaminated soil. Environ Pollut 173:38–46
Tong Z, Bischoff M, Nies L, Applegate B, Turco RF (2007) Impact of fullerene (C_{60}) on a soil microbial community. Environ Sci Technol 41:2985–2991
Torsvik V, Øvreas L (2002) Microbial diversity and function in soil: from genes to ecosystems. Curr Opin Microbiol 5:240–245
Tourinho PS, Van Gestel CAM, Lofts S, Svendsen C, Soares AMVM, Loureiro S (2012) Metal-based nanoparticles in soil: fate, behavior, and effects on soil invertebrates. Environ Toxicol Chem 31:1679–1692
Unrine JM, Hunyadi SE, Tsyusko OV, Rao W, Shoults-Wilson WA, Bertsch PM (2010) Evidence for bioavailability of Au nanoparticles from soil and biodistribution within earthworms (Eisenia fetida). Environ Sci Technol 44:8308–8313
Velegol S, Logan B (2002) Contributions of bacterial surface polymers, electrostatics, and cell elasticity to the shape of AFM force curves. Langmuir 18:5256–5262
Wang H, Wick RL, Xing B (2009) Toxicity of nanoparticulate and bulk ZnO, Al_2O_3 and TiO_2 to the nematode *Caenorhabditis elegans*. Environ Pollut 157:1171–1177
Wu B (2010) Bacterial responses to Cu-doped TiO_2 nanoparticles. Sci Total Environ 408:1755–1758
Yang W, Shen C, Ji Q, An H et al (2009) Food storage material silver nanoparticles interfere with DNA replication fidelity and bind with DNA. Nanotechnology 20:085102
Yang Y, Wang J, Xiu ZM, Alvarez PJJ (2013) Impacts of silver nanoparticles on cellular and transcriptional activity of nitrogen-cycling bacteria. Environ Toxicol Chem 32:1488–1494
Yang Y, Quensen J, Mathieu J et al (2014) Pyrosequencing reveals higher impact of silver nanoparticles than Ag on the microbial community structure of activated sludge. Water Res 48:317–325
Yao X, Walter J, Burke S, Stewart S, Jericho M, Pink D, Beveridge T (2002) Atomic force microscopy and theoretical considerations of surface properties and turgor pressures of bacteria. Colloids Surf B Biointerfaces 23:213–230
Yoon KY, Byeon JH, Park JH, Hwang J (2007) Susceptibility constants of *Escherichia coli* and *Bacillus subtilis* to silver and copper nanoparticles. Sci Total Environ 373:572–575
Yuan SJ, Pehkonen SO (2009) AFM study of microbial colonization and its deleterious effect on 304 stainless steel by *Pseudomonas* NCIMB 2021 and *Desulfovibrio desulfuricans* in simulated seawater. Corrosion Sci 51:1372–1385
Yuan Z, Li J, Cui L, Xu B, Zhang H, Yu CP (2013) Interaction of silver nanoparticles with pure nitrifying bacteria. Chemosphere 90:1404–1411
Zeliadt N (2010) http://www.scientificamerican.com/article/silver-beware-antimicrobial-nanoparticles-in-soil-may-harm-plant-life/
Zhao C, Brinkhoff T, Burchardt M, Simon M, Wittstock G (2009) Surface selection, adhesion, and retention behavior of marine bacteria on synthetic organic surfaces using self-assembled monolayers and atomic force microscopy. Ocean Dynam 59:305–315
Zheng X, Chen YG, Wu R (2011) Long-term effects of titanium dioxide nanoparticles on nitrogen and phosphorus removal from wastewater and bacterial community shift in activated sludge. Environ Sci Technol 45:7284–7290

Chapter 6
Synthesis and Characterization of Pure and Doped ZnO Nanostructures for Antimicrobial Applications: Effect of Dopant Concentration with Their Mechanism of Action

Manika Khanuja, Uma, and Ajit Varma

6.1 Antibiotics and Agriculture

Antimicrobial agents are those materials that have the ability to kill or inhibit the growth of pathogenic microorganisms such as bacteria, pathogens, fungi, or protozoa and have gained immense interest due to their potential use in plant agriculture, production agriculture, sterilization of water, soil sterilization, and food preservation as shown in Fig. 6.1. The use of antibiotics in agriculture is widespread. Antibiotics are used in plant agriculture to control fungal and bacterial infections. Antibiotics have been used in animal agriculture (cattle, swine, and poultry) to increase (1) weight gain, (2) feed utilization, and (3) immunity against infectious diseases. The dose of antibiotics fed to animals varies from 3 to 20 mg depending on the size and type of animal and antibiotic type. Animals are fed with heavy dosage of antibiotic. However, animals do not utilize 90 % of the antibiotic present in their feed, so it is excreted through urine and manure and thus enters into the terrestrial world through land application of these manures as shown in Fig. 6.2 (Kumar et al. 2005). These manures alter the soil microbial ecosystem. The

M. Khanuja (✉)
Centre for Nanoscience and Nanotechnology, Jamia Millia Islamia, New Delhi 110025, India
e-mail: manikakhanuja@gmail.com

Uma
Amity Institute of Microbial Technology, Amity University Uttar Pradesh, E-3 Block, Fourth Floor, Sector 125, Noida, Uttar Pradesh 201303, India

A. Varma
Amity Institute of Microbial Technology, Amity University Uttar Pradesh, E-3 Block, Fourth Floor, Sector 125, Noida, Uttar Pradesh 201303, India

Amity Science, Technology & Innovation Foundation (ASTIF), Noida, Uttar Pradesh, India
e-mail: ajitvarma@amity.edu

M. Ghorbanpour et al. (eds.), *Nanoscience and Plant–Soil Systems*, Soil Biology 48,
DOI 10.1007/978-3-319-46835-8_6

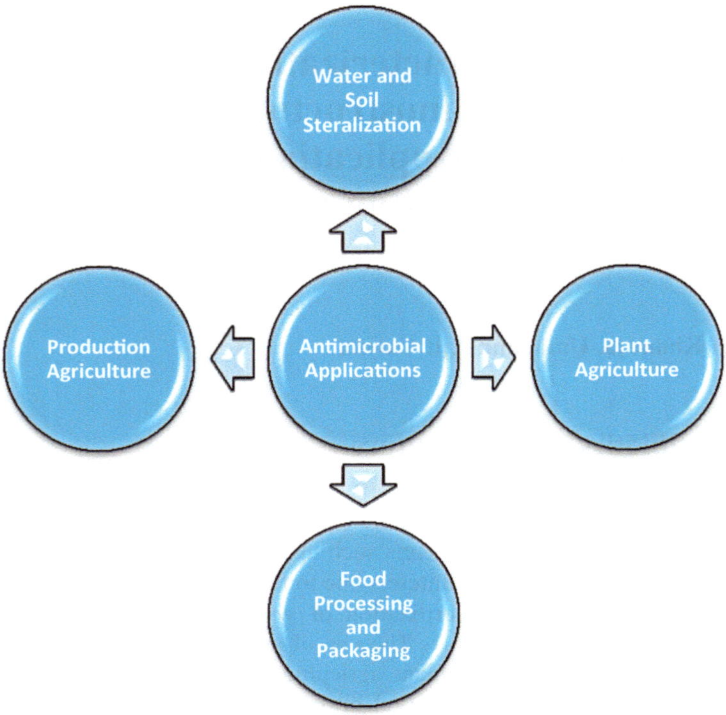

Fig. 6.1 Wide spectrum of antimicrobial applications

biodegradation rate of antibiotics depends on temperature; lower temperature reduces the rate of degradation. The low biodegradation of antibiotics is of great concern in northern world regions as soil freezing at low temperature results in long persistence of antibiotics in soil in deeper soil layer and water leading to its spread in terrestrial environment through snow melt-off. The use of antibiotics for soil sterilization has many benefits. Soil acts as nutrient and water reservoir to support plant life. Soil is the most suitable medium for microbial growth. Certain soil organisms like *Staphylococcus aureus* cause wound inflammation and ulcers (Bowler et al. 2001). Soil sterilization protects plants from harmful organisms such as bacteria, metabolites, virus, fungi, nematodes, and other plant pests. Food antimicrobial agents are used to (1) clean, (2) sanitize, (3) disinfect, and increase the (4) shelf life and (5) nutritional value.

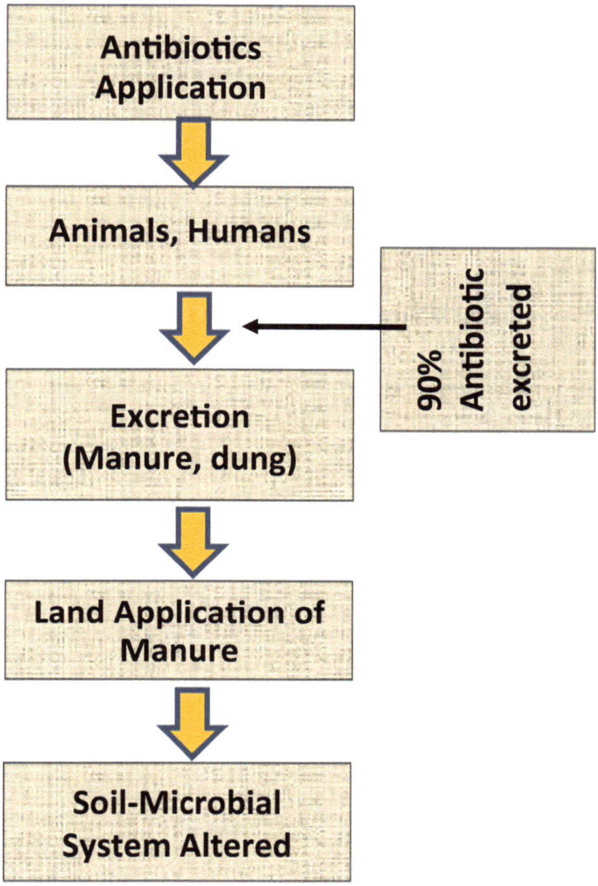

Fig. 6.2 Flowchart to demonstrate relationship between antibiotic application and soil microbial ecosystem

6.2 Antimicrobial Resistance and Nanotechnology

The World Health Day is celebrated with the theme "Antimicrobial Resistance: No action today, No cure tomorrow." Antimicrobial resistance (AMR) is a global threat (Sharma 2011). The ecological principle is "everything is connected to everything else." The use of antibiotics at one place results in development of antimicrobial resistance at another place. A number of strategies are employed throughout the food system, from agriculture to home such that antibiotics are to be used in a proper, appropriate, and judicious manner. The world is entering into the era where "Antibiotics are no longer the future for tomorrow." Microbial world is continuously evolving; bacteria have developed defense mechanisms called "resistant bacteria" and new strains, so microorganisms are becoming resistant to multiple antibiotics. Moreover, conventional antimicrobial agents have the limitation of residual activity.

In this direction, nanotechnology has emerged as a potential candidate to combat increase in microbial resistance against existing antibiotics. Novel nanomaterial-based antimicrobial agents possess high efficiency and selectivity and long time activity and can minimize the environmental problems by reducing residual toxicity of agents. This is due to the novel properties of nanomaterials including small size, large specific surface to volume ratio, their close interaction, and high reactivity with microbial membranes (Prasad et al. 2016).

6.3 Biosynthesized Nanoparticles

Biosynthesized nanoparticles are gaining attention due to their low cost, simplicity, and eco-friendly nature (Bhuyan et al. 2015; Aziz et al. 2015). ZnO has broad range of applications including antimicrobial activity due to its biocompatibility, enhanced surface to volume ratio, wide range of morphologies, surface defects and surface charge, and tunable band gap as summarized in Fig. 6.3. ZnO is considered as safe food (GRAS) by the Food and Drug Administration (Rajiv et al. 2013). Through biosynthesis, i.e., by using eco-friendly, nanotoxic, and safe reagents, controlled and precise synthesis of metal and metal oxide nanoparticles with well-defined diverse sizes and shapes is achievable.

Numerous studies have shown that plants possess antimicrobial activity against, viz., *Pseudomonas aeruginosa*, *Escherichia coli*, *Staphylococcus aureus*, *Salmonella typhi*, *Aspergillus niger*, and *Candida albicans*, tested by disk diffusion method. Researchers have synthesized ZnO using neem and *Trifolium pratense* flower water extract, *T. pratense* L., as summarized in Table 6.1.

Green synthesis of ZnO nanoparticles was done using 25 % (w/v) of *Azadirachta indica* (neem) leaf extract (Bhuyan et al. 2015). 2M Zinc acetate and NaOH were mixed in equal proportions, followed by addition of 1 mL neem leaf extract. The neem mediated biosynthesis of ZnO nanoparticles as shown in Fig. 6.4. The leaf extract consists of a wide variety of metabolites (water-soluble phytochemicals) that helps in reducing Zn ions into nanostructured ZnO. The role of protein molecules is to stabilize the biosynthesized nanoparticles.

The precipitate was filtered and dried and was characterized further using TEM, UV–Vis, XRD, and FTIR. Biosynthesized ZnO has been characterized using transmission electron microscopy for morphology and particle size, band gap using UV–Vis spectroscopy, surface defects using photoluminescence spectroscopy, structural information using X-ray diffraction studies, and elemental composition, stoichiometry, and vacancy using X-ray photoelectron spectroscopy as summarized in Fig. 6.5.

Figure 6.6a shows the transmission electron micrograph (TEM) of biosynthesized ZnO nanoparticles (Bhuyan et al. 2015). As evident, particles are spherical in shape with diameter varying from 9.5 to 25.5 nm. In other studies (as mentioned in Table 6.1), similar size of nanoparticles has been reported. Figure 6.6b shows the EDX (energy dispersive X-ray) micrograph of ZnO

Fig. 6.3 Properties of ZnO nanoparticle contributing for antimicrobial applications

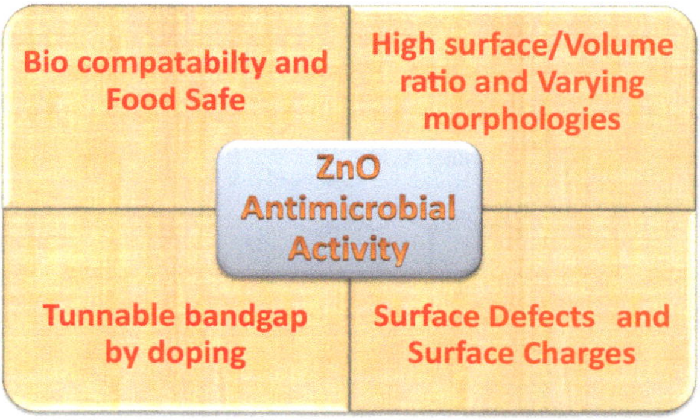

Table 6.1 Biosynthesized ZnO nanoparticles using various plant extracts and fungal biomass (Bhuyan et al. 2015)

Reducing agent	Size (nm)	Shape	Reference
Poncirus trifoliata leaf extract	8.48–32.5	Nearly spherical	Balusamy et al. (2012)
Parthenium hysterophorus L. leaf extract	27 ± 5, 84 ± 2	Spherical, hexagonal	Rajiv et al. (2013)
Aspergillus aeneus	100–140	Spherical	Padmavathy and Vijayaraghavan (2008)
Calotropis procera latex	5–40	Spherical	Rathore and Upadhyay (2013)
Sedum alfredii Hance	53.7	Pseudo-spherical	Yu et al. (2013)
Physalis alkekengi L.	72.5	Triangular, elongated	Qu et al. (2011)

nanoparticles. The strong emission peaks corresponding to Zn and O confirm the presence of zinc and oxygen atoms in the biosynthesized ZnO. No impurity peaks have been observed, indicating that synthesized nanoparticles are pure. Figure 6.6c shows the UV–Vis absorption spectra of biosynthesized ZnO nanoparticles; strong absorption peak at 377 nm confirms the presence of ZnO. Band gap is calculated using Tauc's plot as shown in inset of Fig. 6.6c. The band gap is estimated to be 3.87 eV which is in good match with the earlier studies. Figure 6.6d shows the X-ray diffraction spectra of biosynthesized ZnO nanoparticles. As per Joint Committee on Powder Diffraction Standard (JCPDS, card number 82-1042), the strong and sharp diffraction peaks correspond to crystalline hexagonal wurtzite structure of biosynthesized ZnO.

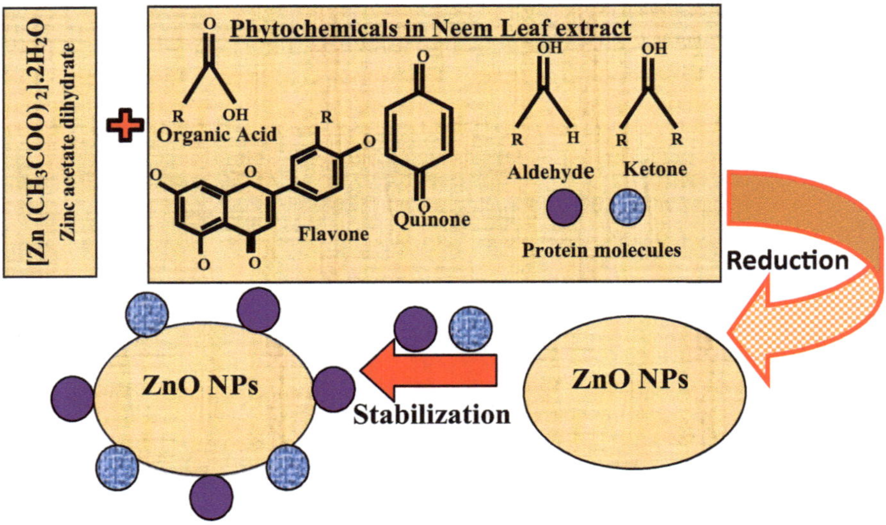

Fig. 6.4 Biosynthesis of ZnO nanoparticles (c.r.f. Bhuyan et al. 2015)

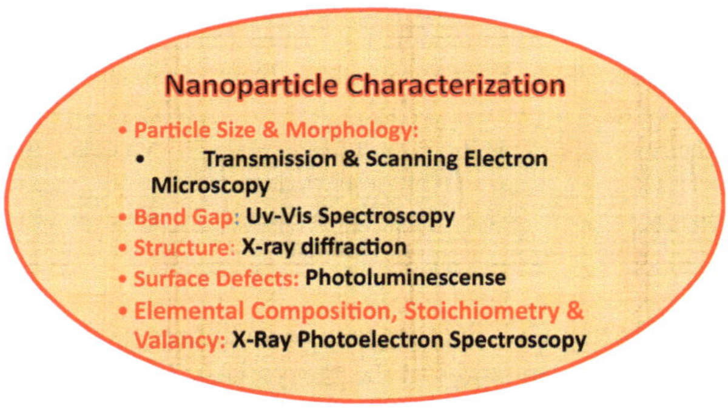

Fig. 6.5 Parameters for nanoparticle characterization

6.4 Copper-Doped ZnO Nanorods

Since ages, copper is known for its antimicrobial action. To date, there are no reports that bacteria have developed antimicrobial resistance against copper. Thus synergetic association of copper and zinc in "copper-doped ZnO nanorods" has the potential to overcome the challenge associated with AMR. This is due to two things: (1) Copper doping tunes the ZnO band gap such that effective utilization of sunlight is more for the generation of superoxide anions, and (2) the changing morphology of ZnO from "spherical" in nanoparticles to 1-D nanostructures in ZnO nanorods results in increase in spatial confinement of electron and hole, thus reducing electron–hole recombination probability, making them available for the formation of superoxide anions.

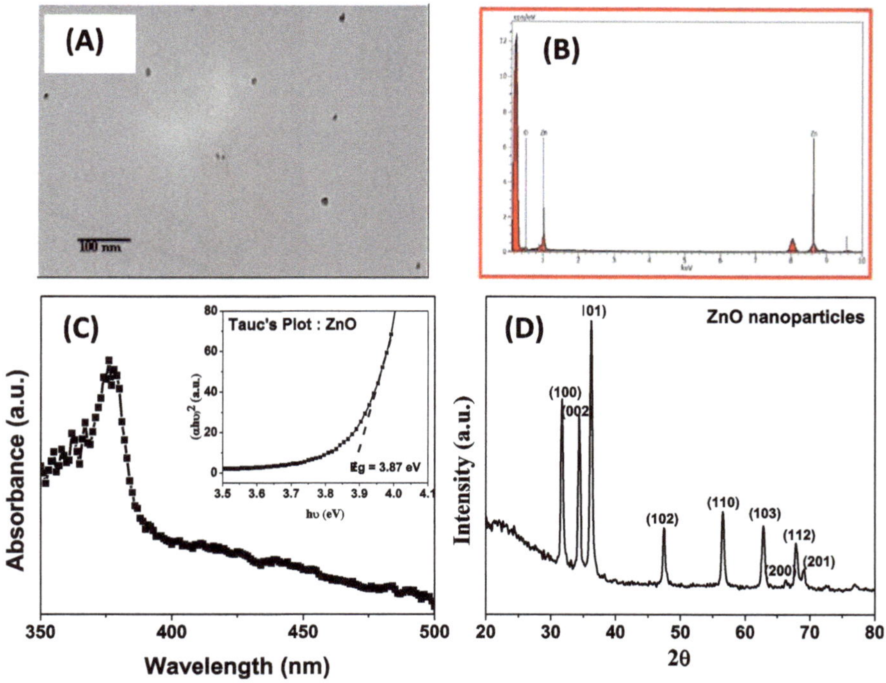

Fig. 6.6 Biosynthesized ZnO characterized by (**a**) transmission electron microscopy for particle size (10–25 nm). (**b**) EDX spectra confirming presence of elements Zn and O. (**c**) UV–Vis absorption spectra and *insert* shows Tauc's plot, estimated band gap (E_g) = 3.87 eV and (**d**) X-ray diffractogram showing (hkl) planes corresponding to hexagonal wurtzite structure (c.r.f. Bhuyan et al. 2015)

6.4.1 Synthesis

Copper-doped ZnO nanostructures have been synthesized using mechanical-assisted thermal decomposition process. Pure zinc acetate (formula) and copper acetate were mixed in mortar pestle for 30 min and then placed in a furnace at 400 °C for 4 h. The mole ratio of Cu:Zn was kept to be 10 % and then mixed in mortar pestle until homogenous mixture is obtained. The obtained powder sample was washed twice with distilled water and then dried in oven at 100 °C for 8 h (Noipa et al. 2014).

6.4.2 Characterization

Figure 6.7a, b shows the SEM image of the pure ZnO nanorods and copper-doped ZnO nanorods. The length of rods is ~500 nm with diameter ~50 nm. Figure 6.7c

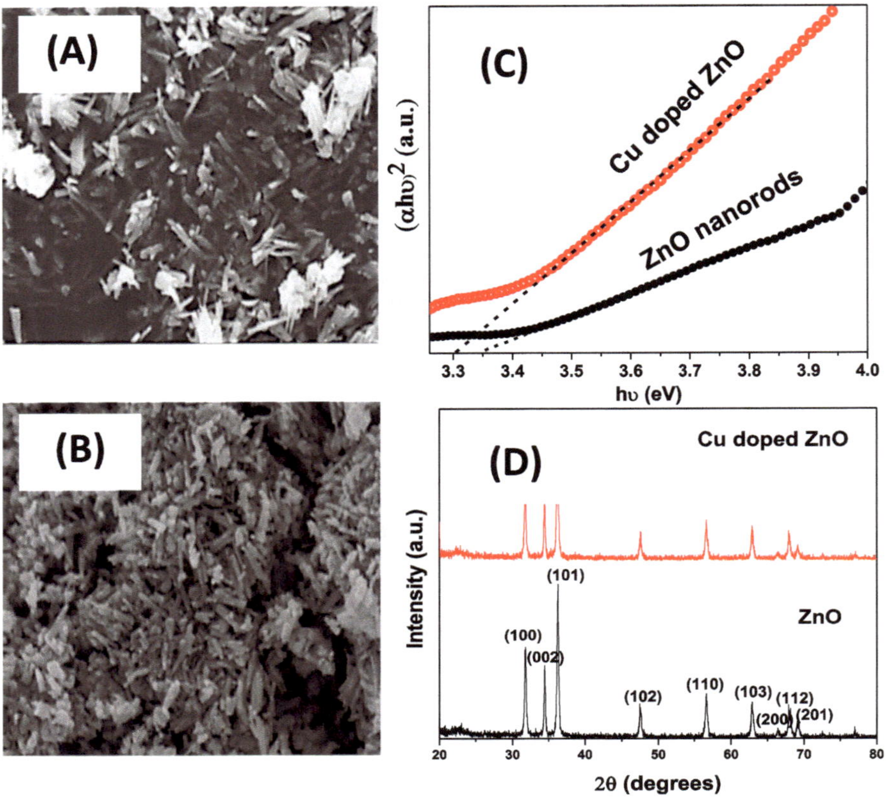

Fig. 6.7 (**a**, **b**) SEM micrograph of ZnO and Cu-doped ZnO nanorods, respectively. (**c**) Tauc's plot of ZnO and Cu-doped ZnO. (**d**) XRD pattern of ZnO and Cu-doped ZnO nanorods

shows the UV spectra depicts that doping plays an important role in varying absorption characteristics of ZnO nanorods. There was a redshift between absorption bands at 374.7 nm of pure ZnO to other absorption bands at 378.2 nm of Cu-doped ZnO. The redshift in absorption peak shows decrease in band gap of ZnO with Cu doping. Thus, redshift in absorption edge indicates that more portion of the solar spectrum can be utilized for electron–hole pair generation (Koziej et al 2014; Jacob et al 2014). Figure 6.7d shows the XRD pattern of pure ZnO and copper-doped ZnO nanorods. All the peaks have been labeled with (hkl) planes after comparing with JCPDS file number 06-2151. XRD pattern shows wurtzite structure with lattice constants $a = 3.247$ Å and $c = 5.203$ Å for ZnO and $a = 3.249$ Å and $c = 5.204$ Å for CZN. No significant changes were observed in lattice constant with copper doping, indicating that all doped copper had gone to substitution sites. No secondary phases, viz., Cu_2O, CuO, or other metallic Cu or Zn phases, were observed. Intensities of XRD peaks of Cu-doped ZnO nanorod sample were found to reduce in comparison to pure ZnO which indicates decrease in crystallinity of ZnO nanorods with Cu doping.

6.5 Antimicrobial Activity

Preparation of Bacterial Cultures

The antibacterial assay of the copper-doped ZnO nanorods was determined against both Gram-positive and Gram-negative bacterial species: *Staphylococcus aureus*, *Streptococcus pyogenes*, and *Escherichia coli*. The stock bacterial cultures were maintained at 37 °C. Sterile Luria–Bertani (LB) broth of 100 mL was prepared in 500 mL Erlenmeyer flasks followed by inoculation of single bacterial colonies from each of the bacterial stock cultures of *Staphylococcus aureus*, *Streptococcus pyogenes*, and *Escherichia coli*. Finally, the prepared bacterial suspensions were placed in an incubator shaker at 37 °C.

6.5.1 SEM Analysis of Bacterial Cells

Fixation of Bacterial Samples

Scanning electron microscopy was employed for examining the morphological changes in the bacterial cells (*Staphylococcus aureus*, *Streptococcus pyogenes*, and *Escherichia coli*) before and after treatment with copper-doped ZnO nanorods. Fixation of the bacterial samples was carried out with 2.5 % glutaraldehyde in 0.1 M sodium phosphate buffer and pH 7.4 for 30 min and then incubated overnight at 4 °C. This is followed by washing the glutaraldehyde-treated samples with 0.1 M sodium phosphate buffer (pH 7.4) thrice and dehydrated with a series of 30, 50, 70, 90, and 100 % ethanol solutions. The prepared samples were then dried for 1 h at 37 °C and finally coated with a thin gold film (<10 nm). The changes in the bacterial morphology were observed under scanning electron microscope operated at a voltage of 5 kV as shown in Fig. 6.8.

6.5.2 Shake-Flask Method

Shake-Flask Test in Luria–Bertani (LB) Broth

The antibacterial activity of biosynthesized ZnO nanoparticles is tested on Gram-negative as well as Gram-positive bacteria: *S. aureus*, *S. pyogenes*, and *E. coli*. The studies have been carried out for different concentrations (μg/ml) of ZnO (0, 20, 40, 60, 80, 100), as summarized in Table 6.2. Thus, growth of bacterial cells were inhibited in the presence of ZnO as compared to ZnO. The inhibition in growth is

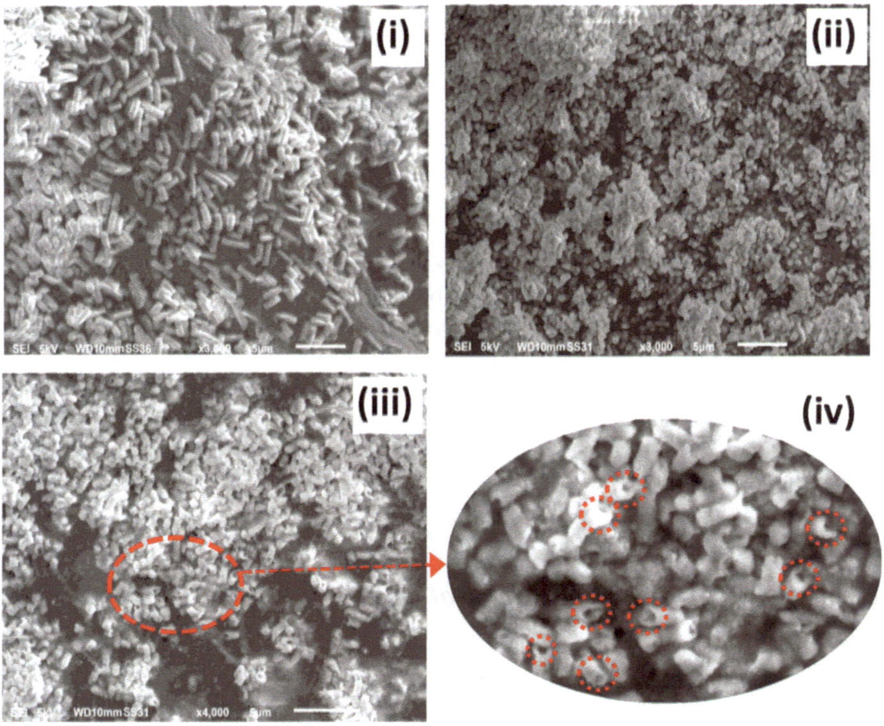

Fig. 6.8 SEM micrograph of *E. coli* in (**i**) control, (**ii**) ZnO nanorods, (**iii**), Cu-doped ZnO nanorods, and (**iv**) magnified view of selected portion of (**iii**) (c.r.f. Bhuyan)

observed to be 70 % for *E. coli*, 57 % for *S. aureus*, and 52 % for *S. pyogenes* as shown in Fig. 6.9. The inhibition in bacterial cells growth is more as concentration of biosynthesized ZnO is increased from 20 to 100 μg/ml.

The antibacterial assay of the copper-doped ZnO nanorods was determined against both Gram-positive and Gram-negative bacterial species: *Staphylococcus aureus*, *Streptococcus pyogenes*, and *Escherichia coli*. Antibacterial activity of CZN sample was carried out using shake-flask method as a function of time (0, 3, 6, 9, and 24 h) against Gram-positive (*S. aureus* and *S. pyogenes*) as well as Gram-negative (*E. coli*) bacteria (Fig. 6.10). The percentage reduction in bacterial growth was found to be $CZN_{E.\ coli}$ (61 %), $CZN_{S.\ aureus}$ (64 %), and $CZN_{S.\ pyogenes}$ (55 %). The doped ZnO samples are effective antibacterial agents on Gram-positive as well as on Gram-negative bacteria as summarized in Table 6.3. The antimicrobial action of sample was found to be highest for *S. aureus* as compared to *E. coli* and *S. pyogenes* due to the differences in (1) cell membrane structure, (2) physiology and metabolic activities of the cell, and (3) degree of contact.

Table 6.2 Antibacterial activity of biosynthesized ZnO nanoparticles (using nutrient broth shake-flask test Bhuyan et al. 2015)

	Test organisms Optical density (O.D) after 24 h (600 nm)					
Concentration of ZnO nanoparticles (μg/ml)	*E. coli* (O.D)	Growth reduction (%)	*S. aureus* (O.D)	Growth reduction (%)	*S. pyogenes* (O.D)	Growth reduction (%)
0	4.6	100.0	5.2	100.0	2.2	100.0
20	4.3	92.4	5.1	97.0	2.1	94.0
40	3.9	83.7	3.4	76.0	1.7	77.4
60	3.7	79.6	3.7	70.4	1.2	56.6
80	3.6	77.7	3.1	60.4	1.1	51.6
100	3.2	69.2	3.0	57.0	1.1	52.0

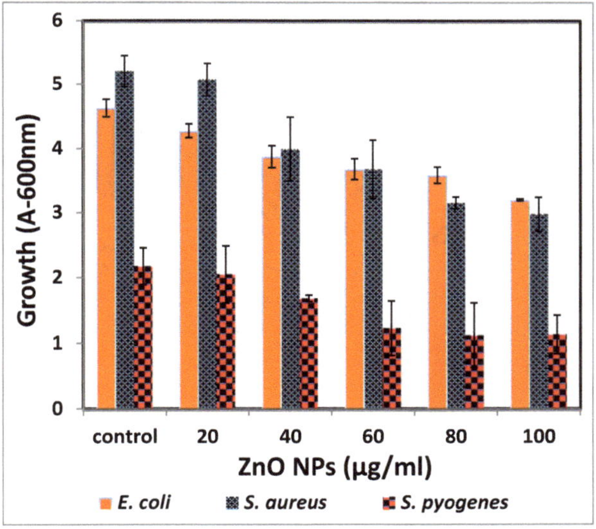

Fig. 6.9 Growth of bacterial strains (*S. aureus*, *S. pyogenes*, and *E. coli*) exposed to various concentrations of ZnO nanoparticles (20–100 μg/mL). Values plotted are mean ± standard deviation (c.r.f. Bhuyan et al. 2015)

6.6 Mechanism of Action

The mechanism of toxicity of the Cu and ZnO nanoparticles mainly involves unfolding of proteins, loss of enzymatic activity, and thiol cross-linking and depends on nanoparticle–biomolecule interaction (Moos et al. 2010). The diffusion of nanoparticles across the cell membrane occurs due to the presence of positive ions or other variables on the surface of nanomaterials and the small size of the Cu and ZnO nanoparticles (Verma et al. 2008; Nel et al. 2009). The precise mechanism revealing the mechanism of toxicity of metal oxide nanoparticles still remains unclear. However, four mechanisms potentially explain the pathways of cellular toxicity of nanostructured Cu and ZnO after their entry into the cell: (1) oxidative

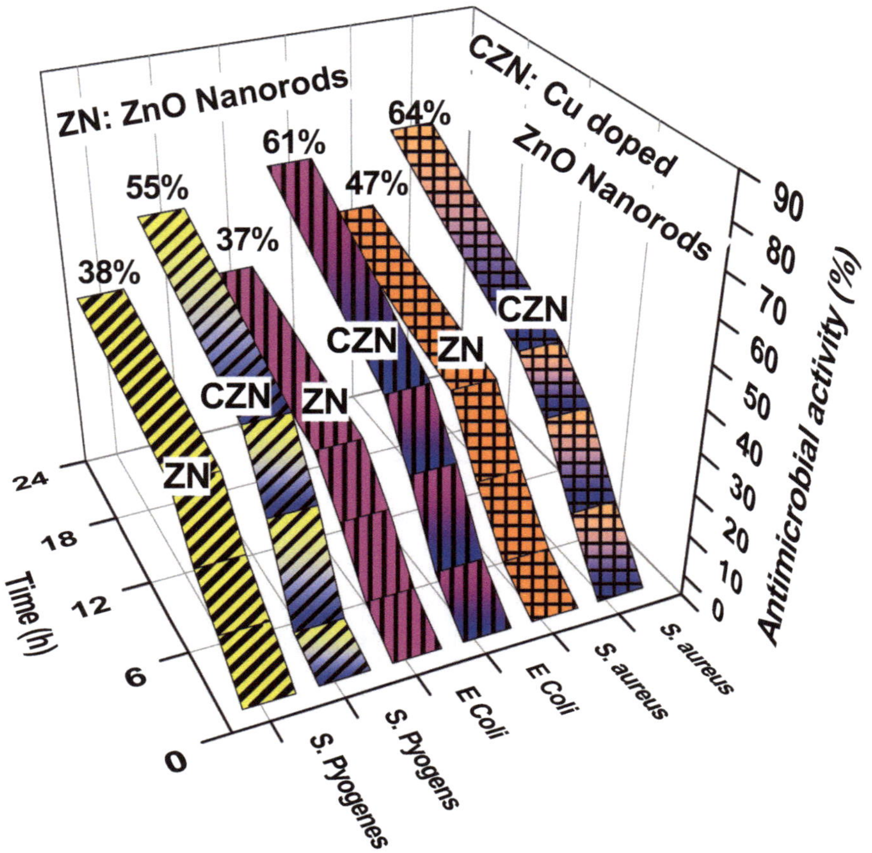

Fig. 6.10 Comparative analysis of the antibacterial efficacy of pure ZnO and Cu-doped ZnO nanorods using shake-flask method

Table 6.3 Antibacterial activity of pure and copper-doped ZnO nanorods

	Antibacterial activity (%)					
	E. coli		*S. aureus*		*S. pyogenes*	
Time (h)	Sample ZN (%)	Sample CZN (%)	Sample ZN (%)	Sample CZN (%)	Sample ZN (%)	Sample CZN (%)
0	0	0	0	0	0	0
3	8.3	6.1	6.2	3.7	8.5	13.4
6	15.5	16.9	17.9	22.5	23.7	28.9
9	27.2	26.4	32.8	36.6	35.5	38.2
24	37.7	36.9	46.6	54.7	60.8	64.2

ZN ZnO nanorods, *CZN* copper-doped ZnO nanorods

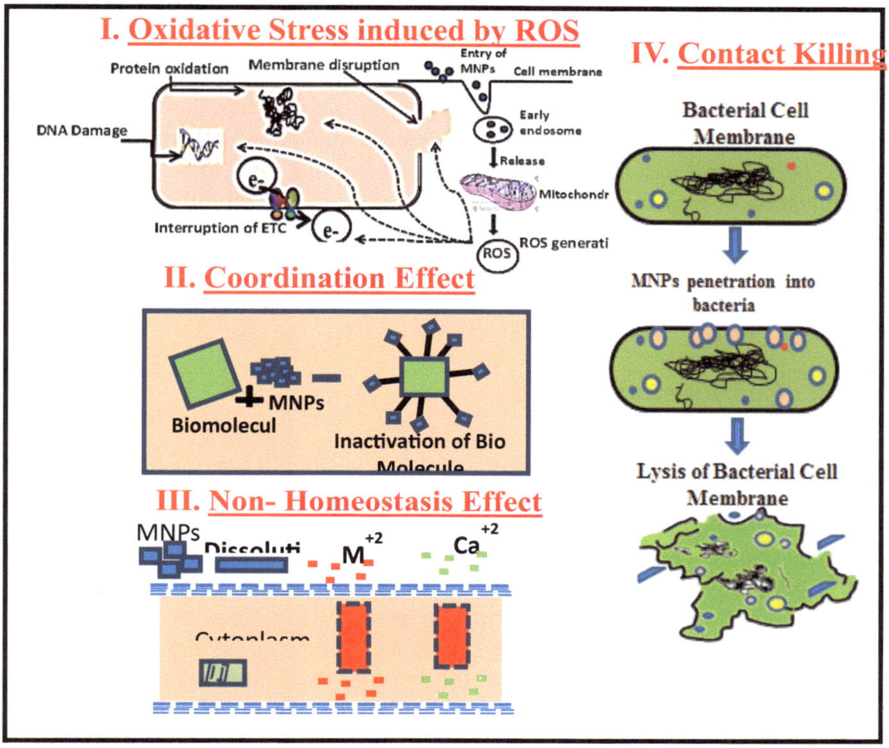

Fig. 6.11 Mechanisms of antimicrobial action of ZnO nanoparticle

stress, (2) coordination effects, (3) contact killing, and (4) non-homeostasis effects (Chang et al. 2012) as shown in Fig. 6.11.

6.6.1 Oxidative Stress

The major toxicological mechanisms of ambient nanoparticles include generation of reactive oxygen species (ROS) and induction of intracellular oxidative stress by creating imbalance between oxidant and antioxidant processes (Chang et al. 2012). Large amounts of ROS are induced when only small amounts of metal oxide nanoparticles interact with oxidative organelles (mitochondria) or get exposed to the acidic environment of lysosomes, causing DNA point mutation or single- or double-strand breaks (Toduka et al. 2012; Singh et al. 2009).

The NP–biomolecule interaction takes place due to large specific surface area, high reactive activity, and electron density of CuO or ZnO nanoparticles (Pisanic et al. 2009). These chemical interactions enhance the formation of superoxide anions (O^{2-}), ultimately leading to ROS accumulation and induction of oxidative stress (Singh et al. 2009).

This results in a variety of biological responses depending upon the relative abundance of ROS generation, the nature of cellular pathways, and the antioxidant response element responsible for oxidative stress induction. In addition, too much of oxidative stress and increased ROS generation might lead to DNA damage and modification of proteins, lipids, and nucleic acids, in turn stimulating the antioxidant defense system or ultimately leading to cell death (Yang et al. 2009; Prasad and Swamy 2013).

6.6.2 Coordination Effects

Coordination and non-covalent interactions occur directly due to the interactions between metal oxide NPs and proteins in vivo or in vitro resulting in major structural changes of proteins and reduction in protein function, decreasing the α-helical content of free protein and many other protein abnormalities (Ueno et al. 2007). Moreover, several biomolecules contain coordination atoms at their active sites (mainly O and N atoms) that form chelates with Cu^{2+} and Zn^{2+} by donating a lone pair of electrons. The resulting NP–biomolecule coordination inactivates the functional biomolecule by affecting the normal physiology process, promoting cellular DNA damage, and then consequently resulting in cellular toxicity (Rao et al. 2010). In addition, metal ions (Cu^{2+} and Zn^{2+}) released by NPs interact with mRNA-stabilizing proteins, leading to degradation of cytoplasmic mRNA and hindering cellular transcription–translation processes (Soenen et al. 2010).

6.6.3 Contact Killing

The accumulation of the metal oxide nanoparticles in the bacterial membrane results in membrane disorganization due to the subsequent release and binding of metal ions to the membrane. However, the toxicity of nanostructured Cu and ZnO is not directly related to their entry inside the cell, but rather, to a certain extent, is related to their close contact onto the cell causing changes in the vicinity of organism–particle contact area. This results in an increase of solubilization of metal and generation of ROS leading to cell death (Walch et al. 2014). Santo et al. 2011 reported that the primary targets for contact killing through surface released copper ions are the cell membranes. The metallic ions cause the inactivation of some enzymes (e.g., hydratase in *E. coli*) necessary for normal cell function and also damage the exposed Fe–S clusters resulting in inhibition of bacterial cell growth (Macomber and Imlay 2009).

6.6.4 *Non-homeostasis Effects*

Metal ions play important roles in maintaining homeostasis of organisms in an independent manner and also keeping the composite functioning (Galhardi et al. 2005). However, any variation in local concentration of these metal ions disrupts metal cation cellular homeostasis ultimately resulting in occurrence of toxicity. It has been reported that the release of Cu^{2+} and Zn^{2+} ions by CuO and ZnO NPs enhances the local concentration of intracellular metal ions accelerating induction of high oxidative stress. This in turn alters the release rate of intracellular Ca^{2+}, which leads to mitochondrial perturbation, imbalance in cellular processes, and ultimately cell death (Xia et al. 2008).

6.7 Conclusions

In the present work, biosynthesized ZnO nanoparticles and copper-doped ZnO nanorods have shown significant antimicrobial activity against Gram-negative as well as Gram-positive bacteria. The antibacterial activity of sample is elucidated using four main mechanisms: oxidative stress, coordination effect, non-homeostasis, and contact killing.

References

Aziz N, Faraz M, Pandey R, Shakir M, Fatma T, Varma A, Barman I, Prasad R (2015) Facile algae-derived route to biogenic silver nanoparticles: synthesis, antibacterial and photocatalytic properties. Langmuir 31:11605–11612

Balusamy B, Kandhasamy YG, Senthamizhan A, Chandrasekaran G, Subramanian MS, Tirukalikundram K (2012) Characterization and bacterial toxicity of lanthanum oxide bulk and nanoparticles. J Rare Earth 30:1298–1302

Bhuyan T, Mishra K, Khanuja M, Prasad R, Varma A (2015) Biosynthesis of zinc oxide nanoparticles from Azadirachta indica for antibacterial and photocatalytic applications. Mater Sci Semicond Process 32:55–61

Bowler PG, Duerden BI, Armstrong DG (2001) Wound microbiology and associated approaches to wound management. Clin Microbiol Rev 14:244–269

Chang YN, Zhang M, Xi L, Zhang J, Xing G (2012) The toxic effects and mechanisms of CuO and ZnO nanoparticles. Materials 5:2850–2871

Galhardi CM, Diniz YS, Rodrigues HG, Faine LA, Burneiko RC, Ribas BO, Novelli ELB (2005) Beneficial effects of dietary copper supplementation on serum lipids and antioxidant defenses in rats. Ann Nutr Metab 49:283–288

Jacob NM, Madras G, Kottam N, Thomas T (2014) Multivalent copper doped ZnO nanoparticles with full solar spectrum absorbance and enhanced photoactivity. Ind Eng Chem Res 53:5895–5904

Koziej D, Lauria A, Niederberger M (2014) Metal oxide nanoparticles in material science. Adv Mater 26:235–257

Kumar KC, Gupta S, Chander Y, Singh AK (2005) Antibiotic use in agriculture and its impact on the terrestrial environment. Adv Agron 87:1–54
Macomber L, Imlay JA (2009) The iron-sulfur clusters of dehydratasesare primary intracellular targets of copper toxicity. Proc Natl Acad Sci USA 106:8344–8349
Moos PJ, Chung K, Woessner D, Honeggar M, Cutler NS, Veranth JM (2010) ZnO particulate matter requires cell contact for toxicity in human colon cancer cells. Chem Res Toxicol 23:733–739
Nel AE, Madler L, Velegol D, Xia T, Hoek EMV, Somasundaran P, Klaessig F, Castranova V, Thompson M (2009) Understanding biophysicochemical interactions at the nano-bio interface. Nat Mater 8:543–557
Noipa K, Rujirawat S, Yimnirun R, Promarak V, Maensiri S (2014) Synthesis, structural, optical and magnetic properties of copper doped ZnO nanorods prepared by a simple direct thermal decomposition route. Appl Phys A 117:927–935
Padmavathy N, Vijayaraghavan R (2008) Enhanced bioactivity of ZnO nanoparticles-an antimicrobial study. Sci Technol Adv Mater 9:035004 (1)–035004 (7)
Pisanic TR, Jin S, Shubayev VI (2009) Nanotoxicity: from in vivo and in vitro models to health risks. Wiley, London, pp 397–425
Prasad R, Swamy VS (2013) Antibacterial activity of silver nanoparticles synthesized by bark extract of Syzygium cumini. J Nanoparticles. doi:10.1155/2013/431218
Prasad R, Pandey R, Barman I (2016) Engineering tailored nanoparticles with microbes: quo vadis. WIREs Nanomed Nanobiotechnol 8:316–330
Qu J, Yuan X, Wang X, Shao P (2011) Zinc accumulation and synthesis of ZnO nanoparticles using Physalis alkekengi L. Environ Pollut 159:1783–1788
Rajiv P, Rajeshwari S, Venckatesh R (2013) Bio-Fabrication of zinc oxide nanoparticles using leaf extract of Parthenium hysterophorus L. and its size-dependent antifungal activity against plant fungal pathogens. Spectrochim Acta A112:384–387
Rao L, Cui Q, Xu X (2010) Electronic properties and desolvation penalties of metal ions plus protein electrostatics dictate the metal binding affinity and selectivity in the copper efflux regulator. J Am Chem Soc 132:18092–18102
Rathore JS, Upadhyay M (2013) Investigation of zinc concentration in some medicinal plant leaves. Res J Pharm Sci 2:15–17
Santo CE, Lam EW, Elowsky CG, Quaranta D, Domaille DW, Christopher CJ, Grass G (2011) Bacterial killing by dry metallic copper surface. Appl Environ Microbiol 77:794
Sharma A (2011) Antimicrobial resistance: no action today, no cure tomorrow. Indian J Med Microbiol 29:91–92
Singh N, Manshian B, Jenkins GJS, Griffiths SM, Williams PM, Maffeis TGG, Wright CJ, Doak SH (2009) NanoGenotoxicology: the DNA damaging potential of engineered nanomaterials. Biomaterials 30:3891–3914
Soenen SJH, Himmelreich U, Nuytten N, Pisanic TR, Ferrari A, De Cuyper M (2010) Intracellular nanoparticle coating stability determines nanoparticle diagnostics efficacy and cell functionality. Small 6:2136–2145
Toduka Y, Toyooka T, Ibuki Y (2012) Flow cytometric evaluation of nanoparticles using side-scattered light and reactive oxygen species-Mediated fluorescence-Correlation with genotoxicity. Environ Sci Technol 46:7629–7636
Ueno T, Yokoi N, Abe S, Watanabe Y (2007) Crystal structure based design of functional metal/protein hybrids. J Inorg Biochem 101:1667–1675
Verma A, Uzun O, Hu Y, Hu Y, Han HS, Watson N, Chen S, Irvine DJ, Stellacci F (2008) Surface-structure regulated cell-membrane penetration by monolayer-protected nanoparticles. Nat Mater 7:588–595
Walch M, Dotiwala F, Mulik S, Thiery J, Kirchhausen T, Clayberger C, Krensky AM, Martinvalet D, Lieberman J (2014) Cytotoxic cells kill intracellular bacteria through granulysin-mediated delivery of granzymes. Cell 157:1309–1323

Xia T, Kovochich M, Liong M, Mädler L, Gilbert B, Shi H, Yeh JI, Zink JI, Nel AE (2008) Comparison of the mechanism of toxicity of zinc oxide and cerium oxide nanoparticles based on dissolution and oxidative stress properties. ACS Nano 2:2121–2134

Yang H, Liu C, Yang DF, Zhang HS, Xi Z (2009) Comparative study of cytotoxicity, oxidative stress and genotoxicity induced by four typical nanomaterials: the role of particle size, shape and composition. J Appl Toxicol 29:69–78

Yu H, Ming H, Gong J, Li H, Huang H, Pan K, Liu Y, Kang Z, Wei J, Wang D (2013) Semiconductor supported gold nanoparticles for photodegradation of rhodamine B. Bull Mater Sci 36:367–372

Chapter 7
Behavior of Nanomaterials in Soft Soils: A Case Study

Zaid Hameed Majeed and Mohd Raihan Taha

7.1 Introduction

There is a severe shortage of desirable soil especially in developing world due to its extensive demand for domestic and industrial developments. As a consequence, undesirable soil must be looked for as alternative and therefore efforts should be concentrated on making it a useful entity and productive enough in the near future, given the fast depletion of desirable soil in nature (Sariosseiri and Muhunthan 2009).

In general, weak soils have caused problems in buildings, embankments, pavements, etc., and hence its engineering properties need to be enhanced.

Soft soil is a type of problematic soil which can be found in areas with high water content, namely, approaching that of the liquid limit, which results in high settlement potential with low shear strength. The construction of infrastructures, such as road embankments and bridge foundations, on soft soils in many civil engineering projects has prompted the introduction of many approaches for soil treatments.

The design solution may include the expensive option of removal and replacement of the undesirable soils. Another design option includes utilizing ground improvement alternatives such as sand drain, grouting, and chemical stabilization.

Soil stabilization is the technique of increasing the strength and durability while decreasing compressibility, permeability, shrinkage limits, and swelling by using mechanical and/or chemical methods (Calik and Sadoglu 2014). The traditional

Z.H. Majeed (✉)
Department of Civil Engineering, College of Engineering, University of Babylon, Babil, Iraq
e-mail: Eng.zaid.hameed@uobabylon.edu.iq

M.R. Taha
Department of Civil and Structural Engineering, Faculty of Engineering and Built Environment, Universiti Kebangsaan Malaysia (UKM), 43600 Bangi, Selangor, Malaysia
e-mail: dr.mrt@eng.ukm.my

M. Ghorbanpour et al. (eds.), *Nanoscience and Plant–Soil Systems*, Soil Biology 48,
DOI 10.1007/978-3-319-46835-8_7

stabilizers such as cement, lime, ash in different sources (fly ash, rice husk ash, and leaf boom ash) have been extensively researched and their stabilization mechanisms have also been discussed extensively (Rahmat and Kinuthia 2011).

During the recent years, there has been a great deal of interest in nanotechnology and nanoparticles (Brar et al. 2009). Some researchers have introduced nanomaterials for use in soil stabilization.

7.2 Soft Soil

Soft soil characteristics include high compressibility, low shear strength, and low permeability. All these characteristics will eventually lead to low bearing capacity and excessive settlement problem. Soft soils are also sensitive and their strength can be reduced by slight disturbances.

According to Kempfert and Gebreselassie (2006), "soft soils" can be defined as clay or silty clay soil which is geologically young and is under a stable condition due its own weight yet has not undergone significant secondary consolidation since its formation. Moreover, the soils are just capable of carrying its own overburden weight and any imposed additional load will result in relatively large deformation.

Clays according to the Unified Soil Classification System (USCS) are fine-grained soils with more than 50 % by weight passing No. 200 US Standard Sieve (0.075 mm) which have much larger surface areas than coarse-grained soils and responsible for the major physical and mechanical differences between coarse-grained soils.

7.2.1 Nanomaterials

Nanotechnology has changed our vision, expectations, and abilities to control the material world (Sobolev et al. 2008). The developments in nanoscience can also have a great impact on the field of construction materials (Pacheco-Torgal and Jalali 2011). Nanotechnology deals with the production and application of physical, chemical, and biological systems at scales ranging from a few nanometers to submicron dimensions, as well as the integration of resulting nanostructures into large systems (Bhushan 2007; Prasad 2014; Prasad et al. 2014, 2016).

Nanomaterials can be metals, ceramics, polymeric materials, or composite materials. Three groups of nanomaterials can be distinguished based on their geometry or shape: quantum well (1 nanosized dimension), quantum wire (2 nanosized dimensions), and quantum dot (3 nanosized dimensions) (Poole and Owens 2003). One of the principal structural units in nanotechnology is quantum dot or nanoparticles, which can be represented as a cluster of tens to thousands of atoms 1–100 nm in diameter (Sobolev and Gutiérrez 2005).

Nanomaterials are poised for widespread use in the construction industry, where they can offer significant advantages for a variety of applications ranging from making more durable concrete to self-cleaning windows (Hosseini et al. 2010; Lee et al. 2010; Givi et al. 2011; Akhnoukh 2013; Ubertini et al. 2014).

Moreover, most properties at the micro scale remain approximately the same as those of the bulk materials. The decrease in one or more geometric dimensions down to the nanoscale completely modifies the behavior of the material. Thus, high surface-to-volume ratio and cation exchange capacity exist at the nanoscale (Olga and Hanna 2014). Nanoparticles interact actively with other particles and solutions, and very minute amounts may lead to considerable effects on the physical and chemical properties of a material. Gravitational force at the nanoscale can be disregarded. Instead, electromagnetic forces are dominant (Mercier et al. 2002).

At nanometer scale, the contacting surfaces between soil particles can be more perfect as the size of particle is smaller. The friction in soil particle at nano scale can change over the class of magnitude (a range of values a designated lower value and an upper value ten times as large) depending on the relative arrangement of the contacting surfaces. Sometimes, the particle movement is restricted in one direction. In other cases the particle rolls or rotates or slides in plane, which can increase shear forces (Falvo and Superfine 2000). In addition, nanoparticles significantly enhance the mechanical properties of different materials, i.e., metals, polymers, ceramic, and concrete composites (Sobolev and Sanchez 2012).

The idea of using nanomaterial to improve soil comes from the inter-particles concept. According to Montesh (2003, 2005) the difference between the particle size contributes to inter-particle filling or interlayer filling, which reduce the void ratio. Moreover, flocculation and dispersion of clay particles can play an important role in hydraulic conductivity. One of the important factors that increase the flocculation is the electrolyte concentration. A small diffuse double layer (DDL) leads to a decrease of electrostatic repulsion, which results to move clay particles toward each other and become flocculation, which causes a increase in hydraulic.

7.3 Materials

7.3.1 Soils Used in the Experimental Work

Three soils were used in this study to evaluate the effects of nanomaterials. Soil samples were obtained from three different locations in Malaysia, shown in Fig. 7.1. Location (A), shown in Fig. 7.2 as S1 is on the Engineering Campus of the UniversitiSains Malaysia, located in Transkrian area is approximately 500 km north of Kuala Lumpur. Location (B), represented by S2, is located in the Sungai Manggis district of the town of Banting, approximately 57 km west of Kuala Lumpur. Location (C), represented by S3, is located in Rengit, Batu Pahat, Johor, approximately 300 km south of Kuala Lumpur. The three soils used were classified

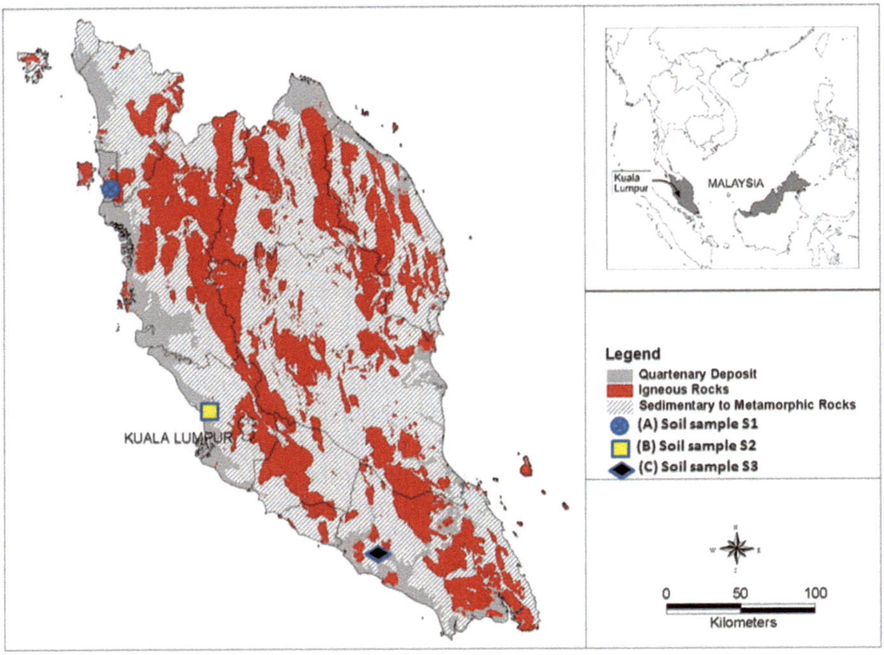

Fig. 7.1 Soil sample collection sites in Malaysia

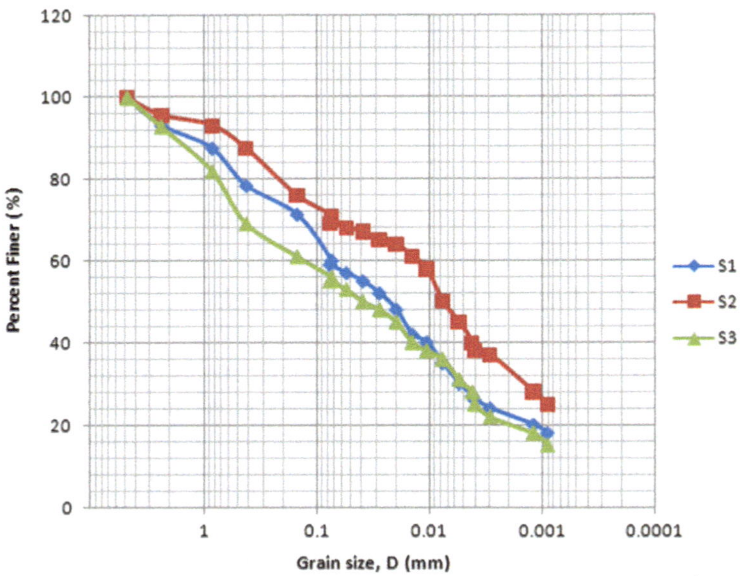

Fig. 7.2 Grain size distributions of the tested soils

Table 7.1 Physical and chemical properties of the soils

Characteristics	Values and descriptions		
	S1	S2	S3
Natural water content (%)	35.53	43.33	29.83
Organic content (%)	12.17	1.31	4.88
Specific gravity	2.42	2.75	2.49
Ph	3.24	4.25	7.50
Clay fraction (%)	29.8	36.2	22.3
Silt fraction (%)	31.3	31.3	33.2
Sand fraction (%)	38.9	32.5	44.5
Liquid limit index (%)	46.35	50.61	52.40
Plasticity index (%)	18.25	25.61	21.1
Linear shrinkage (%)	11.07	8.24	8.96
Unified soil classification (USCS)	OL	CH	MH
Optimum water content (%)	21.60	24.80	17.30
Maximum dry unit weight (kN/m^3)	14.44	15.68	17.46
Unconfined compressive strength (kN/m^2)	90.0	43.0	65.0
Chemical composition			
SiO_2 (%)	61.72	57.03	56.23
Al_2O_3 (%)	17.53	23.63	25.55
Fe_2O_3 (%)	3.61	7.33	4.68
MgO (%)	1.16	0.73	0.40
CaO (%)	0.06	0.04	0.25
TiO_2 (%)	0.89	1.80	1.03
Na_2O (%)	0.40	0.31	0.04
K_2O (%)	2.92	2.51	0.37
Others	11.71	5.31	6.57

according to Unified Soil Classification System (ASTM 2005), respectively, as organic silty soil (OL), clay high plasticity soil (CH), and high plasticity silt soil (MH) as shown in Table 7.1.

All soil samples were disturbed samples. They were collected from a depth of 0–0.75 m below ground level by big plastic bag. After transportation to the laboratory of soil mechanics in UKM University, the soil samples were air-dried and then oven-dried at 100 ± 5 °C over 24 h. Thereafter, pulverization machine was used to crush the soil with proper mixing to obtain uniform soil samples. Thereafter, the soil samples were stored in big plastic drums. Table 7.1 presents the physical and chemical properties of the three soils.

7.3.2 *Nanomaterials*

Three types of nano materials were used in this study, i.e., nano-copper, nano-alumina, and nano-magnesium. Some properties of the nanomaterials are discussed below.

1. *Nano-Copper*
   The copper oxide nanopowder used in this study has a purity of 99.99 %, and supplied by Inframat Advanced Materials, Manchester, CT, USA. The copper oxide nanopowder is insoluble in water. It dissolves slowly in alcohol or ammonia solution. It is also soluble in dilute acids, NH_4Cl, (NH_4) $2CO_3$, potassium cyanide solution under high temperature. Nano-copper oxide is a widely used material and been applied as catalyst for superconducting materials, thermoelectric materials, sensing materials, glass, ceramics, etc. A scanning electronic microscope image (SEM) for copper oxide nanopowder is shown in Fig. 7.3 and an X-ray diffraction (XRD) patterns for nano-copper showed in Fig. 7.4. The average particle diameter for copper oxide nanopowder is about 100 nm. The other properties for copper oxide nanopowder are shown in Table 7.2.
2. *Nano-Alumina*
   The nano-alumina material used in this study is Ultrapure Gamma-Alumina (γ-Al_2O_3) powder with purity of 99.99 %, with high activity and low melting temperature. The nanomaterial has a large surface area and high catalytic activity. It can be made into micro porous spherical structure or honeycomb

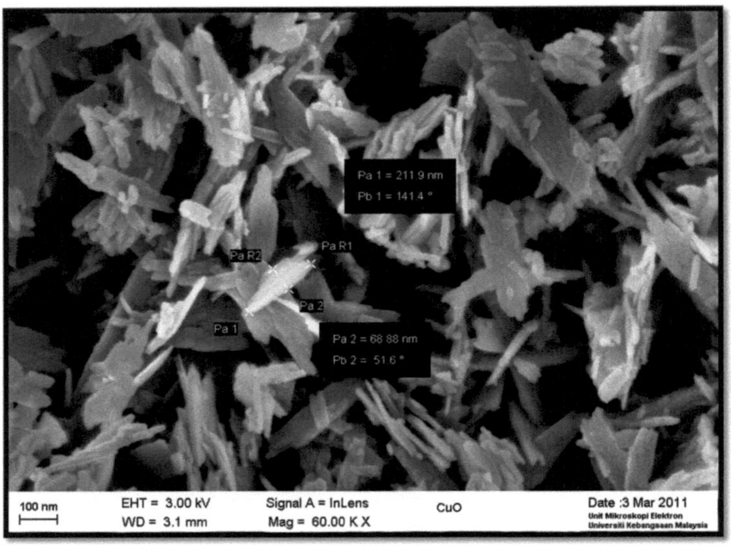

Fig. 7.3 Scanning electronic microscope image for copper oxide nanopowder

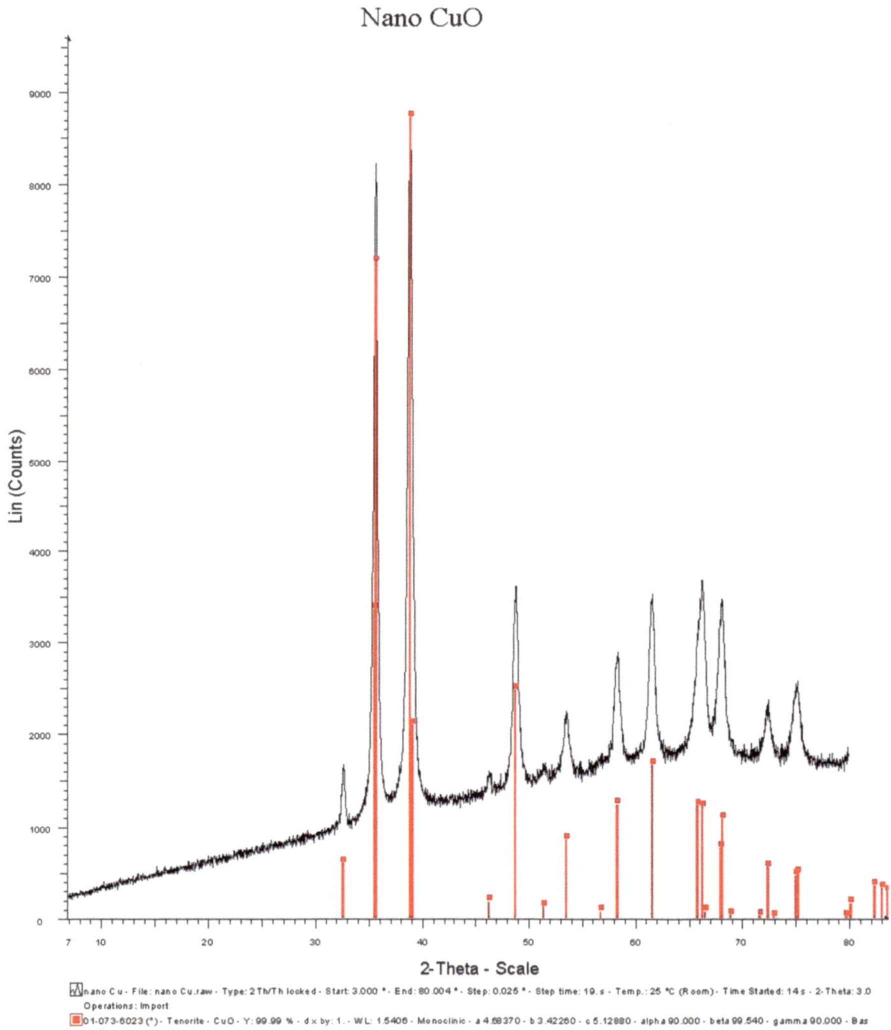

Fig. 7.4 X-ray diffraction (XRD) patterns for copper oxide nanopowder

structure of catalytic materials. These kinds of structures can be excellent catalyst carriers. If used as industrial catalysts, they will be the main materials for petroleum refining, petrochemical and automotive exhaust purification. This powder was supplied by Inframat Advanced Materials, Manchester, CT, USA. A scanning electronic microscope image (SEM) shows the nano-alumina powder in Fig. 7.5, moreover the XRD patterns of nano-alumina shown in Fig. 7.6. The particle size of gamma phase nano-alumina ranged from 20 to 50 nm and the other general properties of nano-alumina powder are shown in Table 7.3.

Table 7.2 Properties of copper oxide (CuO) nano powders

Property	Value
Formula	CuO
Particle density (g/cm^3)	6.3–6.49
Surface area (m^2/g)	N/A
Average particle size (nm)	100
Solubility in water (%)	Insoluble
Melting point (°C)	1326
Composition	
Al_2O_3	N/A
Cu	99.99 %
Appearance and odor	Black powder Odorless
Crystal structure	N/A

Fig. 7.5 Scanning electronic microscope image for nano alumina powder

3. *Nano-Magnesium*
   The nano-magnesium oxide (N-MgO) of high purity was supplied by Inframat Advanced Materials Company, Manchester, USA. The specifications and all the information provided by the company are shown in Table 7.4. Figures 7.7 and 7.8 respectively show the scanning electronic microscope (SEM) and XRD patterns for the nano-magnesium. The average particles size was 100 nm.

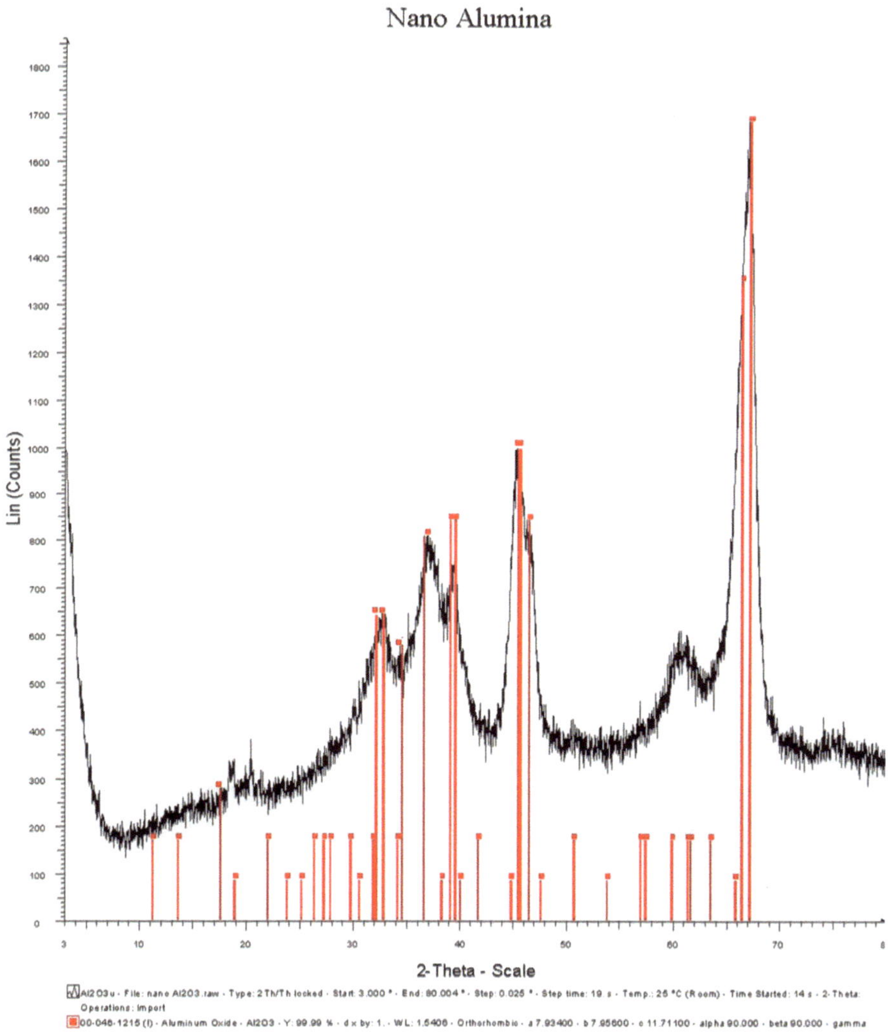

Fig. 7.6 X-ray diffraction (XRD) patterns for nano-alumina nanopowder

7.4 Preparation of Soil: Nanomaterials Mixtures

The amount of nanomaterial required for each specimen was determined by multiplying the nanomaterial percentage by the total dry weight of the soil. The mixture proportions are listed in detail in Table 7.5. Different amounts of the different nanomaterials used.

Taha and Taha (2012) have reported that the best way to obtain homogeneous samples is to mix the nanomaterials with the dry in two stages. This procedure was

Table 7.3 Properties of gamma aluminum oxide (alumina Al_2O_3) nano powders

Property	Value
Formula	Al_2O_3
Particle density (g/cm^3)	3.6
Surface area (m^2/g)	>150
Average particle size (nm)	20–50
Solubility in water (%)	Insoluble
Composition	
Al_2O_3	≥99.99 %
Appearance and odor	White to off-white Odorless
Crystal structure	Transitional gamma

Table 7.4 Properties of magnesium oxide (magnesium MgO) nano powders

Property	Value
Formula	MgO
Particle density (g/cm^3)	3.6
Surface area (m^2/g)	>50
Average particle size (nm)	100
Composition	
MgO	≥99.90 %
Appearance and odor	White to pearl Odorless
Crystal structure	Transitional gamma

Fig. 7.7 Scanning electronic microscope image of nano-magnesium

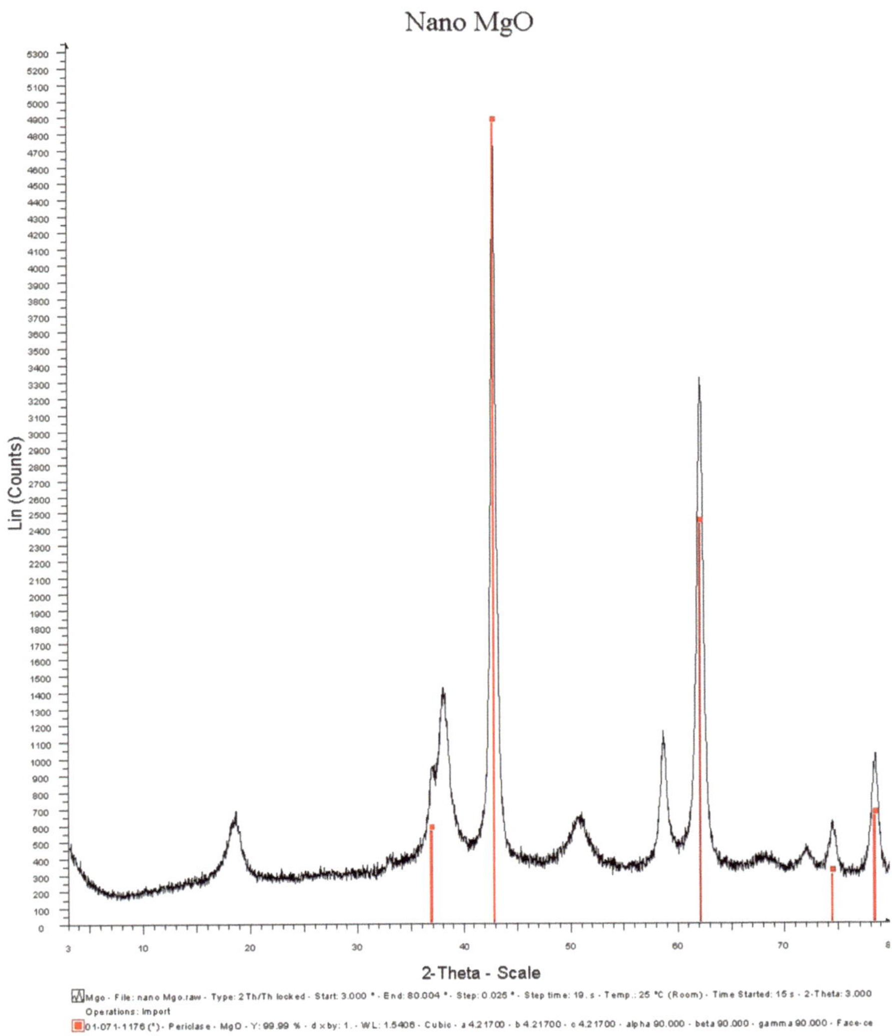

Fig. 7.8 X-ray diffraction (XRD) patterns for nano-magnesium nanopowder

adopted in current study. First, premix or hand mixed the quantity of soil was divided into ten portions and each portion was spread in a square pan (50 × 50 cm), then the required amount of nanomaterial was sprayed using sieve number 200 as shown in Fig. 7.9.

In the second stage, the resulting mixtures were mixed together as shown in Fig. 7.10 for at least 3 h using a horizontal cylindrical mixer to ensure that the samples were homogeneous. In order to avoid the segregation problem, the soil–nanomaterial mixture was mixed with the required water content directly then the

Table 7.5 Mix proportions for nanomaterials (by % weight of dry soil)

Nano-copper (%)	Nano-alumina (%)	Nano-magnesium (%)
0.0	0.0	0.0
0.3	0.05	0.1
0.5	0.075	0.2
0.7	0.1	0.3
0.8	0.15	0.35
1.0	0.2	0.4

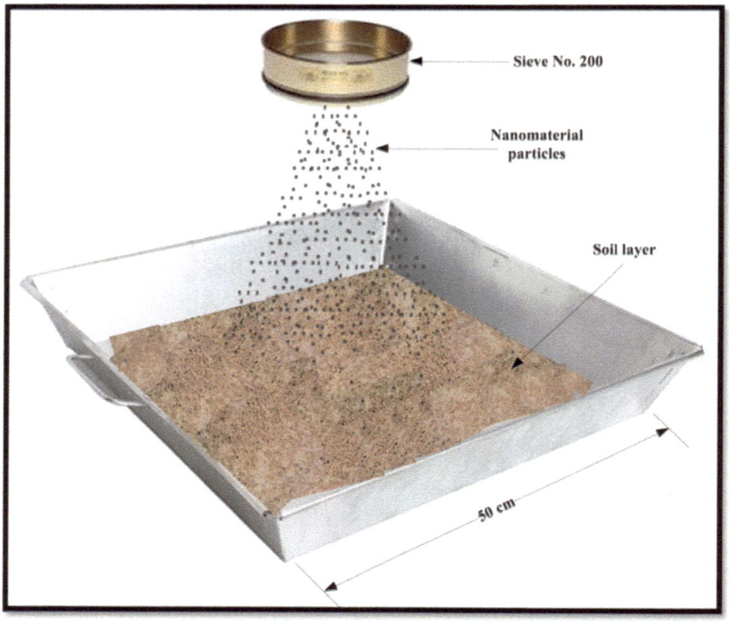

Fig. 7.9 Spraying of nanomaterial on the soil layer

entire mixture was placed in a sealed plastic bag and left for 24 h for hydration. This mixing method was repeated every time needs to prepare a new mixture. This procedure was found to be the best method to obtain homogeneous samples since homogeneous color was obtained after compaction (Fig. 7.11). However, proper care was taken to prepare homogeneous mixtures at each stage.

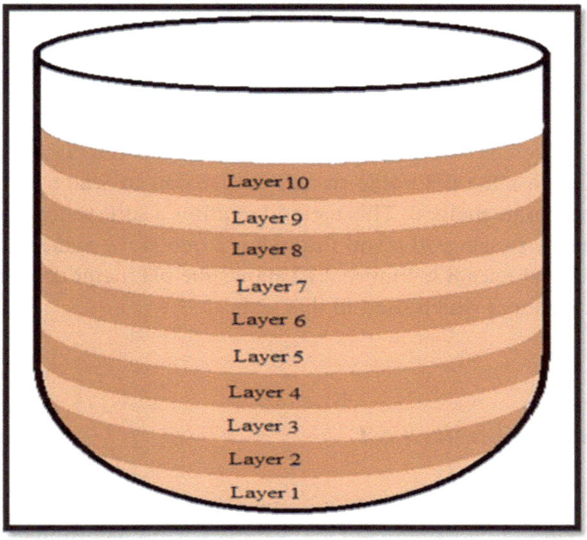

Fig. 7.10 Mixing all quantity in a pot

Fig. 7.11 The homogeneous color of samples obtain after compaction

7.5 Experimental Laboratory Test Program

7.5.1 The pH Test

In this work, the electrometric method (BS 1377: part 3: 1990: clause 9) was utilized to determine the pH value of soil suspension in water and is considered to be the most accurate method. This method requires a soil–water ratio of 1:2.5 and can be obtained by passing 30 g soil through a 2 mm sieve and diluted with 75 ml distilled water for at least 8 h. Figure 7.12 shows the pH meter device with samples preparation procedures which used in this test.

7.5.2 Atterberg Limits Test

The Atterberg limits (i.e., the liquid limit, the plastic limit, and the plasticity index) of each of the natural and stabilized soil samples were determined in accordance with BS 1377, part 2, 1990. The liquid limit tests were performed using a penetration cone assembly, which consisted of a stainless steel cone with a cone angle of $30^\circ \pm 1^\circ$. The plastic limit tests were performed using a manual method. Each sample was rolled at a sufficient pressure on a glass plate to form a thread with a uniform diameter of 3.2 mm along the full length of the sample. The plasticity index

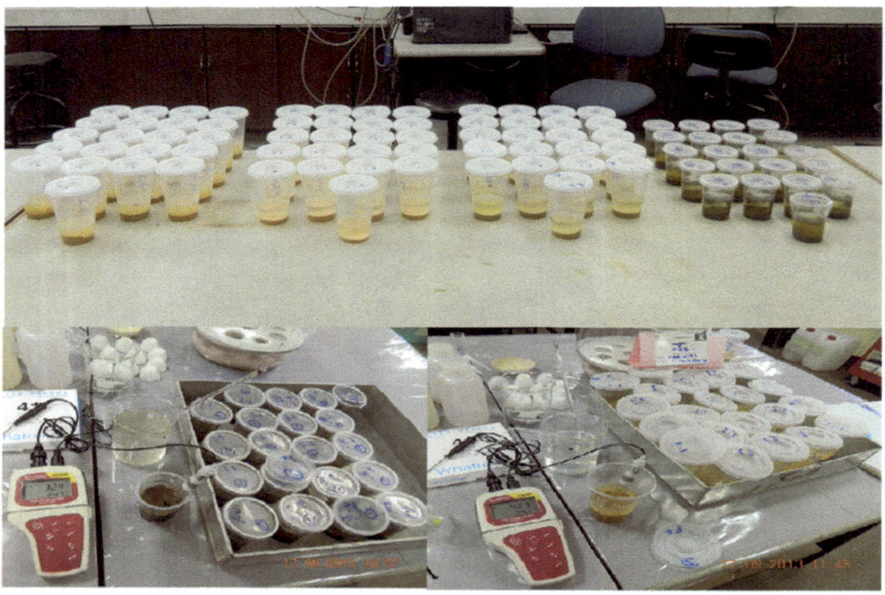

Fig. 7.12 Sample preparation procedures for pH test

was calculated as the difference between the water contents at the liquid and plastic limits.

7.5.3 Fourier Transform Infrared Spectroscopy

The Fourier Transform Infrared (FT-IR) model PerkinElmer System 2000 was used to analyze the possible chemical bonding existing in the untreated and treated soil. Each 25–30 g sample was dried in a microwave oven at the 100 °C for 30 min. The sample was subjected to light grinding in a mortar and the sample was then filtered through a coarse sieve to remove large structures. Approximately 5 g of each filtered sample was ground, passed through a 0.074 mm sieve, and then stored in closed glass vials pending further examination.

The specimens were introduced as pellets prepared from powder mixed with KBr and spectra were recorded in the range 400–4000 cm^{-1}. KBr powder pellets were used as a background.

7.5.4 X-ray Diffraction

X-ray diffraction (XRD) analysis of untreated and treated soft soil was carried out using a BRUKER AXS D8 ADVANCE machine as shown in Fig. 7.13. The XRD patterns were obtained using a Cu-Ka (k = 1.5148 A) X-ray tube with an input voltage of 40 kV and a current of 40 mA. A continuous scan mode and scan rate of 2° per minute was selected. Air-dried powdered samples of treated and untreated soil samples were used as shown in Fig. 7.14. Soil mineralogy obtained from the test provides the basis for understanding the basic mechanisms of chemical stabilization.

7.5.5 Field-Emission Scanning Electron Microscope

Field-emission scanning electron microscope (FESEM) was conducted to disclose the morphology of the native materials and to observe the morphology of the materials after mixing and curing. The change in the soil aggregation and formation of new material can be observed by this analysis, knowing that same specimens were used to perform EDX analysis mentioned previously. Figure 7.15 shows the instrument employed for FESEM and EDX analysis.

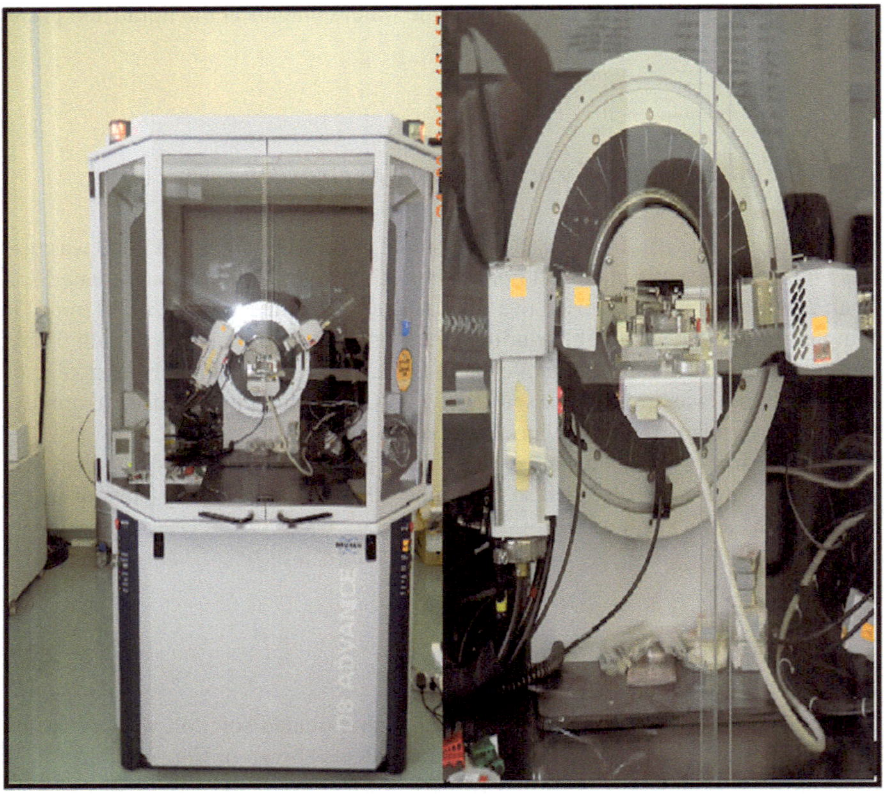

Fig. 7.13 The XRD instrument

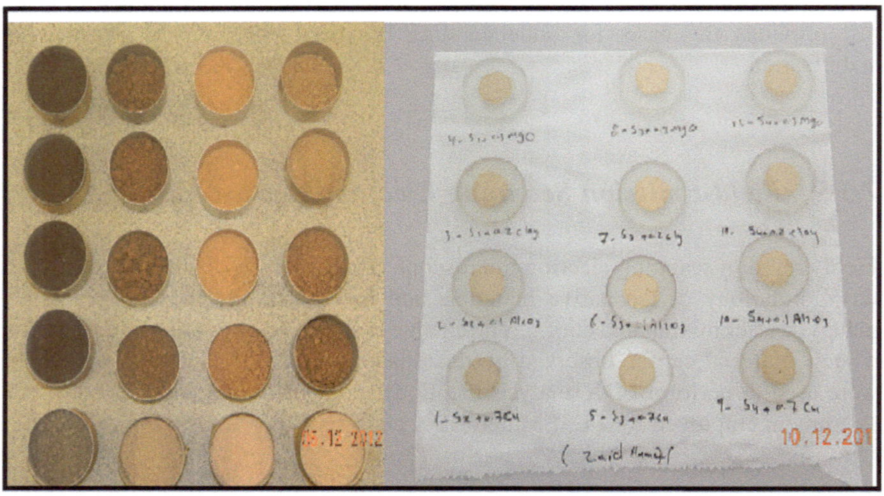

Fig. 7.14 Sample preparation procedures for XRD test

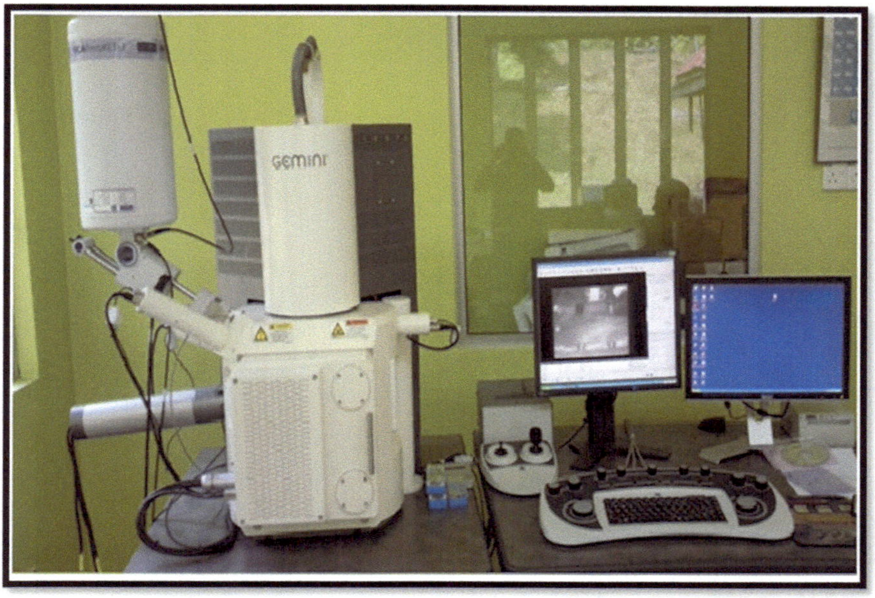

Fig. 7.15 Instrument employed for FESEM and EDX analysis

7.5.6 X-ray Fluorescence Analysis

X-ray fluorescence (XRF) analysis was conducted for the native materials, i.e., virgin soil, and nanomaterials (nano-copper, nano-alumina, and nano-magnesium) to obtain their chemical compositions and to create a perception about the expected reactants. The soil was milled sieved on 425 μm before XRF analysis. For each material, a small compressed disk specimen was prepared as shown in Fig. 7.16. Figure 7.17 presents the machine employed for XRF analysis.

7.6 Physicochemical Aspects of Nanomaterials Treatment

7.6.1 The pH Test

The effect of nanomaterials on pH value were investigated to evaluate the chemical reactions after mixing nanomaterials with the tested soils. The influence of nano-copper, nano-alumina, and nano-magnesium on the pH of different soil samples is shown in Figs. 7.18, 7.19, and 7.20, respectively. The test results showed that the pH of the stabilized soil samples increased with increasing nanomaterials percentage. All nanomaterials had pH higher than the control soils.

Fig. 7.16 Specimens prepared for XRF analysis

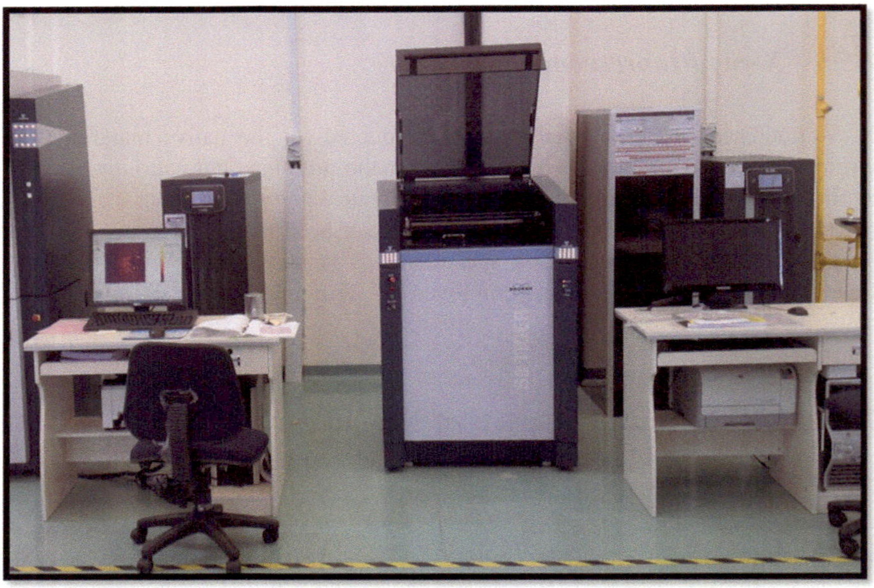

Fig. 7.17 XRF equipment used in this study

Fig. 7.18 pH for different nano-copper percentages after mixing with different soils

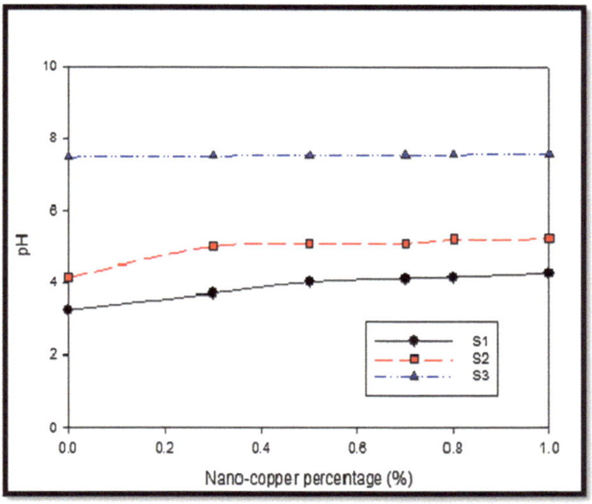

Fig. 7.19 pH for different nano-alumina percentages after mixing with different soils

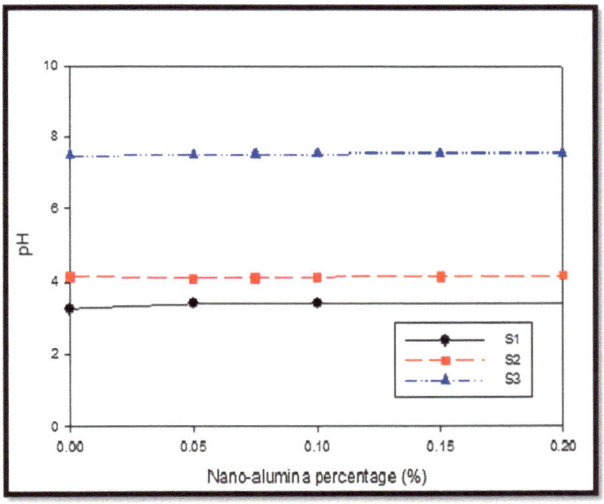

The treatment of the samples with nanomaterial content changed the pH, and the value increased with increasing nanomaterial content for all samples. The pH value of S1, S2, and S3 soil samples treated with nano-copper and nano-alumina increased when the percentage of additive increased due to low pH values of S1, S2, and S3 soil samples (3.24, 4.25, and 7.50, respectively). Moreover, the increase in pH value for samples treated with nano-copper was higher than nano-alumina because the additive content for nano-copper was greater than nano-alumina, also theses changes can be attributed to the properties associated with the original soil sourced from the tropical region. Due to climate effect and the origin of the soil is acidic in nature.

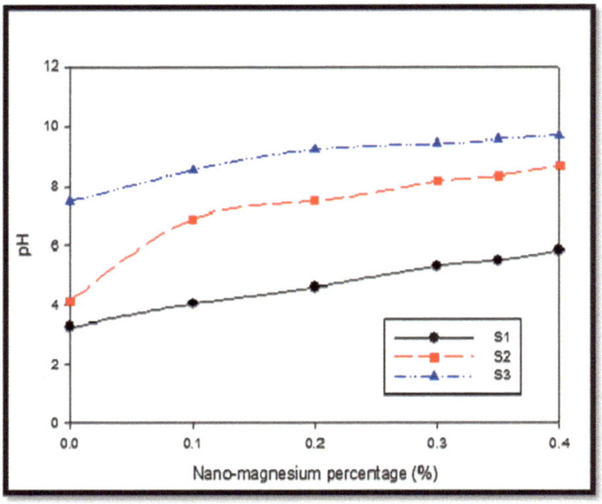

Fig. 7.20 pH for different nano-magnesium percentages after mixing with different soils

Moreover, the chemical action of nanomaterials by formation new hydroxide as nano-magnesium, which is the formation of Mg $(OH)_2$ elucidated above, is followed by the ionization process inducing liberation of hydroxyl ions $(OH)^{-1}$ which is possibly the reason behind the increase the pH value. The following chemical equations can explain the formation and ionization processes (Mo et al. 2012; Shand 2006; Zhang et al. 2011).

$$MgO + H_2O \Rightarrow Mg(OH)_2 \tag{7.1}$$

$$Mg(OH)_2 \Rightarrow Mg^{+2} + 2(OH)^{-1} \tag{7.2}$$

However, the increase in pH value is also one of the indications of increasing the cementation of colloidal particles (Muhunthan et al. 2008).

7.6.2 X-ray Diffraction Analyses

Diffraction analysis of stabilized soil is important to determine the changes that occur in the mineralogical phase reactions. These reactions depend on the chemical and mineralogical composition of each soil and additive. Figures 7.21, 7.22, and 7.23 respectively show the XRD patterns for untreated soil samples S1, S2 and S3. As shown in Figs. 7.21, 7.22, and 7.23, quartz and kaolinite are clearly appearing in the untreated soft soils. In this case, the XRD peaks of quartz are significantly larger than the peaks of other minerals because quartz is better X-ray-scattered than other compounds. The peak located at 2-Theta 26.67° was identified as quartz. The d-spacing shows the change in crystals of materials. It is defined as the distance between two successive structure units and is dependent on the type of the

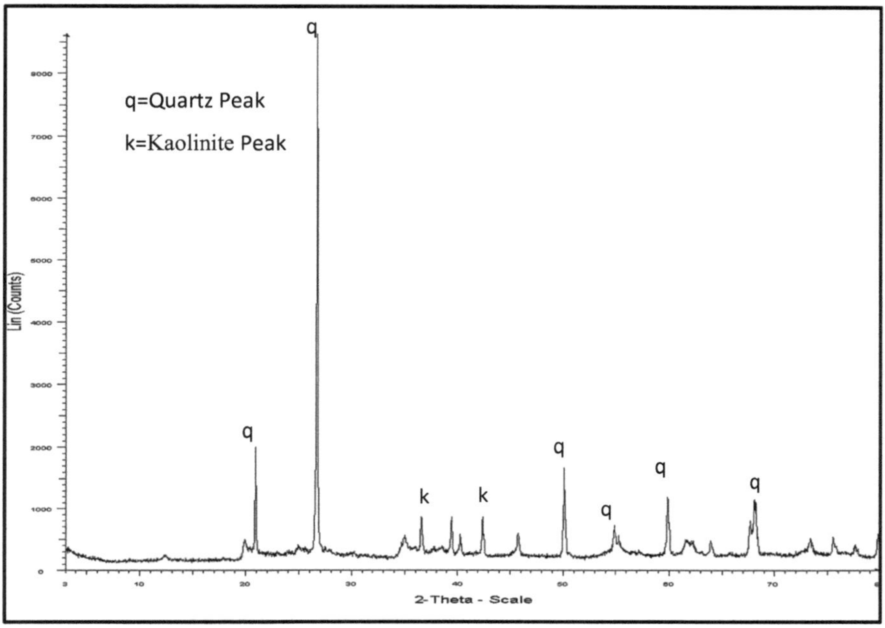

Fig. 7.21 XRD patterns for unstabilized soil sample S1

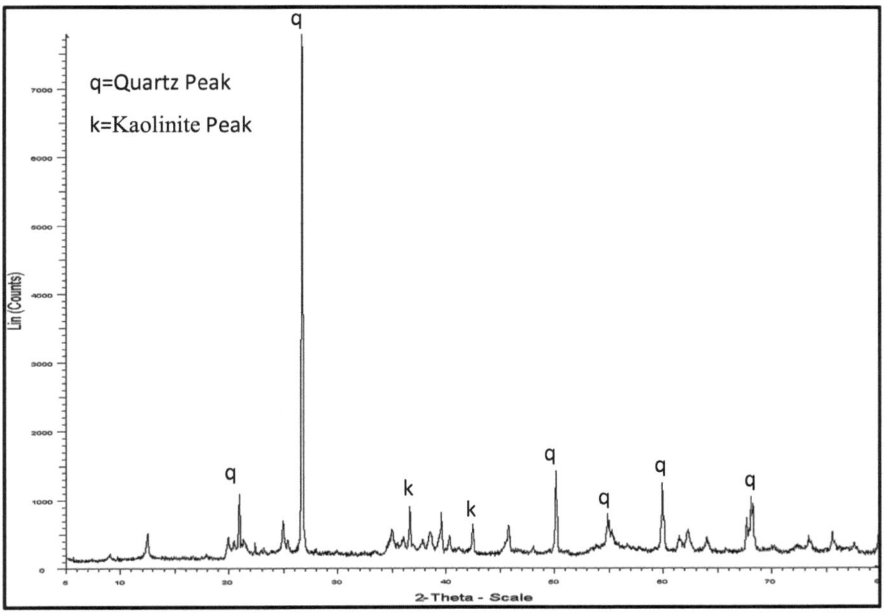

Fig. 7.22 XRD patterns for unstabilized soil sample S2

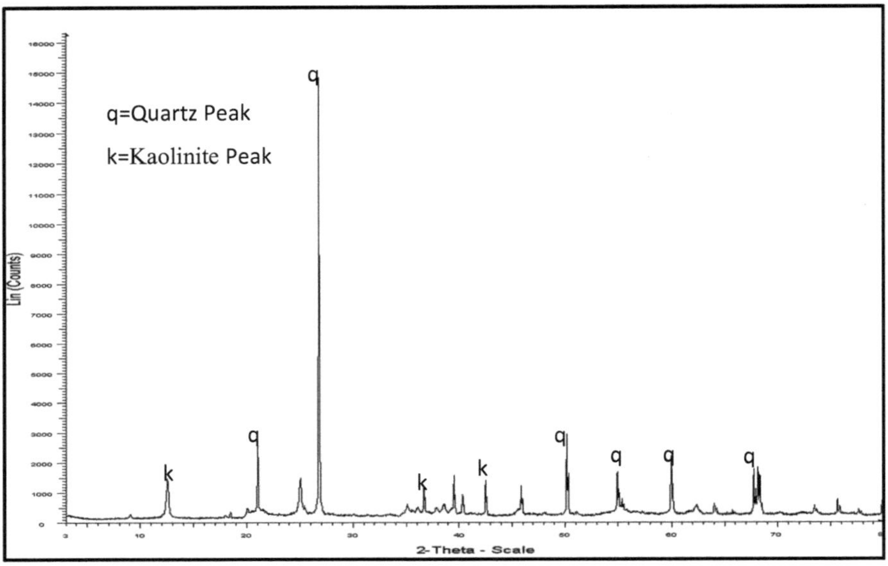

Fig. 7.23 XRD patterns for unstabilized soil sample S3

Table 7.6 The d-spacing values of different soils obtained for the quartz peak

Soil no.	Original soil (Å)	Nano-copper (Å)	Nano-alumina (Å)	Nano-magnesium (Å)
S1	3.3421	3.33562	3.3218	3.32854
S2	3.34305	3.34317	3.32187	3.33656
S3	3.33464	3.32328	3.31686	3.32996

Table 7.7 The intensity values of different soils obtained for the quartz peak

Soil no.	Original soil (Count)	Nano-copper (Count)	Nano-alumina (Count)	Nano-magnesium (Count)
S1	8675	10262	11910	10911
S2	7816	9170	7536	8521
S3	5426	6161	6233	5884

exchangeable cation, the solution composition, and the clay composition (Amorim et al. 2007). The d-spacing for the soils before and after adding the nanomaterials are shown in Table 7.6. For the sake of simplicity, only the optimum nano percentages and the highest quartz peaks for each soil are shown. From Table 7.6, it can be seen that the d-spacing reduced slightly which reflects the effect of the added nanomaterials. To provide further evidence of the chemical reactions of the soils-nanomaterials under consideration, the intensity values for the optimum nanomaterials content and the highest quartz peaks for each soil are tabulated in Table 7.7. It can be seen from the data in Table 7.7 that the intensity increased in all the stabilized soils. This intensity increment again emphasized the effect of the added nanomaterials.

As mentioned before, the XRD patterns of nanomaterials-treated soils reveal several new peaks as compared to the untreated soils. This result may be explained by the fact that chemical action of the nanomaterials agreed with the results obtained from the strength and compressibility test.

7.6.3 Fourier Transform-Infrared Analysis

FT-IR spectroscopy is an important tool for investigating structural changes in soil-stabilized surface. Previous studies used FT-IR to examine the effect of treatment on the changes in the chemical bonds in the stabilizer surface (Millogo et al. 2008).

FT-IR was performed to evaluate the mineral soil bands, which were tested between 400 and 4000 cm^{-1} (Du et al. 2008). The representative FT-IR spectra of the soil samples without nanomaterials are shown in Figs. 7.24, 7.25, and 7.26. The observed wavenumber from all the spectra are given in Tables 7.8, 7.9, and 7.10 along with their corresponding mineral names.

The FT-IR spectra indicated quartz and kaolinite as major constituents and other minerals as minor constituents. Strong bands between 3696 and 3622 cm^{-1} indicated the possibility of hydroxyl linkage. However, a broad absorption band at 3431 cm^{-1} in the clay spectrum suggests the possibility of water hydration in the adsorbent. The bands at 3696, 3625, 1033, and 914 cm^{-1} showed the presence of kaolinite (Manoharan et al. 2012). Kaolinite, the major constituent of clays, provides sharp absorption bands in the 3700–3600 cm^{-1} region. Quartz is a nonclay mineral that is common and invariably present in all samples. Si–O bonds are the

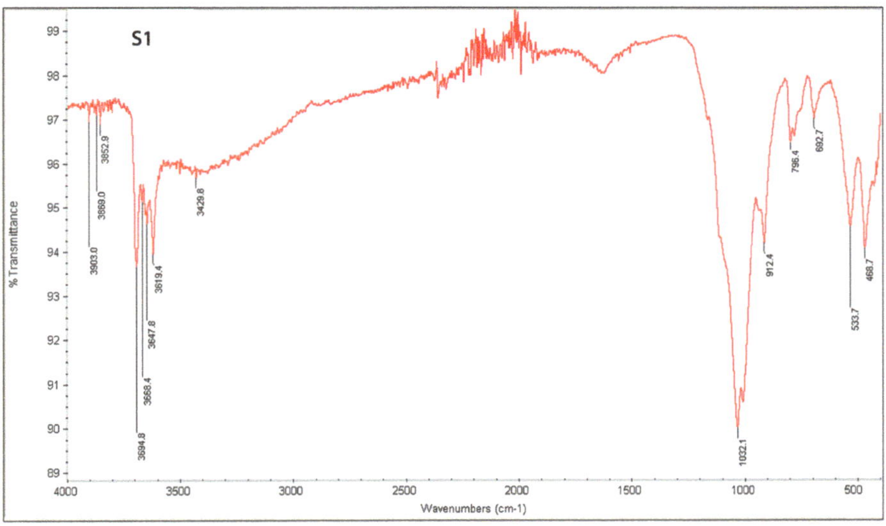

Fig. 7.24 Fourier-transform infrared (FT-IR) spectra of original soil S1

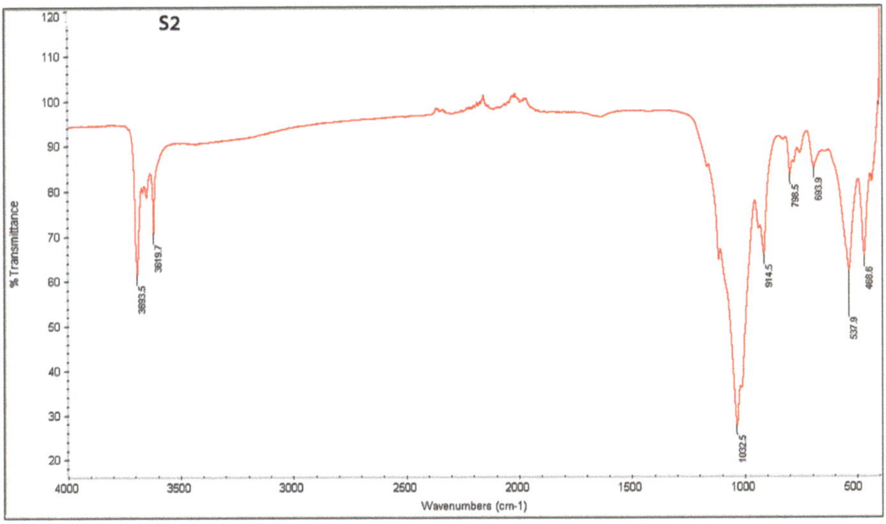

Fig. 7.25 Fourier-transform infrared (FT-IR) spectra of original soil S2

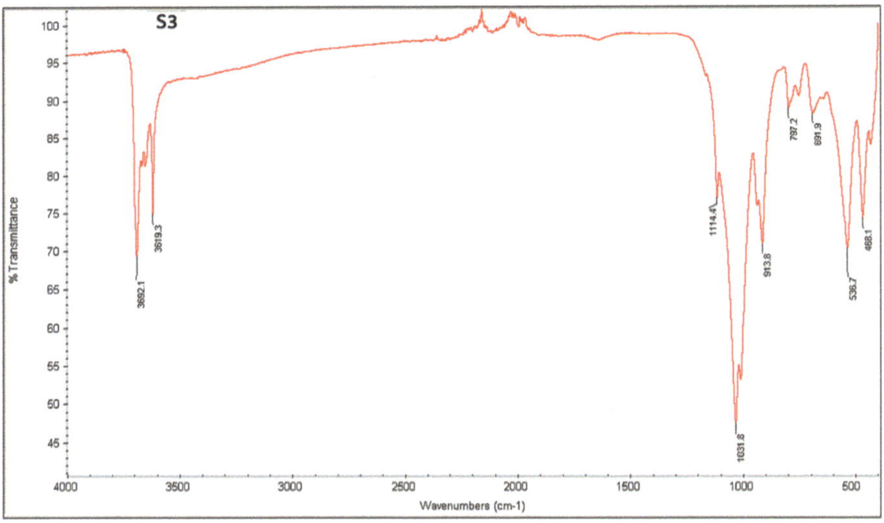

Fig. 7.26 Fourier-transform infrared (FT-IR) spectra of original soil S3

strongest in the silicate structure and can be readily recognized in the infrared spectra of such minerals. The presence of quartz in the samples correspond to Si–O–Si asymmetrical bending vibrations in the 461–467 cm^{-1} region (Manoharan et al. 2012). The strong absorption band at 692 cm^{-1} correspond the Si–O symmetrical bending vibration of quartz. The absorption bands at 779–792 cm^{-1} corresponded to Si–O symmetrical stretching vibrations, whereas those at 1082

Table 7.8 FT-IR spectra (cm^{-1}) data with their corresponding tentative assignments for soil samples without nanomaterial treatment

S1	S2	S3	Tentative assignment	Mineral
3694	3693	3692	In-plane degenerated vibration of the water molecule	Kaolinite
3619	3619	3619	OH stretching of inner-surface hydroxyl groups	Kaolinite
3429	–	–	H–O–H stretching of water molecules	Montmorillonite
–	–	1114	Si–O stretching (longitudinal mode)	Quartz
1032	1032	1031	Si–O stretching clay mineral	Kaolinite
912	914	913	Al_2O–H deformation	Kaolinite
796	798	797	Si–O stretching	Quartz
692	693	691	Si–O stretching	Quartz
533	537	536	Si–O–Al (or) Fe_2O_3	Hematite
468	468	468	Si–O–Si bending	Quartz

Table 7.9 FT-IR spectra (cm^{-1}) data with their corresponding tentative assignments for soil samples with nanomaterial treatment

S1 with			S2 with				
N-Cu	N-Al_2O_3	N-MgO	N-Cu	N-Al_2O_3	N-MgO	Tentative assignment	Mineral
3695	3694	3695	3692	3691	3691	In-plane degenerated vibration of the water molecule	Kaolinite
3620	3620	3619	3619	3651	3651	OH stretching of inner-surface hydroxyl groups	Kaolinite
3397	3382	3382	3415	3422	3619	H–O–H stretching of water molecules	Montmorillonite
–	–	–	–	–	3404	H–O–H stretching of water molecules	Montmorillonite
1640	1630	1640	1642	1640	1644	H–O–H stretching	Illite
–	–	–	–	1114	–	Si–O stretching (longitudinal mode)	Quartz
1030	1029	1030	1031	1029	1030	Si–O stretching of clay mineral	Kaolinite
914	912	913	914	913	913	Al_2 O–H deformation	Kaolinite
795	778	795	798	797	797	Si–O stretching	Quartz
691	692	691	–	692	692	Si–O stretching	Quartz
533	530	532	534	533	534	Si–O–Al (or) Fe_2O_3	Hematite
468	463	464	465	465	467	Si–O–Si bending	Quartz

and 1123 cm^{-1} corresponded to Si–O a symmetrical stretching vibrations (Bertaux et al. 1998). Illite and kaolinite are common main clay minerals in soils and natural aerosols. The observed strong absorption frequencies in the 1620–1646 cm^{-1} region in the clay spectrum suggests the possibility of water hydration in the adsorbent (Gritco et al. 2005). The broad absorption band at 1032 cm^{-1} corresponded to the Si–O stretching of kaolinite which is a clay mineral (Manoharan et al. 2012). The very weak peak at 915 cm^{-1} is due to the $A1_2O$–H

Table 7.10 FT-IR spectra (cm^{-1}) data with their corresponding tentative assignments for soil samples with nanomaterial treatment

S3 with				
N-Cu	N-Al_2O_3	N-MgO	Tentative assignment	Mineral
3690	3691	3690	In-plane degenerated vibration of the water molecule	Kaolinite
3651	3652	3651	O–H stretching of inner hydroxyl group	Kaolinite
3619	3619	3619	H–O–H stretching of water molecules	Montmorillonite
3404	3413	3397	H–O–H stretching of water molecules	Montmorillonite
2117	2114	–	–	Quartz
1641	1640	1642	H–O–H stretching	Illite
1114	1114	1114	Si–O stretching (longitudinal mode)	Quartz
1029	1030	1029	Si–O stretching clay mineral	Kaolinite
1007	1007	1006	In-plane Si–O stretching	Kaolinite
912	912	912	Al_2 O–H deformation	Kaolinite
796	796	796	Si–O stretching	Quartz
689	691	690	Si–O stretching	Quartz
534	534	534	Si–O–Al (or) Fe_2O_3	Hematite
466	466	466	Si–O–Si bending	Quartz

deformation of kaolinite. The strong absorption band at 533 cm^{-1} corresponded to the stretching vibration of Si–O–Al (or) Fe_2O_3 in the presence of hematite (Manoharan et al. 2012).

The FT-IR analysis was performed on treated and untreated soft soil to investigate their functional groups. There is a clear difference in spectra observed in the FT-IR results of treated and untreated samples. The presence of the functional groups of nanomaterials in the soil indicates the possible formation of ionic bonding between the clay mineral lattices and the nanomaterial functional groups, as well as the existence of nanomaterials in the interlayer spacing of soil minerals. These changes were probably caused by the action of stabilizer on the clay structure. In addition, the additional spectra observed in the FT-IR results of nanomaterials-treated soil confirm the presence of new functional groups such as Illite (1644 cm^{-1}) and the kaolinite (1029 cm^{-1}). The presence of the functional groups of nanomaterials-treated soil confirms the formation of ionic bonding between the clay mineral lattices and the nanomaterials functional groups and also the existence of nanomaterials in the interlayer spacing of clay minerals.

7.6.4 Scanning Electron Micrographs (SEM) Observations

The changes of microstructural and microstructural development of soils due to nano-copper, nano-alumina, and nano-magnesium addition play a significant role in the geotechnical properties and the mechanical behavior of these stabilized soils. Figure 7.27a, b illustrate the SEM-micrograph of natural untreated soil S1. The

(a) SEM micrograph of S1@ Mag= 5.00 KX

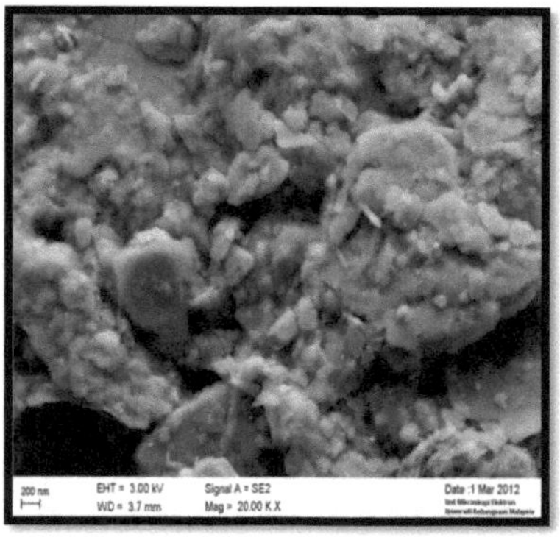

(b) SEM micrograph of S1@ Mag= 20.00KX

(c) SEM micrograph of S2@ Mag= 10.00 KX

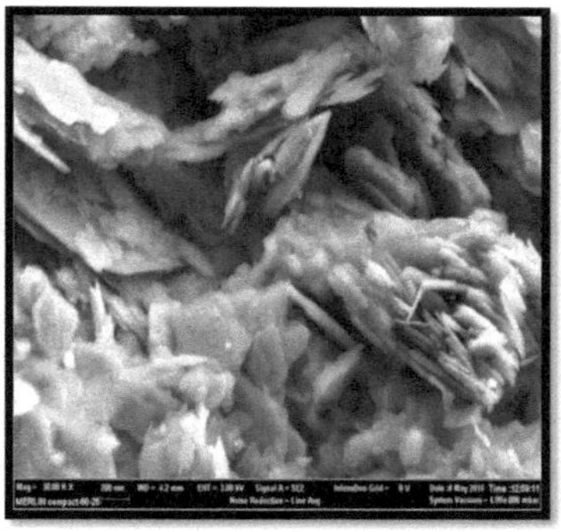

(d) SEM micrograph of S2@ Mag= 30.00 KX

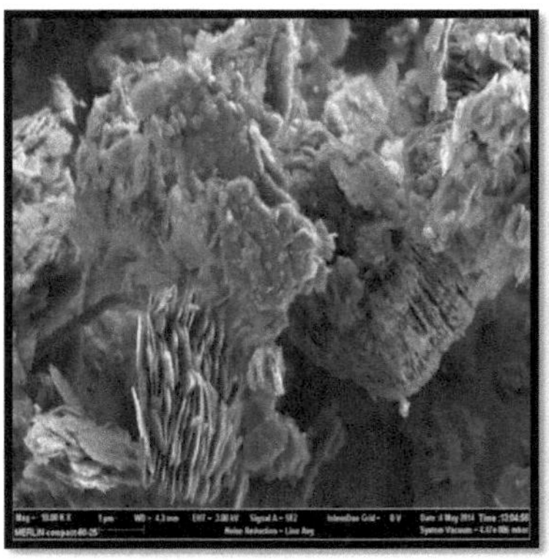

(e) SEM micrograph of S3@ Mag= 10.00 KX

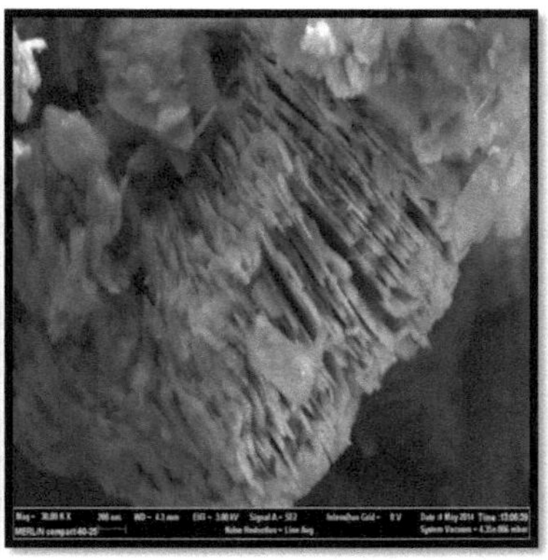

(f) SEM micrograph of S3@ Mag= 30.00 KX

Fig. 7.27 SEM micrograph of the tree natural, untreated soil samples at different magnification. (**a**) SEM micrograph of S1@ Mag = 5.00 KX. (**b**) SEM micrograph of S1@ Mag = 20.00 KX. (**c**) SEM micrograph of S2@ Mag = 10.00 KX. (**d**) SEM micrograph of S2@ Mag = 30.00 KX. (**e**) SEM micrograph of S3@ Mag = 10.00 KX. (**f**) SEM micrograph of S3@ Mag = 30.00 KX

(a) FESEM micrographs and corresponding EDX spectra of elements in S2 soil sample with 0.7% nano-copper

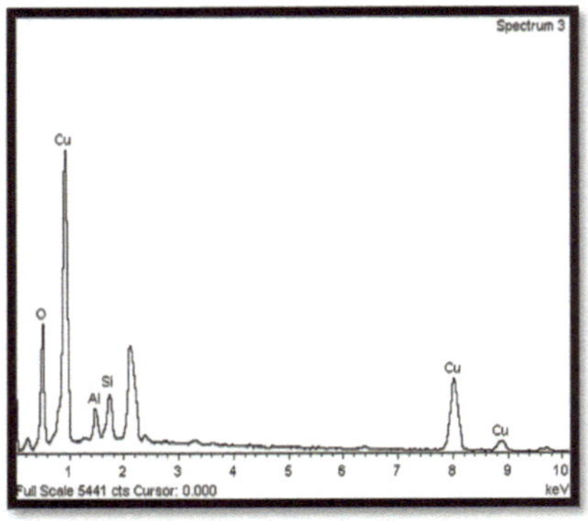

(b) EDX for spectrum which indicate the chemical composition

Fig. 7.28 Soil sample S2 mixed with 0.7 % nano-copper. (**a**) FESEM micrographs and corresponding EDX spectra of elements in S2 soil sample with 0.7 % nano-copper. (**b**) EDX for spectrum which indicate the chemical composition

micrograph shows an occurrence of detrital grains of silt and fine sand fractions and little amount of clay as a matrix between the detrital grains. The specimen has silt-fine sand like structure and characterized by an open fabric system and occurrence of relatively large voids distributed in the specimen. Figure 7.27c, d shows the SEM-micrograph of natural, untreated soil S2 which indicates the sheet-like structure and flaky arrangement of the clay particles. Figure 7.27e, f illustrates the micrograph of natural, untreated soil S3 which shows flaky arrangements of clay particles as matrix between the detrital fine grains. Also the micrograph of untreated soft soil exhibits a fairly open type of microstructure with the platy clay particles assembled in a dispersed arrangement.

Figure 7.28 illustrates the microstructure of S2 soil mixed with 0.7 % nano-copper showing the distribution of the nanomaterial in the soil sample. This image was obtained using a Field Emission Scanning Electron Microscope (FESEM) with an energy-dispersive X-ray (EDX) analysis capability. EDX revealed the energy spectrum of the X-ray character emitted from the element of copper (Cu), oxygen (O), silicon (Si), and aluminum (Al). The percentage of all elements were found in the following order: Cu > O > Si > Al (Fig. 4.7b). Copper and oxygen contents were high and widely distributed in soil. Silicon and aluminum were also found in the soil sample, but significantly lower in the constructions. These showed that the mineral composition of the soils in the study areas were CuO, SiO_2, and Al_2O_3. The distribution of the nanomaterials after mixing in soil samples can be seen in Fig. 7.29. Figure 7.30a, b, c, d shows the micrographs of nano-magnesium treated soft soil specimens. There is a sign of reticulation and the flocculated nature of the structure becomes more evident, with treated clay particle clusters interspersed with large openings and the clay particles were being coated and bound by the

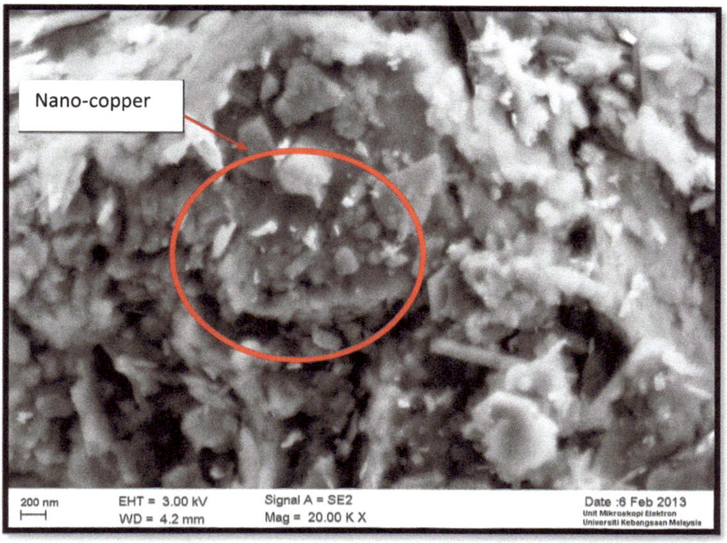

(a) Sample S1 mixed with 0.5% nano-copper

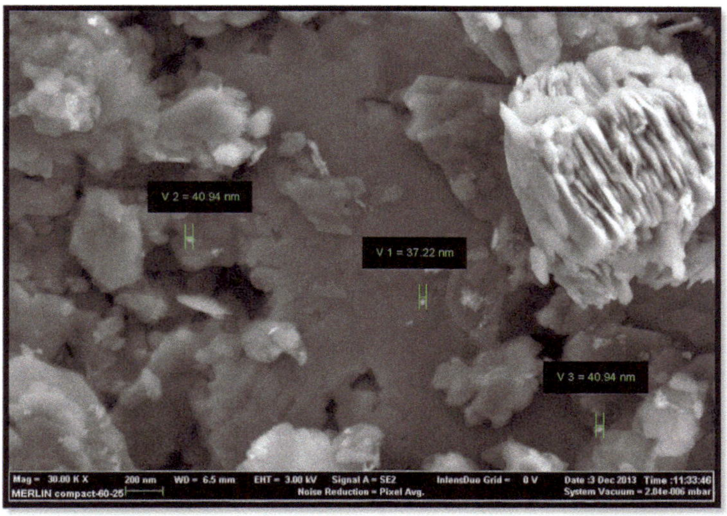

(b) Sample S3 mixed with 0.3% nano-copper

Fig. 7.29 FESEM micrograph of soil samples mixed with nano-copper. (**a**) Sample S1 mixed with 0.5 % nano-copper. (**b**) Sample S3 mixed with 0.3 % nano-copper

magnesium gel, which appear as a kind of spongy gel. At the same time, the particle cluster of the structure becomes less evident and degree of reticulation appears to increase. The increase in the degree of reticulation can be attributed to the chemical

(a) FESEM micrograph of S2 with 0.3% nano-magnesium

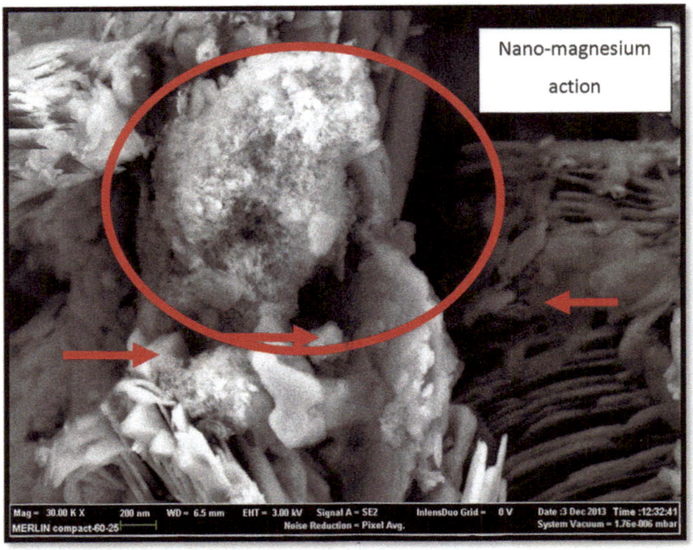

(b) The spongy gel of nano-magnesium

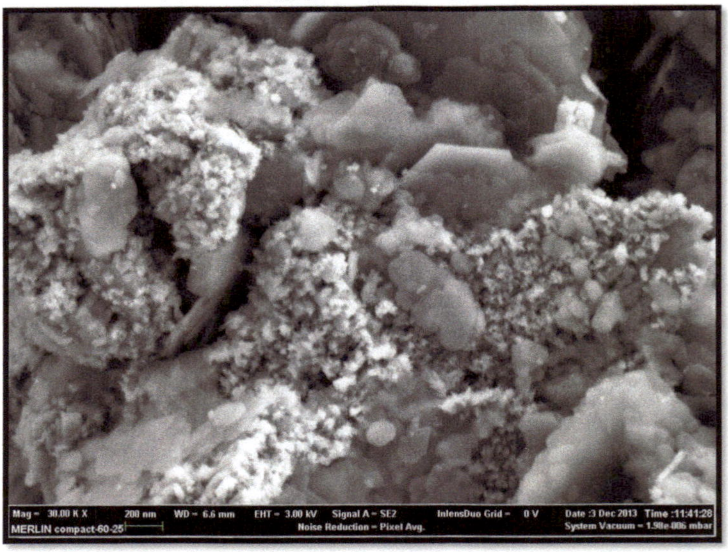

Fig. 7.30 FESEM images of S3 and S2 soil samples treated with nano-magnesium and its chemical action. (**a**) FESEM micrograph of S2 with 0.3 % nano-magnesium. (**b**) The spongy gel of nano-magnesium. (**c**) FESEM micrograph of S3 with 0.3 % nano-magnesium. (**d**) The nano-magnesium action with S3

(c) FESEM micrograph of S3 with 0.3% nano-magnesium

(d) The nano-magnesium action with S3

Fig. 7.30 (continued)

action of the nano-magnesium. This is consistent with the results of the FT-IR analyses discussed above. As shown in Fig. 7.31a, b, c, d, the soil–nano-alumina mixtures micrographs illustrate the new phase consists of an interlocking network of needle like crystals. Subsequently, bridges were formed between adjacent soil particles. These interlocking networks of needle like crystals have grown into the interstices to form a continuous network. This will be further explained in terms of improvement of strength as well as compressibility behavior in later sections.

7.7 Effect of Nanomaterials on the Engineering Properties of the Soils

7.7.1 *Atterberg Limits*

Atterberg limits (Plastic limit "PL," Liquid limit "LL," and Plasticity index "PI" = LL-PL) play an important role in soil identification and classification. These parameters indicate to some of the geotechnical problems such as swell potential and workability.

The results of the liquid limit test of the soil samples containing various percentages of nano-copper, nano-alumina, and nano-magnesium used are shown in Figs. 7.32, 7.33, and 7.34, respectively. The results indicate that for all three soils, the liquid limit decreases with increasing nano-copper, nano-alumina, and nano-magnesium contents especially at optimum nanomaterials contains.

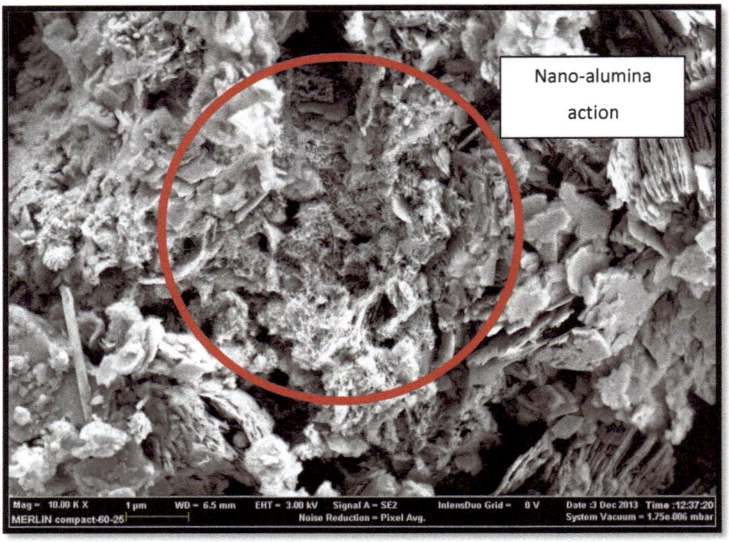

(a) FESEM micrograph of S2 with 0.1% nano-alumina

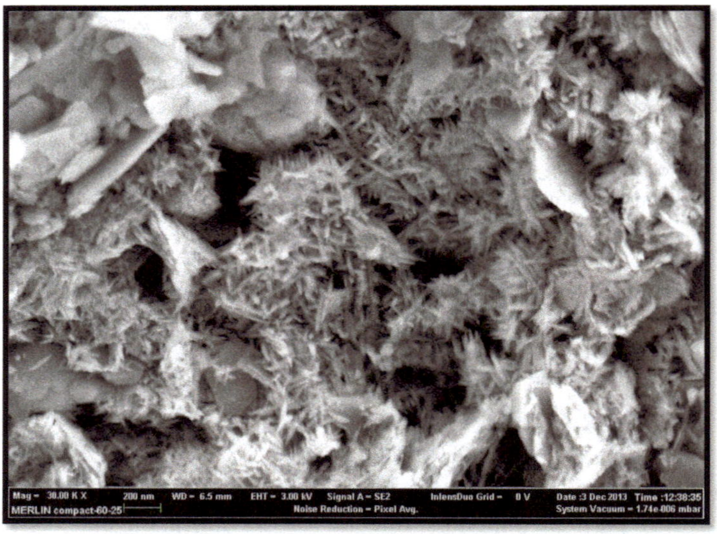

(b) Needle like crystals from the chemical action of nano-alumina

Fig. 7.31 FESEM micrograph of S2 and S3 soil samples treated with nano-alumina. (**a**) FESEM micrograph of S2 with 0.1 % nano-alumina. (**b**) Needle like crystals from the chemical action of nano-alumina. (**c**) FESEM micrograph of S3 with 0.1 % nano-alumina. (**d**) The nano-alumina action with S3

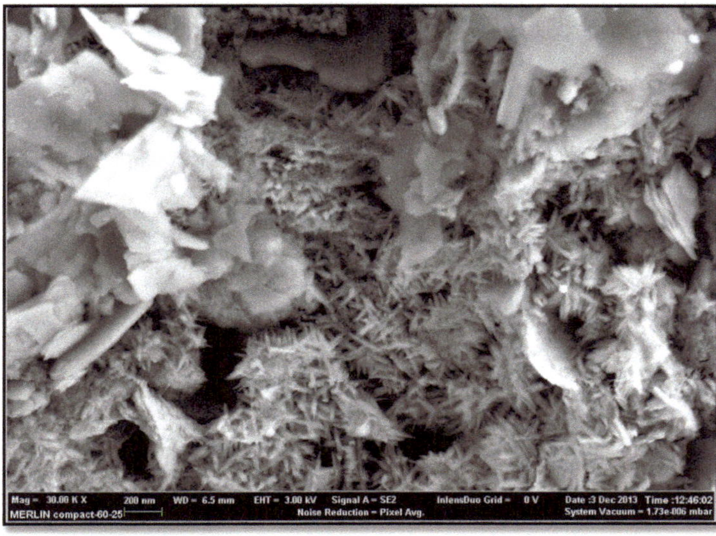

(c) FESEM micrograph of S3 with 0.1% nano-alumina

(d) The nano-alumina action with S3

Fig. 7.31 (continued)

These decreases are obviously due to the change in water content resulting from increasing nanomaterial content. Nanomaterial particles are believed to coat the clay clast, binding them together and filling the clay matrix, thus reducing the voids and the water contained in the voids (Taha and Taha 2012) (Fig. 7.35).

Fig. 7.32 Effects of nano-copper on the liquid limits

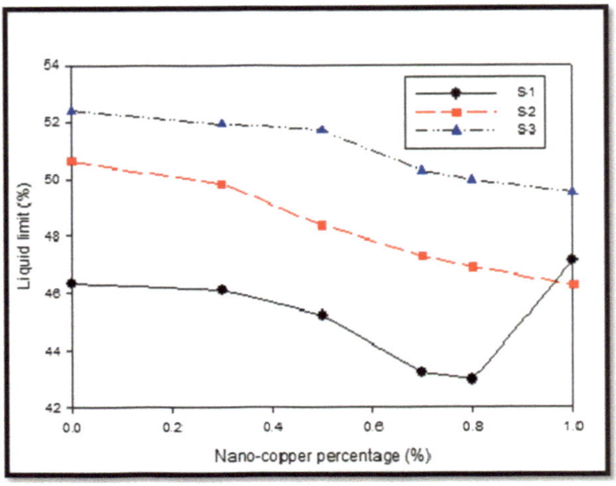

Fig. 7.33 Effects of nano-alumina on the liquid limits

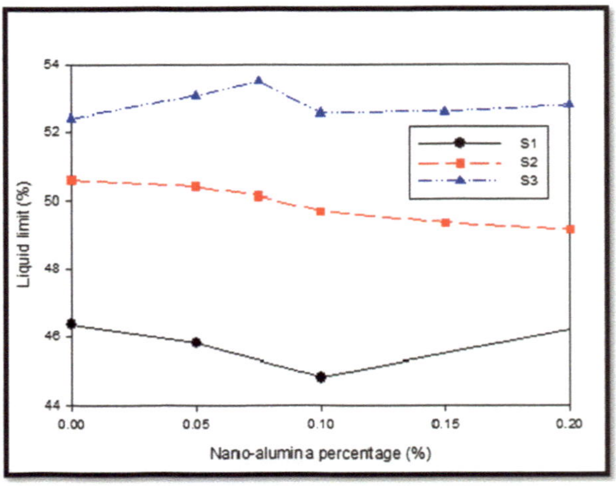

Reductions in the plasticity (Figs. 7.36, 7.37, and 7.38) indices are indicators of soil improvement, because the plasticity index represents the range of water content over which a soil is plastic (Raj 1995). Soils with a high PI may be difficult to work with in construction because of their instability and stickiness when wet. High PI soils also have potential for detrimental volume changes during wetting and drying. Thus, addition of nanomaterials to soil even at low doses can enhance its geotechnical properties. The addition of nano-alumina and nano-magnesium causes some decreases in the plasticity index while an increase in nano-copper content gives a higher decrease in the plasticity index. This is due to the higher density of the nano-alumina, nano-magnesium, and nano-copper particles compared to that of clay

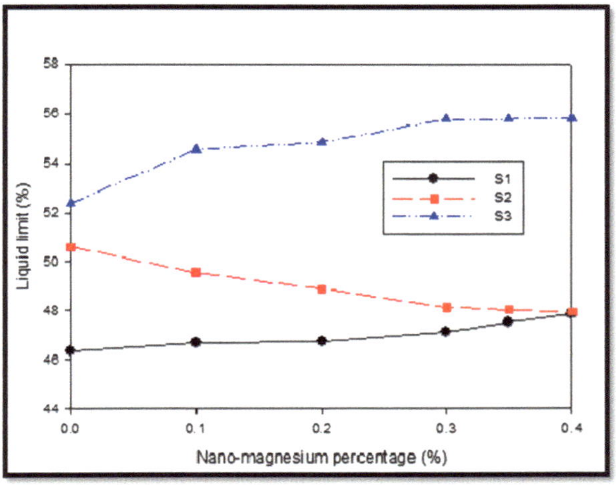

Fig. 7.34 Effects of nano-magnesium on the liquid limits

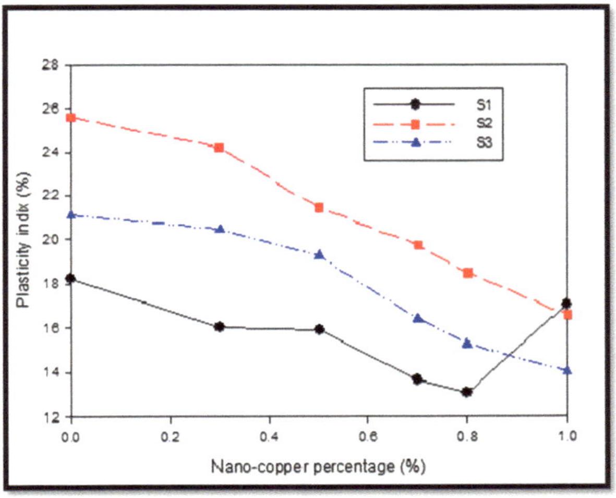

Fig. 7.35 Effects of nano-copper on the plasticity index

particles. Moreover, due to the low content of nano-alumina used with the soil, a slight decrease in plasticity index was observed. However, the reduction in plasticity index is more significant when nano-copper was used due to the high density of nano-copper particle. However, the reduction in plasticity index is more significant for soils with high plasticity index S2 and S3, and especially when nano-copper was used due to the high density of nano-copper particles.

However, the maximum reduction in plasticity index (Fig. 7.38) was noted at 1 % nano-copper content for all types of soils. The reduction in the plasticity index for soil sample S1, S2, and S3 were 1.19, 9.01, and 7.06, respectively, after adding 1 % nano-copper. The reduction in the plasticity index for soil sample S1 was 1.98

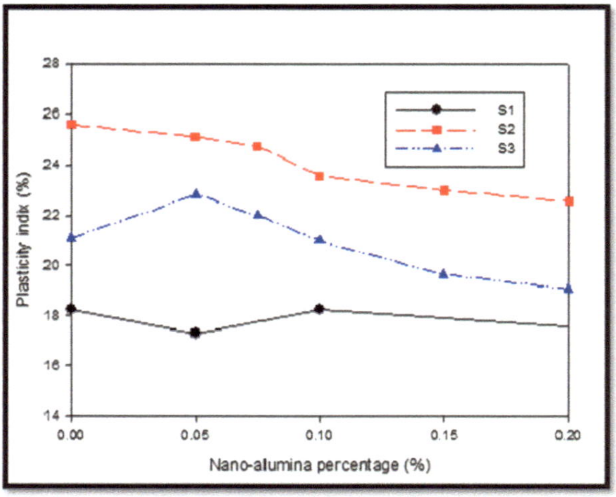

Fig. 7.36 Effects of nano-alumina on the plasticity index

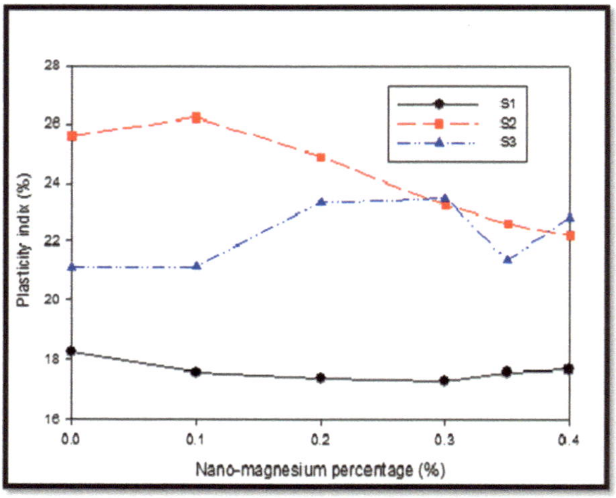

Fig. 7.37 Effects of nanomaterials on the plasticity index

and 0.94 after adding 0.45 % nano-alumina and 0.30 % nano-magnesium, respectively. There was also a reduction in plasticity index for soil sample S2 in which after adding 0.2 % nano-alumina and 0.4 % nano-magnesium, the PI reduced to 3.06 and 3.4, respectively. The reduction in the plasticity index for soil sample S3 was 2.02 after adding 0.2 % nano-alumina.

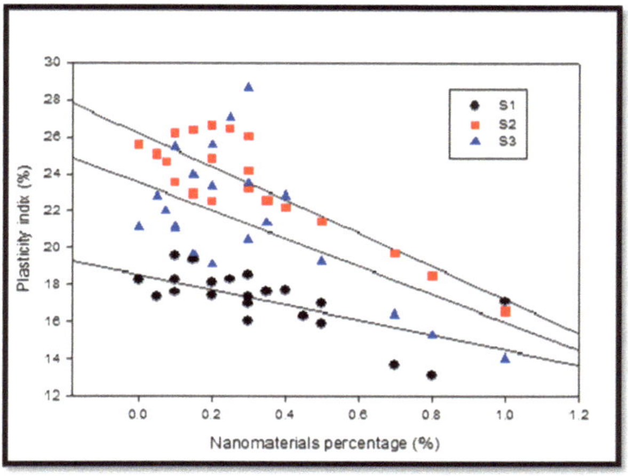

Fig. 7.38 Effects of nanomaterials on the plasticity index tested soils

7.8 Conclusion

The following conclusions are derived from the findings of this study:

1. Evidence from the FT-IR and SEM analyses has shown that the new phase consists of chemical action of the nanomaterials treatment. The success of the treatment process is dependent on the available nanomaterials content, curing time, soil type, soil pH, and clay minerals.
2. The plasticity index exhibits a significant reduction compared with the untreated soil. This reduction is in proportion with and nanomaterials surface area as well as the doses of nanomaterials.

References

Akhnoukh AK (2013) Overview of nanotechnology applications in construction industry in the united states. Micro Nanosyst 5:147–153

Amorim CLG, Lopes RT, Barroso RC, Queiroz JC, Alves DB, Perez CA, Schelin HR (2007) Effect of clay–water interactions on clay swelling by X-ray diffraction. Nucl Inst Methods Phys Res A 580:768–770. doi:10.1016/j.nima.2007.05.103

Bertaux J, Fröhlich F, Ildefonse P (1998) Multicomponent analysis of FTIR spectra: quantification of amorphous and crystallized mineral phases in synthetic and natural sediments. J Sediment Res 68:440–447

Bhushan B (2007) Springer handbook of nanotechnology. Springer, New York

Brar S, Verma M, Tyagi R, Surampalli R (2009) Nanoparticles. Contamin Emerg Environ Concern 416–445. doi:10.1061/9780784410141.ch11

Calik U, Sadoglu E (2014) Classification, shear strength, and durability of expansive clayey soil stabilized with lime and perlite. Nat Hazards 71:1289–1303

Du C, Linker R, Shaviv A (2008) Identification of agricultural Mediterranean soils using mid-infrared photoacoustic spectroscopy. Geoderma 143:85–90. doi:10.1016/j.geoderma.2007.10.012

Falvo M, Superfine R (2000) Mechanics and friction at the nanometer scale. J Nanopart Res 2:237–248

Givi AN, Abdul Rashid S, Aziz FNA, Mohd Salleh MA (2011) Particle size effect on the permeability properties of nano-SiO_2 blended Portland cement concrete. J Compos Mater 45:1173–1180

Gritco A, Moldovan M, Grecu R, Simon V (2005) Thermal and infrared analyses of aluminosilicate glass systems for dental implants. J Optoelectron Adv Mater 7:2845–2847

Hosseini P, Booshehrian A, Farshchi S (2010) Influence of nano-SiO_2 addition on microstructure and mechanical properties of cement mortars for ferrocement.

Kempfert H-G, Gebreselassie B (2006) Excavations and foundations in soft soils. Springer, Berlin

Lee J, Mahendra S, Alvarez PJJ (2010) Nanomaterials in the construction industry: a review of their applications and environmental health and safety considerations. ACS Nano 4:3580–3590

Manoharan C, Sutharsan P, Dhanapandian S, Venkatachalapathy R (2012) Spectroscopic and thermal analysis of red clay for industrial applications from Tamilnadu, India. J Mol Struct 1027:99–103. doi:10.1016/j.molstruc.2012.05.079

Mercier JP, Zambelli G, Kurz W (2002) Introduction to materials science. Elsevier Science, Kidington

Millogo Y, Hajjaji M, Ouedraogo R, Gomina M (2008) Cement-lateritic gravels mixtures: microstructure and strength characteristics. Construct Build Mater 22:2078–2086. doi:10.1016/j.conbuildmat.2007.07.019

Mo L, Deng M, Wang A (2012) Effects of MgO-based expansive additive on compensating the shrinkage of cement paste under non-wet curing conditions. Cem Concr Compos 34:377–383. doi:10.1016/j.cemconcomp.2011.11.018

Montesh G (2005) Swelling-shrinkage measurements of bentonite using coupled environmental scanning electron microscopy and digital image analysis. J Colloid Interface Sci 284:271–277

Montesh G, Duplay J, Martinez L, Mendoza C (2003) Swelling-shrinkage kinetics of Mx80 bentonite. Appl Clay Sci 22:279–293

Muhunthan B, Sariosseiri F, WSDOT Research and Library Services (2008) Interpretation of geotechnical properties of cement treated soils. Washington State Department of Transportation, Office of Research and Library Services

Olga SSK, Hanna D (2014) Nanotechnology based thermosets. Handbook of thermoset plastics, 3rd edn. Elsevier, Amsterdam, pp 623–695

Pacheco-Torgal F, Jalali S (2011) Nanotechnology: advantages and drawbacks in the field of construction and building materials. Construct Build Mater 25:582–590

Poole CP, Owens FJ (2003) Introduction to nanotechnology. Wiley, Hoboken

Prasad R (2014) Synthesis of silver nanoparticles in photosynthetic plants. J Nanopart Article ID 963961. doi:10.1155/2014/963961

Prasad R, Kumar V, Prasad KS (2014) Nanotechnology in sustainable agriculture: present concerns and future aspects. Afr J Biotechnol 13:705–713

Prasad R, Pandey R, Barman I (2016) Engineering tailored nanoparticles with microbes: quo vadis. WIREs Nanomed Nanobiotechnol 8:316–330

Rahmat MN, Kinuthia JM (2011) Effects of mellowing sulfate-bearing clay soil stabilized with wastepaper sludge ash for road construction. Eng Geol 117:170–179

Raj PP (1995) Geotechnical engineering. Tata McGraw-Hill, New Delhi

Sariosseiri F, Muhunthan B (2009) Effect of cement treatment on geotechnical properties of some Washington State soils. Eng Geol 104:119–125

Shand MA (2006) The chemistry and technology of magnesia. Wiley, New York

Sobolev K, Gutiérrez MF (2005) How nanotechnology can change the concrete world. Am Ceram Soc Bull 84:14–18

Sobolev K, Sanchez F (2012) The application of nanoparticles to improve the performance of concrete. Anjuran American Concrete Institute, ACI. Agios Nikolaos, Crete

Sobolev K, Flores I, Hermosillo R, Torres-Martinez L (2008) Nanomaterials and nanotechnology for high-performance cement composites, vol 254. ACI Special Publication, American Concrete Institute, pp 93–120

Taha M, Taha O (2012) Influence of nano-material on the expansive and shrinkage soil behavior. J Nanopart Res 14:1–13. doi:10.1007/s11051-012-1190-0

Ubertini F, Materazzi AL, D'Alessandro A, Laflamme S (2014) Natural frequencies identification of a reinforced concrete beam using carbon nanotube cement-based sensors. Eng Struct 60:265–275

Zhang T, Cheeseman CR, Vandeperre LJ (2011) Development of low pH cement systems forming magnesium silicate hydrate (M-S-H). Cem Concr Res 41:439–442. doi:10.1016/j.cemconres.2011.01.016

Chapter 8
Potentiality of Earthworms as Bioremediating Agent for Nanoparticles

Shweta Yadav

8.1 Introduction

Nanotechnology is a rapidly expanding field of science with development of various nanomaterials. The applications of nanomaterials have gained wide attention because of their novel properties including large surface area and high reaction activity. They are increasingly used in structural engineering, electronics, consumer products, optics, alternative energy, soil and water remediation, diagnostics, and drug delivery devices (Prasad et al. 2014, 2016). Investment in this technology has been increased globally due to their wide potential of application. Advancement in this branch of science promises direct benefits to the society and economy. Despite their bright future outlook, there is growing concern about human health, safety, and environmental impacts including exposure to engineered nanomaterials from different environmental sources. Nanoparticles may be released from various products through normal use and then enter into wastewater stream. These may disposed off into sewage aludge, landfills or applied to agriculture lands. Thus, soil system is an alternative sink for a large portion of nanoparticles (Gottschalk et al. 2009).

However, particles of nanosized range have been present on earth for millions of years and have been used by mankind for thousands of years. But, recently, their production and application have been increased because of our increasing ability to synthesize and manipulate their property. The huge increase in manufacture and application raises concerns about their toxic effects and health implications of NPs (Oberdorster et al. 2005; Kreyling et al. 2006; Nel et al. 2006). Every person is supposed to be exposed with nanometer-sized foreign particles either through air or

S. Yadav (✉)
Department of Zoology, School of Biological Science, Dr HS Gour Vishwavidyalaya (A Central University), Sagar, Madhya Pradesh 470003, India
e-mail: kmshweta@gmail.com

M. Ghorbanpour et al. (eds.), *Nanoscience and Plant–Soil Systems*, Soil Biology 48,
DOI 10.1007/978-3-319-46835-8_8

water. In truth, every organism on earth continuously encounters nanometer-sized entities. However, the vast majority causes little ill affect and goes unnoticed, but occasionally intruder causes harm to organism (Buzea et al. 2007). Free nanoparticles can be easily released into the environment and may pose serious health risk; however, fixed NPs do not show serious ill effects.

Very limited information is available on the fate of nanomaterials in the environment and their potentiality of biotransformation to nullify their toxic forms. Unrine et al. (2010) and Gupta et al. (2014a) reported that intact nanoparticles in the soil have a potential to enter in terrestrial food webs and may also be absorbed by earthworms and biodistributed to tissues remote from portal of entry. Earthworms are known as an ecological receptor that plays an important role in structure and function of terrestrial ecosystem. Therefore, there is a need to investigate accumulation of nanoparticles in food chain and their trophic transfer when assessing the ecological risks of nanomaterials. Earthworms occupy major invertebrate biomass (>80 %) in terrestrial ecosystem and have over 600 million years of experience as environmental managers in the ecosystem. They are known as "waste managers," "soil managers," "fertility improvers," and "plant growth promoters" for a long time. But some comparatively new discoveries about their role in bioremediation of industrial wastes, chemically contaminated soil, dairy industry waste material, and detergent industries have revolutionized the understanding of functioning of this unheralded soldier of mankind. Darwin wrote, "no other creature on earth has done so much for mankind" as the earthworms. This article gives an overview on the behavior of NPs and their potential remediation by earthworm in lines of Dr. Anatoly Igonin who said, "Earthworms create soil & improve soil fertility and provides critical biosphere's functions, disinfecting, neutralizing protective and productive for human welfare."

8.2 Nanoparticles Released in the Environment

Nanoparticles (NPs) may be natural and anthropogenic. They are considered as substances that are less than 100 nm in size in more than one dimension. They can be spherical, tubular, or irregular in shape and can exist in fused, aggregated, or agglomerated forms. They are very small particles and have the ability to enter, translocate, and penetrate physiological barriers and travel within the circulatory system of the host. NPs released into the environment can have a wide range of biological effects. This effect depends not only on the specific chemical nature of nanoparticles but also in the aggregate morphology that the materials may take as they move through in the soil ecosystem. NPs are usually classified on the basis of morphology, dimensions, composition, and agglomeration property.

8.2.1 *Morphology of Nanoparticles*

Buzea et al. (2007) classified NPs in their aspect ratios: high aspect ratio (nanotubes, nanowires with various shapes such as helices, zigzags, and belts) and small aspect ratios (spherical, oval, cubic, prism, helical, or pillar).

8.2.2 *Dimensions of NPs*

Nanomaterials are also classified on the basis of dimensions, viz., *1Ds* (material with one dimension), *2Ds* (two-dimensional nanomaterials firmly attached to substrate or nanopore filters used for small particle separation and filtration (e.g., asbestos), and *3Ds* (three-dimensional nanomaterials that include thin films deposited under conditions that generate atomic-scale porosity, colloids, and free nanoparticles with various morphologies (Buzea et al. 2007).

8.2.3 *Composition of NPs*

NPs can be of single material or be a composite of several materials. Single pure constituent materials are generally anthropogenic and synthesized from various materials, while composite of several materials are found in nature.

8.2.4 *Agglomeration of NPs*

NPs may exist either in dispersed aerosols or in an agglomerate state (as bounded primary particle). In an agglomerate state, NPs may behave as larger particles that also depend on the size of agglomerate.

8.2.5 *Natural Sources of NPs*

Nanoparticles are abundant in nature; they are produced in many natural processes including photochemical reactions, volcanic eruptions, forest fibers, and simple erosion and by plants and animals (shed skin and hair). Dust storms appear to be the longest single source of environmental NPs.

8.2.6 Anthropogenic Sources of NPs

Humans have created nanomaterials for millennia as by-products of simple combustion, food cooking, and, recently, chemical manufacturing, ore refining, welding, combustion in vehicle and airplane engines, and combustion of coal and fuel oil for power generation. Manufactured nanoparticles have been on the market recently and are commonly used in cosmetics, sporting goods, tires, stain-resistant clothings, sunscreens, toothpaste, food additives, etc. The quantity of manmade nanoparticles ranges from multi-ton per year production of carbon black (for car tires) to microgram quantities of fluorescent quantum dots (marker in biological imaging).

8.2.7 Commonly Used Nanoparticles

Metal nanoparticles (MNPs) are largely synthesized to use in wide applications. Silver nanoparticles (Ag-NPs) are largely used nanomaterials among all MNPs due to antibacterial property (Aziz et al. 2015), while gold nanoparticles (Au-NPs) are used in immunochemical studies. Alloy NPs are influenced by both metals and show more advantages over ordinary metallic NPs. Magnetic nanoparticles like Fe_2O_3 (maghemite) are known as very biocompatible nanomaterials.

8.3 Responses of Nanoparticles in Environment

Responses of nanoparticles with life-forms in the environment can take several routes, and their response on organisms against NPs depend on various factors like cellular uptake, receptors and degree of absorbance, aggregation, and cellular interaction. However, specific response to an organism at cellular level may be in terms of the effect of biotic uptake. The uptake of NPs depends on their size, shape, and surface charge that interact with the organism. Their responses depend on cellular endocytosis (depend on acceptor activation) or direct membrane penetration (as reported in nanotubes and fullerenes) and extracellular medium (protein and lipid adsorption pattern). Particles may agglomerate (joined together at the corner or edges) and/or aggregate (total surface area does not differ appreciably) when uptaken by living organism.

8.3.1 Cellular Uptake of NPs

NPs can uptake into cell either by endocytosis or diffusion. Endocytosis is actually the mechanism to uptake particulate matter such as proteins and other nutrients into eukaryotic cells. Cells for the clearance of cell debris and foreign cells from the body also exploit this mechanism. Uptake of opsonized particulate substances and small solute volumes is internalized by phagocytosis, and particles in large amounts of solute are taken up by pinocytosis (either macropinocytosis or receptor-mediated). Kettler et al. (2014) suggested four mechanisms of uptake of NPs by cells: (1) *macropinocytosis* (nonspecific mechanism by which fluid contents are taken up in the same concentration as in the surrounding medium and the vesicle size is 100 nm to 5 μm), (2) *clathrin-mediated endocytosis* (receptor-mediated endocytosis (RME) with vesicle sizes of approximately 120 nm), (3) *caveola-mediated endocytosis* (uptake of RME of vesicles of approximately 80 nm in size), and (4) *clathrin and caveolae* endocytosis (independent endocytosis with vesicles of approximately 50 nm) (Fig. 8.1). Internalization of NPs in cells is influenced by a large number of physical and chemical properties of the specific NP (like shape, surface charge, surface functional groups, and NP hydrophilicity) and also the circumstances in the exposure medium of the cells.

8.3.2 Aggregation of NPs

Aggregation plays an important role in determining the response of NPs, especially in determining their mobility, fate persistence, and toxicity. Shaking of the NPs and cells in solution may increase particle uptake into cells because of a higher hitting frequency.

8.3.3 Accumulation of NPs

The accumulation of NPs in the biological system may be due to strong interactions between NP surfaces and glycoproteins and/or glycans of intra- and extracellular medium. The charge distribution in the polymer, molecular weight, and polymer conformation may also play an important role in accumulation of NPs in organism.

8.3.4 Interaction of NPs in the Biological System

Once NPs have entered in the biological system, they inevitably come into contact with a huge variety of biomolecules including proteins, sugars, and lipids that are

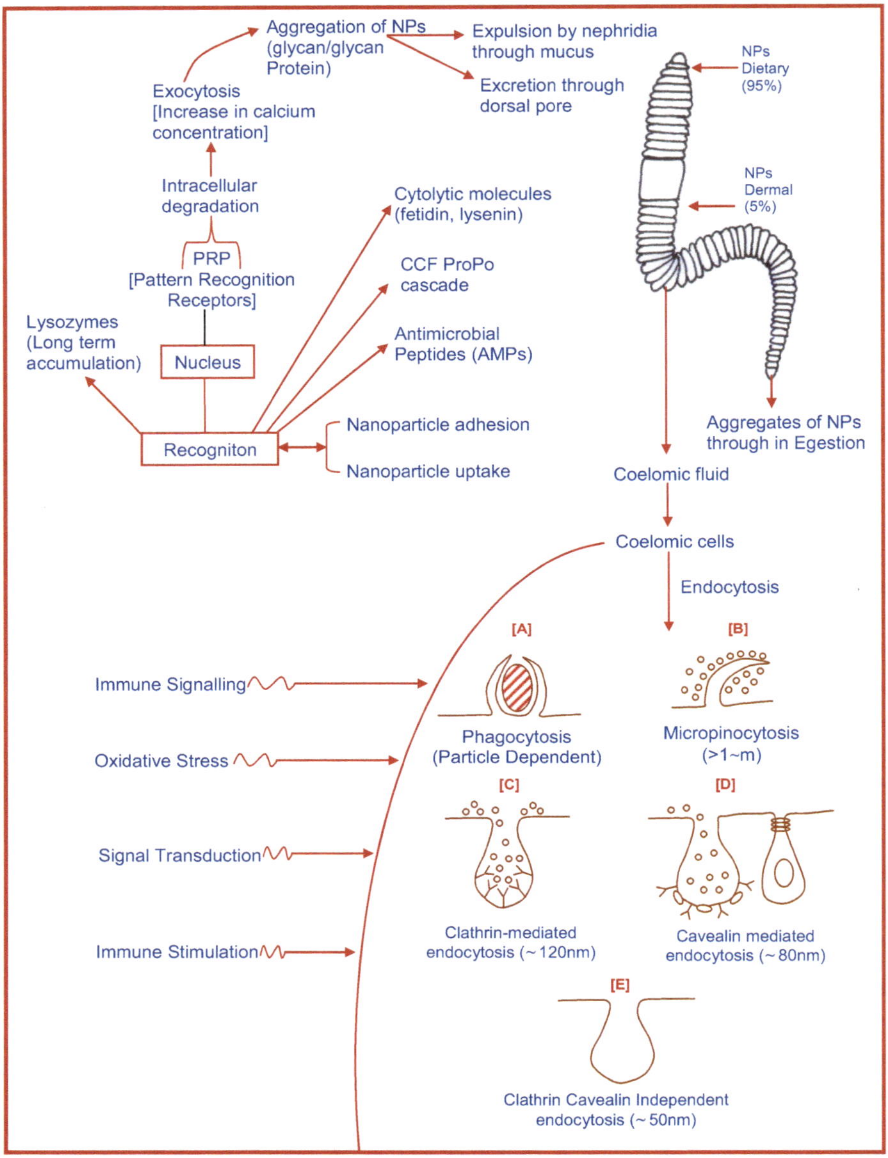

Fig. 8.1 Schematic illustration of potential uptake of NPs by coelomic cells of earthworms and their expulsion in the form of aggregates (in presence of N-linked glycans, high-mannose-type O-linked glycans, mucin-type o-linked glycans) through dorsal pores or nephridia or with egested feed material

dissolved in body fluids. These biomolecules immediately coat the NP surface and form the so-called protein corona (Shang et al. 2014) which determines biological identity of NPs (Lynch et al. 2009). Now they selectively transported into and out of cells via endocytosis and exocytosis. In most of the cells, internalization of NPs occurs via pinocytosis. In this process, an invagination forms in the cell membrane that is finally pinched off so as to generate a vesicle in the cytoplasm that contains the internalized material. The inward budding vesicles contain receptor protein that recognizes specific chemical groups on the molecules to be internalized. Thus, protein absorbed to an NP triggers cell surface receptors and activate cell uptake machinery. Once it entered in cells or tissue, the surface layer including the absorbed biomolecules and also the NP core material will likely be metabolized. Elevated ROS level may also lead to activation of cellular stress-dependent signaling pathways, direct damage of subcellular organelles, and DNA fragmentation in the nucleus that may alter gene expression, apoptosis, and inflammatory response. NPs may also interact with membrane-bound cellular receptors like growth factor receptors and integrins. However, uptake efficiency, internalization pathway selection, intracellular localization, and cytotoxicity of NPs are strongly affected by size of NPs.

8.4 Response of NPs in Soil Environment

Soil contains a variety of inorganic and organic particles of nanoscale dimension including clay minerals, metal hydroxides, humic substances, allophane, and imogolite (in volcanic soils). Organic colloids in soil are largely associated with their inorganic counterparts or form coatings over mineral surfaces. However, the concentration of natural and anthropogenic NPs in the environments still remains unknown; exposure modeling suggests that soil could be a major sink of NPs released into the environment. The concentration of engineered NPs in soil is also recorded higher than water or air by various workers including Gottschalk et al. (2009), Klaine et al. (2008), and Tiede et al. (2009). NPs may affect soil ecosystem via (1) direct effect, (2) changes in the bioavailability of toxins or nutrients, (3) indirect effects resulting from their interaction with natural organic compounds, and (4) interaction with toxic organic compounds which may amplify or alleviate their toxicity (Simonet and Valcarcel 2009).

8.4.1 Effects of NPs on Soil Microorganisms

NPs can be strongly adsorbed to soil surfaces and soil organic matrix making them less mobile or are small enough to be trapped in the interspaces of soil particles and might therefore travel fast than large particles before becoming trapped in the soil matrix. The strength of sorption also depends on size, chemistry of aggregation, and behavioral condition under which it applied. The toxicity of NPs to microorganisms raises serious concern especially because it promotes plant growth and regulates nutrient cycling in soil. Mishra and Kumar (2009) recorded varying degrees of inhibition of plant growth-promoting rhizobacteria (PGPR) especially *Pseudomonas aeruginosa*, *P. putida*, *P. fluorescens*, and *P. subtilis* and soil nitrifying and denitrifying bacteria when exposed to NPs in pure culture and aqueous suspension. Murata et al. (2005) demonstrated the severe effect of Ag on soil dehydrogenase activity and recorded declined growth of bacterial colony. In fact, the degree of hazard of NPs in soil depends not only on its concentration but also on the likelihood of it ever coming into contact with organism as suggested by Simonet and Valcarcel (2009).

8.4.2 Effects of NPs on Soil Invertebrates

Soil invertebrates play an important role in soil ecosystem function (e.g., decomposition and nutrient cycling), and thus addressing NP effects on these organisms is crucial to understand the potential impacts of NPs in the soil environment. It is known that behavior of NPs in soil controls their mobility and bioavailability to soil organisms. The major source of NP deposition in soil is currently through the disposal of wastewater plant sewage sludge where NPs are released from consumer products into wastewater. Thus NPs do not enter in soil in their original form because of the organically rich and reactive environments of wastewater plants. Therefore, soil represents a complex medium of physicochemical behavior of NPs. The effect of NPs on soil fauna is directly affected by dispersibility, agglomeration/aggregation of NPs, soil types, concentration ranges, and frame of exposure. The common defense mechanisms used by most of soil invertebrates against NPs are phagocytosis, production of reactive oxygen (ROS) and nitrogen radicals, synthesis and secretion of antibacterial and antifungal proteins, cytokine-like proteins, hydrolytic enzymes, agglutination and nodule formation, encapsulation of foreign objects, and activation of enzymatic cascades that regulate melanization and coagulation of hemolymph. The hemocytes of insects (Franc et al. 1996) and mollusks (Lu et al. 2011) are known to effect scavenger-mediated uptake of pathogen and apoptotic cells. Their presence suggests an unequivocally conserved role in innate immune recognition that may involve in the uptake of NPs. Gomes et al. (2015) reported a negative effect on survival and reproduction of soil invertebrate on exposure to copper and silver nanoparticles. Such negative impacts are supposed to mediate through oxidative stress and subsequent mitochondrial damage. Thus, NPs affect physicochemical barriers, cellular defenses, and humoral mechanisms of soil invertebrates.

8.4.3 *Effects of NPs on Earthworms*

Earthworms comprises about 80 % of soil biomass and are widely used in standard toxicity tests for studies of soil pollution as recommended by the Organization for Economic Cooperation and Development for standardization. Earthworm appears to be a suitable organism to study nanomaterial interactions with living organisms and in assessing environmental nanosafety. Earthworms are essential for the incorporation and fragmentation of organic debris, mineralization of organic matter, and recycling of mineral nutrients. Their burrowing activity is crucial for water filtration and stabilization of erosion effects. They are considered as bio-indicators of soil health. As different species of earthworms (epigeic, endogeic, anecic) occupy the different ecological niches of soil ecosystem, thus, all are not submitted to the same response on exposure to NPs. McShane et al. (2011) observed earthworms are able to differentiate particles through an unknown mode of action. Microarray analysis revealed altered mRNA levels for specific genes mainly involved in metabolism, transcription, and translation or in the stress response, indicating oxidative stress condition in earthworm exposure to NPs (Gomes-da-Silva et al. 2012).Van der Ploeg et al. (2013) reported a reduction in growth and development and damaged cuticle with underlying pathologies of the epidermis, muscles, and gut barrier on exposure of fullerene NPs (C60) to *Lumbricus rubellus*. However, *Eisenia fetida* showed neither response of antioxidant enzyme expression or activity nor acute toxicity in C60 spiked soil (Li et al. 2011). In our earlier studies, effects of ZnO-NPs on *E. fetida* were recorded in terms of reproductive behavior, antioxidant enzyme activities, and accumulation of Zn^{++} remote from portal of entry (Gupta et al. 2014b). However, the interaction of nanoparticle with earthworm is unpredictable that may result in ecologically significant effects on behavior at environmentally relevant concentrations.

8.4.3.1 Earthworms and TiO_2-NPs

It is known that TiO_2-NPs enter in the composition of several soil layers and considered to be related to the formation of free radicals with water in presence of sunlight. Petkovic et al. (2011) investigated genotoxic responses of TiO_2-NPs to cause persistent increases in DNA strand breaks and oxidized purines. Rahman et al. (2002) observed that 10–20 nm TiO_2-NPs affect enzymatic activities, damage mitochondria, and induce apoptosis in earthworms and observed avoidance of 1000–5000 TiO_2-NPs mg/kg spiked soil. However, soil that contained 200–10,000 TiO_2-NPs mg/kg do not show a significant effect on the activity of worms. Hu et al. (2010) reported an accumulation of Ti in earthworms' tissue and damage of mitochondria at a dose of 5.0 g/kg NPs spiked soil and also concluded harmful effect at exposure to 1.0 g/kg TiO_2-NPs. Schlich et al. (2012) concluded that TiO_2-NPs affect earthworm reproduction activity by abolishing the circannual rhythm that depresses reproduction in winter. Whitfield Aslund et al. (2011)

observed significant changes in their metabolic profile on exposure to 5 nm at 20 and 200 mg/kg TiO_2-NPs. Lapied et al. (2011) noted increased apoptotic frequency in the cuticle and intestinal epithelium of *Lumbricus terrestris* after 7 days of exposure to 100 mg/l nanosized TiO_2 in water, and a similar trend was noted in earthworm exposed to only 15 mg/kg nanosized TiO_2 in soil after 4 weeks of exposure. Several recent studies (Hu et al. 2010; Heckmann et al. 2010) have examined the toxicity of nanosized TiO_2 to earthworms that expressed toxic responses to nanosized TiO_2 at a very high concentration in soil (>1000 mg/kg). The maximum concentrations of TiO_2 are predicted to be 0.3 mg/kg in soil and up to 523 mg/kg in sewage treatment plant sludge. Whitfield Aslund et al. (2011) suggested that release of TiO_2 nanomaterials may be nontoxic to earthworms at environmentally relevant concentrations.

8.4.3.2 Earthworms and AgO-NPs

Coutris et al. (2011) recorded quick excretion of Ag+ ions and Ag-NPs on exposure to 20 nm Ag-NPs and Ag salts in *Eisenia fetida*. Varied activities including expression of several genes of oxidative stress, catalase activity, glutathione reductase inhibitors, phosphatase, and Na^+/K^+ ATPase were observed on exposure to Ag-NPs (30–50 nm) when coated with polyvinylpyrrolidone or to $AgNO_3$. It was found that Ag-NPs were phagocytosed and accumulated by coelomocytes of *E. fetida*. Hayashi et al. (2012) suggested signal transduction, primarily through nitrogen-activated protein kinase (MAPK) pathways that coordinate the cross talk between oxidative stress and immune responses to Ag-NPs.

8.4.3.3 Earthworms and ZnO-NPs

In natural soils, ZnO-NPs were quite toxic in opposition to TiO_2-NP, but in sandy soil it is toxic (Hu et al. 2010). In our earlier studies, the increase in percentage of mortality of earthworms was recorded at exposure to 10 nm ZnO-NPs at 5 mg/kg. However, 10 nm <5 mg/kg were found to be nontoxic to *E. fetida*. Exposure to 50–100 nm at 3 mg/l also does not cause DNA damage (Fig. 8.2). The aggregates of 100–200 nm (Fig. 8.3) were also observed in the cytoplasm on exposure to 50 nm NPs (Gupta et al. 2014a, 2015). That reflects the ability of earthworm to uptake ZnO-NPs from soil system and excrete into the form of aggregates or less toxic form of NPs. The true uptake of ZnO-NP and biodistribution of nanomaterials by earthworms (Fig. 8.4) in an environmentally realistic exposure scenario with potential effects on an ecologically relevant end point have been demonstrated in our earlier studies and shown that the NPs enhance the cellulolytic activity of earthworm. Application of ZnO nanofertilizer may help to increase bioconversion rate including the biotransformation of NPs in the form of aggregates in their tissues and coelomic cavity without causing any genotoxicity to earthworms.

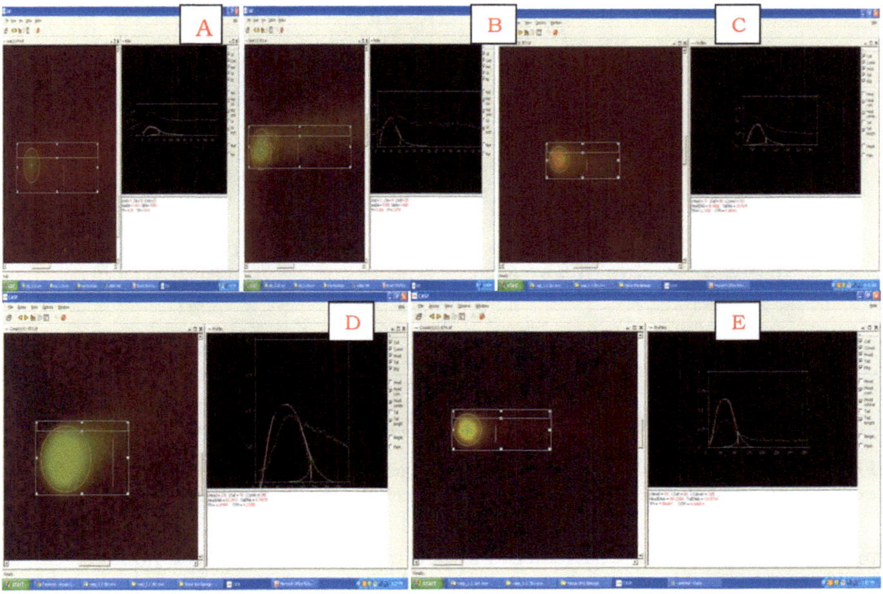

Fig. 8.2 Comet assay *of Eisenia fetida showing no significant genotoxicity after the exposure to NPs* after the exposure of 6 h (**a**) control, (**b**) after exposure to 100 nm, (**c**) after exposure to 50 nm, (**d**) after exposure to 35 nm, and (**e**) after exposure to 10 nm

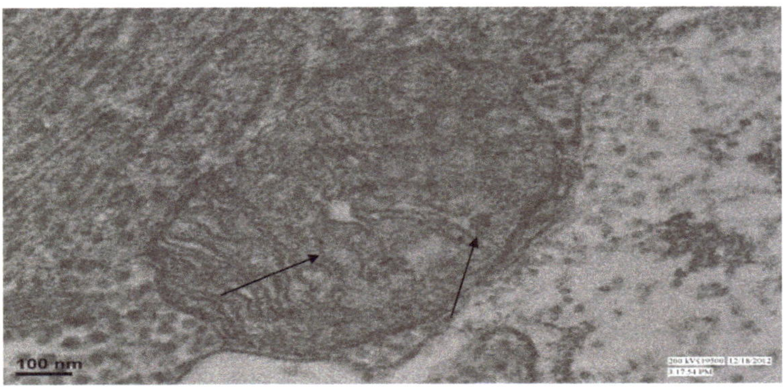

Fig. 8.3 Transmission electron microscopic images of gut cells of *Eisenia fetida* after the exposure to ZnO-NPs (50 nm) at 28 days showing normal appearance of mitochondrial structure and aggregates of NPs

8.4.3.4 Earthworms and C_{60} Fullerene

Van der Ploeg et al. (2010) exposed *L. rubellus* on C_{60} *fullerene* (4 weeks and lifelong) and recorded suppression of heat shock protein 70 (HSP70) gene and coelomic cytolytic factor 1 (CCF1). However, the enzymes involved in antioxidant

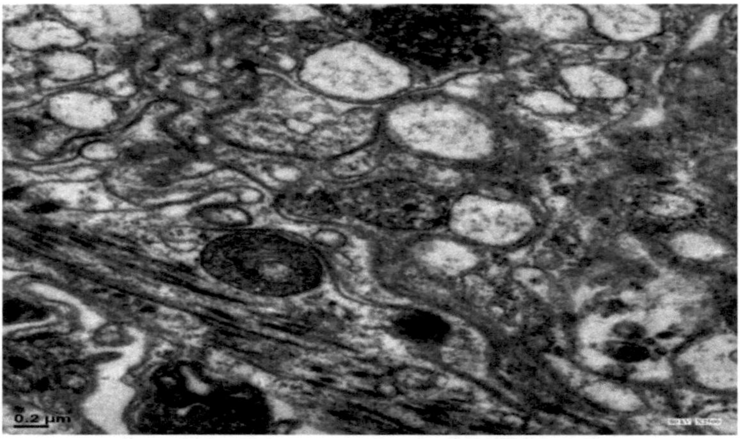

Fig. 8.4 Transmission electron microscopic images of gut wall of *Eisenia fetida* after the exposure to ZnO-NPs (50 nm) at 28 days showing cellular uptake and accumulation of NPs

mechanism were found unaffected. Further, on exposure to a wide range of C_{60} on coelomocytes, CCF1 was found downregulated in concurrence with reduced phagocytic activity.

8.4.3.5 Earthworms and Au-NPs

Unrine et al. (2010) demonstrated the uptake of Au-NPs from soil and biodistribution in tissues of *E. fetida*. They also found that primary particle size (20 or 50 nm) did not consistently influence accumulate concentrations on a mass concentration basis; however, on a particle number basis, the 20 nm particles were more bioavailable. Evidence of biodistribution of NPs to tissues of earthworms remote from the portal of entry highlighted the importance of considering food chain accumulation and trophic transfer when assessing the ecological risks of nanomaterials.

8.4.3.6 Earthworms and Cu-NPs

Alahdadi and Behboudi (2015) studied effects of CuO- and ZnO-NPs on absorption, accumulation, survival, and reproduction of *Eisenia fetida* earthworm in cow manure and spent mushroom compost. They concluded the absorption and accumulation in tissues of earthworm enhanced with increasing NP concentrations. Absalon et al. (2015) recorded negative delayed effect of CuO-NPs on freshwater annelid *Tubifex tubifex*. A higher mortality was recorded, for both the CuO-NPs and Cu-ion solutions, compared to the control solution. And worms were found dead when exposed to the CuO-NP solution; they first started dying after being transferred to a clean sediment. Furthermore, the burrowing behavior was decreased with an average of 15.75 min compared to the 1.75 min in the control solutions and the 4 min in the Cu-ion solution.

8.4.3.7 Earthworms and Quantum Dots

Stürzenbaum et al. (2013) showed exploitation of earthworm's metal detoxification pathway to produce luminescent, water-soluble semiconductor cadmium telluride (CdTe) quantum dots that emit in the green region of the visible spectrum when excited in the ultraviolet region. Standard wild-type *Lumbricus rubellus* earthworms were exposed to soil spiked with $CdCl_2$ and Na_2TeO_3 salts for 11 days. Luminescent quantum dots were isolated from chloragogenous tissues surrounding the gut of the worm and were successfully used in live-cell imaging.

8.5 Potentiality of Earthworms as Bioremediating Agent for NPs

Earthworms are free-living terrestrial animals living in soil, leaf litter under the stones, and mainly in wetter and more heavily vegetated regions. Earthworms can "biotransform" or "biodegrade" chemical contaminants rendering them harmless in their bodies and can bioaccumulate in their tissues. They "absorb" the dissolved chemicals through their moist "body wall" due to interstitial water and also ingest by "mouth" while passing through the gut. Their coelom is filled with coelomic fluid that contains free coelomocytes. Each segment of the body cavity is interfaced with the outer environment by dorsal pores enabling also the entering of microorganisms in coelomic cavity. Therefore, coelomic cavity of earthworms is not aseptic, and it always contains bacteria, protozoa, and fungi from the outer environment. Implications from leading researchers in earthworms revealed the recognition of nanoparticles involved in cellular uptake as well as sub- and intracellular events that further give intriguing insights into earthworm's potentiality as a biotransforming agent.

8.5.1 Uptake of NPs in Coelomic Cells of Earthworms

Earthworms can uptake NPs from soil in two forms, i.e., dietary pathways (~95 % of total uptake) and dermal pathways (~5 % of total uptake). However, their skin with mucus is the first-line defense barrier against foreign invaders, but once they enter the coelom, they are exposed to cellular and humoral responses. Annelids are the first animal in the phylogenetic tree in which not only cellular but also humoral responses are developed. During cellular immune responses, coelomocytes play an important role in phagocytosis, inflammatory processes, graft rejection, and coagulation of coelomic fluid. During the humoral immune response, they secrete lysozyme, agglutinin, peroxidase, phenoloxidases, and antimicrobial factor (fetidin, lysenin, eiseniapore, coelomic cytolytic factor). These cytotoxic molecules increase the intracellular calcium concentration that participates in exocytosis. Exhausted phagocytes with NPs can then be eliminated through the dorsal pores. The internalized NPs can also be excreted by nephridia, while agglomerated/heteroaggregated NPs can be eliminated by process of encapsulation. The internalization of NPs in coelomic cells occurs due to toll like receptors (TLR), and their recognition is mediated by pattern-recognition receptors (PRRs), which lead into various inflammatory cytokines and antimicrobial peptides. The schematic illustration of potential uptake by coelomic cells and their possible expulsion in presence of N-linked glycans and O-linked glycans have been presented in Fig. 8.1.

Coelomic cytolytic factor (CCF) is a well-characterized 42-kDa lytic protein (secreted into coelomic fluid in a stable form) in earthworms that acts as a pattern-recognition molecule and is present on the cells of the mesenchymal lining of the coelomic cavity as well as on free coelomocytes. CCF is formed by two spatially distinct lectin-like domains located in the central part of the molecule which interacts with lipopolysaccharide and β-1,3-glucans, and the second domain is located in the C-terminal part which interacts with peptidoglycan constituents. Upon binding of PAMPs, CCF triggers the activity of prophenoloxidase (ProPO) cascade. The ProPO cascade is sensitive and an efficient defense system consisting of several proteins such as zymogenic proteinases, proteinase inhibitors, ProPO, PO, and PRRs with final product melanin. Melanin exhibits fungal, bacterial, and antiviral properties and is involved in wound healing and defense reaction. This melanization reaction accompanies the cellular defense reaction encapsulation, resulting in the formation of so-called brown bodies of NPs (Fig. 8.5).

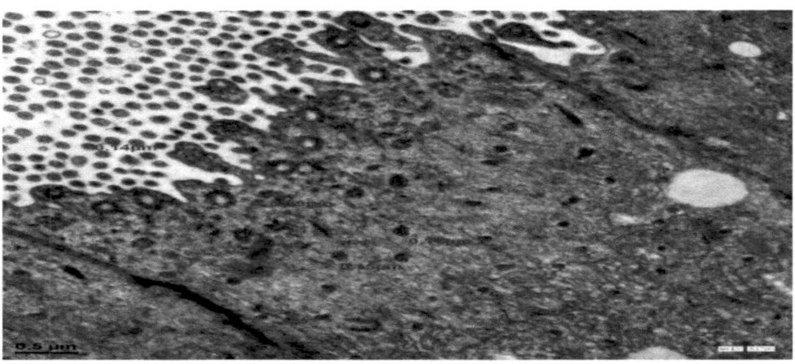

Fig. 8.5 Transmission electron microscopic images *showing uptake of NPs into the body tissue and formation of brown bodies in Eisenia fetida* after the exposure to ZnO-NPs (50 nm)

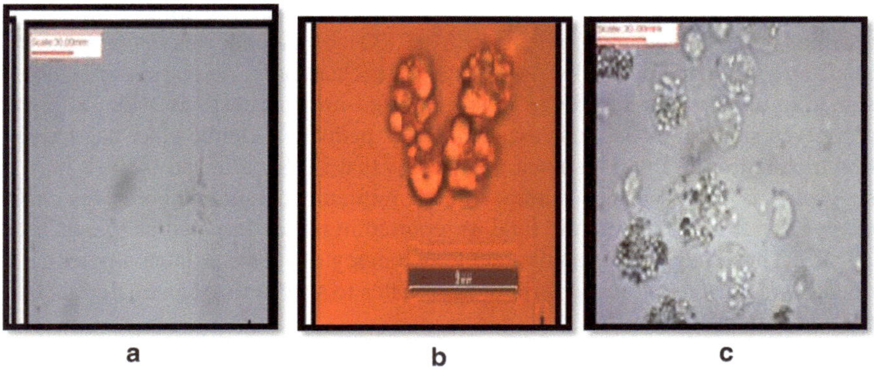

Fig. 8.6 Coelomic cells of *Eisenia fetida*, (**a**) hyaline amoebocyte (×40), (**b**) fluorescent eleocyte (×100), and (**c**) granulocyte and fluorescent eleocyte of *Eudichogaster prashadi*

8.5.2 Coelomic Cells of Earthworms as Nano-scavenger

Earthworm coelomic cells (nonmovable scavenger) can be classified into two basic categories: amebocytes (mainly immune function) and eleocytes (mainly nutritive function). However, a uniform classification of coelomocytes of different species of earthworm is a little difficult (Adamowicz and Wojtaszek 2001) as they exist in various functional states and stages of maturation. The amebocytes moved by pseudopodia devour foreign material and are rich in lysosome (Fig. 8.6), while eleocytes are rich in glycogen particles and lipid droplets and characterized by the presence of distinct yellow granules, or chloragosomes. The origin and relationship of coelomocytes are not yet completely known. It has been assumed that amebocytes are derived from the mesenchymal lining of the coelom, whereas eleocytes originate by detachment of chloragogen cells covering the intestinal tract (Affar et al. 1998; Hamed et al. 2002). Amebocytes participate in the transport and storage of nutritive substances (Valembois et al. 1988), phagocytosis (Stein and Cooper 1981; Dales and Kalac 1992; Ranzelli-Cain and Kaloustian 1995; Cossarizza et al. 1996), while eleocytes play an important role in immune responses producing bacterial substances (Valembois et al. 1982; Ville et al. 1995; Milochau et al. 1997) and also participate in reaction of encapsulation and formation of brown bodies (Cooper and Stein 1981). The number and composition of coelomic cells depend on exogenous (environmental) as well as endogenous (biotic, life cycle) factors. Parry (1975) proved short-term and limited memory in coelomic cells of earthworms in transplantation experiments to autografts and xenografts. Van der Ploeg et al. (2013) reported that enzymes involved in antioxidant mechanism do not affect exposure to C_{60} fullerenes on *L. rubellus*. However, coelomic cytolytic factor 1(CCF1), a pattern-recognition receptor, was found suppressed in lifelong experiment. Clathrin-mediated endocytosis, caveolae-mediated phagocytosis, and micropinocytosis may involve in uptake of NPs by coelomocytes. Scavenger receptor class A (a pattern-recognition receptor) of coelomocytes is a potential pathway to phagocytosis the NPs by amebocytes of earthworms. Macrophage receptor with collagenous structure (MACRO) of amebocytes recognizes and associates with nanoparticles for phagocytic clearance. Activation of calcium signaling on exposure to NPs reflects stress and immune responses like in highly conserved signal transduction cascade, MAPK pathways. On the other hand, secretion/modulation of cytokine from phagocytes that is also affected by NPs may be suggested as indirect effect of intracellular communication of coelomocytes during exocytosis of NPs in coelomic fluid as reported in our earlier studies (Gupta et al. 2014b). Thus phagocytic population of coelomocytes seems to have a potential to uptake and excrete out in the form of aggregates to scavenging nanomaterials from the soil system.

8.5.3 Aggregates/Agglomeration of NPs in Coelomic Fluid

In our earlier studies, we recorded aggregates of ZnO-NPs (50–100 nm) in coelomic fluid and tissues of *Eisenia fetida* (Gupta et al. 2014a). Accumulation of gold and quantum dots has been recently reported in mucus materials of jellyfish by Patwa et al. (2015). Aggregation of Au-NPs in gut tissues of *E. fetida* was also recorded by Unrine et al. (2010). Deposits of 50 nm in HeLa cells (Chitrani et al. 2006) and aggregates of 10–58 nm NPs in mice (Hillyer and Albrecht 2001) were recorded by earlier workers. Aggregation of NPs in coelomic fluid of earthworms may be due to presence of glycoproteins and glycans (high-mannose-type, N-linked glycans, mucin-type O-linked glycans, and high levels of pentose-containing oligomers, Fig. 8.1). This may be due to interactions of electric charges, which allow particles to link together resulting in a zero point of charge. Thus, the size and density of aggregates of particles in coelomic fluid may be due to interactions between NP surfaces and glycoproteins/glycans. The formation of aggregates may be regulated by distribution of charges in polymers, molecular weight (Mabire et al. 1984), polymer conformation (Zeng et al. 2015), hydrogen bonding, ionic interactions, and dehydration of polar groups (Ulrich et al. 2006). This defensive strategy of earthworms may exploit to remediate NPs from soil system.

8.6 Conclusions

Earthworms are significant ecological receptor species that play a key role in the structure and function of soil ecosystems. Evidence that intact nanomaterials can be absorbed and transformed into the form of aggregates by earthworms and biodistributed to tissues remote from the portal of entry highlights the importance of considering food chain accumulation and trophic transfer when assessing the ecological risks of nanomaterials. The strategy of earthworms to accumulate nanoparticles and excrete them in agglutinated/aggregated forms may exploit to remediate nanoparticles from soil system at large scale.

Acknowledgment We acknowledge the financial support of the Department of Biotechnology, Ministry of Science and Technology, Govt. of India, New Delhi, to carry out this study.

References

Absalon L, Charalambous N, Holm A, Jørgensen M (2015) An assessment of the delayed effects of exposure to CuO nanoparticles and $CuCl_2$ spiked sediment on mortality and burrowing behaviour of the freshwater annelid *Tubifex tubifex*. Environmental biology project, Roskilde University, RUC, Denmark

Adamowicz A, Wojtaszek J (2001) Morphological and phagocytotic activity of coelomocytes in *Dendrobaena veneta* (Lumbricidae). Zoologica Poloniae 46:91–104

Affar EB, Dufour M, Poirier GG, Nadeau D (1998) Isolation, purification and partial characterization of chloragocytes from the earthworm species *Lumbricus terrestris*. Mol Cell Biochem 185:123–133

Alahdadi I, Behboudi F (2015) The effects of CuO and ZnO nanoparticles on survival, reproduction, absorption overweight and accumulation in *Eisenia foetida* earthworm tissues in two substrates. Int J Environ Res 9(1):35–42

Aziz N, Faraz M, Pandey R, Sakir M, Fatma T, Varma A, Barman I, Prasad R (2015) Facile algae-derived route to biogenic silver nanoparticles: synthesis, antibacterial and photocatalytic properties. Langmuir 31:11605–11612

Buzea C, Pacheco II, Robbie K (2007) Nanomaterials and nanoparticles: sources and toxicity. Biointerphases 2:17–71

Chitrani BD, Ghazani AA, Chan WCD (2006) Determining the size and shape dependence of gold nanoparticles uptake into mammalian cells. Nano Lett 6:662–668

Cooper EL, Stein EA (1981) Invertebrate blood cells. Academic, London, pp 75–140

Cossarizza A, Cooper EL, Suzuki MM, Salvioli S, Capri M, Gri G, Quaglino D, Franeschi C (1996) Earthworm leucocytes that are not phagocytic and cross-react with several human epitopes can kill human tumor cell lines. Exp Cell Res 224:174–182

Coutris C, Hertel-Aas T, Lapied E, Joner EJ, Oughton DH (2011) Bioavailability of cobalt and silver nanoparticles to the earthworm *Eisenia fetida*. Nanotoxiciology 6:186–195

Dales RP, Kalac Y (1992) Phagocytic defense by the earthworm *Eisenia fetida* against certain pathogenic bacteria. Comp Biochem Physiol 101:487–490

Franc NC, Diarcq J-L, Lagueux M, Hoffmann J, Ezekowitz RAB (1996) Croquemort, a novel *Drosophila* hemocyte/macrophage receptor that recognizes apoptotic cells. Immunity 4:431–443

Gomes SI, Hansen D, Scott-Fordsmand JJ, Amorim MJ (2015) Effects of silver nanoparticles to soil invertebrates: oxidative stress biomarkers in *Eisenia fetida*. Environ Pollut 199:49–55

Gomes-da-Silva LC, Fonseca NA, Moura V, Pedrosa de Lima MC, Simbes S, Moreira JN (2012) Lipid-based nanoparticles or siRNA delivery in cancer therapy: paradigms and challenges. Acc Chem Res 45:1163–1171

Gottschalk F, Sonderer T, Scholz RW, Nowack B (2009) Modeled environmental concentrations of engineered nanomaterials (TiO_2, ZnO, Ag, CNT, Fullerenes) for different regions. Environ Sci Technol 43:9216–9222

Gupta S, Kushwaha T, Yadav S (2014a) Earthworm coelomocytes as nano-scavenger for ZnO NPs. Nanoscale Res Lett 9:259–269

Gupta S, Kushwaha T, Yadav S (2014b) Toxicity of ZnO nanoparticles on earthworm *Eisenia fetida* (Savigny, 1826) and investigating its potential as biotransforming agent. In Pavlicek T,

Cardet P, Almeida MT, Pascoal C, Cassio F (eds) Advances in earthworm taxonomy-VI (Annelida: Oligochaeta), Proceedings of 6th international Oligochaeta taxonomy meeting, Portugal. Kasparek, Heidelberg, pp 158–171

Gupta S, Kushwah T, Vishwakarma A, Yadav S (2015) Optimization of ZnO-NPs to investigate their safe application by assessing their effect on soil nematode *Caenorhabditis elegans*. Nanoscale Res Lett 10:303313. doi:10.1186/s11671-015-1010-4

Hamed SS, Kauchke E, Cooper EL (2002) Cytochemical properties of earthworm coelomocytes enriched by Percoll. In: Beschin A, Bilej M, Cooper EL (eds) A new model for analyzing antimicrobial properties with biomedical applications. IOS Press, Ohmsha, Tokyo, pp 29–37

Hayashi Y, Engelmann P, Foldjerg R, Szabo M, Pollak E, Molnar L, Autru H, Sutherland DS, Scott-Fordsmand J, Heckmann LH (2012) Earthworms and humans *in vitro*: characterizing evolutionarily conserved stress and immune responses to silver nanoparticles. Environ Sci Technol 46:4166–4173

Heckmann LH, Hovgaard MB, Sutherland DS, Autrup H, Besenbacher F, Scott-Fordsmand JJ (2010) Limit-test toxicity screening of selected inorganic nanoparticles to the earthworm *Eisenia fetida*. Ecotoxicology 20:226–233

Hillyer JF, Albrecht RM (2001) Gastrointestinal persorption and tissue distribution of differently sized colloidal gold nanoparticles. J Pharm Sci 90:1927–1936

Hu CW, Li M, Cul YB, Li DS, Chen J, Yang LY (2010) Toxicological effects of TiO2 and ZnO nanoparticles in soil on earthworm *Eisenia fetida*. Soil Biol Chem 42:586–591

Kettler K, Veltman K, van de Meent D, van Wezel A, Hendriks AJ (2014) Cellular uptake of nanoparticles as determined by particle properties, experimental conditions and cell type. Environ Toxicol Chem 33:481–492

Klaine SJ, Alvarez PJJ, Batley GE, Fernandes TF, Handy RD, Lyon DY, Mahendra S, McLaughlin MJ, Lead JR (2008) Nanomaterials in the environment: behavior, fate, bioavailability and effects. Environ Toxicol Chem. doi:10.1897/08-090.1

Kreyling WG, Semmier-Behnke M, Moller W (2006) Health implications of nanoparticles. J Nanopart Res 8:543–562

Lapied E, Nahmani JY, Moudilou E, Cjaurand P, Labille J, Rosae J, Exbrayat JM, Oughton DJ, Joner EJ (2011) Ecotoxicological effects of an aged TiO2 nanocomposite measured as apoptosis in the anecic earthworm *Lumbricus terrestris* after exposure through water, food and soil. Environ Int 37:1105–1110

Li LZ, Zhou DM, Peijnenburg WJ, van Gestel CA, Jin SY, Wang YJ, Wang P (2011) Toxicity of zinc oxide nanoparticles in the earthworm, Eisenia fetida and sub cellular fractionation of Zn. Environ Int 37:1098–1104

Lu X, Qian J, Zhou H, Gan Q, Tang W, Lu J, Yuan Y, Liu C (2011) In vitro toxicity and induction of apoptosis by silica nanoparticles in human HepG2 hepatoma cells. Int J Nanomed 6:1899–9001

Lynch I, Salvat A, Dawson KA (2009) Protein-nanoparticle interactions: what does the cell see? Nat Nanotechnol 4:546–547. doi:10.1038/nnano.2009.248

Mabire F, Audebert R, Quivoron C (1984) Flocculation properties of some water-soluble cationic copolymers toward silica suspensions: a semi-quantitative interpretation of the role of molecular weight and cationicity through a 'patchwork' model. J Colloid Interface Sci 97:120–136

McShane H, Sarrazin M, Whalen JK, Hendershot WH, Sunahara GI (2011) Reproductive and behavioral responses of earthworms exposed to nano-sized titanium oxide in soil. Environ Toxicol 31:184–193

Milochau A, Lassegues M, Valembois P (1997) Purificaton, characterization and activities of two hemolytic and antibacterial proteins from coelomic fluid of the annelid Eisenia fetida andrei. Biochim Biophys Acta 1337:123–132

Mishra VK, Kumar A (2009) Impact of metal nanoparticles on the plant growth promoting rhizobacteria. Dig J Nanomater Biostruct 3:578–592

Murata M, Tanaka Y, Mizukagi T, Ebitani K, Kaneda K (2005) Palladium-platinum bimetallic nanoparticle catalysts using dendron assembly for selective hydrogenation of dienes and their application to thermophilic system. Chem Lett 34:272–273

Nel A, Xia T, Madler L, Li N (2006) Toxic potential of materials at the nanolevel. Science 311:622–627

Oberdorster G, Oberdorster E, Oberdorster J (2005) Nanotoxicology: an emerging discipline evolving from studies of ultrafine particles. Environ Health Perspect 113:825–839

Parry MJ (1975) Evidence of mitotic divisions of coelomocytes in the normal, wounded and grafted earthworm *Eisenia fetida*. Experientia 32:449–451

Patwa A, Thiery A, Lombard F, Lilley MKS, Boisset C, Bramard J-F, Bottero J-Y, Barthelemy P (2015) Accumulation of nanoparticles in jellyfish mucus: a bio-inspired route to decontamination of nano-waste. Sci Rep. doi:10.1038/srep11387

Petkovic J, Zegura B, Stevanovic M, Dmovsek N, Uskokvic D, Novak S, Filipic M (2011) DNA damage and alterations in expressions of DNA damage responsive genes induced by TiO2 nanoparticles in human hepatoma HepG2 cells. Nanotoxicology 5:341–353

Prasad R, Kumar V, Prasad KS (2014) Nanotechnology in sustainable agriculture: present concerns and future aspects. Afr J Biotechnol 13:705–713

Prasad R, Pandey R, Barman I (2016) Engineering tailored nanoparticles with microbes: quo vadis. WIREs Nanomed Nanobiotechnol 8:316–330

Rahman Q, Lohani M, Dopp E, Pemsel H, Jonas L, Weiss DG (2002) Evidence that ultrafine titanium oxide induces micronuclei and apoptosis in Syrian hamster embryo fibroblast. Environ Health Perspect 110:797–800

Ranzelli-Cain R, Kaloustian KV (1995) Evidence for the involvement of opioid peptides in phagocytosis, conformation, granulation and aggregation of immunocompetent Lumbricus *terrestris* amoebocytes. Comp Biochem Physiol 111:205–211

Schlich K, Terytze K, Hund-Rinke K (2012) Effect of TiO2 nanoparticles in earthworm reproduction test. Environ Sci Eur 24:5. doi:10.1186/2190-4715-24-5

Shang L, Nienhaus K, Nienhaus GU (2014) Engineered nanoparticles interacting with cells: size matters. J Nanobiotechnol 12:5. doi:10.1186/1477-3155-12-5

Simonet BM, Valcarcel M (2009) Monitoring nanoparticles in the environment. Anal Bioanal Chem 393:17–21

Stein EA, Cooper EL (1981) Cytochemical observations of coelomocytes the earthworm, Lumbricus terrestris. Dev Comp Immunol 5:15–25

Stürzenbaum SR, Höckner M, Panneerselvam A, Levitt J, Bouillard J-S, Taniguchi S, Dailey L-A, Ahmad KR, Rosca EV, Thanou M, Suhling K, Zayats AV, Green M (2013) Biosynthesis of luminescent quantum dots in an earthworm. Nat Nanotechnol 8:57–60

Tiede K, Tear SP, David H, Boxall AB (2009) Imaging of engineered nanoparticles and their aggregates under fully liquid conditions in environmental matrices. Water Res 43:3335–3343

Ulrich S, Seijo M, Laguecir A, Stoll S (2006) Nanoparticle adsorption on a week polyelectrolyte. Stiffness, pH, charge mobility and ionic concentration effects investigated by Monte Carlo simulations. J Phys Chem B 110:20954–20964

Unrine JM, Hunyadi SE, Tsyusko OV, Rao W, Shoults-Wilson WA, Bertsch PM (2010) Evidence for bioavailability of an nanoparticles from soil and biodistribution with in earthworms (*Eisenia fetida*). Environ Sci Technol 44:8308–8313

Valembois P, Roch P, Lassegues M, Cassand P (1982) Antibacterial activity of the hemolytic system from the earthworm *Eisenia fetida* andrei. J Invertebr Pathol 40:21–27

Valembois P, Roch P, Lassegues M (1988) Evidence of plasma clotting system in earthworms. J Invertebr Pathol 51:221–228

Van der Ploeg MJC, Baveco JM, Van der Hout A, Bakker R, Rietjens IM, Van den Brink NW (2010) Effects of C60 nanoparticle exposure on earthworms (*Lumbricus rubellus*) and implications for population dynamics. Environ Pollut 159:198–203

Van der Ploeg MJ, Handy RD, Heckmann LH, Van der Hout A, Van Den Brink NW (2013) C60 exposure induced tissue damage and gene expression alterations in the earthworm *Lumbricus rubellus*. Nanotoxicology 7:432–440

Ville P, Roch P, Cooper E, Masson P, Narrbomnne J (1995) PCBs increase molecular-related activities (lysosome, antibacterial, hemolysis, proteases) but inhibit macrophage-related functions (phagocytosis, wound healing) in earthworms. J Invertebr Pathol 65:217–224

Whitfield Aslund ML, McShane H, Simpson MJ, Simpson AJ, Whalen JK, Hendershot WH, Sunahara GI (2011) Earthworm sublethal responses to titanium oxide nanomaterial in soil detected by H NMR metabolomics. Environ Sci Technol 46:1111–1118

Zeng Z, Patel J, Lee S-H, McCallum M, Tyagi A, Yan M, Shea KJ (2015) Synthetic polymer nanoparticle-polysaccharide interactions: a systematic study. J Am Chem Soc 134:2681–2690

Chapter 9
Remediation of Environmental Pollutants Using Nanoclays

Mohsen Soleimani and Nasibeh Amini

9.1 Introduction

Nanotechnology and nanomaterials have become more widespread for environmental pollution control over the last decades. Having complex properties, clay minerals are considered very useful in various environmental applications. The terms "nanoclays" and "clay minerals" are mostly used interchangeably because the clay minerals are particles with at least one dimension in the nanometer range (Zhu and Njuguna 2014). Nanoclays are layered silicates clay minerals which can be formed as a result of chemical weathering of other silicate minerals (Bignon 2013; Floody et al. 2009; Choy et al. 2007).Use of nanoclays is increasing in comparison to other nanomaterials because of their specific physicochemical properties (Zhu and Njuguna 2014). Nanoclays have widespread uses in various fields including food industries, agriculture, animal feed as well as environment remediation and medicine (Choy et al. 2007; Newman and Cragg 2007).

Since clay minerals have small particle size, high specific surface area (external and internal surface area up to 100 m^2/g), and a large porosity (Yuan and Wu 2007), they are suitable for pollution control and environmental protection among different nanomaterials. In addition, clay minerals are generally nontoxic and rather inexpensive sorbents (Churchman et al. 2006). Clays and clay minerals are interesting because of their common availability and extraordinary properties (Bergaya and Lagaly 2006). These valuable materials have been used in removal of pollutants from agricultural leachates, soil liners, and water and wastewaters because of their relevant properties mentioned in Table 9.1.

M. Soleimani (✉) • N. Amini
Department of Natural Resources, Isfahan University of Technology, Isfahan 84156-83111, Iran
e-mail: m.soleimani@cc.iut.ac.ir

M. Ghorbanpour et al. (eds.), *Nanoscience and Plant–Soil Systems*, Soil Biology 48,
DOI 10.1007/978-3-319-46835-8_9

Table 9.1 Applications of clays for pollution control and environmental protection (Churchman et al. 2006)

Contaminants for control	Status (actual or potential use)	Relevant clay properties
Heavy metal cations and simple cations	Actual, mainly passive, use (e.g., in soils, liners)	Charge, surface area, reactive surface groups
Organic and biological cations	Potential for water and wastewater treatment, pesticide control	Charge, surface area, especially interlayer
Nonionic organic molecules	Actual, for water and wastewater treatment; potential, for pesticide control, waste liners	Charge, specific surface area, interlayer
Anions	Actual, for water and wastewater treatment; potential, for pesticide and nutrient leaching control	Charge, reactive surface groups
Turbidity and residual treatment chemicals	Actual, for treatment of potable water and some wastewaters and sewage	Colloidal, size and charge; surface area
Leachates	Actual, for waste liners and radioactive waste storage	Swelling, charge, surface area, reactive surface groups

Cation exchange capacity (CEC) of nanoclays is also a positive factor for sorption of various pollutants from contaminated environments (Churchman et al. 2006). This is particularly important for sorption and control of heavy metal cations which can occur at various sites on the surface of clay mineral particles (Inskeep and Baham 1983). Adsorption of heavy metal ions on surface of clay minerals is a complex process indicating formation of covalent bonds. The sorption of heavy metals does not solely depend on CEC of clay minerals. The reason is that heavy metal adsorption takes place due to different processes (Jackson 1998). Adsorption and desorption of heavy metals (e.g., Cu and Cd) were reported on the borders and interlayer sites of montmorillonite as a clay mineral (Undabeytia et al. 1998, 2002). For each type of metal ions, the selected site might depend on various factors such as pH, ionic strength, and the anions that are present in solution (Undabeytia et al. 2002). The heavy metals adsorption capacity of clay minerals is not the same. Furthermore, nanoclays can also react to different types of organic compounds. Since the majority of clay minerals consist of the negatively charged sorbents, they have a strong tendency to adsorb organic cations and they may be limited for adsorption of organic species which are bond to positive charges (Theng 1974).

Considering the importance of clay minerals/nanoclays in environmental issues, in this chapter we present their characteristics and their relevant application in various pollutants removal from contaminated air, water, and soil media.

9.2 Structure, Classification, and Characteristics of Clay Minerals

9.2.1 Structures of Clay Minerals

The unique structure of clay minerals is like the small plates which include several crystal sheets having a replicate atomic structure of alumina and silica sheets. The alumina sheets (i.e., octahedral) consist of a composition of six hydroxyls or oxygen enclosing a metal atom such as aluminum, iron, magnesium, or other atoms (Murray 2006). In the other hand, each silica sheet (i.e., tetrahedral) consists of four oxygen that is linked to adjacent tetrahedral by sharing three corners. The basic building framework of all clay minerals is the same and consists of the tetrahedral and octahedral sheets which linked together by specified ways and create a nano-structure. The fundamental unit of clay mineral particles consists of nanoparticles aluminosilicate with an outer diameter of 3.5–5.0 nm (Brigatti et al. 2006). The primary layers of smectite as a clay mineral have about 1 nm width and less than 100 nm length (Yuan 2004).

9.2.2 Clay Minerals Classification

The first classification of clay minerals at two-class (amorphous and crystalline) was proposed by Grim (1962). This classification becomes the basic of differences between various clay minerals (Murray 2000). Besides, the other classification of clay minerals is based on the layer type and charge per formula unit which has been shown in Table 9.2. In this classification, minerals which have sheets including 1 tetrahedral and 1 octahedral units are considered as 1:1 minerals and those have 2 tetrahedral and 1 octahedral units are 2:1 minerals. If the minerals of 2:1 type include a brucite layer ($Mg(OH)_2$), they may be categorized as 2:1:1 or 2:2 minerals. Furthermore, there is a simple classification of clay minerals in literatures which has classified them into four principal groups including illite, kaolinite, smectite, and vermiculite. The other classification is based on a scheme of hydrous phyllosilicate structures (Guggenheim et al. 2006).

9.2.3 Characteristics of Clay Minerals

The existence of negatively charged surfaces in the clay minerals is an important factor for CEC and swelling properties of the minerals. The charge in clay minerals is caused by their surface and structure. The surface charge usually depends on the value of environment pH while the structural charge originates from the interior of the layers and is permanent which is caused by ion exchange during the crystal

Table 9.2 Classification of clay minerals based on the layer type and charge per formula unit

Layer type	Group	Subgroup	Species
1:1	Kaolin-serpentine $x=0$	Kaolin	Kaolin, dickite, nacrite, halloysite
		Serpentine	Chrysotile, lizardite, amesite
2:1	Pyrophyllite-talc $x=0$	Pyrophyllite	Pyrophyllite
		Talc	Talc
	Smectite $x=0.2$–0.6	Montmorillonite (tioctahedral smectite)	Montmorillonite, beidellite, nontronite
		Saponite (trioctahedral smectite)	Saponite, hectorite
	Vermiculite $x=0.6$–0.9	dioctahedral vermiculite	dioctahedral vermiculite
		trioctahedral vermiculite	trioctahedral vermiculite
	Mica $x=0.5$–1.0	dioctahedral mica	Muscovite, illite, glauconite, paragonite
		trioctahedral mica	Phlogopite, biotite, lepidolite
	Brittle mica $x=2.0$	dioctahedral brittle mica	Margarite
		trioctahedral brittle mica	Clintonite, anandite
	Chlorite $x=$ variable	dioctahedral chlorite	Donbassite
		di- tri-octahedral chlorite	Cookeite, sudoite
		trioctahedral chlorite	Clinochlore, chamosite, nimite
	Palygorskite-sepiolite $x=$ variable	Sepiolite	Sepiolite
		Palygorskite	Palygorskite

Note x: charge per formula unit

formation. In the clay minerals of 2:1 layer type, the surface charge comes from fundamental surface of tetrahedral sheets, and in the clay minerals with 1:1 layer the surface charge exists from both of tetrahedral and octahedral sheets, but the surface charge in the two type clay (1:1 and 2:1) comes from the edges of the sheets (Eslinger and Pevear 1988). The most important characteristic of clay minerals is CEC which is influenced by total layer charge. Since the surface layer charge is affected by pH, usually CEC is measured at neutral pH (Eslinger and Pevear 1988). Ions and water molecules could be sorbed into the space between the sheets of 2:1 clay minerals, while that is impossible in 1:1 minerals in which the sheets are strongly bond together and there is no free space.

9.3 Removal of Pollutants Using Nanoclay Minerals

9.3.1 Gas Pollutants

Industrial activities are the main source of release of toxic gases into the atmosphere. The reduction of gaseous pollutants, particularly volatile organic compounds, using thermal oxidation is expensive due to the need of high amount of energy. In this regard, more efficient processes combined with absorption techniques can be developed to improve the efficiency of catalytic oxidation and finally gas removal (Pires and Pinto 2010). Many studies have shown the absorption of polar and non-polar gases and water vapor molecules by different types of nanoclay minerals, specially pillared nanoclays which have been widely developed for air pollutant abatement nowadays. Nanoclays can be interesting alternatives to reduce and finally remove gaseous pollutants due to having hydrophobic and hydrophilic properties. According to the wide use of nanoclays in control of gaseous pollutants, the researchers have improved sorption processes and control of gaseous pollutants using modified nanoclays. Molina-Sabio et al. (2004) investigated the adsorption of NH_3 and H_2S on activated carbon-sepiolite pellets. Nguyen-Thanh et al. (2005) have studied hydrogen sulfide adsorption using Na-bentonite modified by iron. The iron-doped samples revealed a significant increase in the clay absorption capacity and H_2S removal. Table 9.3 summarizes recent studies regarding the use of clay minerals as the sorbent of various gaseous pollutants. Furthermore, nanoclays have a high potential to remove air pollutants originated from various sources but their efficiency depends on the type of pollutants, the sorbent characteristics, and physicochemical environmental conditions. Besides, using the hybrid sorbents including nanoclays could be a good approach to enhance the pollutants removal efficiency.

9.3.2 Water Pollutants

Industrial, agricultural, and urban wastewaters contain high amounts of heavy metals and organic compounds. These pollutants have hazardous effects on human health and living organisms. Although there are many sorbents used in removal of water pollutants, nanoclays are evaluated as suitable sorbents because of their low cost and high removal efficiency. Therefore, clay minerals and nanoclays have widely used for water and wastewater treatment to remove heavy metals, organic pollutants, and nutrients from water resources (Abdelaal 2004). Clay minerals, such as bentonite and zeolite, having a large specific surface area and negative net charge can adsorb inorganic and organic cations from the environment (Konig et al. 2012; Babel and Kurniawan 2003; Murray 2000).Their capability in adsorption of Cr (VI) from a solution containing 5 ppm Cr and 0.4 g/L sorbent in batch experiments was more than montmorillonite and significantly enhanced when the sorbents were modified by TiO_2 (Fig. 9.1). It seems that we have to modify the

Table 9.3 Removal of various gaseous pollutants using different clays

Pollutant	Clay mineral	Removal (%)	Reference
p-xylene	Na-montmorillonite	–	Cabbar and Cakanyıldırım (2008)
n-Hexane	Montmorillonite		Morozov et al. (2014)
Benzene		–	
SO_2	Palygorskite	–	Zhang et al. (2009)
SO_2	Zeolite	>95 %	Vieira et al. (2011)
SO_2	Zeolite Clinoptilolite	>24 %	Allen et al. (2009)
SO_2	Zeolite Clinoptilolite	–	Ivanova and Koumanova (2009)
H_2S	Zeolite Clinoptilolite	–	Ozekmekci et al. (2015)
H_2S	Kaolinite	–	Batista et al. (2014)
H_2S	Bentonites	–	Stepova et al. (2009)
H_2S	Montmorillonite	–	Nguyen-Thanh et al. (2005)

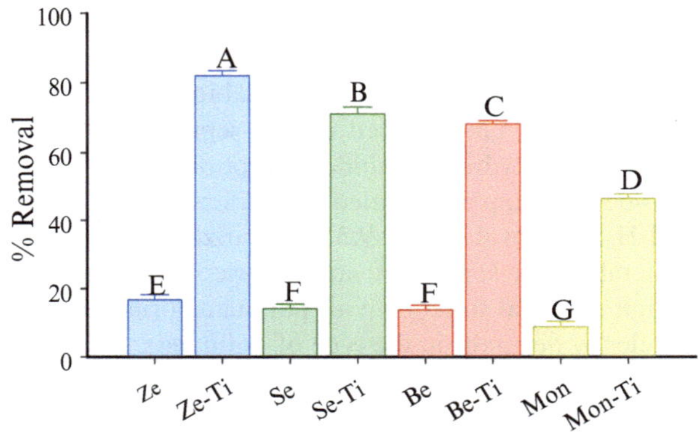

Fig. 9.1 Cr (VI) removal by four nanoclays (*Ze* zeolite, *Se* sepiolite, *Be* bentonite, *Mon* montmorillonite) and their modified types with titanium (Ti) in batch experiments (initial Cr concentration 5 ppm, sorbent dose: 4 g/L, pH: 2, titanium: 13 mmol Ti/g clay, Time: 90 min). The difference letters show the significant difference of means using Tukey test ($p < 0.05$)

clays in some cases in order to sorb a special pollutant and/or to increase their removal efficiency. Gu et al. (2010) compared adsorption of five different metal ions, Cd^{2+}, Cu^{2+}, Pb^{2+}, Ni^{2+}, and Zn^{2+}, by montmorillonite considering pH and ionic strength of the solution. Coppin et al. (2002) studied adsorption of lanthanide on kaolinite and Na-montmorillonite over a pH range and various ionic strengths. Adsorption of Pb onto sepiolite under different pH and temperatures was studied by Bektaş et al. (2004). Table 9.4 summarizes recent studies regarding the use of clay minerals as the sorbents for water pollutant removal.

9.3.3 Soil Pollutants

In recent years, soil pollution has significantly increased due to release of pollutants directly to the environment from various sources such as spills during

Table 9.4 Removal of water pollutants using various clays

Pollutant	Clay mineral	Percent removal	Reference
Cu (II)	Bentonite	48 %	Al-Qunaibit et al. (2005)
Cd (II)–CO(II)	Kaolinite	100 %	Angove et al. (1998)
Fe (II)	Bentonite	>98 %	Tahir and Rauf (2004)
Cu (II)–Zn (II)	Bentonite	99 % Zn (II) 90 % Cu (II)	Veli and Alyüz (2007)
Cd (II)	Montmorillonite	90.2 %	Undabeytia López et al. (1998)
Pb (II)	Zeolite clinoptilolite	–	Inglezakis et al. (2007)
Cu (II)- and Cd (II)	Na-montmorillonite	–	Inskeep and Baham (1983)
Cu (II)- and Cd (II)	Bentonite	84.5 % Cu (II) 87.2 % Cd (II)	Karapinar and Donat (2009)
NH_4	Zeolite clinoptilolite	>90 %	Rahmani et al. (2009)
Phenol	Bentonite	>45 %	Banat et al. (2000)

transportation, leakage from waste disposal or storage sites, and from industrial facilities (Riser-Roberts 1992). Therefore, soil pollution should be controlled via environmental friendly and low cost technologies (Riser-Roberts 1998). Clay is a significant vector transporting pollutants in the soil environment. It's a component of the soil with the ability to adsorb and transport compounds and nutrients, such as phosphorus (Wilkinson et al. 2000; Sharpley et al. 1984), potassium (Petrofanov 2012), metals (Quinton and Catt 2007), and organic pollutants (Wilkinson et al. 2000).

A major part of soil pollutants passes from the soil solution into the groundwater zone. Normally, different adsorbents exist in soil environment which among them clays with high CEC can bind to the positively charged organic materials and metals and make them chemically immobile and consequently reduce their imposed risk to the soil system. The value of soil CEC generally depends on organic substance and clay content of the soil (Pisani and Mirsal 2004). Sorption of ions and other compounds by clay minerals can be performed by two mechanisms: (1) adsorption on planar external surfaces and (2) exchange in the interlayer space. The presence of other minerals (e.g., Al and Fe hydroxides) may considerably increase CEC of clays such as montmorillonite (Terce and Calvet 1977).

Since, clay minerals consist of some silicates and organic components; they are very important sorbents in the soil environment. Due to the presence of clays in the soil, water molecules are adsorbed on their surfaces and this provides sorption sites for pollutants. Therefore, structural properties of individual clay minerals play a main role in determination of selectivity and adsorption mechanism of pollutants. As a consequence, the intensity of sorption in soils will mostly depend on the clay content and clay mineralogy in the soil texture (Pisani and Mirsal 2004).

Clays can be suspended in liquid phase of a porous media and also sorb the contaminants from the solution and therefore may be suitable for a safe infiltration process. Additionally, application of clay minerals in soil to keep the environment safe has been investigated in fundamental problem studies during last decades

(Kühnel 1990). Nowadays, nanoclays are used as a cover for landfills to prevent groundwater contamination. Application of clays in waste disposal as a buffer improved immobilization of hazardous compounds through isolation, stabilization, and fixing (Kühnel 1990). In other words, nanoclays and clay minerals in soil together with metal hydroxides and organic matter control the concentration of heavy metal ions in soil and groundwater. This depends on the type of heavy metal and solution pH and other environmental factors (Churchman et al. 2006). Pedogenic oxide is more effective than organic materials in removing Co^{2+} in the solution (McLaren et al. 1986). Furthermore, combinations of soil components and nanoclays organic matter complexes could enhance heavy metal adsorption. Many reports have mentioned the role of clays in soil stabilization in fields (Morgan 1995). Nanoclays have been used to remove polychlorinated biphenyl (PCB) from the soil (Mobasser and Taha 2013). The results clearly indicated that maximum percentage of PCB adsorption by nanoclays was 77 % (Mobasser and Taha 2013). The removal of diclofenac from soil using nanoclays revealed that the presence of clays in the soil could enhance the pollutant removal efficiency (Santin 2014). Adsorption of organic materials through weak Van der Waals forces on the surfaces of clay minerals can play a significant role in linking organic matter to the surfaces of clays (Stevenson 1994). Organic materials such as fulvic acid could be adsorbed into the space between the layers of montmorillonite (Schnitzer and Kodama 1977). Wei et al. (2015) investigated the removal of pyrene (i.e., one of the polycyclic aromatic hydrocarbons) by kaolin and montmorillonite (Wei et al. 2015). Their results showed that the removal of the pollutant by kaolin increased during the time while desorption from montmorillonite approximately remained stable during the time. The importance of clay minerals and organic matters in sorption of Triton X-100 confirmed the high potential of these compounds by montmorillonite (Zhu et al. 2003). Thus, the clays have a high potential to sorb, remove, and stabilize various pollutants in the soil media.

9.4 Conclusion

Clay minerals and nanoclays as the cheap and environmental friendly sorbents having unique characteristics (e.g., high surface area and porosity, mostly negative charged surface) can be used in various environmental issues. Considering the importance of control and remediation of environmental pollutants throughout the world, especially in developing countries, removal of air, water, and soil pollutants using nanoclays could be considered as a promising approach. Although clays have showed a high potential in removing or stabilizing various organic and inorganic pollutants in the environment, their efficiency depends on their type and structure as well as type of pollutants and the conditions of environment. The use of nanoclays with other sorbents or in combination with other remediation methods may be a promising approach to extend their capability in environmental management issues.

References

Abdelaal A (2004) Using a natural coagulant for treating wastewater. In: 8th international water technology conference, IWTC8, Alexandria, Egypt. Citeseer, pp 781–791

Allen SJ, Ivanova E, Koumanova B (2009) Adsorption of sulfur dioxide on chemically modified natural clinoptilolite. Acid modification. Chem Eng J 152:389–395

Al-Qunaibit M, Mekhemer W, Zaghloul A (2005) The adsorption of Cu (II) ions on bentonite—a kinetic study. J Colloid Interface Sci 283:316–321

Angove MJ, Johnson BB, Wells JD (1998) The influence of temperature on the adsorption of cadmium (II) and cobalt (II) on kaolinite. J Colloid Interface Sci 204:93–103

Babel S, Kurniawan TA (2003) Low-cost adsorbents for heavy metals uptake from contaminated water: a review. J Hazard Mater 97:219–243

Banat F, Al-Bashir B, Al-Asheh S, Hayajneh O (2000) Adsorption of phenol by bentonite. Environ Pollut 107:391–398

Batista LC, de S Dantas D, de Farias RF (2014) Dye adsorption on inorganic matrices as a new strategy to gas capture: hydrogen sulfide adsorption on Rodhamine B modified kaolinite. Synth React Inorg Met Org Nano Met Chem 44:1398–1400

Bektaş N, Ağım BA, Kara S (2004) Kinetic and equilibrium studies in removing lead ions from aqueous solutions by natural sepiolite. J Hazard Mater 112:115–122

Bergaya F, Lagaly G (2006) General introduction: clays, clay minerals, and clay science. Handb Clay Sci 1:1–18

Bignon J (2013) Health related effects of phyllosilicates. Springer Science and Business Media, Heidelberg

Brigatti M, Galan E, Theng B (2006) Structures and mineralogy of clay minerals. Handb Clay Sci 1:19–69

Cabbar HC, Cakanyıldırım C (2008) Adsorption of p-xylene in dry and moist clay. J Int Environ Appl Sci 3:29–36

Choy J-H, Choi S-J, Oh J-M, Park T (2007) Clay minerals and layered double hydroxides for novel biological applications. Appl Clay Sci 36:122–132

Churchman GJ, Gates WP, Theng BKG, Yuan G (2006) Chapter 11.1 Clays and clay minerals for pollution control. In: Faïza Bergaya BKGT, Gerhard L (eds) Developments in clay science, vol 1. Elsevier, Amsterdam, pp 625–675

Coppin F, Berger G, Bauer A, Castet S, Loubet M (2002) Sorption of lanthanides on smectite and kaolinite. Chem Geol 182:57–68

Eslinger E, Pevear DR (1988) Clay minerals for petroleum geologists and engineers. Society of Economic Paleontologists and Mineralogists

Floody MC, Theng B, Reyes P, Mora M (2009) Natural nanoclays: applications and future trends—a Chilean perspective. Clay Miner 44:161–176

Grim R (1962) Applied clay mineralogy. McGraw-Hill, New York

Gu X, Evans LJ, Barabash SJ (2010) Modeling the adsorption of Cd (II), Cu (II), Ni (II), Pb (II) and Zn (II) onto montmorillonite. Geochimica et Cosmochimica Acta 74:5718–5728

Guggenheim S, Adams J, Bain D, Bergaya F, Brigatti MF, Drits V, Formoso ML, Galán E, Kogure T, Stanjek H (2006) Summary of recommendations of nomenclature committees relevant to clay mineralogy: report of the Association Internationale pour l'Etude des Argiles (AIPEA) Nomenclature Committee for 2006. Clay Miner 41:863–877

Inglezakis VJ, Stylianou MA, Gkantzou D, Loizidou MD (2007) Removal of Pb (II) from aqueous solutions by using clinoptilolite and bentonite as adsorbents. Desalination 210:248–256

Inskeep WP, Baham J (1983) Adsorption of Cd (II) and Cu (II) by Na-montmorillonite at low surface coverage. Soil Sci Soc Am J 47:660–665

Ivanova E, Koumanova B (2009) Adsorption of sulfur dioxide on natural clinoptilolite chemically modified with salt solutions. J Hazard Mater 167:306–312

Jackson T (1998) The biogeochemical and ecological significance of interactions between colloidal minerals and trace elements. In: Parker A, Rae JE (eds) Environmental interactions of clays. Springer, Berlin, pp 93–205
Karapinar N, Donat R (2009) Adsorption behaviour of Cu^{2+} and Cd^{2+} onto natural bentonite. Desalination 249:123–129
Konig TN, Shulami S, Rytwo G (2012) Brine wastewater pretreatment using clay minerals and organoclays as flocculants. Appl Clay Sci 67:119–124
Kühnel R (1990) The modern days of clays. Appl Clay Sci 5:135–143
McLaren R, Lawson D, Swift R (1986) Sorption and desorption of cobalt by soils and soil components. J Soil Sci 37:413–426
Mobasser S, Taha MR (2013) Adsorption of PCB from contaminated soil using nano clay particles. J Ind Pollut Contr 29:145–148
Molina-Sabio M, González J, Rodríguez-Reinoso F (2004) Adsorption of NH 3 and H 2 S on activated carbon and activated carbon-sepiolite pellets. Carbon 42:448–450
Morgan RC (1995) Soil erosion and conservation. Longman, London
Morozov G, Breus V, Nekludov S, Breus I (2014) Sorption of volatile organic compounds and their mixtures on montmorillonite at different humidity. Colloids Surfaces A Physicochem Eng Aspect 454:159–171
Murray HH (2000) Traditional and new applications for kaolin, smectite, and palygorskite: a general overview. Appl Clay Sci 17:207–221
Murray HH (2006) Applied clay mineralogy: occurrences, processing and applications of Kaolins, Bentonites, Palygorskitesepiolite, and common clays. Elsevier, Amsterdam
Newman DJ, Cragg GM (2007) Natural products as sources of new drugs over the last 25 years. J Nat Prod 70:461–477
Nguyen-Thanh D, Block K, Bandosz TJ (2005) Adsorption of hydrogen sulfide on montmorillonites modified with iron. Chemosphere 59:343–353
Ozekmekci M, Salkic G, Fellah MF (2015) Use of zeolites for the removal of H2S: a mini-review. Fuel Process Technol 139:49–60. doi:10.1016/j.fuproc.2015.08.015
Petrofanov V (2012) Role of the soil particle-size fractions in the sorption and desorption of potassium. Eurasian Soil Sci 45:598–611
Pires J, Pinto M (2010) Pillared interlayered clays as adsorbents of gases and vapors. In: Gil A, Korili SA, Trujillano R, Vicente MA (eds) Pillared clays and related catalysts. Springer, Heidelberg, pp 23–42
Pisani P, Mirsal I (2004) Soil pollution. Origin, monitoring & remediation. Springer, Heidelberg
Quinton JN, Catt JA (2007) Enrichment of heavy metals in sediment resulting from soil erosion on agricultural fields. Environ Sci Technol 41:3495–3500
Rahmani A, Samadi M, Ehsani H (2009) Investigation of clinoptilolite natural zeolite regeneration by air stripping followed by ion exchange for removal of ammonium from aqueous solutions. J Environ Health Sci Eng 6:167–172
Riser-Roberts E (1992) Bioremediation of petroleum contaminated sites. CK Smoley, Boca Raton, FL, p 197
Riser-Roberts E (1998) Remediation of petroleum contaminated soils: biological, physical, and chemical processes. CRC, Boca Raton, FL
Santin A (2014) Remediation of contaminated soil by nano-to-micro clay particles, case study with diclofenac. Master thesis, University of Padova, Italy
Schnitzer M, Kodama H (1977) Reactions of minerals with soil humic substances. Miner Soil Environ 21:741–770
Sharpley A, Smith S, Stewart B, Mathers A (1984) Forms of phosphorus in soil receiving cattle feedlot waste. J Environ Qual 13:211–215
Stepova KV, Maquarrie DJ, Krip IM (2009) Modified bentonites as adsorbents of hydrogen sulfide gases. Appl Clay Sci 42:625–628
Stevenson FJ (1994) Humus chemistry: genesis, composition, reactions. Wiley, New York

Tahir S, Rauf N (2004) Removal of Fe (II) from the wastewater of a galvanized pipe manufacturing industry by adsorption onto bentonite clay. J Environ Manag 73:285–292
Terce M, Calvet R (1977) Some observations on the role of Al and Fe and their hydroxides in the adsorption of herbicides by montmorillonite. Sonderdruck, Zeitschriftfür Pflanzenkrankheiten und Pflanzenschutz, Sonderheft VIII Stuttgart-Hohenheim
Theng BKG (1974) The chemistry of clay-organic reactions. Wiley, London, p 343
Undabeytia T, Nir S, Rytwo G, Morillo E, Maqueda C (1998) Modeling adsorption–desorption processes of Cd on montmorillonite. Clays Clay Miner 46:423–428
Undabeytia T, Nir S, Rytwo G, Serban C, Morillo E, Maqueda C (2002) Modeling adsorption-desorption processes of Cu on edge and planar sites of montmorillonite. Environ Sci Technol 36:2677–2683
Veli S, Alyüz B (2007) Adsorption of copper and zinc from aqueous solutions by using natural clay. J Hazard Mater 149:226–233
Vieira MGA, Almeida Neto Ad, Gimenes ML, Silva Md (2011) Desulphuration of SO2 by adsorption in fluidized bed with zeolite. Chem Eng Trans 24:1219–1224
Wei Y, Liang X, Lin W, Guo C, Dang Z (2015) Clay mineral dependent desorption of pyrene from soils by single and mixed anionic–nonionic surfactants. Chem Eng J 264:807–814
Wilkinson S, Grunes D, Sumner M (2000) Nutrient interactions in soil and plant nutrition. CRC, Boca Raton, FL
Yuan G (2004) Natural and modified nanomaterials as sorbents of environmental contaminants. J Environ Sci Health Part A 39:2661–2670
Yuan G, Wu L (2007) Allophane nanoclay for the removal of phosphorus in water and wastewater. Sci Technol Adv Mater 8:60–62
Zhang Q, Higuchi T, Sekine M, Imai T (2009) Removal of sulphur dioxide using palygorskite in a fixed bed adsorber. Environ Technol 30:1529–1538
Zhu H, Njuguna J (2014) 7 – nanolayered silicates/clay minerals: uses and effects on health. Health and Environmental Safety of Nanomaterials. Woodhead Publishing, Cambridge, pp. 133–146
Zhu L, Yang K, Lou B, Yuan B (2003) A multi-component statistic analysis for the influence of sediment/soil composition on the sorption of a nonionic surfactant (Triton X-100) onto natural sediments/soils. Water Res 37:4792–4800

Chapter 10
The Role of Nanomaterials in Water Desalination: Nanocomposite Electrodialysis Ion-Exchange Membranes

Sayed Mohsen Hosseini and Sayed Siavash Madaeni

10.1 Introduction

Nowadays, availability of fresh potable water is a fundamental requirement for most aspects of life and key element for all societies (Thampy et al. 2011). More observations are kept on water treatments as demands for drinking water or industrial water increase. Among the various water treatment techniques, membrane processes are more desirable as they are fast, cheap, high selective, and flexible to be integrated with other processes and have low energy consumption and space requirement and no phase change during the separation processes (Mollahosseini et al. 2012). Currently, ion-exchange membranes are being utilized as active separators in divers' electrically driven processes such as electrodialysis for desalting brackish waters, reconcentrating brine from seawater, and production of table salt. Moreover, IEMs are efficient tools in resource recovery, food and pharmacy processing, and environmental protection for treating industrial and biological effluents (Hwang et al. 1999; Nagarale et al. 2004; Volodina et al. 2005; Vyas et al. 2003). In ion-exchange membranes, charged groups are attached to polymer backbone and freely permeable to opposite sign ions under an electrical field influence (Baker 2004). In such processes, ion interactions with membranes, water, and others occur in complex fashions. Knowledge of the electrochemical properties of ion-exchange membranes is a major important factor behind decisions

S.M. Hosseini (✉)
Department of Chemical Engineering, Faculty of Engineering, Arak University, Arak 38156-8-8349, Iran
e-mail: Sayedmohsen_Hosseini@yahoo.com; S-Hosseini@Araku.ac.ir

S.S. Madaeni
Membrane Research Centre, Department of Chemical Engineering, Faculty of Engineering, Razi University, Kermanshah, Iran
e-mail: smadaeni@yahoo.com

M. Ghorbanpour et al. (eds.), *Nanoscience and Plant–Soil Systems*, Soil Biology 48,
DOI 10.1007/978-3-319-46835-8_10

about their applicability in specific separation processes (Gohil et al. 2004; Shahi et al. 2003). The development of ion-exchange membranes with special characteristics such as high selectivity and ionic conductivity, good permeability, and suitable chemical, thermal, and mechanical resistances is highly desired and may be a vital step in future applications (Kerres et al. 1998; Li et al. 2005). The ion-exchange membranes, both homogenous and heterogeneous, supersede each other (Vyas et al. 2001). Homogenous ion-exchange membranes usually show good electrochemical properties but are weak in their mechanical strength, whereas heterogeneous types have acceptable mechanical property but with inadequate electrochemical properties (Kariduraganavar et al. 2006; Tanaka 2007; Vyas et al. 2001, 2003).

10.2 Nanocomposite Ion-Exchange Membranes

Significant researches have already been carried out to improve the physicochemical properties of ion-exchange membranes which may be a vital step in future chemical and treatment applications. During the last decade, utilizing inorganic particles especially nanomaterials into polymeric matrices has been examined in many applications to enhance the mechanical, thermal, and chemical stabilities of membranes in severe conditions such as high temperature and strongly oxidizing environment and also to improve their separation efficiency based on the synergism between the organic-inorganic component properties (Xu 2005). The unique attributes of nanoparticle-incorporated water desalination membranes could open the door to opportunities to overcome the challenges in making clean water easily accessible around the globe (Hegab and Zou 2015).

10.2.1 Electrodialysis Ion-Exchange Membrane Modified by SiO_2 Nanoparticles

PVC-co-silica nanoparticle nanocomposite electrodialysis heterogeneous cation-exchange membranes (Hosseini et al. 2016) were prepared by solution casting technique for the application in water desalination. The effects of SiO_2 nanoparticle concentration in the membrane matrix on physicochemical characteristics of electrodialysis ion-exchange membranes were studied. It was reported that membrane transport number and selectivity were improved initially by increase of SiO_2 nanoparticle concentration up to 1 %wt in prepared membranes and then showed a decreasing trend by more increase in SiO_2 concentration from 1 to 4 %wt. Also ionic flux were also decreased initially by increase of silica nanoparticle concentration up to 0.5 %wt in membrane matrix and then increased again by more nanoparticle concentration from 0.5 to 4 %wt. Moreover, results exhibited that incorporating silica nanoparticles in membrane matrix caused a decrease of areal electrical resistance and increase of membrane mechanical strength obviously.

10.2.2 Nanocomposite Cation-Exchange Membrane Filled with Clay Nanoparticles

Nanocomposite PVC-based heterogeneous ion-exchange membranes were fabricated (Hosseini et al. 2015d) through embedding clay nanoparticles into membrane matrix. It was found that membrane water content was improved initially by increase of clay NP concentration up to 1 %wt in the membrane matrix and then began to decrease. Membrane transport number and permselectivity were improved by increase of clay nanoparticle loading ratio. Utilizing Cloisite nanoparticles up to 1 %wt in membrane matrix also led to increase in ionic flux for prepared membranes initially. The ionic flux was decreased again by more increase in clay NP concentration. The membrane E-conductivity and mechanical strength was enhanced sharply by increase of clay nanoparticle concentration in membrane matrix. Also nanocomposite membranes exhibited good ability in (monovalent/bivalent) ion separation in water desalination.

10.2.3 Mixed Matrix Ion-Exchange Membrane Prepared by TiO_2 Nanoparticles

In another study (Hosseini et al. 2015c), nanocomposite heterogeneous cation-exchange membranes were fabricated by incorporating TiO_2 nanoparticles into PVC matrix by phase inversion method for the application in electrodialysis processes related to water recovery and treatment. Sonication was employed in fabrication of nanocomposite membranes to achieve better homogeneity in membrane matrix and to obtain the balance between electrochemical properties and mechanical integrity. The scanning optical microscopy images showed uniform distribution of nanoparticles and relatively uniform surfaces for the membranes. It was found that membrane ion-exchange capacity was improved initially by increase of TiO_2 nanoparticle content ratio and then decreased. Also membrane transport number and selectivity initially showed a constant trend by increase of TiO_2 nanoparticle content ratio up to 2 %wt and then decreased sharply by more additive concentration. Sodium permeability was enhanced by increase of nanoparticle concentration up to 0.5 %wt and then decreased by more NP percentage from 0.5 to 2 %wt. Sodium permeability was increased another time by more increase in additive concentration from 2 to 16 %wt and then showed a decreasing trend by more nanoparticle content ratio. A different trend was found for barium flux. FTIR spectrum analysis decisively proved stronger affinity of TiO_2 nanoparticles for barium ions compared to sodium ones which caused a different behavior for bivalent ions compared to monovalent ones. Also it was found that decrease of pH enhances positive charge distribution on nanoparticle surface which increases the charge density of membrane and so improves the flux. Membrane mechanical strength was also improved by increase of TiO_2 nanoparticle content ratio in membrane matrix.

10.2.4 Adsorptive Electrodialysis Ion-Exchange Membrane Prepared by Simultaneous Using of Ilmenite-co-Fe_3O_4 Nanoparticles

Also electrodialysis mixed matrix heterogeneous cation-exchange membranes were papered by using ilmenite-co-iron oxide nanoparticles (Hosseini et al. 2015a, b). The membrane potential, transport number, and selectivity were improved in NaCl ionic solution by using $FeTiO_3/Fe_3O_4$ nanoparticles in membrane matrix. Utilizing $FeTiO_3$-co-Fe_3O_4 nanoparticles in the casting solution also led to increase in ionic flux obviously. The modified membranes containing $FeTiO_3/Fe_3O_4$ nanoparticles showed higher transport number, selectivity, and ionic flux compared to prepared membrane including ilmenite clearly. The results showed that membrane areal electrical resistance was declined sharply by using $FeTiO_3$-co-Fe_3O_4 nanoparticles in prepared membranes. Obtained results showed that the membrane electrical resistance decreased initially by increase of electrolyte's pH sharply and then began to increase. Modified membranes containing $FeTiO_3/Fe_3O_4$ nanoparticles showed more appropriate electrochemical properties compared to other prepared membranes. The scanning electron microscopy (SEM) images of prepared nanocomposite membrane are shown in Fig. 10.1.

10.2.5 PVC-co-Zeolite Nanoparticle Ion-Exchange Membrane for Water Desalination

Mixed matrix PVC-based-co-zeolite nanoparticle nanocomposite heterogeneous cation-exchange membranes were prepared by solution casting technique. Hosseini et al. (2014d) reported that membrane water content, membrane potential, transport number, and selectivity were improved initially by utilizing zeolite nanoparticles up to 8 %wt in the casting solution and then began to decrease by more increase in zeolite NP concentration from 8 to 16 %wt. Utilizing zeolite nanoparticles in the membrane matrix also led to increase in membrane electrical conductivity and ionic flux obviously. Furthermore, membrane transport number, membrane selectivity, and membrane electrical conductivity were enhanced by increase of electrolyte concentration. The nanocomposite membrane containing 8 %wt zeolite nanoparticles

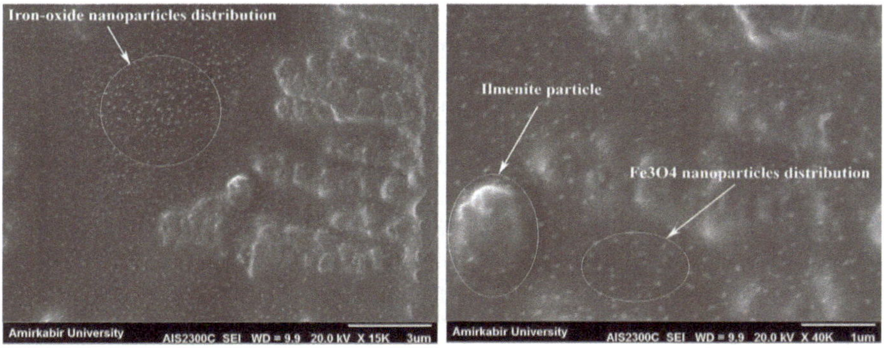

Fig. 10.1 SEM images of nanocomposite heterogeneous cation-exchange membrane prepared by simultaneously using ilmenite-co-Fe_3O_4 nanoparticles

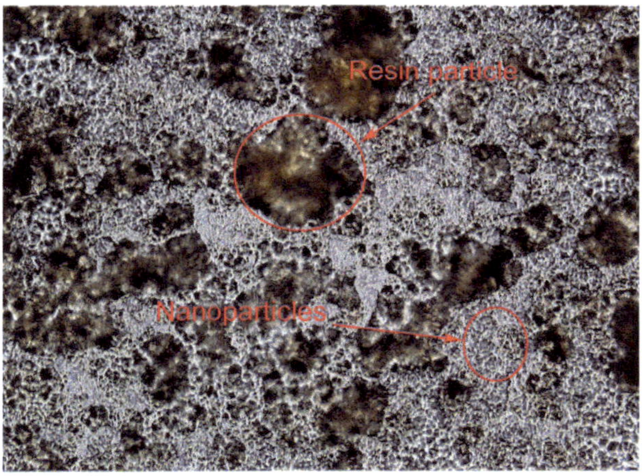

Fig. 10.2 SOM images of PVC-based-co-zeolite nanoparticle nanocomposite heterogeneous cation-exchange membrane

showed more suitable electrochemical properties compared to others. The scanning optical microscopy (SOM) image of PVC-based-co-zeolite nanoparticle nanocomposite heterogeneous cation-exchange membrane is shown in Fig. 10.2.

10.2.6 Mixed Matrix Ion-Exchange Membrane Modified by PANI/MWCNT Composite Nanoparticles

In another research, Hosseini et al. (2014c) used polyaniline (PANI)-co-multiwalled carbon nanotubes (MWCNTs) for modification of polyvinylchloride-based heterogeneous anion-exchange membranes. The effect of PANI/MWCNT nanoparticle concentration in the membrane matrix on electrochemical properties of membranes in water treatment and desalination was studied. Obtained results exhibited that the membrane fixed ionic concentration increased by increase of PANI/MWCNT nanoparticle content ratio in membrane matrix. Membrane transport number and selectivity were also improved in mono- and bivalent ionic solutions by increase of PANI/MWCNT composite nanoparticles. Moreover, membrane flux was enhanced initially in both monovalent and bivalent ionic solutions by increase of NP concentration up to 1 %wt and then showed decreasing trends by more increase of additive content ratio from 1 to 16 %wt. The prepared nanocomposite membranes showed better performance in comparison with pristine membrane. The obtained results are valuable for electromembrane processes especially electrodialysis for water recovery, treatment, and desalination.

10.2.7 Nanocomposite Ion-Exchange Membrane Filled with Al_2O_3 Nanoparticles

Aluminum oxide nanoparticle-incorporated heterogeneous cation-exchange membranes were fabricated by casting technique for the application in electrodialysis processes related to water recovery and treatment. The membrane water content was enhanced by increase of Al_2O_3 nanoparticle concentration in prepared membrane. The obtained results exhibited that membrane permselectivity and transport number were improved initially in sodium and barium chloride ionic solutions by increase of aluminum oxide nanoparticle concentration up to 2 %wt in the membrane matrix and then showed a decreasing trend by more increase in Al_2O_3 content ratio. Furthermore, membrane ionic flux was decreased initially by increase of aluminum oxide nanoparticle loading ratio up to 2 %wt and then increased again by more additive concentration. An opposite trend was found for membrane areal electrical resistance (Hosseini et al. 2014b).

10.2.8 Electrodialysis Ion-Exchange Membranes Modified by Fe_3O_4 Nanoparticles

Nowadays, demands for the superior ion-exchange membranes are increased for separation of dangerous pollutants soluble in water especially heavy metals. Hosseini et al. (2014a) prepared novel polyvinylchloride-based nanocomposite electrodialysis heterogeneous cation-exchange membranes through phase inversion method for the application in water desalination process to Pb removal from wastewater. Results showed that membrane water content and ion-exchange capacity were increased initially by increase of Fe_3O_4 nanoparticle loading ratio to 2 %wt and then showed a decreasing trend by more content from 2 to 8 %wt. Membrane

Fig. 10.3 SEM image of PVC-based nanocomposite membrane filled with Fe_3O_4 nanoparticles

selectivity and transport number were also enhanced by increase of additive concentration in sodium and barium chloride ionic solutions. They showed different behaviors in lead nitrate ionic solution. The sodium and barium flux were increased initially by increase of Fe_3O_4 NPs content ratio up to 2 %wt and then began to decrease by more additive loading. The electrodialysis experiment results showed that dialytic rate of lead ion removal was increased obviously by increase of nanoparticle concentration. Prepared nanocomposite membranes showed better electrochemical properties compared to pristine membrane. The SEM image of prepared nanocomposite membrane modified by Fe_3O_4 NPs is given in Fig. 10.3.

10.2.9 Nanocomposite PVC-Based-co-Multiwalled Carbon Nanotube Mixed Matrix Ion-Exchange Membrane

Hosseini et al. (2013a) reported the effect of multiwalled carbon nanotubes on physicochemical characteristics of heterogeneous cation-exchange membrane for the application in water desalination. This study reported that membrane water content was increased initially by increase of MWCNT concentration up to 4 %wt in membrane matrix and then showed a decreasing trend by more MWCNT percentage. Increase of MWCNT loading ratio in the membrane matrix also caused a decrease in ion-exchange capacity. Membrane permselectivity and transport number were declined initially for monovalent ionic solution by increase of MWCNT concentration up to 4 %wt and then increased by more nanotube concentration. It was found that membrane selectivity and transport number were improved slightly in bivalent ionic solution by increase of MWCNT concentration up to 8 %wt and then decreased sharply by more concentration. The monovalent ionic flux was enhanced by increase of MWCNT content ratio. Also ionic flux was improved initially for bivalent ions by increase of additive concentration up to 2 % wt and then decreased by more MWCNT concentration from 2 to 8 %wt. The bivalent ionic flux showed an increasing trend with further increase in additive concentration. Moreover, membrane areal electrical resistance was decreased by increase of additive concentration.

Moreover, the reported results (Hosseini et al. 2010a) showed that the membrane mechanical strength and thermal stability were improved by utilizing MWCNTs due to the local heterogeneity in prepared nanocomposite membranes with well-dispersed MWCNT particles which make favorable molecular interactions between the surface of MWCNTs and polymer binder which create an interfacial zone of polymer segments.

10.2.10 Cation-Exchange Membrane Coated with Ag Nanoparticles in Water Desalination

Surface modification of electrodialysis heterogeneous ion-exchange membrane was carried out with silver nanoparticles in a vacuum reactor by argon plasma treatment for the application in water treatment. As reported by Hosseini et al. (2010b), scanning optical microscopy images showed uniform nanoparticle distribution in the prepared membranes (Fig. 10.4). Moreover, images revealed that the increment of plasma treatment time resulted in more uniform distribution of silver nanoparticles and its density on the membrane surface. The membrane potential, surface charge density, transport number, and permselectivity of membranes were enhanced initially by increase of deposited Ag nanolayer thickness up to 40 nm on the surface of membranes and then showed a decreasing trend with higher increase in deposited layer thickness from 40 to 60 nm. However, they were increased again with more increase of nanolayer thickness from 60 to 120 nm. The membrane

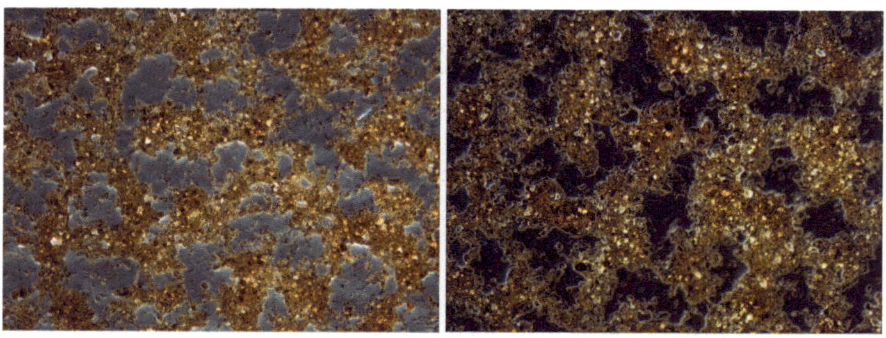

Fig. 10.4 SOM images of Ag nanolayer-coated PVC-based heterogeneous cation-exchange membrane prepared by plasma treatment (Hosseini et al. 2010b)

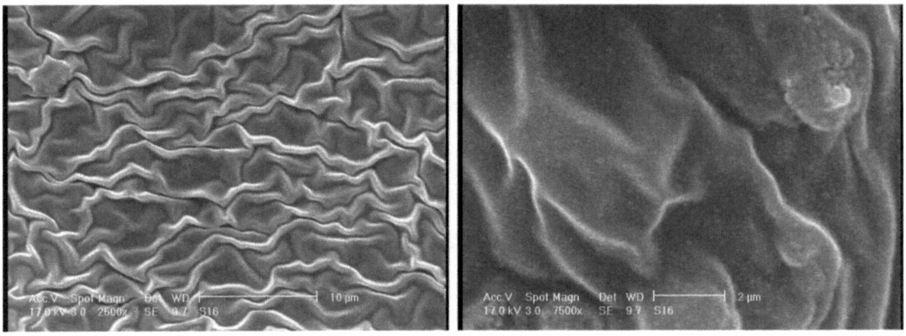

Fig. 10.5 SEM images of Ag nanolayer-coated PVC-based ion-exchange membrane; increase of membrane rugosity by increase of treating temperature (Hosseini et al. 2013b)

electrical conductivity, ionic flux, and current efficiency were improved at first by increase of deposited silver nanolayer thickness up to 40 nm and then decreased by more Ag nanolayer thickness from 40 to 120 nm. Conversely, energy consumption showed an opposite trend. Nanocomposite membrane with 40 nm thickness of Ag nanolayer exhibited more appropriate performance compared to unmodified membrane and the others.

Thermal treatment was also used to improve the Ag nanoparticle stability on membrane surface. Scanning electron microscopy images (Fig. 10.5) showed membrane rugosity increasing by increase of treating temperature. Transport number, selectivity, and flux were enhanced initially by increase of treating temperature up to 80 °C and then showed a decreasing trend. Conversely, membrane resistance showed an opposite trend. Nanocomposite membrane at 80 °C showed better performance compared to others (Hosseini et al. 2013b).

10.2.11 Ion-Selective Cation-Exchange Membrane Prepared by Magnetic Iron-Nickel Oxide Nanoparticles

Demands for the selective ion-exchange membranes are increased for the removal of valuable specific ions from waste solutions, treating effluents and also separation of ions with the same charge from mixed electrolyte solutions. Hosseini et al. (2012) prepared nanocomposite ion-selective polyvinylchloride/styrene-butadiene-rubber blend heterogeneous cation-exchange membranes by incorporating magnetic iron-nickel oxide (Fe_2NiO_4) nanoparticle in membrane matrix. Results showed that increase of Fe_2NiO_4 concentration in casting solution caused a decrease of membrane water content. Membrane ion-exchange capacity initially was increased by increase of additive percentage up to 1 %wt in membrane matrix and then showed a decreasing trend. Membrane potential, charge density, permselectivity, and transport number were enhanced initially by increase of iron-nickel oxide NP loading ratio to 2 %wt, and then they began to decrease by more concentration. Membrane sodium flux and conductivity were also enhanced initially by increase of Fe_2NiO_4 nanoparticle concentration up to 0.5 %wt and then began to decrease. Flux was improved for bivalent ions with increase in additive concentration. The prepared nanocomposite membrane showed good ability for mono- and bivalent ion separation.

10.2.12 Mixed Matrix Nanocomposite Ion-Exchange Membrane Modified by ZnO Nanoparticles

Parvizian et al. (2014) fabricated novel PVC-based nanocomposite ion-exchange membrane by utilizing ZnO nanoparticles. Results revealed that membrane water content, membrane potential, transport number, and selectivity were enhanced initially by increase of zinc oxide nanoparticle concentration up to 10 wt% in membrane matrix and then decreased by more NP concentration. The ionic permeability and flux were also decreased initially by increase of ZnO nanoparticle percentage up to 5 wt%. The permeability and flux were increased another time by more increase in additive content. An opposite trend was observed for membrane electrical resistance.

In another study, Zarrinkhameh et al. (2013) investigated the effect of zinc oxide nanoparticle on physicochemical and antibacterial characteristics of electrodialysis ion-exchange membrane. As reported, scanning electron microscopy images showed compressed structure for the nanocomposite membranes. Also 3D images showed smooth surface for nanocomposite membranes at high ZnO loading ratio content. X-ray diffraction results revealed that membrane crystallinity was improved by increase of nanoparticle concentration. Moreover, prepared nanocomposite membranes containing ZnO nanoparticles showed good antibacterial behavior.

10.2.13 Ion-Exchange Membrane Filled with Synthesized MWCNT-co-Cu Nanolayer Composite Nanoparticles and MWCNT-co-Ag Nanolayer Nanoparticles

Zendehnam et al. (2014, 2015) prepared PVC/MWCNT-co-copper nanolayer composite nanoparticles and PVC/MWCNT-co-silver nanolayer composite nanoparticle heterogeneous cation-exchange membranes for the application in water treatment and desalination process. The results showed crystalline structure for the synthesized composite nanoparticles. Also it was found that membrane transport number, selectivity, and E-conductivity were enhanced by utilizing composite nanoparticles in prepared membranes. The membrane ionic flux were also decreased initially by increase of nanoparticles content ratio up to 0.5 %wt and then increased by more concentration. In addition, mechanical stability of membranes was improved by utilizing synthesized composite nanoparticles.

10.2.14 Nanocomposite Heterogeneous Cation-Exchange Membranes Modified by Silver Nanoparticles

Zarrinkhameh et al. (2014) made nanocomposite heterogeneous cation-exchange membranes by embedding silver nanoparticles in membrane matrix. The analysis images showed compact and crystalline structure for the nanocomposite membranes. The reported results revealed that membrane selectivity and transport number were increased initially by increase of silver nanoparticle percentage up to 4 %wt in membrane matrix and then showed decrease by more NP concentration from 4 to 8 %wt. These parameters were enhanced again by further increase in nanoparticle content ratio from 8 to 16 %wt. The ionic flux was also increased obviously by using Ag nanoparticles. Furthermore, prepared nanocomposite membranes showed good antibacterial ability in *E. coli* removal from water.

10.3 Conclusion

However, preparation of ion-exchange membranes with special physicochemical properties may be a vital step in future applications. The unique characteristics of nanoparticle-incorporated ion-exchange membranes could open the door to chances to overcome the challenges in providing clean water accessible around the world.

References

Baker RW (2004) Membrane technology and applications, 2nd edn. Wiley, West Sussex

Gohil GS, Shahi VK, Rangarajan R (2004) Comparative studies on electrochemical characterization of homogeneous and heterogeneous type of ion-exchange membranes. J Membr Sci 240:211–219

Hegab HM, Zou L (2015) Graphene oxide-assisted membranes: fabrication and potential applications in desalination and water purification. J Membr Sci 484:95–106

Hosseini SM, Madaeni SS, Khodabakhshi AR (2010a) Preparation and characterization of PC/SBR heterogeneous cation exchange membranes filled with carbon nano-tube. J Membr Sci 362:550–559

Hosseini SM, Madaeni SS, Khodabakhshi AR, Zendehnam A (2010b) Preparation and surface modification of PVC/SBR heterogeneous cation exchange membrane with silver nanoparticles by plasma treatment. J Membr Sci 365:438–446

Hosseini SM, Madaeni SS, Heidari AR, Amirimehr A (2012) Preparation and characterization of ion-selective polyvinyl chloride based heterogeneous cation exchange membrane modified by magnetic Iron-nickel oxide nanoparticles. Desalination 284:191–199

Hosseini SM, Koranian P, Gholami A, Madaeni SS, Moghadassi AR, Sakinejad P, Khodabakhshi AR (2013a) Fabrication of mixed matrix heterogeneous ion exchange membrane by multi walled carbon nano tubes: electrochemical characterization and transport properties of mono and bivalent cations. Desalination 329:62–67

Hosseini SM, Madaeni SS, Zendehnam A, Moghadassi AR, Khodabakhshi AR, Sanaeepur H (2013b) Preparation and characterization of PVC based heterogeneous ion exchange membrane coated with Ag nanoparticles by (thermal-plasma) treatment assisted surface modification. Ind Eng Chem 19:854–862

Hosseini SM, Askari M, Koranian P, Madaeni SS, Moghadassi AR (2014a) Fabrication and electrochemical characterization of PVC based electrodialysis heterogeneous cation exchange membranes filled with Fe3O4 nanoparticles. Ind Eng Chem 20(4):2510–2520

Hosseini SM, Gholami A, Koranian P, Nemati M, Madaeni SS, Moghadassi AR (2014b) Electrochemical characterization of mixed matrix heterogeneous cation exchange membrane modified by aluminum oxide nanoparticles: mono/bivalent ionic transportation. J Taiwan Inst Chem Eng 45(4):1241–1248

Hosseini SM, Jeddi F, Nemati M, Madaeni SS, Moghadassi AR (2014c) Electrodialysis heterogeneous anion exchange membrane modified by PANI/MWCNTs composite nanoparticles: preparation, characterization and ionic transport property in desalination. Desalination 341:107–114

Hosseini SM, Rafiei S, Hamidi AR, Moghadassi AR, Madaeni SS (2014d) Preparation and electrochemical characterization of mixed matrix heterogeneous cation exchange membranes filled with zeolite nanoparticles: ionic transport property in desalination. Desalination 351:138–144

Hosseini SM, Hamidi AR, Madaeni SS, Moghadassi AR (2015a) Electrochemical characterization of adsorptive heterogeneous cation exchange membranes modified by simultaneous using ilmenite-co-iron oxide nanoparticles. Kor J Chem Eng 32(3):429–435

Hosseini SM, Hamidi AR, Moghadassi AR, Parvizian F, Madaeni SS (2015b) Preparation and electrochemical characterization of nanocomposite PVC based/FeTiO3-co-Fe3O4 nanoparticles ion exchange membranes: investigation of Concentration and pH effects. Kor J Chem Eng 32(9):1827–1834

Hosseini SM, Nemati M, Jeddi F, Salehi E, Khodabakhshi AR, Madaeni SS (2015c) Fabrication of mixed matrix heterogeneous cation exchange membrane modified by titanium dioxide nanoparticles: mono/bivalent ionic transport property in desalination. Desalination 359:167–175

Hosseini SM, Seidypoor A, Nemati M, Madaeni SS, Parvizian F, Salehi E (2015d) Mixed matrix heterogeneous cation exchange membrane filled with clay nanoparticles: membranes' fabrication and characterization in desalination process. J Water Reuse Desalination 6(2):290–300

Hosseini SM, Ahmadi Z, Nemati M, Parvizian F, Madaeni SS (2016) Electrodialysis heterogeneous ion exchange membranes modified by SiO2 nanoparticles: fabrication and electrochemical characterization. Water Sci Technol 73(9):2074–2084.

Hwang GJ, Ohya H, Nagai T (1999) Ion exchange membrane based on block copolymers. Part III: preparation of cation exchange membrane. J Membr Sci 156:61–65

Kariduraganavar MY, Nagarale RK, Kittur AA, Kulkarni SS (2006) Ion-exchange membranes: preparative methods for electro-dialysis and fuel cell application. Desalination 197:225–246

Kerres J, Cui W, Disson R, Neubrand W (1998) Development and characterization of crosslinked ionomer membranes based upon sulfinated and sulfonated PSU crosslinked PSU blend membranes by disproportionation of sulfinic acid groups. J Membr Sci 139:211–225

Li X, Wang Z, Lu H, Zhao C, Na H, Zhao C (2005) Electrochemical properties of sulfonated PEEK used for ion exchange membranes. J Membr Sci 254:147–155

Mollahosseini A, Rahimpour A, Jahamshahi M, Peyravi M, Khavarpour M (2012) The effect of silver nanoparticle size on performance and antibacteriality of polysulfone ultrafiltration membrane. Desalination 306:41–50

Nagarale RK, Shahi VK, Schubert R, Rangarajan R, Mehnert R (2004) Development of urethane acrylate composite ion-exchange membranes and their electrochemical characterization. J Colloid Interface Sci 270:446–454

Parvizian F, Hosseini SM, Hamidi AR, Madaeni SS, Moghadassi AR (2014) Electrochemical characterization of mixed matrix nanocomposite ion exchange membrane modified by ZnO nanoparticles at different electrolyte conditions "pH/Concentration". J Taiwan Inst Chem Eng 45:2878–2887

Shahi VK, Trivedi GS, Thampy SK, Rangarajan R (2003) Studies on the electrochemical and permeation characteristic of asymmetric charged porous membranes. J Colloid Interface Sci 262:566–573

Tanaka Y (2007) Ion exchange membranes: fundamentals and applications, Membrane science and technology series. Elsevier, Amsterdam

Thampy S, Desale GR, Shahi VK, Makwana BS, Ghosh PK (2011) Development of hybrid electrodialysis-reverse osmosis domestic desalination unit for high recovery of product water. Desalination 282:104–108

Volodina E, Pismenskaya N, Nikonenko V, Larchet C, Pourcelly G (2005) Ion transfer across ion-exchange membranes with homogeneous and heterogeneous surfaces. J Colloid Interface Sci 285:247–258

Vyas PV, Shah BG, Trivedi GS, Ray P, Adhikary SK, Rangarajan R (2001) Characterization of heterogeneous anion-exchange membrane. J Membr Sci 187:39–46

Vyas PV, Ray P, Adhikary SK, Shah BG, Rangarajan R (2003) Studies of the effect of variation of blend ratio on permselectivity and heterogeneity of ion-exchange membranes. J Colloid Interface Sci 257:127–134

Xu T (2005) Ion exchange membrane: state of their development and perspective. J Membr Sci 263:1–29

Zarrinkhameh M, Zendehnam A, Hosseini SM (2013) Electrochemical, morphological and antibacterial characterization of PVC based cation exchange membrane modified by zinc oxide nanoparticles. J Polym Res 20(11):283, 1–9

Zarrinkhameh M, Zendehnam A, Hosseini SM (2014) Preparation and characterization of nanocomposite heterogeneous cation exchange membranes modified by silver nanoparticles. Kor J Chem Eng 31(7):1187–1193

Zendehnam A, Mokhtari S, Hosseini SM, Rabieyan M (2014) Fabrication of novel heterogeneous cation exchange membrane by use of synthesized carbon nanotubes-co-copper nanolayer composite nanoparticles: characterization and performance in desalination process. Desalination 347(15):86–93

Zendehnam A, Rabieyan M, Hosseini SM, Mokhtari S (2015) Novel nanocomposite heterogeneous cation exchange membrane prepared by multi walled carbon nanotubes-co-silver nanolayer composite nanoparticles. Kor J Chem Eng 32(3):501–510

Chapter 11
Nano-fertilizers and Nutrient Transformations in Soil

Kizhaeral S. Subramanian and M. Thirunavukkarasu

11.1 Introduction

Fertilizers are indispensable input in agricultural production systems to sustain soil fertility and productivity. It played a pivotal role in promoting the productivity of a wide array of crops including food grains. The green revolution encouraged the farmers in developing countries, particularly India, to use chemical fertilizers to promote food grain production. The quantity of fertilizers consumed had phenomenally increased from just 0.5 to million tonnes in the 1960s to 25 million tonnes in 2014–2015 that closely coincided with the increase in food grain production from 55 to 256 million tonnes in the same period. The country once referred as a "begging bowl" in the early 1960s has reached the level of "self-sufficiency" in food grain production in early this century and continued to increase production regardless of vagaries of monsoon. Despite the fact that application of chemical fertilizers had consistently increased the food grain production, there were few associated impacts such as less fertilizer response ratio, imbalanced fertilization and multi-micronutrient deficiencies observed that have led to the introduction of nano-fertilizer technology (Subramanian and Tarafdar 2011). Nano-fertilizer technology is very innovative, and scanty reported literature is available in the scientific journals. However, some of the reports and patented products strongly suggest that there is a vast scope for the formulation of nano-fertilizers. In this book chapter, the nutrient dynamics in soil system fertilized with nano-fertilizer formulations are reviewed.

K.S. Subramanian (✉) • M. Thirunavukkarasu
Department of Nano Science & Technology, Tamil Nadu Agricultural University, Coimbatore 641003, Tamilnadu, India
e-mail: kssubra2001@rediffmail.com

M. Ghorbanpour et al. (eds.), *Nanoscience and Plant–Soil Systems*, Soil Biology 48,
DOI 10.1007/978-3-319-46835-8_11

11.2 Nano-fertilizers

Nano-fertilizers are nutrient carriers in the dimension of 1–100 nm. "Nano" refers to one-billionth of a metre or one-millionth of a millimetre. When the size gets reduced, the surface area has tremendously increased (Prasad et al. 2014). For instance, montmorillonite clays have a specific surface area of 15 ± 2 m^2 g^{-1} (Macht et al. 2011), while the surface increased to 1000–1200 m^2 g^{-1} when the clays reduced to nano-dimension (Subramanian and Sharmila Rahale 2009). The high-surface mass ratio is the unique feature of nanotechnology that facilitates atomic scale manipulation to retain and regulate release of nutrients from nano-fertilizers.

A patented nano-composite consists of N, P, K, micronutrients, mannose and amino acids that increase the uptake and utilization of nutrients by grain crops has been reported (Jinghua 2004). Bansiwal et al. (2006) revealed that surface-modified zeolite is used as a carrier of slow release of phosphatic fertilizers. Coating and cementing of nano- and subnano-composites are capable of regulating the release of nutrients from the fertilizer capsule (Liu et al. 2006). Subramanian and Sharmila Rahale (2009) stated that nano-fertilizers are capable of releasing nutrients, especially nitrate nitrogen, for more than 50 days, while nutrient released from conventional fertilizer (urea) ceased to exist beyond 10–12 days and also suggested that nano-fertilizers may be used as a strategy to regulate the smart release of nutrients that commensurate with crop requirement. Ramesh et al. (2011) reported that slow-release fertilizers are excellent alternatives to soluble fertilizers. Nutrients are released at a slower rate throughout the crop growth; plants are able to take up most of the nutrients without waste by leaching. Slow release of nutrients in the environments could be achieved by using zeolites that are a group of naturally occurring minerals having a honeycomb-like layered crystal structure. Its network of interconnected tunnels and cages can be loaded with nitrogen and potassium, combined with other slowly dissolving ingredients containing phosphorous, calcium and a complete suite of minor and trace nutrients. Zeolite acts as a reservoir for nutrients that is slowly released on demand.

De Rosa et al. (2010) defined nano-fertilizer as a product that delivers nutrients to crops encapsulated within a nanoparticle. There are three ways of encapsulation: (a) the nutrient can be encapsulated inside nanomaterials such as nanotubes or nanoporous materials, (b) coated with a thin protective polymer film and (c) delivered as particles or emulsions of nanoscale dimensions. Nano-fertilizers could be used to reduce nitrogen loss due to leaching, emissions and long-term assimilation by soil microorganisms. This suggests that new nutrient delivery systems that exploit the nanoscale porous domains on plant surfaces can be developed. But, the nano-fertilizers should show sustained release of nutrients on demand while preventing them from prematurely converting into chemical/gaseous forms that cannot be absorbed by plants. To achieve this, biosensor could be attached to this nano-fertilizer that allows selective nitrogen release linked to time and environmental and soil nutrient conditions. Slow/controlled release of fertilizers may also improve soil by decreasing toxic effects associated with fertilizer over application (De Rosa 2009). Recently, Subramanian et al. (2015) have reviewed the work on nano-fertilizer research across the globe and narrated how the release of nutrients from various forms of nano-fertilizers carrying macro- (N, P, K, S) and microelements (Fe and Zn) are being regulated. In this book chapter, nutrient dynamics and fate of nutrients in soil system fertilized with nano-forms of nutrients are highlighted.

11.3 Nitrogen Dynamics

Nitrogen is considered as a "kingpin" within the set of essential nutrients required by crops. Plants utilize two distinct forms of nitrogen, namely, NO_3-N (aerobic systems, e.g. maize, wheat) and NH_4-N (rice and aquatic plants). Nitrate-N is a negatively charged ion and is not attracted to soil particles or soil organic matter like NH_4^+-N. Nitrate-N is water-soluble and can move below the crop rooting which may contribute for the groundwater pollution. The losses of N from the soil include leaching, denitrification and volatilization that constitute 50–70 %, and the N use efficiency (NUE) by crops hardly exceeds 30–35 %. In order to improve the N use efficiency of one of the commonly used N fertilizers, urea was coated with neem cake, tar, gypsum and polymers. These coated fertilizers did not appreciably improve the NUE, and thus they were discontinued.

Now, we attempt to use nanotechnology to improve NUE by entrapping N in a clay matrix or intercalation process. One of the natural minerals zeolites possesses extensive surface area and sieving property which make this substrate unique for the development of nano-fertilizers. Ahmed et al. (2009) reported that zeolite amended with urea improved the NUE in maize by 15 % in comparison to the urea without amendment. Komarneni (2009) demonstrated that slow release of NH_4^+ from various zeolites treated with molten NH_4NO_3 and KNO_3 over time. This modified zeolites with occluded ammonium and nitrate showed good promise to be a slow-release N fertilizer.

Fujinuma and Balster (2010) revealed the nitrogen release of the nano-fertilizer from three elevations in Sri Lanka (pH 4.2, 5.2 and 7), and these studies were compared with that of a commercial fertilizer. The nano-fertilizer showed an initial burst and a subsequent slow release even up to 60 days compared to the commercial fertilizer, which released heavily early followed by the release of low and non-uniform quantities until around day 30. Nano-zeolite impregnated with urea causes slow and steady release of nitrogen. NH_4^+ ions occupying the internal channels of zeolite slowly set free N allowing progressive absorption by the crop. Amending sandy soils with NH_4^+-loaded zeolite reduce N leaching while sustaining crop growth. Naturally occurring zeolites are loaded with NH_4NO_3 or KNO_3 by an occlusion process. This was found to more or less double the nitrogen content of the slow-release fertilizer. Nanoporous synthetic layered double hydroxides or anionic clays with nitrate ions in their interlayers also act as slow-release nitrate fertilizers

(Komarneni 2009). Saratin (2010) reported that the rate of nitrogen release from the zeolitic rock is slowed in three ways: (1) by preventing the leaching of urea from the root zone; (2) by slowing the conversion of urea by soil enzymes, thus delaying the formation of NH_4^+ ions; and (3) by taking up ammonium ions onto exchange sites in the zeolite, thus protecting them from nitrifying bacteria. Subramanian et al. (2012) detected that N is released till 1176 h from nano-zeolite-based fertilizer, while conventional fertilizer had detectable amount of N up to 200 h. Kottegoda et al. (2011) reported that the urea-modified hydroxyapatite nanoparticles encapsulated wood-based nano-fertilizer have an initial quick release and subsequently slow and sustained release of nitrogen for more than 60 days in two acidic soils (pH 4.2 and 5.2) and sandy soils (pH 7). Ni et al. (2011) developed that environmentally friendly slow-release formulations with three nitrogen fertilizers (urea, ammonium sulphate and ammonium chloride) on the basis of natural attapulgite clay (APT), ethylcellulose film (EC) and sodium carboxymethyl cellulose (CMC)/ hydroxyl ethylcellulose (HMC) hydrogel as shown in Fig. 11.2. Developed fertilizer release profile contains three stages, namely, slow-release stage with soaking and penetration of water vapour within 24 h, steady-release stage of 5 days and finally concomitant stage luxes of nutrients released for 10 days.

Datta (2011) observed that the use of slow-release fertilizer made out of nano-clay (smectite) polymer composites resulted in 38 % more N recovery than conventional urea fertilizer and 62 % more P recovery than conventional DAP fertilizer, respectively. In a greenhouse experiment, the use of nano-clay polymer composite (NCPC) enhanced N uptake by 32 % and P uptake by 30 % more than conventional urea and DAP fertilizer, respectively. Addition of N and P as NCPC at high dose of 60 and 40 mg kg^{-1} soil of N and P_2O_5 increased biomass yield of pearl millet by 45 % over conventional fertilizer (urea+DAP) at the same dose (Sarkar 2011). Subramanian and Sharmila Rahale (2012) revealed that ^{15}N studies which were taken using maize as a model system have revealed that the N use efficiency from nano-fertilizer was 82 % and the conventional fertilizer (urea) registered 42 % with a net higher nitrogen use efficiency of 40 % which is hardly achievable in the conventional systems.

Kardaya et al. (2012) reported that zeolite or urea-impregnated zeolite as slow-release ammonia was potential in decreasing ruminal ammonia, pH, acetate to propionate ratio, methane and maintaining low-plasma urea within its physiological range. Wanyika et al. (2012) entrapped urea in the mesopores of the siliceous nanomaterial synthesized by liquid crystallization technique and simple immersion for loading of N. About 15.5 % (w/w) of urea was loaded inside the pores mainly by physic-sorption, while the total adsorption capacity of mesoporous silica nanoparticles could reach up to 80 % (w/w). Release process of the urea-loaded mesoporous silica nanoparticles in water and soil indicated a two-stage-sustained slow-release profile. The studies revealed at least fivefold improvements in the release period. By the ability to entrap urea guest molecules into its mesopores and release them in a controlled manner, mesoporous silica nanoparticles demonstrated its great potential as a nano-carrier for agrochemicals.

Subbaiya et al. (2012) reported that the urea-modified hydroxyapatite (HA) particles had higher NUE and slow release of the nitrogen to the soil besides minimizing the adverse effects to the environment. Mohanraj (2013) stated that the slow release of nitrogen from ammonium sulphate-loaded nano-zeolite occurred up to 480 h, but the release of nitrogen from ammonium sulphate was 312 h under percolation reactor study. Further, the study suggests that nano-fertilizer being a slow-release fertilizer circumvents the release of methane in a submerged soil system which may be inferred as a protective factor against global warming.

Manikandan and Subramanian (2014) reported the improvement of NUE using nano-zeolites (80–92 nm) as an adsorbent. Nano-zeolite-fortified urea facilitates adsorption of N in channels and pore spaces. The zeourea (natural zeolite blended) and nano-zeourea contained 18.5 and 28 % of N and capable of releasing N up to 34 and 48 days, respectively, while the N release from conventional urea is just 4 days. The data strongly suggested that zeolite with nano-dimension can help to improve N use efficiency besides sustaining the release of N that may considerably economize the N use in crops with an added advantage of prevention of groundwater contamination.

In all these literatures, it is quite evident that nano-fertilizers are capable of regulating the release of nutrients progressively with the advancement of time. In many cases, there was a burst initially and slow and steady release for an extended period of time which may not commensurate with crop growth. De Rosa et al. (2010) have shown that nanotechnology may assist in developing a customized nano-fertilizer that can release nutrients in accordance with the nutrient requirement of crops which can be achieved through the detection of biochemical signals from the root exudates. This suggests that nano-fertilizers may be served as a strategy to smart release of nutrients in agriculture.

11.4 Phosphorous Dynamics

Phosphorous availability in soil is very enigmatic as it gets fixed in a wide range of pH. Further, a small portion of less than 1 % of P is in an available form which is readily utilized by plants and the rest is in an unavailable form. The mobility of P in soil solution is extremely slow, and the phosphates often get precipitated by the counteracting cations such as Ca, Mg, Fe and Zn. Such precipitation reactions can be overcome by encapsulating the phosphates in a nano-matrix.

Eberl (2008) reported that phosphate (H_2PO_4) can be released to plants from phosphate rock (P-rock) composed largely of calcium phosphate mineral apatite by mixing the rock with zeolite having an exchange ion such as ammonium. Kundu et al. (2010) stated that surface-modified zeolite (SMZ) has been found to be a good sorbent for phosphate, and slow release of P is achievable. In solution culture experiment with maize grown for 30 days and treated with different nanoparticles of phosphates, excellent growth in terms of root length and volume, shoot length, dry matter yield, P content and uptake by shoot and root were observed with hydroxyapatite (200 nm) followed by calcium phosphate (100 nm) and stone rock phosphate (42 nm), respectively.

Zeoponic is a plant demand-driven nutrient delivery system and from this phosphate (PO_4^{3-}) and other nutrients released by controlled dissolution of synthetic apatite. N, P and K were delivered when plant needs them from zeoponics. Basically, the process is a combination of dissolution and ion exchange reactions. The absorption of nutrients from the soil solution by plant roots drives the dissolution and ion exchange reactions, pulling away nutrients as needed. The zeolite is then "recharged" by the addition of more dissolved nutrients.

Recovery of P was higher as compared to that from KH_2PO_4 at lower concentration of P in the form of nano-rock phosphate at sandy soils of Jodhpur, whereas the reverse occurred with higher concentration of P of nanoparticle rock phosphate. Enhancement of soybean crop growth resulted when P was applied to the nutrient solution in the form of nano-rock phosphate. Considerable amount of SiO_2 nanoparticles other than apatite present in nano-rock phosphate could have significant influence on biomass yield of soybean. In maize, relatively higher-yield response to nano-rock phosphate was obtained as compared to micro-sized rock phosphate (Das 2011).

These literatures tend to indicate that phosphate availability can be improved by the positively charged nanoparticles or surface-modified zeolites. The sorption and desorption of P can be modified and regulated to stimulate the availability for a longer period of time. This process may help the plants to be nourished with sufficient quantities of P and sustain crop productivity.

11.5 Potassium Dynamics

Potassium is highly labile in soil, and the availability is primarily controlled by the cation exchange capacity (CEC) of soils. This reflects the soil's ability to hold K and other cations and store them in the soil for crop uptake. Clay minerals and soil organic matter are the two parts of soil that contribute to CEC. In general, the higher the CEC of the soil, the greater the storage capacity and supplying power for K. Recently, it is widely observed that soil is deficient in organic matter content and clays. The lack of adsorptive sites drastically reduced the availability of K.

Hershey et al. (1980) provided the data on the slow release of effect of potassium from K-zeolite. Pino et al. (1995) stated the slow release of potassium from potassium-zeolite. Improvement of nutrient efficiency in crops is an important issue in agriculture for reducing cost in agriculture production and for protecting the environment (Oosterhuis and Howard 2008). Rezaei and Movahedi Naeini (2009) reported that potassium sorbed on zeolite increases with the increase in equilibrium of K concentration. Zhou and Huang (2007) reported the slow and steady release of potassium from nano-zeolite. This may be due to the ion exchangeability of the zeolites with selected nutrient cations; zeolite can become an excellent plant growth medium for supplying plant roots with additional vital nutrient cations and anions. Liu et al. (2005) reported the potassium (K^+)-loaded zeolite (K-Z) as a slow-release fertilizer and investigated the growth characteristics of hot pepper as well as the changes in the nitrogen and potassium contents of soil. Adsorption and desorption of K in nano-fertilizer can be modified to regulate the release of K which in turn facilitates improved use efficiency.

11.6 Dynamics of Sulphur

Sulphur is the fourth major element required for crops, and there is a versatile response to added S due to the increase in deficiencies across India. Despite the importance of S, it is well established particularly for oilseed crops, but the use efficiency remains unchanged (18–20 %) in the past several decades. In order to improve the sulphur use efficiency, nanotechnological approaches may be appropriate.

Sodic zeolite (X type) a value of about 14 sulphur atoms adsorbed per super cage, whereas the sodic zeolite (Y type) adsorb up to 21 sulphur atoms per cavity (Verver and Van Swaaij 1985). Li et al. (1998) indicated that the sorption and desorption of anionic species on surface-modified zeolite (SMZ) surfaces were attributed to surface anion exchange; an increase in ionic strength of the solution will result in competition between background electrolyte chloride against sulphate for sorption sites, thus resulting in more sulphate release. Li (1998) recorded that more sulphate was released when surfactant loading was at 150 % effective cation exchange capacity (ECEC), at which the surfactant bilayer formation was incomplete. A surface adjustment of the sorbed surfactant molecules from patchy bilayer to a monolayer could release the counter ion (sulphate in this case) associated with the patchy bilayer.

The sulphate-loading capacity was lower than that of nitrate on SMZ (Li 2003), comparable to that of phosphate on SMZ (Bansiwal et al. 2006) and higher than chromate sorption (Li and Bowman 1997). Column leaching tests showed that at the same initial sulphate loading of 18 mmol kg^{-1}, the initial sulphate concentration in leachate was about 40 mmol L^{-1} when mixtures of zeolite and soluble sulphate were used. In contrast, the initial sulphate concentration in leachate was only 18 mmol L^{-1} when sulphate-loaded SMZ was used. The column leaching test indicated that sulphate release could be significantly reduced when SMZ was used. Slow release of sulphate could also be achieved as sulphate is the sulphur form ready for plant absorption (Hawkesford and De Kok 2006).

Li and Zhang (2010) stated that at 200 % ECEC, the surfactant bilayer formation is complete and the surface becomes more positively charged, resulting in more sulphate retention, and less release from sulphate-loaded SMZ, noted that the mean percentages of sulphate release were 31 ± 5 % and 15 ± 3 % from SMZ modified to 150 % and 200 % ECEC, respectively, and also reported that surfactant-modified zeolite (SMZ) as fertilizer additives to control sulphate release were tested in batch

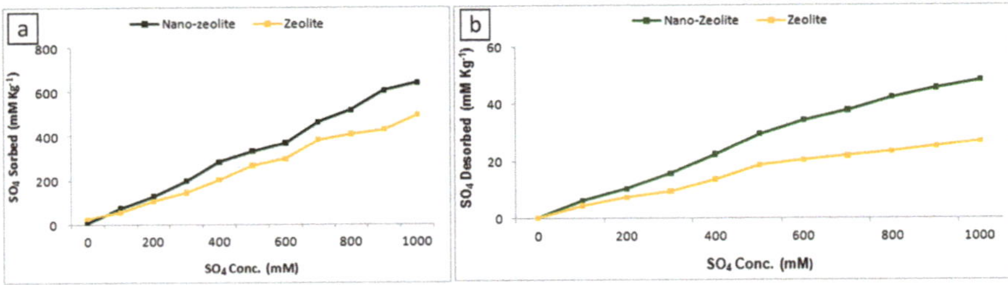

Fig. 11.1 Sulphate sorption (**a**) and desorption (**b**) curves of zeolite and nano-zeolite

and column leaching experiments which indicated that SMZ could be a good carrier for sulphate. Thus, leaching of sulphate can be greatly reduced, and slow release of sulphate can be achieved if SMZ is used as fertilizer additives. Selva Preetha et al. (2014) reported that among the nanoparticles, nano-zeolite response to slow release of fertilizer because of its higher surface area retains and releases anionic (SO_4^{3-}) in a slow and steady state.

The amount of sulphate adsorbed by zeolite and surface-modified nano-zeolite ranged from 24.45 to 495.66 mM and 7.5 to 639.37 mM sulphate at 100–1000 mM concentration, respectively (Fig. 11.1a, b). The nano-zeolite adsorbed 23 % more sulphate ions which may be due to the preferential adsorption at the positive sites of the clays and on surface modification of clay provided positive charge for anionic nutrient adsorption. The quantity of desorbed sulphate ranged from 0.38 to 27.01 mM and 0.39 to 48.17 mM sulphate for zeolite and nano-zeolite at 100–1000 mM concentration, respectively. Desorption of sulphate from zeolite and nano-zeolite has increased linearly to the tune of 77 % when the equilibrium sulphate concentration was maintained at 1000 mM. This may be due to the extensive surface area of nano-zeolite to retain nutrients (Thirunavukkarasu and Subramanian 2015a, b).

A comparative study of the release of sulphate (SO_4^{2-}) from conventional (gypsum) and nano-S was performed using the percolation reactor. The results showed that the SO_4^{2-} supply from nano-S (NS) was available even after 35 days of continuous percolation, whereas SO_4^{2-} from conventional fertilizer (CS) was exhausted within 10 days (Fig. 11.2). These properties suggest that nano-S has a great potential for slow release of SO_4^{2-}, and thereby S use efficiency can be improved (Thirunavukkarasu and Subramanian 2014a, b).

Sulphur fractionation studies was done in soil fertilized with experiment with nano-S and conventional-S and the S fractions include, viz. water-soluble sulphur (WS), exchangeable sulphur (ES), occluded sulphur (Occ. S), organic sulphur (Org. S) and total sulphur (TS) were determined (Fig. 11.3; unpublished data of Thirunavukkarasu and Subramanian). The nano-S fertilized soil (sulphate-loaded surface-modified nano-zeolite) at 40 kg S ha^{-1} registered the highest WS (8.3 ppm), ES (4.4 ppm) and TS (249.4 ppm), whereas conventional-S fertilized soil at same set of treatment recorded lower WS (7.4 ppm), ES (3.9 ppm) and TS (248.6), in comparison to nano-S fertilized soil. The control recorded the lowest water-soluble

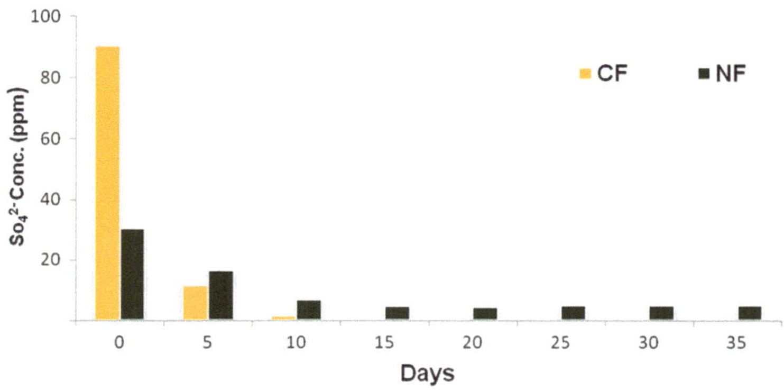

Fig. 11.2 Sulphate release from conventional (CF) and nano-S (NS) fertilizers

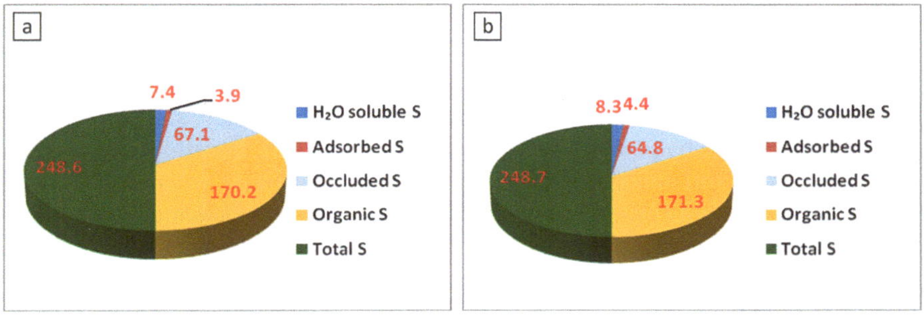

Fig. 11.3 Fractions of S in the soil fertilized with conventional-S (**a**) and nano-S (**b**)

S fraction of 3.5 ppm. The data on Occ. S fraction of soil recorded the highest in conventional-S fertilized at 40 kg S ha^{-1} of 67.1 ppm. The nano-S (sulphate-loaded surface-modified nano-zeolite)-applied soil at similar S dose registered lowest occluded S fraction of 64.2 ppm, in comparison to conventional-S fertilized treatments. The data clearly demonstrated that nano-S fertilization facilitates enhanced availability of sulphur forms (WS and ES), while conventional-S contributes for the increased occluded S. This suggests nano-fertilization assists in enhanced S availability.

The application of sulphur as nano-sulphur at 30 kg S ha^{-1} was recorded highest plant height, number of branches, total chlorophyll content and number of root nodule at various growth stages of groundnut. Nano-sulphur releases nutrients slowly and steadily during critical growth period and thereby improved growth and biochemical parameters since better sulphur uptake of groundnut (Thirunavukkarasu and Subramanian 2014a, b). The soil availability of sulphur was increased due to sulphur application through nano-sulphur at 30 kg ha^{-1} which reduced the leaching losses and nutrient fixation, microbial conversion, oxidation of sulphur and other atmospheric losses that resulted from slow-release mechanism of such fertilizer finally improved the available sulphur status of soil (Thirunavukkarasu and Subramanian 2015a, b).

11.7 Nano-micronutrients

Multi-micronutrient deficiencies are a cause for concern in developing countries particularly Fe and Zn. It has been estimated that at least 3 billion across the globe is affected by Fe and Zn deficiencies, and their impact is alarming in Asian countries where more than 50 % people eat rice as a staple food. There are several strategies have been adopted to improve the Zn use efficiency by crops with relatively little success due to the fact that more 95 % of the added Zn gets fixed. Recently, nanotechnology approaches like "core shell"-fortified Zn appear to improve the Zn nutrition of rice regardless of aerobic or submerged systems of rice cultivation (Yuvaraj and Subramanian 2014).

Broos et al. (2007) reported that the slow release of Zn is attributed to the sparingly solubility of minerals and sequestration effect of exchanger, thereby releasing trace nutrients to zeolite exchange sites where they are more readily available for uptake by plants. The sorption of heavy metals, metalloids and organic contaminants by soil nanoparticles shows a rapid initial phase followed by a slow phase (days) due to diffusion into micropores, structure rearrangement and precipitation. Likewise the rate of desorption is biphasic. As a result, the bioavailability of contaminants decreases with contact time (ageing). However, trace metals, metalloids and radionuclides can move within the soil column and across to adjacent water bodies, hitching onto mobile colloidal particles (Hochella et al. 2008).

Lee et al. (2008) indicated that significant uptake of nano-sized copper by mung bean and wheat was observed. Significant uptake, translocation and accumulation of Fe_2O_3 nanoparticle in the roots and leaves of pumpkin has been reported without any effect on the growth and development of the test species (Zhu et al. 2008). Milani et al. (2010) revealed that nanomaterials could be applied in designing more soluble and diffusible sources of Zn fertiliser for increased plant productivity. The smaller size, higher specific surface area and reactivity of nanoparticulate ZnO compared to bulk ZnO may affect Zn solubility, diffusion and hence availability to plants. However, the unique properties of nanoparticles that make them useful as sources of nutrients could also pose environmental risks. Adhikari (2011) revealed that application of Zn nanoparticles (100 nm) at relatively lower level (0.28 ppm) enhanced the growth of maize plants as compared to normal $ZnSO_4$ (0.5 ppm) and also reported that plant parameters such as plant height, root length and volume and dry matter weight were all improved due to application of zinc oxide nanoparticles.

Das (2011) stated that aluminium and iron hydroxides have a point of zero charge (PZC) at pH 8. They are positively charged in the pH range (4.5–7.5) of most soils. Accordingly, these nanoparticles play an important role in the adsorption and retention of nutrient anions by electrostatic interactions and ligand exchange, and they actively promote clay flocculation and soil aggregate stabilization. Manganese hydroxides have a low PZC (pH 4). Since their surface charge is negative in the pH range of soils, these hydroxides are efficient sorbents and scavengers of heavy metal cations. The adsorption rate obtained with the zeolite seemed to be very satisfactory, and clinoptilolite can be accepted as an efficient adsorbent for zinc removal (Subramanian and Tarafdar 2011).

Subramanian and Sharmila Rahale (2012) recorded that zeolite adsorbed more zinc (220 ppm) than other clays, which was followed by nano-montmorillonite (175 ppm). Shah and Belozerova (2009) have reported improved germination of *Lactuca* seeds treated with metal nanoparticles. The promotional effects of NPs on plant growth, pod yields were significantly increased over control and $ZnSO_4$. Prasad et al. (2012) have shown that the nanoscale ZnO at 1000 ppm proved effective in enhancing both root volume and root dry weight of groundnut seedlings.

Pandey et al. (2010) reported that ZnO NPs increased the level of indole-3-acetic acid (IAA) in roots (sprouts), which in turn indicate the increase in the growth rate of plants as zinc is an essential nutrient for plants. Zhao et al. (2013) observed that ZnO NPs coexisting with Zn dissolved species were continuously released to the soil solution to replenish the Zn ions or ZnO NPs scavenged by roots as compared to soil treated with alginate which promotes the bioaccumulation of Zn in corn plant tissues. Priester (2012) revealed that the effect of ZnO NPs on plants varies with the species, age and characteristics of nanoparticles; the physiological responses of plants need to be studied. It is widely postulated that adverse or beneficial effect of ZnO NPs is closely associated with antioxidant system of plants.

The transformation of applied Zn in submergence and aerobic moisture regimes in soil were studied. The soils fertilized with various nano-forms of Zn along with conventional Zn fertilizer were tested for their Zn fractionation pattern as water-soluble plus exchangeable (WSEX-Zn), organically complexed (OC-Zn), manganese oxide bound (MnOX-Zn), amorphous sesquioxide bound (AMOX-Zn), crystalline sesquioxide bound (CRYOX-Zn) and residual zinc (Res-Zn) in order to determine the mechanism that favours solubilization of Zn towards availability. The data have clearly shown that there is a distinct fraction of Zn among nano-forms of Zn fertilization in soil (Fig. 11.4). Nano-Zn fertilizer prepared by fortifying Zn in nano-sized zeolite appears to be the best which facilitated to accumulate Zn in water-soluble exchangeable and organically bound Zn to the tune of 21–28 % under both submerged and aerobic conditions. Such values were significantly lower for the conventional $ZnSO_4$ fertilization at 16 % and 22 % under submerged and aerobic conditions, respectively. There was a close association between the available pool of Zn and residual Zn. Among the two systems of rice cultivation, submergence of soils retained higher amounts of residual Zn and lower amounts of WS-EX Zn, and the reverse was true with aerobic. A strong negative correlation was established between water-soluble exchangeable Zn and residual forms of Zn (unpublished data of Yuvaraj and Subramanian).

The diethylenetriaminepentaacetic acid (DTPA) extractable (available) Zn can estimate the probability of whether soil can provide sufficient Zn to the roots to meet plant demand (Robson 1993). The nano-fertilizer significantly increased the

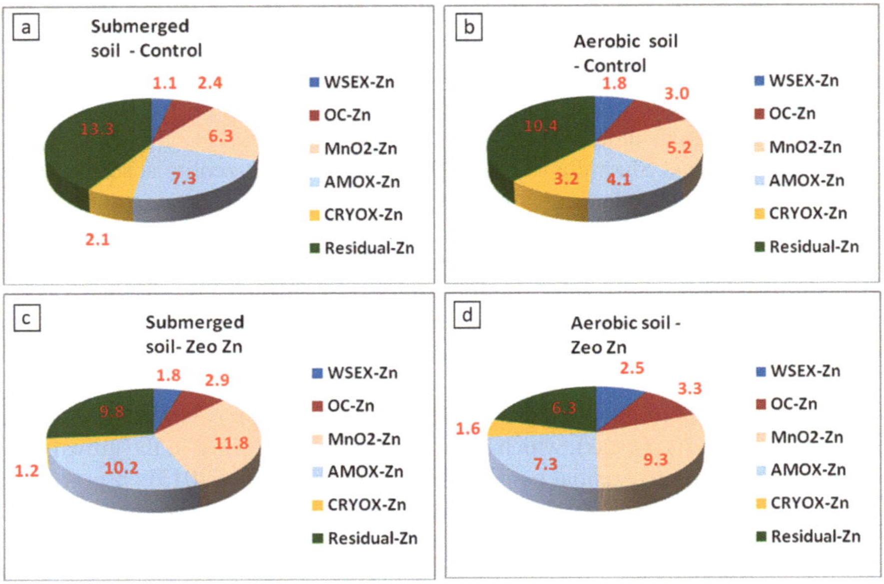

Fig. 11.4 Zn fractions in $ZnSO_4$ fertilized (**a**, **b**) and nano-Zn fertilized soils under submerged (**c**) and aerobic (**d**) conditions

soil available micronutrient status. Zeolites added to fertilizers retained nutrients and long-term soil quality by enhancing its absorption ability. Markus and Othmar (2003) reported on the influence of fertilization with slow-release zeolite-bound zinc uptake of wheat and spinach. Broos et al. (2007) reported that the slow release of Zn is attributed to the sparingly solubility of minerals and sequestration effect of exchange, thereby releasing trace nutrients to zeolite exchange sites where they are more readily available for uptake by plants.

Overall, the data have unequivocally demonstrated that nano-fertilizers are capable of releasing nutrients several days longer than conventional ones. This quite important in tropical agriculture where the availability of moisture is uncertain and controlled release fertilizers are appropriate. Though the release of nutrients is progressive and long lasting, many studies have shown a burst immediately after application, and it turns slow and steady thereafter. More research is needed to evolve customized fertilizers that are capable of releasing nutrients which commensurate with crop demand. Demand-driven supply of nutrients to crops will be a tough task for the scientists involved in nano-fertilizer research. Further, biosafety of nano-fertilizers has to be done as per the stipulated Organization for Economic Co-operation and Development (OECD) protocols to ensure the environmental safety prior the technology hits the farm gate.

References

Adhikari T (2011) Nano-particle research in soil science: micronutrients. In: Proceedings of the national symposium on 'applications of clay science: agriculture, environment and industry', 18–19 February 2011. NBSS & LUP, Nagpur, pp 74–75

Ahmed OH, Hussin A, Mohd Hanif Ahmad H, Boyie Jalloh M, Abd Rahim A, Muhamad Majid N (2009) Enhancing the urea-N use efficiency in maize (*Zea mays*) cultivation on acid soils using urea amended with zeolite and TSP. Am J Appl Sci 6:829–833

Bansiwal AK, Rayalu SS, Labhasetwar NK, Juwarkar AA, Devotta S (2006) Surfactant modified zeolite (SMZ) as a slow release fertilizer for phosphorus. J Agr Food Chem 54:4773–4779

Broos K, Warne J, Heemsbergen DA, Stevens D, Barnes MB, Correll RL, Mclaughlin MJ (2007) Soil factors controlling the toxicity of copper and zinc to microbial processes in Australian soils. Environ Toxicol Chem 26:583–590

Das DK (2011) Application of nanotechnology in soil science. J Indian Soc Soil Sci 59:114–121

Datta SC (2011) Nanoclay research in agriculture environment and industry. In: Proceedings of the national symposium on 'applications of clay science: agriculture, environment and industry', 18–19 February 2011. NBSS&LUP, Nagpur, pp 71–73

De Rosa MC (2009) New opportunities in nanotechnologies. In: Canadian fertilizer products forum, Ottawa, October 14, pp 1–23

De Rosa MC, Monreal C, Schnitzer M, Walsh R, Sultan Y (2010) Nanotechnology in fertilizers. Nat Nanotechnol 5:91

Eberl DD (2008) Controlled release fertilizers using zeolites. USGS science for changing world tech transfer, pp 1–3

Fujinuma R, Balster NJ (2010) Controlled-release nitrogen in tree nurseries. Res Commun 2:123–126

Hawkesford MJ, De Kok LJ (2006) Managing sulfur metabolism in plants. Plant Cell Environ 29:382–395

Hershey DR, Paul JL, Carlson RM (1980) Evaluation of potassium-enriched clinoptilolite as a potassium source for plotting media. Hortscience 15:87–89

Hochella MF, Lower J, Maurice SK, Penn PA, Sahai RL, Sparks DL, Twining BS (2008) Nanominerals, nanoparticles and earth chemistry. Science 21:1631–1635

Jinghua G (2004) Synchrotron radiation, soft X-ray spectroscopy and nano-materials. J Nanotechnol 1:1–21

Kardaya D, Sudrajat D, Dihansih E (2012) Efficacy of dietary urea-impregnated zeolite in improving rumen fermentation characteristics of local lamb. Media Peternakan 35:207–213

Komarneni S (2009) Potential of nanotechnology in environmental soil science. In: Proceedings of 9th international conference of the East and Southeast Asia Federation of Soil Science Societies, 27–30 Oct 2009. Korean Society of Soil Science and Fertilizers, Seoul, pp 16–20

Kottegoda N, Munaweera I, Madusanka N, Karunaratne V (2011) A green slow-release fertilizer composition based on urea-modified hydroxyapatite nanoparticles encapsulated wood. Curr Sci 101:73–78

Kundu S, Adhikari Tapan, Biswas AK, Tarafdar JC, Goswami A, Subba Rao A (2010) Nano-science and nanotechnology in soil fertility and plant nutrition research. Technical bulletin, Indian Institute of Soil Science (ICAR), Bhopal, pp 1–46

Lee WM, An XJ, Yoon H, Kweon HS (2008) Toxicity and bioavailability of copper nano-particles to the terrestrial plants mung beans (*Phaseolus radiata*) and wheat (*Triticum aestivum*): plant auger test for water-insoluble nano-particles. Environ Toxicol Chem 27:1915–1921

Li Z (1998) Chromate extraction from surfactant-modified zeolite surfaces. J Environ Qual 27:240–242

Li Z (2003) Use of surfactant-modified zeolite as fertilizer carriers to control nitrate release. Micropor Mesopor Mater 61:181–188

Li Z, Bowman RS (1997) Effects of the sorption of cationic surfactant and chromate on natural clinoptilolite. Environ Sci Technol 3:2407–2412

Li Z, Zhang Y (2010) Use of surfactant-modified zeolite to carry and slowly release sulfate. Desalination Water Treat 21:73–78

Li Z, Anghel I, Bowman RS (1998) Sorption of oxyanions by surfactant-modified zeolite. J Dispers Sci Technol 19:843–857

Liu X, Zhao-bin F, Fu-dao Z, Shu-qing Z, Xu-sheng HE (2005) Study on adsorption and desorption properties of nano-kaoline to nitrogen phosphorus, potash and organic carbon. Sci Agr 38:102–109

Liu X, Feng Z, Zhang S, Zhang J, Xiao Q, Wang Y (2006) Preparation and testing of cementing nano-subnano composites of slow- or controlled release of fertilizers. Scientia Agricultura Sinica 39:1598–1604

Macht F, Eusterhues K, Pronk GJ, Totsche KU (2011) Specific surface area of clay minerals: comparison between atomic force microscopy measurements and bulk-gas (N_2) and -liquid (EGME) adsorption methods. Appl Clay Sci 53:20–26

Manikandan A, Subramanian KS (2014) Fabrication and characterisation of nanoporous zeolite based N fertilizer. Afr J Agr Res 9:276–284

Markus P, Othmar S (2003) Chemical changes in the rhizosphere of metal hyperaccumulator and excluder Thlaspi species. J Plant Nutr Soil Sci 166:579–584

Milani N, McLaughlin MJ, Hettiaratchchi GM, Beak DG, Kirby JK, Stacey S (2010) Fate of nanoparticulate zinc oxide fertilisers in soil: solubility, diffusion and solid phase speciation. In: 19th world congress of soil science, soil solutions for a changing world, 1–6 August 2010, Brisbane, Australia. Published on DVD: 172–174

Mohanraj J (2013) Effect of nano-zeolite on nitrogen dynamics and green house gas emission in rice soil eco system. M.Tech. (Agricultural Nanotechnology) Thesis, TNAU, Coimbatore, India

Ni B, Liu M, Lu S, Xie L, Wang Y (2011) Environmentally friendly slow release nitrogen fertilizer. J Agr Food Chem 59:10169–10175

Oosterhuis DM, Howard D (2008) Evaluation of slow-release nitrogen and potassium fertilizers for cotton production. Afr J Agr Res 3:068–073

Pandey AC, Sanjay SS, Yadav RS (2010) Ciprofloxacin conjugated zinc oxide nanoparticle: a camouflage towards multidrug resistant bacteria. J Exp Nanosci 5:488–493

Pino N, Arteaga Padron JS, Gonzdlez Martin IJ, Garcfa Herndndez JE (1995) Phosphorus and potassium release from phillipsite based slow release fertilizers. J Contr Release 34:25–29

Prasad TNV, Sudhakar KVP, Sreenivasulu Y, Latha P, Munaswamy V, Raja Reddy K, Sreeprasad TS, Sajanlal PR, Pradeep T (2012) Effect of nanoscale zinc oxide particles on the germination, growth and yield of peanut. J Plant Nutr 356:905–927

Prasad R, Kumar V, Prasad KS (2014) Nanotechnology in sustainable agriculture: present concerns and future aspects. Afr J Biotechnol 13:705–713

Priester JH (2012) Soybean susceptibility to manufactured nanomaterials with evidence for food quality and soil fertility interruption. Proc Natl Acad Sci USA 109:2451–2456

Ramesh K, Reddy DD, Biswas AK, Subba Rao A (2011) Zeolites and their potential uses in agriculture. In: Sparks DL (ed) Advances in agronomy, vol 113. Academic, Burlington, pp 215–236

Rezaei M, Movahedi Naeini SAR (2009) Kinetics of potassium desorption from the losses soil, soil mixed with zeolite and the clinoptilolite zeolite as influenced by calcium and ammonium. J Appl Sci 9:3335–3342

Robson AD (1993) Zinc uptake from soils in zinc in soils and plants. Kluwer, Dordrecht

Saratin JB (2010) Food for turf: slow-release nitrogen. Grounds Maintenance 2:4–6

Sarkar S (2011) Synthesis of clay polymer-nutrient nano-composites and their characterization with respect to water holding capacity and nutrient release behavior. Ph.D. Thesis, IARI, New Delhi

Selva Preetha P, Subramanian KS, Sharmila Rahale C (2014) Characterization of slow release of sulphur nutrient—a zeolite based nano-fertilizer. Int J Dev Res 4:229–233

Shah V, Belozerova I (2009) Influence of metal nanoparticles on the soil microbial community and germination of lettuce seeds. Water Air Soil Pollut 197:143–148
Subbaiya R, Priyanka M, Selvam MM (2012) Formulation of green nano-fertilizer to enhance the plant growth through slow and sustained release of nitrogen. J Pharm Res 5:5178–5183
Subramanian KS, Sharmila Rahale C (2009) Synthesis of nano-fertilizers formulations for balanced nutrition. In: Proceedings of the Indian Society of Soil Science—Platinum Jubilee Celebration, 22–25 December, IARI Campus, New Delhi, p 85
Subramanian KS, Sharmila Rahale C (2012) Ball milled nanosized zeolite loaded with zinc sulphate: a putative slow release Zn fertilizer. Int J Indian Hort 1:33–40
Subramanian KS, Tarafdar JC (2011) Prospects of nanotechnology in Indian farming. Ind J Agr Sci 81:887–893
Subramanian KS, Manikandan A, Praghadeesh M (2012) Smart delivery system—prospects in agriculture. In: Short course on application of nanotechnology in soil science & plant nutrition research, 18 Sept 2012. Indian Institute of Soil Science, Bhopal, pp 122–136
Subramanian KS, Manikandan A, Thirunavukkarasu M, Sharmila Rahale CS (2015) Nano-fertilizers for balanced crop nutrition. In: Rai M et al (eds) Nanotechnologies in food and agriculture. Springer, Cham, pp 1–15
Thirunavukkarasu M, Subramanian KS (2014a) Surface modified nano-zeolite used as carrier for slow release of sulphur. J Appl Nat Sci 6:19–26
Thirunavukkarasu M, Subramanian KS (2014b) Surface modified nano-zeolite based sulphur fertilizer on growth and biochemical parameters of groundnut. Trends Biosci 7:565–568
Thirunavukkarasu M, Subramanian KS (2015a) Synthesis and characterization of surface modified nano-zeolite fortified with sulphate and its sorption and desorption pattern. J Sci Ind Res 74:680–684
Thirunavukkarasu M, Subramanian KS (2015b) Nano-sulphur on biomass, yield attributes, soil microbes and physiological parameters of groundnut. Life Sci Leaflet 63:13–22
Verver AB, Van Swaaij WPM (1985) The rate of oxidation of hydrogen sulphide by oxygen to elemental sulphur over Nax and Nay zeolites and the adsorption of sulphur. Appl Catal 14:185–206
Wanyika H, Gatebe E, Kioni P, Tang Z, Gao Y (2012) Mesoporous silica nanoparticles carrier for urea: potential applications in agrochemical delivery systems. J Nanosci Nanotechnol 12:2221–2228
Yuvaraj M, Subramanian KS (2014) Core shell fortified Zinc as a smart delivery system for rice. J Soil Sci Plant Nutr 1:1–8
Zhao ZJ, Wang YJ, Wang YQ (2013) Absorption properties of ammonia nitrogen and phosphorus by different substrates in constructed wetland. J Zhejiang Forest Sci Technol 31:28–32
Zhou JM, Huang PM (2007) Kinetics of potassium release from illites as influenced by different phosphates. Geoderma 138:221–228
Zhu H, Han J, Xiao JQ, Jin Y (2008) Uptake, transport and accumulation of manufactured iron oxide nano-particles by pumpkin plants. J Environ Monit 10:713–717

Part III
Nanomaterials in Plant Systems

Chapter 12
Nanoparticle Interaction with Plants

Ivan Pacheco and Cristina Buzea

12.1 Introduction

The interaction of plants with nanoparticles is a relatively recent field of study. The effects of nanoparticle exposure on plants may range from subtle changes in the soil environment to changes in plant morphology, physiology, and genetics (Deng et al. 2014). In order to determine the phytotoxicity of nanoparticles some biological end points are measured, such as germination index (comprising germination time and rate), root elongation, shoot and root biomass, and root tip morphology (Deng et al. 2014). The review of Deng et al. (2014) represents a good summary of the current research intended to understand the effects of crop interaction with nanoparticles. The review summarizes nanoparticle uptake and in vivo translocation, as well as physiological, morphological, and genetic consequences and potential trophic transfer (Deng et al. 2014).

In general terms, there is a consensus that the effect of nanoparticles on plants depends on the type of nanoparticle, their physicochemical properties, plant species, and the plant substrate (soil, hydroponics, and culture medium) (Arruda et al. 2015; Aslani et al. 2014; Bakshi et al. 2015; Batley et al. 2013; Bernhardt et al. 2010; Chichiricco and Poma 2015; Deng et al. 2014; Dietz and Herth 2011; Hossain et al. 2015; Ma et al. 2010a; Miralles et al. 2012a; Navarro et al. 2008; Rico et al.

I. Pacheco
Department of Pathology, Grey Bruce Health Services, 1800 8th St East, Owen Sound, ON, Canada, N4K 6M9

Department of Pathology and Laboratory Medicine, Schulich School of Medicine and Dentistry, Western University, London, ON, Canada, N6A 5C1
e-mail: ipacheco@gbhs.on.ca; iblandin@uwo.ca

C. Buzea (✉)
IIPB Medicine Corporation, 2656 8th Ave East, Owen Sound, ON, Canada, N4K 6S5
e-mail: cristinabuzea@mdcorporation.ca

M. Ghorbanpour et al. (eds.), *Nanoscience and Plant–Soil Systems*, Soil Biology 48,
DOI 10.1007/978-3-319-46835-8_12

2011; Yadav et al. 2014; Prasad 2014; Schwab et al. 2015). This was to be expected as science has already a similar understanding involving the toxicity of nanoparticles to humans and animals.

12.2 Nanoparticle–Soil Interaction

Researches have already shown that the toxicity of nanoparticles on plants may be determined by the soil type (Josko and Oleszczuk 2013). This occurs because nanoparticle interaction with the soil will dynamically change their properties: their dispersibility, aggregation, size, surface area, surface charge, and surface chemistry, which will dictate their transport and availability. Furthermore, the physicochemical interaction and transport of nanoparticles in soil are affected not only by the nanoparticle properties but also by the properties of the soil (Bakshi et al. 2015; Prasad et al. 2016).

The soil bacteria are very important for their services to various ecosystems, and several nanoparticles are well known for their antibacterial effects (Aziz et al. 2015; Prasad et al. 2016). Nanoparticle toxicity on bacteria also depends on the soil type (Schlich and Hund-Rinke 2015). The toxicity of silver nanoparticle toward ammonia-oxidizing bacteria varies with the soil type, decreasing with higher clay content and increasing pH. The toxicity was not affected by the organic content of the soil and could not be attributed solely to any soil property. A higher Ag nanoparticle toxicity was observed for soils with a higher sand content or lower clay content; however, the relationship was not linear.

Some natural nanoparticles, such as clay nanoparticles, contribute to soil fertility by preventing the loss of nutrients to groundwater by forming electrostatic bonds with ions, such as NH_4^+, Ca^{2+}, K^+, and Mg^{2+}, due to their charged surface (Bernhardt et al. 2010).

12.3 Nanoparticle Uptake by Plants

12.3.1 Nanoparticle Uptake by Plants: Root Uptake

While the mechanism of nanoparticle interaction with plants is not entirely understood, it is accepted that, depending on their properties, some nanoparticles may form complexes with membrane transporter proteins or root exudates and as a result be translocated into the plant system (Yadav et al. 2014). The surface roughness, hydrophobicity, or surface charge can result in surface binding and cellular uptake.

There are two main ways for nanoparticle uptake by plant roots, the apoplastic and the symplastic route, as shown in Fig. 12.1 (Deng et al. 2014). Plant cell walls consist of a complex matrix with pores between 5 and 20 nm (Deng et al. 2014). The

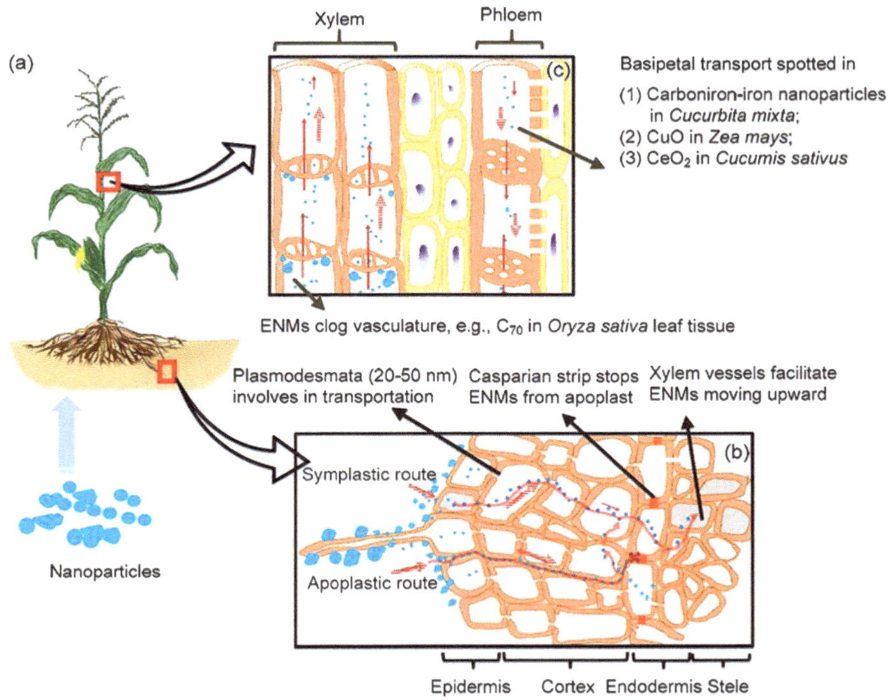

Fig. 12.1 Schematics of nanoparticle uptake: (**a**) in a model crop plant, (**b**) the root uptake and translocation into vasculature, (**c**) the upward movement in xylem and downward transportation in phloem in root and foliar uptake. Journal of Zhejiang University-Science A, 15, 2014, 552–572, DENG, Y. Q. et al, Copyright 2014, with permission of Springer (Deng et al. 2014)

root epidermal cell wall limits the access of particles larger than the size of the pores. Nanoparticles that cross porous cell walls can diffuse between cell walls and plasma membrane and be subjected to osmotic pressure and capillary forces (Lin et al. 2009). Nanoparticles diffusing through the apoplast can reach endodermis (Deng et al. 2014).

The symplastic pathway of nanoparticle entrance is through the inner side of the plasma membrane, which is more important than the apoplastic route. Nanoparticles can enter cells by binding to carrier proteins, through aquaporins which are membrane proteins that operate as water channels, ion channels, and endocytosis, or by piercing the cell membrane and creating new pores (Rico et al. 2011). Endocytic uptake is a pathway of nanoparticle uptake by cells if specific receptor–ligand interactions are taking place. Depending on their morphology, some nanoparticles like carbon nanotubes may pierce their way into the cells and enter cytoplasm (Wild and Jones 2009). When inside the cytoplasm, nanoparticles interact dynamically with their environment via van der Waals, electrostatic, and steric forces. This process results in the binding of proteins on the nanoparticle surface with the formation of a protein corona (Nel et al. 2009). Internalized nanoparticle–endosome or nanoparticle–protein complexes can translocate to neighboring cells through plasmodesmata, which are channels with a diameter of 20–50 nm (Deng et al. 2014). Some nanoparticles (TiO_2) were found to disrupt the cytoskeleton microfilaments that sustain plasmodesmata (Wang et al. 2011b). Therefore, nanoparticle transport through the symplastic pathway may occur via organelles activity, transport proteins, and trans-wall channels (Deng et al. 2014). Once inside the cells, nanoparticles may interact with organelles and disrupt metabolic processes, producing oxidative stress and genetic modifications.

Examples of nanoparticle uptake in root vasculature, root cells, and cytoplasm are shown in Fig. 12.2. The figure shows images of the uptake of gold, ZnO, and silver nanoparticles by roots of *Nicotiana xanthi*, *Allium cepa*, and *Arabidopsis thaliana*, respectively (Sabo-Attwood et al. 2012; Kumari et al. 2011; Geisler-Lee et al. 2013). In Fig. 12.2 (i) (a, b), 3.5 nm-sized gold nanoparticles are seen in tobacco (*Nicotiana xanthi*) root vasculature (400×), while Fig. 12.2 (i) (c) shows nanoparticles in the cell cytoplasm (Sabo-Attwood et al. 2012). In Fig. 12.2 (ii), one can notice ZnO nanoparticles present between cells membrane as well as in the cell interior of *Allium cepa* root cells treated with ZnO nanoparticles (Kumari et al. 2011). Figure 12.2 (iii) demonstrates the presence of Ag nanoparticles in *Arabidopsis* root tips of plants exposed to silver nanoparticles compared to control (Geisler-Lee et al. 2013).

The question still remains of why some plants take up nanoparticles while others do not. Some studies found that nanoparticles will induce plant stress via physical damage without being internalized. For example, MWCNTs adsorb onto root surfaces of alfalfa and wheat, pierce epidermal tissue, and induce plant stress (Miralles et al. 2012b).

12.3.2 Nanoparticle Uptake by Plants: Foliar Uptake

In addition to nanoparticle uptake by roots, foliar uptake is also possible. Nanoparticles are able to penetrate leaf surfaces through stomatal pores (Larue et al. 2014a, b; Hong et al. 2014). For example, *Lactuca sativa* leaves exposed to Ag (Larue et al. 2014a) and TiO_2 nanoparticles (Larue et al. 2014b) show evidence for internalization. *Cucumis sativus* (cucumber) can uptake CeO_2 nanoparticles through the leaves. Particles subsequently translocate to other plant tissue, such as the roots (Hong et al. 2014). Plants that were found to uptake nanoparticles through leaves are *Vicia faba*, *Zea mays*, *Arabidopsis thaliana*, wheat, *Lactuca*, *Cucumis sativus*, and rapeseed (Chichiricco and Poma 2015). Among the studied nanoparticles that penetrate the leaf pores are polystyrene, CeO_2, TiO_2, Fe_2O_3, MgO, ZnO, Zn, Mn, and Ag (Chichiricco and Poma 2015). Their size ranged from a few nanometers to several hundred nanometers. In summary, evidence indicates that several types of nanoparticles penetrate plant leaves and are internalized. Some of these nanoparticles have documented deleterious effects on humans.

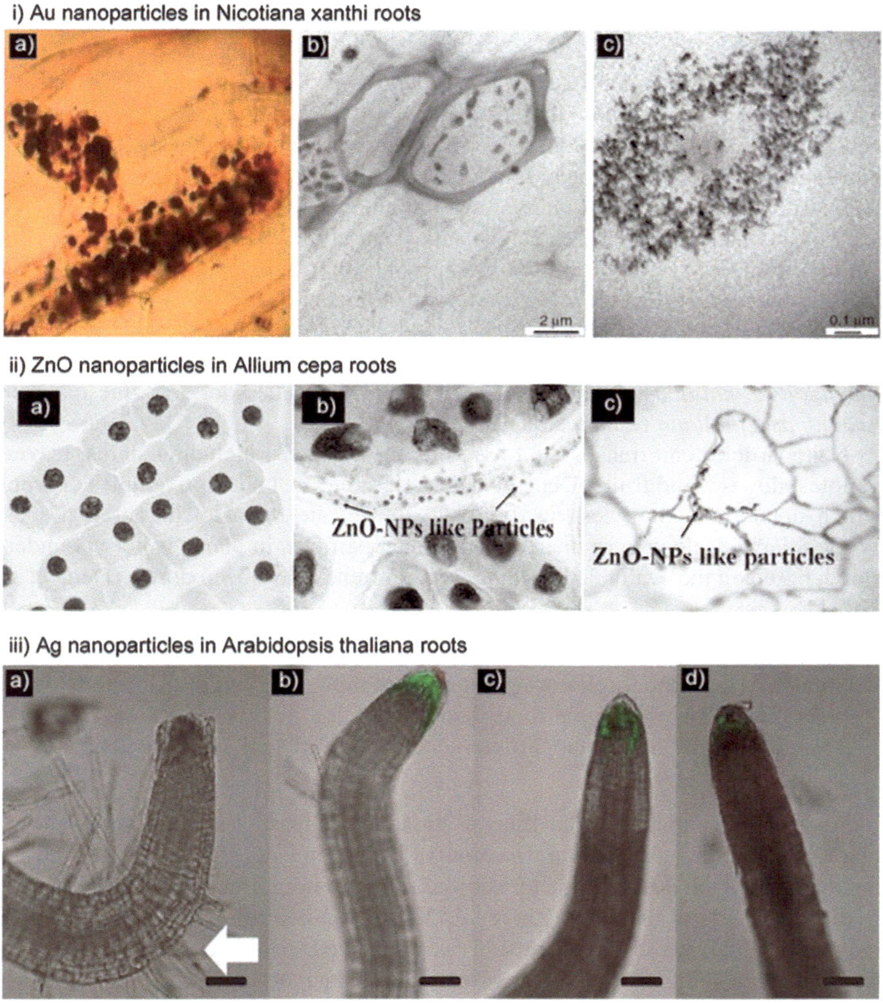

Fig. 12.2 Root uptake of nanoparticles. (i) Gold nanoparticles (3.5 nm diameter) present in tobacco (*Nicotiana xanthi*) seedlings. (**a**) Light microscopy image of nanoparticles present in root vasculature (400×), (**b**) and (**c**) TEM image of root cells where nanoparticle agglomerates can be seen in the vasculature and in the cell cytoplasm. Reprinted from Sabo-Attwood T. et al, Uptake, distribution and toxicity of gold nanoparticles in tobacco (*Nicotiana xanthi*) seedlings. Nanotoxicology, 6, 2012, 353–360. Reprinted by permission of Taylor and Francis Ltd, http://www.tandfonline.com(Sabo-Attwood et al. 2012). (ii) Microscopy images of *Allium cepa* root cells treated with ZnO nanoparticles. (**a**) Control cells (treated with distilled water), (**b**) root cells treated with ZnO nanoparticles exhibiting agglomeration of nanoparticles between the cell membranes. (**c**) Phase contrast image showing particle aggregation in the cell interior. All images have a magnification of ×1000. Reprinted from Kumari M. et al, Cytogenetic and genotoxic effects of zinc oxide nanoparticles on root cells of *Allium cepa*. Journal of Hazardous Materials, 190, 613–621, Copyright (2011), with permission from Elsevier (Kumari et al. 2011). (iii) Confocal laser scanning microscopy images showing the localization of silver after exposure of *Arabidopsis* roots to silver nanoparticles. *Green signals* indicate the presence of silver elements. (**a**) Control roots; (**b**)–(**d**) roots treated with 20-nm silver nanoparticles at different concentrations

12.3.3 Uptake and Translocation of Nanoparticles in Crops

Crops exposed to nanoparticles display adaptive processes in response to stress, which includes but are not limited to nanoparticle endocytosis, modification of genes involved in cell division, as well as decreased photosynthesis, fruit yields, nutritional quality, and nitrogen fixation (Deng et al. 2014).

Research has shown that many crops are able to uptake various nanoparticles, such as Au nanoparticle by tomato plants (Dan et al. 2015); Ag by *Arabidopsis thaliana* (Geisler-Lee et al. 2013); CeO_2 by alfalfa (*Medicago sativa*), corn (*Zea mays*), cucumber (*Cucumis sativus*), and tomato (*Lycopersicon esculentum*) (Lopez-Moreno et al. 2010); MWCNTs and TiO_2 in wheat and rapeseed (Larue et al. 2012b, c); TiO_2 nanoparticles by lettuce leaves (Larue et al. 2014b) and *Arabidopsis thaliana* (Kurepa et al. 2010); and fullerene soot and zinc oxide in *Arabidopsis thaliana* (Landa et al. 2012).

Nanoparticles can translocate to various plant tissues, including stems, leaves, petioles, flowers, and fruits (Deng et al. 2014). Table 12.1 shows examples of crops which uptake nanoparticles via roots and translocate them to the aerial tissues. While a large-scale pattern distribution of nanoparticles in plants is not yet understood, based on the available literature, several trends have been drawn (Deng et al. 2014):

- Nanoparticles in shoots are located mainly near or within the vascular tissue.
- Smaller-size nanoparticles or aggregates are more likely to translocate to aerial parts further away from the roots than larger ones.
- The concentration of nanoparticle accumulated in leaf is larger than that in stems.

An example of uptake and translocation of nanoparticles in plants is shown below. Pumpkin (*Cucurbita maxima*) hydroponically exposed to Fe_3O_4 nanoparticles with 20 nm diameter showed translocation of magnetic nanoparticles in different parts of the plant (Zhu et al. 2008). The control plants did not show a magnetic signal, while most of the tissues analyzed exhibited different levels of magnetization indicative of different amounts of magnetic nanoparticles. This is illustrated in Fig. 12.3, where on a schematic of the pumpkin plant are shown the different levels of magnetization at different sampling locations, measured in units of 10^{-3} emu g^{-1}. One can notice strong magnetic signals in the stem just near the roots in all analyzed leaves irrespective to their distance from the roots and much smaller signal in stem tissue. These results are evidence that magnetic nanoparticles

---

Fig. 12.2 (continued) of 66.8, 133.68, 267.36 mg L^{-1}. *White arrows point* to root hairs. Scale bars = 50 mm. Reprinted from Geisler-Lee, J. et al, 2013, Phytotoxicity, accumulation and transport of silver nanoparticles by *Arabidopsis thaliana*. Nanotoxicology, 7, 323–337, by permission of Taylor and Francis Ltd, http://www.tandfonline.com (Geisler-Lee et al. 2013)

Table 12.1 Translocation and distribution of nanoparticles (NP) in aerial tissues of agricultural crops through roots

Crop name	NP type	NP size (nm)	Location in aboveground parts	References
Bitter melon	$C_{60}(OH)_{20}$	1–10	Petioles, leaves, flowers, fruits	Kole et al. (2013)
Cucumber	CeO_2	7, 25	Leaves	Zhang et al. (2011)
Cucumber	TiO_2	27	Leaves	Servin et al. (2012)
Pea	Carbon–Fe	10	Cortex, leaf petioles, vascular tissue	Cifuentes et al. (2010)
Pumpkin	Fe_3O_4	20	Strong magnetic signals in all leaf specimens	Zhu et al. (2008)
Rapeseed	MWCNTs	41.2	Leaves	Larue et al. (2012b)
Soybean	SPION	9	Stem, leaves, vascular, and parenchyma tissue	Ghafariyan et al. (2013)
Rice	C70	1.19, 17.99	In and near the stem's vascular system, in leaves. Also in leaves of second-generation plants	Lin et al. (2009)
Sunflower	Carbon–Fe	10	In cortex, leaf petioles, internodes, vascular tissue	Cifuentes et al. (2010)
Tobacco	Au	10, 30, 50	Within leaf midrib near petiole	Judy et al. (2011)
Tomato	Carbon–Fe	10	In cortex, leaf petioles, internodes, vascular tissue	Cifuentes et al. (2010)
Tomato	MWCNTs	25	In flower structure	Khodakovskaya et al. (2013)
Tomato	Carbon	10–35 (MWCNTs) 0.86–2.2 (SWCNTs) 2–5 (graphene)	Outside the leaves vascular system	Khodakovskaya et al. (2011)
Wheat	MWCNTs	41.2	Leaves	Larue et al. (2012b)
Wheat	Carbon–Fe	10	In cortex, petioles, internodes, vascular tissue	Cifuentes et al. (2010)

Adapted from Journal of Zhejiang University-Science A, 15, 2014, 552–572, Deng, Y. Q. et al, Copyright 2014, with permission of Springer (Deng et al. 2014)

can undergo uptake by the pumpkin plants and translocation to various tissues (Zhu et al. 2008).

It is important to note that nanoparticle bioaccumulation in plants is species specific. For example, Au nanoparticles bioaccumulate in tobacco but not in wheat (Judy et al. 2012).

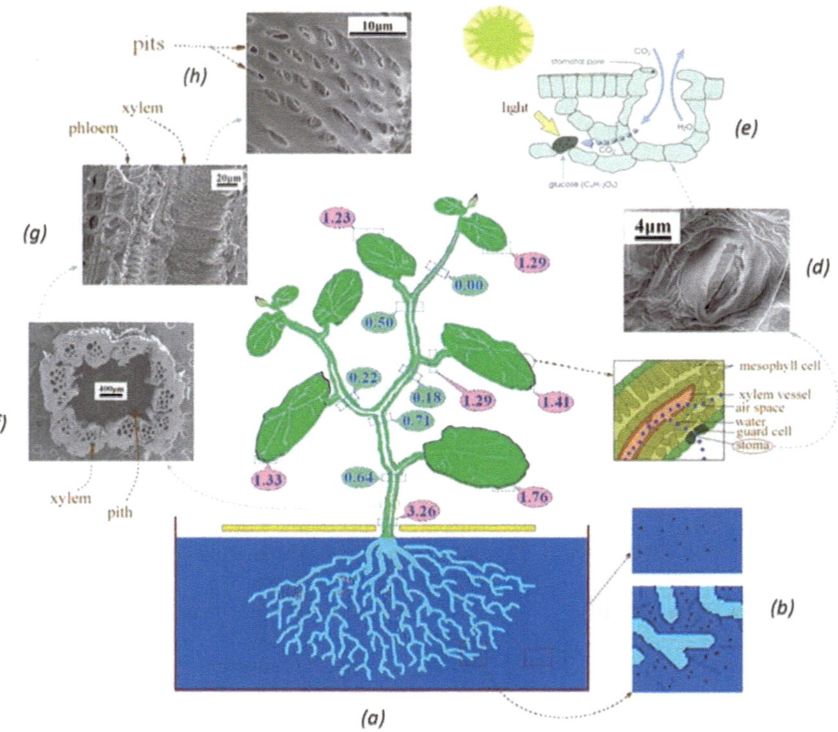

Fig. 12.3 Schematic illustration of Fe_3O_4 nanoparticle uptake and distribution in different tissues of pumpkin. Reproduced from Zhu H. et al., Uptake, translocation, and accumulation of manufactured iron oxide nanoparticles by pumpkin plants. Journal of Environmental Monitoring, 10, 713–717, with permission of The Royal Society of Chemistry (Zhu et al. 2008)

The use of engineered nanoparticles in agriculture might prove a Pandora box which once opened will be very difficult to mitigate. Indeed, it would be ideal to correlate with available research data before potentially toxic nanoparticle that are already patented are released into widespread agricultural use. Indeed, we should assess the risk and potential unwanted effects of potentially harmful nanoparticles in crops due to trophic transfer.

12.4 Plant Uptake and Translocation Dependent on Nanoparticle Properties

Nanoparticle uptake and translocation depends on several factors, such as exposure condition, nanoparticle physicochemical properties, and plant species. Nanoparticle physical properties that are important in plant uptake are size, composition, crystalline state, surface charge, surface functionalization, magnetic properties, hydrophobicity, and/or hydrophilicity.

The translocation time of nanoparticles within plants can be very short. For example, carbon-coated magnetic nanoparticle translocates from roots to shoots in less than 24 h after exposure of sunflower, tomato, pea, and wheat (Cifuentes et al. 2010). Indeed, the rate of translocation could become relevant for harvesting purposes.

12.4.1 Particle Size-Dependent Nanoparticle Uptake and Translocation

Particle size seems to be a very important factor in the uptake of nanoparticles. Smaller nanoparticles are more likely to be internalized by plants than larger ones. For example, 20 nm Fe_3O_4 nanoparticles translocated into pumpkin plants while 25 nm did not (Zhu et al. 2008; Wang et al. 2011a).

A study on the translocation of TiO_2 rutile and anatase with different sizes (between 14 nm and 655 nm) in wheat concluded that the smallest nanoparticles are uptaken by roots and distribute through the entire plant tissue in their original form, without dissolution or crystalline phase change (Larue et al. 2012a). The authors found that only nanoparticles smaller than 140 nm are able to accumulate in wheat roots. Also, nanoparticles larger than 36 nm can accumulate in root parenchyma but do not translocate to the shoot.

A study on the uptake of radioactive ceria nanoparticles of two different sizes (7 and 25 nm) in cucumber seedlings showed that the nanoparticles are absorbed by the plant roots and translocated to leaves (Zhang et al. 2011). A higher ceria content was absorbed when plants were treated with 7 nm nanoparticles when compared to 25 nm ones. The absorption increased with the concentration of nanoparticles. The distribution of nanoparticles in the leaves was similar for both sizes, as can be seen in Fig. 12.4 which shows the location of radioactive ceria nanoparticles in cucumber leaves (Zhang et al. 2011). Young leaves show ceria nanoparticle accumulation exclusively along the leaf margins, while older leaves nanoparticles are spread over the entire surface of the leaves.

12.4.2 Crystalline Structure-Dependent Nanoparticle Uptake and Translocation

The uptake of nanoparticles depends also on their crystalline phase. Cucumber (*Cucumis sativus*) exposed to a mixture of titanium dioxide nanoparticles with anatase and rutile crystalline structure absorbs the nanoparticles in the roots (Servin

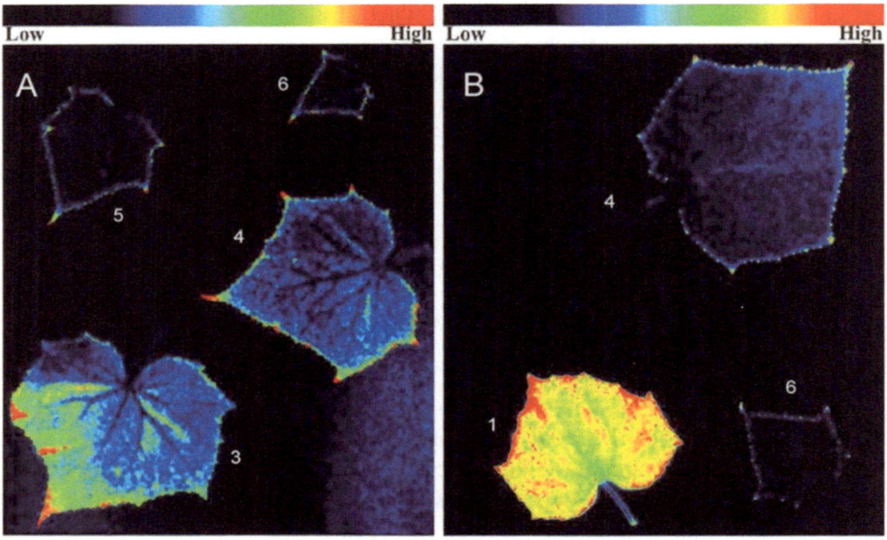

Fig. 12.4 The uptake of radiolabeled ceria nanoparticles in cucumber leaves (**a**) 7 nm ceria, 3rd, 4th, 5th, and 6th leaves; (**b**) 25 nm ceria, 1st, 4th, and 6th leaves. Reproduced from Zhang Z. Y. et al, Uptake and distribution of ceria nanoparticles in cucumber plants. Metallomics, 2011, 3, 816–822 with permission of The Royal Society of Chemistry (Zhang et al. 2011)

et al. 2012). The anatase phase stayed preferentially in the roots, while the rutile phase was transported and accumulated mainly in the aerial tissue of the plant.

12.4.3 Charge-Dependent Nanoparticle Uptake by Plants

The uptake of nanoparticles in plants depends on the nanoparticle surface charge or functionalization. A study showed the effect of cationic (positive) or anionic (negative) charge of nanoparticles in poplar trees (*Populus deltoides* × *nigra*) exposed to quantum dots with different coatings (Wang et al. 2014). The nanoparticles used in the experiment were CdSe/CdZnS quantum dots coated with cationic polyethylenimine (PEI) or poly(ethylene glycol) of anionic poly (acrylic acid) (PAA–EG). As shown in Fig. 12.5 (i) a–b and (ii) a–b, poplar roots showed fluorescence after 2-day exposure, indicative of the presence of both cationic and anionic quantum dots, respectively (Wang et al. 2014). The fluorescence of cationically charged quantum dots inside the plant was fairly constant during the exposure period, as shown in Fig. 12.5 (i) c–d. The absorption of cationic quantum dots was ten times faster than that of anionic quantum dots, probably as a result of the electrostatic attraction between them and the negatively charged root cell wall (Wang et al. 2014). On the other side, the electrostatic repulsion between

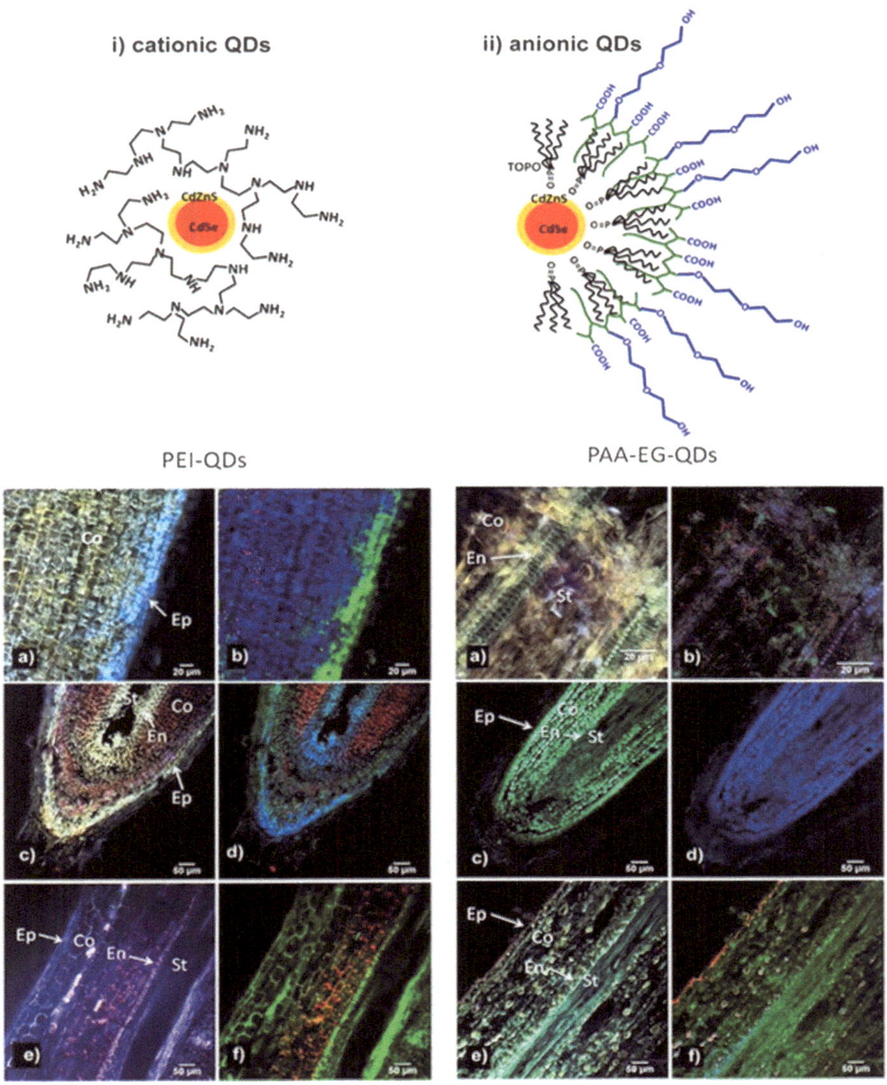

Fig. 12.5 Structure of (i) cationic and (ii) anionic QDs and their coating polymers. QDs have a CdSe core and a CdZnS shell. Coating with polyethylenimine (PEI) or poly(ethylene glycol) confer a cationic charge, while coating with poly(acrylic acid) (PAA–EG) give them an anionic charge. Below (i) fluorescence images showing the distribution of cationic PEI–QDs in poplar root. (**a**) Longitudinal sections of poplar root after 2-day exposure, (**c**) in root tip after 11-day exposure, (**e**) and root after 11-day exposure. (**b**), (**d**), (**f**) are spectral unmixed images of the original fluorescence images (**a**, **c**, **e**). In the spectral unmixed images (**b**, **d**, **f**), PEI-QDs fluorescence is shown in red and plant autofluorescence in *green* and *blue*. Below (ii) fluorescence images show the distribution of anionic PAA-EG-QDs in poplar root. (**a**) Longitudinal sections of poplar root after 2-day exposure, (**c**) in root tip after 11-day exposure, (**e**) and root after 11-day exposure. (**b**), (**d**), (**f**) are spectral unmixed images of the original fluorescence images (**a**, **c**, **e**). In the spectral unmixed images (**b**, **d**, **f**), PAA-EG-QDs fluorescence is shown in *red* and plant autofluorescence in *green* and *blue*. *Ep* epidermis, *Co* cortex, *En*

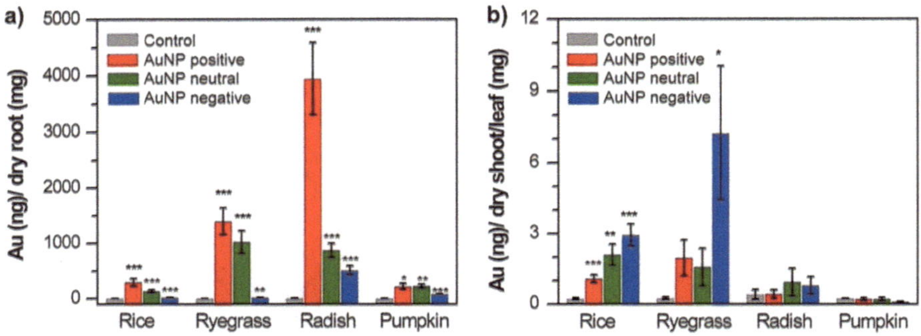

Fig. 12.6 Au nanoparticles uptake by plants (rice, ryegrass, radish, pumpkin) in the roots (**a**) and shoots (**b**). Gold nanoparticles are functionalized with various compounds to induce a positive, neutral, and negative surface charge, respectively. Reprinted with permission from Zhu Z. J. et al, Effect of Surface Charge on the Uptake and Distribution of Gold Nanoparticles in Four Plant Species. Environmental Science and Technology, 2012, 46, 12391–12398. Copyright (2012) American Chemical Society (Zhu et al. 2012)

the negatively charged root surface and the anionic quantum dots resulted in a slower absorption process.

The previous results of charge-dependent nanoparticle uptake in woody plants are in contrast with the results of studies of charged nanoparticle uptake in herbaceous plants (Koelmel et al. 2013; Zhu et al. 2012). More precisely, negatively coated gold nanoparticles had a higher translocation rate in several plant species compared to neutral or positively charged gold nanoparticles (Zhu et al. 2012). Herbaceous plants absorb positively charged nanoparticles in roots, while negatively charged ones are translocated to stems and leaves. This fact is illustrated in Fig. 12.6, where one can see the uptake of nanoparticles with different surface charge by rice, ryegrass, radish, and pumpkin (Zhu et al. 2012).

12.5 Transmission of Nanoparticles to Second-Generation Plants

Nanoparticles can be absorbed by plants and distributed to various plant tissues, including fruits and seeds. The accumulation of nanoparticle in seeds is important in influencing subsequent plant generations. It has been demonstrated that nanoparticles can be transmitted to plant progenies through seeds, even though

---

Fig. 12.5 (continued) endodermis, *St* stele. Reprinted with permission from Wang J. et al, Uptake, Translocation, and Transformation of Quantum Dots with Cationic versus Anionic Coatings by Populus deltoides × nigra Cuttings. Environmental Science and Technology 2014, 48, 6754–6762. Copyright (2014) American Chemical Society (Wang et al. 2014)

they are not exposed to external nanoparticles from soil, water, or air. To our knowledge, there are only two studies that researched upon this topic (Lin et al. 2009; Wang et al. 2013).

Lin et al. studied generational transmission of C70 from seeds of rice (Lin et al. 2009). Seeds from control plants and plants treated with C70 were harvested and then planted in a media without C70 nanoparticles. These germinated plants, called the second-generation plants, were studied, and the authors found C70 black aggregates in progenies, more exactly near the stem's vascular system and even in leaf tissue of the second-generation rice plants, as seen in Fig. 12.7 (Lin et al. 2009). These findings are extremely important for the negative impact that nanoparticles might pose for food safety and environment.

Second-generation tomato seedling plants grown from seeds of plants previously exposed to ceria nanoparticles have a different development compared to plants grown from seeds of plants that were not exposed to nanoparticles (Wang et al. 2013). Indeed, the benefits related to first-generation seedling exposed to CeO_2 nanoparticles were not present in second-generation seedlings. Second-generation seedlings treated with ceria nanoparticles had a much diminished biomass than seedlings grown from seeds of untreated plants. While ceria nanoparticles lead to a minor growth improvement for first-generation seedlings, it had a detrimental effect

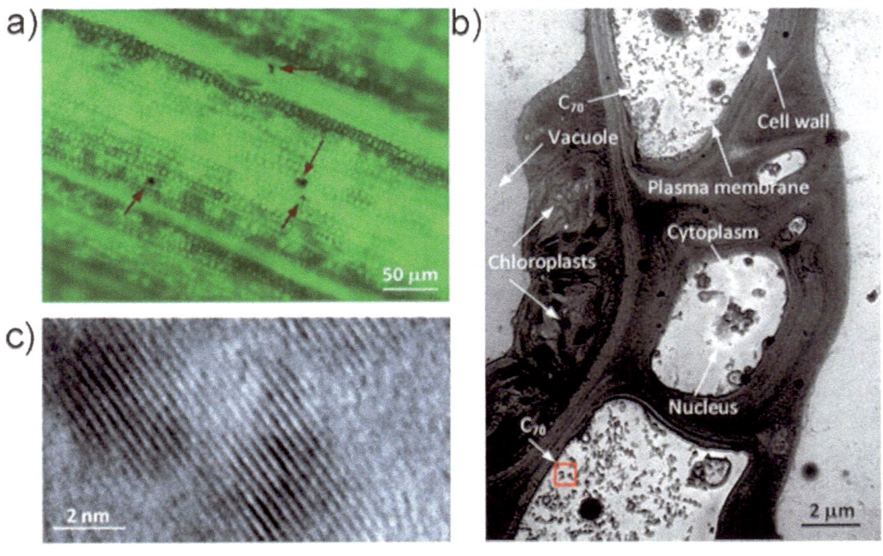

Fig. 12.7 The presence of C70 nanoparticles in second-generation rice plants. (**a**) Bright-field image of rice leaf. *Arrows* show C70 aggregates in or near the leaf vascular system. (**b**) TEM image showing C70 nanoparticles in the leaf cells, plant cell walls, and organelles. (**c**) High-magnification TEM image of the C70 nanoparticles shown in (**b**) in a *red square*. Reproduced from Lin S. J. et al. Uptake, Translocation, and Transmission of Carbon Nanomaterials in Rice Plants. Small, 5, 2009, 1128–1132 Copyright (2009), with permission from John Wiley and Sons (Lin et al. 2009)

for the response of second-generation plants to the fertilization effects of ceria nanoparticles (Wang et al. 2013). This emphasizes the existence of long-term multigenerational effects of nanoparticles on the growth of plants and implies that a buildup of nanoparticles in the environment may have a detrimental effect.

12.6 Adverse Effects of Nanoparticle Interaction with Plants

12.6.1 Shoot and Root Biomass Modification of Plants Exposed to Nanoparticles

The assessment of nanoparticles toxicity on plants usually involves measurements of germination index (time and rate), root elongation, shoot and root biomass, and root tip morphology, among others (Deng et al. 2014).

Shoot and root biomass is an important indicator of nanoparticle toxicity. While exposure time and dose varies among different studies, one can conclude that the negative effect on root and shoot elongation and biomass are plant and nanoparticle specific. While some nanoparticle might be toxic due to the release of ions and subsequent accumulation of these ions in plant tissue, other nanoparticles are toxic upon their uptake and translocation (Deng et al. 2014). Nanoparticles that were found to have a negative effect on seedling root and shoot elongation of different plants are made of materials such as Au, ZnO, cobalt oxide, CuO, Ag, Al_2O_3, and MWCNTs (Deng et al. 2014; Feichtmeier et al. 2015; Burklew et al. 2012; Dimkpa et al. 2013b; Begum and Fugetsu 2012; Ghodake et al. 2011). The observed phytotoxicity was attributed to massive adsorption of nanoparticles into the root system.

Figure 12.8 shows the negative effect of seedling and plant exposure to increasing concentration of various nanoparticles. One can notice a general trend of decreased shoot length and root length with increasing concentrations of nanoparticles for (a) barley exposed to Au nanoparticles (Feichtmeier et al. 2015), (b) tobacco treated with Al_2O_3 (Burklew et al. 2012), (c) wheat exposed to Ag nanoparticles (Dimkpa et al. 2013b), and (e)–(g) red spinach treated with MWCNTs (Begum and Fugetsu 2012). Figure 12.8d shows that ionic gold alone exhibits a negative effect, similar to other nanoparticles described above, to *Arabidopsis* (Taylor et al. 2014). More on the effect of ionic versus nanoparticles on plants will be elaborated in the next subchapter.

Some nanoparticles with beneficial effects on plant growth are MWCNTs, Au, TiO_2, Pd, Cu, Si, Fe_2O_3, CeO_2, and fullerol $C_{60}(OH)_{20}$ (Khodakovskaya et al. 2009; Zheng et al. 2005; Arruda et al. 2015; Deng et al. 2014; Shah and Belozerova 2009; Kole et al. 2013; Siddiqui and Al-Whaibi 2014; Chichiricco and Poma 2015).

It is interesting to note that many of the nanoparticles that show negative effects on some plant biomass have a beneficial effect for the growth of other plants. For

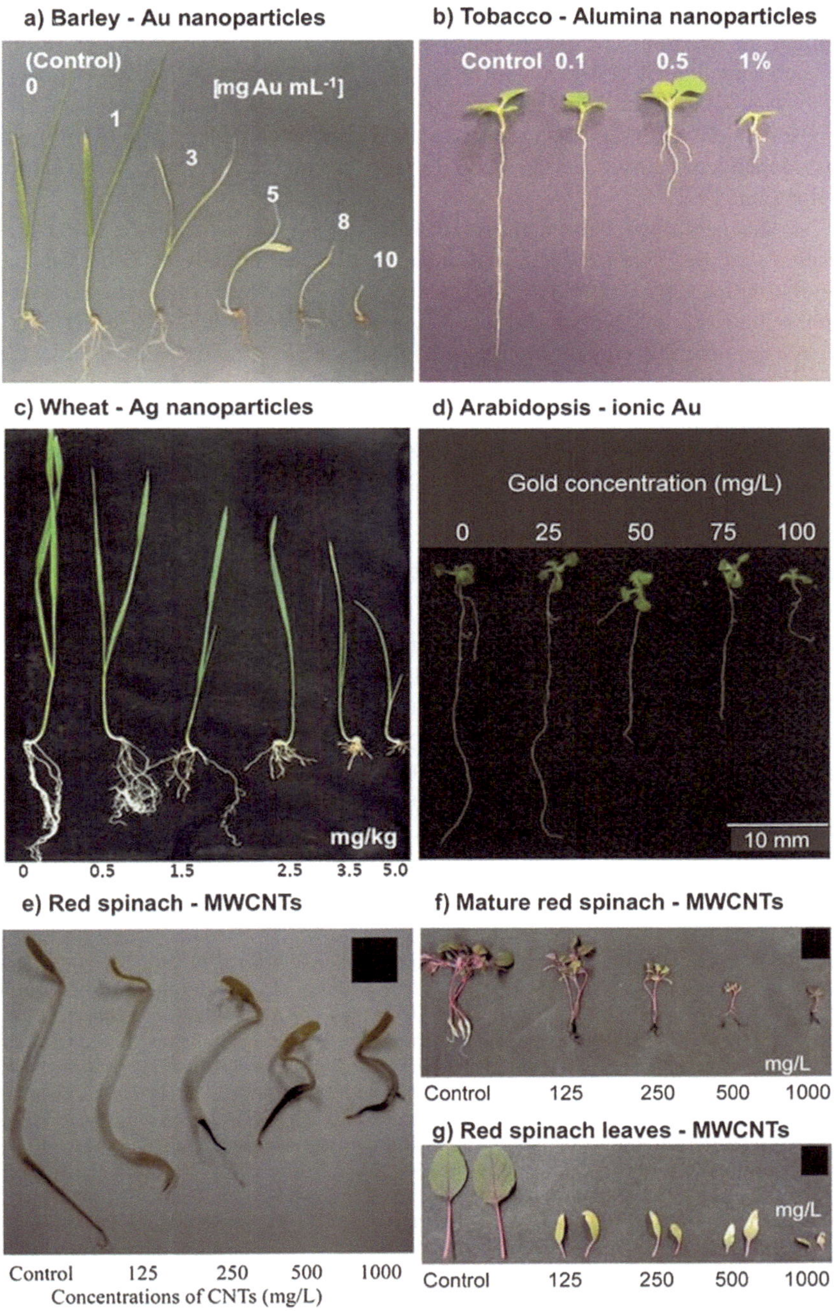

Fig. 12.8 Photographs showing the effect of exposure to different concentrations of different nanoparticles or ionic metals of seedlings and plants. (**a**) Barley seedlings exposed to different concentrations of gold nanoparticles after 2 weeks of exposure. Reprinted from Feichtmeier, N. S. et al, Uptake, effects, and regeneration of barley plants exposed to gold nanoparticles.

example, while MWCNTs were found to increase seed germination rate and plant biomass in tomatoes (Khodakovskaya et al. 2009), they inhibit the growth of red spinach and induce cell death after 15 days of hydroponic culture (Begum and Fugetsu 2012). Figure 12.8e–g shows red spinach seeds, mature plants, and leaves exposed to MWCNTs. One can see a decrease in shoot and plant growth following a dose-dependent manner. See more on the topic of beneficial effects of nanoparticles in the Sect. 12.7.

As we mentioned before, the toxic effects of nanoparticles are material and species specific. This is illustrated by the phytotoxicity study of rare earth oxide nanoparticles, CeO_2, La_2O_3, Gd_2O_3, and Yb_2O_3 on several plant species (radish, rape, tomato, lettuce, wheat, cabbage, and cucumber) (Ma et al. 2010b). The effect of each nanoparticle type on root growth is also species specific. For example, CeO_2 had an effect on the root elongation of lettuce alone, while La_2O_3, Gd_2O_3, and Yb_2O_3 nanoparticles resulted in severely reduced root elongations for all studied plants. This is illustrated in Fig. 12.9 (Ma et al. 2010b).

12.6.2 Ionic Effects Versus Nanoparticle Effects

Some nanoparticles are toxic to plants not only due to their small size but due to the release of ions that interact with their cellular mechanisms (Arruda et al. 2015). In general, the toxicity of nanoparticles of these materials is higher than the toxicity of the ions alone, and nanoparticle toxicity cannot be solely explained due to the release of the dissolved components. Some nanoparticles might be toxic due to the release of ions and their subsequent accumulation of these ions in plant tissue as nanoparticles (Deng et al. 2014).

Fig. 12.8 (continued) Environmental Science and Pollution Research, 22, 2015, 8549–8558, with permission from Springer (Feichtmeier et al. 2015). (**b**) Tobacco seedlings treated with aluminum oxide, Al_2O_3. From *left* to *right*, control, 0.1 %, 0.5 %, and 1 % Al_2O_3. Reproduced from (Burklew et al. 2012). (**c**) Wheat exposed to silver nanoparticles for 14 days in a sand matrix. Reprinted with permission from Dimkpa C. O. et al, Silver Nanoparticles Disrupt Wheat (*Triticum aestivum* L.) Growth in a Sand Matrix. Environmental Science and Technology, 47, 1082–1090. Copyright (2013) American Chemical Society (Dimkpa et al. 2013b). (**d**) *Arabidopsis thaliana* L. (*Arabidopsis*) seedlings in the presence of different concentrations of ionic gold with subsequent formation of nanoparticles inside the shoots. Reprinted from Taylor et al., Investigating the Toxicity, Uptake, Nanoparticle Formation and Genetic Response of Plants to Gold. Plos One, 2014, 9. Copyright (2014) Taylor et al. (Taylor et al. 2014). (**e–g**) Red spinach untreated and treated with multi-walled carbon nanotubes (MWCNTs) at different concentrations: (**e**) seeds after incubation with and without MWCNTs for 5 days; (**f**) mature plants; and (**g**) leaves exposed to different concentrations of MWCNTs. Images (**e–g**) reprinted from Begum P and Fugetsu B, 2012, Phytotoxicity of multi-walled carbon nanotubes on red spinach (Amaranthus tricolor L.) and the role of ascorbic acid as an antioxidant, Journal of Hazardous Materials, 243, 212–222, Copyright (2012), with permission from Elsevier (Begum and Fugetsu 2012)

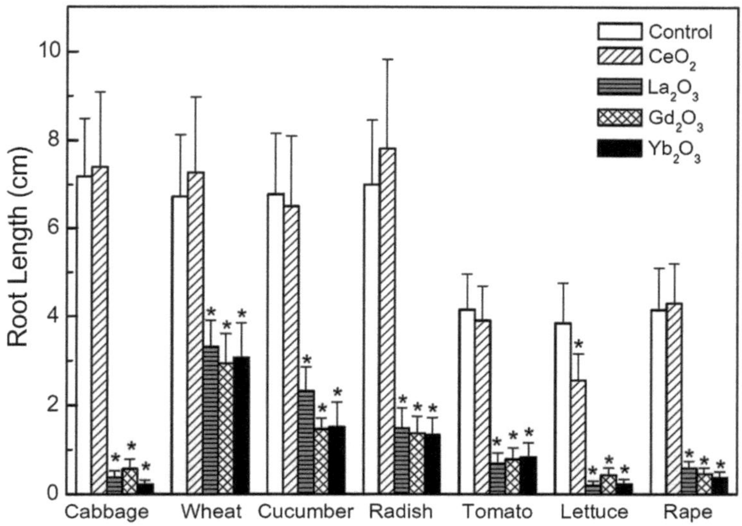

Fig. 12.9 The effect of root lengths of the seeds of cabbage, wheat, cucumber, radish, tomato, lettuce, and rape soaked and incubated in 2000 mg L^{-1} nanoparticle of CeO_2, La_2O_3, Gd_2O_3, and Yb_2O_3. Significant difference was indicated by "*asterisk*." Reprinted from Ma Y. et al, Effects of rare earth oxide nanoparticles on root elongation of plants. Chemosphere, 78, 273–279, Copyright (2010), with permission from Elsevier (Ma et al. 2010b)

Sometimes ionic metals will be absorbed by plants, and they will agglomerate as nanoparticle in plant tissue. Figure 12.10 shows an example of uptake of ionic gold and its accumulation as gold nanoparticles by *Arabidopsis* (Taylor et al. 2014). It was observed that ionic gold can enter the shoot tissue directly via passive uptake and accumulate as nanoparticles. Also root uptake of ionic gold results in gold nanoparticle formation in the root cell.

A study on the effect of ionic and nanoparticulate Zn on *Allium sativum* (garlic) showed that the root length Zn nanoparticles alone are responsible for the severe root length inhibition with increasing concentrations of Zn, as seen in Fig. 12.11a (Shaymurat et al. 2012). Also, the mitosis index values diminished in a concentration- and time-dependent manner. The treatment of garlic with ionic zinc alone had no effect on root length, as shown in Fig. 12.11b.

The comparative effect of ZnO nanoparticles and ionic zinc on the internalization and translocation in *Lolium perenne* (ryegrass) in a hydroponic culture was investigated (Lin and Xing 2008). In order to verify if the dissolution of ZnO nanoparticles has a contribution to the toxicity of ryegrass, the effect of Zn ions alone was also studied. ZnO nanoparticle exposure resulted in a significant reduction of ryegrass biomass, shrinking of the root tips, and high vacuolization of root epidermal and cortical cells. The reduction on root and shoot biomass is illustrated in Fig. 12.11c. Both treatments lead to an increased content of Zn in roots and shoots, however at different rates, as shown in Fig. 12.11d. The Zn content in shoots for plants exposed to ZnO nanoparticles was much lower than in the shoots of plants

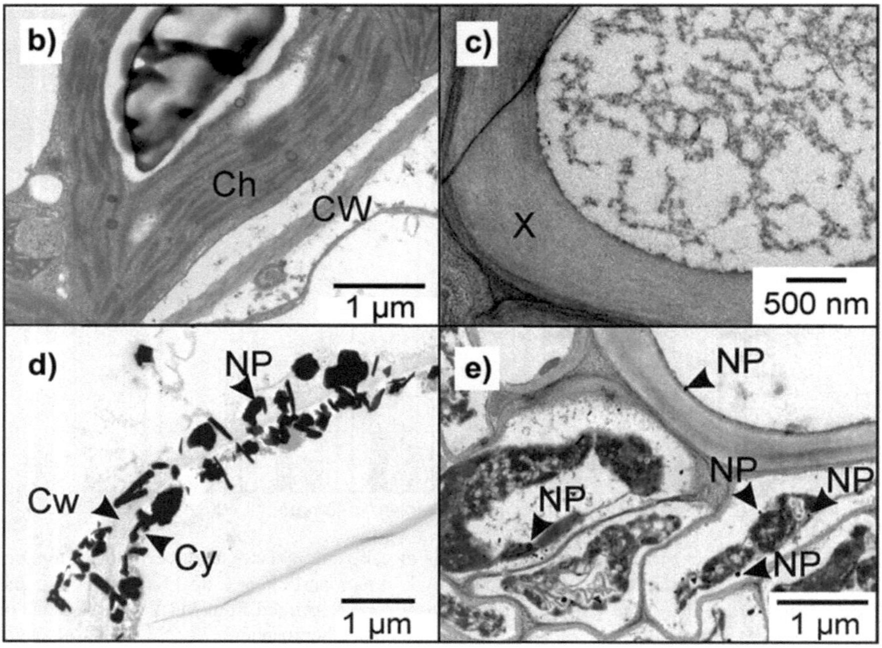

Fig. 12.10 Growth of *Arabidopsis thaliana* L. (*Arabidopsis*) seedlings in the presence of gold nanoparticles. (**a**) Photographs of seedlings at different gold concentrations. TEM images of plant tissue showing gold nanoparticles: (**b**) Leaf mesophyll, (**c**) leaf vascular tissue, (**d**) root cortex, (**e**) root vascular tissue. *NP* nanoparticle, *Cy* cytoplasm, *Cw* cell wall, *X* xylem, *Ch* chloroplast. Reprinted from Taylor et al, Investigating the Toxicity, Uptake, Nanoparticle Formation and Genetic Response of Plants to Gold. Plos One, 2014, 9. Copyright (2014) Taylor et al. (2014)

treated with ionic zinc. In contrast, the amount of Zn in roots treated with nanoparticles was much higher than in roots treated with ionic zinc. The amount of soluble Zn available in the ZnO nanoparticle-treated nutrient solutions was smaller than the toxic threshold of Zn ions to the ryegrass. Therefore, the authors concluded that the phytotoxicity of ZnO nanoparticles was not related to their dissolution in the nutrient solution (Lin and Xing 2008). A large amount of nanoparticles adhered onto roots exposed to ZnO nanoparticles, some being localized in apoplast and protoplast of the root endodermis and stele. Zinc translocation to shoots for plants exposed to nanoparticles was much lower than those exposed to ionic zinc. The toxicity on root tips of ionic zinc and ZnO nanoparticles compared to control plants is shown in Fig. 12.11e–g. Both ZnO nanoparticles and ionic Zn had a negative effect on the root tips, with a shrank vascular cylinder, showing a broken epidermis and root cap, with highly vacuolated cortical cells.

Many types of nanoparticles are absorbed and translocated by plants either in nanoparticle form or in their ionic form (Arruda et al. 2015). Plants can absorb a significant amount of metals, and during this process, they can synthesize nanoparticles. This aspect can be used in phytoremediation, a process when plants

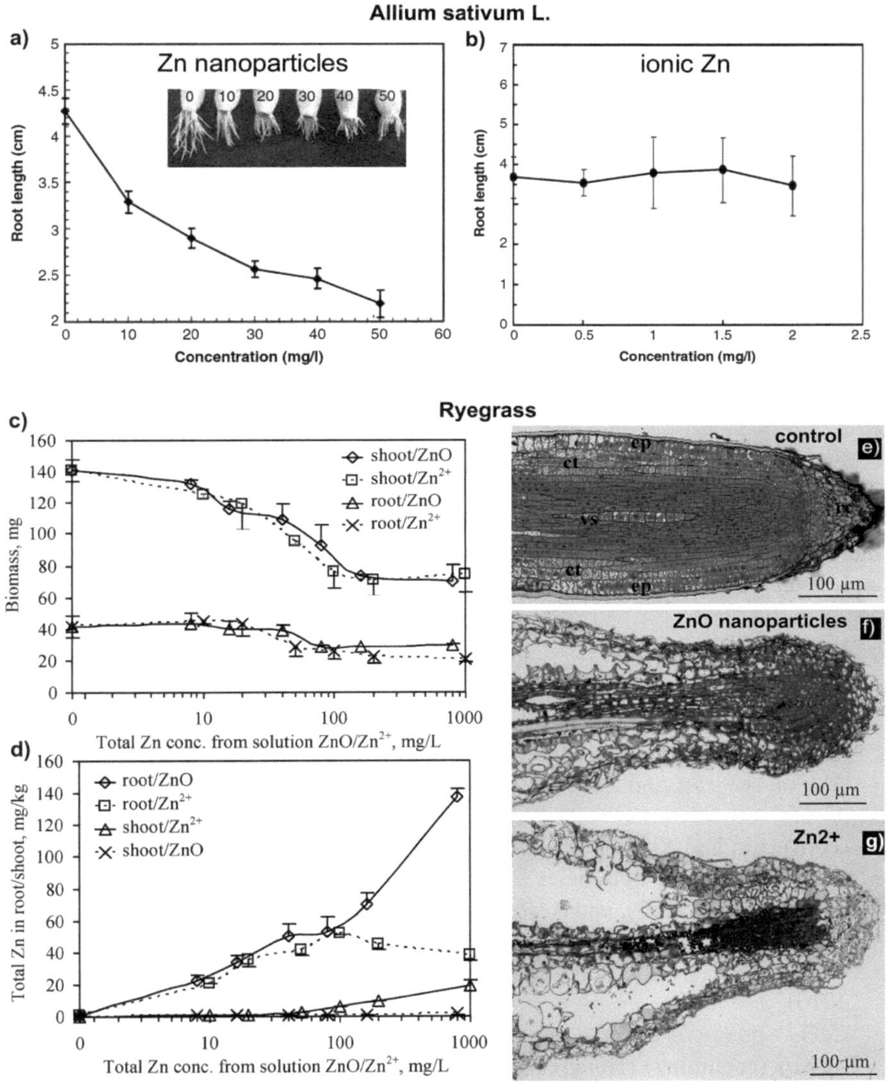

Fig. 12.11 The effects of exposure of plants to materials as nanoparticle and in ionic form. (**a**) ZnO nanoparticles and (**b**) Zn^{2+} with different concentrations on root length of ZnO nanoparticles on garlic (*Allium sativum* L.). Figures (**a**) and (**b**) reprinted from Shaymurat T. et al, Phytotoxic and genotoxic effects of ZnO nanoparticles on garlic (*Allium sativum* L.): A morphological study. Nanotoxicology, 6, 2012, 241–248, reprinted by permission of the publisher (Taylor and Francis Ltd, http://www.tandfonline.com) (Shaymurat et al. 2012). (**c**)–(**g**) The effects of exposure of ryegrass to ZnO nanoparticles and ionic zinc Zn^{2+}. (**c**) Biomass reduction and (**d**) the amount of Zn accumulated in root and shoot after treatment with ZnO nanoparticles or Zn^{2+}. (**e**), (**f**), (**g**) Light microscopic images of longitudinal sections of ryegrass root tips treated with ZnO nanoparticles: (e) control, (**f**) 1000 mg L^{-1} ZnO nanoparticles, (**g**) 1000 mg L^{-1} Zn^{2+}. Figures (**c**)–(**g**) reprinted with permission from Lin D. H. and Xing B. S., Root uptake and phytotoxicity of ZnO nanoparticles. Environmental Science and Technology, 42, 5580–5585. Copyright (2008) American Chemical Society (Lin and Xing 2008)

are used to decontaminate soils and aquatic environments. For example, plants that can be used in phytoremediation for Pt are *L. sativum* and *S. alba*, which tolerate relatively high concentrations of platinum nanoparticles (Asztemborska et al. 2015b). They uptake Pt nanoparticles and translocate it to aboveground organs, the majority remaining in the roots (about 90 %). Another example of plant suitable for the phytoremediation of Al_2O_3 is *Zea mays* (Asztemborska et al. 2015a). Heavy metals are accumulated in most plants mainly in the roots, with only a small percentage of translocation to aboveground parts (Arruda et al. 2015).

12.6.3 Phytotoxicity of Nanoparticles of Various Compositions

A review of nanoparticle interaction with edible plants indicates that plants show a heterogeneous behavior when exposed to nanoparticles, some exhibiting positive effects and other toxicity after nanoparticle exposure (Rico et al. 2011). However, there are many studies that show a negative impact of nanoparticles on plant growth and development. Examples of nanoparticles that show adverse effects on plants are:

- Ag—on *Triticum aestivum* (wheat) (Dimkpa et al. 2013b), *Arabidopsis thaliana* (Geisler-Lee et al. 2013), *Zea mays L.* (corn) and *Brassica oleracea* (cabbage) (Pokhrel and Dubey 2013), *Phaseolus radiatus* (mung bean), *Sorghum bicolor* (sorghum) (Lee et al. 2012), and *Oryza sativa* L. (rice) (Mirzajani et al. 2013)
- Al—on ryegrass and *Zea mays L.* (corn) (Lin and Xing 2007)
- Al_2O_3—on *Nicotiana xanthi* (tobacco) (Burklew et al. 2012) and *Zea mays L.* (corn) (Lin and Xing 2007)
- Au—on barley (Feichtmeier et al. 2015), *Nicotiana xanthi* (tobacco) (Sabo-Attwood et al. 2012), and *Arabidopsis thaliana* (Taylor et al. 2014)
- C_{60}—on *Allium cepa* (onion) (Chen et al. 2010)
- $C_{70}(C(COOH)_2)_{4-8}$ water-soluble fullerene—on *Arabidopsis thaliana* (Liu et al. 2010)
- CeO_2—on *Lactuca sativa* (lettuce) (Gui et al. 2015; Zhang et al. 2015), *Cucumis sativus* (cucumber) (Hong et al. 2014; Lopez-Moreno et al. 2010; Wang et al. 2013), *Medicago sativa* (alfalfa), *Zea mays* (corn), and *Lycopersicon esculentum* (tomato) (Lopez-Moreno et al. 2010; Wang et al. 2013)
- CdSe quantum dots (QD)—on *Oryza sativa* (rice) (Nair et al. 2011)
- CoO_x and ZnO—on *Allium cepa* (onion) (Ghodake et al. 2011)
- CuO—on *Triticum aestivum* (wheat) (Dimkpa et al. 2013a)
- La_2O_3, Gd_2O_3, and Yb_2O_3—on *Brassica napus* (rape), *Raphanus sativus* (radish), *Triticum aestivum* (wheat), *Lactuca sativa* (lettuce), *Brassica oleracea* (cabbage), *Lycopersicon esculentum* (tomato), and *Cucumis sativus* (cucumber) (Ma et al. 2010b)
- Mo—on rapeseed (Aubert et al. 2012)

- MWCNT—on *Allium cepa* (onion) (Ghosh et al. 2015)
- Nanoclay—on *Zea mays L.* (corn) (Asli and Neumann 2009)
- TiO_2—on *Allium cepa* (onion) and *Nicotiana tabacum* (tobacco) (Ghosh et al. 2010), *Zea mays* (corn) (Asli and Neumann 2009), *Triticum aestivum* (wheat) (Du et al. 2011), and *Arabidopsis thaliana* (Wang et al. 2011b)
- Zn—on ryegrass, *Raphanus sativus* (radish), rape, *Lactuca sativa* (lettuce), *Zea mays L.* (corn) (Pokhrel and Dubey 2013; Lin and Xing 2007), *Cucumis sativus* (cucumber) (Lin and Xing 2007), and *Brassica oleracea* (cabbage) (Pokhrel and Dubey 2013)
- ZnO—on ryegrass, *Raphanus sativus* (radish), rape, *Lactuca sativa* (lettuce), *Zea mays L.* (corn), *Cucumis sativus* (cucumber) (Lin and Xing 2007), *Triticum aestivum* (wheat) (Dimkpa et al. 2013a; Du et al. 2011), *Lepidium sativum* (cress) (Josko and Oleszczuk 2013), *Allium cepa* (onion) (Kumari et al. 2011), and *Allium sativum L.* (garlic) (Shaymurat et al. 2012)

Ag nanoparticles are used extensively in agriculture for their antibacterial and antifungal activity. Uncontrolled disposal of Ag nanoparticles in the environment can affect the balance of the symbiotic relationship involving fungi and nitrogen-fixing bacteria that may affect the physicochemical characteristics of soil (Anjum et al. 2013). Soil microbiota perturbation can have severe consequences on ecosystems and plant growth. Nanoparticles may be accumulated in their biomass and enter the food chains. Silver can enter the soil via agricultural application of organic waste and sewage sludge as a fertilizer, accidental spills, the use of silver nanoparticles as pesticide, waste incineration, and from consumer products that contain silver nanoparticles (Anjum et al. 2013). The exposure of some plants to silver nanoparticles resulted in reduced biomass and transpiration; reduced germination; a dose-dependent reduction in shoot and root length; change in gene expression, cytotoxicity, and reactive oxygen species (ROS)-induced cell death; decrease in mitosis; and increase in frequency of chromosomal abnormalities (Anjum et al. 2013; Arruda et al. 2015). For example, a study investigating the phytotoxicity of silver nanoparticles on mung bean (*Phaseolus radiates*) and sorghum (*Sorghum bicolor*) indicated a concentration-dependent growth inhibition effect (Lee et al. 2012). Another study on the effect of Ag nanoparticles on rice (*Oryza sativa* L.) indicated that nanoparticles penetrated the cell wall and damaged cells (Mirzajani et al. 2013). Treatment with Ag nanoparticle concentrations of less than 30 $\mu g\ mL^{-1}$ accelerated root growth, and at 60 $\mu g\ mL^{-1}$, it inhibited root growth (Mirzajani et al. 2013). The induction of root branching were ascribed to stress due to the presence of nanoparticles as a result of reactive oxygen species and local root tissue death.

Plants exposed to Cu nanoparticles show in general an inhibition of seed germination and seedling growth, oxidative damage, loss of root cell viability, and concentration-dependent bioaccumulation of copper, while the soil microbial community is negatively impacted (Anjum et al. 2015). Zinc oxide nanoparticles reduce alfalfa plant growth and dry biomass production (Bandyopadhyay et al.

2015). Barley can uptake palladium nanoparticles and suffer a significant effect on leaf length growth (Battke et al. 2008).

12.6.4 Nutrient Depletion in Nanoparticle-Contaminated Plants

Plants and fruits are an important source of mineral and nutrients (such as K and mineral salts) for humans. Exposure to nanoparticles was shown to influence the amount of these nutrients in plants and their fruits (Rico et al. 2013; Zhao et al. 2014; Antisari et al. 2015; Petersen et al. 2014; Deng et al. 2014). The results of the studies testing nutritional contents of plants exposed to nanoparticles suggest that nanoparticles may alter nutritional values, fruit flavor, antioxidant content, and growth performance. As a result, the application of nanoparticles as agrichemicals becomes a main area of concern.

Several studies found that CeO_2 nanoparticles affect the nutrient levels of crops, such as rice, cucumber, corn, soybean and tomato (Rico et al. 2013; Zhao et al. 2014, 2015; Peralta-Videa et al. 2014; Antisari et al. 2015). *Oryza sativa* (rice) grains from plants treated with CeO_2 nanoparticles showed compromised nutritional quality, having lower amounts of Fe, S, glutelin, lauric and valeric acids, starch, and antioxidants (except flavonoids) (Rico et al. 2013). CeO_2 nanoparticles were also found to change Mo micronutrient content in *Cucumis sativus* fruit, altered sugar content, phenolic content, and fractionation of proteins (Zhao et al. 2014). Ceria and zinc oxide nanoparticles alter the nutritional element composition in *Glycine max* (soybean) plants cultivated in soil amended with nanoparticles (Peralta-Videa et al. 2014). The exposure to ceria nanoparticles resulted in modified allocation of calcium in corn kernels as well as reduced Ca translocation from the cob to the kernels compared to control and a reduced yield by 38 % (Zhao et al. 2015).

Zhao et al. (2015) showed that zinc oxide nanoparticles alter the quality of nutrients in *Zea mays* (corn), which is a source of many nutrients, such as K, Ca, Mg, Zn, and Cu (Zhao et al. 2015). Zinc oxide nanoparticles were found to reduce net photosynthesis, relative chlorophyll content, and yield by 49 % and accumulated in corn cobs. The amount of Zn and Ca in fully developed cobs was significantly smaller than in undeveloped corn.

In addition to the direct effect of nanoparticles on plants, of concern is their antibacterial effect on soils (Dinesh et al. 2012) and negative impact upon endophytic bacteria symbionts (Deng et al. 2014). Legume nitrogen fixation is altered and severely diminished by legumes exposure to nanoparticles. This was the case for *Glycine max* (soybean) exposed to ceria nanoparticles whose diminished nitrogen fixation was correlated to nearly absent bacteroids in nodules (Priester et al. 2012). Also, *Pisum sativum* (peas) exhibited a disrupted *Rhizobium*–legume symbiosis with subsequent delayed nitrogen fixation (Fan et al. 2014).

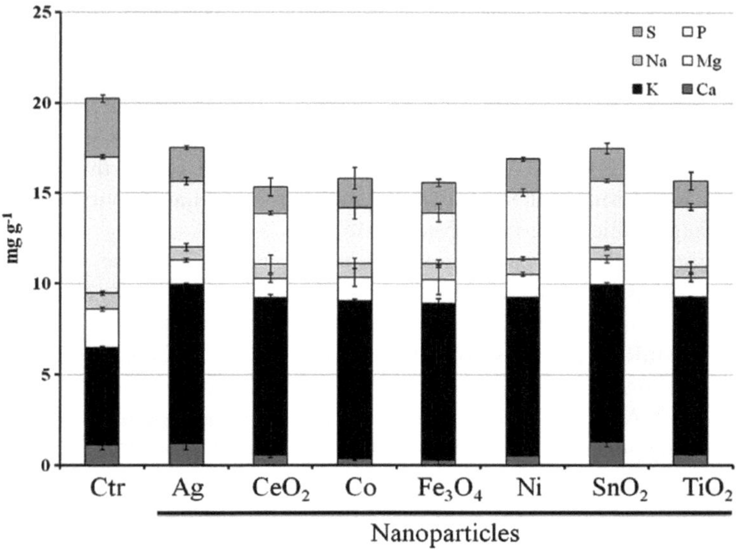

Fig. 12.12 Concentration of nutrients (mg/g) in tomato fruits of plants grown in soil containing metal oxide and metallic nanoparticle. Reproduced from Antisari L. V. et al, Uptake and translocation of metals and nutrients in tomato grown in soil polluted with metal oxide (CeO_2, Fe_3O_4, SnO_2, TiO_2) or metallic (Ag, Co, Ni) engineered nanoparticles. Environmental Science and Pollution Research, 2015, 22, 1841–1853. Copyright (2015) with permission of Springer (Antisari et al. 2015)

It was found that nanoparticles, depending on their type, can have different effects on the root growth, accumulation rate, and fruit yield of tomato plants exposed to metal oxides (CeO_2, Fe_3O_4, SnO_2, TiO_2) and metallic (Ag, Co, Ni) nanoparticles (Antisari et al. 2015). Iron oxide nanoparticles promoted the root growth, while tin oxide resulted in decreased root growth. Metal nanoparticles accumulated mostly in tomato roots, except Ag and Co which had higher concentrations in both aboveground and belowground organs. Higher silver content was also found in the tomato fruits. In addition, the fruits of the plants exposed to nanoparticles showed a modified composition of different elements, a depletion of Mg, P, and S with a general enrichment in K (Antisari et al. 2015). This is shown in Fig. 12.12.

12.6.5 Nanoparticle-Induced Genetic Alterations in Plants

A more accurate determination of phytotoxicity, better than measuring biological end points (such as germination index, root elongation, shoot and root biomass, and root tip morphology), is perhaps the genetic response of plants to nanoparticles.

However, there are very few studies assessing molecular changes in plants exposed to nanoparticles.

Several nanoparticles (CuO, Ag, ZnO, CeO_2, TiO_2, CNTs) were analyzed for genotoxicity in various plants (such as *Raphanus sativus*, *Lolium perenne*, *Lolium rigidum*, *Allium cepa*, *Glycine max*, *Nicotiana tabacum*, etc.) (Atha et al. 2012; Chichiricco and Poma 2015; Ghosh et al. 2015). The plants that showed root inhibition had errors in the cell division and chromosomal activity and oxidative DNA damage (Chichiricco and Poma 2015).

The root length of tobacco seedlings was found to decrease with increasing concentrations of alumina nanoparticles (Burklew et al. 2012). As a result to stress generated by alumina nanoparticles, there is an increase in the expression of several microRNA (miRNA) in tobacco plants (Burklew et al. 2012). miRNA is a class of posttranscriptional gene regulators that change gene expression by acting upon messenger RNA (mRNA) and mediate gene expression of more than 30 % of protein-coding genes (Burklew et al. 2012). miRNA are mediators of abiotic stress responses and regulate leaf and root development and cell proliferation among others.

While a type of nanoparticle is associated with beneficial effects in one plant, it may also generate genotoxic effects in other plants. For example, carbon nanotubes play a factor in the activation of stress-related genes in roots and leaves of tomato seedlings, resulting in increased germination and growth (Khodakovskaya et al. 2011). Tobacco cells cultured with MWCNTs show upregulation of genes controlling the cell division, cell wall formation, and water flow through the membrane (Khodakovskaya et al. 2012). This suggests that the observed enhanced growth of cells was due to water uptake and cell division. As opposed to the beneficial effects from above, CNTs are associated with DNA damage and apoptosis in roots of *Allium cepa* (Ghosh et al. 2015).

12.7 Nanoparticles as Promoters of Plant Growth

Some nanoparticles have a beneficial effect on some plant species manifested by enhancing seed germination, enhancing crop yield, or suppressing plant disease (Servin et al. 2015; Zheng et al. 2005; Arruda et al. 2015). Some examples are shown in Fig. 12.13 (Servin et al. 2015). However, we cannot predict what effect may arise from their progressive accumulation in soil.

Beneficial effects were observed specially for seeds with low germination when treated with TiO_2 and CNTs (Zheng et al. 2005; Miralles et al. 2012b; Feizi et al. 2013). It is believed that TiO_2 nanoparticles induced reactive oxygen species with subsequent enhancement of stress resistance and facilitation of capsule penetration for water and oxygen (Khot et al. 2012). This in turn, enabled faster germination of the seeds. In the case of CNTs, the impurities that served as catalysts for the growth of nanotubes (Fe or Al_2O_3) might be responsible for enhanced germination rather than CNT due hormesis (Deng et al. 2014). Hormesis is a term that refers to a

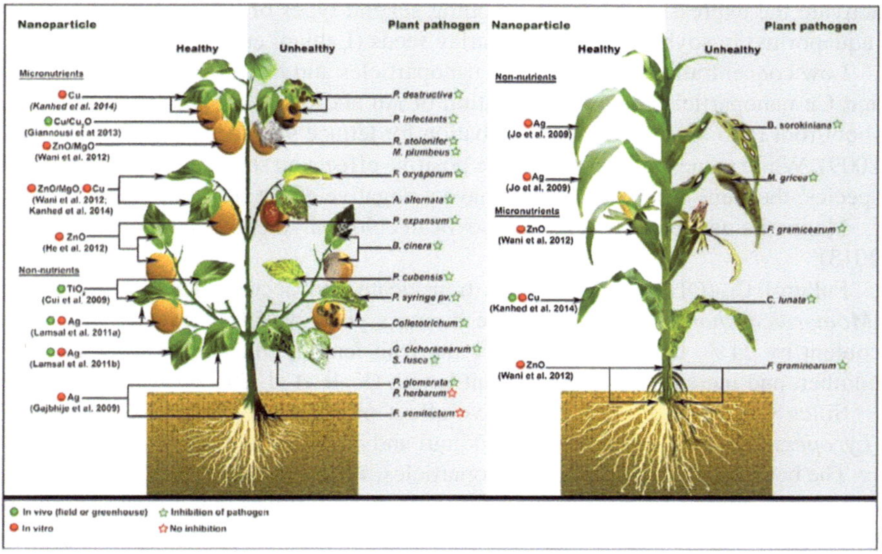

Fig. 12.13 Effect of nanoparticle nutrients and non-nutrients on crop disease. Reproduced from Journal of Nanoparticle Research, A review of the use of engineered nanomaterials to suppress plant disease and enhance crop yield, Servin A. et al, 17, 2015. Copyright (2015) with permission of Springer (Servin et al. 2015)

beneficial effect for a low dose of an agent which at a higher dose has a toxic or inhibitory effect.

TiO_2 nanoparticles increase seed germination of fennel (Feizi et al. 2013). Rutile TiO_2 nanoparticles enhance the growth of spinach when administered to seeds or when sprayed onto leaves (Zheng et al. 2005). They promoted nitrate absorption and the activity of several enzymes. Seeds treated with titania nanoparticles had 73 % more dryweight and increased photosynthesis and chlorophyll formation.

It was suggested that root elongation is a developmental adaptation of plants whose roots become clogged by nanoparticles (Asli and Neumann 2009). This study showed that clay and TiO_2 nanoparticles present in the external water supply of hydroponic maize accumulate at the cell wall surfaces of the primary root and block pores, lower water transport capacity, leaf growth, and transpiration. As a response to the effect of root clogging, the plants develop a large root system as a plant adaptation to nanoparticles.

Carbon nanotubes were found to have a promoting effect on seedling root elongation and seed germination of some plants, such as onion, cucumber, wheat, mustard, and tomatoes (Chichiricco and Poma 2015). MWCNTs were found to increase seed germination rate and plant biomass in tomatoes (Khodakovskaya et al. 2009). Increased germination rate for tomato seeds exposed to MWCNTs is assumed to be due to the fact that carbon nanotubes penetrate the seeds (Khodakovskaya et al. 2009). Subsequent research found that MWCNTs may activate the expression of genes encoding several types of water channel proteins (aquaporins) in soybean, corn, and barley seeds (Lahiani et al. 2013).

Low concentrations of Pd and Au nanoparticles and higher concentrations of Si and Cu nanoparticles and a combination of Au and Cu nanoparticles increase the shoot/root ratio after 15 days of incubation for lettuce seeds (Shah and Belozerova 2009). While some nanoparticles have positive effects on seed germination of some species, the same nanoparticles can have a negative effect on other plant species.

Magnetite nanoparticles enhance soybean chlorophyll content (Ghafariyan et al. 2013).

Fullerol $C_{60}(OH)_{20}$ nanoparticles translocate and accumulate in bitter melon (*Momordica charantia*), improve the biomass yield by 54 %, increase the water content by 24 %, increase up to 20 % the fruit length, enhance by 59 % the fruit number, and increase by 70 % the fruit length (Kole et al. 2013).

Silica was found to have a positive effect on the seed germination of tomato (*Lycopersicum esculentum* Mill.) (Siddiqui and Al-Whaibi 2014).

The beneficial effects of some nanoparticles, such as Ag, ZnO, Mg, Si, and TiO_2, manifest by inhibiting plant pathogens due to their antimicrobial properties (Servin et al. 2015).

Some studies showed a positive impact of some nanoparticles on crop growth and/or pathogen inhibition (Servin et al. 2015). Some nanoparticles have shown to inhibit fungal pathogens, among them being Cu (Kanhed et al. 2014; Giannousi et al. 2013), ZnO (Wani and Shah 2012; He et al. 2011), MgO (Wani and Shah 2012), TiO_2 (Cui et al. 2009), and Ag (Lamsal et al. 2011a, b; Gajbhiye et al. 2009; Jo et al. 2009).

Rare earth additives have been used extensively in fertilizers due to increase yield, darker green foliage, higher production of roots and rate of development, and better fruit color (Yuan et al. 2001).

One must emphasize that some authors are promoting the use of specific nanoparticles in agriculture despite the fact that the same nanoparticles are extremely toxic to humans and animals. This is a shortcoming of the regulatory bodies that did not pass yet laws and regulations regarding the safe handling and limit the use of these nanoparticles in plants for consumption. For example, it has been suggested that carbon nanotubes, which are extremely toxic to humans, can be used to increase fruit and crop production in edible plants and vegetables (Husen and Siddiqi 2014). It is likely that many authors working in the agricultural field are not aware of the current knowledge on nanoparticle toxicity to humans and animals and are conducting experiments expected to improve plant yield and promote their growth.

There is the need for environmental monitoring of anthropogenic nanoparticles as well as for regulations and laws regarding the production, use, and safe handling of nanomaterials (Paterson et al. 2011). Currently there is no specific legislation dedicated to regulate nanomaterial and nanoparticle use anywhere in the world. European Union and Switzerland seem to be the only regions where provisions regarding nanomaterials have been incorporated in existing legislation (Amenta et al. 2015).

12.8 Conclusions

There is a large amount of newly discovered nanoparticles from different natural but mostly newly generated anthropogenic sources that are becoming available for interaction with plants. The field of science investigating the interaction between plants and nanoparticles, more precisely plant growth, development, and gene expression, is still in its infancy. Plants have been exposed to nanoparticle since their appearance on Earth, and, as a result, some plants developed adaptive mechanisms to some of the natural occurring nanoparticles. However, the large number of engineered nanoparticles intentionally and unintentionally released into the environment will most likely pose a novel threat on plant physiology.

Up-to-date studies have found that depending on the type of nanoparticles, plants, and soil, nanoparticles can have negative, insignificant, or positive effects on plants. Starting with plant morphology, nanoparticles were found to alter morphological features of plants in vital organs such as roots and leaves. Also they can influence seed germination. A few studies report on genetic alterations due to plant–nanoparticle interactions. Nanoparticle bioaccumulation in plants is species specific and depends on the nanoparticle physicochemical properties. While some studies report beneficial effects on some plant species, the overall negative effect of the accumulation of these nanoparticles in the soil and plants might exceed the minor beneficial temporary effects. The main negative effects uncovered up to date involve growth inhibition, oxidative stress, and genetic alteration, among others. Many nanoparticles are translocated within plants and are likely to enter the food chain, be available for trophic transfer, and become available in food for humans and animals. Many nanoparticles are already shown to be toxic to humans, and uptake of nanoparticles in plants poses major safety concerns. If these safety concerns are not properly addressed now, nanomaterials can become an environmental pollutant that might be conducive to irreversible or undesirable modifications with potentially harmful consequences on plants, animals, and humans alike.

References

Amenta V, Aschberger K, Arena M, Bouwmeester H, Moniz FB, Brandhoff P, Gottardo S, Marvin HJP, Mech A, Pesudo LQ, Rauscher H, Schoonjans R, Vettori MV, Weigel S, Peters RJ (2015) Regulatory aspects of nanotechnology in the agri/feed/food sector in Eu and non-Eu countries. Regul Toxicol Pharmacol 73:463–476

Anjum NA, Gill SS, Duarte AC, Pereira E, Ahmad I (2013) Silver nanoparticles in soil-plant systems. J Nanopart Res 15:1–26

Anjum NA, Adam V, Kizek R, Duarte AC, Pereira E, Iqbal M, Lukatkin AS, Ahmad I (2015) Nanoscale copper in the soil-plant system – toxicity and underlying potential mechanisms. Environ Res 138:306–325

Antisari LV, Carbone S, Gatti A, Vianello G, Nannipieri P (2015) Uptake and translocation of metals and nutrients in tomato grown in soil polluted with metal oxide (CeO_2, Fe_3O_4, SnO_2, TiO_2) or metallic (Ag, Co, Ni) engineered nanoparticles. Environ Sci Pollut Res 22:1841–1853

Arruda SCC, Silva ALD, Galazzi RM, Azevedo RA, Arruda MAZ (2015) Nanoparticles applied to plant science: a review. Talanta 131:693–705

Aslani F, Bagheri S, Julkapli NM, Juraimi AS, Hashemi FSG, Baghdadi A (2014) Effects of engineered nanomaterials on plants growth: an overview. Sci World J 64:165–177

Asli S, Neumann PM (2009) Colloidal suspensions of clay or titanium dioxide nanoparticles can inhibit leaf growth and transpiration via physical effects on root water transport. Plant Cell Environ 32:577–584

Asztemborska M, Steborowski R, Kowalska J, Bystrzejewska-Piotrowska G (2015a) Accumulation of aluminium by plants exposed to nano- and microsized particles of Al_2O_3. Int J Environ Res 9:109–116

Asztemborska M, Steborowski R, Kowalska J, Bystrzejewska-Piotrowska G (2015b) Accumulation of platinum nanoparticles by *Sinapis alba* and *Lepidium sativum* plants. Water Air Soil Pollut 226:1–7

Atha DH, Wang HH, Petersen EJ, Cleveland D, Holbrook RD, Jaruga P, Dizdaroglu M, Xing BS, Nelson BC (2012) Copper oxide nanoparticle mediated DNA damage in terrestrial plant models. Environ Sci Technol 46:1819–1827

Aubert T, Burel A, Esnault MA, Cordier S, Grasset F, Cabello-Hurtado F (2012) Root uptake and phytotoxicity of nanosized molybdenum octahedral clusters. J Hazard Mater 219:111–118

Aziz N, Faraz M, Pandey R, Sakir M, Fatma T, Varma A, Barman I, Prasad R (2015) Facile algae-derived route to biogenic silver nanoparticles: synthesis, antibacterial and photocatalytic properties. Langmuir 31:11605–11612

Bakshi S, He ZLL, Harris WG (2015) Natural nanoparticles: implications for environment and human health. Crit Rev Environ Sci Technol 45:861–904

Bandyopadhyay S, Plascencia-Villa G, Mukherjee A, Rico CM, Jose-Yacaman M, Peralta-Videa JR, Gardea-Torresdey JL (2015) Comparative phytotoxicity of ZnO NPs, Bulk ZnO, and ionic zinc onto the alfalfa plants symbiotically associated with Sinorhizobium meliloti in soil. Sci Total Environ 515:60–69

Batley GE, Kirby JK, Mclaughlin MJ (2013) Fate and risks of nanomaterials in aquatic and terrestrial environments. Acc Chem Res 46:854–862

Battke F, Leopold K, Maier M, Schmidhalter U, Schuster M (2008) Palladium exposure of barley: uptake and effects. Plant Biol 10:272–276

Begum P, Fugetsu B (2012) Phytotoxicity of multi-walled carbon nanotubes on red spinach (Amaranthus Tricolor L) and the role of ascorbic acid as an antioxidant. J Hazard Mater 243:212–222

Bernhardt ES, Colman BP, Hochella MF, Cardinale BJ, Nisbet RM, Richardson CJ, Yin LY (2010) An ecological perspective on nanomaterial impacts in the environment. J Environ Qual 39:1954–1965

Burklew CE, Ashlock J, Winfrey WB, Zhang BH (2012) Effects of aluminum oxide nanoparticles on the growth, development, and microrna expression of tobacco (*Nicotiana tabacum*). PLoS One 7:e34783

Chen R, Ratnikova TA, Stone MB, Lin S, Lard M, Huang G, Hudson JS, Ke PC (2010) Differential uptake of carbon nanoparticles by plant and mammalian cells. Small 6:612–617

Chichiricco G, Poma A (2015) Penetration and toxicity of nanomaterials in higher plants. Nanomaterials 5:851–873

Cifuentes Z, Custardoy L, De La Fuente JM, Marquina C, Ibarra MR, Rubiales D, Perez-De-Luque A (2010) Absorption and translocation to the aerial part of magnetic carbon-coated nanoparticles through the root of different crop plants. J Nanobiotechnol 8:26

Cui HX, Zhang P, Gu W, Jiang JF (2009) Application Of Anatase TiO(2) Sol derived from peroxotitannic acid in crop diseases control and growth regulation. Boca Raton, Crc Press-Taylor & Francis Group 2:286–289

Dan YB, Zhang WL, Xue RM, Ma XM, Stephan C, Shi HL (2015) Characterization of gold nanoparticle uptake by tomato plants using enzymatic extraction followed by single-particle inductively coupled plasma-mass spectrometry analysis. Environ Sci Technol 49:3007–3014

Deng YQ, White JC, Xing BS (2014) Interactions between engineered nanomaterials and agricultural crops: implications for food safety. J Zhejiang Univ Sci A 15:552–572
Dietz KJ, Herth S (2011) Plant nanotoxicology. Trends Plant Sci 16:582–589
Dimkpa CO, Latta DE, Mclean JE, Britt DW, Boyanov MI, Anderson AJ (2013a) Fate of CuO and ZnO nano- and microparticles in the plant environment. Environ Sci Technol 47:4734–4742
Dimkpa CO, Mclean JE, Martineau N, Britt DW, Haverkamp R, Anderson AJ (2013b) Silver nanoparticles disrupt wheat (*Triticum aestivum* L.) growth in a sand matrix. Environ Sci Technol 47:1082–1090
Dinesh R, Anandaraj M, Srinivasan V, Hamza S (2012) Engineered nanoparticles in the soil and their potential implications to microbial activity. Geoderma 173:19–27
Du WC, Sun YY, Ji R, Zhu JG, Wu JC, Guo HY (2011) TiO_2 and ZnO nanoparticles negatively affect wheat growth and soil enzyme activities in agricultural soil. J Environ Monit 13:822–828
Fan RM, Huang YC, Grusak MA, Huang CP, Sherrier DJ (2014) Effects of Nano-TiO_2 on the agronomically-relevant rhizobium-legume symbiosis. Sci Total Environ 466:503–512
Feichtmeier NS, Walther P, Leopold K (2015) Uptake, effects, and regeneration of barley plants exposed to gold nanoparticles. Environ Sci Pollut Res 22:8549–8558
Feizi H, Kamali M, Jafari L, Moghaddam PR (2013) Phytotoxicity and stimulatory impacts of nanosized and bulk titanium dioxide on fennel (*Foeniculum vulgare* Mill). Chemosphere 91:506–511
Gajbhiye M, Kesharwani J, Ingle A, Gade A, Rai M (2009) Fungus-mediated synthesis of silver nanoparticles and their activity against pathogenic fungi in combination with fluconazole. Nanomed Nanotechnol Biol Med 5:382–386
Geisler-Lee J, Wang Q, Yao Y, Zhang W, Geisler M, Li KG, Huang Y, Chen YS, Kolmakov A, Ma XM (2013) Phytotoxicity, accumulation and transport of silver nanoparticles by *Arabidopsis thaliana*. Nanotoxicology 7:323–337
Ghafariyan MH, Malakouti MJ, Dadpour MR, Stroeve P, Mahmoudi M (2013) Effects of magnetite nanoparticles on soybean chlorophyll. Environ Sci Technol 47:10645–10652
Ghodake G, Seo YD, Lee DS (2011) Hazardous phytotoxic nature of cobalt and zinc oxide nanoparticles assessed using *Allium cepa*. J Hazard Mater 186:952–955
Ghosh M, Bandyopadhyay M, Mukherjee A (2010) Genotoxicity of titanium dioxide (Tio2) nanoparticles at two trophic levels plant and human lymphocytes. Chemosphere 81:1253–1262
Ghosh M, Bhadra S, Adegoke A, Bandyopadhyay M, Mukherjee A (2015) Mwcnt uptake in *Allium cepa* root cells induces cytotoxic and genotoxic responses and results in DNA hypermethylation. Mutat Res 774:49–58
Giannousi K, Avramidis I, Dendrinou-Samara C (2013) Synthesis, characterization and evaluation of copper based nanoparticles as agrochemicals against phytophthora infestans. RSC Adv 3:21743–21752
Gui X, Zhang ZY, Liu ST, Ma YH, Zhang P, He X, Li YY, Zhang J, Li HF, Rui YK, Liu LM, Cao WD (2015) Fate and phytotoxicity of CeO_2 nanoparticles on lettuce cultured in the potting soil environment. PLoS One 10:e0134261
He LL, Liu Y, Mustapha A, Lin MS (2011) Antifungal activity of zinc oxide nanoparticles against *Botrytis cinerea* and *Penicillium expansum*. Microbiol Res 166:207–215
Hong J, Peralta-Videa JR, Rico C, Sahi S, Viveros MN, Bartonjo J, Zhao L, Gardea-Torresdey JL (2014) Evidence of translocation and physiological impacts of foliar applied CeO_2 nanoparticles on cucumber (*Cucumis sativus*) plants. Environ Sci Technol 48:4376–4385
Hossain Z, Mustafa G, Komatsu S (2015) Plant responses to nanoparticle stress. Int J Mol Sci 16:26644–26653
Husen A, Siddiqi KS (2014) Carbon and fullerene nanomaterials in plant system. J Nanobiotechnol 12:16
Jo YK, Kim BH, Jung G (2009) Antifungal activity of silver ions and nanoparticles on phytopathogenic fungi. Plant Dis 93:1037–1043
Josko I, Oleszczuk P (2013) Influence of soil type and environmental conditions on ZnO, TiO_2 And Ni nanoparticles phytotoxicity. Chemosphere 92:91–99

Judy JD, Unrine JM, Bertsch PM (2011) Evidence for biomagnification of gold nanoparticles within a terrestrial food chain. Environ Sci Technol 45:776–781

Judy JD, Unrine JM, Rao W, Wirick S, Bertsch PM (2012) Bioavailability of gold nanomaterials to plants: importance of particle size and surface coating. Environ Sci Technol 46:8467–8474

Kanhed P, Birla S, Gaikwad S, Gade A, Seabra AB, Rubilar O, Duran N, Rai M (2014) In vitro antifungal efficacy of copper nanoparticles against selected crop pathogenic fungi. Mater Lett 115:13–17

Khodakovskaya M, Dervishi E, Mahmood M, Xu Y, Li ZR, Watanabe F, Biris AS (2009) Carbon nanotubes are able to penetrate plant seed coat and dramatically affect seed germination and plant growth. ACS Nano 3:3221–3227

Khodakovskaya MV, De Silva K, Nedosekin DA, Dervishi E, Biris AS, Shashkov EV, Galanzha EI, Zharov VP (2011) Complex genetic, photothermal, and photoacoustic analysis of nanoparticle-plant interactions. Proc Natl Acad Sci USA 108:1028–1033

Khodakovskaya MV, De Silva K, Biris AS, Dervishi E, Villagarcia H (2012) Carbon nanotubes induce growth enhancement of tobacco cells. ACS Nano 6:2128–2135

Khodakovskaya MV, Kim BS, Kim JN, Alimohammadi M, Dervishi E, Mustafa T, Cernigla CE (2013) Carbon nanotubes as plant growth regulators: effects on tomato growth, reproductive system, and soil microbial community. Small 9:115–123

Khot LR, Sankaran S, Maja JM, Ehsani R, Schuster EW (2012) Applications of nanomaterials in agricultural production and crop protection: a review. Crop Prot 35:64–70

Koelmel J, Leland T, Wang HH, Amarasiriwardena D, Xing BS (2013) Investigation of gold nanoparticles uptake and their tissue level distribution in rice plants by laser ablation-inductively coupled-mass spectrometry. Environ Pollut 174:222–228

Kole C, Kole P, Randunu KM, Choudhary P, Podila R, Ke PC, Rao AM, Marcus RK (2013) Nanobiotechnology can boost crop production and quality: first evidence from increased plant biomass, fruit yield and phytomedicine content in bitter melon (*Momordica charantia*). BMC Biotechnol 13:37

Kumari M, Khan SS, Pakrashi S, Mukherjee A, Chandrasekaran N (2011) Cytogenetic and genotoxic effects of zinc oxide nanoparticles on root cells of *Allium cepa*. J Hazard Mater 190:613–621

Kurepa J, Paunesku T, Vogt S, Arora H, Rabatic BM, Lu JJ, Wanzer MB, Woloschak GE, Smalle JA (2010) Uptake and distribution of ultrasmall anatase TiO_2 Alizarin red S nanoconjugates in *Arabidopsis thaliana*. Nano Lett 10:2296–2302

Lahiani MH, Dervishi E, Chen JH, Nima Z, Gaume A, Biris AS, Khodakovskaya MV (2013) Impact of carbon nanotube exposure to seeds of valuable crops. ACS Appl Mater Interfaces 5:7965–7973

Lamsal K, Kim S-W, Jung JH, Kim YS, Kim KS, Lee YS (2011a) Inhibition effects of silver nanoparticles against powdery mildews on cucumber and pumpkin. Mycobiology 39:26–32

Lamsal K, Kim SW, Jung JH, Kim YS, Kim KS, Lee YS (2011b) Application of silver nanoparticles for the control of Colletotrichum species in vitro and pepper anthracnose disease in field. Mycobiology 39:194–199

Landa P, Vankova R, Andrlova J, Hodek J, Marsik P, Storchova H, White JC, Vanek T (2012) Nanoparticle-specific changes in *Arabidopsis thaliana* gene expression after exposure to ZnO, TiO_2, and fullerene soot. J Hazard Mater 241:55–62

Larue C, Laurette J, Herlin-Boime N, Khodja H, Fayard B, Flank AM, Brisset F, Carriere M (2012a) Accumulation, translocation and impact of TiO_2 nanoparticles in wheat (*Triticum aestivum* Spp.): influence of diameter and crystal phase. Sci Total Environ 431:197–208

Larue C, Pinault M, Czarny B, Georgin D, Jaillard D, Bendiab N, Mayne-L'hermite M, Taran F, Dive V, Carriere M (2012b) Quantitative evaluation of multi-walled carbon nanotube uptake in wheat and rapeseed. J Hazard Mater 227:155–163

Larue C, Veronesi G, Flank AM, Surble S, Herlin-Boime N, Carriere M (2012c) Comparative uptake and impact of TiO_2 nanoparticles in wheat and rapeseed. J Toxicol Environ Health Part A Curr Iss 75:722–734

Larue C, Castillo-Michel H, Sobanska S, Cecillon L, Bureau S, Barthes V, Ouerdane L, Carriere M, Sarret G (2014a) Foliar exposure of the crop *Lactuca sativa* to silver nanoparticles: evidence for internalization and changes in Ag speciation. J Hazard Mater 264:98–106
Larue C, Castillo-Michel H, Sobanska S, Trcera N, Sorieul S, Cecillon L, Ouerdane L, Legros S, Sarret G (2014b) Fate of pristine TiO_2 nanoparticles and aged paint-containing TiO_2 nanoparticles in lettuce crop after foliar exposure. J Hazard Mater 273:17–26
Lee WM, Kwak JI, An YJ (2012) Effect of silver nanoparticles in crop plants *Phaseolus radiatus* and *Sorghum bicolor*: media effect on phytotoxicity. Chemosphere 86:491–499
Lin DH, Xing BS (2007) Phytotoxicity of nanoparticles: inhibition of seed germination and root growth. Environ Pollut 150:243–250
Lin DH, Xing BS (2008) Root uptake and phytotoxicity of ZnO nanoparticles. Environ Sci Technol 42:5580–5585
Lin SJ, Reppert J, Hu Q, Hudson JS, Reid ML, Ratnikova TA, Rao AM, Luo H, Ke PC (2009) Uptake, translocation, and transmission of carbon nanomaterials in rice plants. Small 5:1128–1132
Liu QL, Zhao YY, Wan YL, Zheng JP, Zhang XJ, Wang CR, Fang XH, Lin JX (2010) Study of the inhibitory effect of water-soluble fullerenes on plant growth at the cellular level. ACS Nano 4:5743–5748
Lopez-Moreno ML, De La Rosa G, Hernandez-Viezcas JA, Peralta-Videa JR, Gardea-Torresdey JL (2010) X-ray absorption spectroscopy (Xas) corroboration of the uptake and storage of CeO_2 nanoparticles and assessment of their differential toxicity in four edible plant species. J Agric Food Chem 58:3689–3693
Ma XM, Geiser-Lee J, Deng Y, Kolmakov A (2010a) Interactions between engineered nanoparticles (ENPs) and plants: phytotoxicity, uptake and accumulation. Sci Total Environ 408:3053–3061
Ma YH, Kuang LL, He X, Bai W, Ding YY, Zhang ZY, Zhao YL, Chai ZF (2010b) Effects of rare earth oxide nanoparticles on root elongation of plants. Chemosphere 78:273–279
Miralles P, Church TL, Harris AT (2012a) Toxicity, uptake, and translocation of engineered nanomaterials in vascular plants. Environ Sci Technol 46:9224–9239
Miralles P, Johnson E, Church TL, Harris AT (2012b) Multiwalled carbon nanotubes in alfalfa and wheat: toxicology and uptake. J R Soc Interface 9:3514–3527
Mirzajani F, Askari H, Hamzelou S, Farzaneh M, Ghassempour A (2013) Effect of silver nanoparticles on *Oryza sativa* L. and its rhizosphere bacteria. Ecotoxicol Environ Saf 88:48–54
Nair R, Poulose AC, Nagaoka Y, Yoshida Y, Maekawa T, Kumar DS (2011) Uptake of FITC labeled silica nanoparticles and quantum dots by rice seedlings: effects on seed germination and their potential as biolabels for plants. J Fluoresc 21:2057–2068
Navarro E, Baun A, Behra R, Hartmann NB, Filser J, Miao AJ, Quigg A, Santschi PH, Sigg L (2008) Environmental behavior and ecotoxicity of engineered nanoparticles to algae, plants, and fungi. Ecotoxicology 17:372–386
Nel AE, Madler L, Velegol D, Xia T, Hoek EMV, Somasundaran P, Klaessig F, Castranova V, Thompson M (2009) Understanding biophysicochemical interactions at the nano-bio interface. Nat Mater 8:543–557
Paterson G, Macken A, Thomas KV (2011) The need for standardized methods and environmental monitoring programs for anthropogenic nanoparticles. Anal Methods 3:1461–1467
Peralta-Videa JR, Hernandez-Viezcas JA, Zhao LJ, Diaz BC, Ge Y, Priester JH, Holden PA, Gardea-Torresdey JL (2014) Cerium dioxide and zinc oxide nanoparticles alter the nutritional value of soil cultivated soybean plants. Plant Physiol Biochem 80:128–135
Petersen EJ, Henry TB, Zhao J, Maccuspie RI, Kirschling TL, Dobrovolskaia MA, Hackley V, Xing B, White JC (2014) Identification and avoidance of potential artifacts and misinterpretations in nanomaterial ecotoxicity measurements. Environ Sci Technol 48:4226–4246
Pokhrel LR, Dubey B (2013) Evaluation of developmental responses of two crop plants exposed to silver and zinc oxide nanoparticles. Sci Total Environ 452:321–332

Prasad R (2014) Synthesis of silver nanoparticles in photosynthetic plants. J Nanopart, 963961. doi:10.1155/2014/963961
Prasad R, Pandey R, Barman I (2016) Engineering tailored nanoparticles with microbes: quo vadis. Wires Nanomed Nanobiotechnol 8:316–330
Priester JH, Ge Y, Mielke RE, Horst AM, Moritz SC, Espinosa K, Gelb J, Walker SL, Nisbet RM, An YJ, Schimel JP, Palmer RG, Hernandez-Viezcas JA, Zhao LJ, Gardea-Torresdey JL, Holden PA (2012) Soybean susceptibility to manufactured nanomaterials with evidence for food quality and soil fertility interruption. Proc Natl Acad Sci USA 109:E2451–E2456
Rico CM, Majumdar S, Duarte-Gardea M, Peralta-Videa JR, Gardea-Torresdey JL (2011) Interaction of nanoparticles with edible plants and their possible implications in the food chain. J Agric Food Chem 59:3485–3498
Rico CM, Morales MI, Barrios AC, Mccreary R, Hong J, Lee WY, Nunez J, Perata-Videa JR, Gardea-Torresdey JL (2013) Effect of cerium oxide nanoparticles on the quality of rice (*Oryza sativa* L.) grains. J Agric Food Chem 61:11278–11285
Sabo-Attwood T, Unrine JM, Stone JW, Murphy CJ, Ghoshroy S, Blom D, Bertsch PM, Newman LA (2012) Uptake, distribution and toxicity of gold nanoparticles in tobacco (*Nicotiana xanthi*) seedlings. Nanotoxicology 6:353–360
Schlich K, Hund-Rinke K (2015) Influence of soil properties on the effect of silver nanomaterials on microbial activity in five soils. Environ Pollut 196:321–330
Schwab F, Zhai G, Kern M, Turner A, Schnoor JL, Wiesner MR (2015) Barriers, pathways and processes for uptake, translocation and accumulation of nanomaterials in plants – critical review. Nanotoxicology 10:257–278
Servin AD, Castillo-Michel H, Hernandez-Viezcas JA, Diaz BC, Peralta-Videa JR, Gardea-Torresdey JL (2012) Synchrotron micro-Xrf and micro-xanes confirmation of the uptake and translocation of TiO_2 nanoparticles in cucumber (*Cucumis sativus*) plants. Environ Sci Technol 46:7637–7643
Servin A, Elmer W, Mukherjee A, De La Torre-Roche R, Hamdi H, White JC, Bindraban P, Dimkpa C (2015) A review of the use of engineered nanomaterials to suppress plant disease and enhance crop yield. J Nanopart Res 17:1–21
Shah V, Belozerova I (2009) Influence of metal nanoparticles on the soil microbial community and germination of lettuce seeds. Water Air Soil Pollut 197:143–148
Shaymurat T, Gu JX, Xu CS, Yang ZK, Zhao Q, Liu YX, Liu YC (2012) Phytotoxic and genotoxic effects of ZnO nanoparticles on garlic (*Allium sativum* L.): a morphological study. Nanotoxicology 6:241–248
Siddiqui MH, Al-Whaibi MH (2014) Role of Nano-SiO_2 in germination of tomato (*Lycopersicum esculentum* seeds Mill.). Saudi J Biol Sci 21:13–17
Taylor AF, Rylott EL, Anderson CWN, Bruce NC (2014) Investigating the toxicity, uptake, nanoparticle formation and genetic response of plants to gold. PLoS One 9:e93793
Wang HH, Kou XM, Pei ZG, Xiao JQ, Shan XQ, Xing BS (2011a) Physiological effects of magnetite (Fe_3O_4) nanoparticles on perennial ryegrass (*Lolium perenne* L.) and pumpkin (*Cucurbita mixta*) plants. Nanotoxicology 5:30–42
Wang SH, Kurepa J, Smalle JA (2011b) Ultra-small TiO_2 nanoparticles disrupt microtubular networks in *Arabidopsis thaliana*. Plant Cell Environ 34:811–820
Wang Q, Ebbs SD, Chen YS, Ma XM (2013) Trans-generational impact of cerium oxide nanoparticles on tomato plants. Metallomics 5:753–759
Wang J, Yang Y, Zhu HG, Braam J, Schnoor JL, Alvarez PJJ (2014) Uptake, translocation, and transformation of quantum dots with cationic versus anionic coatings by *Populus deltoides* × *Nigra cuttings*. Environ Sci Technol 48:6754–6762
Wani AH, Shah MA (2012) A unique and profound effect of MgO and ZnO nanoparticles on some plant pathogenic fungi. J Appl Pharm Sci 2:40–44
Wild E, Jones KC (2009) Novel method for the direct visualization of in vivo nanomaterials and chemical interactions in plants. Environ Sci Technol 43:5290–5294

Yadav T, Mungray AA, Mungray AK (2014) Fabricated nanoparticles: current status and potential phytotoxic threats. In: Whitacre DM (ed) Reviews of environmental contamination and toxicology. Springer, Cham

Yuan DG, Shan XQ, Huai Q, Wen B, Zhu XR (2001) Uptake and distribution of rare earth elements in rice seeds cultured in fertilizer solution of rare earth elements. Chemosphere 43:327–337

Zhang ZY, He X, Zhang HF, Ma YH, Zhang P, Ding YY, Zhao YL (2011) Uptake and distribution of ceria nanoparticles in cucumber plants. Metallomics 3:816–822

Zhang P, Ma YH, Zhang ZY, He X, Li YY, Zhang J, Zheng LR, Zhao YL (2015) Species-specific toxicity of ceria nanoparticles to lactuca plants. Nanotoxicology 9:1–8

Zhao LJ, Peralta-Videa JR, Rico CM, Hernandez-Viezcas JA, Sun YP, Niu GH, Servin A, Nunez JE, Duarte-Gardea M, Gardea-Torresdey JL (2014) CeO_2 and ZnO nanoparticles change the nutritional qualities of cucumber (*Cucumis sativus*). J Agric Food Chem 62:2752–2759

Zhao LJ, Sun YP, Hernandez-Viezcas JA, Hong J, Majumdar S, Niu GH, Duarte-Gardea M, Peralta-Videa JR, Gardea-Torresdey JL (2015) Monitoring the environmental effects of CeO_2 and ZnO nanoparticles through the life cycle of corn (*Zea mays*) plants and in situ Mu-Xrf mapping of nutrients in kernels. Environ Sci Technol 49:2921–2928

Zheng L, Hong FS, Lu SP, Liu C (2005) Effect of Nano-TiO_2 on strength of naturally and growth aged seeds of spinach. Biol Trace Elem Res 104:83–91

Zhu H, Han J, Xiao JQ, Jin Y (2008) Uptake, translocation, and accumulation of manufactured iron oxide nanoparticles by pumpkin plants. J Environ Monit 10:713–717

Zhu ZJ, Wang HH, Yan B, Zheng H, Jiang Y, Miranda OR, Rotello VM, Xing BS, Vachet RW (2012) Effect of surface charge on the uptake and distribution of gold nanoparticles in four plant species. Environ Sci Technol 46:12391–12398

Chapter 13
Stimulatory and Inhibitory Effects of Nanoparticulates on Seed Germination and Seedling Vigor Indices

Mehrnaz Hatami

13.1 Introduction

Nanotechnology application is now widely distributed throughout life, and especially in plant-soil systems. The development of nanotechnology and nanoscience in agricultural section has spread out the domain area of nano-structured materials in various fields because of their unique physiochemical traits (e.g., large surface area-to-volume ratio, ability to engineer electron exchange, and highly surface reactive capabilities) (Scrinis and Lyons 2007). Subsequently, different types of nanomaterials including metal- and carbon-based materials have been produced.

The effects of nanosized materials on plant performance are complex; even the same type of these materials may have various biological impacts (positive, negative, and or inconsequence) on various plant species; also, the same nanoscale materials that can be toxic in high concentrations may have positive role in low levels. Up to now, wide kinds of impacts of nanomaterials on seed germination and plant growth have been reported (Hatami et al. 2013, 2014; Hatami and Ghorbanpour 2013; Feizi et al. 2013; Ghorbanpour and Hatami 2014, 2015; Hatami et al. 2016; Larue et al. 2011; Khodakovskaya et al. 2009; Zhang et al. 2015; Kim et al. 2015; Rico et al. 2013; Hong et al. 2014; Zhao et al. 2012; Ushahra et al. 2014; Mohammadi et al. 2013; Wang et al. 2015; Corral-Diaz et al. 2014; Yasur and Rani 2014; Hernandez-Viezcas et al. 2011; Lee et al. 2013; Siddiqui et al. 2014).

In recent years, many scientists have studied the effects of nanomaterials on seed germination and plant growth with the aim to promote its use for agricultural productions. Most of these studies are focused on the potential toxicity of nanoparticles in higher plants and both positive and subsequently negative or

M. Hatami (✉)
Department of Medicinal Plants, Faculty of Agriculture and Natural Resources, Arak University, 38156-8-8349 Arak, Iran
e-mail: hatamimehrnaz@yahoo.com; m-hatami@araku.ac.ir

M. Ghorbanpour et al. (eds.), *Nanoscience and Plant–Soil Systems*, Soil Biology 48,
DOI 10.1007/978-3-319-46835-8_13

inconsequential effects were presented. Most researches were mainly studied in laboratory conditions in vitro, and the mechanisms of the interactions between nanoparticles and whole plants are not well understood. Also, the mechanism by which nanoparticulates affect plant growth and productivity requires further investigation and elucidation.

13.2 Germination Indices of Seeds in Response to Different Nanomaterial Treatments

Because of their physicochemical characteristics, nanoparticles are among the potential candidates for modulating the redox status and changing the seed germination, growth, performance, and quality of plants (Mukherjee and Mahapatra 2009).

Seed germination is a critical phase of the plant life cycle resulting in many biological processes. Seed germination combined with various catabolic and anabolic processes is sensitive to various internal and external stimuli. Consequently, the regulation of seed germination by nanomaterials has been performed abundantly in the research.

In a study, we investigated the influence of different concentrations (0, 10, 20, 40, and 80 mg/L) of nanosized TiO_2 [10–25 nm, TEM image of the TiO_2 NPs is shown in Fig. 13.1, and the XRD measurement showed that applied TiO_2 NPs were all present in the anatase form)] on seed germination parameters including germination percentage (GP), mean germination time (MGT), germination rate (GR), germination index (GI), and seedling vigor index (SVI) of five different plant species namely *Salvia mirzayanii* Rech. F. and Esfand (Labiatae), *Alyssum homolocarpum* (Brassicaceae), *Sinapis alba* L. (Brassicaceae), *Carum copticum* L. (Umbelliferae), and *Nigella sativa* (Ranunculaceae). Results showed that in *Salvia mirzayanii* plants, the highest and the lowest GP (81 and 47 %), GR (0.39 and 0.28), and VI (46.62 and 12.3) were obtained in 80 mg/L concentration of TiO_2

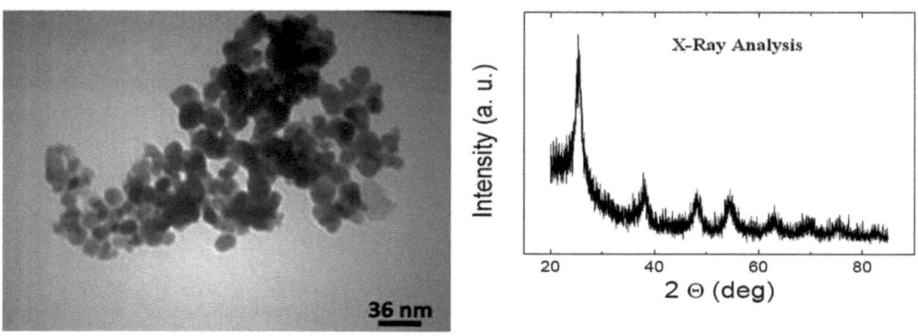

Fig. 13.1 Transmission electron microscopy, TEM (*left*) image, and XRD pattern of TiO_2 NPs (*right*) (Hatami et al. 2014)

NPs and control, respectively. However, the lowest MGT (2.55 days) was observed in the highest TiO_2 NP concentration. Moreover, GI of the seeds was not significantly affected by treatments. Our findings possibly revealed that the use of TiO_2 NPs increased the seed germination parameters of *S. mirzayanii* mucilage's seeds (Table 13.1). In *Alyssum homolocarpum*, the final GP was significantly affected by the employed treatments; however, increasing the concentration of TiO_2 NPs up to 40 mg/L caused an increase in seed germination and after that it declined. The lowest MGT (1.55 days) and the highest GR (0.62) were obtained in 10 mg/L concentration of TiO_2 NP treatment. However, the maximum seedling SVI (14.27) was observed in the medium concentration (40 mg/L).

Table 13.1 Seed germination responses and seedling vigor index values of the five plant species exposed to different nanosized TiO_2 concentrations (Hatami et al. 2014)

Traits						
	Treatment					
Plant species	TiO_2 NPs (mg/L)	GP (%)	MGT (day)	GR (seed/day)	SVI	GI
Salvia mirzayanii	0	47.0 ± 1.2	4.58 ± 0.3	0.28 ± 0.01	12.30 ± 0.1	5.26 ± 0.9
	10	58.0 ± 1.4	2.99 ± 0.3	0.33 ± 0.01	24.19 ± 0.2	5.15 ± 1.0
	20	68.0 ± 1.9	2.85 ± 0.2	0.36 ± 0.02	34.43 ± 0.3	5.53 ± 0.8
	40	70.0 ± 1.6	3.10 ± 0.2	0.36 ± 0.01	35.23 ± 0.1	5.81 ± 0.4
	80	81.0 ± 2.1	2.55 ± 0.1	0.39 ± 0.03	46.62 ± 0.1	6.08 ± 0.5
Alyssum homolocarpum	0	68.3 ± 1.4	2.36 ± 0.4	0.44 ± 0.02	9.15 ± 0.3	5.62 ± 0.2
	10	76.6 ± 1.2	1.55 ± 0.1	0.62 ± 0.03	11.94 ± 0.2	5.84 ± 0.3
	20	77.6 ± 1.6	1.62 ± 0.2	0.54 ± 0.03	11.98 ± 0.4	6.62 ± 0.1
	40	83.3 ± 2.3	1.89 ± 0.1	0.55 ± 0.02	14.27 ± 0.1	7.36 ± 0.2
	80	71.6 ± 1.3	2.25 ± 0.2	0.57 ± 0.04	14.0 ± 0.5	6.38 ± 0.3
Sinapis alba	0	71.3 ± 2.3	8.10 ± 0.2	0.11 ± 0.05	45.0 ± 1.5	5.47 ± 0.2
	10	76.6 ± 2.1	7.95 ± 0.1	0.10 ± 0.03	52.17 ± 2.4	5.65 ± 0.5
	20	87.0 ± 2.5	7.90 ± 0.4	0.14 ± 0.02	102.67 ± 5.7	5.68 ± 0.4
	40	66.6 ± 1.8	8.86 ± 0.2	0.12 ± 0.01	88.33 ± 3.1	5.91 ± 0.2
	80	48.3 ± 1.5	9.46 ± 0.2	0.10 ± 0.03	22.0 ± 1.8	5.71 ± 0.2
Carum copticum	0	65.0 ± 1.4	9.64 ± 1.5	0.52 ± 0.04	13.70 ± 1.3	2.64 ± 0.4
	10	75.0 ± 2.2	8.45 ± 1.2	0.81 ± 0.07	29.12 ± 1.5	3.82 ± 0.3
	20	72.9 ± 1.9	7.07 ± 1.1	0.64 ± 0.02	24.34 ± 1.2	3.65 ± 0.3
	40	51.0 ± 1.3	9.84 ± 1.6	0.46 ± 0.04	12.70 ± 0.9	2.0 ± 0.2
	80	42.5 ± 1.1	9.91 ± 1.4	0.41 ± 0.01	4.19 ± 0.3	1.3 ± 0.1
Nigella sativa	0	68.0 ± 1.5	8.47 ± 0.9	0.72 ± 0.03	22.73 ± 0.3	6.12 ± 0.2
	10	79.0 ± 2.1	7.78 ± 0.8	0.75 ± 0.01	34.49 ± 0.6	8.62 ± 0.1
	20	84.6 ± 2.2	6.43 ± 0.5	0.83 ± 0.04	53.20 ± 0.2	9.29 ± 0.6
	40	80.0 ± 1.8	5.95 ± 0.2	0.86 ± 0.02	42.76 ± 0.7	8.66 ± 0.3
	80	70.0 ± 1.3	9.46 ± 1.1	0.77 ± 0.01	23.45 ± 0.1	6.21 ± 0.2

GP germination percentage, *MGT* mean germination time, *GR* germination rate, *SVI* seedling vigor index, *GI* germination index. Values represent mean ± standard deviation

In *Sinapis alba* plants, the maximum GR (0.14) with the lowest MGT (7.9 days) was obtained in 20 mg/L of TiO_2 NP concentration. Moreover, the highest (102.6) and the lowest (22) SVI were observed in the 20 and 80 mg/L of the TiO_2 NP treatment, respectively (Fig. 13.2). For *Carum copticum*, the highest (75 %) and lowest (42.5 %) germination percentage was obtained at 10 and 80 mg/L of TiO_2 NP concentration, respectively. Results also showed that the maximum GR (0.81) and the lowest MGT (8.45 days) were obtained in 10 mg/L concentration of employed treatment. However, the lowest SVI (4.19) and GI (1.3) were observed in seeds treated with the highest TiO_2 NPs. In *Nigella sativa*, the lowest and the highest SVI (6.12 vs. 9.29) were obtained in 20 mg/L concentration of TiO_2 NPs and control, respectively. However, the maximum (84.6 %) and the minimum GP (68 %) were observed in 20 mg/L TiO_2 NPs and in control treatments, respectively.

Employing TiO_2 NPs in appropriate concentration could promote the seed germination features and early growth of plants in comparison to control. Generally, nanoparticles are materials that are small enough to fall within the nanometric range, with at least one of their dimensions being less than a few hundred nanometers. This reduction in size brings about significant changes in their physical properties with respect to those observed in bulk materials. Most of these changes are related to the appearance of quantum effects as the size decreases (Joseph and Morrison 2006). The engineered carbon nanotubes at the range of 10–40 mg/L dramatically enhanced the seed germination and growth of tomato plants (Khodakovskaya et al. 2009).

The positive effects of nanomaterials arose from the capability of them to penetrate seed coat and therefore promote water uptake. Water uptake in seed germination is critical because mature seeds are relatively dry and need a substantial amount of water to initiate cellular metabolism and growth. The measured water moisture content of seeds and the detection of nanomaterials inside seeds supported the above-mentioned hypothesis.

According to Lu et al. (2002) a mixture of TiO_2 and SiO_2 nanoparticles at low concentrations increased nitrate reductase activity in the rhizosphere of soybean and consequently expedited soybean germination and growth. Also, it has been reported that SiO_2 NPs increased seed germination by providing better nutrient availability to the maize seeds, and pH and conductivity to the growing medium (Suriyaprabha et al. 2012).

Metallic oxide nanoparticles such as ZnO were shown to be inhibitive at different developmental stages of plants such as seed germination and root elongation (Lin and Xing 2007). Seed germination and root elongation are two standard indictors of phytotoxicity suggested by US Environmental Protection Agency; yet several researches have indicated the insensitivity of seed germination for nanoparticles (Stampoulis et al. 2009).

Different responses of the plant species to nanomaterials could be due to differences in concentration, particle size and specific surface area, physicochemical properties of nanoparticles, plant species, plant age/life cycle stage, growth media conditions, nanoparticle stability, and dilution agent.

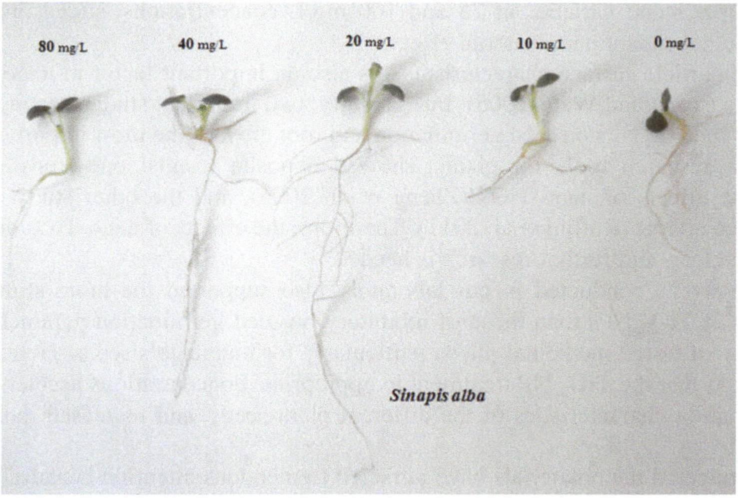

Fig. 13.2 Effect of different concentrations of TiO_2 nanoparticles on seed germination morphology in *Sinapis alba* (Hatami et al. 2014)

The size of seeds could render more sensitivity to nanoparticle exposure. This is because a large seeded species (e.g., *S. mirzayanii*) in our study has a lower surface-to-volume ratio than a small-seeded species (e.g., *S. alba*). However, a clear effect of the size of seeds on the toxicity of nanoparticles in plants cannot be confirmed at this time (Lee et al. 2008; Lin and Xing 2007).

Lee et al. (2008) reported that the mung bean plant was more sensitive than the wheat plant to Cu NP toxicity, probably due to differences in root anatomy because xylem structures determine the speed of water transport and different xylem structures may demonstrate different uptake kinetics of nanoparticles. Mung bean is a dicot with one large primary root and several smaller lateral roots, whereas wheat is a monocot with numerous small roots without a primary root.

However, generalization on whether the toxicity is based on dicot or monocot classification cannot be made.

Nanoparticles show different effects on seed germination of different plants, suggesting species-specific-dependant nanosized material effect. Most recently Karami-Mehrian et al. (2015) studied the effects of synthesis silver nanoparticles (Ag NPs, 50 nm) at five different concentrations including 0, 25, 50, 75, and 100 mg/L, on the seed germination, germination percentage (GP), seedling vigor index (VI), germination index (GI), tolerance index (TI), root and shoot length (RL and SL), and silver content in seven varieties of *Lycopersicon esculentum* Mill (tomato) plants, namely Peto early CH, Primo early, Cal.j.n3, Early urbanay VF, King stone, Super stone, and Super strain B. They reported that Ag NP-treated seeds sprouted within first 3 days, whereas the control seeds (deionized water) took longer time to sprout, exhibiting an increase in GI, in Early urbanay VF, Super strain B, and Primo early varieties. Also, Ag NPs decreased in GP of Super strain B, and Super stone varieties at 75 and 100 mg/L concentrations, suggesting dose-specific-dependant nanomaterial effect.

The particle surface characteristic was also an important factor in nanoparticle toxicity (Yang and Watts 2005). Interestingly, two important studies testing on the effects of TiO_2 NPs on seed germination and root growth, the most important basic toxicity research tools for plants, showed opposite results: one study showed positive effects of nano-TiO_2 (Zheng et al. 2005), and the other study showed negative effects (Ruffini et al. 2011). Therefore, the effects of nano-TiO_2 on plants are not clear, and further research is needed.

The works conducted in our laboratory also supported the more stimulatory effects of TiO_2 NPs than those of inhibitory on seed germination parameters and vigority of tested medicinal plants particularly for mucilage's seeds. Here, results indicated that the TiO_2 NP treatment in appropriate concentrations accelerated the germination characteristics of the different plant seeds, and increased their vigor as well.

Engineered nanomaterials have attracted tremendous attention because of their positive effects in consumer products, pharmaceutics, transportation, cosmetics, energy, plant sciences, and agriculture. However, the unique properties of nanomaterials could lead to unpredicted biological effects, such as toxicity. The phytotoxicity profile of nanoparticulates has also been investigated by researchers via seed germination and root elongation tests which evaluate the acute effects of nanoparticulates on plant physiologies.

In recent years, most of the studies are focused on the potential toxicity of nanoparticles in higher plants and both positive and subsequently negative or inconsequential effects were presented (Table 13.2). Moreover, many studies have found that although having obviously negative effect on root elongation, nanoparticles did not affect seed germination (Lin and Xing 2007; Wang et al. 2012).

The reported data from different studies suggested that the effect of nanoparticulates on seed germination was concentration dependent. Khodakovskaya et al. (2009) reported that the same nano-structured materials that can be toxic for plants in high concentrations may have stimulatory impact on the physiological processes at low concentrations. Also, Raskar and Laware (2014) suggested that lower concentration of ZnO NPs showed positive effect on onion seed germination. However, higher dose of ZnO NPs impaired seed germination.

In different studies, researchers have reported that multi-walled carbon nanotubes (MWCNTs) have a great ability to affect the seed germination and plant growth. MWCTs induce the water and essential nutrient (including Ca and Fe) uptake efficiency that could improve the seed germination and seedling growth (Tiwari et al. 2014). It has been reported that MWCNTs supplemented to the agar medium stimulate seed germination of three plant species (barley, soybean, corn) due to the ability of MWCNTs to penetrate the seed coats, as the nanotube agglomerates were detected inside the seed coats using Raman spectroscopy and transmission electron microscopy analysis (Lahiani et al. 2013). Also, they reported

Table 13.2 The effects of different types of nanomaterials on seed germination and growth of various plant species

Nanoparticle	Particle size (nm)	Concentration	Plant species	Growth medium/condition	Observed effects	References
TiO_2	10–25 nm	0, 10, 20, 40 and 80 mg/L	*Salvia mirzayanii*, *Alyssum homolocarpum*, *Sinapis alba*, *Carum copticum*, *Nigella sativa*	Filter paper in Petri dishes	Among the plant species tested, the highest germination percentage was observed in *S. alba* seeds at 20 mg/L TiO_2 NP concentration. Also, the lowest mean germination time was observed in *A. homolocarpum* seeds at 10 mg/L TiO_2 NP concentration. Moreover, the highest seedling vigor index was observed in the 20 mg/L of the TiO_2 NPs for *S. alba* seeds	Hatami et al. (2014)
Ag	5–35 nm	0, 20, 40, 80 mg/L in presence or absence of thidiazuron (0, 50, 75, and 100 μM)	*Pelargonium graveolens*	Potting mix	Increased plant height and the number of branches with increasing concentrations of both thidiazuron (up to 75 μM) and NS (up to 40 mg/L); however, TDZ was found to be more efficient	Ghorbanpour and Hatami (2015)
MWCNT	5–15 nm	0, 25, 50, 100, 250, 500 μg/mL	*Satureja khuzestanica*	Gamborg B5, the effects of MWCNTs on callus induction and their growth parameters were studied	Improved calli growth with the increase of MWCNT concentration, peaked at 50 μg/mL, and then followed a rapid decrease at 500 μg/mL	Ghorbanpour and Hadian (2015)
TiO_2	10–15 nm	0, 20, 40, 80 mg/L	*Hyoscyamus niger*	Potting mix	Improved the plant dry weight to 42 % at 40 mg/L TiO_2 NPs compared to the unexposed control plants	Ghorbanpour et al. (2015)

(continued)

Table 13.2 (continued)

Nanoparticle	Particle size (nm)	Concentration	Plant species	Growth medium/condition	Observed effects	References
TiO_2	10–15 nm	0, 10, 50, 100, 200, 1000 mg/L	*Salvia officinalis*	Potting mix	The highest and the lowest dry weights of both root and shoot tissues were obtained from plants exposed to 100 and 1000 mg/L TiO_2 NPs, respectively	Ghorbanpour (2015)
TiO_2 and CeO_2	33 and 21 nm, respectively	0 and 500 mg/L	*Arabidopsis thaliana*	Whatman GF/A paper in Petri dishes	Increased the percentage of seeds observed with an emergent radical exposed to both ENPs, and also both ENPs increased the percentage of seeds exhibiting hypocotyls and cotyledons. The two ENPs affected leaf development compared with control. While both ENPs decreased the percentages of germinants without leaves, the proportions of germinants with fully grown leaves increased	Hatami et al. (2016)
MWCNTs	50, 100, and 200 μg/mL	14–40 nm	Barley, soybean, and corn	MS medium	The exposure of seeds of all tested crops to MWCNTs resulted in the acceleration of the process of seed germination. Significant acceleration of germination was observed when MWCNTs were used in doses of 100 and 200 μg/mL; however, no significant effect was observed for the lowest MWCNT concentration (50 μg/mL). The study proved the significant potential of carbon nanotubes as regulators of germination and plant growth	Lahiani et al. (2013)

MWCNTs	13 nm	0, 20, 200, 1000, and 2000 mg/L	*Lactuca sativa*, *Oryza sativa*, *Cucumis sativus*, *Amaranthus tricolor*, *Abelmoschus esculentus*, *Capsicum annuum*, and *Glycine max*	Modified Hoagland media in hydroponic culture	Reduced the root and shoot lengths of *A. tricolor*, *L. sativa*, and *C. sativus* following exposure to 1000 mg/L and 2000 mg/L MWCNTs. However, *A. tricolor* and *L. sativa* were most sensitive to MWCNTs, followed by *O. sativa* and *C. sativus*	Begum et al. (2014)
Ag (Colloidal, Ag_5 and Ag_{20})	2–20 nm, 5 nm and 20 nm, respectively	0, 100, 250, 500, 1000, 2000, and 5000 mg/kg	*Linum usitatissimum*, *Lolium perenne* and *Hordeum vulgare*	Filter paper and soil in Petri dishes	The three types of Ag NPs affected seed germination differently for the all plant species examined. The smallest particle type, Ag_{coll}, had an inhibitory effect at a concentration as low as 10 mg/ L, but only with ryegrass. Ag NPs inhibited seed germination at lower concentrations, but showed no clear size-dependant effects, and never completely impeded germination. The intermediately sized particle type (Ag_5) also had weak inhibitory effect at 10 mg/ L, which increased at higher concentrations, but this effect was only observed with barley. Barley was also the only plant species affected by the largest sized type of Ag NPs (Ag_{20}), with a weak inhibition observed at 10 mg/L, increasing slightly at higher concentrations. No effect on germination percentage of flax was observed for any types of Ag NPs, even at the highest concentrations	El-Temsah and Joner (2012)

(continued)

Table 13.2 (continued)

Nanoparticle	Particle size (nm)	Concentration	Plant species	Growth medium/condition	Observed effects	References
CeO_2, La_2O_3, Gd_2O_3, and Yb_2O_3	7.2 ± 0.7, 21.8 ± 2.7, 23.4 ± 2.9, and 11.7 ± 0.6 nm, respectively	2000 mg/L	*Brassica napus*, *Raphanus sativus*, *Triticum aestivum*, *Lactuca sativa*, *Brassica oleracea*, *Lycopersicon esculentum*, and *Cucumis sativus*	Filter paper in Petri dish	A suspension of 2000 mg/LCeO_2 NPs had no effect on the root elongation of six plants, except lettuce. On the contrary, 2000 mg/L suspensions of La_2O_3, Gd_2O_3, and Yb_2O_3 NPs severely inhibited the root elongation of all the seven species. Inhibitory effects of La_2O_3, Gd_2O_3, and Yb_2O_3 NPs also differed in the different growth process of plants. For wheat, the inhibition mainly observed during the seed incubation process, while lettuce and rape were inhibited on both seed soaking and incubation process. The 50 % inhibitory concentrations (IC_{50}) for rape were about 40 mg/L of La_2O_3, 20 mg/L of Gd_2O_3, and 70 mg/L of Yb_2O_3 NPs, respectively	Yuhui et al. (2010)
Ag	40 nm	0, 100, 200, 400, and 800 μM	*Brassica nigra*	Petri dish	Inhibited seed germination, lipase activity, soluble and reducing sugar contents in germinating seeds and seedlings. Application of 200–400 mg/L Ag NPs increased transcription of hemeoxygenase-1. However, at 800 mg/L, Ag NPs suppressed HO-1 expression. At 400 mg/L, Ag NPs induced inhibitory effects on seed germination and were ameliorated by the HO-1 inducer, hematin, or NO donor, sodium nitroprusside	Amooaghaie et al. (2015)

Ag	<100 nm	0, 100, 200, 500, 1000, 2000, and 4000 mg/L	*Ricinus communis*	Filter paper in Petri dishes	Ag NPs at all the concentrations tested and even at the highest dose employed did not disturb the germination of the seeds or the seedling growth of *R. communis*	Yasur and Rani (2014)
Ag_2O	218.34 ± 2.21 nm	0.1, 1, 10, 100, and 1000 mg/L	*Vigna radiata*	Petri dishes	Decreased root length, shoot length, and dry weight of seedlings. Silver content in seedlings increased with increasing concentrations of Ag NPs. Brown tips and necrosis were detected in the roots of *V. radiata* seedlings indicating toxic effects of Ag NPs. The trends of toxic effects of Ag NPs on shoot growth were found to be similar to those of root growth	Singh and Kumar (2015)
Ag	20, 30–60, 70–120, and 150 nm	0, 0.1, 1, 10, 100, and 1000 mg/L	*Oryza sativa*	Hydroponic system	The level of seed germination and subsequent growth of those seedlings that germinated were both decreased with increasing sizes and concentrations of Ag NPs. Higher uptake was found when the seeds were treated with the smaller Ag NPs (20 nm), but it was trapped in the roots rather than transported to the leaves. These resulted in the less negative effects on seedling growth, when compared to the seed soaking with the larger Ag NPs (150 nm). The negative effects of Ag NPs were supported by leaf cell deformation when rice seeds were treated with 150-nm-diameter Ag NP at the concentration of 10 or 100 mg/L during seed germination	Thuesombat et al. (2014)

(continued)

Table 13.2 (continued)

Nanoparticle	Particle size (nm)	Concentration	Plant species	Growth medium/condition	Observed effects	References
Ag	25 nm	0, 20, 40, and 60 ppm	*Ocimum basilicum*	Sprayed whole plant at seed growth stage	Improved seed yield with increasing concentration of nano silver from 20 to 60 ppm. Also, the lowest amount of seed yield was found with control	Nejatzadeh-Barandozi et al. (2014)
CeO_2	231 ± 16 nm	0, 62.5, 125, 250, and 500 mg/kg	*Raphanus sativus*	Potting mix	At 500 mg/kg, CeO_2 NPs significantly retarded seed germination, but did not reduce the number of germinated seeds. In addition, cerium accumulation in tubers of plants treated with 250 and 500 mg/kg reached 72 and 142 mg/kg dwt, respectively. None of the treatments affected plants' length or biomass accumulation	Corral-Diaz et al. (2014)
TiO_2	21 nm	0, 5, 20, 40, 60, and 80 ppm	*Agropyron desertorum*	Filter paper in Petri dishes	Improved germination percentage by 9 % following exposure to 5 ppm TiO_2 NP treatment comparing to control. Also, exposure of *A. desertorum* seeds to 5 and 60 ppm TiO_2 NPs showed the lowest mean germination time. Application of TiO_2 NPs at 40 ppm demonstrated the highest shoot and seedling length	Azimi et al. (2013)

TiO_2	8–15 nm	0, 100, 200, and 400 mg/L	*Lycopersicum esculentum*, *Allium Cepa*, and *Raphanus sativus*	Peat: perlite (1:1, v/v) in Petri dishes (laboratory test) or pots (greenhouse test)	TiO_2 NPs at 100 and 200 mg/L had the most positive effect on germination. In the laboratory, the highest germination percentage of *L. esculentum* and *A. cepa* was observed at 100 mg/L (100 and 30 %, respectively), and in *R. sativus*, 100 % germination was obtained with 400 mg/L. In the greenhouse, seedlings were tallest after exposure to 400 and 200 mg/L for *L. esculentum* and *A. cepa*, respectively, and 400 and 100 mg/L for *R. sativus*	Haghighi and Teixeira da Silva (2014)
ZnO	30 ± 12 nm	0, 10, 100, and 1000 mg/L	*Zea mays* and *Cucumis sativus*	Filter paper in Petri dishes	ZnO NPs at 1000 mg/L reduced root length of *Z. mays* and *C. sativus* by 17 % and 51 %, respectively, but exhibited no effects on germination. Root growth was clearly reduced with increasing concentrations	Zhang et al. (2015)
Four ZnO NPs (spheric ZnO-30, spheric ZnO-50, columnar ZnO-90, and hexagon rodlike ZnO-150)	30 ± 12.5, 50 ± 20, 90 ± 45, 150 ± 100 nm, respectively	0, 1, 5, 10, 20,40, and 80 mg/L	*Brassica pekinensis*	Filter paper in Petri dishes	ZnOs NPs did not affect germination rates at concentrations of 1-80 mg/L but significantly inhibited the root and shoot elongation of *B. pekinensis* seedlings, with the roots being more sensitive. ZnO-30 NPs with the smallest primary size caused the greatest inhibition of root elongation, and vice versa (ZnO-150)	Xiang et al. (2015)

(continued)

Table 13.2 (continued)

Nanoparticle	Particle size (nm)	Concentration	Plant species	Growth medium/condition	Observed effects	References
TiO_2	<100 nm	0.2, 1.0, 2.0, and 4.0 %	*Vicia narbonensis* and *Zea mays*	Petri dishes	In *V. narbonensis* no germination process was detected following TiO_2 NP treatments, while more than 20 % of control seeds were germinated within 24 h. In *Z. mays* control seeds reached a value of about 50 % of germination; following 0.2 % TiO_2 NP treatment, the germination rate was almost halved, and at higher concentrations it further tended to slightly decrease. Root elongation was affected only after treatment with the higher TiO_2 NP concentration. Further significant effects were detected showing mitotic index reduction and concentration-dependent increase in the aberration emergence that evidenced a TiO_2 NP-induced genotoxic effect for both species	Castiglione et al. (2011)
MWCNTs	6–9 nm	0, 5, 10, 20, 40, and 60 mg/L	*Zea mays*	BA gel medium	The MWCNTs increased the water content of the root but less unequivocally for the whole seedling. For the shoot however, after a spurt in the %water at 10 mg/L of MWCNT, the water content decreased gradually. The fresh weight followed the trends of the water contents. The dry weight for each morphological part was not much different at the low MWCNT concentration, but showed	Tiwari et al. (2014)

					a sudden spurt at 20 mg/L MWCNT particularly for the root, while slowly declined at the higher MWCNT dose	
Ag	13 nm	0, 20, and 50 mg/L	*Pennisetum glaucum*	Filter paper in Petri dish	Enhanced seed germination compared to the control; however, the highest percentage of seed germination was 93.33 % at 50 mg/L of Ag NPs. The Ag NPs at 50 mg/L decreased the root and shoot lengths followed by 20 mg/L	Parveen and Rao (2015)
ZnO	<50 nm	0, 5, 10, 25, 50, 75, 100, 125, 250, and 500 mg/L	*Brassica napus*	Paper filters in the Petri dishes	Inhibited root elongation at Ag NP concentrations of 250 and 500 mg/L; however, inhibitory effects of Ag NPs on shoot elongation were less severe	Mousavi Kouhi et al. (2014)
TiO_2	21 nm	0, 1, 2, 10, 100, and 500 ppm	*Triticum aestivum*	Paper in Petri dishes	Among the seed germination indices, only mean germination time was affected by treatments. The lowest and the highest mean germination time (0.89 vs. 1.35 days) was obtained in 10 ppm concentration of TiO_2 NPs and control treatments, respectively. Shoot and seedling lengths at 2 and 10 ppm concentrations of TiO_2 NPs were higher than those of the untreated control and bulk TiO_2 at 2 and 10 ppm concentrations	Feizi et al. (2012)

(continued)

Table 13.2 (continued)

Nanoparticle	Particle size (nm)	Concentration	Plant species	Growth medium/condition	Observed effects	References
ZnO, TiO_2, and Ni	<100, <21, and <100 nm, respectively	0, 10, 100, and 1000 mg/L	*Lepidium sativum*	Seeds on the filter paper were placed directly on the ENP-contaminated soil. Different methods of application of nanoparticles to the soil were also investigated	No effect of the studied ENPs and their bulk counterparts on the germination of seeds was observed. Root elongation was reported as a more sensitive indicator. The root growth inhibition of *L. sativum* depended on the kind of test applied. The trend between concentration of ENPs and effect depended on the method used and kind of ENPs	Josko and Oleszczuk (2014)
MWCNT and MWCNT-COOH	10–30 nm	0.01, 1.0, or 100 µg/mL (positive control: activated carbon at 100 mg/mL, negative control: water treatment)	*Parmotrema tinctorum*	Open top chamber	Reduced cell viability of *P. tinctorum*. The treatment with 100 µg/mL of MWCNT-COOH resulted in intracellular ion leakage, probably due to changes in membrane permeability. Carbon nanotube entrapment and internalization into the lichen thallus were observed. Short-term exposition of CNT produced measurable physiological changes in *P. tinctorum* lichen	Viana et al. (2015)
TiO_2		0, 0.25, 0.5, 1.0, 1.5, 2.0, 2.5, 4.0, and 6.0 %	*Spinacia oleracea*	Moistened gauze in pearlite-containing Petri dishes with Hoagland culture solution	Increased germination, germination rate, and vigor indices of aged spinach seeds at 0.25–4 % TiO_2 (rutile) NPs treatment; however, the best results were found at 2.5 % TiO_2 NPs	Zheng et al (2005)

SiO_2	50 nm	10 mg/L	*Zea mays*	Pot under laboratory condition	Nanosilica promoted seed germination percentage (100 %) in maize than conventional Si sources	Karunakaran et al. (2013)
TiO_2 (anatase/rutile = 80:20)	27 nm	0, 100, 500, 1000, 2500, and 5000 mg/L	*Brassica campestris*, *Lactuca sativa* L., and *Phaseolus vulgaris*	Seed germination and root elongation test were done in Petri dish, and commercial soil and circular hydroponic system used for testing the toxicity of NPs	The various treatments did not affect the germination rates of the three species tested. However, TiO_2 NPs had positive effects on root elongation in some species, with the longest root length in the 500 mg/L treatment. The Ti content was three times higher in seeds treated with 1000 mg/L TiO_2 NPs than in seeds treated with a higher concentration of 5000 mg/L. No physiological differences in enzyme activities were detected, even though the plants absorbed TiO_2 NPs	Song et al. (2013)
GO	0.7 ± 0.2 nm	0 and 40 μg/mL	Tomato	Glass bottles	GO accelerated the process of seed germination and significantly shortened the germination time. GO significantly enhanced the water uptake inside tomato seeds by 17.5 %. Seedlings germinated from GO-treated seeds had slightly lower biomass accumulation than the control, but exhibited significantly longer stems and roots than the control	Zhang et al. (2015)

(continued)

Table 13.2 (continued)

Nanoparticle	Particle size (nm)	Concentration	Plant species	Growth medium/condition	Observed effects	References
ZVI	<100 nm	0, 0.005–0.5 g/L, and 5 and 10 mg/L of Fe^{2+} ions	*Arabidopsis thaliana*	1/2 MS medium into a Petri dish	Exposure of plants to 0.5 g/L ZVI NPs enhanced root elongation by 150–200 % over that in the control, and this occurred via ZVI NP-mediated OH radical-induced cell wall loosening. All of the ZVI NP-treated seedlings grew normally without severe damages in phenotype, whereas those exposed to Fe^{2+} ions (except the 0.005 g/L-treated group) failed to grow after germination	Kim et al. (2014)
TiO_2		0, 10, 20, 30, and 40 mg/mL	*Petroselinum crispum*	MS medium into glass bottles	Increased percentage of germination, germination rate index, root and shoot length, fresh weight, vigor index, and chlorophyll content of the seedlings with increasing TiO_2 NP concentration; however, the best concentration was 30 mg/mL	Hashemi-Dehkourdi and Mousavi (2013)
Yb_2O_3, bulk, and $YbCl_3.6H_2O$	12 ± 2 nm	0, 0.32, 0.8, 2, 5, 20, 200, 2000 mg/L	*Cucumis sativus*	Growth conditions were according to the SI	Decreased biomass accumulation at the lowest dose (0.32 mg/L) of Yb_2O_3 NPs, while at the highest concentration, the most severe inhibition was from $YbCl_3$	Zhang et al. (2012)

Au	10–20 nm	0, 10, 25, 50, and 100 ppm	*Brassica juncea*	MS growth media	Percentage germination of *B. juncea* seeds was positively affected by Au NP treatment up to 25 ppm concentration. A maximum increase of 25 % in leaf number was recorded in seedlings treated with 10 ppm Au NPs. However, Au NP treatment did not increase the average leaf area of the treated seedlings	Arora et al. (2012)
Au	6–10 nm	0 and 31.25 nM	*Oryza sativa*, *Lolium perenne*, *Raphanus Sativus*, *Cucurbita mixta*	Exposure of the Au NPs to the seedlings under hydroponic conditions for a 5-day period was investigated	*R. Sativus* and *L. perenne* roots generally accumulated higher amounts of the Au NPs than *O. sativa* and *C. mixta* roots. Each of the Au NPs was found to accumulate to statistically significant extents in *O. sativa* shoots, while none of the Au NPs accumulated in the shoots of *R. Sativus* and *C. mixta*	Zhu et al. (2012)
TiO_2 and bulk		0, 5, 20, 40, 60, and 80 mg/L	*Foeniculum vulgare*	Filter paper in Petri dishes	Improved germination percentage following exposure to 60 ppm TiO_2 NPs. Similar positive effects occurred in terms of shoot dry weight and germination rate. Enhanced mean germination time by 31.8 % in comparison to the untreated control at 40 ppm TiO_2 NPs. In addition, low and moderate concentrations of TiO_2 NPs improved indices such as germination value, vigor index, and mean daily germination	Feizi et al. (2013)

Yb_2O_3 ytterbium oxide, $YbCl_3$ ytterbium chloride, TiO_2 titanium dioxide, *Au* gold, *ZVI* zero-valent iron, *GO* graphene oxide, SiO_2 silicon dioxide, *SWCNTs* single-walled carbon nanotubes, *MWCNTs* multi-walled carbon nanotubes, *MWCNT* functionalized multi-walled carbon nanotube-*COOH*, *ZnO* zinc oxide, *Ni* nickel, *Ag* silver, CeO_2 cerium oxide, Ag_2O silver oxide, La_2O_3 lanthanum oxide, Gd_2O_3 gadolinium oxide

that MWCNTs regulated gene expression encoding several types of water channel proteins in seed coat. As we know, aquaporins (water channel) play a key role in plant–water relations, seed germination, cell growth, and development. Furthermore, it has been reported that a high degree of porosity and a partial dismemberment of the surface structure are seen upon introduction of the MWCNT to maize (*Zea mays*). Such pores would greatly facilitate the entry of water, nutrients, and oxygen as well as the dispersed aqueous phase MWCNT themselves into the germinating seed. It is well known that scarification aids the germination of seeds. Visually, the appearance of the effect of the MWCNT at the black-layer region seems akin to scarification at the microscopic scale (Tiwari et al. 2014).

On the contrary, many researchers confirmed the negative effects of different nanomaterials on seed germination. For example, El-Temsah and Joner (2012) reported that zero-valent iron nanoparticles and silver nanoparticles inhibited seed germination at different concentrations. Lin and Xing (2007) also found that Zn and ZnO nanoparticles have significant inhibition effect on seed germination and root growth. Also, Wu et al. (2012) studied the effects of the metal oxide nanoparticles including CuO, NiO, TiO_2, Fe_2O_3, and Co_3O_4 on germination index of lettuce, radish, and cucumber and it was found that CuO and NiO showed the strongest inhibitory effects on these seed germination. Nair et al. (2011) placed rice seeds in CdSe quantum dots and observed that the quantum dots inhibited seed germination.

However, the mechanism of the phytotoxicity of engineered nanomaterials is unclear, with factors such as chemical composition and reactive oxygen species (ROS) derived from surface-catalyzed reactions considered potential causes of engineered nanomaterial toxicity, especially for metal-based nanoparticles (Lin and Xing 2007; Yang and Watts 2005).

13.3 Changes in Plant Growth and Development Exposed to Different Nanomaterials

Growth can be defined as an irreversible permanent increase in size of an organ or its parts or even of an individual cell. Generally, growth is accompanied by metabolic processes (both anabolic and catabolic) that occur at the expense of energy. However, development is a term that includes all changes that an organism goes through during its life cycle from seed germination to senescence. Plant growth and development are affected by internal and external factors.

Given the wide-range commercial, environmental, medical, and agricultural application of nanosized materials, their production has reached the highest industrial scale, and accordingly different types of nanomaterials, such as metal-based materials and carbon-based materials, have been produced. These materials possess distinctive characteristics provided by their high surface area-to-volume ratio, surface charge, and size.

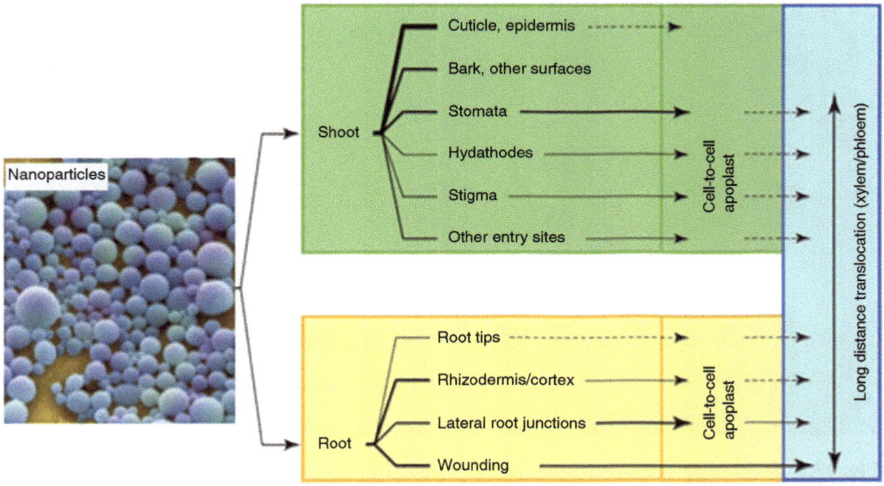

Fig. 13.3 A schematic model for nanoparticle uptake and distribution in plants. The thickness of *line* relates significance of the route. *Broken lines* represent very low rates of nanoparticle translocation (adapted from Dietz and Herth 2011, with slight modification)

As plants contribute the largest interface between the environment and the biosphere, there is a chance that they encounter the released nanomaterials. It is unclear what impact these nanomaterials will have on plant's life, and it is essential to understand the effects of nanomaterials on plant growth. Such studies are important not only from the point of view of the application of nanomaterials in plants, but also for understanding presumed effects on plants and causes for bioaccumulation. As of yet, the risks of nanomaterials in environment have not yet been fully characterized.

Nanomaterials can be absorbed by plants via many entry routes and translocated within the plants body (Fig. 13.3). Therefore, it is important to understand the processes of plant growth and development in relation to nanomaterials.

However, only a limited number of studies on the effects of nanoparticles on higher plants are available, with varied conclusions (negative, positive, and even protective effects) reported (Hatami et al. 2015; Feizi et al. 2012; Lin and Xing 2007; Stampoulis et al. 2009), although a relatively broad range of species have been tested.

In a study, we performed the effects of different concentrations of nano-silver particles (NS, 20, 40, and 80 mg/L) and thidiazuron (TDZ, 50, 75, and 100 μM) and their combinations on variations of growth indices in geranium (*Pelargonium graveolens* L.) plant (Ghorbanpour and Hatami 2015). Specific surface area and purity of NS particles were 210 m^2/g and 99 %, respectively. Average size of NS was determined using scanning electron microscope (SEM) and transmission electron microscope (TEM), and estimated to be 5–35 nm in diameter (Fig. 13.4). We observed that the plant height and the number of branches increased with increasing

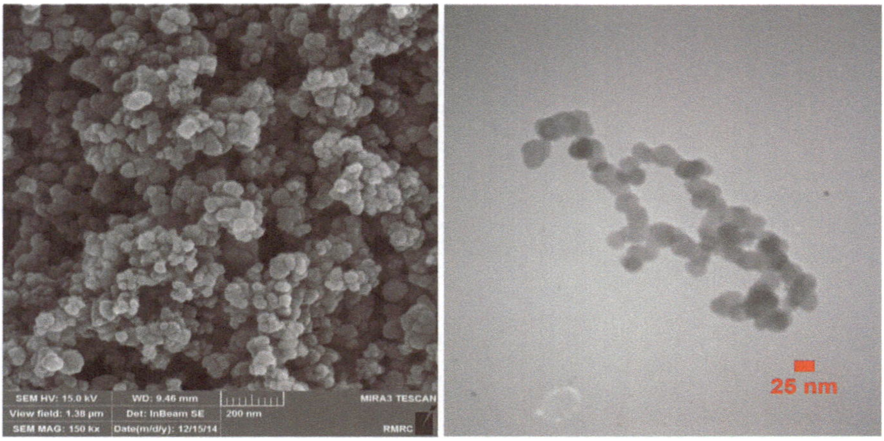

Fig. 13.4 Scanning electron microscopy, SEM (*left*) and transmission electron microscopy, TEM (*right*) images of nano-silver particle (Ghorbanpour and Hatami 2015)

concentrations of both NS (up to 40 mg/L) and TDZ (up to 75 μm), and TDZ was found to be more efficient for increasing the plant height and the number of its branches compared to NS. Also, the fresh and dry weight of aerial parts of the plant increased with application of NS and TDZ at the same concentrations as mentioned above. However, dry mass of the plant significantly ($P \leq 0.05$) reduced at the highest NS and TDZ concentrations compared to the other employed levels of NS and TDZ, but the extent of reduction only in TDZ-treated pants was not lower than that of the control. The highest fresh weight (238.3 g/plant) and dry weight (24.4 g/plant) were obtained at 75 μm TDZ; the untreated control fresh and dry weights were 196.7 and 15.1 g/plant, respectively. Also, plant height, the number of branches, and their fresh and dry weights were improved by all employed concentrations of NS + TDZ except when supplied at 80 mg/L NS + 100 μm TDZ. This study approved that application of NS at certain levels could improve growth parameters of *P. graveolens* plants. It has been reported that NS, due to high capacity to support electron exchange with iron (Fe^{2+}) and copper (Co^{3+}) ions, is one of the most potential candidates for modulating the biological redox reactions of the plants (Mukherjee and Mahapatra 2009). Our previous studies on different ornamental cultivars of *Pelargonium* showed that plant physiological and biochemical characteristics were significantly affected by various concentrations of silver nanoparticles (Hatami and Ghorbanpour 2013, 2014).

In another study, we carried out an experiment to evaluate the potential effects of different concentrations (0, 10, 20, 40, and 80 mg/L) of TiO_2 nanoparticles and bulk on growth parameters of *Salvia mirzayanii* plant at different stages of growth (Figs. 13.5 and 13.6) (Hatami et al. 2015). Results showed that at vegetative stages (45 days after planting) the maximum and the minimum growth indices were obtained at 10 and 80 mg/L concentrations of TiO_2 NPs, respectively. However, at flowering and reproductive stages the highest growth and shoot yield were

Fig. 13.5 Effect of different concentrations of TiO_2 nanoparticles and bulk on seedling growth of *Salvia mirzayanii* at early vegetative stage (Hatami et al. 2015)

observed at 20 mg/L TiO_2 NP concentration. However, we did not observe statistically significant differences among the other treatments and control plants based on growth indices.

The effects of different concentrations (0, 10, 100, and 500 mg/L) of nanosized (10–25 nm) titanium dioxide (TiO_2) on growth parameters and seed yield of

Fig. 13.6 Effect of different concentrations of TiO_2 nanoparticles on *Salvia mirzayanii* growth at late vegetative and flowering stages (Hatami et al. 2015)

Linum usitatissimum plant under sufficient and scarce water conditions were studied by Baiazidi-Aghdam et al. (2016). They found that the plant height and the number of subsidiary branches per plant were not significantly changed under employed treatments. However, the maximum plant height (49.8 cm) and the number of subsidiary branches per plant (8) were observed in plants exposed to

10 and 500 mg/L TiO_2 NPs, respectively. At drought stress, the number of capsules per plant decreased up to 33.35 compared to well-watered conditions. Moreover, the number of capsules per plant significantly changed in plants exposed to different concentrations of TiO_2 NPs. However, exposure to 10 and 500 mg/L TiO_2 NPs increased the number of capsules per plant by 22.5 and 9.98 % over untreated control plants under drought stress. Also they observed that TiO_2 NPs generally enhanced the seed weight under both normal and stress conditions. According to De Rosa et al. (2010) nano-silica particles absorbed by roots have been shown to form films at the cell walls, which can enhance the plant's resistance to stress and lead to improved yields.

Studies on the toxicity (phyto-cyto-geno- and ecotoxicity) of different nanomaterials are still emerging and have shown negative effects on the growth and development of various plants.

Ecotoxicity of silver nanoparticles is reported in a few studies as well. They are known to reduce cell growth, photosynthesis, and chlorophyll production of a marine diatom (*Thalassiosira weissflogii*) and in freshwater alga (*Chlamydomonas reinhardtii*), and these toxic effects are implied to be due to the release of dissolved silver (Miao et al. 2009; Navarro et al. 2008a, b).

Ma et al. (2010) reported that nano-CeO_2, nano-La_2O_3, and nano-Gd_2O_3 could significantly inhibit root elongation of seven plant species (radish, rape, tomato, lettuce, wheat, cabbage, and cucumber), but no inhibitions of root elongation in the six plants (except lettuce) were observed in nano-CeO_2 treatment. However, toxicity studies of several nanomaterials such as TiO_2, ZnO, MWCNTs, Cu, Si, and C60 fullerenes showed both negative and positive effects on plant growth on certain higher plants (Table 13.2).

Different studies showed contradictory results on the phytotoxicity of carbon-based nanomaterials in plants. The effects of functionalized and nonfunctionalized single-walled carbon nanotubes on root elongation of six different crop species [cabbage (*Brassica oleracea*), carrot (*Daucus carota*), cucumber (*Cucumis sativus*), lettuce (*Lactuca sativa*), onion (*Allium* sp.), and tomato (*Lycopersicon esculentum*)] have been extensively studied (Buzea et al. 2007). Also, inhibitory effect of MWCNTS on plants growth has been reported by many researchers (Tiwari et al. 2014; Ikhtiar et al. 2013; Begum et al. 2014). Thus, the effect of nanomaterials on plants varies from plant to plant, their growth stages, and the nature of nanoparticles.

References

Amooaghaie R, Tabatabaei F, Ahadi AM (2015) Role of hematin and sodium nitroprusside in regulating *Brassica nigra* seed germination under nano silver and silver nitrate stresses. Ecotoxicol Environ Saf 113:259–270

Arora S, Sharma P, Kumar S, Nayan R, Khanna PK, Zaidi MGH (2012) Gold-nanoparticle induced enhancement in growth and seed yield of *Brassica juncea*. Plant Growth Regul 66:303–310

Azimi R, Feizi H, Can K-HM (2013) Bulk and nanosized titanium dioxide particles improve seed germination features of wheatgrass (*Agropyron desertorum*). Not Sci Biol 5:325–331

Baiazidi-Aghdam MT, Mohammadi H, Ghorbanpour M (2016) Effects of nanoparticulate anatase titanium dioxide on physiological and biochemical performance of *Linum usitatissimum* (Linaceae) under well watered and drought stress conditions. Braz J Bot 39:139–146

Begum P, Ikhtiari R, Fugetsu B (2014) Potential impact of multi-walled carbon nanotubes exposure to the seedling stage of selected plant species. Nanomaterials 4:203–221

Buzea C, Blandino IIP, Robbie K (2007) Nanomaterials and nanoparticles: sources and toxicity. Biointerphases 2:MR17–MR172

Castiglione MR, Giorgetti L, Geri C, Cremonini R (2011) The effects of nano-TiO_2 on seed germination, development and mitosis of root tip cells of *Vicia narbonensis* L. and *Zea mays* L. J Nanopart Res 13:2443–2449

Corral-Diaz B, Peralta-Videa JR, Alvarez-Parrilla E, Rodrigo-Garcia J, Morales MI, Osuna Avila P, Niu G, Hernandez-Viezcas JA, Gardea-Torresdey JL (2014) Cerium oxide nanoparticles alter the antioxidant capacity but do not impact tuber ionome in *Raphanus sativus* (L.). Plant Physiol Biochem 84:277–285

De Rosa MC, Monreal C, Schnitzer M, Walsh R, Sultan Y (2010) Nanotechnology in fertilizers. Nat Nanotechnol 5:91

Dietz KJ, Herth S (2011) Plant nanotoxicology. Trends Plant Sci 16:582–589

El-Temsah YS, Joner EJ (2012) Impact of Fe and Ag nanoparticles on seed germination and differences in bioavailability during exposure in aqueous suspension and soil. Environ Toxicol 27:42–49

Feizi H, Rezvani-Moghaddam P, Shahtahmassebi N, Fotovat A (2012) Impact of bulk and nanosized titanium dioxide (TiO_2) on wheat seed germination and seedling growth. Biol Trace Elem Res 146:101–106

Feizi H, Kamali M, Jafari L, Rezvani MP (2013) Phytotoxicity and stimulatory impacts of nanosized and bulk titanium dioxide on fennel (*Foeniculum vulgare* Mill). Chemosphere 91:506–511

Ghorbanpour M (2015) Major essential oil constituents, total phenolics and flavonoids content and antioxidant activity of *Salvia officinalis* plant in response to nano-titanium dioxide. Indian J Plant Physiol 20:249–256

Ghorbanpour M, Hadian J (2015) Multi-walled carbon nanotubes stimulate callus induction, secondary metabolites biosynthesis and antioxidant capacity in medicinal plant *Satureja khuzestanica* grown in vitro. Carbon 94:749–759

Ghorbanpour M, Hatami M (2014) Spray treatment with silver nanoparticles plus thidiazuron increases anti-oxidant enzyme activities and reduces petal and leaf abscission in four cultivars of geranium (*Pelargonium zonale*) during storage in the dark. J Hortic Sci Biotechnol 89:712–718

Ghorbanpour M, Hatami H (2015) Changes in growth, antioxidant defense system and major essential oils constituents of *Pelargonium graveolens* plant exposed to nano-scale silver and thidiazuron. Indian J Plant Physiol 20:116–123

Ghorbanpour M, Hatami M, Hatami M (2015) Activating antioxidant enzymes, hyoscyamine and scopolamine biosynthesis of *Hyoscyamus niger* L. plants with nano-sized titanium dioxide and bulk application. Acta Agric Slov 105:23–32

Haghighi M, Teixeira da Silva JA (2014) The effect of N-TiO_2 on tomato, onion, and radish seed germination. J Crop Sci Biotechnol 17:221–227

Hashemi-Dehkourdi E, Mousavi MM (2013) Effect of anatase nanoparticles (TiO_2) on parsley seed germination (*Petroselinum crispum*) in vitro. Biol Trace Elem Res 155:283–286

Hatami M, Ghorbanpour M (2013) Effect of nanosilver on physiological performance of *Pelargonium* plants exposed to dark storage. J Hortic Res 21:15–20

Hatami M, Ghorbanpour M (2014) Defense enzymes activity and biochemical variations of *Pelargonium zonale* in response to nanosilver particles and dark storage. Turk J Biol 38:130–139

Hatami M, Hatamzadeh A, Ghasemnezhad M, Ghorbanpour M (2013) The comparison of antimicrobial effects of silver nanoparticles (SNP) and silver nitrate ($AgNO_3$) to extend the vase life of 'Red Ribbon' cut rose flowers. Trakia J Sci 2:144–151
Hatami M, Ghorbanpour M, Salehiarjomand H (2014) Nano-anatase TiO_2 modulates the germination behavior and seedling vigority of the five commercially important medicinal and aromatic plants. J Biol Environ Sci 8:53–59
Hatami M, Ghorbanpour M, Salehiarjomand H (2015) Evaluation of nanosized titanium dioxide (TiO_2) on primary growth parameters and secondary metabolites production in *Salvia mirzayanii* plants. Research project (contract number: 92. 13497), Arak University (In Persian)
Hatami M, Kariman K, Ghorbanpour M (2016) Engineered nanomaterial-mediated changes in the metabolism of terrestrial plants. Sci Total Environ 571:275–291
Hernandez-Viezcas JA, Castillo-Michel H, Servin AD, Peralta-Videa JR, Gardea orresdey JL (2011) Spectroscopic verification of zinc absorption and distribution in the desert plant *Prosopis julif loravelutina* (velvet mesquite) treated with ZnO nanoparticles. Chem Eng J 170:346–352
Hong J, Peralta-Videa JR, Rico CM, Sahi S, Viveros MN, Bartonjo J, Zhao L, Gardea-Torresdey JL (2014) Evidence of translocation and physiological impacts of foliar applied CeO_2 nanoparticles on cucumber (*Cucumis sativus*) plants. Environ Sci Technol 48:4376–4385
Ikhtiar R, Begum P, Watari F, Fugetsu B (2013) Toxic effect of multiwalled carbon nanotubes on lettuce (*Lactuca Sativa*). Nano Biomed 5:18–24
Joseph T, Morrison M (2006) Nanotechnology in agriculture and food. Institute of Nanotechnology, Nanoforum Organization. Available: http://www.nanoforum.org
Josko I, Oleszczuk P (2014) Phytotoxicity of nanoparticles—problems with bioassay choosing and sample preparation. Environ Sci Pollut Res 21:10215–10224
Karami-Mehrian S, Heidari R, Rahmani F (2015) Effect of silver nanoparticles on free amino acids content and antioxidant defense system of tomato plants. Indian J Plant Physiol 20:257–263
Karunakaran G, Suriyaprabha R, Manivasakan P, Yuvakkumar R, Rajendran V, Prabu P, Kannan N (2013) Effect of nanosilica and silicon sources on plant growth promoting rhizobacteria, soil nutrients and maize seed germination. IET Nanobiotechnol 7:70–77
Khodakovskaya M, Dervishi E, Mahmood M, Xu Y, Li Z, Watanabe F, Alexandru SB (2009) Carbon nanotubes are able to penetrate plant seed coat and dramatically affect seed germination and plant growth. ACS Nano 3:3221–3227
Kim JH, Lee Y, Kim EJ, Gu S, Sohn EJ, Seo YS, An HJ, Chang YS (2014) Exposure of iron nanoparticles to *Arabidopsis thaliana* enhances root elongation by triggering cell wall loosening. Environ Sci Technol 48:3477–3485
Kim JH, Oh Y, Yoon H, Hwang I, Chang YS (2015) Iron nanoparticle-induced activation of plasma membrane H^+-ATPase promotes stomatal opening in *Arabidopsis thaliana*. Environ Sci Technol 49:1113–1119
Lahiani MH, Dervishi E, Chen J, Nima Z, Gaume A, Biris AS, Khodakovskaya MV (2013) Impact of carbon nanotube exposure to seeds of valuable crops. ACS Appl Mater Interfaces 5:7965–7973
Larue C, Khodja H, Herlin-Boime N, Brisset F, Flank AM, Fayard B, Chaillou S, Carrier M (2011) Investigation of titanium dioxide nanoparticles toxicity and uptake by plants. J Phys 304, 012057
Lee WM, An YJ, Yoon H, Kwbon HS (2008) Toxicity and bioavailability of copper nanoparticles to the terrestrial plants mung bean (*Phaseolus radiatus*) and wheat (*Triticum aestrivum*): plant agar test for water-insoluble nanoparticles. Environ Toxic Chem 27:1915–1921
Lee S, Kim S, Kim S, Lee I (2013) Assessment of phytotoxicity of ZnO NPs on a medicinal plant, *Fagopyrum esculentum*. Environ Sci Pollut Res 20:848–854
Lin D, Xing B (2007) Phytotoxicity of nanoparticles: inhibition of seed germination and root growth. Environ Pollut 150:243–250
Lu CM, Zhang CY, Wu JQ, Tao MX (2002) Research of the effect of nanometer on germination and growth enhancement of *Glycine max* and its mechanism. Soybean Sci 21:168–172

Ma Y, Kuang L, He X, Bai W, Ding Y, Zhang Z, Zhao Y, Chai Z (2010) Effects of rare earth oxide nanoparticles on root elongation of plants. Chemosphere 78:273–279

Miao AJ, Schwehr K, Xu C, Zhang AJ, Luo Z, Quigg A (2009) The algal toxicity of silver engineered nanoparticles and detoxification by exopolymeric substances. Environ Pollut 157:3034–3041

Mohammadi R, Maali-Amiri R, Abbasi A (2013) Effect of TiO_2 nanoparticles on chickpea response to cold stress. Biol Trace Elem Res 152:403–410

Mousavi Kouhi SM, Lahouti M, Ganjeali A, Entezari MH (2014) Comparative phytotoxicity of ZnO nanoparticles, ZnO microparticles, and Zn^{2+} on rapeseed (*Brassica napus* L.): investigating a wide range of concentrations. Toxicol Environ Chem 96:861–868

Mukherjee M, Mahapatra A (2009) Effect of coinage metal nanoparticles and zwitterionic surfactant on reduction of [Co(NH3)5Cl](NO3)2 by iron(III). Colloid Surf 350:1–7

Nair R, Poulose AC, Nagaoka Y, Yoshida Y, Maekawa T, Kumar DS (2011) Uptake of FITC labeled silica nanoparticles and quantum dots by rice seedlings: effects on seed germination and their potential as biolabels for plants. J Fluoresc 21:2057–2068

Navarro E, Baun A, Behra R, Hartmann NB, Filser J, Miao AJ, Quigg A, Santschi PH, Sigg L (2008a) Environmental behavior and ecotoxicity of engineered nanoparticles to algae, plants, and fungi. Ecotoxicology 17:372–386

Navarro E, Piccapietra F, Wagner B, Marconi F, Kaegi R, Odzak N, Sigg L, Behra R (2008b) Toxicity of silver nanoparticles to *Chlamydomonas reinhardtii*. Environ Sci Technol 42:8959–8964

Nejatzadeh-barandozi F, Darvishzadeh F, Aminkhani A (2014) Effect of nano silver and silver nitrate on seed yield of (*Ocimum basilicum* L.). Org Med Chem Lett 4:11

Parveen A, Rao S (2015) Effect of nanosilver on seed germination and seedling growth in *Pennisetum glaucum*. J Clust Sci 26:693–701

Raskar SV, Laware SL (2014) Effect of zinc oxide nanoparticles on cytology and seed germination in onion. Int J Curr Microbiol App Sci 3:467–473

Rico CM, Hong J, Morales MI, Zhao L, Barrios AC, Zhang JY, Peralta-Videa JR, Gardea Torresdey JL (2013) Effect of cerium oxide nanoparticles on rice: a study involving the antioxidant defense system and in vivo fluorescence imaging. Environ Sci Technol 47:5635–5642

Ruffini Castiglione M, Giorgetti L, Geri C, Cremonini R (2011) The effects of nano-TiO_2 on seed germination, development and mitosis of root tip cells of *Vicia narbonensis* L. and *Zea mays* L. J Nanopart Res 1:2443–2449

Scrinis G, Lyons K (2007) the emerging nano-corporate paradigm: nanotechnology and the transformation of nature, food and agri-food systems. Int J Soc Agric Food 15:22–44

Siddiqui MH, Al-Whaibi MH, Faisal M, Al Sahli AA (2014) Nano-silicon dioxide mitigates the adverse effects of salt stress on *Cucurbita pepo* L. Environ Toxicol Chem 33:2429–2437

Singh D, Kumar A (2015) Effects of nano silver oxide and silver ions on growth of *Vigna radiate*. Bull Environ Contam Toxicol 95:379–384

Song U, Shin M, Lee G, Roh J, Kim Y, Lee EJ (2013) Functional analysis of TiO_2 nanoparticle toxicity in three plant species. Biol Trace Elem Res 155:93–103

Stampoulis D, Sinha SK, White JC (2009) Assay-dependent phytotoxicity of nanoparticles to plants. Environ Sci Technol 43:9473–9479

Suriyaprabha R, Karunakaran G, Yuvakkumar R, Rajendran V, Kannan N (2012) Silica nanoparticles for increased silica availability in maize (*Zea mays* L.) seeds under hydroponic conditions. Curr Nanosci 8:902–908

Thuesombat P, Hannongbua S, Akasit S, Chadchawan S (2014) Effect of silver nanoparticles on rice (*Oryza sativa* L. cv. KDML 105) seed germination and seedling growth. Ecotoxicol Environ Saf 104:302–309

Tiwari DK, Dasgupta-Schubert N, Cendejas LMV, Villegas J, Montoya LC, Garcia SEB (2014) Interfacing carbon nanotubes (CNT) with plants: enhancement of growth, water and ionic nutrient uptake in maize (*Zea mays*) and implications for nanoagriculture. Appl Nanosci 4:577–591

Ushahra J, Bhati-Kushwaha H, Malik CP (2014) Biogenic nanoparticle-mediated augmentation of seed germination, growth, and antioxidant level of *Eruca sativa* Mill. varieties. Appl Biochem Biotechnol 174:729–738

Viana Cde O, Vaz RP, Cano A, Santos AP, Cançado LG, Ladeira LO, Junior AC (2015) Physiological changes of the lichen *Parmotrema tinctorum* as result of carbon nanotubes exposition. Ecotoxicol Environ Saf 120:110–116

Wang ZY, Xie XY, Zhao J, Liu XY, Feng WQ, White JC, Xing BS (2012) Xylem- and phloem based transport of CuO nanoparticles in maize (Zea mays L.). Environ Sci Technol 46:4434–4441

Wang S, Wang F, Gao S (2015) Foliar application with nano-silicon alleviates Cd toxicity in rice seedlings. Environ Sci Pollut Res 22:2837–2845

Wu SG, Huang L, Head J, Chen DR, Kong IC, Tang YJ (2012) Phytotoxicity of metal oxide nanoparticles is related to both dissolved metals ions and adsorption of particles on seed surfaces. J Pet Environ Biotechnol 3:712–749

Xiang L, Zhao HM, Li YW, Huang XP, Wu XL, Zhai T, Yuan Y, Cai QY, Mo CH (2015) Effects of the size and morphology of zinc oxide nanoparticles on the germination of Chinese cabbage seeds. Environ Sci Pollut Res 22:10452–10462

Yang L, Watts DJ (2005) Particle surface characteristics may play an important role in phytotoxicity of alumina nanoparticles. Toxicol Lett 158:122–132

Yasur J, Rani PU (2014) Environmental effects of nanosilver: impact on castor seed germination, seedling growth, and plant physiology. Environ Sci Pollut Res Int 20:8636–8648

Zhang P, Ma Y, Zhang Z, He X, Guo Z, Tai R, Ding Y, Zhao Y, Chai Z (2012) Comparative toxicity of nanoparticulate/bulk Yb_2O_3 and $YbCl_3$ to cucumber (*Cucumis sativus*). Environ Sci Technol 46:1834–1841

Zhang M, Gao B, Chen J, Li Y (2015) Effects of graphene on seed germination and seedling Growth. J Nanopart Res 17:1–8

Zhao L, Peng B, Hernandez-Viezcas JA, Rico C, Sun Y, Peralta-Videa JR, Tang X, Niu G, Jin L, Varela-Ramirez A, Zhang JX, Gardea-Torresdey JL (2012) Stress response and tolerance of *Zea mays* to CeO_2 nanoparticles: cross talk among H_2O_2, heat shock protein, and lipid peroxidation. ACS Nano 6:9615–9622

Zheng L, Hong F, Lu S, Liu C (2005) Effect of nano-TiO_2 on strength of naturally aged seeds and growth of spinach. Biol Trace Elem Res 104:83–91

Zhu ZJ, Wang H, Yan B, Zheng H, Jiang Y, Miranda OR, Rotello VM, Xing B, Vachet RW (2012) Effect of surface charge on the uptake and distribution of gold nanoparticles in four plant species. Environ Sci Technol 46:12391–12398

Chapter 14
Role of Nanoparticles on Plant Growth with Special Emphasis on *Piriformospora indica*: A Review

Ajit Varma, Uma, and Manika Khanuja

14.1 Introduction

Currently, the wide spectrum of challenges is faced by world agriculture scientist including climate change, urbanization, and environmental issues: accumulation of insecticides and pesticides, decay in soil organic matter, and sustainable use of natural resources. These challenges are going to be further intensified due to increase in food demand. Nanotechnology has significant benefits on food and agriculture system. Through nanotechnology, optimization of agriculture inputs (viz. nano-pesticides, nano-herbicides) to enhance the effectiveness of the active ingredients including targeted delivery and release and less dosage per application and to reduce by-products that otherwise degrade ecosystem can be achieved.

14.1.1 Arbuscular Mycorrhiza "Piriformospora indica"

Piriformospora indica is an arbuscular mycorrhizae and was discovered from Thar Desert of western India (Jaisalmer, Rajasthan) by Prof. Dr. Ajit Varma in 1992 (Fig. 14.1). *Piriformospora indica* (*Hymenomycetes*, *Basidiomycota*), a novel cultivable, can colonize monocot as well as dicot, is a root endophyte fungus, and

A. Varma (✉) • Uma
Amity Institute of Microbial Technology, Amity University Uttar Pradesh, E-3 Block, Fourth Floor, Sector 125, Noida, Uttar Pradesh 201303, India
e-mail: ajitvarma@amity.edu; uma.singhal87@gmail.com

M. Khanuja
Centre for Nanoscience and Nanotechnology, Jamia Millia Islamia,
New Delhi 110025, India
e-mail: manikakhanuja@gmail.com

M. Ghorbanpour et al. (eds.), *Nanoscience and Plant–Soil Systems*, Soil Biology 48,
DOI 10.1007/978-3-319-46835-8_14

Fig. 14.1 Gold mine for microorganisms (*P. indica* was discovered from Thar Desert of India by Prof. Dr. Ajit Varma in 1992)

possesses multidimensional actions like plant growth promoter, biofertilizer, immunomodulator, obviates biotic (pathogens, hyphal feeders, and organic matter) and abiotic (water stress, heavy metal toxicity, soil structure, pH, salt) stresses, bio-herbicide, and phytoremediator (Varma et al. 2012a; Prasad et al. 2013; Gill et al. 2016). *P. indica* is an axenically cultivable phytopromotional endosymbiont, which mimics the capabilities of arbuscular mycorrhizal fungi (AMF) (Fig. 14.2). *P. indica* belongs to Sebacinaceae; its relation to other symbiotic arbuscular mycorrhizal fungi has been identified by molecular methods (Fig. 14.3). The entire genome of the fungus has been sequenced.

14.1.2 Nanoparticles

In nanotechnology, ultrafine particles of size range 1–100 nm are termed as nanoparticles. Nanoparticle research is an area of intense scientific research as these particles possess novel, improved, and tunable properties with potential applications in multiple domains including physical, chemical, biological, health

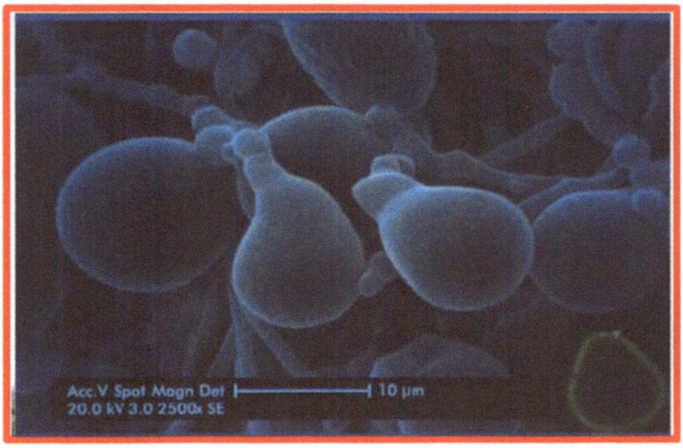

Fig. 14.2 Typical pear-shaped spores of model organism: *Piriformospora indica* (c.f. Varma et al. 2013)

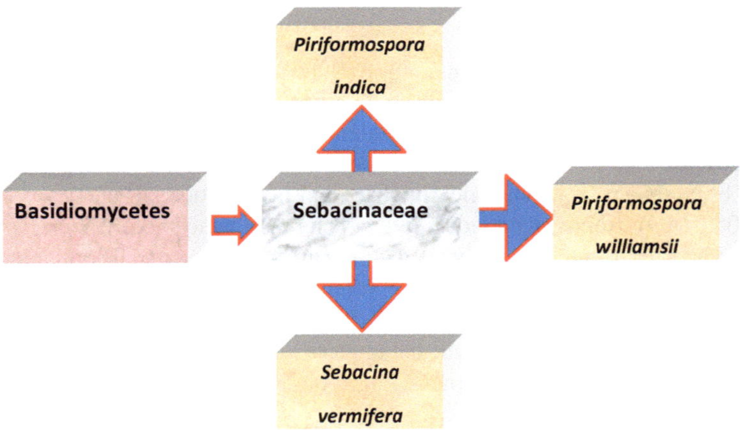

Fig. 14.3 Molecular taxonomic positions of *P. indica*

sciences, and agriculture. The recent usage of nanomaterials in agricultural crops has already been reported to (1) enhance plant growth; (2) deliver (a) DNA to plants called genetic engineering (Chang et al. 2013), (b) hormones, and (c) vaccines (Makidon et al. 2010); and (3) detect pathogens using nanosensors (Inbaraj and Chen 2015). Thus, agricultural global challenges such as food scarcity, climate change, and the limited availability of important plant nutrients such as phosphorus and potassium can be successfully addressed by nanotechnology-based nanotools (Prasad et al. 2014).

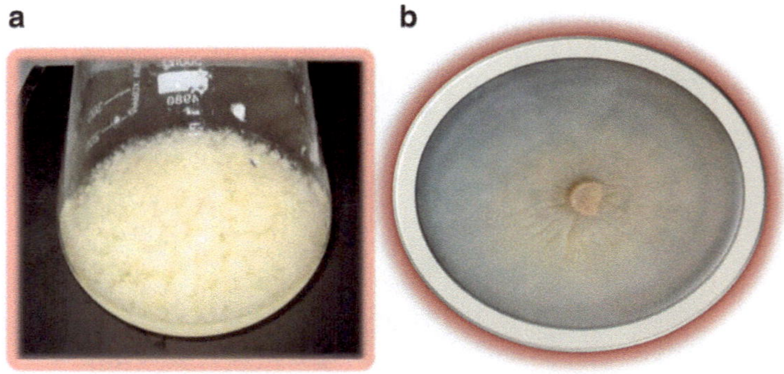

Fig. 14.4 *P. indica* grown on (**a**) HK broth, (**b**) HK plate at 28 ± 2 °C (photographed after 10 days of inoculation)

14.1.3 Nanotechnology and **Piriformospora indica** *"Nanoembedded* **Piriformospora indica***"*

In the present chapter, the novel nanotool "Nanoembedded *Piriformospora indica*" and its role in increasing agricultural productivity and restoring ecosystem balance are described. Research studies revealed that the synergetic association of nanoparticles with *P. indica* called "Nanoembedded *P. indica*" results in enhanced fungal biomass, spore count, thick hyphae, and less vacuoles (Patent no. 14/DEL/2003; 08.01.09). Nanoembedded *P. indica* is a novel nano-biofertilizer, has shown early germination, and has enhanced plant growth (root and shoot length, leaf count) on wide variety of plants. The effect of Nanoembedded *Piriformospora indica* on plant growth is compared with *Piriformospora indica* alone (referred as control). The various plant growth parameters were analyzed including seed germination index, root and shoot length, chlorophyll content, flavonoid (Mishra et al. 2014).

14.2 *Piriformospora indica*

14.2.1 Growth Conditions of **P. indica**

The root-colonizing fungus *P. indica* (Fig. 14.4) is cultured in modified Hill and Kaefer medium. It is shown that the fungus can grow axenically on different synthetic media. Among the tested media, the best growth is reported to be on Hill and Kaefer medium. Circular agar disks (4 mm diameter) infested with chlamydospores and actively growing hyphae of *P. indica* are placed onto petri dishes containing solidified Hill and Kaefer medium. Incubation is carried out at 25 °C in dark for 7–10 days. Broth jars are constantly shaken at 80 rpm. After 7–10

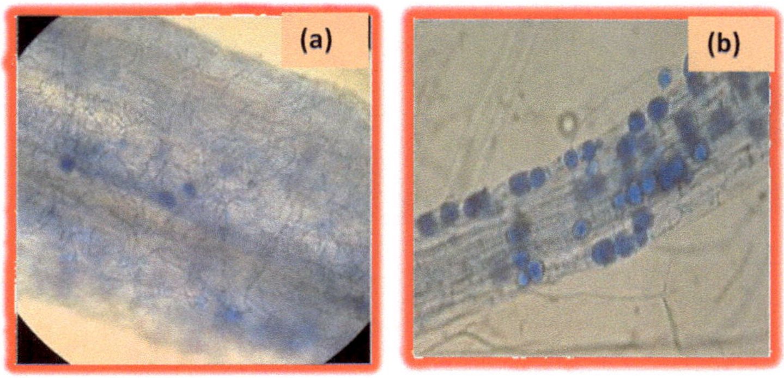

Fig. 14.5 Root colonization of *A. thaliana* after 18 days co-cultivation with *P. indica* on PNM medium (**a**) hyphae and young chlamydospores; (**b**) mature chlamydospores inside the root cells (Ref, Irena Sherameti et al., Chapter 20, Sebacinales)

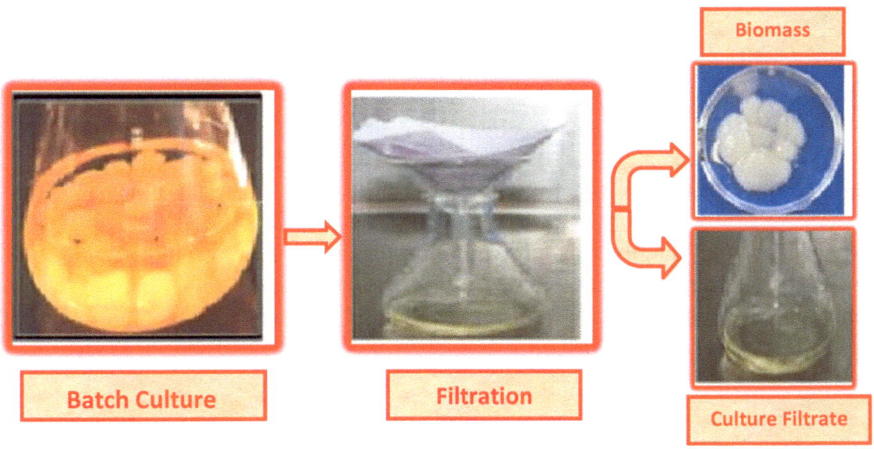

Fig. 14.6 Separation of *P. indica* biomass and culture filtrate

days, the petri plate is completely filled up with biomass. In broth jars, small and large colonies appear which consist of hyphae and chlamydospores. Spores are extrametrical and intercellular (Fig. 14.5).

14.2.2 Separation of P. indica *Biomass and Culture Filtrate*

One of the unique features of the fungus is that the culture filtrate also acts as an excellent source for plant promotion. Culture filtrate was separated from fungal biomass using simple filtration procedure (Fig. 14.6). In an independent

Fig. 14.7 Protocol for seed treatment with fungal biomass formulated with talcum powder

experiment, the fungus was grown in broth. After 10 days the biomass was removed. The culture filtrate caused early seed germination and flowering. In the long run, the culture filtrate may serve as liquid biofertilizer (Uttamkumar et al. 2013).

14.2.3 Formulation of Rootonic for Field Application

The Rootonic biofertilizer was formulated to enhance its handling, storage, propagation, and overall convenience of use by a common farmer. To enhance the usage so that the benefits of the fungus are used by common farmer, it was formulated with magnesium silicate which acts as a carrier, and quantity for field trial was optimized. For this, 2 % (w/w) formulation served as effective and stable carrier. On an average the colony forming unit (CFU) count was maintained as 108 and moisture 20 %. Protocol for seed treatment for field trial has been given in Fig. 14.7.

14.3 Applications of *Piriformospora indica*

14.3.1 Plant Growth Promotion

P. indica is a wide-host symbiotic fungus and colonizes members of bryophytes, pteridophytes, gymnosperms, and angiosperms, monocots and dicots including orchids and members of the Brassicaceae (e.g., *Arabidopsis thaliana*) (Peskan-Berghofer et al. 2004). Plants colonized by *P. indica* display a wide range of beneficial effects including enhanced host growth and resistance to biotic and abiotic stresses, promotion of adventitious root formation in cuttings, and enhanced nitrate and phosphate assimilation (Zuccaro et al. 2011). They not only act as a plant promoter but also bio-protectant against pathogens (Waller et al. 2005; Deshmukh et al. 2006). Baltruschat et al. in 2008 studied biochemical mechanisms underlying *P. indica*-mediated salt tolerance in barley with special focus on anti-oxidants. *P. indica*-colonized barley roots in salt stress conditions had increased plant growth, elevated amount of ascorbic acid, and increased activities of antioxidant. These findings have suggested that antioxidants might play a role in both inherited and endophyte-mediated plant tolerance to salinity as reported in *Brassica napus* L. (Chen et al. 2012; Varma et al. 2012a, b). The fungus-treated *Brassica* plants showed significant increase in the size and numbers of their leaves and the weights of their fresh roots, dry roots and shoots, and early flowering and increased seed yield and oil content. Nutritional analysis revealed that fungus-treated plants had reduced erucic acid and glucosinolate contents and increased accumulation of N, P, K, S, and Zn. Also, RT-PCR results showed that the expression of Bn-FAE1 and BnECR genes, encoding enzymes responsible for regulating erucic acid bio-synthesis, was downregulated at mid- and late-life stages during seed development in colonized plants (Binggen Lou-personal communication). Thus, the results confirmed that *P. indica* plays an important role in enhancing growth, seed yield, and seed quality of *Brassica napus*.

A large number of medicinal plants like *Spilanthes calva*, *Withania somnifera*, *Bacopa monnieri*, *Coleus forskohlii*, and others were inoculated with the *P. indica* in pots as well as in fields to study its influence on the host plants. Pictures of growth promotional effect of *P. indica* on few plants have been given in (Table 14.1). The effect of *P. indica* on growth of *Bacopa monnieri*, *A. vasica*, and *S. calva* (Fig. 14.8) has been given.

14.3.2 Protection Against Pathogens and Insects

In addition to plant promotion, this fungus also protects the plant against pathogens and insects. Some of the plants' growth promotional factors involved in *Arabidopsis* protection are glucosinolates, ethylene, etc. (Fig. 14.9A). *P. indica*

Table 14.1 Beneficial effect of *Piriformospora indica* on crop plant seeds for better yield and value addition

Crop type	*P. indica*-mediated improved productivity	References
Stevia rebaudiana	Improved vegetative growth	Rai et al. (2001)
Spilanthes calva and *Withania somnifera*	Increased growth	Varma et al. (2013)
Artemisia annua	Increased leaf area fresh biomass	Rai et al. (2001)
Coleus forskohlii	Modulation of secondary metabolites production, flowering and growth performance	Das et al. (2012)
Chickpea and *black lentil*	Improved endogenous NPK and growth performance	Nautiyal et al. (2010)
Phaseolus sp.	Improved root length and root dry weight	Tuladhar et al. (2013)
Phaseolus sp.	Elevated shoot length and shoot dry weight	Tuladhar et al., (2013)
Linum album	Enhanced growth	Kumar et al. (2013)
Piper nigrum	Increased leaf number and improved fresh weight	Anith et al. (2011)
Lycopersicon esculentum	Better growth and development	Fakhro et al. (2010)
Cicer arietinum	Increased total dry weight	Nautiyal et al. (2010)
Brassica rapa	Auxin-mediated plant growth and development	Michal-Johnson et al. (2013)
Arabidopsis thaliana	Enhanced seed production	Shahollari et al. (2007)
Helianthus annuus	Higher seed yield with increased oil content	Bagde et al. (2010)
Jatropha and Populus	Early seed germination	Varma et al. (2013)
Triticum aestivum	Growth sustainability under salt	Zarea et al. (2012)
H. vulgare	Relieve the plants from the attack of leaf pathogens to maintain plant growth performance	Molitor et al. (2011)
Wheat (Triticum sp.)	Stable growth performance of plant against biotic stress	Serfling et al. (2007)
Brassica campestris sp. Chinensis	Consistent growth under drought	Sun et al. (2010)
L. esculentum	Defend plants to sustain plant productivity under biotic stress	Andrade-Linares et al. (2013)
Arabidopsis sp.	Retention plant productivity under water deficit condition	Sherameti et al. (2008)
Hordeum vulgare cv. Ingrid	Salinity tolerance to maintain proper vegetative growth	Waller et al. (2005)
Hordeum vulgare	Salinity leads to increased productivity	Baltruschat et al. (2008)

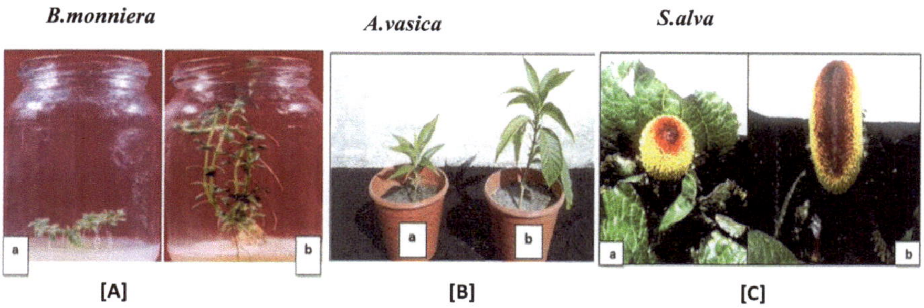

Fig. 14.8 Effect of *P. indica* on (**A**) in vitro grown *Bacopa monnieri*, (**a**) control (without *P. indica*) and (**b**) inoculated with *P. indica*; (**B**) *A. vasica*, (**a**) control (without *P. indica*) and (**b**) inoculated with *P. indica*; (**C**) pronounced growth response and flowering in *S. calva* after inoculation with *P. indica*, (**a**) control (without *P. indica*) and (**b**) inoculated with *P. indica*

co-inoculation protected bottle gourd and *Nicotiana tabacum* from other fungal and viral infection leading to healthy growth of plant (Fig. 14.9B).

14.3.2.1 Glucosinolates

A second class of plant glycosides, called the glucosinolates or mustard oil glycosides, break down to release defensive substances. Found principally in the Brassicaceae and related plant families, glucosinolates break down to produce the compounds responsible for the smell and taste of vegetables such as cabbage, broccoli, and radishes. Glucosinolate breakdown is catalyzed by a hydrolytic enzyme, called a thioglucosidase or myrosinase that cleaves glucose from its bond with the sulfur atom. These defensive products function as toxins and herbivore repellents. Like cyanogenic glycosides, glucosinolates are stored in the intact plant separately from the enzymes that hydrolyze them, and they are brought into contact with these enzymes only when the plant is crushed. Several studies have reported that glucosinolates exhibit growth inhibition or feeding deterrence to a wide range of general herbivores such as birds, slugs, and generalist insects (Giamoustaris and Mithen 1996; Giamoustaris and Mithen 1995). It was also found that plants respond to herbivore or insect damage by systematically accumulating higher levels of glucosinolates and thus presumably increasing their resistance (Martin and Müller 2006). Usually it is the indole glucosinolates which become induced.

14.3.2.2 Ethylene

The plant response to damage by insect herbivores involves both a wound response and the recognition of certain insect-derived compounds referred to as elicitors. Although repeated mechanical wounding can induce responses similar to those

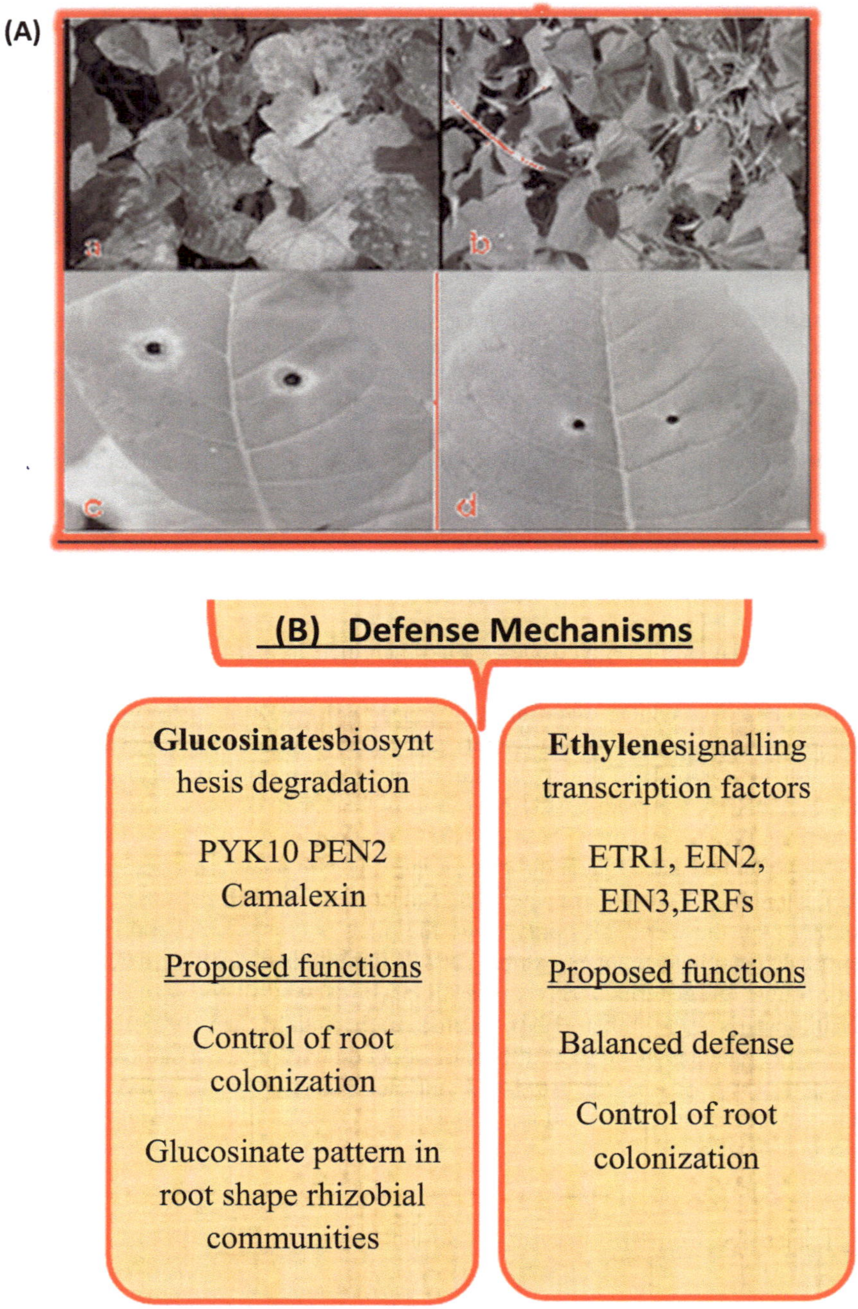

Fig. 14.9 (**A**) *P. indica* protects against plant pathogens like fungi and viruses. (**a**) Bottle gourd infested with insects and virus in the field. (**b**) Bottle gourd plants treated by *P. indica* are healthy. (**c**) *Alternaria longipes* infection status of untreated *Nicotiana tabacum* and (**d**) *P. indica*-

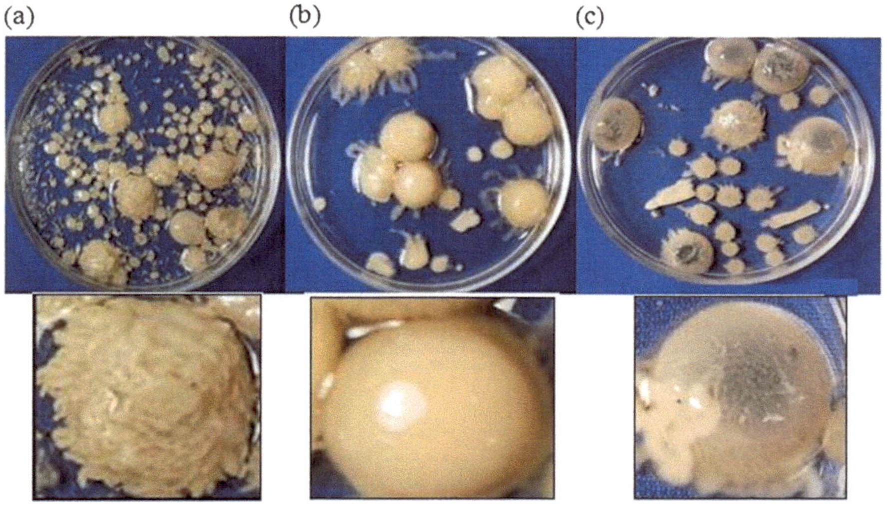

Fig. 14.10 The fungus *Piriformospora indica* interacting with the nanomaterials: control, titanium dioxide nanoparticles (TiO_2), carbon nanotubes (CNT), and activated charcoal. (**a**) Control without nanoparticles—fungal surface is rough and overall size of the colonies is not very large. (**b**) TNP environment—colonies are larger in size, more smooth, and spherical. (**c**) CNT environment—colonies are bigger, and morphologically it is not smooth but protruding outwards (Ref: Ajit Varma et al. 2013, Chapter 21)

caused by insect herbivorous in some plants, certain molecules in insect saliva can serve as enhancers of this stimulus. In addition, such insect-derived elicitors can trigger signaling pathways systemically, thereby initiating defensive responses in distant regions of the plant in anticipation of further damage. After being regurgitated by an insect, elicitors become part of its saliva and are thus applied to the feeding site during herbivory. Plants then recognize these elicitors and activate a complex signal transduction pathway that induces their defenses. Ethylene is one of the signaling compounds induced by insect herbivory. In many cases, the concerted action of ethylene is necessary for the full activation of induced defenses.

14.3.3 Stress Tolerance

P. indica also shows tolerance toward stresses of extremes of climate that is very high and very low temperature and also salt stress (Frank et al. 2005; Gill et al. 2016).

Fig. 14.9 (continued) colonized plants (Courtesy of Amit Kharkwal and Bingganlau). (**B**) Plant defense compounds identified in the beneficial interaction between *P. indica* and *Arabidopsis*

Table 14.2 Percentage increase in growth of *P. indica* after co-inoculation with nanoparticles (Ajit Varma, V.K. Jain, Suman Ram Prasad, Patent no. 14/DEL/2003; 08.01.09) entitled A Nanomaterial Based Culture Medium For Microbial Growth Enhancement

Nanoparticles	Fresh biomass (g/100 ml)	Percent increase over control
Control (no addition)	2.98	–
Titanium dioxide (TiO_2)	4.12	38.25
Carbon nanotubes (CNT)	3.86	29.53
Silver (Ag)	3.48	16.77
Zinc oxide (ZnO)	4.61	56

It also shows synergism with other plant growth bacterium like *Azospirillum brasilense*.

14.4 Effect of Nanoparticles on *P. indica*

Previous research in application of nanotechnology in agriculture has inspired the author to work with this novel organism expecting enhancement in growth promotion capacity. Changes in growth pattern of *P. indica* after interaction with nanoparticles have been demonstrated in picture (Fig. 14.10). Interaction with nanoparticles showed percentage enhancement in growth (Table 14.2). Nanomaterials enter into fungal cells and act as carrier to nutrient molecules, thus enhancing growth (Fig. 14.11). Co-inoculation with nanoparticles resulted in enhanced growth of *P. indica*.

Fungal biomass interacted with nanomaterial by introduction of nanoparticles into liquid nutrient medium. After optimum incubation, nanomaterial-embedded *P. indica* biomass and culture filtrate were separated by the protocol given in Fig. 14.12.

Nanomaterial-embedded *P. indica* biomass and culture filtrate were used to treat seeds of plant in a similar manner as that of control organism, and its effect on growth promotional quality was studied for few plants. Test trials on broccoli seeds have shown remarkable enhancement in growth (Fig. 14.12 and Table 14.3).

The nanomaterial-treated *P. indica* culture filtrate has also shown to be helpful in seed germination and growth of seedlings in plants. In future, it can be utilized not only as a microbial growth enhancer but also as a potential tool for early diagnosis of diseases.

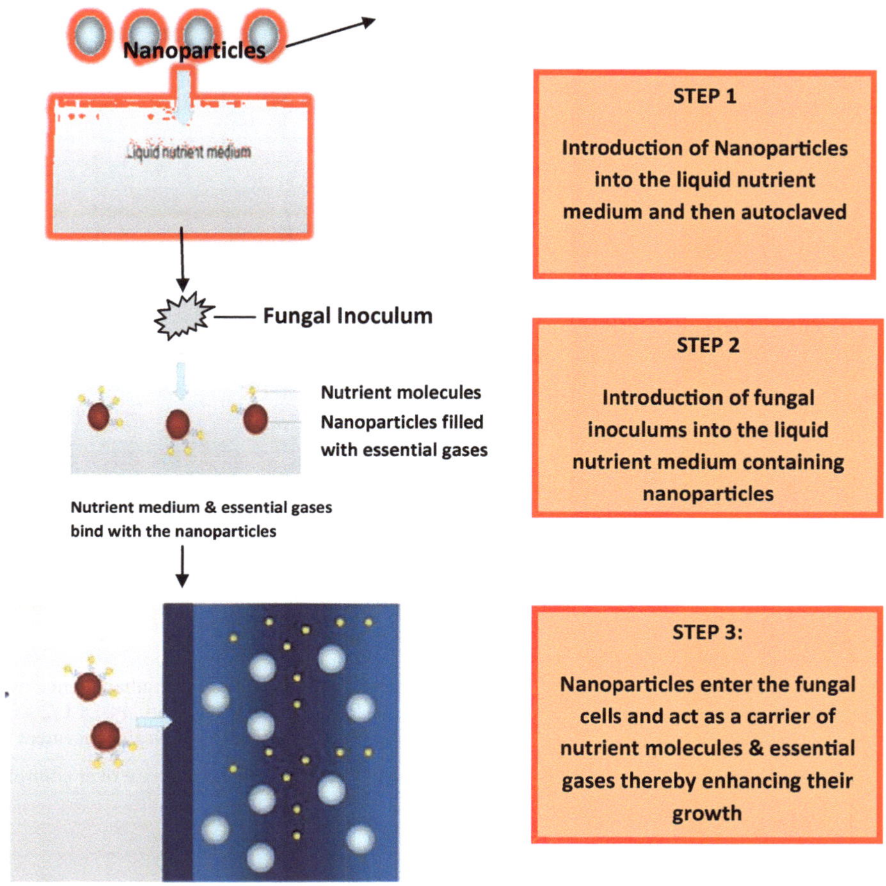

Fig. 14.11 Schematic to illustrate mechanism of fungal interaction with nanoparticles

14.5 Conclusion

In summary, Nanoembedded *P. indica* has emerged as a novel nanotool that has the potential to overcome the existing agricultural challenges including low productivity and ecosystem imbalance through its potential use as plant growth promoter, biofertilizer, and protector against biotic and abiotic stress. The synergetic association of *P. indica* with nanoparticles has shown encouraging effects like enhanced fungal biomass, increased spore count, thick hyphae, and less vacuoles; thus, nanotechnology and *P. indica* together have shown promising effects on plants in terms of the betterment of agricultural production quality as well as quantity.

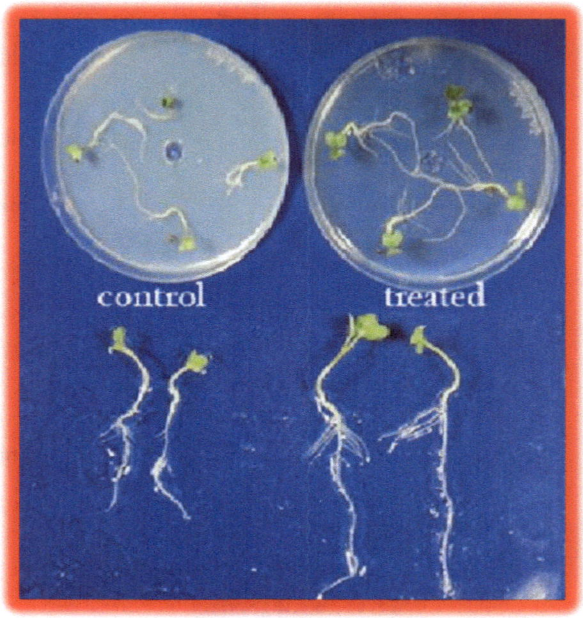

Fig. 14.12 Test trial results on interaction of broccoli seeds with nanomaterial-embedded *P. indica* showing improved growth (Varma et al. 2013)

Table 14.3 Data giving details of result of co-inoculation with broccoli seeds resulted in increase in the seedling length (Ajit Varma, V.K. Jain, Suman Ram Prasad, Patent no. 14/DEL/2003; 08.01.09) entitled A Nanomaterial Based Culture Medium for Microbial Growth Enhancement

Nanoparticles	Seedling length (cm)	Percentage increase over control
Control (no addition)	5.25	–
Titanium dioxide (TiO_2)	8.50	61.90
Carbon nanotubes (CNT)	6.87	30.85
Silver (Ag)	6.35	20.95
Zinc oxide (ZnO)	10.5	50

14.6 Future Prospects

Nanotechnology has significant benefits on food and agriculture system. The development of potential metal oxide nanostructured materials along with plant growth promoting properties contributes to enhanced agro-horticulture productivity and reduces by-products that harm environment or human health.

Acknowledgment Authors (Ajit Varma and Uma) are thankful to ICAR, BIRAC, DST, and DBT for partial funding.

References

Andrade-Linares DR, Müller A, Fakhro A, Schwarz D, Franken P (2013) Impact of *Piriformospora indica* on tomato. Soil Biol 33:107–117

Anith KN, Faseela KM, Archana PA, Prathapan KD (2011) Compatibility of *Piriformospora indica* and *Trichoderma harzianum* as dual inoculants in black pepper (*Piper nigrum* L.). Symbiosis 55:11–17

Bagde US, Ram P, Varma A (2010) Mass cultivation of *Piriformospora indica* in New Brunswick Fermenter and its formulation as biofertilizer. Asian J Microbiol Biotechnol Environ Sci 12:911–916

Baltruschat H, Fodor J, Harrach BD, Niemczyk E, Barna B, Gullner G, Janeczko A, Kogel KH, Schaefer P, Schwarczinger I, Zuccaro A, Skoczowski A (2008) Salt tolerance of barley induced by the root endophyte *Piriformospora indica* is associated with a strong increase in antioxidants. New Phytol 180:501–510

Chang F-P, Kuang L-Y, Huang C-A, Jane W-N, Hung Y, Hsingie Y, Mou C-Y (2013) A simple plant gene delivery system using mesoporous silica nanoparticles as carriers. J Mater Chem B 1:5279–5287

Chen HJ, Wu SD, Huang GJ, Shen CY, Afiyanti M, Li WJ, Lin YH (2012) Expression of a cloned sweet potato catalase SPCAT1 alleviates ethephon-mediated leaf senescence and H_2O_2 elevation. J Plant Physiol 169:86–97

Das A, Kamal S, Shakil Najam A, Sherameti I, Oelmuller R, Dua M, Tuteja N, Johri AK, Varma A (2012) The root endophyte fungus *Piriformospora indica* leads to early flowering, higher biomass and altered secondary metabolites of the medicinal plant, *Coleus forskohlii*. Plant Signal Behav 7:1–10

Deshmukh S, Hueckelhoven R, Schaefer P, Imani J, Sharma M, Weiss M, Waller F, Kogel KH (2006) The root endophytic fungus *Piriformospora indica* requires host cell death for proliferation during mutualistic symbiosis with barley. Proc Natl Acad Sci USA 103:18450–18457

Fakhro A, Andrade-Linares DR, Bargen S, Bandte M, Buttner C, Grosch R, Schwarz D, Franlen P (2010) Impact of *Piriformospora indica* on tomato growth and on interaction with fungal and viral pathogens. Mycorrhiza 20:191–200

Frank W, Beate A, Helmut B, Jozsef F, Katja B, Marina F, Tobias H, Ralph H, Christina N, von Diter W, Philipp F, Karl-Heinz K (2005) The endophytic fungus *Piriformospora indica* reprograms barley to salt-stress tolerance, disease resistance, and higher yield. Proc Natl Acad Sci USA 102:13386–13391

Giamoustaris A, Mithen R (1995) The effect of modifying the glucosinolate content of leaves of oilseed rape (*Brassica napus* ssp. Oleifera) on its interaction with specialist and generalist pests. Ann Appl Biol 126:347–363

Giamoustaris A, Mithen R (1996) The effect of flower colour and glucosinolates on the interaction between oilseed rape and pollen beetles. Entomol Exp Appl 80:206–208

Gill SS, Gill R, Trivedi DK, Anjum NA, Sharma KK, Ansari MW, Johri AK, Prasad R, Pereira E, Varma A, Tuteja N (2016) *Piriformospora indica*: potential and significance in plant stress tolerance. Front Microbiol. doi:10.3389/fmicb.2016.00332

Inbaraj BS, Chen BH (2015) Nanomaterial-based sensors for detection of foodborne bacterial pathogens and toxins as well as pork adulteration in meat products. J Food Drug Anal 24:15–28

Kumar V, Sahai V, Bisaria VS (2013) Effect of *Piriformospora indica* on enhanced biosynthesis of anticancer drug, podophyllotoxin, in plant cell cultures of *Linum album*. Soil Biol 33:119–137

Makidon PE, Nigavekar SS, Bielinska AU, Mank N, Shetty AM, Suman J et al (2010) Characterization of stability and nasal delivery systems for immunization with nanoemulsion-based vaccines. J Aerosol Med Pulm Drug Deliv 23:77–89

Martin N, Müller C (2006) Induction of plant responses by a sequestering insect: relationship of glucosinolate concentration and myrosinase activity. Basic Appl Ecol 8:13–25

Michal-Johnson J, Lee YC, Camehl I, Sun C, Yeh KW, Oelmuller R (2013) *Piriformospora indica* promotes growth of Chinese cabbage by manipulating auxin homeostasis – role of auxin in symbiosis. In: Varma A (ed) *Piriformospora indica*. Soil biology. Springer, Berlin

Mishra M, Prasad R, Varma A (2014) Rootonic with bio-zinc to accelerate *Pennisetum glaucum* seed germination and plant growth. Int J Plant Anim Environ Sci 4:552–561

Molitor A, Zajic D, Voll L, Pons-Kuehnemann J, Samans B, Kogel KH, Waller F (2011) Barley leaf transcriptome and metabolite analysis reveals new aspects of compatibility and *Piriformospora indica*-mediated systemic induced resistance to powdery mildew. Mol Plant Microbe Interact 24:1427–1439

Nautiyal CS, Chauhan PS, DasGupta SM, Seem K, Varma A, Staddon WJ (2010) Tripartite interactions among *Paenibacillus lentimorbus* NRRL B-30488, *Piriformospora indica* DSM 11827, and *Cicer arietinum* L. World J Microbiol Biotechnol 26:1393–1399

Peskan-Berghofer T, Shahollari B, Giong PH, Hehl S, Markert C, Blanke V, Kost G, Varma A, Oelmuller R (2004) Association of *Piriformospora indica* with *Arabidopsis thaliana* roots represents a novel system to study beneficial plant–microbe interactions and involves early plant protein modifications in the endoplasmic reticulum and at the plasma membrane. Physiol Plant 122:465–477

Prasad R, Kamal S, Sharma PK, Oelmueller R, Varma A (2013) Root endophyte *Piriformospora indica* DSM 11827 alters plants morphology, enhances biomass and antioxidant activity of medicinal plant *Bacopa monniera*. J Basic Microbiol 53:1016–1024

Prasad R, Kumar V, Prasad KS (2014) Nanotechnology in sustainable agriculture: present concerns and future aspects. Afr J Biotechnol 13:705–713

Rai M, Acharya D, Singh A, Varma A (2001) Positive growth responses of the medicinal plants *Spilanthes calva* and *Withania somnifera* to inoculation by *Piriformospora indica* in a field trial. Mycorrhiza 11:123–128

Serfling A, Wirsel SG, Lind V, Deising HB (2007) Performance of the biocontrol fungus *Piriformospora indica* on wheat under greenhouse and field conditions. Phytopathology 97:523–531

Shahollari B, Vadassery J, Varma A, Oelmüller R (2007) A leucine-rich repeat protein is required for growth promotion and enhanced seed production mediated by the endophytic fungus *Piriformospora indica* in *Arabidopsis thaliana*. Plant J 50:1–13

Sherameti I, Tripathi S, Varma A, Oelmueller R (2008) The root-colonizing endophyte *Piriformospora indica* confers drought tolerance in Arabidopsis by stimulating the expression of drought stress–related genes in leaves. Mol Plant-Microbe Interact 21:799–807

Sun C, Johnson JM, Cai D, Sherameti I, Oelmüller R, Lou B (2010) *Piriformospora indica* confers drought tolerance in Chinese cabbage leaves by stimulating antioxidant enzymes, the expression of drought-related genes and the plastid-localized CAS protein. J Plant Physiol 167:1009–1017

Tuladhar R, Shrestha J, Singh A, Varma A (2013) Enhanced productivity associated with tripartite symbiosis between phaseolus, rhizobia, and *Piriformospora indica*: in presence of vermin compost. In: Varma A, Kost G, Oelmüller R (eds) *Piriformospora indica*. Springer, Berlin, pp 191–199

Uttamkumar SB, Ram P, Ajit V (2013) Impact of culture filtrate of *Piriformospora indica* on biomass and biosynthesis of active ingredient aristolochic acid in *Aristolochia elegans* Mart. Int J Biol 6:29

Varma A, Jain VK, Prasad SR, Patent no. (14/DEL/2003; 08.01.09) entitled *A* Nanomaterial Based Culture Medium for Microbial Growth Enhancement

Varma A, Bakshi M, Lou B, Hartmann A, Oelmueller R (2012a) *Piriformospora indica*: a novel plant growth promoting mycorrhizal fungus. Agric Res 1:117–131

Varma A, Tripathi S, Prasad R et al (2012b) The symbiotic fungus *Piriformospora indica*: review. In: Hock B (ed) The mycota, vol 32. Springer, Berlin, pp 231–254

Varma A, Bajaj R, Agarwal A, Asthana S, Rajpal K, Das A, Prasad R, Kharkwal AC (2013) Memoirs of 'Rootonic' – the magic fungus. Amity University Press, Noida

Waller F, Achatz B, Baltruschat H, Fodor J, Becker K, Fischer M, Heier T, Hückelhoven R, Neumann C, von Wettstein D, Franken P, Kogel KH (2005) The endophytic fungus *Piriformospora indica* reprograms barley to salt stress tolerance, disease resistance and higher yield. Proc Natl Acad Sci USA 102 :13386-13391**

Zarea MJ, Hajinia S, Karimi N, Goltapeh EM, Rejali F, Varma A (2012) Effect of *Piriformospora indica* and Azospirillum strains from saline or non-saline soil on mitigation of the effects of NaCl. Soil Biol Biochem 45:139–146

Zuccaro A, Lahrmann U, Güldener U, Langen G, Pfiffi S, Biedenkopf D et al (2011) Endophytic life strategies decoded by genome and transcriptome analyses of the mutualistic root symbiont *Piriformospora indica*. PLoS Pathog 7(10):e1002290. doi:10.1371/journal.ppat.1002290

Chapter 15
Application of Nanofertilizer and Nanopesticides for Improvements in Crop Production and Protection

Mujeebur Rahman Khan and Tanveer Fatima Rizvi

15.1 Introduction

There is a growing pressure on the agricultural resources throughout the world due to burgeoning human population. The green revolution introduced during 1970s solved the problem of hunger, by and large, in many developing countries. Since then, the global human population has doubled, and stagnation in the agricultural productivity has been experienced in many food crops. This has necessitated the need of innovative technologies for crop improvement and resource conservation. The nanotechnology is one of the recent innovative sciences that has tremendous potential to revolutionize agriculture and allied fields, such as crop production and protection. Nanoagriculture focuses on target farming that involves the use of nanosized particles with unique properties to boost crop productivity and pest suppression (Scott and Chen 2012; Batsmanova et al. 2013). Nanotechnology application in the agriculture and food sectors is relatively recent compared with its use in drug delivery and pharmaceuticals (Garcia et al. 2010). Nanotechnology has tremendous potential to improve crop productivity (Gruère et al. 2011), protect plants (Pérez-de-Luque and Hermosín 2013), monitor/detect plant diseases (Frewer et al. 2011), increase global food production (Biswal et al. 2012), enhance food quality (Sonkaria et al. 2012), and minimize wastage of resources for sustainable intensification (Prasad et al. 2014). Food and agricultural production are among the most important fields of nanotechnology application (Coles and Frewer 2013; Chen et al. 2014).

M.R. Khan (✉) • T.F. Rizvi
Department of Plant Protection, Faculty of Agricultural Sciences, Aligarh Muslim University, Aligarh 202002, India
e-mail: mrkhan777in@yahoo.co.in; tanveer.124000@gmail.com

M. Ghorbanpour et al. (eds.), *Nanoscience and Plant–Soil Systems*, Soil Biology 48,
DOI 10.1007/978-3-319-46835-8_15

15.2 Application of Nanotechnology in Crop Production

Nanoscale refers to a billionth of a meter (1–100 nm or 0.1–99.0 nm). At this scale the properties of materials differ with respect to their physical, chemical, and biological properties from those at a larger scale. Nanotechnology may prove highly useful in improving current agriculture practices through the enhancement of management and conservation of inputs in crop production systems (Thornton 2010). The potential for improving the effectiveness of agricultural active ingredients using nanosized particles is one of the commonest applications (Zhao et al. 2012). Application of nanotechnology in agriculture and crop production mainly includes nanotechnology-enabled delivery of agriculture chemicals (Fig. 15.1) (Manimegalai et al. 2011), field-sensing systems to monitor the environmental stresses (Scott 2007) and crop conditions (Knauer and Bucheli 2009), and improvement of plant traits against environmental stress (Owolade et al. 2008) and diseases

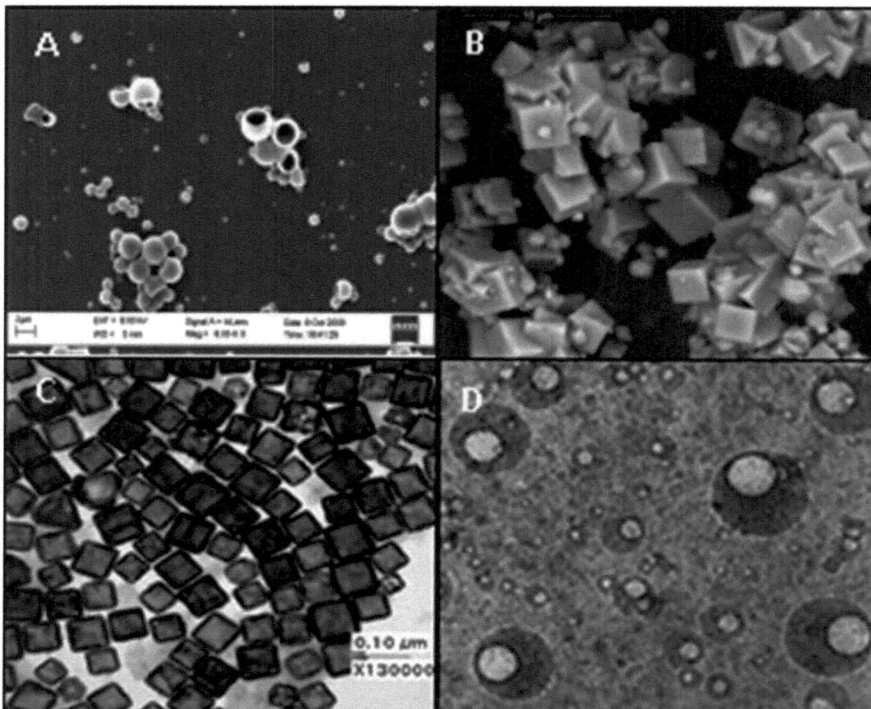

Fig. 15.1 Nano-delivery materials: SEM image of nanocapsules, nanocontainers (Source: http://www.itc.uji.es/notDestacada/PublishingImages/Nanocontainers.jpg), nanocages (Source: http://bit.bme.jhu.edu/showimg/nanocage.jpg) and nanoemulsions (Source: https://encrypted-tbn3.gstatic.com/images?q=tbn:ANd9GcSpWbJHE3yCQMIK2fsgAtu_VSfBRRiG77nchzu1LNLOyjvc2VDO0w) (clockwise)

(Forsberg and de Lauwere 2013; Khan and Rizvi 2014). However, in the present chapter discussion is confined to nanotechnology based fertilizers and pesticides.

15.2.1 Nanofertilizers

Fertilizers are chemical compounds applied to promote plant growth and yield (Behera and Panda 2009). Fertilizers are usually applied either through the soil (for uptake by plant roots) or by foliar spray. Inorganic fertilizers constitute a huge proportion of fertilizers used to provide additional nutrition to plants. The inorganic fertilizers are artificially synthesized and are formulated in appropriate concentrations and the combinations that usually supply three main nutrients: nitrogen, phosphorus, and potassium (N, P, and K) for various crops and growing conditions. Nitrogen promotes leaf growth and forms proteins and chlorophyll. Phosphorus contributes to root, flower and fruit development. Potassium contributes to stem and root growth and the synthesis of proteins (Mandal et al. 2009; Gu et al. 2009). About 30–60 % of N, 10–20 % P, and 30–50 % K of the applied dose is utilized by plants and the rest is lost to the environment. This causes substantial economic and resource loss as well as serious soil and water contamination. With the application of nanotechnology, these demerits of conventional fertilizers can be minimized, so as to utilize the major proportion of the applied dose of the chemical. This can be achieved by encapsulating the nutrients by nanomaterials, coated with a thin protective film, or delivered as emulsions or nanoparticles (de la Rosa et al. 2010).

The nanobased slow-release or controlled release of fertilizers have the potential to increase the efficiency of nutrient uptake and to significantly reduce their wastage. The nanotechnology may be applied in the soil nutrition by developing the formulations in two ways, i.e., fertilizers coated, encapsulated, or buried in the nanomaterials; and nanoforms of fertilizers and other growth promoting materials.

15.2.2 Fertilizers Coated or Encapsulated with Nanoparticles

To prevent wastage of fertilizer, reduce the dose, and increase efficiency, the fertilizer can be coated, binded or encapsulated by some specific nanomaterials. Coating and binding of nano and subnano-composites help to regulate the release of nutrients from the fertilizer capsule (Fig. 15.1) (Liu et al. 2001). In this context, researches show that application of a nanocomposite consisting of nitrogen, phosphorus, potassium, micronutrients, mannose, and amino acids enhanced the uptake and use of nutrients by grain crops (Guo 2011). Besides, zinc–aluminum layered double-hydroxide nanocomposites have been employed for the controlled release of chemical compounds that act as plant growth regulators. Urea-modified hydroxyapatite nanoparticle-encapsulated *Gliricidia sepium* nanocomposite displayed a slow and sustained release of nitrogen over time at three different pH values (Kottegoda et al. 2011). Nanoporous zeolite base on nitrogen fertilizer can be used as an alternate strategy to improve the efficiency of nitrogen use in crop production systems (Manikandan and Subramanian 2014). As a super fertilizer, carbon nanotubes were found to penetrate tomato seeds and improved their germination and growth rates. Analytical methods indicated that the carbon nanotubes penetrated the thick seed coat and supported water uptake inside seeds (Khodakovskaya et al. 2009). Encapsulation of fertilizers within a nanoparticle is done by encapsulating the nutrient by a nanoporous material. The nutrient may also be coated with thin polymer film, or be delivered as particles or emulsions of nanoscale dimensions (Rai et al. 2012).

In recent years, the use of slow release fertilizers has become one of the important innovative technologies to save fertilizer consumption and to minimize environmental pollution (Guo et al. 2005). In this technology the fertilizers are entrapped within nanoparticles (Teodorescu et al. 2009). Consequently, the fertilizers are protected by the nanoparticles for longer presence in the soil, allowing their controlled release into the soil (Saigusa 2000). The application of nanofertilizers increases efficiency of the elements, minimizes their toxicity in the soil, and reduces the frequency of application. The nanofertilizer application leads to a gradual and controlled release of nutrients in the soil, and prevents eutrophication and pollution of water resources.

Chitosan nanoparticles have been largely investigated as a carrier for drug delivery (Fig. 15.2). This nanomaterial may also prove efficient for the controlled release of NPK fertilizers. Chitosan (CS) is a biodegradable, bioabsorbable, and bactericidal polymer (Coma et al. 2002; No et al. 2007). Due to its polymeric cationic characteristics, chitosan nanoparticles may interact with negatively charged molecules and polymers, showing a favorable interaction. The chitosan nanoparticles obtained by polymerizing methacrylic acid were used to incorporate into NPK fertilizers (Corradini et al. 2010).

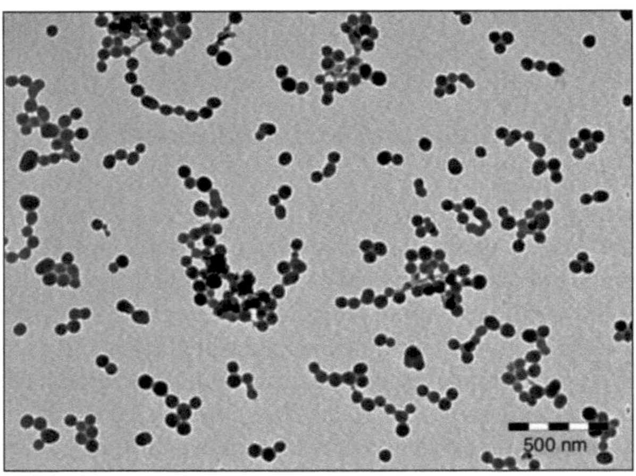

Fig. 15.2 TEM microphotograph obtained for chitosan nanoparticles (CS-PMAA) at pH 4 (Source: Corradini et al. 2010)

15.2.3 HA Nanoparticles

A nano strategy involving a slow-release fertilizer composition based on urea-modified hydroxyapatite (HA) nanoparticles encapsulated into the cavities present in soft wood (Kottegoda et al. 2011). HA (Ca_{10} $(PO_4)_6(OH)_2$) nanoparticles are rated as one of the prominent candidates to supplement the phosphorous nutrition in agricultural soils. Much of the current literature on HA, is however, focused on its biomedical applications due to its excellent biocompatibility and bioactivity, while potential agricultural applications have not been adequately addressed (Liu 2008; Cao et al. 2010). The HA nanoparticles can be synthesized by different methods such as wet chemical precipitation (Han et al. 2008; Poinern et al. 2009), hydrothermal method (Manafi and Joughehdoust 2009; Montazeri et al. 2010), sol–gel method (Liu et al. 2001), and spray pyrolysis process (Cho and Kang 2008). Morphology and the level of the crystallinity of the nanoparticles highly depend on the preparation method and the experimental conditions. Wet chemical precipitation of HA nanoparticles is the simplest available method for the synthesis with a high yield. The HA nanoparticles have been found quite suitable for ready surface modification due to their rich surface chemistry. The chitosan–nanohydroxyapatite composites, have also been found suitable for tissue engineering applications (Katti et al. 2008; Han and Misra 2009; Li et al. 2010). The biodegradable and biocompatible nanohydroxyapatite polyvinyl alcohol composites (Poursamar et al. 2009), and amino acid-functionalized hydroxyapatite nano rods, have potential prospects for being used to prepare the bioinorganic nanocomposites (Mcqquire et al. 2004).

The HA nanoparticles were synthesized by wet chemical methods (Mateus et al. 2007) and surface modified with urea (water-soluble plant nitrogen nutrient source), and a fertilizer composition was prepared by encapsulation of urea-modified HA nanoparticles into micro/nano porous cavities of the young stem of *Glyricidia sepium* under pressure (Kottegoda et al. 2011). These cavities are made up of cellular polymers such as cellulose, hemicellulose, and lignin. The bionanocomposite was then dried and its pellets were prepared. Upon soil application of the pellets, it was stipulated that this nanofertilizer formulation will absorb the moisture, leading to a slow release of nitrogen into the soil due to diffusion and microbial degradation, of the biopolymeric matrix.

15.2.4 Fertilizers as Nanoparticles

15.2.4.1 Nitrogen Nanoparticles as a Fertilizer

Nitrogen is a key nutrient source for food, fiber and biomass production in plants. However, considering the energy required in its synthesis and the large tonnage required, the nitrogen fertilizer has a high monetary value. The 50–70 % of the nitrogen applied using conventional fertilizers, is lost to the soil due to leaching and lower nitrogen utilization efficiency (NUE) by plants. Attempts to increase the NUE in conventional fertilizer formulations have not been much effective. On the other hand, the emerging nano strategies indicate that, due to the high surface area to volume ratio, nano nitrogen is expected to be far more effective than even polymer-coated conventional slow-release N fertilizers (Hossain et al. 2008; De Rosa et al. 2010).

15.2.4.2 Phosphorus Nanoparticles as a Fertilizer

Agriculture is the major user of mined phosphorus (P), accounting for 80–90 % of the world demand for P (Childers et al. 2011). Burgeoning human population, growing preferences towards meat-based diets and rising demands for bio-energy crops will increase the future demand for P fertilizers. However, application of P fertilizers leads to eutrophication problem in surface waters (Correll 1998; Carpenter 2005; Conley 2009). Numerous regulations (Litke 1999), best managements practices (BMPs) (Hoffmann 2009), and remediation technology have been proposed to reduce P fertilizer application and to prevent the applied P from entering into water bodies (De-Bashana and Bashana 2004; Buda et al. 2012). However, there have been a few attempts to solve the eutrophication problem through modifications of the chemical properties of a P fertilizer to reduce the P mobility in the soil or decreasing bioavailability of P to the algae. Liu and Lal (2014) hypothesized that use of P nanoparticle fertilizer, as an alternative to the regular P fertilizers on agricultural lands, would enhance agronomic production, use efficiency of P, and improve the surface-water quality.

Generally, commercially available P fertilizers such as MAP (mono ammonium phosphate, $NH_3H_2PO_4$), DAP (diammonium phosphate, $(NH_3)_2HPO_4$), or TSP (triple superphosphate, Ca $(H_2PO_4)_2$) are water soluble phosphate salts, which are easily dissolved in the soil solution and re-available for plant uptake, and thus, are regarded as high quality fertilizers. However, these soluble phosphates are also very mobile in the soil and large portion often ends up in surface-water bodies through runoff or seepage, causing eutrophication. On the other hand, solid forms of P such as naturally occurring phosphate rocks, and apatites ($Ca_5(PO_4)_3X$, X_5F, Cl, Br, or OH) have also been attempted as P fertilizers where the phosphate is locked in a solid form and is less easily available to the alga and also less easily being transported by runoff or soil erosion (Fageria 2009). However, these solids are less effective in providing nutrient P at the time when the plants are in need (Fageria, 2009). In addition, application of solid phosphates in agriculture is hindered by the large size of the particles, which limits phosphate mobility in the soil, and prevents phosphate from reaching the root zone and nurturing the crops in a timely fashion. For these limitations of the conventional P application, the nano-sized apatite particles could be as effective in providing the nutrient P as the commonly used soluble P fertilizers, and shall also minimize the eutrophication and the delivery problem associated with the later. Rather, application of nano-sized solid P as fertilizer would be a good compromise between agricultural benefits and the environmental hazards. It had been found that the P nanoparticle suspension and an aqueous solution of P have same mobility rate in the soil columns. The P nanoparticles can be easily delivered to the root zones with conventional methods like spray or irrigation. Moreover, the nanoparticles are environmentally benign because the P in solid form is much less bioavailable to the algae than those in soluble forms (Reynolds and Davies 2001). The algae-bioavailable P is primarily responsible for the eutrophication in fresh surface-waters (Carpenter 2008; Conley 2009; Childers et al. 2011).

Liu and Lal (2014) reported that in greenhouse test, the height of soybean plants applied with nano-sized hydroxyapatite (nHA, Figs. 15.2, 15.3, and 15.4) increased by 30 % over regular P fertilizer ($Ca(H_2PO4)_2$) treatment. However, the control soybean plants that were grown under tap water without any fertilizer application were much shorter. The P is a limiting and indispensable nutrient for healthy growth of soybean and other legumes, and depends much lesser on the outer inputs.

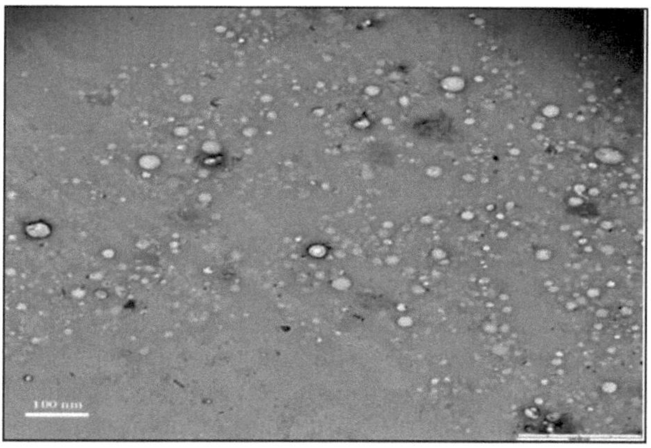

Fig. 15.3 A TEM image of nano-sized hydroxyapatite (nHA; Source: Liu and Lal 2014)

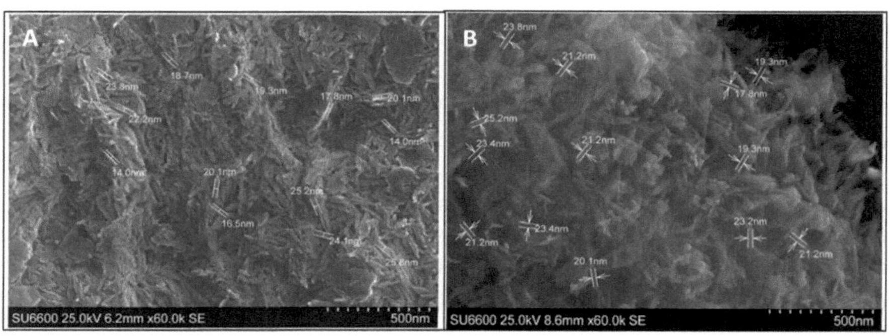

Fig. 15.4 Scanning electron microscope images of synthesized HA nanoparticles (**a**) and urea surface-modified HA nanoparticles (**b**) (Source: Kottegoda et al. 2011)

The nHA treatment also enhanced the growth, dry biomass (aboveground and underground) and the soybean seed yield greater than regular P applied soybean plants. The 20-week-long greenhouse test in an inert growing medium showed that application of nHA as a P source promoted the soybean growth 32.6 % higher than regular P fertilizer treatment. This study has demonstrated that nHA could be used as a P fertilizer in enhancing crop yields and biomass production.

15.2.4.3 Titanium Nanoparticles as a Fertilizer

The treatment with TiO_2 nanoparticles promoted the plant growth of maize, however, the effect of TiO_2 bulk treatment was negligible. Titanium nanoparticles increased light absorption and photo energy transmission (Sekhon 2014). In another experiment, a compound of SiO_2 and TiO_2 nanoparticles increased the activity of nitrate reductase in soybeans and intensified plant absorption capacity, making its use of water and fertilizer more efficient (Lu et al. 2002).

15.2.4.4 Zinc Nanoparticles as a Fertilizer

Zinc deficiency is one of the most widely distributed micronutrient problems that limits the crop productivity. Approximately 49 % of the arable soils of the world are Zn deficient (Sillanpaa 1982, 1990). Typically, solid Zn fertilizers are blended with, incorporated into, or coated onto macronutrient fertilizer to maintain a more uniform distribution of Zn in the field and to provide a cost-effective delivery of the small amounts of Zn required. The effectiveness of Zn fertilizers for providing plants with Zn in Zn-deficient soils mainly depends on the solubility of the Zn source in soil. Mortvedt and Giordano (1969) found a significant correlation between water-soluble fractions of Zn and Zn availability to crops from several macronutrient fertilizers with zinc oxide (ZnO) or zinc sulfate.

Researchers have ascertained that water-soluble Zn, (not the total Zn concentration), is the major parameter controlling the effectiveness of Zn-enriched fertilizers for plant growth and development (Westfall et al. 1991; Amrani et al. 1999; Gangloff et al. 2002). The ZnO (Inorganic sources of Zn) is a most commonly used Zn fertilizer which is applied to the crops in Zn-deficient regions (Martens and Westermann 1991). Incorporation of ZnO nanoparticles (ZnO NPs, Fig. 15.5) into fertilizers as a source of Zn might be a promising approach which can exploit novel solubility features of ZnO NPs to improve the efficiency of Zn fertilizers. Application of ZnO NPs as a source of Zn in Zn fertilizers may improve the efficiency of the fertilizer and Zn availability to plants by enhancing the rate and extent of Zn dissolution. The Zn NPs may be applied as a foliar spray. This treatment may potentially enhance uptake and the penetration of zinc oxide nanoparticles in the plant leaves. Pot studies with foliar spray have demonstrated that plants sprayed with 20 mg mL^{-1} ZnO NPs solution showed improved growth and biomass production over control plants (Panwar et al. 2012; de la Rosa et al. 2013).

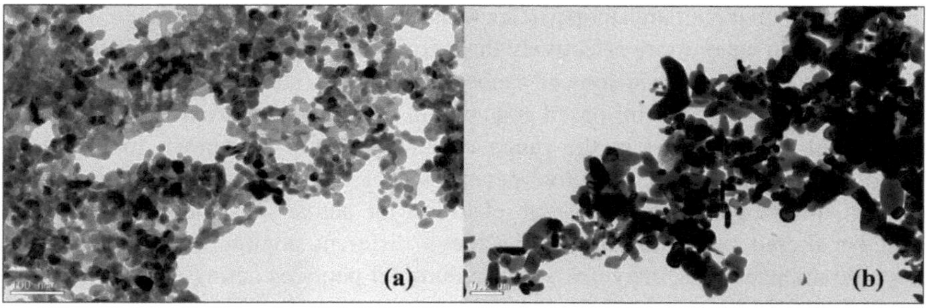

Fig. 15.5 Transmission electron microscopy image of the ZnO powders used in the experiments: (**a**) ZnO NPs (nominal diameter of 20 nm), (**b**) bulk ZnO (nominal diameter <1 μm) (Source: Milani et al. 2012)

15.3 Nanopesticides

Pesticides are commonly used in agriculture to protect plants from the attack of pests and diseases so as to get higher crop yields. Although pesticides are highly effective and reliable means of pest and disease control, but their application presents several ill effects such as environmental contamination and residual toxicity in the food. These ill effects have resulted due to indiscriminate and over use of pesticides. Hardly 1 % of the applied pesticide kills the target pest/pathogen, while the rest 99 % affects the non-target organisms, and accumulate in the plant body. Numerous studies have shown that pesticide application is safer if applied at a lower dose, but at the lower doses, effectiveness of pesticides is severely affected. At the present state of our knowledge, the nanotechnology with new innovative strategies may address the limitations associated with lesser effectiveness of lower doses of pesticides and environmental contamination (Sasson et al. 2007). Nanopesticides may offer a way to control delivery of pesticide and to achieve greater effectiveness of much smaller dose of a chemical. The current research trend in the agrochemical companies is to reduce the particle size of existing chemical emulsions to the nanoscale, or to encapsulate the active ingredients in the nanocapsules designed to split open, for example, in response to sunlight, heat, or the alkaline conditions in an insect's stomach. Microencapsulation may be used as a versatile tool for hydrophobic pesticides, enhancing their dispersion in aqueous media and allowing a controlled release of the active compound. Polymer NPs have a potential use in the nanopesticide production (Perlatti et al. 2013). Besides, there have been numerous studies on the toxic effects of nanoparticles on bacteria, fungi, and insects (Feng et al. 2000; Samuel and Guggenbichler 2004; Khan and Rizvi 2014). The nanoformulations are generally found to increase the solubility of poorly soluble active ingredients, to release the active ingredient in a slow/targeted manner, and/or to protect against premature degradation (Kah et al. 2013). The agrochemical companies are planning to produce the formulation having the nanoparticles and emulsions to make them more potent. Many companies make formulations that contain nanoparticles within the 100–250 nm size range which are dissolvable in water more effectively than existing ones (Kumar et al. 2010). Other companies employ suspensions of nanoscale particles (nanoemulsions), which can be either water-based or oil-based and contain uniform suspensions of pesticidal or herbicidal nanoparticles in the range of 200–400 nm. As shown in Fig. 15.1, nanocapsules can enable effective penetration of herbicides through cuticles and tissues, allowing slow and constant release of the active substances. Viral capsids can be altered by mutagenesis to achieve different configurations and deliver specific nucleic acids, enzymes, or antimicrobial peptides acting against the parasites (Pérez-de-Luque and Rubiales 2009).

Nanoparticles have been used as a physical approach to alter and improve the effectiveness of properties of synthetic chemical pesticides or in the production of bio-pesticides directly. Nanomaterials serve equally as additives (mostly for controlled release) and active constituents (Gogos et al. 2012). Controlled-release (CR) formulations of imidacloprid (1-(6 chloro-3-pyridinyl methyl)-*N*-nitro imidazolidin-2-ylideneamine), synthesized from polyethylene glycol and various aliphatic diacids using encapsulation techniques, have been used for efficient pest management in different crops.

In addition, nanoparticles of some metals have also been found to suppress the pests and pathogens (Khan and Rizvi 2014). Silver nanoparticles at 100 mg/kg inhibited mycelial growth and conidial germination of powdery mildew fungus on cucurbits and pumpkins. Silver nanoparticles have received significant attention as a pesticide for agricultural applications (Afrasiabi et al. 2012). The potential of nanomaterials in pest and disease management as a modern approach of nanotechnology has been realized in recent years (Rai and Ingle 2012; Khan and Rizvi 2014). Broadly, nanotechnology can be used in the crop protection in two following ways, i.e., nanoparticles toxic to pests and pathogens, and nanomaterials as a carrier of pesticides.

15.3.1 Nanoparticles Toxic to Pests and Pathogens

15.3.1.1 Nanoparticles Toxic to Pathogens

Silver Nanoparticles

The use of nano-sized silver particles as antimicrobial agents have become more common with the technological advances in the production techniques. Silver displays multiple modes of inhibitory action to microorganisms (Clement and Jarret 1994), hence, it may be used for controlling various plant pathogens in a relatively safer way compared to synthetic fungicides (Park et al. 2006). Various forms of silver ions and nanoparticles were tested for the antifungal activity against *Bipolaris sorokiniana* and *Magnaporthe grisea* (Jo et al. 2009) (Table 15.1). Nanoparticle treatments significantly suppressed the colony formation of the above two pathogenic fungi. Effective concentrations of the Ag NPs in inhibiting the

Table 15.1 Effect of nanoformulations/nanoparticles on some pest and pathogens

Nanoparticles and its formulations	Pest/pathogen	Effect	References
Ag NPs (suspension, 20–30 nm)	*Bipolaris sorokiniana* and *Magnaporthe grisea*	Inhibited colony formation (In vitro)	Jo et al. (2009)
Ag NPs (solution)	*Spaherotheca fusca*	Inhibited growth (*In vitro* and *vivo*)	Lamsal et al. (2010)
Ag NPs (solution)	*Fusarium culmorum*	Inhibited growth (In vitro)	Kasprowicz et al. (2010)
Ag NPs	Bacteria	(In vitro)	Kamran et al. (2011)
Ag NPs (solution)	*Pediculus humanus*, *Anopheles subpictus* and *Culex-quinquefasciatus*	Pediculocidal and larvicidal (In vitro)	Jayaseelan et al. (2011)
Ag NPs (suspension)	*Aphis nerii*	Insecticidal	Rouhani et al. (2012)
Au NPs (DNA tagged)	*Spodoptera litura*	Insecticidal	Chakravarthy et al. (2012)
Zn NPs (ZnO, suspension)	*Salmonella typhimurium* and *Staphylococcus aureus*	Antibacterial (In vitro)	Ahamed et al. (2011)
S NPs (35 nm)	*Fusarium solani* and *Venturia inaequalis*	Fungicidal	Rao and Paria (2013)
Cu NPs (encapsulated in soda lime glass powder, 30 ± 5 nm)	*Escherichia coli* (gram-negative bacteria), *Micrococcus luteus* (gram-positive bacteria) and *Issatchenkia orientalis* (yeast)	Antibacterial and antifungal (In vitro)	Esteban-Tejeda et al. (2009)
Al NPs (Inorganic)	*Sarocladium oryzae* and *Rhyzopertha dominica*	Insecticidal	Stadler et al. (2010)
Al-Si NPs (suspension)	*Bombyxmori*	Insecticidal	Goswami et al. (2010)
Si-Ag NPs	*Botrytis cinerea*, *Rhizoctonia solani*, and *Colletotrichum gloeosporioides*	Fungicidal	Park et al. (2006)
Carbofuran and imidacloprid (CR formulation)	*Aphis gossypii* and *Amrasca biguttula biguttula*	Insecticidal (In vivo)	Kumar et al. (2011)
Zineb and mancozeb encapsulated into carbon nanotubes	*Alternaria alternata*	Inhibited growth (In vitro)	Sarlak et al. (2014)
Nano-silicon as carrier of diatom frustules	Weeds	Herbicidal	Lodriche et al. (2013)
Zinc-layered hydroxide nanohybridas a carrier of [4-(2,4-dichlorophenoxy)	Weeds	Herbicidal	Hussein et al. (2012)

(continued)

Table 15.1 (continued)

Nanoparticles and its formulations	Pest/pathogen	Effect	References
butyrate (DPBA) and 2-(3-chlorophenoxy) propionate (CPPA)] (CR formulation)			
Manganese carbonate as a carrier of pendimethalin	Weeds	Herbicidal	Kanimozhi and Chinnamuthu (2012)

colonization by 50 % (EC_{50}) were higher for *B. sorokiniana* than for *M. grisea*. Growth chamber inoculation assays further confirmed that both ionic and nanoparticles of silver significantly reduced the diseases caused by the above two fungi on perennial ryegrass (*Lolium perenne*). Nanoparticles effectively reduced disease severity with an application at 3 h before spore inoculation, but their efficacy significantly diminished when applied at 24 h after inoculation. Kasprowicz et al. (2010) recorded a significant reduction in the mycelial growth when spores were incubated with silver nanoparticles. The sporulation test showed that relative to control samples, the number of spores formed by mycelia on the nutrient-poor PDA medium increased in the culture after contact with silver nanoparticles. The 24 h incubation of FC spores with a 2.5 ppm solution of silver nanoparticles greatly reduced the number of germinating fragments and sprout length relative to the control. Kamran et al. (2011) reported that nano silver had a good potential for removing of the bacterial contaminants in tobacco plant tissue culture procedures.

Application of silver nanoparticles (WA-CV-WA13B) at different concentrations was investigated on cucurbits under field condition. The NP treatments were given before and after disease outbreak. The treatment with 100 ppm silver nanoparticles caused greater fungal disease suppression when given before and after the outbreak of disease on cucumbers and pumpkins. The treatment also caused maximum inhibition in the growth of fungal hyphae and conidial germination in in vivo tests. The SEM results indicated that the silver nanoparticles caused detrimental effects on both mycelial growth and conidial germination (Lamsal et al. 2010).

Zinc Nanoparticles

The antibacterial potential of zinc oxide nanoparticles (ZnO NPs), compared with conventional ZnO powder, against nine bacterial strains, mostly food-borne including pathogens, was evaluated using qualitative and quantitative assays. ZnO NPs were more efficient as antibacterial agent than powder. Gram-positive bacteria were generally more sensitive to ZnO than Gram negative. The exposure of *Salmonella typhimurium* and *Staphylococcus aureus* to ZnO NP reduced the cell number to zero within 8 and 4 h of application, respectively. The SEM of the treated bacteria with NPs show that the disruptive effect of ZnO on *S. aureus* was vigorous as all treated cells were completely exploded or lysed after only 4 h of exposure. Promising results of ZnO NP antibacterial activity suggest its usage in food systems as preservative agent after further required investigations and risk assessments (Ahamed et al. 2011).

Sulfur Nanoparticles

Rao and Paria (2013) studied fungicidal efficiency of sulfur nanoparticles (SNPs) against two phytopathogens, *Fusarium solani* (early blight and wilt diseases) and *Venturia inaequalis* (apple scab disease). The 35 nm sized particles were found more effective than the bigger particles in preventing the fungal growth. Microscopic study confirmed that the fungicidal effect is mainly because of the deposition of particles on the cell wall and subsequent damage. The NP deposition caused an imbalance in the cell wall structure as supported by a Biuret assay test. Hence, the fungicides containing sulfur NPs can effectively control the fungal diseases of crops under organic as well as conventional farming (Rao and Paria 2013).

Copper and Silica Nanoparticles

Nanocopper particles suspended in water have been used since at least 1931, in a product known as Bouisol as fungicide in grapes and fruit trees (Hatschek 1931). Antifungal activities of polymer-based copper nanocomposites against pathogenic fungi (Cioffi et al. 2004) and silica–silver nanoparticles against *Botrytis cinerea*, *Rhizoctonia solani*, and *Colletotrichum gloeosporioides* (Park et al. 2006) have been reported. Copper nanoparticles in soda lime glass powder showed efficient antimicrobial activity against gram-positive and gram-negative bacteria and fungi (Esteban-Tejeda et al. 2009).

15.3.1.2 Nanoparticles Toxic to Insects

Silver Nanoparticles

The silver nanoparticles (AgNPs) have also been found a potential candidate for the management of insect pests. The pediculocidal and larvicidal activity of synthesized silver nanoparticles using an aqueous leaf extract of *Tinospora cordifolia* showed maximum mortality against the head louse, *Pediculus humanus* and fourth instar larvae of *Anopheles subpictus* and *Culex-quinque fasciatus*. Synthesized silver nanoparticles possessed excellent antilice and mosquito larvicidal activity (Jayaseelan et al. 2011) (Table 15.1). The insecticidal activity of AgNPs against the

Aphis nerii has also been reported (Rouhani et al. 2012). Nanoparticles of Ag and Ag–Zn were synthesized through a solvothermal method, and using them, insecticidal solutions of different concentrations were prepared and tested on *A. nerii*. For comparison purposes, imidacloprid was used as a conventional insecticide. The LC_{50} value for imidacloprid, Ag, and Ag–Zn nanoparticles were calculated to be 0.13 $\mu L\ mL^{-1}$, 424.67 $mg\ mL^{-1}$, and 539.46 $mg\ mL^{-1}$, respectively. Overall, imidacloprid at 1 $\mu L\ mL^{-1}$ and nanoparticles at 700 $mg\ mL^{-1}$ had the highest insect mortality effect.

Other Nanoparticles

Treatment of *Bombyx mori* infested leaves with grasserie disease with ethanolic suspension of hydrophobic alumina–silicate nanoparticles significantly reduced the viral load (Goswami et al. 2010). DNA-tagged gold nanoparticles have been found effective against *Spodoptera litura* and may serve as a useful component of an integrated pest-management strategy (Chakravarthy et al. 2012). Development of nano-based viral diagnostics can help to detect the exact strain of virus and identify different proteins in healthy and diseased plants during the infection cycle to stop disease (Prasanna 2007).

Nanosilica has been successfully employed to control a range of agricultural insect pests and ectoparasites in animals. The silica nanoparticles are absorbed into cuticular lipids by physisorption. The cuticular lipids are used by insects to prevent death from desiccation. The absorption of silica NPs with lipids leads to the insect death. To target insects, the NPs may be applied on leaves and stem surfaces of plants (Ulrichs et al. 2005). A novel photodegradable insecticide involving nanoparticles has been reported. The toxicity of the photodegradable insecticide was evaluated against the adult stage of *Martianus dermestoides*. The results showed that thermodynamically stable IMI (imidacloprid) microcrystals were obtained by association and had a mean length of 7 μm and a ζ-potential of −37.5. The drug loading and encapsulation efficiency were $56.15 \pm 0.96\,\%$ and $81.57 \pm 0.96\,\%$, respectively. The polysaccharide capsules prolonged the release time of the encapsulated IMI crystals. Among the photocatalysts, SDS/Ag/TiO_2 had the highest photocatalytic activity. Toxicity of the novel 50 % nano-SDS/Ag/TiO_2—IMI was higher in the adult stage compared to the 95 % IMI as indicated by the lower LC_{50} value (Guan et al. 2008).

The insecticidal activity of nanostructured alumina against two insect pests, *Sarocladium oryzae* (L.) and *Rhyzopertha dominica* (F.), which are major insect pests in stored food supplies throughout the world has also been recorded. Significant mortality was observed after 3 days of continuous exposure to nanostructured alumina-treated wheat. The inorganic nanostructured alumina is a low cost material and may provide a reliable management of insect pests (Stadler et al. 2012).

15.3.2 Nanomaterials as a Carrier of Pesticides

15.3.2.1 Nanoinsecticides

Nanoencapsulation helps slow release of a chemical to the particular host for insect pest control through release mechanisms that include dissolution, biodegradation, diffusion, and osmotic pressure with specific pH (Vidyalakshmi et al. 2009). Nanoparticles loaded with garlic essential oil proved effective against *Tribolium castaneum* (Barik et al. 2008). The nano-encapsulated pesticides, generally, have the ability to target a specific insect, thereby reducing the amount of the dose when compared to traditional pesticides. The nano-pesticides are absorbed on the surface of the plant, facilitating a prolonged release that lasts for a longer time compared to conventional pesticides that wash away in the rain (Scrinis and Lyons 2007). Significant mortality of two insect pests, *Sarocladium oryzae* and *Rhyzopertha dominica*, after 3 days of exposure to nanostructured alumina-treated wheat was reported (Stadler et al. 2010). Halloysite nanotube has potential to be applied as a nanocontainer for encapsulation of chemically and biologically active agents such as agromedicines and pesticides (Abdullayev and Lvov 2011; Murphy 2008).

Polycaprolactone and poly(lactic) acid nanospheres were used for encapsulation of the insecticide ethiprole. The examination of the nanoformulation revealed that nanospheres did not provide a controlled release of agrochemical active ingredients, but, due to their small size, they enhanced the penetration in the plant body compared to the classical suspension (Boehm et al. 2003). In vivo experiments carried out with Egyptian cotton leafworm, *Spodoptera littoralis* larvae have indicated that the toxicity of nanoparticles of novaluron resembled with the conventional commercial formulation (Elek et al. 2010). The bioefficacy of the nanoformulations and a commercial formulation were evaluated against major pests of soybean, namely stem fly, *Melanagromyza sojae* and white fly, *Bemisia tabaci*. Most of the controlled release (CR) formulations of imidacloprid exhibited better control of the pests compared with its commercial formulations. The poly [poly(oxyethylene-1000)-oxy suberoyl] amphiphilic polymer-based formulation performed better than rest of the CR formulations against the target pests. Application of some of the CR formulations resulted in a higher yield over commercial formulation and control (Adak et al. 2012). The CR formulations of carbofuran and imidacloprid provided better or equal control against the aphid, *Aphis gossypii* and leafhopper, *Amrasca biguttula biguttula* Ishida on potato crop, than conventional formulations. The residues of carbofuran and imidacloprid in potato tuber and soils were not detectable at the time of harvesting in any one of the formulations (Kumar et al. 2011).

Pheromones are naturally occurring volatile semiochemicals and are considered ecofriendly biological control agents. Pheromones immobilized in a nanogel exhibited high residual activity and excellent efficacy in an open orchard. Environment-friendly management of fruit flies involving pheromones for the reduction of pest populations has been reported. A nanogel of the methyl eugenol, pheromone using a low-molecular-mass gelator such as all-trans tri (*p*-phenylene vinylene) bis-aldoxime was prepared. The nanogel offered stability at open ambient conditions, reduced evaporation and sustained release of the pheromone (Bhagat et al. 2013). This formulation required easy handling and transportation without refrigeration, and lesser frequency of pheromone recharging in the orchard. This nanopheromone also offered an easy sampling technique for the trapping of the pests in mango and guava orchards. Notably the involvement of the nanogelled pheromone brought about an effective management of *Bactrocera dorsalis*, a prevalent harmful pest for a number of fruits, including mango and guava (Bhagat et al. 2013).

15.3.2.2 Nanofungicides

In a study by Sarlak et al. (2014), zineb and mancozeb were encapsulated into multiwall carbon nanotubes-graft-poly (citric acid) (MWCNT-g-PCA) hybrid material. This process successfully converted bulk pesticide into nanofungicide. Polymerization of citric acid onto the surface of oxidized multiwall carbon nanotubes led to MWCNT-g-PCA hybrid materials. Because of the presence of conjugated citric acid branches, synthesized MWCNT-g-PCA hybrid materials were not only soluble in water but also able to trap water-soluble chemical species and metal ions. Trapping of pesticides such as zineb and mancozeb in aqueous solution by MWCNT-g-PCA hybrid materials led to encapsulated pesticide (EP) in the polycitric acid shell. Effective parameters in the encapsulation process, such as pH, temperature and time of stirring were optimized via the UV-vis spectroscopic method. The effect of encapsulated pesticide was studied on *Alternaria alternata* on potato dextrose agar. Results showed that nanofungicide in contrast with bulk pesticide had extraordinarily much superior suppressive effect on *A. alternata* fungi. This study was the first that reported the use of water soluble polymerized carbon nanotubes for encapsulation of pesticides for better management of plant diseases (Sarlak et al. 2014). This study has demonstrated that the Pesticide encapsulated CNT-g-PCA hybrid material is more stable and effective than bulk pesticide; nanopesticide usage will decrease the amount of pesticide.

15.3.2.3 Nanoherbicides

Among all kinds of pests, weeds are responsible for greater loss in crop productivity (Khan and Jairajpuri 2012). Hence, their management becomes essential for reducing the crop losses. There is lot of scope for application of nanotechnology in the development of novel herbicides. The weed management using nanoherbicides is considered as an economically viable alternative to the conventional herbicides. The herbicides are quite effective in controlling weeds, but their effectiveness under rain-fed areas depends on the moisture availability. Lack of moisture limits the use and efficiency of the herbicide treatment. The nano-silicon carrier comprising diatom frustules (pore size 1–100 nm) has been used for delivery of pesticides and herbicides in plants as well as in hormonal waste-water treatment (Lodriche et al. 2013). The CR formulation was found superior to its counterpart and resulted in a higher yield and better crop quality. Such a formulation may also be used in herbicides, pesticides, and plant growth regulators (Sarijo et al. 2010; Hussein et al. 2010).

The potential application of a layered single-metal hydroxide, particularly zinc-layered hydroxide, as the host for the preparation of a nanohybrid compound with a tunable CR property containing two herbicides simultaneously has been demonstrated (Hussein et al. 2012). In this context, a nanohybrid containing both herbicides [4-(2,4-dichlorophenoxy) butyrate (DPBA) and 2-(3-chlorophenoxy) propionate (CPPA)] labeled as ZCDX was found a suitable host for the CR formulation of two herbicides, namely DPBA and CPPA, simultaneously. The zinc-layered hydroxide nanohybrid containing two herbicides, CPPA and DPBA, were found to be composed of a much greater loading of DPBA (83.8 %) compared to CPPA (16.2 %) between the zinc-layered hydroxide inorganic interlayers (Hussein et al. 2012).

Researchers have reported a functional hybrid nanocomposite based on the intercalation of two herbicides' anions (2,4-dichlorophenoxy acetate and 4-chlorophenoxy acetate) with zinc–aluminum-layered double hydroxide (Bashi et al. 2011). The CR formulations of nanocomposites such as 4-chlorophenoxy acetate–zinc–aluminum-layered double hydroxide and 4-dichlorophenoxy acetate–zinc–aluminum-layered double hydroxide have also been developed (Hussein et al. 2005, 2007; Ghazali et al. 2013). Another nanoherbicides based on manganese carbonate core–shell nanoparticles loaded with pendimethalin (pre-emergence herbicide) has been developed. This nanoherbicide is programmed to release the chemical smartly as per the requirements (Kanimozhi and Chinnamuthu 2012). Recent advances in nanoscale engineering have created a new class of particulate bionanotechnology that uses biomimicry to better integrate adjuvant and antigen. These pathogen-like particles originate from a variety of sources, ranging from fully synthetic platforms to biologically derived, self-assembling systems (Rosenthal et al. 2014), and can be used to deliver the herbicides.

15.4 Conclusion and Future Prospects

Nanotechnology is one of the most important tools in modern agriculture and agri-food production, and this tool is anticipated to become a driving economic force in the near future. Agri-food planning focuses on sustainability of crop production and protection. Nanotechnology provides new agrochemicals and delivery tools to improve crop productivity, and it promises to reduce doses of chemicals. The available agrochemicals are quite effective in enhancing crop productivity and suppressing pests and diseases. However, the main problem of contamination and toxicity lies with their high dose of effectivity and indiscriminate application.

Development and application of nanofertilizers and pesticides is one of the potentially effective options of enhancing the global agricultural productions and reducing the chemical inputs. Application of nanotechnology in the direction of reduction in the dose of agrochemicals through coating, encapsulation, etc. may greatly minimize the adverse effects of synthetic chemicals on crop production. However, coating, encapsulation, nanoemulsions are in the developing stage, and it is probably a long way in reaching nanoproducts in the farms in developing and underdeveloped countries. Hence accelerated research efforts are needed towards mass production and commercial development of nano fertilizers and pesticides. Further, concerns on the biosafety of nanomaterials/nanoproducts need to be examined very critically on priority.

References

Abdullayev E, Lvov Y (2011) Halloysite clay nanotubes for controlled release of protective agents. J Nanosci Nanotechnol 11:10007–10026

Adak T, Kumar J, Dey D, Shakil NA, Walia S (2012) Residue and bio-efficacy evaluation of controlled release formulations of imidacloprid against pests in soybean (*Glycine max*). J Environ Sci Health B 47:226–231

Afrasiabi Z, Eivazi F, Popham H, Stanley D, Upendran A, Kannan R (2012) Silver nanoparticles as pesticides. National Institute of Food and Agriculture 1890 Capacity Building Grants Program Project Director's Meeting. Huntsville, AL

Ahamed AT, Wael FT, Shaaban MA, Mohammed FS (2011) Antibacterial action of zinc oxide nanoparticle agents' foodborne pathogens. J Food Saf 31:211–218

Amrani M, Westfall DG, Peterson GA (1999) Influence of water solubility of granular zinc fertilizers on plant uptake and growth. J Plant Nutr 22:1815–1827

Barik TK, Sahu B, Swain V (2008) Nanosilica-from medicine to pest control. Parasitol Res 103:253–258

Bashi AM, Haddawi SM, Dawood AH (2011) Synthesis and characterizations of two herbicides with Zn/Al layered double hydroxide nano hybrides. J Kerbala Univ 9:9–16

Batsmanova LM, Gonchar LM, Taran NY, Okanenko AA (2013) Using a colloidal solution of metal nanoparticles as micronutrient fertiliser for cereals. Proc Int Conf Nanomater 2:14 (2 pp)

Behera SK, Panda RK (2009) Integrated management of irrigation water and fertilizers for wheat crop using field experiments and simulation modeling. Agric Water Manage 96:1532–1540

Bhagat D, Samanta SK, Bhattacharya S (2013) Efficient management of fruit pests by pheromone nanogels. Sci Rep 3:1294

Biswal SK, Nayak AK, Parida UK, Nayak PL (2012) Applications of nanotechnology in agriculture and food sciences. IJSID 2:21–36

Boehm AL, Martinon I, Zerrouk R, Rump E, Fessi H (2003) Nanoprecipitation technique for the encapsulation of agrochemical active ingredients. J Microencapsul 20:433–441

Buda AR, Koopmans GF, Bryant RB, Chardon WJ (2012) Emerging technologies for removing nonpoint phosphorus from surface water and groundwater: introduction. J Environ Qual 41:621–627

Cao H, Zhang L, Zheng H, Wang Z (2010) Hydroxyapatite nanocrystals for biomedical applications. J Phys Chem C 114:18352–18357

Carpenter SR (2005) Eutrophication of aquatic ecosystems: biostability and soil phosphorus. Proc Natl Acad Sci USA 102:9999–10001

Carpenter SR (2008) Phosphorus control is critical to mitigating eutrophication. Proc Natl Acad Sci USA 105:11039–11040
Chakravarthy AK, Chandrashekharaiah, Kandakoor SB (2012) Bio efficacy of inorganic nanoparticles CdS, Nano-Ag and Nano-TiO_2 against *Spodoptera litura* (Fabricius) (Lepidoptera: Noctuidae). Curr Biotica 6:271–281
Chen H, Seiber JN, Hotze M (2014) ACS select on nanotechnology in food and agriculture: a perspective on implications and applications. J Agri Food Chem 62:1209–1212
Childers DL, Corman J, Edwards M, Elser JJ (2011) Sustainability challenges of phosphorus and food: solutions from closing the human phosphorus cycle. Bioscience 61:117–124
Cho JS, Kang YC (2008) Nano-sized hydroxyapatite powders prepared by flame spray pyrolysis. J Alloy Compd 464:282–287
Cioffi N, Torsi L, Ditaranto N (2004) Antifungal activity of polymer-based copper nanocomposite coatings. Appl Phys Lett 85:2417–2419
Clement JL, Jarret PS (1994) Antimicrobial silver. Metal-Based Drugs 1:467–482
Coles D, Frewer LJ (2013) Nanotechnology applied to European food production: a review of ethical and regulatory issues. Trends Food Sci Technol 34:32–43
Coma V, Martial-Gros A, Garreau S, Copinet A, Salin F, Deschamps A (2002) Edible antimicrobial films based on chitosan matrix. J Food Sci 67:1162–1169
Conley DJ (2009) Controlling eutrophication: nitrogen and phosphorus. Science 323:1014–1015
Corradini E, de Moura MR, Mattoso LHC (2010) A preliminary study of the incorporation of NPK fertilizer into chitosan nanoparticles. Polym Lett 4:509–515
Correll DL (1998) The role of phosphorous in the eutrophication of receiving waters: a review. J Environ Qual 27:261–266
De la Rosa G, Lopez-Moreno ML, De Haro D, Botez CE, Peralta-Videa JR, Gardea-Torresdey J (2013) Effects of ZnO nanoparticles in alfalfa, tomato, and cucumber at the germination stage: root development and X-ray absorption spectroscopy studies. Pure Appl Chem 85:2161–2174
De Rosa MC, Monreal C, Schnitzer M, Walsh R, Sultan Y (2010) Nanotechnology in fertilizers. Nat Nanotechnol 5:91
De-Bashana LE, Bashana Y (2004) Recent advances in removing phosphorus from wastewater and its future use as fertilizer (1997–2003). Water Res 38:4222–4246
Elek N, Hoffman R, Raviv U, Resh R, Ishaaya I, Magdassi S (2010) Novaluron nanoparticles: formation and potential use in controlling agricultural insect pests. Colloids Surf A Physicochem Eng Asp 372:66–72
Esteban-Tejeda L, Malpartida F, Esteban-Cubillo A, Pecharromán C, Moya JS (2009) Antibacterial and antifungal activity of a soda-lime glass containing copper nanoparticles. Nanotechnology 20:505701
Fageria NK (2009) The use of nutrient in crop plants. CRC Press, Boca Raton, FL, p 430
Feng QL, Wu J, Chen GQ, Cui FZ, Kim TN, Kim JO (2000) A mechanism study of the antibacterial effect of silver ions on *Escherichia coli* and *Staphylococcus aureus*. J Biomed Mater Res 52:662–668
Forsberg EM, de Lauwere C (2013) Integration needs in assessments of nanotechnology in food and agriculture. Etikk. i. Praksis 1:38–54
Frewer LJ, Norde W, Fischer ARH, Kampers FWH (eds) (2011) Nanotechnology in the agri-food sector: implications for the future. Wiley, Weinheim
Gangloff WJ, Westfall DG, Peterson GA, Mortvedt JJ (2002) Relative availability coefficients of organic and inorganic Zn fertilizers. J Plant Nutr 25:259–273
Garcia M, Forbe T, Gonzalez E (2010) Potential applications of nanotechnology in the agro-food sector. Food Sci Technol (Campinas) 30:573–581
Ghazali SMSAI, Hussein MZ, Sarijo SH (2013) 3,4-Dichlorophenoxyacetate interleaved into anionic clay for controlled release formulation of a new environmentally friendly agrochemical. Nanoscale Res Lett 8:362
Gogos A, Knauer K, Bucheli TD (2012) Nanomaterials in plant protection and fertilization: current state, foreseen applications, and research priorities. J Agric Food Chem 60:9781–9792

Goswami A, Roy I, Sengupta S, Debnath N (2010) Novel applications of solid and liquid formulations of nanoparticles against insect pests and pathogens. Thin Solid Films 519:1252–1257
Gruère G, Narrod C, Abbott L (2011) Agriculture, food, and water nanotechnologies for the poor: opportunities and constraints. Policy Brief 19, vol 19. International Food Policy Research Institute, Washington, DC
Gu YF, Zhang ZP, Tu SH, Lindström K (2009) Soil microbial biomass, crop yields, and bacterial community structure as affected by long-term fertilizer treatments under wheat-rice cropping. Eur J Soil Biol 45:239–246
Guan H, Chi D, Yu J, Li X (2008) A novel photodegradable insecticide: preparation, characterization and properties evaluation of nano-Imidacloprid. Pestic Biochem Physiol 92:83–91
Guo JS (2011) Synchrotron radiation, soft x-ray spectroscopy and nanomaterials. Int J Nanotechnol 1:193–225
Guo MY, Liu MZ, Zhan FL, Wu L (2005) Preparation and properties of a slow-release membrane-encapsulated urea fertilizer with super absorbent and moisture preservation. Ind Eng Chem Res 44:4206–4211
Han WWT, Misra RDK (2009) Biomimetic chitosan–nanohydroxyapatite composite scaffolds for bone tissue engineering. Acta Biomater 5:1182–1197
Han Y, Wang X, Li S (2008) A simple route to prepare stable hydroxyapatite nanoparticles suspension. J Nanopart Res 11:1235–1240
Hatschek E (1931) Electro Chem. Processes, Ltd, assignee. Brouisol. British patent No 392,556
Hoffmann CC (2009) Phosphorus retention in riparian buffers: review of their efficiency. J Environ Qual 38:1942–1955
Hossain KZ, Monreal CM, Sayari A (2008) Adsorption of urease on PE-MCM-41 and its catalytic effect on hydrolysis of urea. Colloid Surf B 62:42–50
Hussein MZ, Yahya AH, Zainal Z, Kian LH (2005) Nanocomposite-based controlled release formulation of herbicide 2,4-dichlorophenoxy acetate encapsulated in Zn/Al layered double hydroxide. Sci Tech Adv Mater 6:956–962
Hussein MZ, Sarijo SH, Yahaya AH, Zainal Z (2007) Synthesis of 4-chlorophenoxyacetate-zinc-aluminium-layered double hydroxide nanocomposite: physico-chemical and controlled release properties. J Nanosci Nanotechnol 7:2852–2862
Hussein MZ, Hashim N, Yahaya AH, Zainal Z (2010) Synthesis and characterization of [4-(2, 4-dichlorophenoxybutyrate)-zinc layered hydroxide] nanohybrid. Solid State Sci 2:770–775
Hussein MZ, Rahman NS, Sarijo SH, Zainal Z (2012) Herbicide-intercalated zinc layered hydroxide nanohybrid for a dual-guest controlled release formulation. Int J Mol Sci 13:7328–7342
Jayaseelan C, Rahuman AA, Rajakumar G (2011) Synthesis of pediculocidal and larvicidal silver nanoparticles by leaf extract from heartleaf moonseed plant, *Tinospora cordifolia* Miers. Parasitol Res 109:185–194
Jo YK, Kim BH, Jung G (2009) Antifungal activity of silver ions and nano-particles on phytopathogenic fungi. Plant Dis 93:1037–1043
Kah M, Beulke S, Tiede K, Hofmann T (2013) Nanopesticides: state of knowledge, environmental fate, and exposure modeling. Crit Rev Environ Sci Technol 43:1823–1867
Kamran S, Forogh M, Mahtab E, Mohammad (2011) In vitro antibacterial activity of nanomaterials for using in tobacco plants tissue culture. World Acad Sci Eng Technol 79:372–373
Kanimozhi V, Chinnamuthu CR (2012) Engineering core/hallow shell nanomaterials to load herbicide active ingredient for controlled release. Res J Nanosci Nanotechnol 2:58–69
Kasprowicz MJ, Kozioł M, Gorczyca A (2010) The effect of silver nanoparticles on phytopathogenic spores *of Fusarium culmorum*. Can J Microbiol 56:247–253
Katti KS, Katti DR, Dash R (2008) Synthesis and characterization of a novel chitosan/montmorillonite/hydroxyapatite nanocomposite for bone tissue engineering. Biomed Mater 3:034122

Khan MR, Jairajpuri MS (2012) Nematode infestation in horticultural crops, national scenario. In: Khan MR, Jairajpuri MS (eds) Nematode infestation Part III: horticultural crops. National Academy of Sciences, Allahabad, pp 1–30
Khan MR, Rizvi TF (2014) Nanotechnology: scope and application in plant disease management. Plant Pathol J 13:214–231
Khodakovskaya M, Dervishi E, Mahmood M (2009) Carbon nanotubes are able to penetrate plant seed coat and dramatically affect seed germination and plant growth. ACS Nano 3:3221–3227
Knauer K, Bucheli TD (2009) Nano-materials: research needs in agriculture. Revue Suisse d'Agric 41:337–341
Kottegoda N, Munaweera I, Madusanka N, Karunaratne V (2011) A green slow-release fertilizer composition based on urea-modified hydroxyapatite nanoparticles encapsulated wood. Curr Sci 101:73–78
Kumar R, Sharon M, Choudhary AK (2010) Nanotechnology in agricultural diseases and food safety. J Phytol 2:83–92
Kumar J, Shakil NA, Khan MA, Malik K, Walia S (2011) Development of controlled release formulations of carbofuran and imidacloprid and their bioefficacy evaluation against aphid, Aphis gossypii and leafhopper, *Amrasca biguttula biguttula* Ishida on potato crop. J Environ Sci Health B 46:678–682
Lamsal K, Sang-Woo K, Jung JH, Kim YS, Kim KU, Lee YS (2010) Inhibition effects of silver nanoparticles against powdery mildews on cucumber and pumpkin. Microbiology 39:26–32
Li J, Shu D, Yin J, Liu Y, Yao F, Yao V (2010) Formation of nano-hydroxyapatite crystal *in situ* in chitosan–pectin polyelectrolyte complex network. Mater Sci Eng C 30:795–803
Litke DW (1999) Review of phosphorus control measures in the United States and their effects on water quality. U.S. Geological Survey Water-Resources Investigations Report 99–4007 (1999)
Liu C (2008) Biomimetic synthesis of collagen/nano-hydroxyapatite scaffold for tissue engineering. J Bionic Eng 5:1–8
Liu R, Lal R (2014) Synthetic apatite nanoparticles as a phosphorus fertilizer for soybean (*Glycine max*). Environ Sci Environ Chem 4:5686
Liu DM, Troczynski T, Tseng WJ (2001) Water-based sol-gel synthesis of hydroxyapatite: process development. Biomaterials 22:1721–1730
Lodriche SS, Soltani S, Mirzazadeh R (2013) Inventors and assignees. Silicon nanocarrier for delivery of drug, pesticides and herbicides, and for waste water treatment. United States patent US 20130225412 A1. August 29, 2013
Lu CM, Zhang CY, Wen IQ, Wu GR (2002) Effects of nano materials on germination and growth of soybean. Soybean Sci 21:168–171
Manafi SA, Joughehdoust S (2009) Synthesis of hydroxyapatite nanostructure by hydrothermal condition for biomedical application. Iran J Pharm 5:89–94
Mandal KG, Hati KM, Misra AK (2009) Biomass yield and energy analysis of soybean production in relation to fertilizer-NPK and organic manure. Biomass Bioenergy 33:1670–1679
Manikandan A, Subramanian KS (2014) Fabrication and characterisation of nanoporous zeolite based N fertilizer. Afr J Agric Res 9:276–284
Manimegalai G, Kumar SS, Sharma C (2011) Pesticide mineralization in water using silver nanoparticles. Int J Chem Sci 9:1463–1471
Martens DC, Westermann DT (1991) Fertilizer applications for correcting micronutrient deficiencies. In: Mortvedt JJ, Cox FR, Shuman LM, Welch RM (eds) Micronutrients in agriculture. Soil Science Society of America, Madison, pp 549–592
Mateus AYP, Barrias CC, Ribeiro C, Ferraz MP, Monteiro FJ (2007) Comparative study of nanohydroxyapatite microspheres for medical applications. J Biomed Mater Res A 86:483–493
Mcqquire RG, Ching JYC, Vignaud E, Lebuge A, Mann S (2004) Synthesis and characterization of amino acid-functionalized hydroxyapatite nanorods. J Mater Chem 14:2277–2281
Milani N, Mc Laughlin MJ, Stacey SP, Kirby JK, Hettiarachchi GM, Beak DG (2012) Dissolution kinetics of macronutrient fertilizers coated with manufactured zinc oxide nanoparticles. J Agric Food Chem 60:3991–3998

Montazeri L, Javadpour J, Shokrgozar ML, Bonakdar S, Javadian S (2010) Hydrothermal synthesis and characterization of hydroxyapatite and fluorohydroxyapatite nano-size powders. Biomed Mater 5:7
Mortvedt JJ, Giordano PM (1969) Extractability of zinc granulated with macronutrient fertilizers in relation to its agronomic effective mess. J Agric Food Chem 17:1272–1275
Murphy K (2008) Nanotechnology: agriculture's next "industrial" revolution. Financial Partner, Yankee Farm Credit, ACA, Williston, pp 3–5
No HK, Meyers SP, Prinyawiwatkul W, Xu Z (2007) Applications of chitosan for improvement of quality and shelf life of foods: a review. J Food Sci 72:87–100
Owolade OF, Ogunleti DO, Adenekan MO (2008) Titanium dioxide affects diseases, development and yield of edible cowpea. EJEAF Che 7:2942–2947
Panwar J, Jain N, Bhargaya A, Akhtar MS, Yun YS (2012) Positive effect of zinc oxide nanoparticles on tomato plants: a step towards developing "Nano-fertilizers" 01/2012. In: Proceeding of 3rd international conference on environmental research and technology (ICERT); May 30–June 1, 2012, Penang
Park HJ, Kim SH, Kim HJ, Choi SH (2006) A new composition of nanosized silica-silver for control of various plant diseases. Plant Pathol J 22:295–302
Pérez-de-Luque A, Hermosín MC (2013) Nanotechnology and its use in agriculture. In: Bagchi D, Bagchi M, Moriyama H, Shahidi F (eds) Bio-nanotechnology: a revolution in food, biomedical and health sciences. Wiley, West Sussex, pp 299–405
Pérez-de-Luque A, Rubiales D (2009) Nanotechnology for parasitic plant control. Pest Manage Sci 65:540–545
Perlatti B, de Souza, Bergo PL, da Silva MF (2013) Polymeric nanoparticle-based insecticides: a controlled release purpose for agrochemicals, insecticides. In: Trdan S (ed) Insecticides: development of safer and more effective technologies. InTech, Croatia, pp 523–550
Poinern GE, Brundavanam RK, Mondinos N, Jiang ZT (2009) Synthesis and characterization of nanohydroxyapatite using an ultrasound assisted method. Ultrason Sonochem 16:469–474
Poursamar SA, Rabiee M, Samadikuchaksaraei V, Karimi M, Azami M (2009) Influence of the value of the pH on the preparation of nano hydroxyapatite polyvinyl alcohol composites. J Ceram Process Res 10:679–682
Prasad R, Kumar V, Prasad KS (2014) Nanotechnology in sustainable agriculture: present concerns and future aspects. Afr J Biotechnol 13:705–713
Prasanna BM (2007) Nanotechnology in agriculture. Indian Agricultural Statistics Research Institute, New Delhi
Rai M, Ingle A (2012) Role of nanotechnology in agriculture with special reference to management of insect pests. Appl Microbiol Biotechnol 94:287–293
Rai V, Acharya S, Dey N (2012) Implications of nanobiosensors in agriculture. J Biomater Nanobiotechnol 3:315–324
Rao KJ, Paria S (2013) Use of sulfur nanoparticles as a green pesticide on *Fusarium solani* and *Venturia inaequalis* phytopathogens. RSC Adv 3:10471–10478
Reynolds CS, Davies PS (2001) Sources and bioavailability of phosphorus fractions in freshwaters: a British perspective. Biol Rev 76:27–64
Rosenthal JA, Chen L, Baker JL, Putnam D, DeLisa MP (2014) Pathogen-like particles: biomimetic vaccine carriers engineered at the nanoscale. Curr Opin Biotechnol 28:51–58
Rouhani M, Samih MA, Kalantari S (2012) Insecticide effect of silver and zinc nanoparticles against *Aphis nerii* Boyer De Fonscolombe (Hemiptera: Aphididae). Chilean J Agric Res 72:590–594
Saigusa M (2000) Broadcast application versus band application of polyolefin-coated fertilizer on green peppers grown on andisol. J Plant Nutri 23:1485–1493
Samuel U, Guggenbichler JP (2004) Prevention of catheterrelated infections: the potential of a new nano-silver impregnated catheter. Int J Antimicrob Agents 23S1:S75–S78

Sarijo SH, bin Hussein MZ, Yahaya AH, Zainal Z, Yarmo MA (2010) Synthesis of phenoxyherbicides-intercalated layered double hydroxide nanohybrids and their controlled release property. Curr Nanosci 6:199–205

Sarlak N, Taherifar A, Salehi F (2014) Synthesis of nanopesticides by encapsulating pesticide nanoparticles using functionalized carbon nanotubes and application of new nanocomposite for plant disease treatment. J Agric Food Chem 62:4833–4838

Sasson Y, Levy-Ruso G, Toledano O, Ishaaya I (2007) Nanosuspensions: emerging novel agrochemical formulations. In: Ishaaya I, Nauen R, Horowitz AR (eds) Insecticides design using advanced technologies. Springer, Berlin, pp 1–39

Scott NR (2007) Nanotechnology opportunities in agriculture and food systems. Biological and Environmental Engineering, Cornell University NSF Nanoscale Science and Engineering Grantees Conference, Arlington. http://www.nseresearch.org/2007/overviews/Day3_Scott.pdf. Accessed 19 Apr 2014

Scott N, Chen H (2012) Nanoscale science and engineering for agriculture and food systems. National Planning Workshop, Washington, DC. http://www.nseafs.cornell.edu/web.roadmap.pdf. Accessed 18 Apr 2014

Scrinis G, Lyons K (2007) The emerging nano-corporate paradigm: nanotechnology and the transformation of nature, food and agri-food systems. Int J Sociol Food Agric 15:22–44

Sekhon BS (2014) Nanotechnology, science and applications. Dovepress 7:31–53

Sillanpaa M (1982) Micronutrients and the nutrient status of soils: a global study. Food and Agriculture Organization of the United Nations (FAO), Rome, p 444

Sillanpaa M (1990) Micronutrient assessment at country level: an international study. FAO, Rome, p 208

Sonkaria S, Ahn SH, Khare V (2012) Nanotechnology and its impact on food and nutrition: a review. Recent Pat Food Nutr Agric 4:8–18

Stadler T, Buteler M, Weaver DK (2010) Novel use of nanostructured alumina as an insecticide. Pest Manage Sci 66:577–579

Stadler T, Buteler M, Weaver DK, Sofie S (2012) Comparative toxicity of nanostructured alumina and a commercial inert dust for *Sitophilus oryzae* (L.) and *Rhyzopertha dominica* (F.) at varying ambient humidity levels. J Stored Product Res 48:81–90

Teodorescu M, Lungu A, Stanescu PO, Neamtu C (2009) Preparation and properties of novel slow-release NPK agrochemical formulations based on poly(acrylic acid) hydrogels and liquid fertilizer. Ind Eng Chem Res 48:6527–6534

Thornton PK (2010) Livestock production: recent trends, future prospects. Philos Trans R Soc B 365:2853–2867

Ulrichs C, Mewis I, Goswami A (2005) Crop diversification aiming nutritional security in West Bengal: biotechnology of stinging capsules in nature's water-blooms. Ann Tech Issue State Agric Technol Ser Assoc 1–18

Vidyalakshmi R, Bhakyaraj R, Subhasree RS (2009) Encapsulation "the future of probiotics" – a review. Adv Biol Res 3:96–103

Westfall DG, Mortvedt JJ, Peterson GA, Gangloff WJ (1991) Efficient and environmentally safe use of micronutrients in agriculture. Commun Soil Sci Plant Anal 36:169–182

Zhao M, Liu L, Her R (2012) Nano-sized delivery for agricultural chemicals. In: Tiddy G, Tan R (eds) NanoFormulation. Royal Society of Chemistry, Cambridge, pp 256–265

Chapter 16
Engineered Nanomaterials and Their Interactions with Plant Cells: Injury Indices and Detoxification Pathways

Mansour Ghorbanpour and Javad Hadian

16.1 Introduction

Although nanotechnology is a fairly new science, nanomaterials are not. Gold and silver nanoparticles were used in Persia in the tenth century BC to fabricate ceramic glazes to provide a lustrous or iridescent effect (Brayner 2008).

Nanoparticles (also known as particulate nanomaterials) include natural or engineered particles, with at least one dimension in the nanoscale (<100 nm) that are utilized for many purposes including medicine, biomedical engineering, material sciences, electronics, magnetic energy resources and agricultural researches (Prasad 2014; Prasad et al. 2014, 2016). Because of their particular properties such as small size, high surface-area-to-volume ratio, ability to engineer electron exchange and highly surface reactive capabilities (Scrinis and Lyons 2007), the use of nanomaterials in industries and a wide range of consumer products are increasing extremely.

However, the extensive use of various nanomaterials (Fig. 16.1) such as carbon-based materials and metal-based materials for multipurpose applications can result in their diffusion in biosphere, which has also raised concerns about the fate and effects of these substances on human health and the environment (Kahru and Dubourguier 2010).

Terrestrial plants, the most important components of ecosystem, interact directly with the soil, water and atmospheric environmental compartments, all of which can

M. Ghorbanpour (✉)
Department of Medicinal Plants, Faculty of Agriculture and Natural Resources, Arak University, 38156-8-8349 Arak, Iran
e-mail: m-ghorbanpour@araku.ac.ir

J. Hadian
Medicinal Plants and Drug Research Institute, Shahid Beheshti University, G.C. Evin, 1483963113 Tehran, Iran

M. Ghorbanpour et al. (eds.), *Nanoscience and Plant–Soil Systems*, Soil Biology 48,
DOI 10.1007/978-3-319-46835-8_16

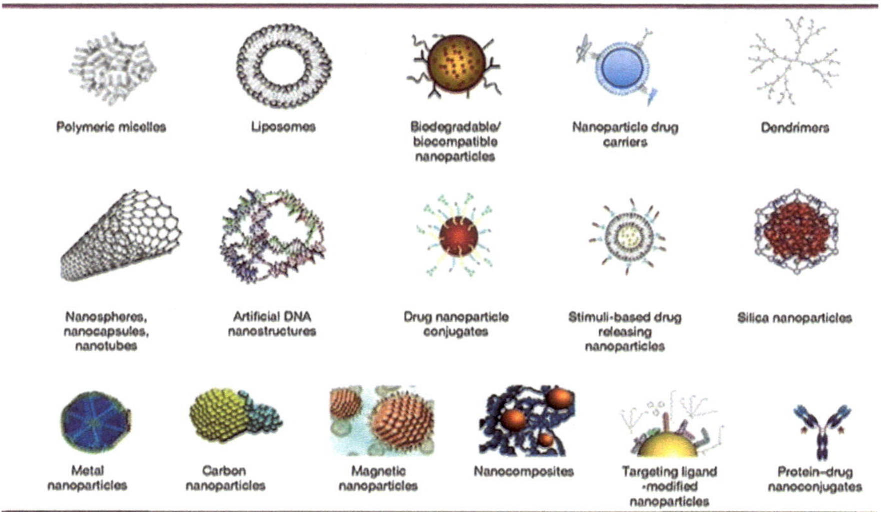

Fig. 16.1 Different types of engineered nanomaterials

be routes of engineered nanomaterials distribution. Plants are also subject to extensive human manipulation and are thus potentially subject to engineered nanomaterials exposure from different sources (Miralles et al. 2012). Therefore, the interactions between vascular plants and ultrafine materials can shed light on the environmental consequences of nanotechnology.

The effects of nanoscale materials on plant physiological and biochemical processes are complex; even the same type of these materials may have both positive and negative effects on various plant species (Baiazidi-Aghdam et al. 2016; Ghorbanpour 2015; Hatami et al. 2014, 2015; Yasur and Rani 2013; Siddiqui et al. 2014; Hatami and Ghorbanpour 2013; Mohammadi et al. 2013; Hatami et al. 2013).

Several studies have shown that engineered nanomaterials induce oxidative stress or produce reactive oxygen species (ROS) (Ghorbanpour and Hadian 2015; Oukarroum et al. 2012; Begum et al. 2011) and influence the plant defense mechanisms through antioxidative enzyme activities including superoxide dismutase (SOD), peroxidases (POX), ascorbic peroxidase (APX), catalase (CAT) and glutathione peroxidase (GPX) and nonenzymatic antioxidants such as carotenoids, glutathione, ascorbic acid, alpha-tocopherol and proline (Song et al. 2012; Ghorbanpour and Hatami 2015; Zhao et al. 2012; Rico et al. 2013a, b; Hong et al. 2014; Ghorbanpour et al. 2015; Hatami and Ghorbanpour 2014; Hernandez-Viezcas et al. 2011; Ghorbanpour and Hatami 2014; Zhao et al. 2012; Hatami et al. 2016) in plants. However, further studies are necessary in order to draw a comprehensive picture of the plant-engineered nanomaterials interactions at the cellular level.

16.2 Cellular Damages Induced by Nanomaterials

Cell membranes are one of the first targets of many plant stresses, and it is generally accepted that the maintenance of their integrity and stability under stress conditions is a major component of stress tolerance in plants. The degree of cell membrane injury induced by various stress may be easily estimated through measurements of electrolyte leakage from the cells. On the other hand, electrolyte leakage is a hallmark of stress response in intact plant cells and is ubiquitous among different species, tissues and cell types. This phenomenon is widely used as a test for the oxidative-induced injury of plant tissues (Lee and Zhu 2010). Membrane lipid peroxidation, which is normally associated with natural course of ageing, senescence and environmental stresses, is mechanistically very important from the perspective of generation of ROS and considered as one of the few examples of carbon-centred radical production in cells (Winston 1990).

The lipid peroxidation can be caused due to the accumulation of the ROS which are the principal causes of oxidative stress-related membrane damage (Maaouia Houimli et al. 2010). H_2O_2 is the major ROS of the oxidative burst in plants, because it is the most long-lived and able to cross plant cell membranes and thereby act as a diffusible and relatively lasting signal (Karuppanapandian et al. 2011). In addition to the formation of malondialdehyde (MDA), induction of antioxidant mechanisms may be a sign of ROS overproduction and thereby of oxidative stress. Plasma membrane senses different environmental stimuli and transduces them to downstream intracellular and intercellular signalling networks. Exposure to both abiotic and biotic stresses causes changes in membrane architecture. In fact, membranes must respond to various environmental stresses.

The widespread application of engineered nanoparticles reveals all organisms on contact with them. It is acknowledged that nanoparticles can significantly influence cell components. The effects of nanomaterials on the oxidative stress in plants have been widely investigated using techniques that measure cellular injury indices including production of H_2O_2 (or ROS in general), membrane electrolyte leakage and MDA content. Studies have shown that engineered nanomaterials are able to induce stress, generating excess ROS with the potential to affect cell organelles and structures, proteins, carbohydrates, lipids and DNA in plants. Figure 16.2 schematically illustrates potential oxidative stress injury in different organelles within plant cells through interactions with engineered nanoparticles (Zahed et al. 2015).

In an in vitro study, the effects of multiwalled carbon nanotubes (MWCNTs, 5–15 nm, scanning electron microscopy (SEM) and transmission electron microscopy (TEM) images and Raman spectra of the nanotubes are shown in Fig. 16.3) on callus induction and H_2O_2 production in *Satureja khuzestanica* were investigated (Ghorbanpour and Hadian 2015). Leaf segments were aseptically cultured in Gamborg's B-5 medium with various MWCNTs concentrations (0, 25, 50, 100, 250 and 500 μg/mL). Results showed that application of 250 and 500 μg/mL MWCNTs significantly increased the H_2O_2 generation compared to untreated controls. However, no significant changes were observed in H_2O_2 content between

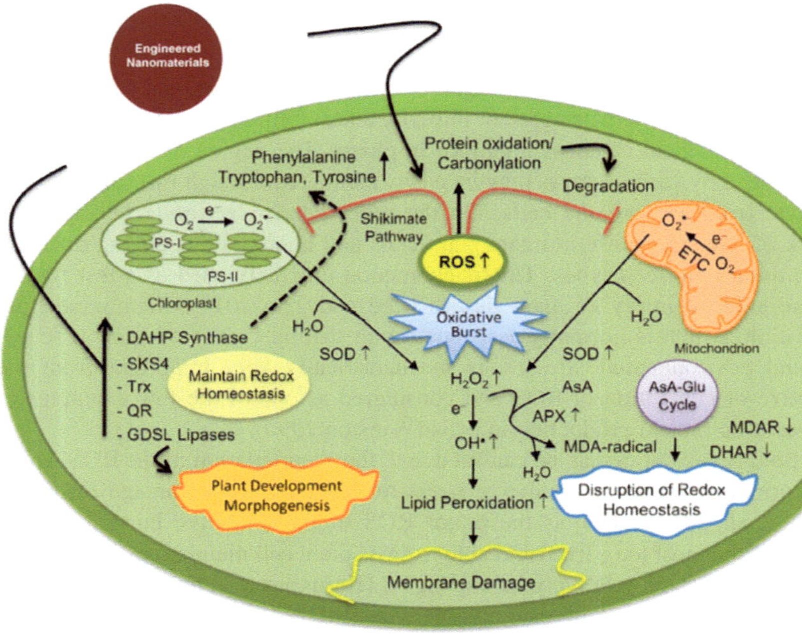

Fig. 16.2 Schematic model for cellular damages induced by nanomaterials. When nanomaterials entered the plant cell, they can transport apoplastically or symplastically. Exposure to nanomaterials cause toxic effects including increased generation of reactive oxygen species (ROS), disruption of redox homeostasis, peroxidation of lipid, impaired mitochondrial function and membrane damage. Upward and downward arrows indicate increased and decreased protein abundance in response to nanomaterials exposure, respectively. The shikimate pathway (*dotted arrow*) consists of several enzymatic reactions whose end product chorismate is the precursor for the synthesis of the aromatic amino acids. Abbreviations: *APX* ascorbate peroxidase, *SOD* superoxide dismutase, *AsA* reduced ascorbate, *DAHP* 3-deoxy-*D*-arabino-heptulosonate-7-phosphate, *DHAR* dehydroascorbate reductase, *ETC* electron transport chain, H_2O_2 hydrogen peroxide, *MDA* malondialdehyde, *MDAR* monodehydroascorbate reductase, *PS* photosystem, *QR* quinone reductase, *SKS4*, *SKU5* similar 4 protein, *Trx* thioredoxin (adapted from Zahed et al. 2015)

the control treatment and those supplemented with MWCNTs up to 100 μg m/L (Table 16.1).

It has been reported that carbon nanotubes could induce ROS accumulation enhancing lipid peroxidation in cell culture (Liu et al. 2010) and seedling root tips (Liu et al. 2013). According to Rico et al. (2013a, b), metal-based nanoparticles and/or the liberated ions from the nanoparticles may produce oxidative stress inducing ROS accumulation in plants. The stress induced by nanoscale ZnO and CuO has been linked to the nanoparticles and liberated Zn and Cu ions (Rico et al. 2013a, b; Nair and Chung 2014).

Baiazidi-Aghdam et al. (2016) evaluated the effects of various concentrations (0, 10, 100 and 500 mg/L) of nano-sized (10–25 nm) titanium dioxide (TiO_2) on H_2O_2 and in *Linum usitatissimum* under optimal and water deficit stress conditions.

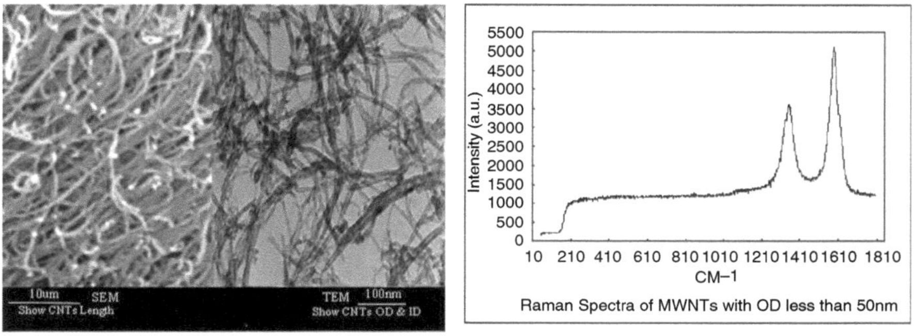

Fig. 16.3 SEM and TEM images (*left*) and Raman spectra (*right*) of MWCNTs (Ghorbanpour and Hadian 2015)

Table 16.1 The effect of different concentrations of MWCNTs on H_2O_2 content and enzymatic activities of polyphenol oxidase (PPO), phenylalanine ammonia-lyase (PAL) and peroxidase (POD) in callus cell of *Satureja khuzestanica* grown in vitro (Ghorbanpour and Hadian 2015)

MWCNTs Concentration (μg/mL)	H_2O_2 content (μmol g^{-1} FW)	Enzyme activities		
		PPO (Unit g^{-1} FW min^{-1})	PAL (Unit g^{-1} FW)	POD (Unit mg^{-1} protein min^{-1})
0	2.76 ± 0.18	18.04 ± 2.8	15.12 ± 1.6	0.24 ± 0.014
25	2.78 ± 0.15	11.72 ± 2.3	21.46 ± 2.3	0.22 ± 0.015
50	2.83 ± 0.13	14.55 ± 1.9	32.75 ± 1.7	0.28 ± 0.018
100	2.89 ± 0.14	17.76 ± 1.4	53.42 ± 2.8	0.31 ± 0.014
250	4.52 ± 0.19	21.43 ± 1.5	56.94 ± 2.9	0.32 ± 0.012
500	6.16 ± 0.21	28.13 ± 2.1	38.36 ± 1.4	0.44 ± 0.012

The results showed that application of nanoscale TiO_2 at low concentration better improved the morphological and physiological traits of a plant compared to other doses particularly under water scare conditions, leading to better plant performance. The levels of H_2O_2 and MDA in plants exposed to nanoscale TiO_2 at 10 mg/L were lower than that of other treatments; therefore, lipid peroxidation was less pronounced in such plants.

In another study, the effects of TiO_2 nanoparticles (2, 5 and 10 ppm) on physiological and biochemical responses were studied in two chickpea (*Cicer arietinum* L.) genotypes differing in cold sensitivity (tolerant Sel11439 and sensitive ILC533) during cold stress (4 °C) (Mohammadi et al. 2013). They found that H_2O_2 and MDA contents and electrolyte leakage index (ELI) increased under cold stress conditions in both genotypes and that these damage indices were higher in ILC533 than in Sel11439 plants. In plants treated with TiO_2 nanoparticles, a decreased H_2O_2 level was accompanied by a decrease in the MDA content and ELI compared to control plants, and these changes occurred more effectively in Sel11439 than in ILC533 plants (Mohammadi et al. 2013).

Also, investigations have shown that TiO_2 nanoparticles (both rutile and anatase forms) produced ROS in spinach plant (Fenoglio et al. 2009). Begum et al. (2011) studied the toxicity impacts of graphene at different concentrations (00, 1000 and 2000 mg/L) in cabbage, tomato and red spinach plants. Results exhibited a graphene dose-dependent increase in H_2O_2 generation, cell death and electrolyte leakage in graphene-treated leaves.

Anjum et al. (2013, 2014) investigated the *Vicia faba* seedlings tolerance to different concentrations (0, 100, 200, 400, 800 and 1600 mg/L) of single-bilayer graphene oxide (GO) sheet and underlying potential mechanisms. They found both positive and negative dose-dependent GO impacts on *V. faba*. Significant negative effects of GO concentrations (ranked of effects: $1600 > 200 > 100$ mg/L) were indicated by decreases in the levels of H_2O_2, electrolyte leakage and lipid and protein oxidation. The positive impacts of 400 mg/L of GO included significant improvement of the seedling health status indicated by decreased levels of above-mentioned cell injury indices. Prakash et al. (2016) investigated the effects of copper oxide (CuO) nanoparticles at different concentrations (0, 50, 100, 200, 400 and 500 mg/L) on biochemical, anatomical and molecular changes in cucumber (*Cucumis sativus* L.) seedlings. A dose-dependent increase in ROS generation and MDA content were observed.

In another study, the effects of copper oxide (CuO) nanoparticles at morphological, physiological and molecular levels were studied in *Brassica juncea* (Nair and Chung 2015). The seedlings were exposed to 0, 20, 50, 100, 200, 400 and 500 mg/L of CuO nanoparticles in semisolid half-strength MS medium for 14 days. Exposure to CuO nanoparticles caused increases in H_2O_2 production and lipid peroxidation level, and lignification of shoots and roots organs was observed in *B. juncea* seedlings.

The effects of cerium oxide (CeO_2) nanoparticles and bulk at different concentrations (0, 10 and 100 mg/L) on physiological and biochemical attributes of *Brassica rapa* plants revealed that while the bulk CeO_2 treatment resulted in significantly higher concentration of H_2O_2 in plant tissues at the vegetative growth stage, CeO_2 nanoparticles led to significantly higher H_2O_2 levels in plant tissues at the flowering phase. According to Franklin et al. (2007) and Navarro et al. (2008), the toxicity of the nanoparticles may be associated with the possible release of toxicants from their surface, such as metal ions or residues after synthesis.

16.3 Detoxification Pathways and Defense Mechanisms

Under the optimal environmental conditions, ROS can be by-products of plant normal metabolic pathways localized in cell organelles including chloroplasts, mitochondrion and peroxisomes (Møller et al. 2007). However, the metabolism of plants under stress conditions is biochemically characterized as an increase in the production of ROS (Foyer and Noctor 2000). To scavenge ROS, plants employ specific mechanisms, including activation of antioxidant enzymes such as SOD,

POX, APX, CAT and GPX, and nonenzymatic antioxidants such as carotenoids, glutathione, ascorbic acid, alpha-tocopherol and proline (Mittler 2002). ROS produced under normal conditions of the environment are balanced and/or removed by aforesaid specific antioxidant enzymes, whose activity helps to avoid oxidative damages in plants.

However, overproduced ROS under various stressful conditions of the environment can react with proteins, lipids and DNA molecules, which can then cause severe oxidative damage to plant biomolecules through electron transfer, resulting in a number of metabolic disorders, destruction of cell membranes and in consequence cell death (Lushchak 2011; Saibo et al. 2009). Although the high level of ROS is potentially harmful to plant cells, its production during oxidative stress could have a role in stress perception and protection (Suzuki and Mittler 2006). Therefore, study of antioxidative capacities under oxidative stress compared to indicators of nanotoxicity may reflect some mechanisms underlying nanomaterials tolerance. The content of H_2O_2 can be assayed to determine ROS generation as a role of metal-based nanoparticles exposure in plant cells.

Rico et al. (2013a, b) reported that changes in H_2O_2 content between two rice cultivars exposed to the same concentration of CeO_2 nanoparticles were varied. They also found that the H_2O_2 content in the cultivar 'Neptune' was approximately three times higher than the control group upon exposure to 500 mg/L CeO_2 nanoparticles, whereas cultivar 'Cheniere' showed no significant difference upon exposure to the same concentration of nano CeO_2. However, at low concentration (62.5 mg/L), the H_2O_2 was scavenged in both cultivars.

It has previously been reported that the same nanostructured materials that can be toxic for plants in high concentrations may have stimulatory impact on the physiological processes at low concentrations (Khodakovskaya et al. 2009). According to Lee et al. (2013a, b), CeO_2 nanoparticles at low concentration (50 mg/L) could eliminate ROS through a Fenton-type reaction. Zhao et al. (2012) assayed H_2O_2 content in *Zea mays* grown in a soil amended with CeO_2 nanoparticles and reported effective antioxidant defense through CAT and APX activities, both of which converted ROS to water molecule. Arora et al. (2012) suggested that exposure to 100 ppm Au nanoparticles increased the ROS levels by 29 % compared to control in *Brassica juncea*; however, there were no any negative effects observed on growth and seed yield. There are several antioxidant enzymes that may efficiently detoxify ROS in plants. In a study, we investigated the effects of various concentrations (0, 25, 50, 100, 250 and 500 μg/mL) of MWCNTs on enzyme activities in *Satureja khuzestanica* grown in vitro Ghorbanpour and Hadian (2015). Results showed that significant differences ($p < 0.05$) in enzyme activities were observed in calli grown in the culture media (B5) fortified with various MWCNTs concentrations (Table 16.1). The polyphenol oxidase (PPO) activity showed a significant decrease in treatments with the two lowest MWCNTs concentrations (25 and 50 μg/mL), while it was significantly enhanced by the two highest MWCNTs concentrations (250 and 500 μg/mL). An increase in phenylalanine ammonia-lyase (PAL) activity was observed in callus extracts under all employed treatments compared to control and peaked at 500 μg/mL MWCNTs treatment. The peroxidase (POD) activity was unchanged where the lowest MWCNTs concentration applied (25 μg/mL), whereas all other treatments had significantly higher activity compared to the untreated control peaking at 500 μg/mL MWCNTs treatment.

In another study, the impacts of different concentrations (0, 20, 40 and 80 mg/L) of nanosilver particles (5–35 nm, SEM and TEM images are given in Fig. 16.4) were investigated on H_2O_2 content and antioxidant enzymes including SOD, POD and CAT activities in *Pelargonium graveolens* (Ghorbanpour and Hatami 2015). Results showed that except in plants treated with 80 mg/L nanosilver, the H_2O_2 generation of the leaves was sharply decreased with other nanosilver supply compared to control (Table 16.2). The application of nanosilver at 40 mg/L reduced significantly H_2O_2 accumulation rate by 20.8 % than control. In the current study, significant differences ($p \leq 0.05$) in antioxidant enzyme activities were observed in

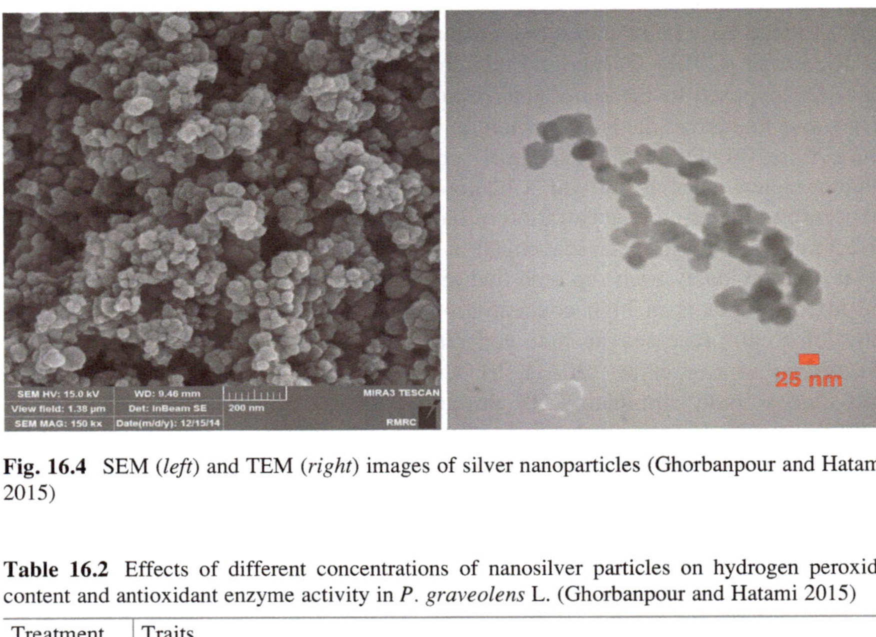

Fig. 16.4 SEM (*left*) and TEM (*right*) images of silver nanoparticles (Ghorbanpour and Hatami 2015)

Table 16.2 Effects of different concentrations of nanosilver particles on hydrogen peroxide content and antioxidant enzyme activity in *P. graveolens* L. (Ghorbanpour and Hatami 2015)

Treatment	Traits			
Nanosilver (mg/L)	H_2O_2 (μmol g^{-1} fw)	SOD (U min^{-1} mg $protein^{-1}$)	CAT (μmol min^{-1} mg $protein^{-1}$)	POD (μmol min^{-1} mg $protein^{-1}$)
0	4.8^{ab}	56.4^{c}	2.1^{b}	0.15^{c}
20	4.4^{b}	64.5^{b}	3.4^{a}	0.19^{b}
40	3.8^{c}	71.2^{a}	2.2^{b}	0.23^{a}
80	5.1^{a}	53.0^{c}	1.75^{b}	0.10^{d}
LSD ($p \leq 0.05$)	0.20	2.4	0.31	0.011

The values followed by different letter(s) are significantly different using least significant difference (*LSD*) test at $p \leq 0.05$

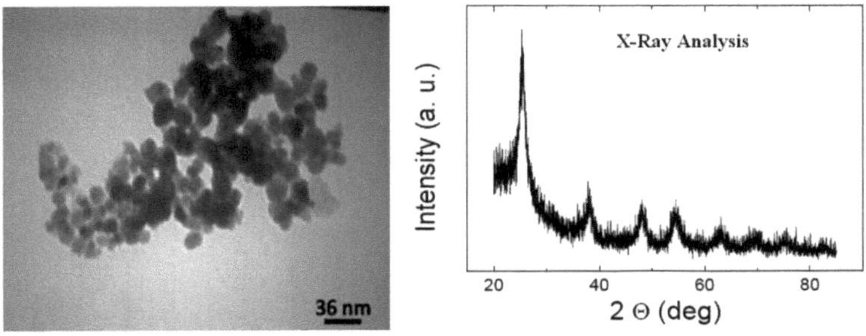

Fig. 16.5 TEM image (*left*) and XRD (X-ray diffraction) pattern (*right*) of TiO_2 (anatase) nanoparticles (Ghorbanpour et al. (2015)

leaves of the plants with different treatments used. However, the highest POD activity was observed in plants treated with 40 mg/L nanosilver compared to the plants treated with other employed treatments and to untreated control plants. Similarly, application of nanosilver (at 40 mg/L) led to an enhanced SOD activity when compared to control. In contrast to POD and SOD, CAT activity was strongly increased to maximum value upon exposure to 20 mg/L nanosilver.

Ghorbanpour et al. (2015)studied the effects of different concentrations (0, 20, 40 and 80 mg/L) of nano-sized titanium dioxide (NT, 10–15 nm, SEM and TEM images of TiO_2 nanoparticles are shown in Fig. 16.5) and bulk (BT) on antioxidant enzyme activities including SOD, POX and CAT in *Hyoscyamus niger*. We found noticeable differences in antioxidant enzyme activities under employed treatments. SOD activity increased with NT and BT levels and application of a significant role in adjusting the enzyme activity (Fig. 16.6). SOD activity increased with increasing TiO_2 concentration in both nano-sized and bulk forms. However, the highest SOD activity was observed at maximum NT and BT supply. Moreover, CAT activity increased with NT application up to 20 mg/L and then decreased compared to other NT level, whereas BT at all concentrations enhanced the CAT activities. The maximum CAT activity was observed in BT at 80 mg/L treatment. With regard to the effects of NT and BT on adjusting CAT activity, low NT and high BT application significantly increased CAT activity up to 50 % compared to untreated control plants. However, POX activity significantly increased under employed NT up to 40 mg/L and then decreased with NT concentration (Fig. 16.6). The high concentrations of NT and BT significantly decreased POX activity; however, the final value was not lower than that of control in NT-treated plants. Generally, all tested enzyme activities were higher in NT-treated plants that those of BT except CAT activity at 80 mg/L (Ghorbanpour et al. 2015).

In a study, the post-storage influence of nanosilver particles (10–20 nm) at different concentrations (0, 20, 40, 60 and 80 mg/L) was tested on cell injury indices and enzyme activity in two *Pelargonium* pot plants of the cultivars ‘Flower fairy’ and ‘Foxi’ during dark storage period (Hatami and Ghorbanpour 2014) as shown in Table 16.3. In both cultivars, MDA content decreased significantly with the increase of nanosilver concentration up to 60 mg/L and then followed a rapid

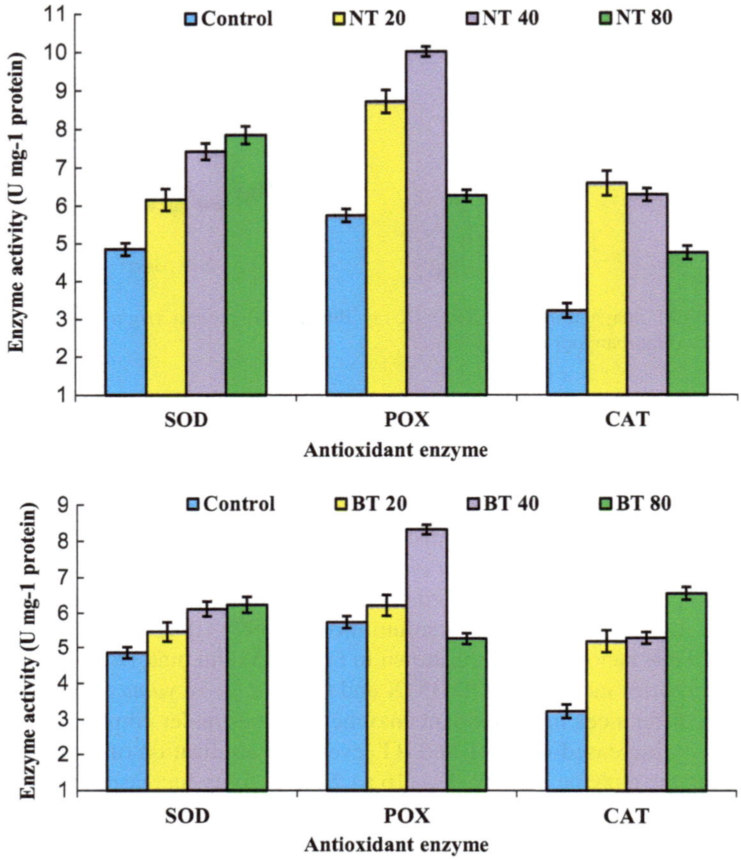

Fig. 16.6 The effects of nano-sized titanium dioxide (*NT*) and bulk (*BT*) concentrations (0, 20, 40 and 80 mg/L) on superoxide dismutase (*SOD*), peroxidase (*POX*) and catalase (*CAT*) activities in *Hyoscyamus niger* plant (Ghorbanpour et al. 2015)

increase at 80 mg/L (Fig. 16.7). After 5 days of dark storage marked differences in enzyme activity were found between two *Pelargonium* cultivars exposed to different nanosilver concentrations. The highest APX activity was observed at 40 mg/L nanosilver concentration, but specific activities of SOD and POD increased in a maximum level at 60 mg/Lin both cultivars. CAT activity was pronounced at 20 mg/Land 40 mg/L nanosilver treatments in Flower fairy and Foxi, respectively. Both under control conditions as well as the highest nanosilver concentration, ß-glucosidase (as a glycoside hydrolyse enzyme) showed lower activity than other nanosilver levels.

In plant cells, SOD catalyses the conversion of the O^{-2} to O_2 and H_2O_2 (Hafis et al. 2011). Enhanced SOD activity of leaves under nanosilver treatments may be interpreted as a direct response to augmented O^{-2} formation. Also, reduced SOD

Table 16.3 The effects of different types of nanomaterials on cell injury indices and defense enzyme activities in various plant species

Nanoparticle	Particle size (nm)	Concentration	Plant species	Growth medium/condition	Observed effects	Reference
ZnO	<50 nm	0, 1, 5, 100, 1000, and 2000 mg/L	*Fagopyrum esculentum*	Hoagland solution in hydroponic culture	Zn bioaccumulation in plant increased with increasing treatment concentrations. Also, ROS generation by ZnO NPs was estimated as the reduced glutathione level and CAT activity. At the high NP doses (1000 or 2000 mg/L), the amount of GSH was lower than that of low NP doses. This phenomenon was also observed for CAT activity after treatments of 1000 and 2000 mg/L of ZnO NPs. Increased CAT activity at all treatment concentrations	Lee et al. (2013a, b)
SiO_2	12 nm	1.5–3.0, 4.5, 6 and 7.5 g/L	*Cucurbita pepo*	Filter paper in Petri dishes	Improved germination and growth parameters by reducing cell injury indices (e.g. MDA, H_2O_2 and ELI), exhibiting reduction in oxidative damage as a result of the expression of antioxidant enzymes such as CAT, POD, SOD, GR and APX. Also, the concentration of 6 g/L of SiO_2 NPs proved to be the best for alleviating salt stress	Siddiqui et al. (2014)

(continued)

Table 16.3 (continued)

Nanoparticle	Particle size (nm)	Concentration	Plant species	Growth medium/condition	Observed effects	Reference
ZnO	8–10 nm	500 – 4000 mg/L	*Proposis julif lora-velutina*	Whole plant exposed for 15 days	CAT activity increased in the whole plant; however, APX activity decreased in roots, and increased in leaves and stem	Hernandez-Viezcas et al. (2011)
Ag	<100 nm	0 , 100, 200, 500, 1000, 2000, and 4000 mg/L	*Ricinus communis*	Filter paper in Petri dishes	The Ag NPs application to *R. communis* seeds enhanced antioxidant enzymatic activities. Increased POD activity in Ag NPs-exposed *R. communis* seedlings, but no significant difference in the activity between the doses of 500 and 1000 mg/L was observed. Increased SOD activity till 1000 mg/L and then decreased up to 2000 mg/L; however, there was not much difference between 2000 and 4000 mg/L concentrations with no significant change	Yasur and Rani (2013)
CeO_2	231 ± 16 nm	0, 62.5, 125, 250 and 500 mg/kg	*Raphanus sativus*	Potting mix	Increased CAT activity at 125 mg CeO_2 NPs in tubers, but it decreased in leaves compared to control. APX activity also increased in tubers of plants treated with	Corral-Diaz et al. (2014)

					125 and 500 mg/kg treatments	
Si	60 nm	0 and 2.5 mM	*Oryza sativa*	Plastic box with 1/4 strength Hoagland nutrient solution, (sprayed whole seedlings)	Foliar application of Si NPs improved the growth, Mg, Fe, and Zn nutrition of the rice seedlings under Cd stress, and decreased Cd accumulation and translocation of Cd from root to shoot. Also, nanosilver-treated seedlings had lower MDA but higher glutathione content and different antioxidant enzyme activities	Wang et al. (2015)
TiO_2	7–40 nm	0, 2, 5 and 10 ppm	*Cicer arietinum*	Filter paper in Petri dishes The effect of TiO_2 NPs concentrations was assessed on physiological performance of *C. arietinum* genotypes during cold stress	TiO_2 NPs at 5 ppm caused a decrease in ELI during thermal treatments, whereas ELI content under cold stress (4 °C) treatment increased at 0 ppm TiO_2 NPs in both sensitive and tolerant *C. arietinum* genotypes. Under thermal treatments, although the tolerant genotype showed lower accumulation of MDA than sensitive genotype, a significant decrease was observed in MDA content at 5 ppm TiO_2 NPs. Results showed that TiO_2 NPs treatments not only did not induce	Mohammadi et al. (2013)

(continued)

Table 16.3 (continued)

Nanoparticle	Particle size (nm)	Concentration	Plant species	Growth medium/condition	Observed effects	Reference
					oxidative damage in sensitive and tolerant chickpea genotypes but also alleviated membrane damage indexes under cold stress treatment	
Biogenic NPs (synthesized from *Tridax procumbens*)		30 ppm	*Eruca sativa*	Filter paper	Decreased in ELI value and level of MDA in biogenic NPs treated seeds compared to control. Enhanced the levels of proline and ascorbic acid and stimulated the antioxidant enzyme activities and reduced level of ROS. Biogenic NPs promoted seed germination in *E. sativa* by overcoming the detrimental effects of ROS and improving the antioxidative defense system which finally result in increased seedling growth	Ushahra et al. (2014)

CeO_2	10 ± 1 nm	400 and 800 mg/kg soil	*Zea mays*	Magenta boxes filled with soil. Stress-related parameters were analysed at 10, 15, and 20 days post germination	Increased CAT and APX activities in shoot, concomitant with the H_2O_2 levels. Both 400 and 800 mg/kg CeO_2 NPs triggered the upregulation of the heat shock protein 70 in roots. None of the CeO_2 NPs increased the level of thiobarbituric acid reacting substances, indicating that no lipid peroxidation occurred. CeO_2 NPs, at both concentrations, did not induce ion leakage in either roots or shoots, suggesting that membrane integrity was not compromised. The results suggest that the CAT, APX, and heat shock protein 70 might help the plants defend against CeO_2 NP-induced oxidative injury and survive NPs exposure	Zhao et al. (2012)

(continued)

Table 16.3 (continued)

Nanoparticle	Particle size (nm)	Concentration	Plant species	Growth medium/condition	Observed effects	Reference
CeO_2	Primary size 8 ± 1 nm, particle size 231 ± 16 nm	powder at 0, 0.98 and 2.94 g/m^3 or suspensions at 0, 20, 40, 80, 160, and 320 mg/L	*Cucumis sativus*	Germination paper towels, and hydroponic jars (magenta boxes) containing a modified Hoagland nutrient solution	Decreased APX activity in stems and increased in roots as the external CeO_2 NPs increased. The activity of CAT significantly increased in leaves and roots at the higher NP concentration, but no changes were observed in stems. The activity of dehydroascorbate reductase showed an increase in leaves only in plants treated with 80 mg/L, but it increased in stems and roots of plants treated with 80–320 mg/L	Hong et al. (2014)
CeO_2	Primary size 8 ± 1 nm, particle size 231 ± 16 nm	0, 62.5, 125, 250, and 500 mg/L	*Oryza sativa*	Petri dishes	Relative to the control, the 62.5 and 125 mg CeO_2 NPs/L treatments significantly reduced the H_2O_2 generation in both shoots and roots. Enhanced electrolyte leakage and lipid peroxidation were found in the shoots of seedlings grown at 500 mg CeO_2 NPs/L. All enzyme activities in rice shoots, except for SOD and GPOX, exhibited a u-shaped trend with the lowest activity	Rico et al. (2013a)

					at 125 mg CeO_2 NPs/L. Altered enzyme activities and levels of ascorbate and free thiols resulting in enhanced membrane damage and photosynthetic stress in the shoots were observed at 500 mg CeO_2 NPs/L	
ZVI	<100 nm	0 and 0.1 g/L	*Arabidopsis thaliana*	Hydroponic and soil culture	Triggered high plasma membrane H^+-ATPase activity, leading to the possibility of increased CO_2 uptake. The increase in activity caused a decrease in apoplastic pH, an increase in leaf area, and also wider stomatal opening	Kim et al. (2015)

Abbreviations: *ZVI* zero valent iron, *CeO_2* cerium oxide, *TiO_2* titanium dioxide, *SiO_2* silicon dioxide, *Si* silicon, *ZnO* zinc oxide, *Ag* silver

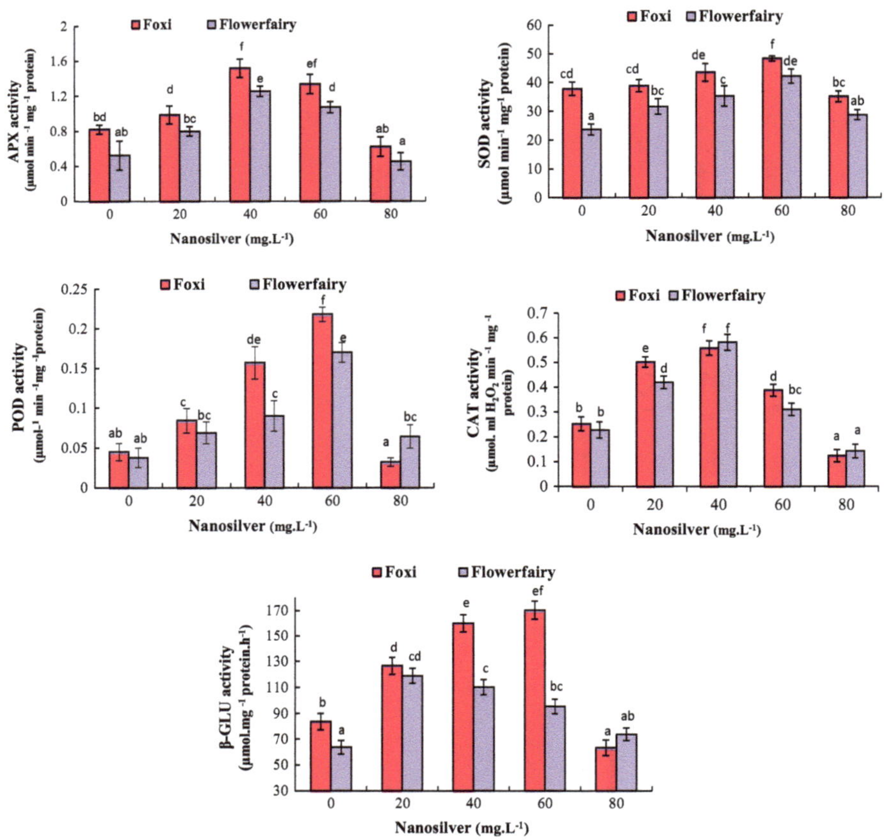

Fig. 16.7 Ascorbate peroxidase (*APX*), superoxide dismutase (*SOD*), peroxidase (*POD*), catalase (*CAT*) and β-Glucosidase (*β-GLU*) activities of two *Pelargonium* cultivars, Foxi and Flower fairy, in response to different nanosilver concentrations after 5 days of dark storage. The error bars represent the standard deviation for three replications. Different *letters* indicate significant differences in each treatment as determined by Tukey's HSD test at $P = 0.05$ (Hatami and Ghorbanpour 2014)

activity in both cultivars (Foxi and Flower fairy) under control and the highest nanosilver concentrations may reflect the low ROS scavenging capacity and increased damage to the plant parts and structure. We found that certain nanosilver concentrations upregulated the activity of POD in *Pelargonium* cultivars. This might be an important protection mechanism in plants against the excessive increase of H_2O_2 during dark-induced stress.

It is well known that the overexpression of SOD, if joined with increment of H_2O_2 scavenging mechanisms like POD and CAT, has been considered as a strategy to cope with oxidative damage (Kohler et al. 2006). Our results also indicated significant role of nanosilver; particularly the application of 40 and

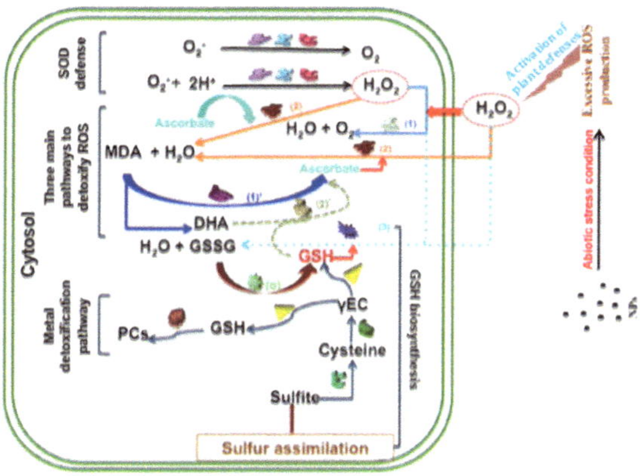

Possible locations of listed antioxidant enzymes in plant cell

Symbol	Abbreviation of antioxidant enzymes	Possible locations in plant cell
	Fe-SOD	
	Mn-SOD	Chl, Mit, Per, Nu, Cy
	Cu-Zn-SOD	
	CAT	Mit, Chl, Per, CW, Nu, Cy
	APX	Mit, Chl, Per, Go, PM, Cy
	MDAR	Chl, Cy, Per
	DHAR	Chl, Per, PM, Cy
	GPx	Cy, Per, Nu, Chl, Go
	GR	Chl, Cy, Per
	CS	V, Chl
	γECS	Chl, CW
	GS	Chl, CW
	PCS	Chl

'Abbreviations: Chl, chloroplast; Mit, mitochondrion; Per, peroxisome; V, vacuole; Nu, nucleus; PM, plasma membrane; Go, Golgi; Cy, cytosol

Fig. 16.8 A schematic model of antioxidant enzymes functions in plants to scavenge excessive ROS generation induced by metal-based nanoparticles. SOD is the first line of cellular defense against the oxidative stress and plays an important role in modulating the relative amount of $O2^{\bullet-}$ and H_2O_2 in plants and, hence, performs a key role in the defense mechanism against ROS toxicity. However, CAT is one of the important antioxidant enzymes that convert H_2O_2 to H_2O and O_2. Excessive amounts of H_2O_2 can be removed and converted to H_2O by CAT (1), APX (2) and GPx (3). In reaction 2, both oxidized ascorbate (monodehydroascorbate, MDA) and dehydroascorbate (*DHA*) can be recycled to ascorbate with the participation of MDA reductase (*MADR*) (1) and DHA reductase (DHAR) (2), respectively. In reaction 3, DHA is reduced to ascorbate by DHAR at the expense of glutathione (*GSH*), yielding oxidized glutathione (*GSSG*). Finally GSSG is reduced by glutathione reductase (*GR*) using NADPH as electron donor. Another reaction is termed the ascorbate-glutathione cycle, which is a metabolic pathway that detoxifies H_2O_2. The cycle involves the antioxidant metabolites: ascorbate, glutathione and NADPH and the enzymes linking these metabolites (Noctor and Foyer 1998). Also, GSH is a key antioxidant agent that can directly transform free radical H_2O_2. GSH, a key pathway for oxidative stress tolerance, can scavenge ROS and be oxidized to GSSG, which can then be recycled by GR. The third reaction is catalysed by GPX with products as GSSG, which can be subsequently reduced GR. The reduction of dehydroascorbate may be nonenzymatic or catalysed by proteins with dehydroascorbate reductase (DHAR) activity. Since glutathione, ascorbate and NADPH are present in high concentrations in plant cells, it is assumed that the ascorbate-glutathione cycle plays a key role for H_2O_2 detoxification (Rouhier et al. 2008). Nevertheless, other enzymes (peroxidases) including peroxiredoxins and glutathione peroxidases, which use thioredoxins or glutaredoxins as reducing substrates, also contribute to H_2O_2 removal in plants (Reproduced from Ma et al. 2015, with permission from American Chemical Society)

60 mg/L provides a protective mechanism by increasing the activity of defense enzymes against dark-induced oxidative damage.

However, APX and CAT activities showed uniform increase from low to moderate (20 and 40 mg/L) nanosilver concentration. Similar result was reported by Krishnaraj et al. (2012), who found that high levels of CAT and POD activities were recorded from leaf samples of plants exposed to nanosilver treatment, resulting in less ROS formation and subsequently less toxicity to the plants. They also reported that CAT and POD are enzymes that play major role in ensuring protection against oxidative damage in plants exposed to nanosilver particles treatments. CAT is an important antioxidant defense enzyme because it converts free radical H_2O_2 in water and oxygen; therefore, it protects plant cells against oxidative stress damage. According to Lei et al. (2008), TiO_2 nanoparticles declined oxidative damage in spinach chloroplast through enhancing APX, SOD, POD and CAT activities.

In order to defend excessive amounts of ROS induced by nanoparticles, different enzymes can protect plants from further damages caused by ROS and nanoparticles. In our aforesaid study, responses of ß-GLU enzyme to nanosilver application in two *Pelargonium* cultivars were completely different (Fig. 16.7). In plants, ß-GLU exhibits several important functions, such as bioactivation of defense compounds, cell wall degradation, activation of phytohormones and lignifications and abscisic acid liberation (Suzuki et al. 2006). These defense compounds are regarded as a protective mechanism against the toxicity of its own chemical defense system, to add to solubility and to facilitate storage (Lee et al. 2006).

Priyadarshini et al. (2012) reported that nanosilver particles decreased H_2O_2 production and increased the efficiency of redox reactions. Also, they reported that higher concentration of nanosilver enhanced the activity of H_2O_2 metabolizing enzymes. It can be noted that nanosilver releases Ag^+ ion, which enables to interact with cytoplasmic organelles and nucleic acids to inhibit respiratory enzymes and to interfere with cellular functions such as membrane leakage (Lu et al. 2010).

Exposure to CeO_2 nanoparticles at 400 and 800 mg/kg soil caused significant increases in H_2O_2 level, as well as significantly higher CAT and APX activities in *Zea mays* plant (Zhao et al. 2012). It has been reported that extra ROS produces changes in the redox status of ascorbate and/or glutathione, as well as in the activity of their redox enzymes (De-Gara 2003).

Rico et al. (2013b) reported that CeO_2 nanoparticles did not change APX activity in the root of the rice (cultivar Cheniere) until the concentration extend to 500 mg/L. Even though exposure to 500 mg/L CeO_2 nanoparticles exhibited high activities of APX and GPX in the roots, no variations in enzyme status were observed in the shoots. Moreover, they found significant differences in CAT activity of the roots, indicating that the ROS detoxification pathway may be through the ascorbate-glutathione cycle (reaction 2) and glutathione peroxidase cycle (reaction 3) (Rico et al. 2013b). However, application of CeO_2 nanoparticles at 62.5 mg/L increased root CAT activity of the cultivar Neptune (Rico et al. 2013a). The changing trend of APX and GPX activities were similar in both cultivars.

The physiological processes under which plant cells may scavenge ROS induced by metal-based nanoparticles are shown in Fig. 16.8 (Ma et al. 2015).

Servin et al. (2013) reported that TiO_2 nanoparticles (at concentrations of 250–750 mg/kg) induced CAT activity of cucumber leaves when compared to control; however, APX activity was not significantly changed except a remarked decrease at 500 mg/kg treatment. ROS production and antioxidant activity may vary with type of nanoparticles, exposure conditions and plant species (Ma et al. 2015). In a study, Dimkpa et al. 2012 found that CAT activity was significantly increased in wheat roots exposed to 500 mg/kg CuO nanoparticles, whereas it inhibited at 500 mg/kg ZnO nanoparticles treatment, implying that alternative detoxification pathways may be involved in ROS scavenging.

In a research, the toxic effects of silver (Ag) nanoparticles and silver ions (Ag^+) at different concentrations (0, 20,40, 60 ppm) were studied on callus cells of two varieties of wheat plant: stress tolerant (Parabola) and stress sensitive (Raweta) (Barbasz et al. 2016). Results showed that silver had no significant effect on the SOD activity, while it significantly reduced the specific activity of CAT. The changes in the activity of POX in both varieties were opposite. However, the highest content of intracellular GSH was noticed at a concentration of 20 ppm of both Ag nanoparticles and silver ions.

Determination of antioxidant enzymes as well as the regulation of genes encoding antioxidant enzymes could provide key information about the molecular mechanisms involving antioxidant response to oxidative stress from application of nanomaterials (Ma et al. 2015). In this respect gene expression studies should also be considered. According to Burklew et al. (2012) Al_2O_3 nanoparticles upregulated two abiotic stress-related genes encoding APX and alcohol dehydrogenase (ADH) in tobacco cell. In another study, the effects of copper oxide (CuO) nanoparticles at morphological, physiological and molecular levels were studied in *Brassica juncea* (Nair and Chung 2015). The seedlings were exposed to 0, 20, 50, 100, 200, 400 and 500 mg/L of CuO nanoparticles in semisolid half-strength MS medium for 14 days. The gene expression results showed significant activation of CuZn superoxide dismutase (CuZn SOD) in roots at all CuO nanoparticles concentrations tested. In shoots, significant upregulation of CuZn SOD gene was observed under 100, 200 and 400 mg/L of CuO nanoparticles treatment. However, no change in the expression levels of Mn SOD gene was observed in both stem and roots of the seedlings. Also, they reported that the expression of CAT and APX genes was not changed in shoots. However, CAT and APX genes were significantly inhibited in roots of plants exposed to 100, 200, 400 and 500 mg/L of CuO nanoparticles. In roots and shoots, SOD-specific activities increased under exposure to 50–500 mg/L and 100–500 mg/L of CuO nanoparticles, respectively. The APX activity reduced in roots and roots at 50–500 mg/L and 200–500 mg/L of CuO nanoparticles, respectively.

The activities of antioxidant enzymes DHAR and GR were assayed in CeO_2 nanoparticles-treated rice plants (Rico et al. 2013a, b). They found that GR activity in rice (cultivar Neptune) was significantly increased at 500 mg/L CeO_2 nanoparticles treatment, allowing for more GSSG conversion to GSH for ROS deactivation. However, the activity of DHAR was low at the same abiotic stress, exhibiting that the main detoxification pathway in this case might be GSH peroxidase cycle instead of ascorbate-glutathione cycle (Rico et al. 2013b).

Response to oxidative stress involves the induction of antioxidant compounds and detoxification enzymes. Thus, the determination of antioxidant enzyme activities as plant defense response is important in understanding the interaction mechanisms between the nanoparticles and plant systems.

References

Arora S, Sharma P, Kumar S, Nayan R, Khanna PK, Zaidi MGH (2012) Gold-nanoparticle induced enhancement in growth and seed yield of *Brassica juncea*. Plant Growth Regul 66:303–310

Anjum NA, Singh N, Singh MK, Sayeed I, Duarte AC, Pereira E, Ahmad I (2013) Single-bilayer graphene oxide sheet tolerance and glutathione redox system significance assessment in faba bean (*Vicia faba* L.). J Nanopart Res 15:1–12

Anjum NA, Singh N, Singh MK, Sayeed I, Duarte AC, Pereira E, Ahmad I (2014) Single-bilayer graphene oxide sheet impacts and underlying potential mechanism assessment in germinating faba bean (*Vicia faba* L.). Sci Total Environ 472:834–841

Baiazidi-Aghdam MT, Mohammadi H, Ghorbanpour M (2016) Effects of nanoparticulate anatase titanium dioxide on physiological and biochemical performance of *Linum usitatissimum* (Linaceae) under well watered and drought stress conditions. Braz J Bot 39:139–146

Barbasz A, Kreczmer B, Oc wieja M (2016) Effects of exposure of callus cells of two wheat varieties to silver nanoparticles and silver salt ($AgNO_3$). Acta Physiol Plant 38:76–87

Begum P, Ikhtiari R, Fugetsu B (2011) Graphene phytotoxicity in the seedling stage of cabbage, tomato, red spinach, and lettuce. Carbon 49:3907–3919

Brayner R (2008) The toxicological impact of nanoparticles. Nanotoday 3:48–55

Burklew CE, Ashlock J, Winfrey WB, Zhang B (2012) Effects of aluminum oxide nanoparticles on the growth, development, and microRNA expression of tobacco (*Nicotiana tabacum*). PLoS One 7, e34783

Corral-Diaz B, Peralta-Videa JR, Alvarez-Parrilla E, Rodrigo-Garcia J, Morales MI, Osuna-Avila P, Niu G, Hernandez-Viezcas JA, Gardea-Torresdey JL (2014) Cerium oxide nanoparticles alter the antioxidant capacity but do not impact tuber ionome in *Raphanus sativus*(L). Plant Physiol Biochem 84:277–285

De-Gara L (2003) Ascorbate metabolism and plant growth-from germination to cell death. In: Asard H, May J, Smirnoff N (eds) Vitamin C: its function and biochemistry in animals and plants. BIOS Scientific Publishers Ltd, Oxford, pp 83–95

Dimkpa C, McLean J, Latta D, Manangon E, Britt D, Johnson W, Boyanov M, Anderson A (2012) CuO and ZnO nanoparticles: phytotoxicity, metal speciation, and induction of oxidative stress in sand grown wheat. J Nanopart Res 14:1–15

Fenoglio I, Greco G, Livraghi S, Fubini B (2009) Non-UV-induced radical reactions at the surface of TiO_2 nanoparticles that may trigger toxic responses. Chemistry 15:4614–4621

Foyer CH, Noctor G (2000) Oxygen processing in photosynthesis: regulation and signalling. Tansley Review No. 112. New Phytol 146:359–388

Franklin NM, Rogers NJ, Apte SC, Batley GE, Gadd GE, Casey PS (2007) Comparative toxicity of nanoparticulate ZnO, bulk ZnO, and ZnCl to a freshwater microalga (*Pseudokirchneriella subcapitata*): the importance of particle solubility. Environ Sci Technol 41:8484–8490

Ghorbanpour M (2015) Major essential oil constituents, total phenolics and flavonoids content and antioxidant activity of Salvia officinalis plant in response to nano-titanium dioxide. Ind J Plant Physiol 20:249–256

Ghorbanpour M, Hadian J (2015) Multi-walled carbon nanotubes stimulate callus induction, secondary metabolites biosynthesis and antioxidant capacity in medicinal plant *Satureja khuzestanica* grown *in vitro*. Carbon 94:749–759

Ghorbanpour M, Hatami M (2014) Spray treatment with silver nanoparticles plus thidiazuron increases anti-oxidant enzyme activities and reduces petal and leaf abscission in four cultivars of geranium (*Pelargonium zonale*) during storage in the dark. J Hort Sci Biotechnol 89: 712–718

Ghorbanpour M, Hatami H (2015) Changes in growth, antioxidant defense system and major essential oils constituents of *Pelargonium graveolens* plant exposed to nano-scale silver and thidiazuron. Ind J Plant Physiol 20:116–123

Ghorbanpour M, Hatami M, Hatami M (2015) Activating antioxidant enzymes, hyoscyamine and scopolamine biosynthesis of *Hyoscyamus niger* L. plants with nano-sized titanium dioxide and bulk application. Acta Agric Slov 105:23–32

Hafis C, Romero-Puertas MC, Rio LA, Abdelly C, Sandalio LM (2011) Antioxidative response of Hordeum maritimum L. to potassium deficiency. Acta Physiol Plant 33:193–202

Hatami M, Ghorbanpour M (2013) Effect of nanosilver on physiological performance of *Pelargonium* plants exposed to dark storage. J Hort Res 21:15–20

Hatami M, Ghorbanpour M (2014) Defense enzymes activity and biochemical variations of *Pelargonium zonale* in response to nanosilver particles and dark storage. Turk J Biol 38: 130–139

Hatami M, Hatamzadeh A, Ghasemnezhad M, Ghorbanpour M (2013) The comparison of antimicrobial effects of silver nanoparticles (SNP) and silver nitrate ($AgNO_3$) to extend the vase life of 'Red Ribbon' cut rose flowers. Trakia J Sci 2:144–151

Hatami M, Ghorbanpour M, Salehiarjom H (2014) Nano-anatase TiO_2 modulates the germination behavior and seedling vigority of the five commercially important medicinal and aromatic plants. J Biol Environ Sci 8:53–59

Hatami M, Ghorbanpour M, Salehiarjom H (2015) Evaluation of nanosized titanium dioxide (TiO_2) on primary growth parameters and secondary metabolites production in *Salvia mirzayanii* plants. Research project (contract number: 92. 13497), Arak University, (In Persian)

Hatami M, Kariman K, Ghorbanpour M (2016) Engineered nanomaterial-mediated changes in the metabolism of terrestrial plants. Sci Total Environ 571:275–291

Hernandez-Viezcas JA, Castillo-Michel H, Servin AD, Peralta-Videa JR, Gardea-orresdey JL (2011) Spectroscopic verification of zinc absorption and distribution in the desert plant *Prosopis julif loravelutina* (velvet mesquite) treated with ZnO nanoparticles. Chem Eng J 170:346–352

Hong J, Peralta-Videa JR, Rico CM, Sahi S, Viveros MN, Bartonjo J, Zhao L, Gardea-Torresdey JL (2014) Evidence of translocation and physiological impacts of foliar applied CeO_2 nanoparticles on cucumber (*Cucumis sativus*) plants. Environ Sci Technol 48:4376–4385

Kahru A, Dubourguier HC (2010) From ecotoxicology to nano ecotoxicology. Toxicology 269: 105–119

Karuppanapandian T, Wang HW, Prabakaran N, Jeyalakshmi K, Kwon M, Manoharan K, Kim W (2011) 2,4-dichlorophenoxyacetic acid-induced leaf senescence in mung bean (*Vigna radiate* L. Wilczek) and senescence inhibition by co-treatment with silver nanoparticles. Plant Physiol Biochem 49:168–177

Khodakovskaya M, Dervishi E, Mahmood M, Xu Y, Li Z, Watanabe F, Alexandru SB (2009) Carbon nanotubes are able to penetrate plant seed coat and dramatically affect seed germination and plant growth. ACS Nano 3:3221–3227

Kim JH, Oh Y, Yoon H, Hwang I, Chang YS (2015) Iron nanoparticle-induced activation of plasma membrane H^+-ATPase promotes stomatal opening in *Arabidopsis thaliana*. Environ Sci Technol 49:1113–1119

Kohler J, Caravaca F, Carrasco L, Roldan A (2006) Contribution of Pseudomonas mendocina and Glomus intraradices to aggregates stabilisation and promotion of biological properties in rhizosphere soil of lettuce plants under field conditions. Soil Use Manage 22:298–304

Krishnaraj C, Jagan EG, Ramachandran R, Abirami SM, Mohan N, Kalaichelvan PT (2012) Effect of biologically synthesized silver nanoparticles on Bacopa monnieri (Linn.) Wettst. plant growth metabolism. Process Biochem 47:651–658

Lee B, Zhu JK (2010) Phenotypic analysis of Arabidopsis mutants: electrolyte leakage after freezing stress. Cold Spring Harbour Protocols 2010, pdb.prot4970. doi: 10.1101/pdb.prot4970

Lee KH, Piao HL, Kim HY, Chio SM, Jiang F, Harting W, Hwang I, Kwak JM, Lee IJ (2006) Activation of glucosidase via stress-induced polymerization rapidly increases active pools of abscisic acid. Cell 126:1109–1120

Lee S, Kim S, Kim S, Lee I (2013a) Assessment of phytotoxicity of ZnO NPs on a medicinal plant, *Fagopyrum esculentum*. Environ Sci Pollut Res 20:848–854

Lee SS, Song W, Cho M, Puppala HL, Nguyen P, Zhu H, SegatoriL CVL (2013b) Antioxidant properties of cerium oxide nanocrystals as a function of nanocrystal diameter and surface coating. ACS Nano 7:9693–9703

Lei Z, Mingyu S, Xiao W, Chao L, Chunxiang Q, Liang C, Hao H, Xiao-qing L, Fashui H (2008) Antioxidant stress is promoted by nano-anatase in spinach chloroplasts under UV-B radiation. Biol Trace Elem Res 121:69–79

Liu Q, Zhao Y, Wan Y, Zheng J, Zhang X, Wang C, Fang X, Lin J (2010) Study of the inhibitory effect of water-soluble fullerenes on plant growth at the cellular level. ACS Nano 4:5743–5748

Liu Q, Zhang X, Zhao Y, Lin J, Shu C, Wang C, Fang X (2013) Fullerene-induced increase of glycosyl residue on living plant cell wall. Environ Sci Technol 47:7490–7498

Lu J, Cao P, He S, Liu J, Li H, Cheng G, Ding Y, Joyce DC (2010) Nano-silver pulse treatments improve water relations of cut rose cv. Movie Star flowers. Postharvest Biol Technol 57: 196–202

Lushchak VI (2011) Environmentally induced oxidative stress in aquatic animals. Aquat Toxicol 101:13–30

Ma X, Wang Q, Rossi L, Zhang W (2015) Cerium oxide nanoparticles and bulk cerium oxide leading to different physiological and biochemical responses in *Brassica rapa*. Environ Sci Technol 50(13):6793–802

Maaouia Houimli SI, Denden M, Mouhandes BD (2010) Effects of 24-epibrassinolide on growth, chlorophylii, electrolyte leakage and proline by pepper under NaCl-stress. EurAsia J BioSci 4:96–104

Miralles P, Church TL, Harris AT (2012) Toxicity, uptake, and translocation of engineered nanomaterials in vascular plants. Environ Sci Technol 46:9224–9239

Mittler R (2002) Oxidative stress, antioxidants and stress tolerance. Trends Plant Sci 7:405–410

Mohammadi R, Maali-Amiri R, Abbasi A (2013) Effect of TiO_2 nanoparticles on chickpea response to cold stress. Biol Trace Elem Res 152:403–10

Møller IM, Jensen PE, Hansson A (2007) Oxidative modifications to cellular components in plants. Annu Rev Plant Biol 58:459–481

Nair PMG, Chung IM (2014) Impact of copper oxide nanoparticles exposure on *Arabidopsis thaliana* growth, root system development, root lignification, and molecular level changes. Environ Sci Pollut Res Int 21:12709–12722

Nair PM, Chung IM (2015) Study on the correlation between copper oxide nanoparticles induced growth suppression and enhanced lignification in Indian mustard (*Brassica juncea* L.). Ecotoxicol Environ Saf 113:302–313

Navarro E, Piccapietra F, Wagner B, Marconi F, Kaegi R, Odzak N, Sigg L, Behra R (2008) Toxicity of silver nanoparticles to Chlamydomonas reinhardtii. Environ Sci Technol 42: 8959–8964

Noctor G, Foyer CH (1998) Ascorbate and glutathione: keeping active oxygen under control. Annu Rev Plant Physiol Plant Mol Biol 49:249–279

Oukarroum A, Bras S, Perreault F, Popovic R (2012) Inhibitory effects of silver nanoparticles in two green algae, Chlorella vulgaris and Dunaliella tertiolecta. Ecotoxicol Environ Safe 78: 80–85

Prakash M, Nair G, Chung M (2016) Biochemical, anatomical and molecular level changes in cucumber (*Cucumis sativus*) seedlings exposed to copper oxide nanoparticles. Biologia 70: 1575–1585

Prasad R (2014) Synthesis of silver nanoparticles in photosynthetic plants. J Nanopart 2014: Article ID 963961

Prasad R, Kumar V, Prasad KS (2014) Nanotechnology in sustainable agriculture: present concerns and future aspects. Afr J Biotechnol 13:705–713

Prasad R, Pandey R, Barman I (2016) Engineering tailored nanoparticles with microbes: quo vadis. WIREs Nanomed Nanobiotechnol 8:316–330

Priyadarshini S, Deepesh B, Zaidi MGH, Pardha saradhi P, Khanna PK, Arora S (2012) Silver nanoparticle-mediated enhancement in growth and antioxidant status of *Brassica juncea*. Appl Biochem Biotechnol 167:2225–2233

Rico CM, Morales MI, McCreary R, Castillo-Michel H, Barrios AC, Hong J, Tafoya A, Lee WY, Varela-Ramirez A, Peralta-Videa JR, Gardea-Torresdey JL (2013a) Cerium oxide nanoparticles modify the antioxidative stress enzyme activities and macromolecule composition in rice seedlings. Environ Sci Technol 47:14110–14118

Rico CM, Hong J, Morales MI, Zhao L, Barrios AC, Zhang JY, Peralta-Videa JR, Gardea-Torresdey JL (2013b) Effect of cerium oxide nanoparticles on rice: a study involving the antioxidant defense system and in vivo fluorescence imaging. Environ Sci Technol 47:5635–5642

Rouhier N, Lemaire SD, Jacquot JP (2008) The role of glutathione in photosynthetic organisms: emerging functions for glutaredoxins and glutathionylation. Annu Rev Plant Biol 59:143–66

Saibo NJ, Lourenco T, Oliveira MM (2009) Transcription factors and regulation of photosynthetic and related metabolism under environ stresses. Ann Bot 103:609–623

Scrinis G, Lyons K (2007) the emerging nano-corporate paradigm: nanotechnology and the transformation of nature, food and agri-food systems. Int J Soc Agr Food 15:22–44

Servin AD, Morales MI, Castillo-Michel H, Hernandez-Viezcas JA, Munoz B, Zhao L, Nunez JE, Peralta-Videa JR, Gardea-Torresdey JL (2013) Synchrotron verification of TiO_2 accumulation in cucumber fruit: a possible pathway of TiO_2 nanoparticle transfer from soil into the food chain. Environ Sci Technol 47:11592–11598

Siddiqui MH, Al-Whaibi MH, Faisal M, Al Sahli AA (2014) Nano-silicon dioxide mitigates the adverse effects of salt stress on *Cucurbita pepo* L. Environ Toxicol Chem 33:2429–2437

Song G, Gao Y, Wu H, Hou W, Zhang C, Ma H (2012) Physiological effect of anatase TiO_2 nanoparticles on Lemna minor. Environ Toxicol Chem 31:2147–2152

Suzuki N, Mittler R (2006) Reactive oxygen species and temperature stresses: a delicate balance between signaling and destruction. Physiol Plantarum 126:45–51

Suzuki H, Takahashi S, Watanabe R, Fukushima Y, Fujita N, Noguchi A, Yokoyama R, Nishitani K, Nishino T, Nakayama T (2006) An isoflavone conjugate-hydrolyzing beta-glucosidase from the roots of soybean (*Glycine max*) seedlings: purification, gene cloning, phylogenetics, and cellular localization. J Biol Chem 28:30251–30259

Ushahra J, Bhati-Kushwaha H, Malik CP (2014) Biogenic nanoparticle-mediated augmentation of seed germination, growth, and antioxidant level of *Eruca sativa* Mill. Varieties. Appl Biochem Biotechnol 174:729–738

Wang S, Wang F, Gao S (2015) Foliar application with nano-silicon alleviates Cd toxicity in rice seedlings. Environ Sci Pollut Res 22:2837–2845

Winston GW (1990) Physicochemical basis of free radical formation in cells: production and defenses. In: Smallwood W (ed) Stress responses in plants: adaptation and acclimation mechanisms. Willey Liss Inc, UK, pp 57–86

Yasur J, Rani PU (2013) Environmental effects of nanosilver: impact on castor seed germination, seedling growth, and plant physiology. Environ Sci Pollut Res 20:8636–8648

Zahed H, Ghazala M, Setsuko K (2015) Plant responses to nanoparticle stress. Int J Mol Sci 16: 26644–26653

Zhao L, Peng B, Hernandez-Viezcas JA, Rico C, Sun Y, Peralta-Videa JR, Tang X, NiuG JL, Varela-Ramirez A (2012) Stress response and tolerance of Zea mays to CeO_2 nanoparticles: cross talk among H_2O_2, heat shock protein, and lipid peroxidation. ACS Nano 6:9615–9622

Chapter 17
Gold Nanoparticles from Plant System: Synthesis, Characterization and their Application

Azamal Husen

17.1 Introduction

Nanotechnology is science, engineering and technology conducted at the nanoscale (at least one dimension between 1 and 100 nm). The unique physicochemical feature of these nanomaterials is not observed in the corresponding bulk materials (Nel et al. 2006). Hence, they have gained huge attention in industry and technology. Over the past few years, synthesis of nanomaterials and their characterization has accelerated due to huge applications in various fields of biology, chemistry, physics and medicine. The main concerns with chemical synthetic routes are environmental contamination, and physical methods need enormous amount of energy to maintain the high pressure and temperature. Moreover, chemical and physical methods are usually expensive processes. Many researchers have diverted their interest to biological synthesis of nanoparticles. Plants and plant-related product synthesis of nanomaterials are generally simple, inexpensive, available and eco-friendly (Husen and Siddiqi 2014a, b, c; Yasmin et al. 2014; Pasca et al. 2014; Prasad 2014; Khan et al. 2015; Yu et al. 2016; Tripathi et al. 2016; Siddiqi and Husen 2016a). It is understood that in comparison to microorganism-mediated synthesis of nanoparticles, the use of plants and plant-related products is more advantageous due to the ease of scaling up, less biohazards on production and elimination of the elaborate process of maintaining cell cultures. Thus, exploring plants with high metal accumulation capacity/phytomining, as well as its engineering, is a need of the hour (Husen and Siddiqi 2014b; Iqbal et al. 2015).

Plants or their extracts/products have been extensively used to produce a range of metal nanoparticles with well-defined size and shape (Husen and Siddiqi 2014b).

A. Husen (✉)
Department of Biology, College of Natural and Computational Sciences, University of Gondar, P.O. Box 196 Gondar, Ethiopia
e-mail: adroot92@yahoo.co.in; adroot92@gmail.com

M. Ghorbanpour et al. (eds.), *Nanoscience and Plant–Soil Systems*, Soil Biology 48,
DOI 10.1007/978-3-319-46835-8_17

Gold acquires elite properties, namely, high free electron density, malleability and conductivity, and favours opportunities to produce stable and adjustable gold nanoparticles for potential applications, for instance, in diagnostics, biological imaging, biosensors, therapeutic agent delivery, photodynamic therapy, electronics, catalytic activity, antioxidant, antibacterial, larvicidal activity, environmental monitoring/cleanup, etc. (Castaneda et al. 2007; Chen et al. 2008; Baruah and Dutta 2009; Yeh et al. 2012; Spivak et al. 2013; Kesik et al. 2013; Kumar et al. 2013; Husen and Siddiqi 2014b; Li et al. 2014; Kuppusamy et al. 2015; Yu et al. 2016).

It is well known that in the biological process, extracts from the living system served as reducing and capping agents. Numerous routes have been developed for the biological or biogenic synthesis of gold nanoparticles from the corresponding salts. In this connection, plants have been proven to be capable for the rapid intra- or extracellular synthesis of gold nanoparticles such as *Acacia nilotica* (Majumdar et al. 2013), *Achyranthes aspera* (Tripathi et al. 2016), *Azadirachta indica* (Ramezani et al. 2008), *Beta vulgaris* (97), *Brassica juncea* (Arora et al. 2012), *Camellia sinensis* (Vilchis-Nestor et al. 2008), *Cicer arietinum* (Ghule et al. 2006), *Cinnamomum camphora* (Huang et al. 2007), *Citrus maxima* (Yu et al. 2016), *Cymbopogon flexuosus* (Shankar et al. 2004a), *Euphorbia hirta* (Annamalai et al. 2013), *Hamamelis virginiana* (Pasca et al. 2014), *Madhuca longifolia* (Fayaz et al. 2011), *Salicornia brachiata* (Ahmed et al. 2014), *Vitis vinifera* (Ismail et al. 2014), *Zingiber officinale* (Kumar et al. 2011) and so on (Fig. 17.1 and Table 17.1). Hence based on this information, the present review was focused on the plant-mediated syntheses of gold nanoparticles, possible mechanisms, characterization as well as the potential applications in various fields, including medicine, industry, agriculture and pharmaceuticals.

17.2 Phytosynthesis of Gold Nanoparticles

Phytosynthesis of gold nanoparticles depends on several important factors such as concentration of plant extract or biomass, concentration of metal salt, incubation/reaction time, temperature and pH of the solution (Fig. 17.2). Thus, by establishing the relationship of these factors with size and shape of the concerned nanoparticles, it is possible to obtain nanoparticles of the desired properties in a controlled way. Plant extract or biomass can be prepared from various parts of plant (leaves, stems, roots, shoots, barks, seeds, flowers or floral parts) or the whole plant. Usually, for the extraction procedure, the desired plant parts are soaked in a solvent. The obtained plant extract contained the reducing and capping agents needed to reduce metallic ions. The advantage of using dried desired plant parts is that these can be stored at room temperature for a longer period of time if at all required, while to prevent deterioration, the fresh plant/plant parts should be stored at −20 °C. In addition, the variation due to seasonal fluctuations which lead to the variations in plant constituents is eliminated by using dried materials (Huang et al. 2007; Sheny et al. 2011). It is well known that plants or their extracts contain different

Fig. 17.1 Plants and their parts used for fabrication of gold nanoparticles

biomolecules such as proteins, sugars, amino acids, enzymes and other traces of metals. These metabolites are strongly involved in the bioreduction process.

The main idea behind the nanoparticle formation is the reduction of metal ion to elemental metal. It has been reported that due to the limited capacity of plants for reducing metal ions, the biosynthetic process usually works well for metal ions with large positive electrochemical potential such as Au and Ag ions (Haverkamp and Marshall 2009). Synthesis of gold nanoparticles was carried out by Shankar et al. (2003) using geranium (*Pelargonium graveolens*) leaf extract. The shape of the gold nanoparticles was spherical, triangular, decahedral and icosahedral. This reaction was completed within 48 h. Authors proposed that the terpenoids in the leaf extract may be responsible for the reduction of gold ions and formation of gold nanoparticles. Synthesis of gold nanoparticles using Fourier transform infrared (FTIR) spectroscopy exhibited that the flavanones and terpenoids which are abundant in *Azadirachta indica* leaf broth have probably been adsorbed on the surface of the nanoparticles and led to their stability for 4 weeks (Shankar et al. 2004b). In this

Table 17.1 Phytosynthesis of gold nanoparticles with their size and shape

Plant	Part used	Size	Shape	References
Acacia nilotica	Leaves	6–12 nm	Spherical	Majumdar et al. (2013)
Aegle marmelos	Leaves	38.2 ± 10.5 nm	Spherical	Rao and Paria (2014)
Aerva lanata	Leaves	17.97 nm	–	Joseph and Mathew (2015)
Aloe vera	Leaves	–	Crystalline	Chandran et al. (2006)
Angelica archangelica	Root	3–4 nm	Spherical, ovals, polyhedral	Pasca et al. (2014)
Azadirachta indica	Leaves	5.5−7.5 nm	Crystalline	Ramezani et al. (2008)
Beta vulgaris	Sugar beet pulp		Spherical, rod-shaped nanowires	Castro et al. (2011)
Brassica juncea	Leaves	10−20 nm	Near spherical	Arora et al. (2012)
Cacumen platycladi	Leaves	2.2−42.8 nm	Face-centred cubic (fcc) crystalline	Zhan et al. (2011)
Camellia sinensis	Leaves (tea bag)	40 nm	Spherical, triangular, irregular	Vilchis-Nestor et al. (2008)
Chenopodium album	Leaves	10–30 nm	Quasi-spherical	Dwivedi and Gopal (2010)
Cicer arietinum	Bean	–	Triangular	Ghule et al. (2006)
Cinnamomum camphora	Leaves	80, 23.4, 21.5 nm	Spherical, triangular	Huang et al. (2007)
Cinnamomum zeylanicum	Leaves	25 nm	Spherical, prism	Smitha et al. (2009)
Coleus amboinicus	Leaves	4.6–55.1 nm	Spherical, triangular, truncated triangular, hexagonal, decahedral	Narayanan and Sakthivel (2010)
Coriandrum sativum	Leaves	6.7−57.9 nm	Spherical, triangular, truncated, triangular, decahedral	Narayanan and Sakthivel (2008)
Cuminum cyminum	Seed	1−10 nm	Spherical	Krishnamurthy et al. (2011)
Cymbopogon flexuosus	Leaf	12–30 nm	Triangular	Shankar et al. (2004a)
Diopyros kaki	Leaves	5−300 nm	Spherical, triangular, pentagonal, hexagonal	Song et al. (2009)
Emblica officinalis	Fruit	15−25 nm	–	Ankamwar et al. (2005a)

(continued)

Table 17.1 (continued)

Plant	Part used	Size	Shape	References
Eucalyptus camaldulensis	Leaves	5.5–7.5 nm	Crystalline	Ramezani et al. (2008)
Euphorbia hirta	Leaves	6–71 nm	Spherical	Annamalai et al. (2013)
Hypericum perforatum	Bark	4–6 nm	Spherical, polyhedral	Pasca et al. (2014)
Hamamelis virginiana	Bark	4–6 nm	Spherical, polyhedral	Pasca et al. (2014)
Madhuca longifolia	Leaves	–	Triangular, spherical, hexagonal nanoplates	Fayaz et al. (2011)
Magnolia kobus	Leaves	5–300 nm	Spherical, triangular, pentagonal, hexagonal	Song et al. (2009)
Mangifera indica	Leaves	17–20 nm	Spherical	Phillip (2010)
Memecylon edule	Leaves	10–45 nm	Circular, triangular, hexagonal	Elavazhagan et al. (2011)
Menta piperita	Leaves	150 nm	Spherical	Ali et al. (2011)
Momordica charantia	Fruit	500–600 nm	–	Pandey et al. (2012)
Morinda citrifolia	Root	12.17–38.26 nm	Cubic	Suman et al. (2014)
Murraya koenigii	Leaves	20 nm	Spherical, triangular	Philip et al. (2011)
Nyctanthes arbor-tristis	Flower	19.8 nm	Spherical, triangular, hexagonal	Das et al. (2011)
Pelargonium graveolens	Leaves	20–40 nm	Decahedral, icosahedral	Shankar et al. (2003)
Pelargonium roseum	Leaves	5.5–7.5 nm	Crystalline	Ramezani et al. (2008)
Pistacia integerrima	Gall	20–200 nm	–	Islam et al. (2015a)
Psidium guajava	Leaves	25–30 nm	Spherical	Raghunandan et al. (2009)
Punica granatum	Juice	23–36 nm	Triangular, pentagonal, hexagonal, spherical	Dash and Bag (2014)
Rosa hybrida	Petal	~10 nm	Spherical, triangular, hexagonal	Noruzi et al. (2011)
Rosa rugosa	Leaves	11 nm	Triangular and hexagonal	Dubey et al. (2010a)
Salix alba	Leaves	50–80 nm	–	Islam et al. (2015b)
Sesbania drummondii	Seed	6–20 nm	Spherical	Sharma et al. (2007)
Sphaeranthus amaranthoides	Leaves	39 nm	Spherical	Nellore et al. (2012)

(continued)

Table 17.1 (continued)

Plant	Part used	Size	Shape	References
Stevia rebaudiana	Leaves	8–20 nm	Octahedral	Mishra et al. (2010)
Salicornia brachiata	Plant	22–35 nm	Spherical	Ahmed et al. (2014)
Tanacetum vulgare	Fruit	11 nm	Triangular	Dubey et al. (2010b)
Terminalia catappa	Leaves	10–35 nm	Spherical	Ankamwar (2010)
Terminalia arjuna	Fruit	20–50 nm	Spherical	Gopinath et al. (2014)
Terminalia arjuna	Bark	15–20 nm	Triangular, tetragonal, pentagonal, hexagonal, rod-like, spherical	Majumdar and Bag (2012)
Trigonella foenum-graecum	Seeds	15–25 nm	Spherical	Aromal and Philip (2012)
Vitis vinifera	Leaves	18–25	Triangular, pentagonal, spherical	Ismail et al. (2014)
Zingiber officinale	Roots	5–15 nm	Spherical	Kumar et al. (2011)

study, the morphology of the gold nanoparticles was predominantly planar (triangular and a few hexagonal) along with spherical shapes. Control over shape and size of gold nanoparticles has been achieved using *Cymbopogon flexuosus* extract. Anisotropic gold nanotriangles have been synthesized by the reaction of lemongrass extract with aqueous gold ions. In this study, 45 % of the population of total gold nanoparticles was triangular in shape in the range of 0.05–1.8 μm, while other shapes were spherical, hexagonal and cubic. Triangle size was controlled by varying the concentration of the lemongrass extract in the reaction medium. Authors have claimed that with increasing amounts of extract added to the $HAuCl_4$ solution, the average size of the triangular and hexagonal particles decreased, while the ratio of the number of spherical nanoparticles to triangular/hexagonal particles increased (Shankar et al. 2004a). Transmission electron microscopy (TEM) analysis revealed that the most polar fraction produces only triangular shapes similar to that produced by the total extract, while the most non-polar fraction produces only cubic shapes. The study of the FTIR and nuclear magnetic resonance (NMR) spectroscopy of the most polar reaction exhibited that aldehydes and ketones were responsible for the stabilization and formation of gold nanoparticles. These gold nanotriangles might be building blocks for the synthesis of electrically conductive thin films (coatings), which can be used effectively in vapour sensing. Bioreduction of $HAuCl_4$ using tamarind leaf extract led to the formation of flat and thin single crystalline gold nanotriangles with unique and highly anisotropic planar shapes. These gold nanotriangles may find application in

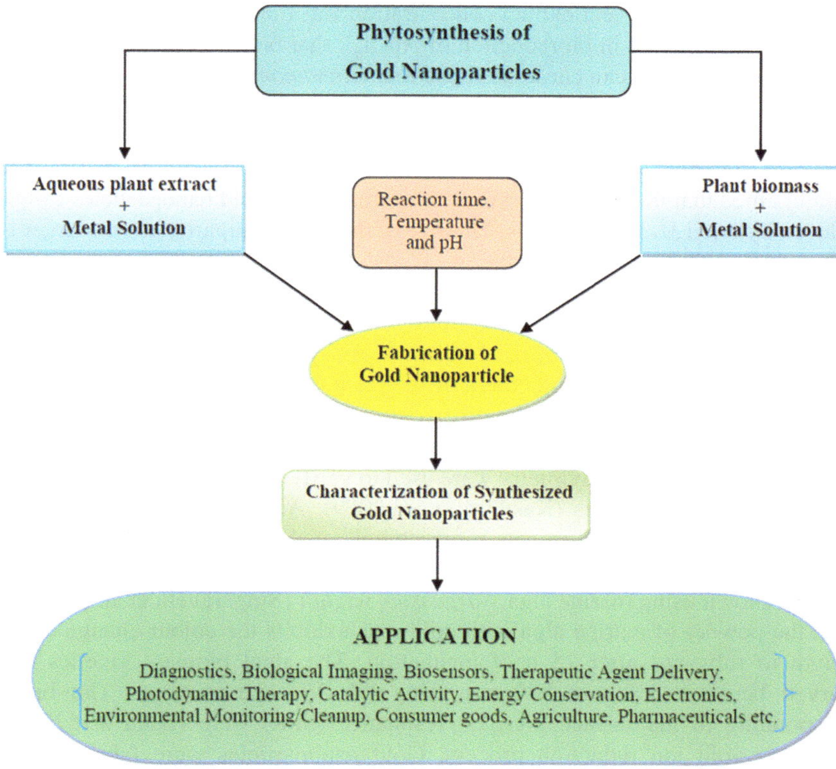

Fig. 17.2 Phytosynthesis of gold nanoparticles from aqueous plant extract/biomass and their application

photonics, optoelectronics and optical sensing (Ankamwar et al. 2005b). Proteins and other biomolecules from *Cicer arietinum* mediate the bioreduction of aqueous Au (III) ions directing the formation of microscale triangular gold prisms (Ghule et al. 2006). Control of the morphology of gold nanoparticles has been achieved by varying compositions of *C. arietinum* extract and aqueous Au (III) solution.

The shape and size of the synthesized gold nanoparticles were modulated and produced by *Aloe vera* leaf extract (Chandran et al. 2006). The gold nanoparticles were triangular and ranged from 50 to 350 nm, which was dependent on the quantity of leaf extract. When a low amount of leaf extract was added to $HAuCl_4$ solution, the fabrications of nanogold triangles were larger in sizes. Moreover, when the quantity of leaf extract was increased, the ratio of nanogold triangles to spherical was decreased. In this study, carbonyl functional groups were found to be responsible for the reduction of gold ions and production of nanoparticles. As progress is made in green synthesis, instead of using the plant extract by boiling, the sun-dried leaf powder in water at ambient temperature is used in some studies. In this type of nanoparticle fabrication, a moderator and accelerator like ammonia was not needed, but the leaf extract concentration was the rate-determining step. It was an important move in bioreduction of $AuCl_4^-$ that biomolecules of molecular weight less than 3 kDa can cause its reduction. *Cinnamomum camphora* sun-dried powder of leaves was used for the production of gold and silver nanoparticles (Huang et al. 2007). *Mangifera indica* leaf (extract and dried powder) was also used to produce spherical gold nanoparticles at room temperature (Philip 2010). The author claimed that the smaller and more uniform size of gold nanoparticles can be produced by dried *M. indica* leaves. The size of the gold nanoparticles was larger in lower extract quantities perhaps due to lack of stabilizing biomolecules in small quantities. These gold nanoparticles were stable for more than 5 months. The FTIR study exhibited the role of water-soluble compounds, for instance, flavonoids, terpenoids and thiamine as stabilizing agents in the synthesis of gold nanoparticles.

Phyllanthin an important ingredient of the plant was separated from the extract of *Phyllanthus amarus* by liquid–liquid extraction and chromatography. Thereafter, a purified component was used to synthesize gold nanoparticles. Further, cyclic voltammetry and thermogravimetry were used to verify the conversion of gold ions to zero-valent nanoparticles (Kasthuri et al. 2009). Although pure nanoparticles were obtained, this procedure was more complicated than the traditional plant-mediated methods. Rapid gold nanoparticle synthesis within a short duration has also been shown using marine alga, *Sargassum wightii* (Singaravelu et al. 2007), in which the powder of marine alga with gold ions exhibits the colour change of the medium to ruby red after 15 h of incubation. The algal biomass kinetics was observed between 300 and 800 nm using UV–Vis spectroscopy. The bands corresponding to the surface plasmon resonance (SPR) were found at 527 nm during this gold ion reduction process. Gold nanoparticles were fabricated by using *Gnidia glauca* flower extract and were used as a chemocatalytic agent in the reduction of 4-nitrophenol to 4-aminophenol in the presence of sodium borohydride (Ghosh et al. 2012). The formation of gold nanoparticles was observed by the change in colour from yellow to dark red in the visible range of the spectrum from 450 to 600 nm (Fig. 17.3). Further, the UV–Vis spectrum of gold nanoparticles as a function of time showed that the reaction was completed within 20 min. It has been observed that the fabrication of gold nanoparticles starts 2 min after the interaction of *G. glauca* flower extract with $HAuCl_4$. This method of gold nanoparticle synthesis (Ghosh et al. 2012) was faster and more efficient than that reported earlier (Vankar and Bajpai 2010) which took about 2 h for the completion of the reaction.

Freshly cut leaves of *Hibiscus rosa-sinensis* were exposed to microwave heating for 3 min, and rapidly gold nanoparticles were synthesized. The SPR at 520 nm confirmed the synthesis of gold nanoparticles. TEM study exhibits the spherical-shaped nanoparticles in the size range of 16–30 nm. The nanoparticles' stability was proved during in vitro stability tests. It was found that alkaloids and flavonoids played a crucial role in the nanoparticle synthesis which was identified using the FTIR (Yasmin et al. 2014). In an experiment, leaf extract of two plants (*Magnolia kobus* and *Diopyros kaki*) were used to fabricate gold nanoparticles (Song et al. 2009). They observed that the reaction temperature and the leaf extract

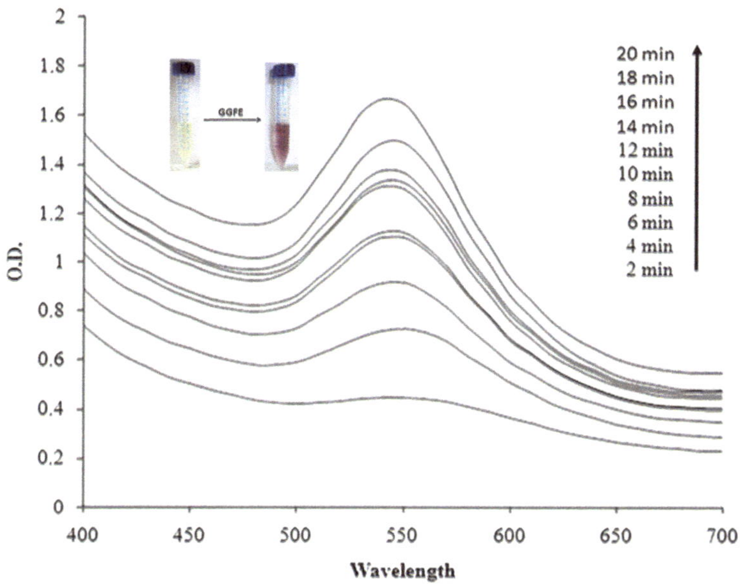

Fig. 17.3 UV–Vis spectra recorded as a function of reaction time of 1 mM chloroauric acid solution with *Gnidia glauca* flower extract at 40 °C (Ghosh et al. 2012)

concentration influenced the shape and the size of the gold nanoparticle formation. At higher temperatures and extract concentrations, smaller and mainly spherical nanoparticles were fabricated, whereas a variety of other morphologies in larger sizes were achieved at lower temperatures and extract concentrations. When an aqueous extract of *Dysosma pleiantha* was added to a solution of gold ions, it gave gold nanoparticles in a spherical shape of 127 nm in size (Karuppaiya et al. 2013). Interestingly, in this investigation gold nanoparticles were produced at boiling temperature. The average particle size of gold nanoparticles was inversely dependent on temperature in the range of 30–60 °C. Further, the antimetastatic activity of nanoparticles against human fibrosarcoma cancer cell line HT-1080 was tested. These gold nanoparticles had no toxic effect on cell proliferation. Moreover, these gold nanoparticles exhibited a high potential for the inhibition of cell migration of human fibrosarcoma cancer cell line HT-1080.

Pasca et al. (2014) reported that the reduction of Au (III) to Au(0) with plant extracts of *Angelica archangelica*, *Hamamelis virginiana* and *Hypericum perforatum* was rapid at room temperature, and for high dilutions of the plant extract, indicating the presence of a reducing substance in large amounts. In this study, the stability of the gold nanoparticles was more at high pH (8–10). It was suggested that the stabilization of nanoparticles was due to the adsorption of stabilizing substances present in the same plant extracts or perhaps also due to their oxidation products (quinones). Moreover, besides this extract, other substances are capable to mediate the self-assembly of nanoparticles. The best stability was found by using *Angelica* extract, whereas self-aggregation tendency was higher in the presence of the *Hypericum* extract. Here a common trend was also observed; as such at a lower concentration of the plant extract, larger particles were formed. The tendency to self-aggregation was increased as a result of the dilution of protecting substances.

In several other studies, the size and shape of fabricated nanoparticles can be manipulated by adjusting the pH of the reaction mixtures. The main effect of the pH was in its ability to change the electrical charges of biomolecules which might have affected their capping and stabilizing abilities and subsequently the growth and production rate of nanoparticles. This may resulted in the favourable formation of nanoparticles of certain shapes at a specific pH range so that a greater stability can be obtained. For instance, in another study the pH-dependent fabrication of gold NPs by *Avena sativa* biomass was also performed (Armendariz et al. 2004) where face-centred cubic (fcc), tetrahedral, hexagonal, decahedral, icosahedral and irregular rod-shaped gold nanoparticles were produced. The yield was more at low pH (3). It was found that at higher pH, the gold nanoparticles of small size are produced. On the other hand, the rod-shaped gold nanoparticles were produced at all pH which has been reported to be formed mainly by electrodeposition in the presence of $KAuCl_4$ which produced $AuCl_4^-$ anion in water. Oat biomass has exhibited the ability to bind $AuCl_4^-$ and its subsequent reduction to gold nanoparticles. An available protein from *Macrotyloma uniflorum* was used to produce gold nanoparticles as capping and stabilizing agents (Aromal et al. 2012). UV–Vis spectrum of the produced gold nanoparticles in different extract quantities from 0.5 to 2 mL demonstrated a shift to shorter wavelengths, which shows the decrease in particle size, when the temperature was increased from room temperature to the 373 K at the same extract quantity SPR bands became broader and shifted to the longer wavelengths; this indicated the increment of particle size. Similarly, gold nanoparticles were synthesized using *Anacardium occidentale* extract where the pH of the reaction mixture was varied from 3 to 8 at room temperature. At pH 5 and 8, the SPR bands were broad, which shows that polydispersed nanoparticles were synthesized, while at pH 4, 6 and 7, the SPR bands were sharper; and at pH 6 a narrow band was recorded which was the characteristic of monodispersed spherical nanoparticles that was confirmed by TEM images (Sheny et al. 2011).

In a recent study, an enzymatic digestion process was developed for the simultaneous determination of nanoparticle size, distribution, particle concentration, and dissolved gold concentration in tomato plant tissues (Dan et al. 2015). The authors suggested that tomato plants can uptake gold nanoparticles of 40 nm diameter and transport them to various parts of the plant. It was suggested that the macerozyme R-10 enzyme can be used to extract the gold nanoparticles from the plant tissue system. Plant organ-dependent yield of gold nanoparticles has been also reported in *Cucurbita pepo* (Gonnelli et al. 2015). Gold nanoparticle one-pot synthesis was carried out at 40 °C for 30 min with diluted $HAuCl_4$. Shoot extracts produced a high number of spherical nanoparticles with lower size than gold nanoparticles as obtained from root extracts of *C. pepo*. Further, when root extracts grown in the presence of Cu (II), Ag (I) or Au (III) produced nanoparticles with treatment-dependent shape, while using shoot extracts, this phenomenon was not recorded. This may be due to metal-imposed specific changes in the cell antioxidant pool, as the total polyphenol concentration in the extracts was correlated with the differences achieved in nanoparticle production. In another recent study, Tetgure et al. (2015) have reported the synthesis and properties of silver and gold nanoparticles using *Ficus racemosa* latex as reducing agent. The colloidal solutions of the nanoparticles exhibited characteristic absorption peaks in the UV-vis region of spectra containing a mixture of nanoparticles of varying size. As suggested by Tetgure et al. (2015) that under acidic condition COOH and NH_3+ groups of amino acids binds with nanoparticles but under basic conditions the COO^- and NH_2 of the same acids cannot bind the nanoparticles. It is very strange obsession of these workers who have hypothesized such imaginary chemical binding of nanoparticles with amino acids under acidic condition. First the nanoparticles are neutral atoms which can associate themselves under both acidic and basic conditions. Second only two charged species may be bonded such as a metal ion and an electron donor. They have mistaken the agglomeration with complexation.

17.3 Characterization of Synthesized Gold Nanoparticles

Characterization of nanoparticles is an important process to understand the reaction mechanism and its subsequent applications. Quite often used techniques for the characterization of nanoparticles, viz. UV–Vis spectroscopy, TEM, scanning electron microscopy (SEM), X-ray diffraction (XRD), FTIR spectroscopy, atomic force microscopy (AFM), energy-dispersive X-ray spectroscopy (EDX), dynamic light scattering (DLS) and zeta potential, are used. These techniques are useful for the determination of size, shape, surface modification, surface area, crystallinity and dispersity of nanoparticles. Generally, UV–visible spectroscopy analysis is used initially for the characterization of noble metallic nanoparticles including gold. Gold nanoparticles have strong absorption in the visible region with the maximum in the range of 500–600 nm due to the SPR phenomenon. This is attributed to the collective oscillation of free conduction electrons induced by an interacting electromagnetic field with the concerned metallic nanoparticles. The appearance of extract colour in red, purple, violet or pink-ruby red due to excitation of SPR vibration in the above-mentioned wavelength confirms the production of gold nanoparticles. For instance, *Solanum melongena* leaf extract leads to the production of a higher quantity of gold nanoparticles followed by *Datura metel* and *Carica papaya* (Rajasekharreddy et al. 2010). When $HAuCl_4$ solution was exposed to *Tridax procumbens*, *Jatropha curcas*, *Calotropis gigantea*, *S. melongena*, *D. metel*, *C. papaya* and *Citrus aurantium* leaf extract solutions exhibited purple colours in the reaction mixture which indicated the formation of gold nanoparticles (Fig. 17.4a–g). For the size and shape of the synthesized nanoparticles, SPR band is able to provide useful information. SPR wavelength variations with the variations

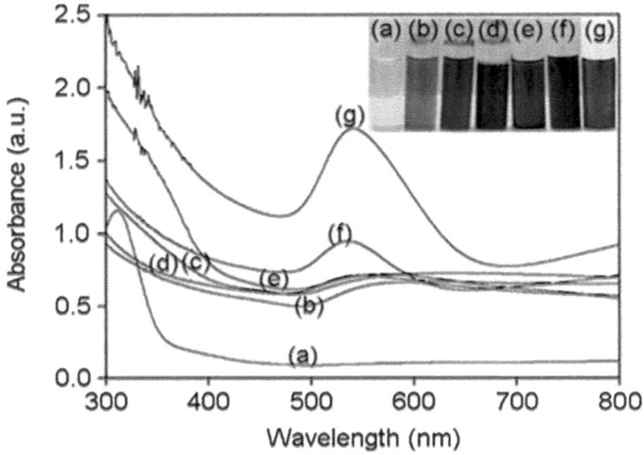

Fig. 17.4 UV–Vis absorption spectra of colloidal gold nanoparticles synthesized using (**a**) *Calotropis gigantea*, (**b**) *Jatropha curcas*, (**c**) *Tridax procumbens*, (**d**) *Citrus aurantium*, (**e**) *Carica papaya*, (**f**) *Datura metel* and (**g**) *Solanum melongena* leaf extracts (the *inset* of the figure shows glass vials of the gold nanoparticle solution formed at the end of the reaction) (Rajasekharreddy et al. 2010)

in particle size in the extract of various plant species have already been established by different workers as with *Cinnamomum camphora* (Huang et al. 2007), *Camellia sinensis* (Vilchis-Nestor et al. 2008), cypress (Noruzi et al. 2012), *Artocarpus heterophyllus* (Jiang et al. 2013), *Hibiscus rosa-sinensis* (Yasmin et al. 2014) and *Citrus maxima* (Yu et al. 2016).

Microscopic techniques, for example, SEM, TEM and AFM, are mostly used for morphological studies of the concerned nanoparticles. The use of these microscopic techniques in morphological studies of nanoparticles has been mentioned previously. TEM is used for greater magnification and resolution than SEM. TEM is also used to differentiate the crystalline structures from amorphous structures using the electron diffraction pattern for a selected area (SAED) (Kasthuri et al. 2009; Tripathi et al. 2016) (Fig. 17.5). AFM is used to study the shape of gold nanoparticles (Ghodake et al. 2010; Pasca et al. 2014).

The XRD technique is used to establish the structural information of crystalline metallic nanoparticles and confirms the formation of zero-valent nanoparticles (Jun et al. 2014). X-rays are electromagnetic radiation with typical photon energies in the range 100 eV–100 keV. Only short-wavelength X-rays in the range of a few angstroms to 0.1 Å (1–120 keV) are used for diffraction applications. Since the X-ray wavelength is comparable to the atom size, they are perfectly suitable for probing the structural arrangement of atoms and molecules in a wide range of materials. The energetic X-rays are able to penetrate deep into the materials and provide valuable information about the bulk structure (Putnam et al. 2007). The XRD technique is also used to calculate the crystallite sizes by the use of the Debye–Scherrer equation (Dubey et al. 2010b). Reports are available on the use of

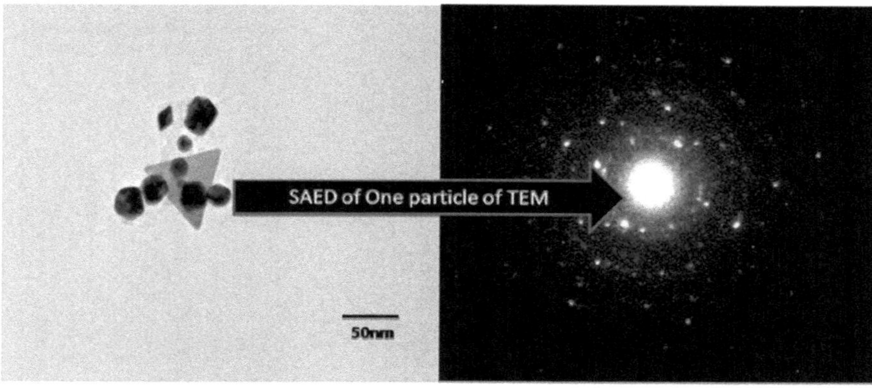

Fig. 17.5 TEM and SAED image showing the size, morphology and texture of gold nanoparticles as obtained from the aqueous leaf extract of *Achyranthes aspera* and gold chloride solution (Tripathi et al. 2016)

XRD pattern/peaks during fabrication of gold nanoparticles (Rajasekharreddy et al. 2010; Jun et al. 2014; Yu et al. 2016).

FTIR spectroscopy is used to measure infrared intensity against wavelength (wave number) of light. It is used to identify the biomolecules involved in the reduction and formation of the concerned nanoparticles. On the basis of wave number, infrared light can be considered as far infrared (4–400 cm^{-1}), mid infrared (400–4000 cm^{-1}) and near infrared (4000–14,000 cm^{-1}). Several studies have compared the FTIR spectrum during fabrication of gold nanoparticles and produced the information about the reducing and capping agents such as proteins, polysaccharides, flavonoids, terpenoids, phenols, ascorbic acids and so on. For instance, FTIR pattern of the gold nanoparticles synthesized using *C. maxima* fruit extracts exhibited bands at 617, 1125, 1376, 1658 and 3278 cm^{-1} (Yu et al. 2016). Aromal et al. (2012) have reported the presence of intense band at 1729 and 1642 cm^{-1} which indicates that the gold nanoparticles are probably bound to proteins present in the aqueous extract of *M. uniflorum* through amine group. The FTIR spectra comparison of plants before and after reaction, the functional groups, such as—OCH_3 of phyllanthin (Kasthuri et al. 2009) and polyols of the *C. camphora* leaf extract (Huang et al. 2007) were identified. In another study, Jiang et al. (2013) demonstrated that the FTIR spectra (Fig. 17.6a) of the biomass of *A. heterophyllus* leaf extract before reduction exhibits bands at 3402, 2931, 1606, 1518, 1402, 1260 and 1076 cm^{-1}. The bands at 3402, 1606 and 1518 cm^{-1} were given to the aromatic hydroxyl and benzene ring, which indicated that there are phenols in the extract (Andrei et al. 2012). Bands at 2931 and 1402 cm^{-1} are stretching vibrations of the methylene and deformation vibration of the methyl, while bands at 1260, 1113 and 1076 cm^{-1} are assigned to epoxy bond, semi-acetal and primary alcohol indicated the presence of sugars in the extract. Further, FTIR spectra of the biomass *A. heterophyllus* leaf extract after reduction (Fig. 17.6b) exhibits analogous bands at 2931, 1615, 1509, 1385 and 1068 cm^{-1}. The main variation between the two

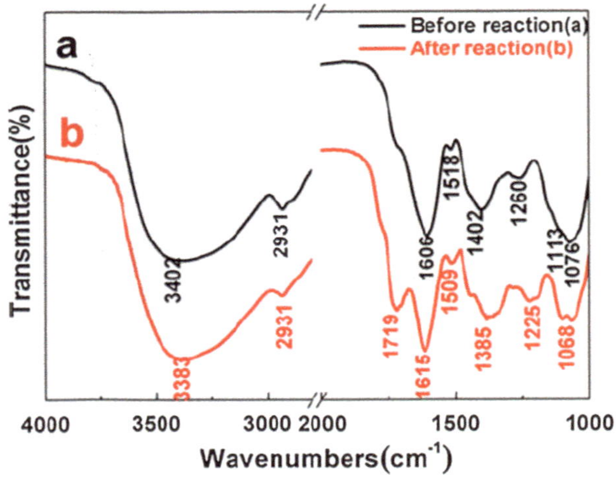

Fig. 17.6 FTIR spectra of the biomass *Artocarpus heterophyllus* leaf extract before (**a**) and after (**b**) reaction (biomass, 2.965 mg mL^{-1}; HAuCl4: 1 mM; 30 °C; 180 min) (Jiang et al. 2013)

waves lies in the fact that, original peaks at 3402 and 1260 cm^{-1} disappeared, whereas new peaks at 3383, 1719 and 1225 cm^{-1} appeared after the reaction, which meant that the epoxy band was broken and aromatic hydroxyl was oxidized to carbonyl. Hence, these FTIR studies suggested that the reducing sugars and the phenols are responsible for Au(III) reduction and gold nanoparticle stabilization.

Raman spectroscopy is a molecular spectroscopy that is observed as inelastically scattered light that allows for the interrogation and identification of vibrational (phonon) states of molecules. Thus, this spectroscopy provides an invaluable analytical tool for molecular finger printing as well as monitoring changes in molecular bond structure. In this technique, very little sample preparation and a rapid, non-destructive optical spectrum are easily achieved. Raman spectra are normally carried out with green, red or near-infrared lasers. Gold nanoparticles enhance the intensity of Raman scattering of adjacent molecules. Thus, they are usually employed in surface-enhanced Raman scattering (SERS) for the detection and quantitative study of Raman active materials such as some organic and inorganic species at low concentration. The uses of SERS in new material characterization, identification and their applications have been already reported (Dieringer et al. 2006; Alvarez-Puebla et al. 2007; Kalmodia et al. 2013; Sun et al. 2014; Prasad et al. 2016). NMR spectroscopy has also been used to confirm the functionalization of gold nanoparticles (Shankar et al. 2004a; Das et al. 2011).

Zeta potential is a potential difference between the two suspended particles present in colloidal suspension. It is a physical property which confirms the stability of nanoparticles. Zeta potential values may be positive or negative but values above than −30 mV or +30 mV favour the good quality and stability of nanoparticles and such nanoparticles can be stored for a longer period of time. Zeta potential strongly depends on the pH of the solution. For instance, in a fixed temperature (25 °C) at pH 6, zeta potential is −9.62 mV that exhibits poor quality, unstable gold nanoparticles. At pH 7 zeta potential value is −25.7 mV, with zeta deviation

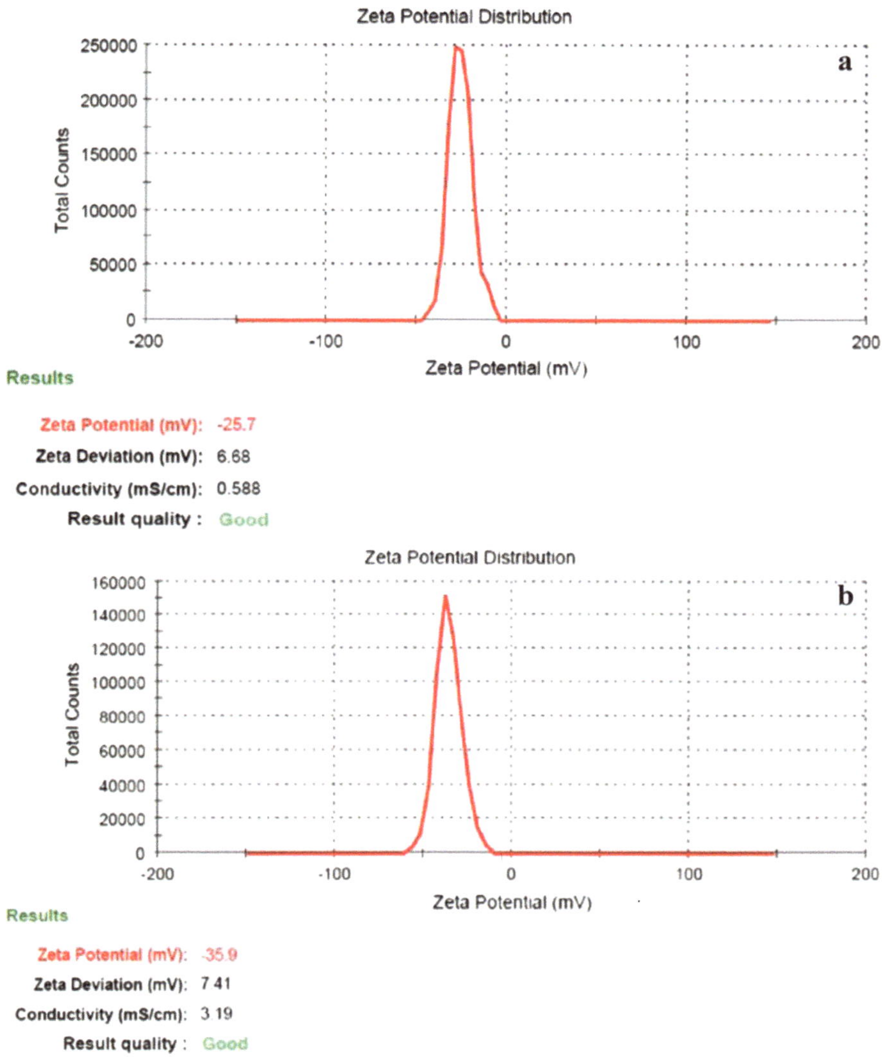

Fig. 17.7 Zeta potential value and zeta potential distribution graph of *Achyranthes aspera* gold nanoparticles at pH 7 (**a**) and 10 (**b**) (Tripathi et al. 2016)

6.68 mV and conductivity 0.588 mS cm^{-1} which represents good quality but unstable nanoparticles (Fig. 17.7a). Whereas at pH 10 zeta potential value is –35.9 mV, with zeta deviation 7.41 mV and conductivity 3.19 mS cm^{-1} which exhibits good quality and comparatively better stability of gold nanoparticles (Fig. 17.7b). Tripathi et al. (2016) claimed that the gold nanoparticles that have zeta potential of −35.9 mV can be stored for up to 2 months without compromising their quality and stability.

17.4 Applications of Gold Nanoparticles

Surprisingly, the applications of nanoparticles have a long history. Though, in recent years different metal nanoparticles with unique properties have been synthesized and applied in many research areas. Among these, gold nanoparticles have attracted intense interests due to their unique optical and electrical properties, high stability and good biocompatibility. Since the ancient civilizations, gold has been used to treat various types of diseases such as smallpox, skin ulcers, syphilis and measles (Tanaka 1999; Huaizhi and Yuantao 2001; Richards et al. 2002; Gielen and Tiekink 2005; Kumar 2007; Chen et al. 2008). Moreover, the Romans in the fourth century used gold for adding a striking red colour to glass (Lycurgus Cup); it appears with a green colour in daylight but changes to red, when illuminated from the inside (Leonhardt 2007; Freestone et al. 2007) (Fig. 17.8). Currently, advances in synthesis and surface functionalization of nanoparticles (effective manipulation) have led to numerous promising and diverse applications of gold nanoparticles as mentioned in Fig. 17.2. Gold nanoparticles are competent of delivering large biomolecules, without restricting themselves as carriers of only small molecular drugs. Facile tunable size and functionality make them a valuable scaffold for efficient recognition and delivery of biomolecules. Gold nanoparticles have shown success in the delivery of peptides, proteins or nucleic acids like DNA or RNA (Verma et al. 2004; Bhumkar et al. 2007; Ghosh et al. 2008; Rana et al. 2012; Ding et al. 2014). They were used as a carrier in the preparation of the anticancer agent, paclitaxel (Gibson et al. 2007), and attached with vascular endothelial growth factor antibodies which are employed in treating B-chronic lymphocytic leukaemia (Mukherjee et al. 2007). Biosensors are generally defined as sensors that consist of biological recognition elements, often called bioreceptors or transducers (Vo-Dinh and Cullum 2000). SERS by gold nanoparticles has been used to identify tumours in cancer research (Huang and El-Sayed 2010), immunoassays (Grubisha et al. 2003; Neng et al. 2010), study of living cells (Kneipp et al. 2002), detection of Alzheimer's disease markers (Neely et al. 2009), determination

Fig. 17.8 The Lycurgus Cup in reflected (*left*) and in transmitted (*right*) light. © Trustees of the British Museum

of protease activity (Guarise et al. 2006) and several other purposes (Dykman and Khlebtsov 2012).

Gold nanoparticles possess catalytic activity and are thus widely used for selective reactions at low temperature such as the water–gas shift reaction and selective oxidation of carbon monoxide (Andreeva 2002; Grisel et al. 2002; Hutchings and Haruta 2005), methanol (Hernández et al. 2006), glycerol (Carrettin et al. 2002), hydrogenation of unsaturated materials (Claus et al. 2000), reduction of aromatic nitro compounds (Kundu et al. 2009) and a toxic pollutant 4-nitrophenol to 4-aminophenol (Sharma et al. 2007; Ghosh et al. 2012; Yu et al. 2016).

Gold nanoparticles have shown antimicrobial properties. Cationic and hydrophobic functionalized gold nanoparticles are found to effectively suppress the growth of 11 clinical multidrug-resistant isolates, including both Gram-negative and Gram-positive bacteria (Li et al. 2014). The nanogold bioconjugate exhibits antimicrobial activity against several Gram-negative and Gram-positive pathogenic bacteria as well as *Saccharomyces cerevisiae* and *Candida albicans* (Das et al. 2009). The synthesized silver and gold nanoparticles from *Mentha piperita* exhibited a strong antibacterial activity against *Staphylococcus aureus* and *Escherichia coli* (Ali et al. 2011). In addition, the antibacterial activities of honey-mediated gold and silver nanoparticles have been observed (Sreelakshmi et al. 2011).

Several uses of gold nanoparticles, as biosensor substrates, have been reported (Liu and Lu 2003; Luo et al. 2004; Li et al. 2010). For instance, they have been utilized in the biosensor design to improve the performance for the detection of infectious diseases and biothreats (Lin et al. 2013). Gold nanoparticle-based sensors are useful to detect toxins, heavy metals and inorganic and organic pollutants in water rapidly with high sensitivity, and thus they are believed to play an important role in environmental cleaning and monitoring. They have been used in chemical sensing such as potassium (Lin et al. 2002), lithium (Obare et al. 2002) and toxic heavy metals like mercury, lead and cadmium (Kim et al. 2001). This was demonstrated by using surface-engineered gold nanoparticles. Gold nanoparticles are also useful for the removal of heavy metals by the formation of alloys with varying composition, for instance, Au_3Hg, AuHg and $AuHg_3$ phases, and therefore they can be utilized for the removal of Hg ions from the contaminated water (Pradeep and Anshup 2009). Gold nanoparticle-based sensor for the selective detection of Cr^{3+} in aqueous solution, in the presence of 15 other metal ions, has been demonstrated (Dang et al. 2009). They can also be used for the detection and removal of organic compounds such as pesticides (Han et al. 2003) endosulfan, malathion and chlorpyrifos (Nair et al. 2003). Sulphur-containing compounds bind with gold nanoparticles, causing a change in the suspension colour (to purple). Aggregation of the endosulfan bond gold nanoparticles ultimately occurs, which basically removes the endosulfan into a concentrated, solid form. In the availability of methanol co-solvent, Bootharaju and Pradeep (2012) found the decomposition of chlorpyrifos at room temperature in the presence of gold nanoparticles. Salt-induced aggregation of gold nanoparticles was also used for the detection of pesticides in drinking water at low concentration (Burns et al. 2006). Moreover,

Gupta and Kulkarni (2011) have shown the removal of diesel oil droplets floating on water through swelling and absorption of the gold nanoparticle composite. Application of fabricated gold nanoparticles in plant growth and production has shown significant promising potential. Both beneficial and harmful response of gold nanoparticles in plant system has been reported (Husen and Siddiqi 2014b; Siddiqi and Husen 2016b). Gold nanoparticles enter into plant system by a size-dependent mechanism where they may trigger the growth/biomass or inhibit the growth by leading an imbalance in physiological, biochemical and molecular processes; and producing oxidative stress. They exhibited inhibition of reactive oxygen species which suggests the free radical scavenging ability of gold nanoparticle. In addition, their exposure has also altered microRNA and gene expression in plants. It has been suggested that the gold nanoparticles may be applied in fruiting plants to increase the quality and quantity of the fruits and vegetable (Siddiqi and Husen 2016b).

17.5 Conclusion

Owing to reach plant biodiversity, the phytosynthesis of gold nanoparticles is able to create facile, eco-friendly, inexpensive and stable nanoparticles in comparison to physical and chemical methods. Available biomolecules in plant systems played a significant role during bioreduction process. Some studies have shown control over the shape and size of gold nanoparticles by adjustment of concentration of plant extract or biomass, concentration of metal salt, incubation/reaction time, temperature and pH of the solution. However, this fabrication mechanism is not fully understood and is still in its infancy stage. Development of a highly controllable fabrication approach is desirable. The stability of synthesized gold nanoparticles is another concern to achieve a longer time/duration and maximum practical application. Thus, more experiments are anticipated to elucidate the reduction mechanism to control well-defined size and shape and the stability of gold nanoparticles. The high potential applications of gold nanoparticles in various sectors are worth exploring specially in biomedicine and catalysis. The recent use of engineered gold nanoparticles in drug and gene delivery, catalytic activity, infectious diseases control and environmental monitoring is an established fact. At the same time, the unique physicochemical properties of gold and other metallic nanoparticles are of great concern regarding the potential adverse effects as arisen with their increasing production, use and disposal which unavoidably lead to ecological risk. In addition to this, another key issue to consider in the use of gold nanoparticles is to calculate the cost to see as if it is economically viable.

Acknowledgment The author is thankful to the publisher and the Trustees of the British Museum for the permission to adopt their figures for this review.

References

Ahmed KBA, Subramanian S, Sivasubramanian A, Veerappan G, Veerappan A (2014) Preparation of gold nanoparticles using *Salicornia brachiata* plant extract and evaluation of catalytic and antibacterial activity. Spectro Acta A Mol Biomol Spectrosc 130:54–58

Ali MD, Thajuddin N, Jeganathan K, Gunasekaran M (2011) Plant extract mediated synthesis of silver and gold nanoparticles and its antibacterial activity against clinically isolated pathogens. Colloids Surf B Biointerfaces 85:360–365

Alvarez-Puebla RA, Dos Santos DS Jr, Aroca RF (2007) SERS detection of environmental pollutants in humic acid-gold nanoparticle composite materials. Analyst 132:1210–1214

Andreeva D (2002) Low temperature water gas shift over gold catalysts. Gold Bull 35:82–88

Andrei AB, Hassan YAE, Serban F (2012) FTIR spectrophotometric methods used for antioxidant activity assay in medicinal plants. Appl Spectros Rev 47:245–255

Ankamwar B, Damle C, Ahmad A, Satry M (2005a) Biosynthesis of gold and silver nanoparticles using *Emblica officinalis* fruit extract, their phase transfer and transmetallation in an organic solution. J Nanosci Nanotechnol 5:1665–1671

Ankamwar B, Chaudhary M, Sastry M (2005b) Gold nanotriangles biologically synthesized using tamarind leaf extract and potential application in vapor sensing. Synth React Inorg Metal-Org Nano-Metal Chem 35:19–26

Ankamwar B (2010) Biosynthesis of gold nanoparticles (green-gold) using leaf extract of *Terminalia catappa*. E-J Chem 7:1334–1339

Annamalai A, Christina VL, Sudha D, Kalpana M, Lakshmi PT (2013) Green synthesis, characterization and antimicrobial activity of Au NPs using *Euphorbia hirta* L. leaf extract. Colloids Surf B Biointerfaces 108:60–65

Armendariz V, Herrera I, Peralta-Videa JR, Jose-Yacaman M, Troiani H, Santiago P, Gardea-Torresdey JL (2004) Size controlled gold nanoparticle formation by *Avena sativa* biomass: use of plants in nanobiotechnology. J Nano Res 6:377–382

Aromal SA, Philip D (2012) Green synthesis of gold nanoparticles using *Trigonella foenum-graecum* and its size dependent catalytic activity. Spectrochim Acta Part A 97:1–5

Aromal SA, Vidhu VK, Philip D (2012) Green synthesis of well-dispersed gold nanoparticles using *Macrotyloma uniflorum*. Spectrochim Acta A Mol Biomol Spectrosc 85:99–104

Arora S, Sharma P, Kumar S, Nayan R, Khanna PK, Zaidi MGH (2012) Gold-nanoparticle induced enhancement in growth and seed yield of *Brassica juncea*. Plant Growth Regul 66:303–310

Baruah S, Dutta J (2009) Nanotechnology applications in pollution sensing and degradation in agriculture: a review. Environ Chem Lett 7:191–204

Bhumkar DR, Joshi HM, Sastry M, Pokharkar VB (2007) Chitosan reduced gold nanoparticles as novel carriers for transmucosal delivery of insulin. Pharm Res 24:1415–1426

Bootharaju MS, Pradeep T (2012) Understanding the degradation pathway of the pesticide, chlorpyrifos by noble metal nanoparticles. Langmuir 28:2671–2679

Burns C, Spendel WU, Puckett S, Pacey GE (2006) Solution ionic strength effect on gold nanoparticle solution color transition. Talanta 69:873–876

Carrettin S, McMorn P, Johnston P, Griffin K, Hutchings GJ (2002) Selective oxidation of glycerol to glyceric acid using a gold catalyst in aqueous sodium hydroxide. Chem Commun 7:696–697

Castaneda MT, Merkoci A, Pumera M, Alegret S (2007) Electrochemical genosensors for biomedical applications based on gold nanoparticles. Biosens Bioelectron 22:1961–1967

Castro L, Blazquez ML, Munoz JA, Gonzalez F, Garcia-Balboa C, Ballester A (2011) Biosynthesis of gold nanowires using sugar beet pulp. Process Biochem 46:1076–1082

Chandran SP, Chaudhary M, Pasricha R, Ahmad A, Sastry M (2006) Synthesis of gold nanotriangles and silver nanoparticles using *Aloe vera* plant extract. Biotechnol Prog 22:577–583

Chen PC, Mwakwari SC, Oyelere AK (2008) Gold nanoparticles: from nanomedicine to nanosensing. Nanotechnol Sci Appl 1:45–66

Claus P, Brückner A, Mohr C, Hofmeister H (2000) Supported gold nanoparticles from quantum dot to mesoscopic size scale: effect of electronic and structural properties on catalytic hydrogenation of conjugated functional groups. J Am Chem Soc 122:11430–11439

Dan Y, Zhang W, Xue R, Ma X, Stephan C, Shi H (2015) Characterization of gold nanoparticle uptake by tomato plants using enzymatic extraction followed by single-particle inductively coupled plasma–mass spectrometry analysis. Environ Sci Technol 49:3007–3014

Dang YQ, Li HW, Wang B, Li L, Wu Y (2009) Selective detection of trace Cr^{3+} in aqueous solution by using 5,5′-dithiobis (2-nitrobenzoic acid)-modified gold nanoparticles. ACS Appl Mater Interfaces 1:1533–1538

Das RK, Gogoi N, Bora U (2011) Green synthesis of gold nanoparticles using *Nyctanthes arbortristis* flower extract. Bioprocess Biosyst Eng 34:615–619

Das SK, Das AR, Guha AK (2009) Gold nanoparticles: microbial synthesis and application in water hygiene management. Langmuir 25:8192–8199

Dash SS, Bag BG (2014) Synthesis of gold nanoparticles using renewable *Punica granatum* juice and study of its catalytic activity. Appl Nanosci 4:55–59

Dieringer JA, McFarland AD, Shah NC, Stuart DA, Whitney AV, Yonzon CR, Young MA, Zhang X, Van Duyne RP (2006) Surface enhanced Raman spectroscopy: new materials, concepts, characterization tools, and applications. Faraday Discuss 132:9–26

Ding Y, Jiang Z, Saha K, Kim CS, Kim ST, Landis RF, Rotello VM (2014) Gold nanoparticles for nucleic acid delivery. Mol Ther 22:1075–1083

Dubey SP, Lahtinen M, Sillanpää M (2010a) Green synthesis and characterizations of silver and gold nanoparticles using leaf extract of *Rosa rugosa*. Colloids Sur A Physicochem Eng Aspect 364:34–41

Dubey SP, Lahtinen M, Sillanpää M (2010b) Tansy fruit mediated greener synthesis of silver and gold nanoparticles. Process Biochem 45:1065–1071

Dwivedi AD, Gopal K (2010) Biosynthesis of silver and gold nanoparticles using *Chenopodium album* leaf extract. Colloids Sur A Physicochem Eng Aspect 369:27–33

Dykman L, Khlebtsov N (2012) Gold nanoparticles in biomedical applications: recent advances and perspectives. Chem Soc Rev 41:2256–2282

Fayaz MA, Girilal M, Venkatesan R, Kalaichelvan P (2011) Biosynthesis of anisotropic gold nanoparticles using *Maduca longifolia* extract and their potential in infrared absorption. Colloids Surf B 88:287–291

Freestone I, Meeks N, Sax M, Higgitt C (2007) The Lycurgus Cup—a Roman nanotechnology. Gold Bull 40:270–277

Ghodake G, Deshpande N, Lee Y, Jin E (2010) Pear fruit extract-assisted room-temperature biosynthesis of gold nanoplates. Colloids Surf B 75:584–589

Ghosh P, Han G, De M, Kim CK, Rotello VM (2008) Gold nanoparticles in delivery applications. Adv Drug Del Rev 60:1307–1315

Ghosh S, Patil S, Ahire M, Kitture R, Gurav DD, Jabgunde AM, Kale S, Pardesi K, Shinde V, Bellare J, Dhavale DD, Chopade BA (2012) *Gnidia glauca* flower extract mediated synthesis of gold nanoparticles and evaluation of its chemocatalytic potential. J Nanobiotechnol 10:17

Ghule K, Ghule AV, Liu JY, Ling YC (2006) Microscale size triangular gold prisms synthesized using Bengal gram beans (*Cicer arietinum* L.) extract and $HAuCl_4.3H_2O$: a green biogenic approach. J Nanosci Nanotechnol 6:3746–3751

Gibson JD, Khanal BP, Zubarev ER (2007) Paclitaxel-functionalized gold nanoparticles. J Am Chem Soc 129:11653–11661

Gielen M, Tiekink ERT (2005) Metallotherapeutic drugs and metal-based diagnostic agents. The use of metals in medicine. Wiley, Hoboken, New York

Gonnelli C, Cacioppo F, Cristiana G, Capozzoli L, Salvatici C, Salvatici MC, Colzi I, Bubba MD, Ancillotti C, Ristori S (2015) *Cucurbita pepo* L. extracts as a versatile hydrotropic source for the synthesis of gold nanoparticles with different shapes. Green Chem Lett Rev 8:39–47

Gopinath K, Gowri S, Karthika V, Arumugam A (2014) Green synthesis of gold nanoparticles from fruit extract of *Terminalia arjuna*, for the enhanced seed germination activity of *Gloriosa superb*. J Nanostruct Chem 4:115

Grisel R, Weststrate KJ, Gluhoi A, Nieuwenhuys BE (2002) Catalysis by gold nanoparticles. Gold Bull 35:39–45

Grubisha DS, Lipert RJ, Park H-Y, Driskell J, Porter MD (2003) Femtomolar detection of prostate-specific antigen: an immunoassay based on surface-enhanced Raman scattering and immunogold labels. Anal Chem 75:5936–5943

Guarise C, Pasquato L, De Filippis V, Scrimin P (2006) Gold nanoparticles-based protease assay. Proc Natl Acad Sci USA 103:3978–3982

Gupta R, Kulkarni GU (2011) Removal of organic compounds from water by using a gold nanoparticle-poly(dimethylsiloxane) nanocomposite foam. Chemsuschem 4:737–743

Han A, Dufva M, Belleville E, Christensen CB (2003) Detection of analyte binding to microarrays using gold nanoparticle labels and a desktop scanner. Lab Chip 3:329–332

Haverkamp RG, Marshall AT (2009) The mechanism of metal nanoparticle formation in plants: limits on accumulation. J Nanopart Res 11:1453–1463

Hernández J, Solla-Gullón J, Herrero E, Aldaz A, Feliu JM (2006) Methanol oxidation on gold nanoparticles in alkaline media: unusual electrocatalytic activity. Electrochim Acta 52:1662–1669

Huaizhi Z, Yuantao N (2001) China's ancient gold drugs. Gold Bull 34:24–9

Huang J, Li Q, Sun D, Lu Y, Su Y, Yang X, Wang H, Wang Y, Shao W, He N, Hong J, Chen C (2007) Biosynthesis of silver and gold nanoparticles by novel sundried *Cinnamomum camphora* leaf. Nanotechnology 18:105104

Huang X, El-Sayed MA (2010) Gold nanoparticles: optical properties and implementations in cancer diagnosis and photothermal therapy. J Adv Res 1:13–28

Husen A, Siddiqi KS (2014a) Carbon and fullerene nanomaterials in plant system. J Nanobiotechnol 12:16

Husen A, Siddiqi KS (2014b) Phytosynthesis of nanoparticles: concept, controversy and application. Nano Res Lett 9:229

Husen A, Siddiqi KS (2014c) Plants and microbes assisted selenium nanoparticles: characterization and application. J Nanobiotechnol 12:28

Hutchings GJ, Haruta M (2005) A golden age of catalysis: a perspective. Appl Catal A 291:2–5

Iqbal M, Ahmad A, Ansari MKA, Qureshi MI, Aref IM, Khan PR, Hegazy SS, El-Atta H, Husen A, Hakeem KR (2015) Improving the phytoextraction capacity of plants to scavenge metal(loid)-contaminated sites. Environ Rev 23:44–65

Islam NU, Jalil K, Shahid M, Muhammad N, Rauf A (2015a) *Pistacia integerrima* gall extract mediated green synthesis of gold nanoparticles and their biological activities. Arab J Chem. doi:10.1016/j.arabjc.2015.02.014

Islam NU, Jalil K, Shahid M, Rauf A, Muhammad N, Khan A, Shah MR, Khan MA (2015b) Green synthesis and biological activities of gold nanoparticles functionalized with *Salix alba*. Arab J Chem. doi:10.1016/j.arabjc.2015.06.025

Ismail EH, Khalil MMH, Al Seif FA, El-Magdoub F (2014) Biosynthesis of gold nanoparticles using extract of grape (*Vitis vinifera*) leaves and seeds. Prog Nanotechnol Nanomat 3:1–12

Jiang X, Sun D, Zhang G, He N, Liu H, Huang J, Odoom-Wubah T, Li Q (2013) Investigation of active biomolecules involved in the nucleation and growth of gold nanoparticles by *Artocarpus heterophyllus* Lam leaf extract. J Nanopart Res 15:1741

Joseph S, Mathew B (2015) Microwave assisted facile green synthesis of silver and gold nanocatalysts using the leaf extract of *Aerva lanata*. Spectrochim Acta Part A Mol Biomol Spectrosc 136:1371–1379

Jun SH, Kim HS, Koo YK, Park Y, Kim J, Cho S, Park Y (2014) Root extracts of *Polygala tenuifolia* for the green synthesis of gold nanoparticles. J Nanosci Nanotechnol 14:6202–6208

Kalmodia S, Harjwani J, Rajeswari R, Yang W, Barrow CJ, Ramaprabhu S, Krishnakumar S, Elchuri SV (2013) Synthesis and characterization of surface-enhanced Raman-scattered gold nanoparticles. Int J Nanomed 8:4327–4338
Karuppaiya P, Satheeshkumar E, Chao WT, Kao LY, Chen ECF, Tsay HS (2013) Anti-metastatic activity of biologically synthesized gold nanoparticles on human fibrosarcoma cell line HT-1080. Colloids Surf B 110:163–170
Kasthuri J, Kathiravan K, Rajendiran N (2009) Phyllanthin assisted biosynthesis of silver and gold nanoparticles: a novel biological approach. J Nanopart Res 11:1075–1085
Kesik M, Ekiz Kanik F, Hızalan G, Kozanoglu D, Nalbant Esenturk E, Timur S, Toppare L (2013) A functional immobilization matrix based on a conducting polymer and functionalized gold nanoparticles: synthesis and its application as an amperometric glucose biosensor. Polymer 54:4463–4471
Khan M, Al-Marri AH, Khan M, Shaik MR, Mohri N, Adil SF, Kuniyil M, Alkhathlan MZ, Al-Warthan A, Tremel W, Tahir MN, Siddiqui MRH (2015) Green approach for the effective reduction of graphene oxide using *Salvadora persica* L. root (Miswak) extract. Nano Res Lett 10:1–9
Kim Y, Johnson RC, Hupp JT (2001) Gold nanoparticle-based sensing of "Spectroscopically Silent" heavy metal ions. Nano Lett 1:165–167
Kneipp K, Haka AS, Kneipp H, Badizadegan K, Yoshizawa N, Boone C, Shafer-Peltier KE, Motz JT, Dasari RR, Feld MS (2002) Surface-enhanced Raman spectroscopy in single living cells using gold nanoparticles. Appl Spectrosc 56:150–154
Krishnamurthy S, Sathishkumar M, Lee SY, Bae MA, Yun YS (2011) Biosynthesis of Au nanoparticles using cumin seed powder extract. J Nanosci Nanotechnol 11:1811–1814
Kumar AN, Jeyalalitha T, Murugan K, Madhiyazhagan P (2013) Bioefficacy of plant mediated gold nanoparticles and *Anthocepholus cadamba* on filarial vector, *Culex quinquefasciatus* (Insecta: Diptera: Culicidae). Parasitol Res 112:1053–1063
Kumar CSSR (2007) Nanomaterials for cancer diagnosis. Wiley, Weinheim, Germany
Kumar KP, Paul W, Sharma CP (2011) Green synthesis of gold nanoparticles with *Zingiber officinale* extract: characterization and blood compatibility. Process Biochem 46:2007–2013
Kundu S, Lau S, Liang H (2009) Shape-controlled catalysis by cetyltrimethylammonium bromide terminated gold nanospheres, nanorods, and nanoprisms. J Phys Chem C 113:5150–5156
Kuppusamy P, Yusoff MM, Maniam GP, Govindan N (2015) Biosynthesis of metallic nanoparticles using plant derivatives and their new avenues in pharmacological applications—An updated report. Saudi Pharma J 24:473–484. doi:10.1016/j.jsps.2014.11.013
Leonhardt U (2007) Optical metamaterials: invisibility cup. Nat Photon 1:207–208
Li X, Robinson SM, Gupta A, Saha K, Jiang Z, Moyano DF, Sahar A, Riley MA, Rotello VM (2014) Functional gold nanoparticles as potent antimicrobial agents against multi-drug-resistant bacteria. ACS Nano 8:10682–10686
Li Y, Schluesener HJ, Xu S (2010) Gold nanoparticle-based biosensors. Gold Bull 43:29–41
Lin M, Pei H, Yang F, Fan C, Zuo X (2013) Applications of gold nanoparticles in the detection and identification of infectious diseases and biothreats. Adv Mater 25:3490–3496
Lin SY, Liu SW, Lin CM, Chen CH (2002) Recognition of potassium ion in water by 15-crown-5 functionalized gold nanoparticles. Anal Chem 74:330–335
Liu J, Lu Y (2003) A colorimetric lead biosensor using DNAzyme-directed assembly of gold nanoparticles. J Am Chem Soc 125:6642–6643
Luo X-L, Xu J-J, Du Y, Chen H-Y (2004) A glucose biosensor based on chitosan–glucose oxidase–gold nanoparticles biocomposite formed by one-step electrodeposition. Anal Biochem 334:284–289
Majumdar R, Bag BG (2012) *Terminalia arjuna* bark extract mediated size controlled synthesis of polyshaped gold nanoparticles and its application in catalysis. Int J Res Chem Environ 2:338–344

Majumdar R, Bag BG, Maity N (2013) *Acacia nilotica* (Babool) leaf extract mediated size-controlled rapid synthesis of gold nanoparticles and study of its catalytic activity. Int Nano Lett 3:53
Mishra AN, Bhadauria S, Gaur MS, Pasricha R, Kushwah BS (2010) Synthesis of gold nanoparticles by leaves of zero-calorie sweetener herb (*Stevia rebaudiana*) and their nanoscopic characterization by spectroscopy and microscopy. Int J Green Nanotechnol Phys Chem 1:118–124
Mukherjee P, Bhattacharya R, Bone N, Lee YK, Patra C, Wang S, Lu L, Secreto C, Banerjee PC, Yaszemski MJ, Kay NE, Mukhopadhyay D (2007) Potential therapeutic application of gold nanoparticles in B-chronic lymphocytic leukemia (BCLL): enhancing apoptosis. J Nanobiotechnol 5:4
Nair AS, Tom RT, Pradeep T (2003) Detection and extraction of endosulfan by metal nanoparticles. J Environ Monitor 5:363–365
Narayanan KB, Sakthivel N (2010) Phytosynthesis of gold nanoparticles using leaf extract of *Coleus amboinicus* Lour. Mat Charac 61:1232–1238
Narayanan KB, Sakthivel N (2008) Coriander leaf mediated biosynthesis of gold nanoparticles. Mater Lett 62:4588–4590
Neely A, Perry C, Varisli B, Singh AK, Arbneshi T, Senapati D, Kalluri JR, Ray PC (2009) Ultrasensitive and highly selective detection of Alzheimer's disease biomarker using two-photon Rayleigh scattering properties of gold nanoparticle. ACS Nano 3:2834–2840
Nel A, Xia T, Mädler L, Li N (2006) Toxic potential of materials at the nanolevel. Science 622:622–627
Nellore J, Pauline PC, Amarnath K (2012) Biogenic synthesis of *Sphearanthus amaranthoids* towards the efficient production of the biocompatible gold nanoparticles. Dig J Nanomater Biostruct 7:123–133
Neng J, Harpster MH, Zhang H, Mecham JO, Wilson WC, Johnson PA (2010) A versatile SERS-based immunoassay for immunoglobulin detection using antigen-coated gold nanoparticles and malachite green-conjugated protein A/G. Biosens Bioelectron 26:1009–1015
Noruzi M, Zare D, Davoodi D (2012) A rapid biosynthesis route for the preparation of gold nanoparticles by aqueous extract of cypress leaves at room temperature. Spectrochim Acta Part A Mol Biomol Spectrosc 94:84–88
Noruzi M, Zare D, Khoshnevisan K, Davoodi D (2011) Rapid green synthesis of gold nanoparticles using *Rosa hybrida* petal extract at room temperature. Spectrochim Acta Part A Mol Biomol Spectrosc 79:1461–1465
Obare SO, Hollowell RE, Murphy CJ (2002) Sensing strategy for lithium ion based on gold nanoparticles. Langmuir 18:10407–10410
Pandey S, Oza G, Mewada A, Madhuri S (2012) Green synthesis of highly stable gold nanoparticles using Momordica charantia as nano fabricator. Arch Appl Sci Res 4:1135–1141
Pasca RD, Mocanu A, Cobzac SC, Petean I, Horovitz O, Tomoaia-Cotisel M (2014) Biogenic syntheses of gold nanoparticles using plant extracts. Part Sci Technol 32:131–137
Philip D (2010) Rapid green synthesis of spherical gold nanoparticles using *Mangifera indica* leaf. Spectro Acta A Mol Biomol Spectrosc 77:807–810
Philip D, Unni C, Aromal SA, Vidhu VK (2011) *Murraya koenigii* leaf-assisted rapid green synthesis of silver and gold nanoparticles. Spectrochim Acta Part A 78:899–904
Phillip D (2010) Rapid green synthesis of spherical gold nanoparticles using *Mangifera indica* leaf. Spectrochim Acta Part A 77:807–810
Pradeep T, Anshup A (2009) Noble metal nanoparticles for water purification: a critical review. Thin Solid Film 517:6441–6478
Prasad R (2014) Synthesis of silver nanoparticles in photosynthetic plants. J Nanopart 2014, 963961
Prasad R, Pandey R, Barman I (2016) Engineering tailored nanoparticles with microbes: quo vadis. WIREs Nanomed Nanobiotechnol 8:316–330

Putnam CD, Hammel M, Hura GL, Tainer JA (2007) X-ray solution scattering (SAXS) combined with crystallography and computation: defining accurate macromolecular structures, conformations and assemblies in solution. Q Rev Biophys 40:191–285

Raghunandan D, Basavaraja S, Mahesh B, Balaji S, Manjunath SY, Venkataraman A (2009) Biosynthesis of stable polyshaped gold nanoparticles from microwave-exposed aqueous extracellular antimalignant Guava (*Psidium guajava*) leaf extract. Nanobiotechnology 5:34–41

Rajasekharreddy P, Rani PU, Sreedhar B (2010) Qualitative assessment of silver and gold nanoparticle synthesis in various plants: a photobiological approach. J Nanopar Res 12:1711–1721

Ramezani N, Ehsanfar Z, Shamsa F, Amin G, Shahverdi HR, Esfahani HRM, Shamsaie A, Bazaz RD, Shahverdi AR (2008) Screening of medicinal plant methanol extracts for the synthesis of gold nanoparticles by their reducing potential. Z Naturforsch 63:903–908

Rana S, Bajaj A, Mout R, Rotello VM (2012) Monolayer coated gold nanoparticles for delivery applications. Adv Drug Del Rev 64:200–216

Rao KJ, Paria S (2014) Green synthesis of gold nanoparticles using aqueous *Aegle marmelos* leaf extract and their application for thiamine detection. RSC Adv 4:28645–28652

Richards DG, McMillin DL, Mein EA, Nelson CD (2002) Gold and its relationship to neurological/glandular conditions. Int J Neurosci 112:31–53

Shankar SS, Ahmad A, Pasricha R, Sastry M (2003) Bioreduction of chloroaurate ions by geranium leaves and its endophytic fungus yields gold nanoparticles of different shapes. J Mater Chem 13:1822–1826

Shankar SS, Rai A, Ahmad A, Sastry M (2004a) Rapid synthesis of Au, Ag, and bimetallic Au core–Ag shell nanoparticles using neem (*Azadirachta indica*) leaf broth. J Colloid Interf Sci 275:496–502

Shankar SS, Rai A, Ankamwar B, Singh A, Ahmad A, Sastry M (2004b) Biological synthesis of triangular gold nanoprisms. Nat Mater 3:482–488

Sharma NC, Sahi SV, Nath S, Parsons JG, Gardea-Torresdey JL, Tarasankar P (2007) Synthesis of plant-mediated gold nanoparticles and catalytic role of biomatrix-embedded nanomaterials. Environ Sci Technol 41:5137–5142

Sheny D, Mathew J, Philip D (2011) Phytosynthesis of Au, Ag and Au–Ag bimetallic nanoparticles using aqueous extract and dried leaf of *Anacardium occidentale*. Spectrochim Acta Part A Mol Biomol Spectrosc 79:254–262

Siddiqi KS, Husen A (2016a) Green synthesis characterization and uses of palladium/platinum nanoparticles. Nano Res Lett. doi:10.1186/s11671-016-1695-z

Siddiqi KS, Husen A (2016b) Engineered gold nanoparticles and plant adaptation potential. Nano Res Lett 11:400

Singaravelu G, Arockiamary JS, Ganesh Kumar V, Govindaraju K (2007) A novel extracellular synthesis of monodisperse gold nanoparticles using marine alga, *Sargassum wightii* Greville. Colloids Surf B Biointerfaces 57:97–101

Smitha SL, Philip D, Gopchandran KG (2009) Green synthesis of gold nanoparticles using *Cinnamomum zeylanicum* leaf broth. Spectrochim Acta Part A 74:735–739

Song JY, Jang HK, Kim BS (2009) Biological synthesis of gold nanoparticles using *Magnolia kobus* and *Diopyros kaki* leaf extracts. Process Biochem 44:1133–1138

Spivak MY, Bubnov RV, Yemets IM, Lazarenko LM, Tymoshok NO, Ulberg ZR (2013) Development and testing of gold nanoparticles for drug delivery and treatment of heart failure: a theranostic potential for PPP cardiology. EPMA J 4:20

Sreelakshmi C, Datta K, Yadav J, Reddy B (2011) Honey derivatized Au and Ag nanoparticles and evaluation of its antimicrobial activity. J Nanosci Nanotechnol 11:6995–7000

Suman TY, Rajasree SRR, Ramkumar R, Rajthilak C, Perumal P (2014) The Green synthesis of gold nanoparticles using an aqueous root extract of *Morinda citrifolia* L. Spectrochim Acta Part A Mol Biomol Spectrosc 118:11–16

Sun D, Zhang G, Huang J, Wang H, Li Q (2014) Plant-mediated fabrication and surface enhanced Raman property of flower-like Au@Pd nanoparticles. Materials 7:1360–1369

Tanaka K (1999) Nanotechnology towards the 21st century. Thin Solid Film 341:120–125
Tetgure SR, Borse AU, Sankapal BR, Garole VJ, Garole DJ (2015) Green biochemistry approach for synthesis of silver and gold nanoparticles using Ficus racemosa latex and their pH-dependent binding study with different amino acids using UV/Vis absorption spectroscopy. Amino Acids 47:757–765
Tripathi A, Kumari S, Kumar A (2016) Toxicity evaluation of pH dependent stable *Achyranthes aspera* herbal gold nanoparticles. Appl Nanosci 6:61–69
Vankar PS, Bajpai D (2010) Preparation of gold nanoparticles from *Mirabilis jalapa* flowers. Ind J Biochem Biophys 47:157–160
Verma A, Simard JM, Worrall JWE, Rotello VM (2004) Tunable reactivation of nanoparticle-inhibited β-galactosidase by glutathione at intracellular concentrations. J Am Chem Soc 126:13987–13991
Vilchis-Nestor AR, Sánchez-Mendieta V, Camacho-López MA, Gómez-Espinosa RM, Arenas-Alatorre JA (2008) Solvent less synthesis and optical properties of Au and Ag nanoparticles using *Camellia sinensis* extract. Mater Lett 62:3103–3105
Vo-Dinh T, Cullum B (2000) Biosensors and biochips: advances in biological and medical diagnostics. Fresenius J Anal Chem 366:540–551
Yasmin A, Ramesh K, Rajeshkumar S (2014) Optimization and stabilization of gold nanoparticles by using herbal plant extract with microwave heating. Nano Converg 1:12
Yeh YC, Creran B, Rotello VM (2012) Gold nanoparticles: preparation, properties, and applications in bionanotechnology. Nanoscale 4:1871–1880
Yu J, Xu D, Guan HN, Wang C, Huang LK, Chi DF (2016) Facile one-step green synthesis of gold nanoparticles using *Citrus maxima* aqueous extracts and its catalytic activity. Mat Lett 166:110–112
Zhan G, Huang J, Lin L, Lin W, Emmanuel K, Li Q (2011) Synthesis of gold nanoparticles by *Cacumen platycladi* leaf extract and its simulated solution: toward the plant-mediated biosynthetic mechanism. J Nanopart Res 13:4957–4968

Chapter 18
Encapsulation of Nanomaterials and Production of Nanofertilizers and Nanopesticides: Insecticides for Agri-food Production and Plant Disease Treatment

Nahid Sarlak and Asghar Taherifar

18.1 Introduction

Organic pesticides including insecticides and herbicides are widely used in agriculture because of their powerful biological activity (Al-Degs et al. 2009). Mancozeb and Zineb were first introduced during the 1940s and widely used. Mancozeb as a pesticide protects many fruit, vegetable, nut, and agricultural crops against a wide range of fungal diseases. It is a grayish-yellow powder, and its chemical name is manganese ethylene bis (dithiocarbamate) (polymeric). Zineb is a light-colored powder or crystal, the chemical name of which is zinc ethylene bis (dithiocarbamate) Dithiocarbamates (A. Matthews 2000; Honeycutt Richard et al. 1985; Peter E Rijtema et al. 1999; Scher Herbert 1984; Tordoir 1994). Fig. 18.1a, b shows the chemical structure of Mancozeb and Zineb molecules, respectively.

Benomyl is a general use pesticide (GUP). It is used against a wide range of fungal diseases of field crops, fruits, nuts, ornamentals, mushrooms, and turf (McCarroll et al. 2002; Suwalsky et al. 2000). Trichlorfon is a pale clear, white, or yellow crystalline solid with an ethyl ether odor. It is an organophosphate insecticide used to control cockroaches, crickets, silverfish, bedbugs, fleas, cattle grubs, flies, etc. It is also used for treating domestic animals for the control of internal parasites (Burnett 2005). Pymetrozine belongs to the class of pyridine azomethine. It is used to control aphids and whiteflies in vegetables, potatoes, tobacco, citrus, fruit, hops, and ornamentals (Brück et al. 2009). Figure 18.1c–e shows the shape of Benomyl, Trichlorfon, and Pymetrozine (chess) molecules, respectively.

Encapsulation is a process of surrounding or enveloping one substance within another material on a very small scale. Another definition of encapsulation is the process by which a material is surrounded and protected from extraneous conditions

N. Sarlak (✉) • A. Taherifar
Department of Chemistry, Lorestan University, khorram Abad, Iran
e-mail: sarlak.n@gmail.com

M. Ghorbanpour et al. (eds.), *Nanoscience and Plant–Soil Systems*, Soil Biology 48,
DOI 10.1007/978-3-319-46835-8_18

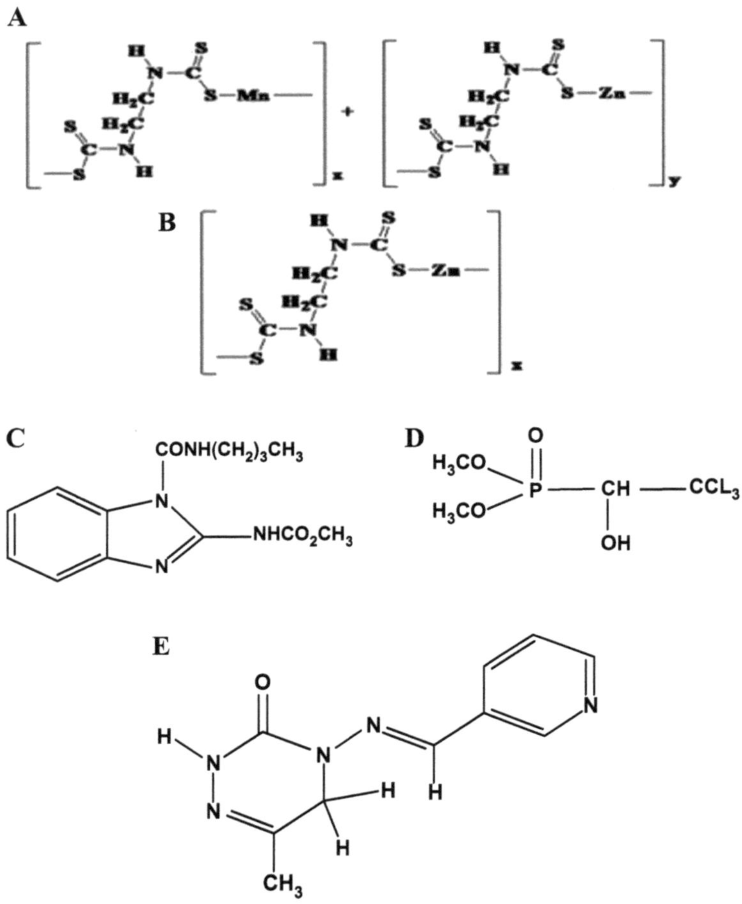

Fig. 18.1 Shape of pesticides molecule: (**a**) Mancozeb, (**b**) Zineb, (**c**) Benomyl, (**d**) Trichlorfon (**e**) Pymetrozine

that would otherwise break it down, yielding capsules ranging from less than one micron to several hundred microns in size (Del Carmen Giménez-López et al. 2011). Encapsulation can be achieved by several techniques with different purposes in mind. Encapsulation of materials may occur with the intention that the core material be confined within capsule walls for a specific period of time. Alternatively, core substances can be encapsulated so that the core material will be released either gradually through the capsule walls, known as controlled release or diffusion, or when external conditions activate the capsule walls to break, melt, or dissolve (Doane 1992). There are many applications for encapsulated materials in various fields such as agriculture, pharmaceuticals, foods, cosmetics and fragrances, textiles, paper, paints, coatings and adhesives, printing applications, and many other industries (Prasad et al. 2014).

Pesticides that are encapsulated to be released over time allow farmers to apply the pesticides less often rather than requiring very highly concentrated and perhaps toxic initial applications followed by repeated applications to battle the loss of efficacy due to leaching, evaporation, and degradation (Lewis and Cowsar 1977). Protection of pesticides from full contact with the elements lessens the risk to the environment and those that might be exposed to the chemicals and provides a more efficient strategy for pest control (Ulbricht and Hertel 2003).

Since carbon nanotubes (CNTs) were discovered by Iijima in 1991, they have come under intense multidisciplinary study because of their unique physical and chemical properties and their possible applications. One of the disadvantages of CNTs is their toxic effect on environment, but there are some reports which demonstrate that immobilization of chemical agents on the surface of CNTs reduces its toxic effects remarkably (Sayes et al. 2006; Tian et al. 2006).

In the past decade, there has been an increasing interest in the studies of polymer/CNT nanocomposites due to the unique combination of promising properties and construction of multifunctional structures of each component (Ajayan 1999).

Previously, (Trimnell et al. 1982) investigated the encapsulation of pesticide within a starch–borate complex. In this work, the starch, pesticide, and water were mixed and alkali was added to gelatinize the starch. Then the mixture was treated with boric acid. As reported in their paper, pesticides had been only encapsulated within a starch–borate in which decomposition of pesticide had decreased; however, the effect of capsulated pesticides wasn't studied on malady agent.

In this work, we synthesized polymerized CNTs using citric acid which is an environmental friendly material. The obtained hybrid material is completely water soluble. On the other hand, the amount of CNT used for polymerization reaction is about 0.03 g/L which in comparison with toxic dosage of CNTs in soil that was reported previously (Chung et al. 2011; Lin and Xing 2007; Petersen et al. 2011) is very little. Then, Zineb, Mancozeb, Benomyl, Trichlorfon, and Pymetrozine were encapsulated into MWCNT-g-PCA hybrid material. This process leads to conversion of bulk pesticide to pesticide nanoparticles. Effective parameters in encapsulation process, such as pH, temperature, and stirrer time, were optimized via UV-Vis spectroscopic method. The influence of encapsulated pesticide was studied on malady Alternaria alternate, so for this purpose Potato Dextrose Agar (PDA) was used as a growth medium. Results show that pesticide nanoparticles in contrast with bulk pesticide have extraordinary function. To the best of our knowledge, this paper is the first report about using water-soluble polymerized carbon nanotubes for encapsulation of pesticides and application of the new nanopesticides in malady agent treatment.

In comparison with previous reports, there are some novel and specific advantages:

(a) Encapsulated pesticide into CNT-g-PCA hybrid material is more stable and effective than bulk pesticide.
(b) Pesticide usage will be decreased in agriculture.

(c) Water solubility of encapsulated pesticides will be increased.
(d) Multiwall carbon nanotubes are modified to nontoxic form and will be soluble in water.

This product can be used extensively in several greenhouse and field studies and Agriculture.

18.2 Experimental

18.2.1 Reagents

The MWCNT was purchased from Nutrino. The outer diameter of MWCNT was between 20 and 40 nm. Monohydrate citric acid, sulfuric and nitric acid, tetrahydrofurane, and cyclohexane were purchased from Merck. Pesticides standard was purchased from Delta Green South Corporation of Iran.

18.2.2 Apparatus

Fourier transform infrared (FT-IR) spectroscopic measurements were performed using an FT-IR spectrometer (Thermo Nicolet Magna-IR 560 spectroscopy, USA). The FTIR spectra of acid-modified CNTs and CNT-g-PCA were obtained in transmittance mode by placing a small amount of the materials in KBr pellets. The UV-Vis measurements were obtained using a double beam spectrophotometer (Shimadzu 1650 UV–Vis Spectrophotometer, Japan) with 1-cm quartz cells. In order to well disperse pesticide in the polymeric shell of hybrid material, an ultrasonic bath (22 KHz) was used. The TEM study was carried out using a microscope operating at 100 kV. The samples for the TEM analysis were prepared by dilution of nanocomposite with distilled water. A drop of the suspension was applied onto a carbon-coated copper grid and was dried in air. A centrifuge (Eppendorf 5810) was used for separation of CNT-g-PCA-EP from pesticide molecules.

18.2.3 Preparation of MWCNT-g-PCA Hybrid Materials

MWCNTs were opened according to reported procedures in literatures (Stobinski et al. 2010). Briefly, MWCNT (2 g) was added to 40 ml of sulfuric and nitric acid mixture (3/1) in a reaction flask and refluxed for 24 h at 120 °C. The mixture was cooled and diluted with distilled water; then it was filtrated and washed with distilled water till the pH reaches to 5; finally it was dried by vacuum oven. MWCNT-g-PCA hybrid materials were prepared according to the reported procedure (Adeli et al. 2008).

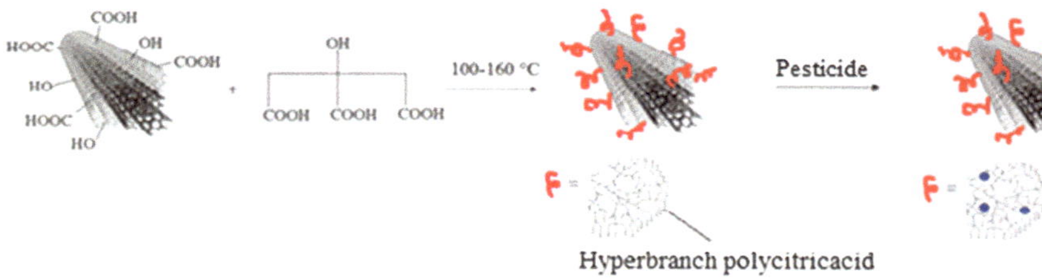

Fig. 18.2 Synthetic scheme of CNT-g-PCA-EP

18.2.4 Preparation of MWCNT-g-PCA-EP

Water solution of MWCNT-*g*-PCA separately was mixed with each of pesticides (Mancozeb, Zineb, Benomyl, Trichlorfon, and Pymetrozine) with appropriate concentrations and placed in an ultrasonic bath for 10 min in order to well disperse pesticides in the polymer shell of hybrid material. Then the mixtures were stirred for complete encapsulation of pesticides into CNT-g-PCA. Encapsulated pesticide nanoparticles were purified by centrifugation. Pesticide molecules which were not reacted with CNT-g-PCA were separated from encapsulated pesticide nanoparticles using a centrifuge with rotating speed of 2000 r/min within 7 min. During the process of centrifugation, the sample is separated into two phases, a pellet consisting of MWCNT-g-PCA-EP and a supernatant that is unreacted pesticide molecules. Then the supernatant was separated and wasted. Figure 18.2 shows the synthetic scheme of the CNT-g-PCA-EP.

18.2.5 Preparation of Medium Potato Dextrose Agar (PDA)

Usually for PDA medium preparation, 250 g of potatoes was boiled for 20 min and the sap of potato was taken. Afterwards, 20 g dextrose agar was added to the sap under continuous stirring and slowly warmed until it was well solved. Then the medium was diluted to 1 l by warmed distilled water and autoclaved for 15 min at 121 °C and pressure 1.5 kg/cm^2. Finally, PDA medium was chilled until 45 °C and was poured into a Petri dish (Adeli et al. 2008; Kuo et al. 2009; Xu et al. 2009; Zheng and Shetty 1998).

18.3 Findings

18.3.1 Nanohybrid Characterization

The FTIR spectrum shown in Fig. 18.3 confirms the presence of functional groups in MWCNT-COOH and MWCNT-g-PCA hybrid materials. The carbonyl stretching frequency in MWCNT-COOH was seen at 1701 cm^{-1} (COOH). The

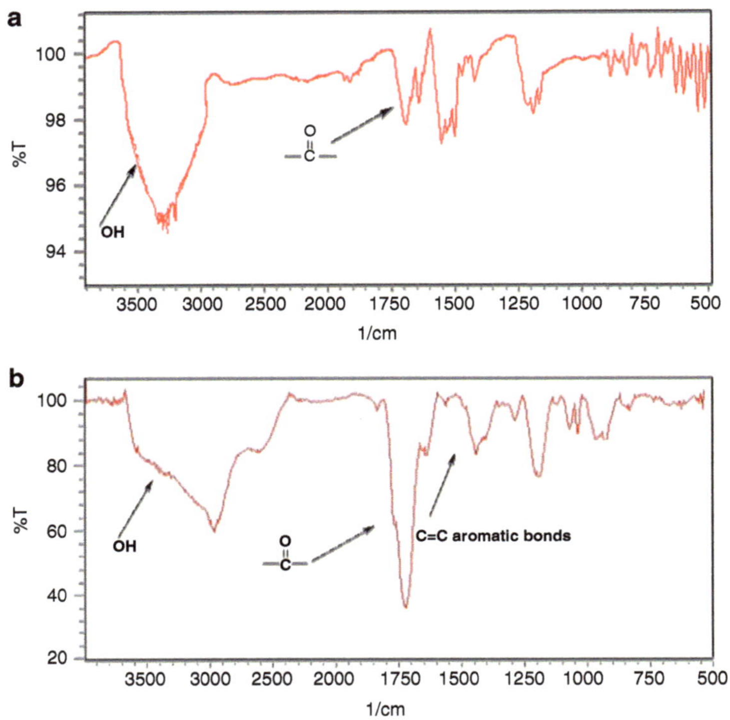

Fig. 18.3 FTIR data for (**a**) MWCNT-COOH and (**b**) MWCNT-g-PCA hybrid material

3200–3500 cm^{-1} band present in MWCNT-COOH spectrum was attributed to the hydroxyl vibration of the carboxylic acid group attached through the opening.

The peak around 1400–1500 cm^{-1} was assigned to the C=C stretching of the carbon skeleton. As shown in Fig. 18.3b, a broad absorbance band at 2600–3573 cm $^{-1}$ was attributed to hydroxyl functional groups of grafted PCA. In this spectrum, a band of carbonyl groups of citric acid was seen at 1718 cm^{-1}. Also the C=C band of MWCNT appeared at 1400–1500 cm^{-1}.

18.3.2 Optimization of Effective Parameters on Synthesis of CNT-g-PCA-EP

The effect of various parameters on encapsulation of pesticides into CNT-g-PCA was studied. Figure 18.4 shows the role of pH in the encapsulation of pesticides by MWCNT-g-PCA hybrid material. The maximum encapsulation of pesticide molecules into polymeric shell corresponds to maximum absorption in UV-Vis spectra. Maximum encapsulation for Mancozeb, Zineb, Benomyl, Trichlorfon, and Pymetrozine according to Fig. 18.4 takes place at pH 3,8,7, 2, and 3, respectively. Furthermore, temperature had no significant effect on the encapsulation reaction.

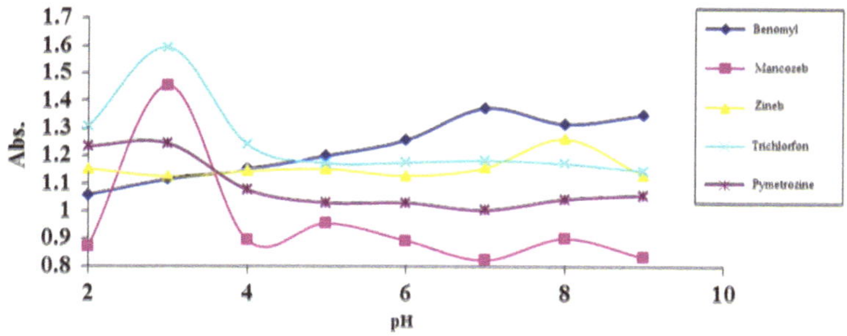

Fig. 18.4 Effect of pH on the encapsulation of pesticides with CNT-g-PCA

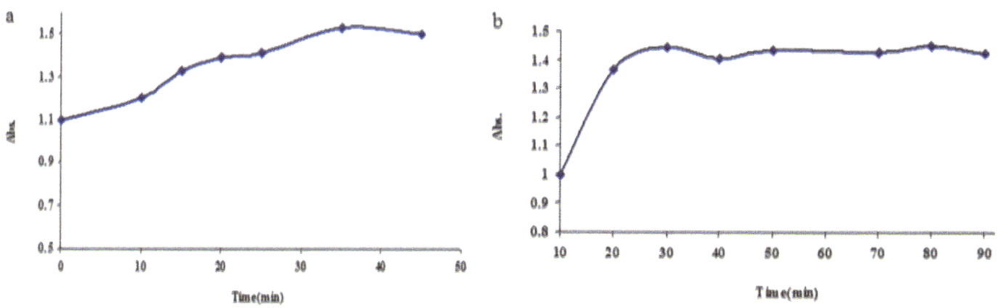

Fig. 18.5 The effect of stirring time on encapsulation of pesticides into CNT-g-PCA hybrid material. Mancozeb (**a**) and Zineb (**b**)

The required time for complete encapsulation of pesticides in CNT-g-PCA is very important. Figure 18.5 shows the effect of stirrer time on the reaction between pesticides and CNT-g-PCA. As shown in Fig. 18.5a,b with the addition of pesticide aqueous solutions to CNT-g-PCA, the respective absorption peak increases. In other words, maximum encapsulation for Mancozeb and Zineb in hyper branch polycitric acid shell takes place after 35 min and 80 min, respectively.

As can be seen in Fig. 18.6, the amount of encapsulated Trichlorfon and Pymetrozine into hyper branched polycitric acid increases with increasing stirrer time, but it has no significant effect on encapsulation of Benomyl. Encapsulation of Trichlorfon and Pymetrozine was completed after 50 and 60 min, respectively, whereas encapsulation of Benomyl was completed immediately, after sonication of Benomyl with CNT-g-PCA.

Figure 18.7a, b, c shows the color change during the progress of the complexation reaction between Zineb and CNT-g-PCA hybrid material under optimum conditions. As shown in Fig. 18.7 for Zineb, the light yellow color of solution turned gradually to pale yellow after 2 h. The darkening of the solution continued and fixed after 8 h which indicated the encapsulation process was completed. Also the encapsulation of Pymetrozine into the CNT-g-PCA was accompanied with color change during the time. The color change through the progression of the encapsulation reaction between Pymetrozine and CNT-g-PCA hybrid material was shown in Fig. 18.7d, e, f.

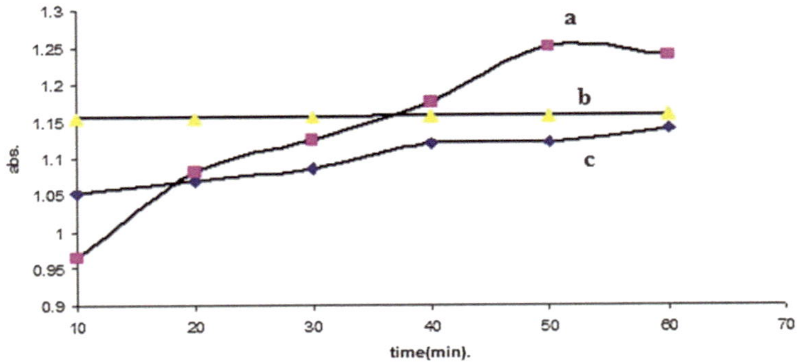

Fig. 18.6 Effect of stirrer time on encapsulation of (**a**) Trichlorfon, (**b**) Benomyl, and (**c**) Pymetrozine into CNT-g-PCA hybrid

Fig. 18.7 The color change during time progression of the encapsulation between Zineb and MWCNT-g-PCA (**a**) initiated (**b**) after 2 h and (**c**) after 8 h and pymetrozine and MWCNT-g-PCA (**d**) initiated (**e**) after 2 h and (**f**) after 8 h

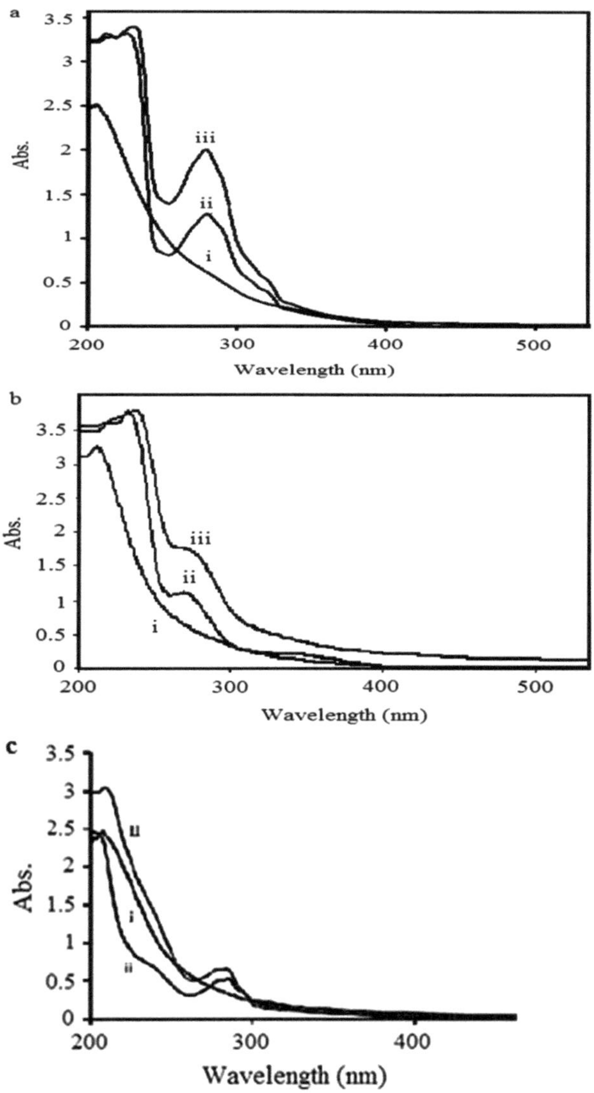

Fig. 18.8 UV spectra for encapsulation of (**a**) Zineb, (**b**) Mancozeb, (**c**) Benomyl, (**d**) Trichlorfon, (**e**) Pymetrozine by CNT-g-PCA. (i) Aqueous solution of CNT-g-PCA (ii) aqueous solution of pesticide (iii) after addition of pesticide to CNT-g-PCA

18.3.3 Encapsulation of the Pesticides by CNT-g-PCA: UV-Vis Study

The encapsulation of the Zineb, Mancozeb, Benomyl, Trichlorfon, and Pymetrozine by polymeric shell of CNT-g-PCA was investigated spectrophtometrically. Fig. 18.8a (iii), b (iii), c (iii), d (iii), e (iii) shows the UV–Vis spectra for encapsulated of pesticides by CNT-g-PCA, respectively.

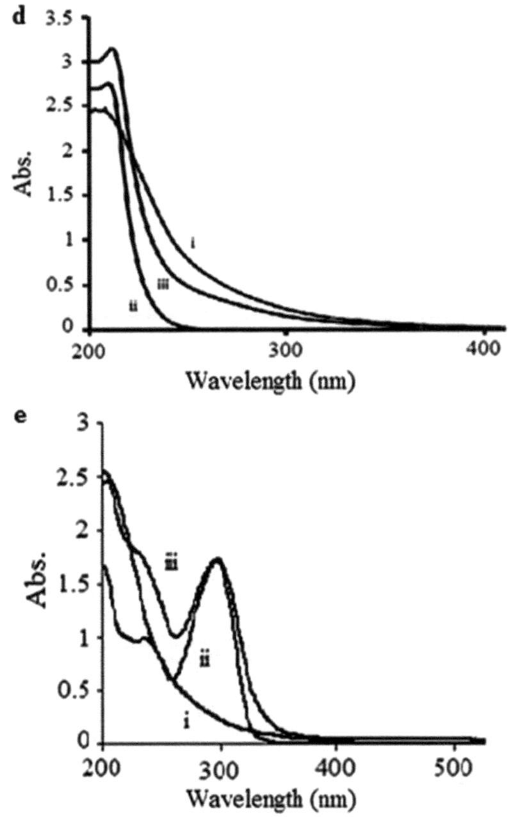

Fig. 18.8 (continued)

A simple route which clearly approves the encapsulation of pesticides in hyper branched polycitric acid shell was attained with the addition of various amounts of CNT-g-PCA to each pesticide and monitoring the UV spectra. As shown in Fig. 18.9, addition of CNT-g-PCA led to increasing the peak intensity while the fine structures of pesticide spectra follow a same pattern. These behaviors confirm the encapsulation of pesticides into polycitric acid hyper branched shell.

18.3.4 TEM Images of Synthesized CNT-g-PCA-EP Pesticide

Finally, encapsulation of pesticides by the CNT-g-PCA was confirmed by TEM analysis. Figure 18.10 shows TEM images of Zineb and Mancozeb encapsulation

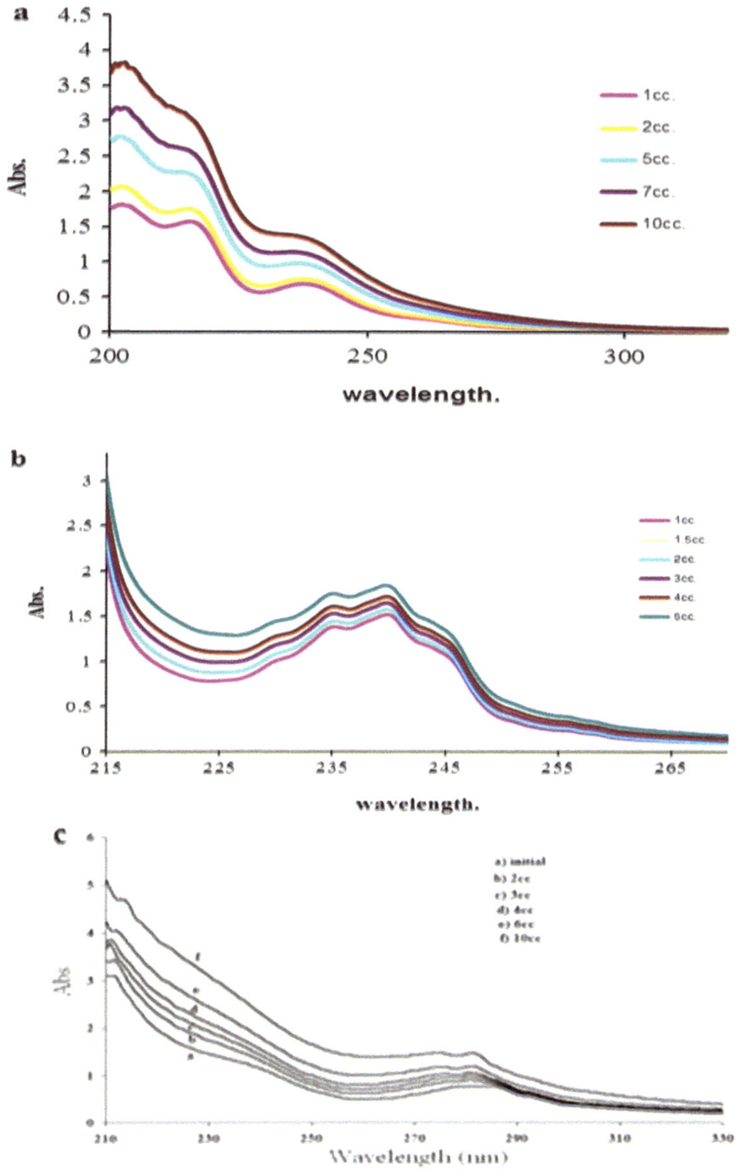

Fig. 18.9 Addition of various amounts of a water solution of CNT-g-PCA to pesticides (**a**) Mancozeb, (**b**) Zineb, (**c**) Benomyl, (**d**) Trichlorfon, (**e**) Pymetrozine

by the CNT-g-PCA hybrid material which clearly indicates polycitric acid (PCA) shell around CNTs and encapsulated pesticide nanoparticles. Figure 18.11 shows more TEM images of Zineb encapsulation into polymeric shell of CNT-g-PCA.

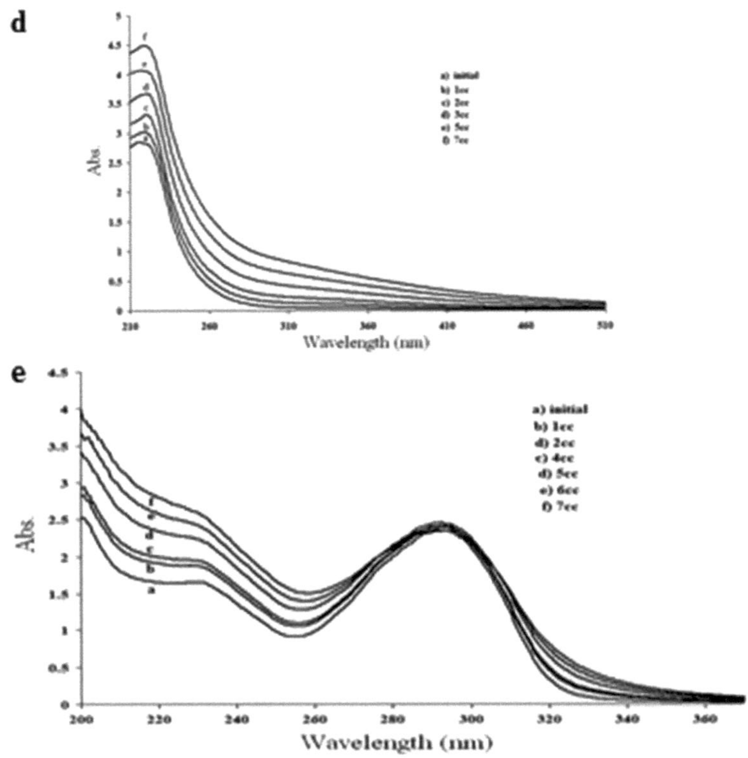

Fig. 18.9 (continued)

Figure 18.12 shows TEM images of encapsulated Benomyl (a, b), Trichlorfon (c, d), and Pymetrozine (e, f). This picture clearly indicates polycitric acid (PCA) shell around the CNTs and encapsulated pesticide.

Also distribution of Pymetrozine in the polycitric acid shell can be seen in Dark field image, in which bright points correspond to encapsulated Pymetrozine (Fig. 18.13).

18.3.5 Application of Encapsulated Pesticide Nanoparticles on Alternaria alternate Fungi

Encapsulation of pesticide nanoparticles in hyper branched polycitric acid results in the appearance of extraordinary characteristics. In this study, the influence of

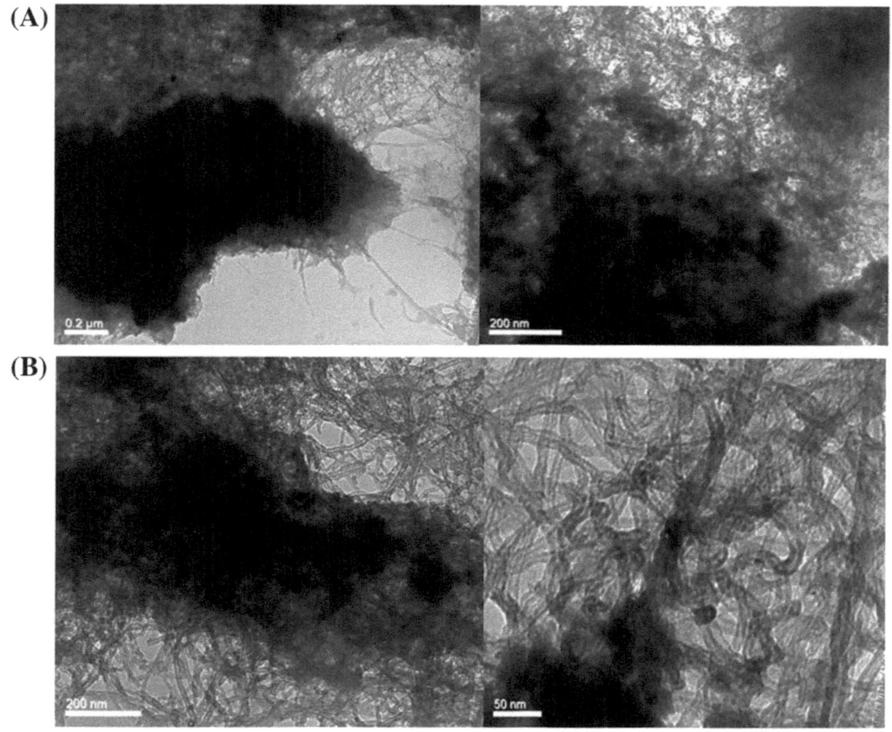

Fig. 18.10 TEM images of encapsulation of Zineb (**a**) and Mancozeb (**b**) with MWCNT-g-PCA

CNT-g-PCA-EP on malady Alternaria alternate fungi was studied. For this purpose, Potato dextrose agar (PDA) medium was used to grow malady Alternaria alternate. Bulk and encapsulated Zineb were added to the control medium, and their effects on growth of colonies were studied after 3 and 10 days (Fig. 18.14). The results show that encapsulated Zineb nanoparticles have more effect on growth restriction in comparison to the bulk Zineb. This effect after 10 days reaches the maximum intensity. Probably, this phenomenon is attributed to the increase of the contact area which was formed between encapsulated pesticide and fungi.

Also, the effect of encapsulated Benomyl was studied on malady Alternaria alternata fungi. Bulk and encapsulated Benomyl were added to the control mediums separately, and their effects on growth of colonies were studied after 10 days. Results showed that encapsulated Benomyl had more significant effect on growth restriction of malady in comparison with the bulk Benomyl. This effect after 10 days reached the maximum intensity as shown in Fig. 18.15. As can be seen in Fig. 18.15c, malady has largely disappeared. Probably, this phenomenon is due to

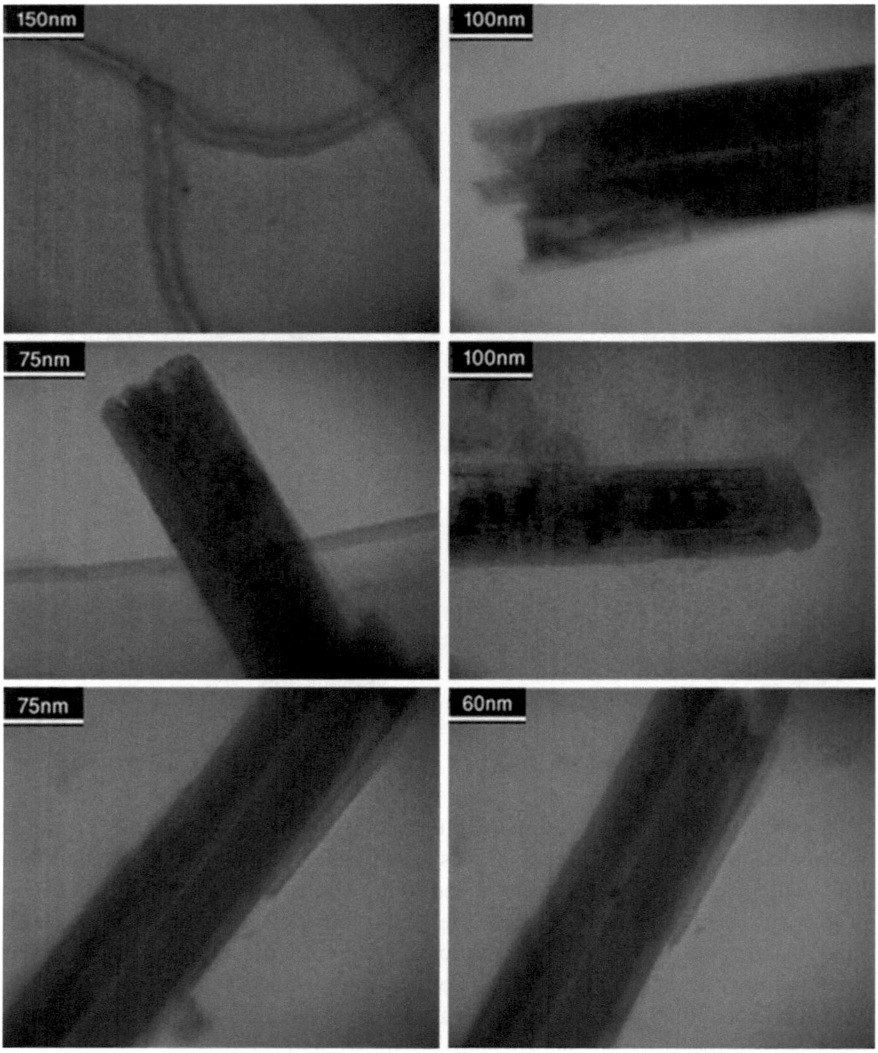

Fig. 18.11 TEM images of encapsulation of Zineb with MWCNT-g-PCA

the high surface area of Bbenomyl nanoparticles which increases the contact area between pesticide and fungi.

In summary, pesticides were successfully encapsulated into CNT-g-PCA hybrid material. Results showed that encapsulation of pesticides into hybrid material is dependent on the pH and time of stirring, but temperature has no significant effect

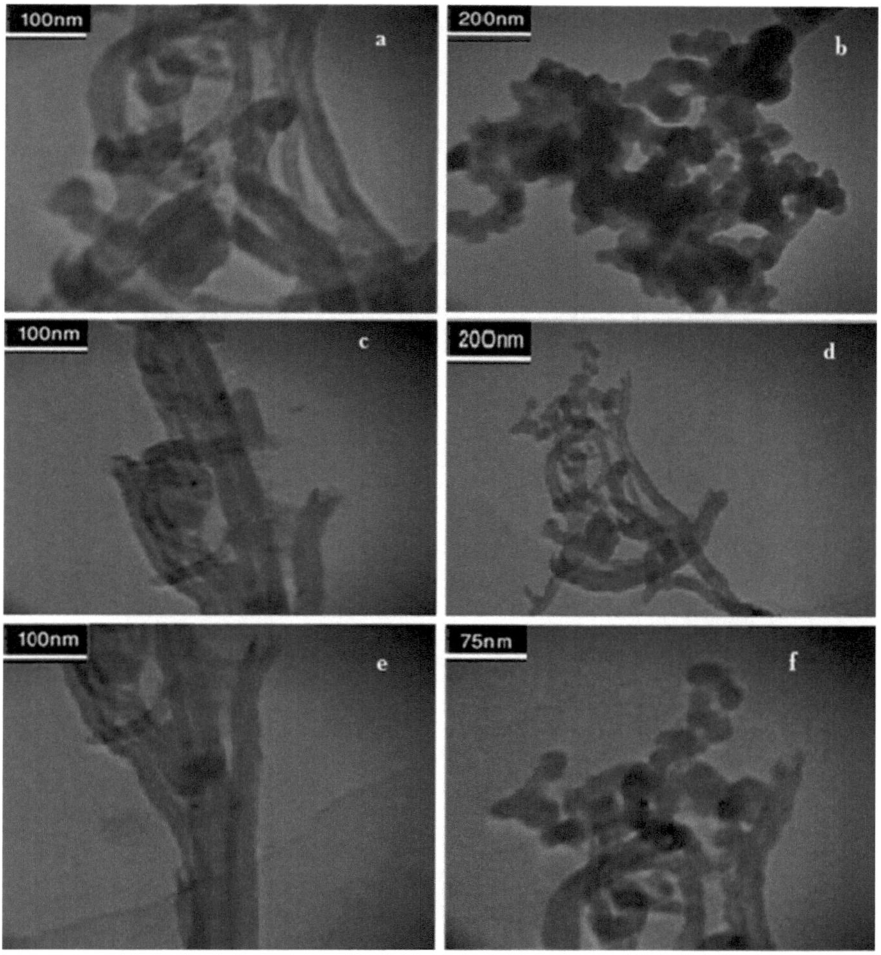

Fig. 18.12 TEM images of encapsulation of (**a**, **b**) Benomyl (**c**, **d**), Trichlorfon, and (**e**, **f**) Pymetrozine into MWCNT-g-PCA

on encapsulation process. Experiments show the optimum pH for encapsulation of Mancozeb and Zinebare 3 and 8, respectively.

The optimum pHs for encapsulation of Benomyl, Trichlorfon, and Pymetrozine are 7, 2, and 3, respectively. Furthermore, the time of stirring for complete encapsulation of Mancozeb and Zineb was 35 and 80 min, respectively. The stirring time for complete encapsulation of Trichlorfon and Pymetrozine was 50 and 60 min, respectively, while encapsulation of Benomyl was constant during stirrer time. As a result, in this work we could produce pesticide nanoparticles. So encapsulated

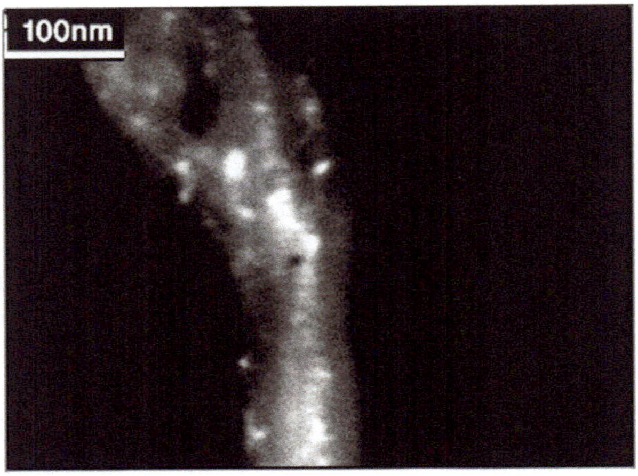

Fig. 18.13 Dark field image of encapsulation of Pymetrozine with MWCNT-g-PCA

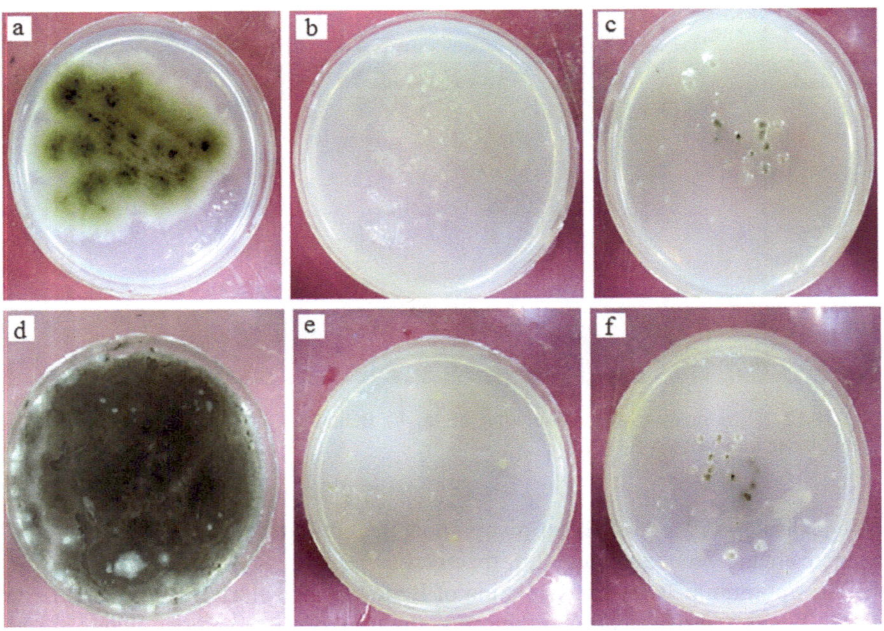

Fig. 18.14 The influence of Zineb encapsulation on Alternaria alternate fungi. Medium control (**a** and **d**), medium control with encapsulated Zineb (**b** and **e**), and medium control with bulk Zineb (**c** and **f**). *Top* after 3 and *down* after 10 days

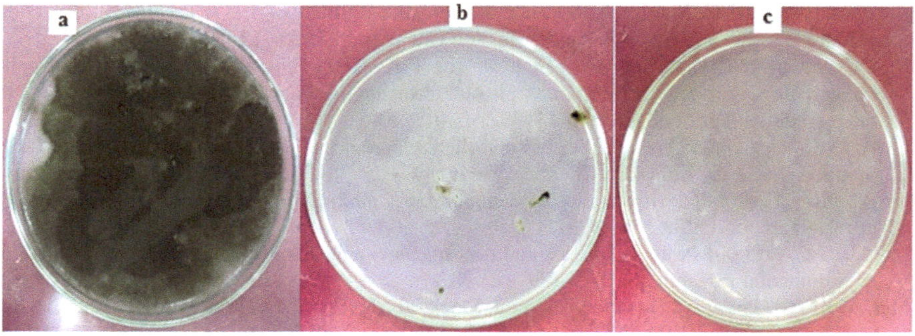

Fig. 18.15 The effect of Benomyl on Alternaria alternate fungi. (**a**) The growth of malady Alternaria alternate fungi in Potato Dextrose Agar (PDA) medium, (**b**) the growth of malady after addition of bulk Benomyl, and (**c**) the growth of malady after addition of encapsulated Benomyl after 10 days

Zineb as pesticide nanoparticles in comparison with bulk Zineb showed an extraordinary effect on malady Alternaria alternata fungi and its effect reached to maximum after 10 days. Also, encapsulated Benomyl as pesticide nanoparticles in comparison with bulk Benomyl showed an extraordinary effect on malady Alternaria alternata fungi and its effect reached to maximum after 10 days. As an important result, using pesticides nanoparticles can help farmers to use less pesticide with much more effect in eliminating disease in agricultural applications and increases environmental safety.

References

Adeli M, Bahari A, Hekmatara H (2008) Carbon nanotube-graft-poly (citric acid) nanocomposites. Nano 3:37–44
Ajayan P (1999) Nanotubes from carbon. Chem Rev 99:1787–1800
Al-Degs YS, Al-Ghouti MA, El-Sheikh AH (2009) Simultaneous determination of pesticides at trace levels in water using multiwalled carbon nanotubes as solid-phase extractant and multivariate calibration. J Hazard Mater 169:128–135
Brück E, Elbert A, Fischer R, Krueger S, Kühnhold J, Klueken AM, van Waetermeulen X (2009) Movento®, an innovative ambimobile insecticide for sucking insect pest control in agriculture: biological profile and field performance. Crop Protect 28:838–844
Burnett KG (2005) Chapter 8 Impacts of environmental toxicants and natural variables on the immune system of fishes. In: Mommsen TP, Moon TW (eds) Biochemistry and molecular biology of fishes, vol 6. Elsevier, St.Louis, pp 231–253
Chung H, Son Y, Yoon TK, Kim S, Kim W (2011) The effect of multi-walled carbon nanotubes on soil microbial activity. Ecotoxicol Environ Saf 74:569–575
Del Carmen Giménez-López M, Moro F, La Torre A, Gómez-García CJ, Brown PD, van Slageren J, Khlobystov AN (2011) Encapsulation of single-molecule magnets in carbon nanotubes. Nat Commun 2:407. doi:10.1038/ncomms1415
Doane WM (1992) Encapsulation of pesticides in starch by extrusion. Ind Crops Prod 1:83–87

Honeycutt Richard C, Zweig G, Ragsdale N, Durham William F (1985) Preface, introduction dermal exposure related to pesticide use. Am Chem Soc 273:11–14
Kuo C-F, Chyau C-C, Wang T-S, Li C-R, Hu T-J (2009) Enhanced antioxidant and anti-inflammatory activities of Monascus pilosus fermented products by addition of turmeric to the medium. J Agr Food Chem 57(11397):11405
Lewis D, Cowsar D (1977) Principles of controlled release pesticides. Paper presented at the ACS Symposium Series American Chemical Society
Lin D, Xing B (2007) Phytotoxicity of nanoparticles: inhibition of seed germination and root growth. Environ Pollut 150:243–250
Matthews GA (2000) Pests, pesticides and pest management. In: Mason SJ (Ed.), Highlights in environmental research. pp. 165–189
McCarroll NE, Protzel A, Ioannou Y, Frank Stack HF, Jackson MA, Waters MD, Dearfield KL (2002) A survey of EPA/OPP and open literature on selected pesticide chemicals. III. Mutagenicity and carcinogenicity of benomyl and carbendazim. Mutat Res 512:1–35
Petersen EJ, Pinto RA, Zhang L, Huang Q, Landrum PF, Weber WJ Jr (2011) Effects of polyethyleneimine-mediated functionalization of multi-walled carbon nanotubes on earthworm bioaccumulation and sorption by soils. Environ Sci Technol 45:3718–3724
Prasad R, Kumar V, Prasad KS (2014) Nanotechnology in sustainable agriculture: present concerns and future aspects. Afr J Biotechnol 13:705–713
Rijtema PE, Groenendijk P, Kroes JG (1999) Environmental influences on processes. In: Rijtema PE, Groenendijk P, Kroes JG (eds) Environmental impact of land use in rural regions, vol 1. Imperial College Press, London, pp 167–203
Sayes CM, Liang F, Hudson JL, Mendez J, Guo W, Beach JM, Billups WE (2006) Functionalization density dependence of single-walled carbon nanotubes cytotoxicity in vitro. Toxicol Lett 161:135–142
Scher Herbert B (1984) Advances in pesticide formulation technology advances in pesticide formulation technology. Am Chem Soc 254:1–7
Stobinski L, Lesiak B, Kövér L, Tóth J, Biniak S, Trykowski G, Judek J (2010) Multiwall carbon nanotubes purification and oxidation by nitric acid studied by the FTIR and electron spectroscopy methods. J Alloys Compound 501:77–84
Suwalsky M, Benites M, Norris B, Sotomayor P (2000) Toxic effects of the fungicide benomyl on cell membranes. Compar Biochem Physiol Part C: Pharmacol Toxicol Endocrinol 125:111–119
Tian F, Cui D, Schwarz H, Estrada GG, Kobayashi H (2006) Cytotoxicity of single-wall carbon nanotubes on human Fibroblasts. Toxicol In Vitro 20:1202–1212
Tordoir WF (1994) The development of safe chemical pesticides occupational health in national development. World Scientific, pp 27–42
Trimnell D, Shasha B, Wing R, Otey F (1982) Pesticide encapsulation using a starch–borate complex as wall material. J Appl Polym Sci 27:3919–3928
Ulbricht H, Hertel T (2003) Dynamics of C60 encapsulation into single-wall carbon nanotubes. J Phys Chem B 107:14185–14190
Xu M-J, Yang Z-L, Liang Z-Z, Zhou S-N (2009) Construction of a Monascus purpureus mutant showing lower citrinin and higher pigment production by replacement of ctnA with pks1 without using vector and resistance gene. J Agr Food Chem 57:9764–9768
Zheng Z, Shetty K (1998) Solid-state production of beneficial fungi on apple processing wastes using glucosamine as the indicator of growth. J Agr Food Chem 46:783–787

Chapter 19
Simultaneous Determination of Pesticides at Trace Levels in Water Using Functionalized Multiwalled Carbon Nanotubes as Solid-Phase Extractant and Partial Least-Squares (PLS) Method

Nahid Sarlak and Asghar Taherifar

19.1 Introduction

Organic pesticides including insecticides and herbicides are widely used in agriculture. Because of their powerful biological activity, analytical procedures have been developed to determine and control pesticides in surface and ground waters. Pymetrozine first registered in 1999 by EPA has a unique mode of action that is not fully understood. Pymetrozine is used to control aphids and whiteflies in vegetables, potatoes, tobacco, citrus, fruit, hops, and ornamentals; although of low toxicity it did produce some evidence of tumorigenicity in both rats and mice. Pymetrozine belongs to the class of pyridine azomethine (Tomlin 2003). Carbon nanotubes (CNTs) were discovered by Iijimain 1991 (Iijima 1991); they come under intense multidisciplinary study because of their unique physical and chemical properties and their possible applications. CNTs include single-walled carbon nanotubes (SWCNTs) and multiwalled carbon nanotubes (MWCNTs) are dependent on the number of layers comprising them. Because of their structure properties with nanometer-order size and pseudo-graphite layers, MWCNTs have been expected to apply for the electrochemical storage of hydrogen (Frackowiak and Beguin 2002) and a promising sorbent of heavy metal ions and radionuclides (Dong et al. 2012) and herbicides (Zhou et al. 2006). MWCNTs have great adsorption capacity and show great potential application in preconcentration for environmental pollutants (Cai et al. 2003); although sorption of lead by carbon nanotubes and clay minerals has been studied extensively, the sorption mechanism of lead by nanotubes is still ambiguous, especially in the presence of different anion or cation ions.

N. Sarlak (✉) • A. Taherifar
Department of Chemistry, Faculty of Sciences, Lorestan University, khorram Abad, Iran
e-mail: sarlak.n@gmail.com

M. Ghorbanpour et al. (eds.), *Nanoscience and Plant–Soil Systems*, Soil Biology 48,
DOI 10.1007/978-3-319-46835-8_19

In this study, separation and simultaneous determination of pesticides by functionalized MWCNTs under ambient conditions was investigated. Based on it, a novel and simple method was developed for the simultaneous determination of mancozeb and pymetrozine at trace level in water.

A certain number of these calibration methods are available as affordable commercial software is used with existing instruments. The most popular among them include multiple linear regression (MLR) (Blanco et al. 1989; Hamilton and Gemperline 1990), principal component regression (PCR) (Msimanga et al. 1997; Rius et al. 1997), target factor analysis (TFA), partial least-squares (PLS) regression (Afkhami et al. 2007; Martens and Naes 1992; Sarlak and Anizadeh 2011; Zupan and Gasteiger 1993), and artificial neural networks (ANN) (Zupan and Gasteiger 1993).

19.2 Experimental

19.2.1 Instrumentation and Software

The absorbance measurements were obtained using a double beam spectrophotometer (Shimadzu, 1650 model UV–Vis Spectrophotometer, Japan) with 1-cm quartz cells, The data treatment and chemometric calculations were carried out using MATLAB® (version 6.0). The pH measurements were made with a Hana (Germany) pH meter using a companied glass electrode. The surface functional groups of oxidized MWCNTs were characterized by Fourier transform infrared spectra (FT-IR). The sample for the FT-IR measurement was placed shaped like a film on KBr pellet at room temperature (Shimadzu, FT-IR Spectrophotometer, Japan).

19.3 Pesticides and Their Stock Solutions

All chemicals used in the experiments were purchased in analytical purity and used without any purification. The MWCNT was purchased from Nutrino. The outer diameter of CNT was between 20 and 40 nm. Standard pesticides were purchased from Kingquenson Corporation of China. Stock standard solutions containing 1000 mgL^{-1} of pymetrozine and 1000 mgL^{-1} of mancozeb were prepared by diluting commercial standard solutions with distilled water. Solutions and working standards were prepared daily by appropriate dilution of corresponding standard solutions.

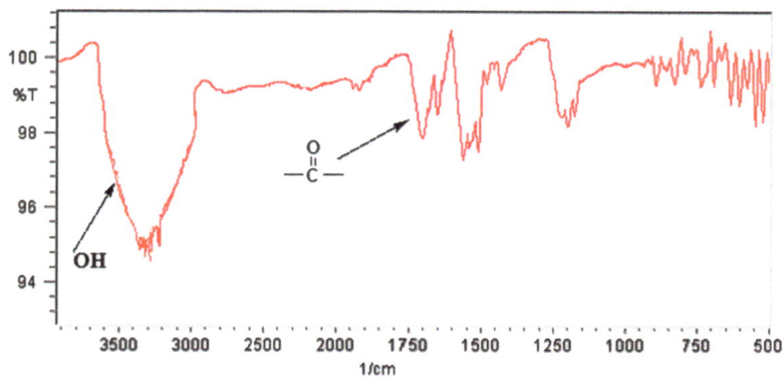

Fig. 19.1 IR spectra of opened MWCNT

19.3.1 Characterization of Oxidized MWCNTs

MWCNTs were opened according to reported procedures in literatures (Hamilton and Gemperline 1990). Briefly, MWCNTs (2 g) were added to 40 ml of sulfuric and nitric acid mixture (3/1) in a reaction flask and refluxed for 24 h at 120 °C. The mixture was cooled and diluted by distilled water and it was filtrated; then product was washed by distilled water. Figure 19.1 shows the FT-IR spectra of opened MWCNTs. Absorbance band of alcoholic hydroxyl functional groups, carbonyl groups, and C=C aromatic appeared at 3200–3500, 1701, and 1590 and 1400 cm^{-1}, respectively. These surface functional groups provide active sites. MWCNT-COOH was not soluble in water and any organic solvent.

19.3.2 Solid-Phase Extraction Procedures

Oxidized MWCNT-packed cartridges were prepared by empty polypropylene cartridges; the adsorbent was washed many times with double distilled water and then dried at 50 °C. 500 mg of functionalized MWCNTs weighted exactly and loaded in a typical 6.0 ml polyethylene extraction tube. The adsorbent was then conditioned by washing with 5 ml HCL+HNO_3 diluted mixture followed by double distilled water. Then several condition parameters, such as sample loading time, eluent species and eluent volume, pH of solution, and concentration of acid, were optimized to achieve good sensitivity and precision for the extraction and simultaneous determination of analyst.

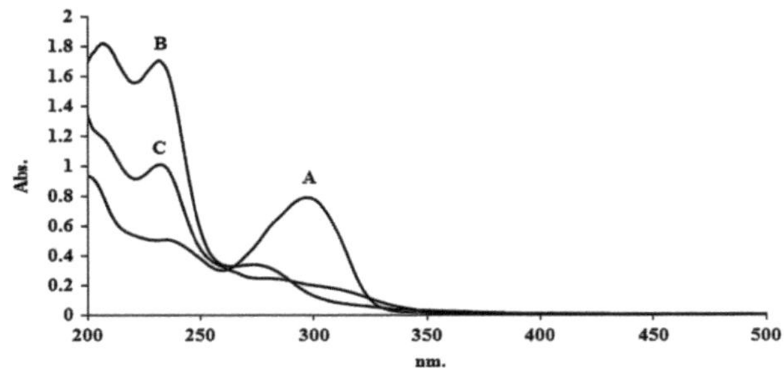

Fig. 19.2 Absorption spectra of pesticides. Absorption spectra of (**a**) Pymetrozine, (**b**) Mancozeb, and (**c**) mixture of two pesticides

19.4 Findings

19.4.1 Spectral Overlap and Importance of Multivariate Calibration

The absorption spectra of pesticide and their mixture are given in Fig. 19.2. As indicated in Fig. 19.2, the pesticides were active within the spectral region 200–350 nm. As can be seen in Fig. 19.2, the spectra of pesticides were overlapped within the spectral region (200–350 nm) and no wavelength can be found where any of the solutes is the only absorber. This indicates that conventional calibration procedures would have a limited application for quantitative determination.

19.5 Optimization of Variables

The effect of reaction variables was studied for separation of mancozeb and pymetrozine individually, and their optimum values were selected. There are many chemical variables that affect the absorption characteristics, stability, and hydrolysis of pesticides in water. Among these variables, pH and eluent are the most influential and, therefore, should be investigated. The influence of pH on the absorption separation of pesticides was studied in the pH range 2–8. Maximum adsorption was obtained in pH = 2, so it was selected as optimum pH for both pesticides. The initial pH was adjusted using 0.01 M HCl or 0.01 M NaOH solutions. Also, eluent type is very important in solid-phase extraction. The effect of different eluents was studied. Acetone, methanol, acetonitrile, n-hexane, acetone +HCL, acetone+ n-hexane, acetonitrile+methanol were used as eluent. Adsorption increased significantly with acetone + HCL as eluent. According to Fig. 19.3, extraction reaches to maximum using acetone + HCL as eluent.

The scheme of extraction process is shown in Fig. 19.4 in three steps. As can be seen, in the first step, pesticide adsorbs on the functionalized CNTs as solid-phase extractant, in the next step, it is desorbed and ion exchange occurs, and then in final step, pesticide elution and separation is completed.

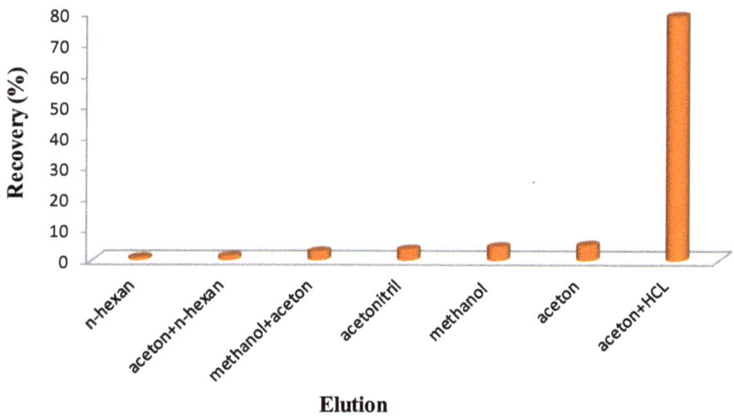

Fig. 19.3 Effect of different eluents of extraction

The effect of acid concentration was studied on extraction process. As can be seen in Fig. 19.5, the concentration of 3 M of HCl is appropriate for extraction process. Optimized conditions are shown in Table 19.1.

19.6 Univariate Calibration

Under the optimum conditions, calibration graphs were constructed for mancozeb and pymetrozine individually by plotting absorbance change values at wavelength function of the analyte concentration. The calibration graphs were linear in the range of 10–600 and 3–168 ($\mu g.L^{-1}$) for pymetrozine and mancozeb, respectively (Fig. 19.6). The detection limit for mancozeb and pymetrozine were 2.3 and 8.2 ($\mu g.L^{-1}$), respectively. RSD% of method was less than 2 % for mancozeb and 3.2 % and for pymetrozine.

19.7 Multivariate Calibration

The first step in the simultaneous determination of mancozeb and pymetrozine by PLS involves constructing the calibration set for the binary mixtures of them. Quantity of 22 binary mixtures was selected as calibration set. Their composition was randomly designed to obtain more information from the calibration procedure. Under these conditions, the calibration models were obtained. The obtained model was validated with eight synthetic mixture sets containing mancozeb and pymetrozine in different proportions (prediction set) that were randomly selected (Table 19.2).

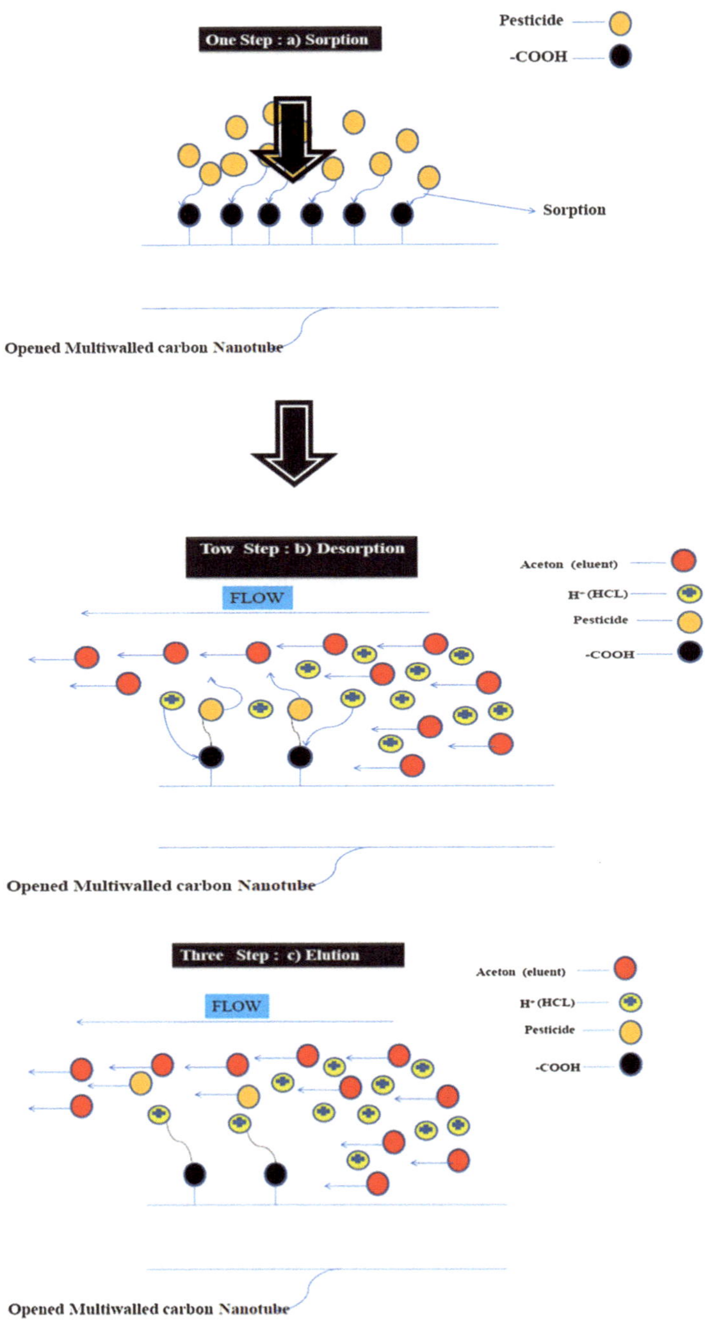

Fig. 19.4 Extraction process of pesticide particles

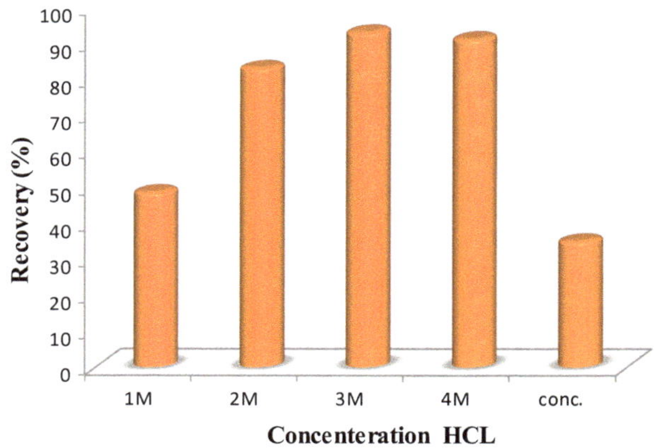

Fig. 19.5 Effect of different concentration of HCL on extraction

Table 19.1 Optimization condition of separation pesticides by solid-phase extraction

Pesticide	Transmittal time (min)	Eluent	Eluent volume (ml)	pH
Mancozeb	16	Acetone+HCL	7	2
Pymetrozine	16	Acetone +HCL	7	2

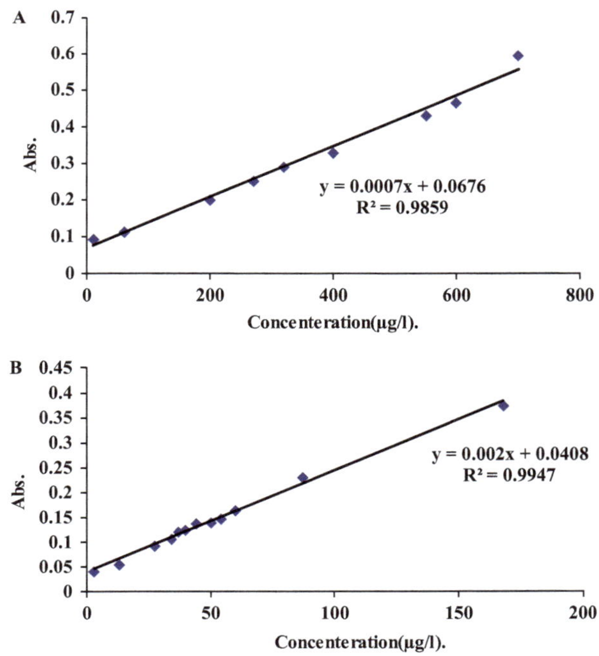

Fig. 19.6 Calibration curve of pesticides (**a**) Pymetrozine and (**b**) Mancozeb

Table 19.2 Concentration data for the different mixtures used in the calibration set and Prediction set for the simultaneous determination of pymetrozine and mancozeb

Calibration set (μg/l)		Prediction set (μg/l)	
Mancozeb	Pymetrozine	Mancozeb	Pymetrozine
20	100	20	200
20	150	40	300
20	200	60	250
20	300	60	150
40	50	80	150
40	100	100	50
40	200	100	200
40	250	120	300
60	50		
60	100		
60	200		
60	300		
80	50		
80	100		
80	200		
80	250		
100	100		
100	150		
100	250		
120	50		
120	150		
120	200		

19.8 Selection of Optimal Number of Factors

To select the number of factors in the PLS algorithm, a cross-validation method, leaving out one sample at a time, was employed (Abdollahi 2001). The prediction error was calculated for each component for the prediction set, which are the samples not Participating in the construction of the model. The optimum number of factors (latent variables) to be included in the calibration model was determined by computing the predication error sum of squares (PRESS) for the first variable, which built the PLS-1 modeling in the calibration step, and then, another latent variable was added for the model building and the PRESS was calculated again. This process was repeated for one to 8 latent variables, which were used in the PLS-1 modeling. The predicted concentrations of the compounds in each sample were compared with the already known concentration, and the prediction error sum of squares (PRESS) was calculated. The optimal number of factors for pymetrozine and mancozeb was obtained 5 and 4, respectively (Fig. 19.7). The results obtained are given in Table 19.3; the prediction error of a single component in the mixture was calculated as the relative standard error (R.S.E) of the prediction concentration.

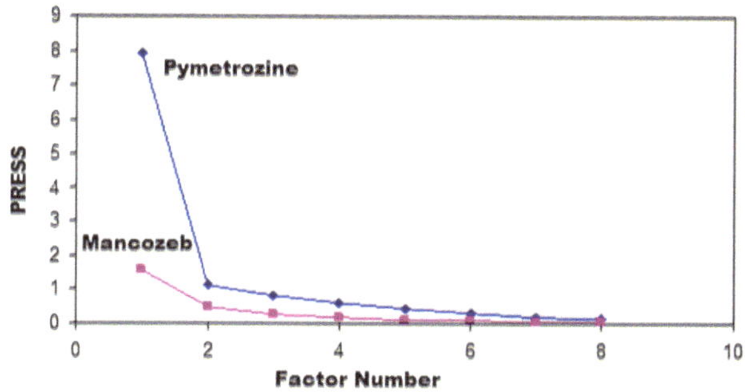

Fig. 19.7 Plot of PRESS against the number of factors for pymetrozine and mancozeb

Table 19.3 Composition of synthetic samples, their predictions by PLS-1 model, and statistical parameters for system

Pesticides (μg/l)		Prediction (μg/l)		Recovery (%)	
Mancozeb	Pymetrozien	Mancozeb	Pymetrozine	Mancozeb	Pymetrozine
20	200	21/78	200/57	108/9	100/28
40	300	43/02	275/36	107/55	91/78
60	250	61/83	249/01	103/05	99/6
60	150	52/69	153/49	78/81	102/32
80	150	86/38	155/08	107/97	103/38
100	50	95/34	58/54	95/34	117/08
100	200	107/11	215/32	107/11	107/66
120	300	133/3	268/05	111/08	89/35
Mean recovery (%)				103/6	101/43
R.S.E.$_{\cdot single}$				8/4	7/3
R.S.E.$_{\cdot total}$				7/4	7/4

$$\text{R.S.E}(\%) = \left[\sum\nolimits_{J-1}^{N} \left(\hat{C} - C_J\right)^2 / \sum \left(C_J\right)^2\right]^{1/2} \times 100 \tag{1}$$

where N is the number of samples, Cj the concentration of the component in the jth mixture, and $\hat{Cj}$ is the estimated concentration. The total prediction error of N samples is calculated as follows:

$$\text{R.S.E.}_{\cdot t}(\%) = \left[\frac{\sum\nolimits_{i=1}^{M} \sum\nolimits_{J-1}^{N} \left(\hat{C} - C_J\right)^2}{\sum\nolimits_{i=1}^{M} \sum\nolimits_{J-1}^{N} \left(C_{iJ}\right)^2}\right]^{1/2} \times 100 \tag{2}$$

where M is the number of components, Cij is the concentration of the ith component in the jth sample, and $\hat{Cij}$ its estimation. Table 19.3 also shows reasonable single and total relative errors for such a system.

Table 19.4 Determination of pymetrozine and mancozeb in water samples of khoramabad city of lorestan

Compounds (μg/l)		Prediction (μg/l)		Recovery (%)	
Mancozeb	Pymetrozine	Mancozeb	Pymetrozine	Mancozeb	Pymetrozine
20	200	21.95	218.65	109.75	109.32
40	300	43.65	270.52	109.12	90.17
100	200	108.25	208.78	108.25	104.39
Average (%)				109.04	101.29

19.9 Application

To evaluate the analytical applicability of the proposed method, it was applied to the simultaneous determination of pesticides in water samples, so they were prepared by adding known amounts of pymetrozine and mancozeb. The water samples were found to be free from pymetrozine and mancozeb. The results are given in Table 19.4. The results show that the PLS-1 model is able to predict the simultaneous determination of pymetrozine and mancozeb concentrations in such samples.

The results show that SPE-PLS-1 is an appropriate method for separation and simultaneous determination of pesticides in water samples. The obtained results in this work allow us to conclude that both components of the binary mixture are accurately determined by spectroscopy with PLS-1 calibration. Experimental design and the nature of the validation set are seen to be of major importance when assessing the quality of the model. If a calibration data set is correlated, it may predict itself fairly well. The results in Table 19.4 show that PLS-1 can appropriately model multicomponent systems and predict unknown analyte concentrations with satisfactory results. Accuracy, precision, reproducibility, sensitivity, and line arrange for the proposed method are satisfactory.

References

Abdollahi H (2001) Simultaneous spectrophotometric determination of chromium (VI) and iron (III) with chromogenic mixed reagents by H-point standard addition method and partial least squares regression. Analytica Chimica Acta 442:327–336

Afkhami A, Sarlak N, Zarei AR (2007) Simultaneous kinetic spectrophotometric determination of cyanide and thiocyanate using the partial least squares (PLS) regression. Talanta 71:893–899

Blanco M, Iturriaga H, Maspoch S, Tarin P (1989) A simple method for spectrophotometric determination of two-components with overlapped spectra. J Chem Educ 66:178–180

Cai Y, Jiang G, Liu J, Zhou Q (2003) Multiwalled carbon nanotubes as a solid-phase extraction adsorbent for the determination of bisphenol A, 4-n-nonylphenol, and 4-tert-octylphenol. Anal Chem 75:2517–2521

Dong Y, Liu Z, Chen L (2012) Removal of Zn (II) from aqueous solution by natural halloysite nanotubes. J Radioanal Nucl Chem 292:435–443

Frackowiak E, Beguin F (2002) Electrochemical storage of energy in carbon nanotubes and nanostructured carbons. Carbon 40:1775–1787

Hamilton JC, Gemperline PJ (1990) Mixture analysis using factor analysis. II: Self-modeling curve resolution. J Chemometrics 4:1–13

Iijima S (1991) Helical microtubules of graphitic carbon. J Nature 354:56–58

Martens H, Naes T (1992) Multivariate calibration. Wiley, Chichester

Msimanga HZ, Charles MJ, Martin NW (1997) Simultaneous determination of aspirin, salicylamide, and caffeine in pain relievers by target factor analysis. J Chem Educ 74:1114

Rius A, Callao M, Ferré J, Rius F (1997) Assessing the validity of principal component regression models in different analytical conditions. Analytica Chimica Acta 337:287–296

Sarlak N, Anizadeh M (2011) Simultaneous kinetic spectrophotometric determination of sulfide and sulfite ions by using an optode and the partial least squares (PLS) regression. Sens Actuat B: Chem 160:644–649

Tomlin CT (2003) The pesticide manual. The British Crop Protection Council, Farnham

Zhou Q, Xiao J, Wang W, Liu G, Shi Q, Wang J (2006) Determination of atrazine and simazine in environmental water samples using multiwalled carbon nanotubes as the adsorbents for preconcentration prior to high performance liquid chromatography with diode array detector. Talanta 68:1309–1315

Zupan J, Gasteiger J (1993) Neural networks for chemists: an introduction. Wiley, New York

Chapter 20
Nanomaterials–Plant–Soil System: Challanges and Threats

Jośko Izabela, Stefaniuk Magdalena, and Oleszczuk Patryk

20.1 Transfer Routes of NMs to Plants

NMs' intra-plant transfer might occur via various paths, which is conditioned by both NM type and plant species (Rico et al. 2011) (Fig. 20.1). Just like other compounds, NMs may penetrate cell wall pores to enter plants. Only the smallest NMs may penetrate this natural tunnel as average diameter of cell wall pores is estimated at $\leq$10 nm (Carpita et al. 1979; Chesson et al. 1997). Nevertheless, NMs can induce creating new pores or enlarging those that already exist in cell walls, which leads the way for larger NMs. Such changes may be triggered by NMs' interaction with proteins or polysaccharides (Tan et al. 2009). Additionally, NMs may be transported into plant cells through endocytosis, i.e., via channels or using carrier proteins (aquaporins) (Gardea-Torresdey et al. 2014).

A plant growth in the soil basis also determines the ability of NMs to penetrate inside plants together with the bound natural organic matter (NOM) (Rico et al. 2011). On the other hand, NOM may reduce the availability of NMs for plants. Studies by Zhu et al. (2008) proved there was no uptake of nano-Fe_2O_3 by pumpkin growing in soil or sand, whereas the accumulation of those NMs was discovered in hydroponic cultivation. NMs' hampered transport into plants might also be connected with the plant physiology. Roots of some species produce mucilage (Schwab et al. 2015), which protects a plant against numerous stress factors, such as a presence of pollutants, including NMs. Studies by Kurepa et al. (2010) show

J. Izabela (✉)
Institute of Plant Genetics, Breeding and Bionanotechnology, University of Life Sciences in Lublin, Lublin, Poland
e-mail: izabela.josko@up.lublin.pl

S. Magdalena • O. Patryk
Department of Environmental Chemistry, University of Marie Skłodowska-Curie, Lublin, Poland

M. Ghorbanpour et al. (eds.), *Nanoscience and Plant–Soil Systems*, Soil Biology 48,
DOI 10.1007/978-3-319-46835-8_20

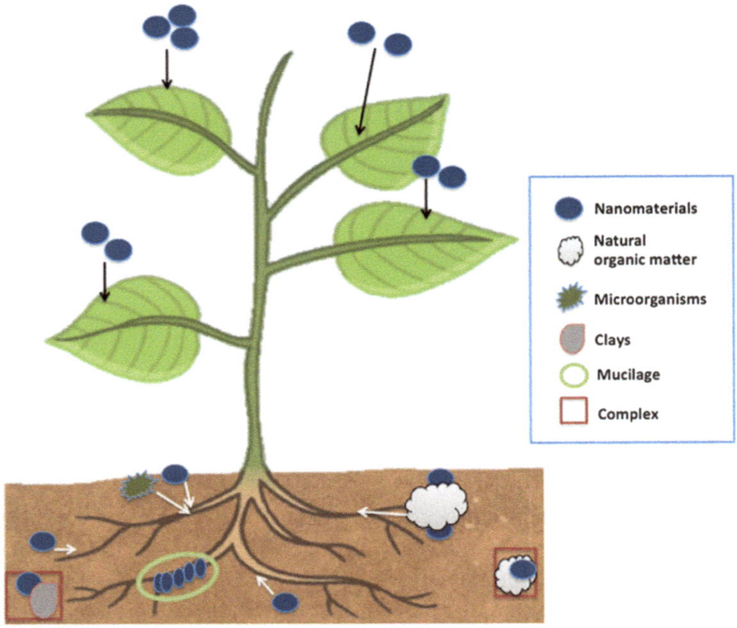

Fig. 20.1 Scheme of different transfer routes of NMs to plant

Arabidopsis thaliana roots to have stopped the uptake of nano-TiO_2 inside root cells by producing mucilage. Furthermore, investigation by Watanabe et al. (2008) demonstrated mucilage produced by *Melastoma malabathricum* to be conducive to accumulating nano-Al_2O_3. In a soil environment, soil flora determined by soil properties could also be relevant to NM transfer. A presence of some symbiotic microorganisms might support NM accumulation, probably through perforate cuticule and cell walls (Whiteside et al. 2009). However, Feng et al. (2013) proved arbuscular mycorrhizal fungi to cause a decrease in the nano-Ag and nano-Fe_2O_3 uptake through clover, *Trifolium repens*. Studies explaining NM transfer into plants in soils are still scarce, which makes it difficult to specify mycorrhiza mechanisms in response to stress caused by NM presence.

In agricultural cultivation, NMs (after intra-foliage application of NM-including fertilizers and pesticides) may penetrate into plants through stomal openings (their diameters might even reach 40 nm, depending on a plant species) (Eichert and Goldbach 2008) or through the bases of trichomes (Corredor et al. 2009). NMs' properties (type, degree of aggregation and dissolution, surface charge) affect their potential for permeating into plants. A strong tendency for NMs to aggregate results in their lesser mobility and the size of aggregates prevents them from breaching the pore barrier in cell walls. Root surface of most plants is characterized by negative charge, while NMs typically have positive charge, which results in NMs' adhesion to root surfaces of plants (Zhu et al. 2012). NMs' adsorption on plant surfaces may block cell pores, which hampers transport of nutrients and water, as well as pollutants, including NMs. In soil conditions, NM adhesion to root surfaces can be reinforced by the presence of humic acids (Navarro et al. 2012). Interestingly, research by Schwabe et al. (2013) proved fulvic acid to be conducive to CeO_2 dispersion, which resulted in a decreased adhesion to plant roots.

20.2 The Impact of NM on the Growth and Development of Plants

The risk of plants contacting NMs requires an assessment of potential toxicity of NMs (Rico et al. 2011). Studies that have been conducted to date into NM phytotoxicity mainly focused on hydroponic cultures, where toxic potential was assessed for carbon-based NMs (fullerenes, single-walled carbon nanotubes [SWCNTs], multi-walled carbon nanotubes [MWCNTs]), and inorganic NMs (metal oxides and metals). There is even less data concerning the influence of NMs on plants in soil environment, where a biological reaction of plant depends not only on concentration and type of NMs and the plant species but also on physical and chemical properties of soils. Seed germination and inhibition of root/shoot growth are the tests that to date have been most commonly used in investigations (Table 20.1) aimed at assessing the impact of NMs on plants in soil environment.

20.2.1 Toxicity of Nano-metals

Research on nano-metal toxicity for plants in soil environment is rather scarce. Literature merely features reports on the influence of few nano-metals (Fe, Ag, and Ni) on higher plants. Negative effect of nano-Ag on *Phaseolus radiatus* and *Sorghum bicolor* was proved by Lee et al. (2008), who observed slight inhibition in root growth in plants tested depending on the nano-Ag (100–2000 mg/kg) dose and failed to observe any correlation between the toxicity level and NMs' dose being used. Additionally, these researchers observed considerably lower nano-Ag toxicity for plants tested in soil matrix than in agar, which was probably connected with a decrease in the bioavailability of these particles in the soil matrix. A considerably greater inhibition of the root growth (up to 80 %), despite significantly lower doses (0.5–5 mg/kg) of nano-Ag in *Triticum aestivum,* was reported by Dimkpa et al. (2013b). Discrepancies between the findings by Lee et al. (2008) and Dimkpa et al. (2013b) probably resulted from researchers using different soils (OECD and sand, respectively), different lengths of test periods (5 and 14 days), and different sensitivity of plants tested for the presence of nano-Ag. In case of nano-Fe, various plant reactions were also observed in different soils (El-Temsah and Joner 2012). El-Temsah and Joner (2012) reported greater inhibition of seed

Table 20.1 Influence of NMs on different plants in soil environment

NMs	Size (nm)	Plant	Concentration	Exposure time	Type of soil	Effects	References
Nano-metals							
Fe	–	Ryegrass *(Lolium perenne)*, barley (*Hordeum vulgare)*,flax *(Linum usitatissimum)*	100–5000 mg/kg	5, 7 days	Clay loam, sandy loam	Seed germination was observed up to 500 mg/kg nZVI; at higher concentrations, no seed germination was observed for any of the species	El-Temsah and Joner (2012)
Ag	5–25	Mung bean *(Phaseolus radiates)*,sorghum *(Sorghum bicolor)*	100–2000 mg/kg	5 days	OECD	The growth of *P. radiates* and *S. bicolor* showed an approximate 20 % inhibition under the maximum exposure concentration of 2000 mg/kg nano-Ag	Lee et al. (2012)
	10	Wheat *(Triticum aestivum)*	0.5–5.0 mg/kg	14 days	Sand	Nano-Ag reduced the length of shoots and roots of wheat in a dose-dependent manner	Dimkpa et al. (2013a)
Ni	<100	Cress *(Lepidium sativum)*	10–1000 mg/kg	3 days	OECD, silt, sand	Toxicity of NMs was significantly determined by the soil type	Jośko and Oleszczuk (2013)
Nano-oxides of metals							
CeO_2	–	Radish *(Raphanus sativus)*	0–500 mg/kg	4 days 12 days	Loamy sand	at 500 mg/kg nano-CeO_2 significantly retarded seed germination but did not reduce the number of germinated seeds	Corral-Diaz et al. (2014a)
	8	Cilantro *(Coriandrum sativu)*	0–500 mg/kg	30 days	Organic potting soil	At 125 mg/kg, plants produced the longest roots, in other concentrations were also observed stimulation	Morales et al. (2013)
	8	Soybean (*Glycine max*)	0.1–1.0 g/kg	48 days	Organic farm soil	Nano-CeO_2 substantially entered the roots and root nodules and nearly paralyzing nodule-associated N_2 fixation	Priester et al. (2012)

	8	Cucumber (*Cucumis sativus*)	0–800 mg/kg	53 days	Loam sand:	Nano-ZnO and nano-CeO_2 did not show toxicity to cucumber plants	Zhao et al. (2013a)
ZnO	10						
	<50 <100	Cress (*Lepidium sativum*)	10–10,000 mg/kg	3 days	OECD, silt, sand	Toxicity of nanoparticles was significantly determined by the soil type	Jośko and Oleszczuk (2013)
	10	Green peas (*Pisum sativum*)	0–500 mg/kg	25 days	Sandy loam	Nano-ZnO increased root elongation	Mukherjee et al. (2014a))
	<50	Buck wheat (*Fagopyrum esculentum*)	10, 100, 1000 mg/kg	5 days	Loamy sand	Seedlings growth was significantly reduced in response to nano-ZnO	Lee (2012)
	<100	Wheat (*Triticum aestivum*)	125, 250, 500 mg/kg	7 days	Calcareous agricultural soil, acidic soil	Phytotoxicity was observed in acid but not in alkaline soils	Watson et al. (2015)
	<50	Soybean (*Glycine max*)	0, 50, 500 mg/kg	57, 65 days	OECD soil	Nano-ZnO negatively affected the developmental stages and reproduction of soybean plants in a soil microcosm	Yoon et al. (2014)
	<100	Cowpea (*Vigna unguiculata*)	500 mg/kg	4 weeks	Sandy clay	There was no significant difference in plant growth and accumulation	Wang et al. (2013)
SiO_2	40–150	Maize (*Zea mays*)	0–20 kg/ha	20, 40days	Sandy loam	Growth characteristics were influenced with increasing concentration of nano-SiO_2 up to 15 kg/ha^{-} whereas at 20 kg/hano significant increments were noticed	Suriyaprabha et al. (2012)
TiO_2	<21	Cress (*Lepidium sativum*)	10–10,000 mg/kg	3 days	OECD soil, silt, sand	The toxicity of nanoparticles was significantly determined by the soil type	Jośko and Oleszczuk (2013)
	20–100	Wheat (*Triticum aestivum*)	–	7 months	Loamy clay	Nano-TiO_2 were retained in the soil for long periods and primarily adhered to cell walls of wheat, additionally soil protease, catalase, and peroxidase activities were inhibited	

(continued)

Table 20.1 (continued)

NMs	Size (nm)	Plant	Concentration	Exposure time	Type of soil	Effects	References
	27	Cucumber (*Cucumis sativus*)	0–750 mg/kg	150 days	Sandy loam	Activity catalase in leaves increased and ascorbate peroxidase decreased compared to control, additionally observed increased total chlorophyll content in leaves	Servin et al. (2013)
CuO	<50	Wheat *(Triticum aestivum)*	500 mg/kg	14 days	Commercial white silica sand	Shoot and root length was reduced significantly compared the control	Dimkpa et al. (2012)
Nanocomposites							
CMC/Fe	–	Flax *(Linum usitatissimum)*, barley (*Hordeum vulgare*)	1 g/L	5 days	Sandy loam	Inhibition of germination and root growth of the tested plants was observed relative to control	El-Temsah and Joner (2013)
Fe/ZnO	10	Green peas *(Pisum sativum)*	0–500 mg/kg	5 days	Sandy loam	Nano-Fe/ZnO did not show visible signs of toxicity; iron doping may be considered as an effective approach to reduce the toxicity of nano-ZnO in higher terrestrial plants	Mukherjee et al. (2014b)

germination in sand soil than in clay soil during nano-Fe application. Additionally, diversified tolerance of tested plants (*Lolium perenne, Hordeum vulgare, Linum usitatissimum*) on nano-Fe presence was observed. *L. usitatissimum* proved to suffer the greatest germination inhibition, which might have been caused by its weakest shoot-producing potential.

On the other hand, root growth stimulation of *Lepidium sativum* was observed in case of nano-Ni applied to sand soil, irrespectively of the concentration used. Furthermore, having used reference soil OECD, researchers subsequently obtained stimulation for the lowest and the highest nano-Ni concentration (Jośko and Oleszczuk 2013). Diversified impact of NMs on plants in different soil matrices is probably connected with different proportions of nutrients present and soil colloids in soil environment, which could affect the behavior of NMs. These elements may specially impact the aggregation of NMs. The susceptibility of NMs to aggregation results in decreasing NMs mobility, consequently decreasing their bioavailability, which can finally reduce their toxicity (Hotze et al. 2010).

20.2.2 Phytotoxicity of Nano-Oxides of Metals

Literature of the subject also features reports concerning phytotoxicity of nano-oxides of metals in soil environment. Nano-ZnO and nano-CeO_2 appear to be most frequently tested NMs from this group. Yet, only few studies also present behavior of nano-SiO_2, nano-TiO_2, and nano-CuO in various plants. Phytotoxicity of nano-ZnO in soil depends on a number of soil properties, such as pH, organic matter content, and cation-exchange capabilities. The study demonstrates that nano-ZnO application into acidic soil triggers negative effects for various test plants, e.g., for *Fagopyrum esculentum* (Lee 2012), *T. Aestivum* (Watson et al. 2015), and *Glycine max* (Yoon et al. 2014). Negative influence of nano-ZnO on plants in acidic soil may be connected not only with its direct effect but also with the indirect effect of decreasing soil condition (Kim et al. 2011), including a decrease in the activity of soil enzymes such as dehydrogenase (Watson et al. 2014), urease, or phosphatase (Jośko et al. 2014). Preserving microbiological balance of soil may indirectly contribute to restricting plant growth, which in this case correlated with a length reduction of *Fagopyrum esculentum* roots (Lee et al. 2012). In case of nano-ZnO application into alkaline soil, no significant changes were observed, nor any stimulation of root length growth in comparison with control (Wang et al. 2013a; Mukherjee et al. 2014a; Watson et al. 2015). Additionally, an increase was reported in producing side roots in comparison with control (Watson et al. 2014) and a decrease in toxic effect on soil toxicity in comparison with hydroponic culture (Wang et al. 2013a). Mukherjee et al. (2013) reported a stimulation of *Pisum sativum's* root length growth. However, after 25 days, in contrast to results after 15 days, a decrease in chlorophyll content in leaves was observed. This means that Zn started to induce toxicity when the plants were at middle age (Mukherjee et al. 2014a).

Nano-CeO_2 is yet another nano-oxide widely tested in soil environment plants. The majority of published studies prove that the presence of nano-CeO_2 in soil fails to inhibit plant growth (Morales et al. 2013; Zhao et al. 2013a; Corral-Diaz et al. 2014a), and it is only with an increase in NM dose in soil that a delay in germination (Corral-Diaz et al. 2014b) and a decrease in crop output occurs (Morales et al. 2013; Zhao et al. 2013a). Soybean is an exception in whose case Priester et al. (2012) recorded not only a decrease in the crop yield but also an inhibition in plants' growth. This phenomenon was probably caused by blocking the N_2 bond by plants, especially with high concentration of nano-CeO_2.

Additionally, literature features only scarce information concerning toxicity of nano-SiO_2, nano-TiO_2, or nano-CuO for various test plants. The study shows that an addition of nano-SiO_2 into soils fails to exert a considerable influence of *Zea mays* germination and causes stimulation (with the dose of 15 and 20 kg/ha) of root growth and an increase in the leaf surface in comparison to control, which might promote photosynthesis activity (Suriyaprabha et al. 2012). However, a varied response of *Lepidium sativum* was reported in case of nano-TiO_2 depending on the type of soil. The greatest toxicity was observed for silt soil, while root growth stimulation occurred in case of sand soil (Jośko and Oleszczuk 2013). In case of nano-CuO (500 mg/kg) application, a growth inhibition was reported both for wheat root and stem in comparison with control (Dimkpa et al. 2012). The research proved that after the application of nano-CuO a release of dissolved Cu ions occurred into matrix, which was probably responsible for the inhibited root growth.

20.2.3 Toxicity of Nanocomposites

It is not only nano-metals and oxides that potentially pose environmental threat but also nanocomposites, which are created through modifying/improving NMs. Recent literature features information concerning phytotoxicity of modified zero-valence nano-iron (CMC/Fe) (El-Temsah and Joner 2013) and widely used zinc oxide (Fe/ZnO) (Mukherjee et al. 2014b).

Application of nano-iron into soils is a promising technology aimed at degrading numerous pollutants in this matrix; however, its influence on live organisms has not been fully understood. El-Temsah and Joner (2013) demonstrated that adding fresh CMC/Fe to soil in the form of suspension (with the condensation of 1 g/L) completely stops germination of plants such as barley or flax. Additionally, having used the column test, these researchers found that the upper (the closest to the place of application) layer of soil proved most toxic and the bottom layer least toxic after 5 days of NM application into soil. Root growth inhibition of flax and barley in soil taken from various layers of the column ranged from 17 to 45 %. However, one must account for the fact that soil in the experiment was additionally contaminated with DDT; thus, the toxicity effect might result not only from the operation of modified nano-iron. On the other Mukherjee et al. (2014b) observed that adding Fe/ZnO to soils did not cause visible signs of toxicity for *Pisum sativum*, such as necrosis, chlorosis, or withering but might affect physiological and biochemical processes occurring in plants (plant growth, chlorophyll content, and the number of free radicals). Nevertheless, the study proved that ZnO modification could bring about a decrease in toxic influence of NMs on plants by lessening their bioavailability to plants through a reduction of their solubility.

20.3 Accumulation and Translocation NM and Metal Components Inside Plants

Numerous potential ways for NMs permeating into plants can result in their bioaccumulation, intensity of which is determined by the NM type, plant species, and test design, in particular the type of contact and properties of the matrix soil (Priester et al. 2012; Morales et al. 2013; Wang et al. 2013a; Zhao et al. 2013a; Hernandez-Viezcas et al. 2013; Mukherjee et al. 2014b). Most of the studies have so far involved a content analysis of the metal component of NMs in plant tissues, without a distinction between nanoparticulate form and a dissolved ion. Nevertheless, both types of analyses are extremely useful, especially in the context of the quality of plant resources for the food industry (Gardea-Torresdey et al. 2014).

Investigations into the NM uptake have been so far conducted mainly in hydroponic cultures, where the behavior and bioaccessibility of NMs are not distorted by the influence of ingredients and matrices. Thus, NMs' uptake rates by plants growing in culture medium are considerably higher than by those living in soils (Ma et al. 2015). NMs are subject to complexation by NOM as well as clays, which might reduce their bioavailability (Dimkpa et al. 2012). However, Zhao et al. (2012) found considerably higher concentration of Ce in organic soil solution in comparison with unenriched soil. Recorded increase in Ce accumulation in roots of corn cultivated in organic soil might have resulted from Ce adsorption on root surfaces, which humic acid was supposed to reinforce. Furthermore, soil matrix may precipitate NMs' solubility, and then a plant is exposed to metal ions that make up NMs. Nano-ZnO added to acidic soil was not detected (Wang et al. 2013a) after one hour in the soil and cowpea (*Vigna unguiculata*). NM bioavailability and consequently its uptake may also by determined by pH. 200 times greater concentration of soluble Zn was recorded in acid soil than in alkaline soil, which translated into the 10 times greater Zn content in shoots in acid soil. Soil constituents may considerably affect the rate of NMs uptake and metal component of NMs. Alignates, which are constituents of organic matter, increased Ce accumulation in roots and shoots of corn cultivated in soil with nano-CeO_2 (Zhao et al. 2014a). Greater accumulation of Ce in corn tissues was observed with lower concentration of alginate. Soil macroelements can also determine NMs' transfer into plants. Study by Steward et al. (Stewart et al. 2015) also demonstrated the type of ingredient to be relevant, because whereas N and K contributed to Zn uptake by wheat, Ca reduced Zn accumulation in wheat shoots under nano-ZnO treatment. In case of nano-

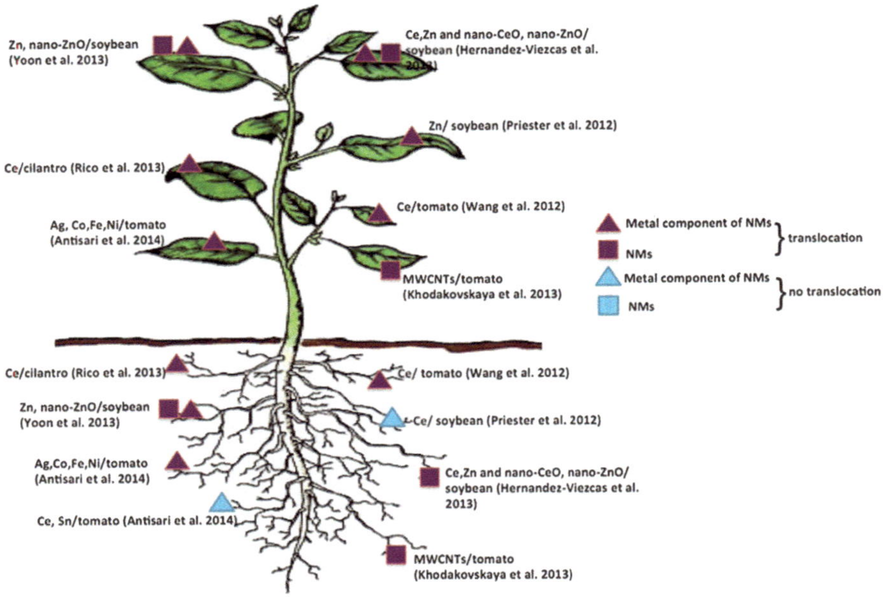

Fig. 20.2 Translocation of NM to above-ground parts of plants under root exposure

CuO, N, K, Ca salts also similarly influenced Cu loading in shoots, and these macroelements reduced Cu concentration in wheat tissues. A decrease in Zn and Cu concentration in plant tissue in the presence of macroelements might have resulted from the competition for binding and transport (Stewart et al. 2015). Furthermore, the species of a plant, and in particular the type of root system, is important for the potential NM absorption. Studies conducted into hydroponic cultures proved that nano-Cu accumulation in *Triticum aestivum* with the fibrous root system (thin and numerous roots—greater potential for NM penetration) was greater than in those with the taproot system (Lee et al. 2008).

Research into accumulation of NMs and their components in plants requires specifying NMs' accumulation places in individual parts of a plant, which is particularly important in the context of food transfer (Fig. 20.2). Most of the research that has been so far conducted on NM accumulation showed exposition to NMs to be directed to plant roots. Further transport of NMs into plants with the symplastic or the apoplastic channel may result in NM accumulation in above-ground plant parts (Fig. 20.2) (Morales et al. 2013). It is mostly NM type and its concentration that proves decisive for NM translocation in plants. Research by Larue et al. (2012) proved that Ti accumulation (with nano-TiO_2 with the 14 nm diameter) was mostly placed with the root parenchyma of rice and seeds of rape; furthermore, following the exposition to 25 nm nano-TiO_2, Ti accumulated in the root vascular cylinder. Moreover, Priester et al. (2012) proved Ce (with nano-CeO_2) to be taken up by soybean; however, no relevant translocation was observed into their above-ground parts. On the other hand, Zn (with nano-ZnO) taken up by plants was translocated and accumulated in stems and leaves. What is more, Zn concentration was higher in the above-ground parts than in roots. Concentration of both metals in respective plant parts was greater in case of higher NMs' doses (10 and 50 g/kg nano-ZnO; 50–100 g/kg nano-CeO_2) in comparison with the lowest dose (5 and 10 g/kg, respectively, for nano-ZnO and nano-CeO_2). Antisari et al. (2014) also observed considerable differences in the translocation rate of metal component of NMs (Ag, Co, Ni, CeO_2, Fe_3O_4, SnO_2, TiO_2) in tomato *Lycopersicon esculentum* cultivated in natural soil with peat (1:4 *v*/*v*): Ag (accumulated in stem, leaves, and fruits), Co (in stem and leaves), Fe (in fruits), and Ni (stem and fruits). Nevertheless, no translocation of Ce and Sn was observed. There are still merely scarce reports on the accumulation and translocation of carbon-based NMs in plants. Yet, Khodakovskaya et al. (2013) observed that MWCNTs to translocate inside plants as Raman spectroscopy confirmed their presence in tomato flowers.

Soil cultivated plants may be exposed to NMs not only through roots but also through seeds (having been sown) or through leaves (being treated with NM-component-based foliar fertilizers or plant protection products). Studies into hydroponic cultures concerning the influence of foliar application of NMs to plants exhibited accumulation of nano-CeO_2 in *C. Sativus* leaves; furthermore, NMs were translocated into other plant parts, including roots. Unfortunately, similar soil-involving studies in laboratory or natural conditions are rather scarce (Arora et al. 2012; Larue et al. 2014a; b). The research confirmed accumulation of nano-Ag (Larue et al. 2014a) and nano-TiO_2 (Larue et al. 2014b) in *Lactuca sativa* leaves, nano-Ag in *C. Sativus* leaves and fruits (Shams et al. 2013), and nano-Au in *Brassica juncea* rape (Arora et al. 2012). Moreover, Kole et al. (Kole et al. 2013) observed accumulation and translocation of fullerol to petiole, leaves, flowers, and fruits of bitter melon (*Momordica charantia*) after the seeds have been exposed to carbon-based NMs.

NM bioaccumulation presents the risk of food transfer. Nonetheless, research into this problem has not been conducted with the use of plants. Solely available findings including food chain stimulation confirmed Ce transfer from the producer (zucchini leaves) to primary consumers (crickets, *Acheta domesticus*); however, Ce content in secondary consumer (wolf spiders) was nonquantifiable (Hawthorne et al. 2014).

20.4 Biotransformation of NM

NMs transfer into plants determines their contact with various plant metabolites, which can lead to NM biotransformation. Synchrotron (XAS—X-ray absorption spectroscopy) and electron microscopy (TEM—transmission electron microscopy, SEM—scanning electron microscopy with EDS—energy-dispersive X-ray spectroscopy) analysis is most often employed to determine speciation of NMs in plants and consequently to verify NM biotransformation. NM biomodifications mostly depend on the NM type as well as the plant species and its physiology. Some NMs may be resistant to phytomodifications. Despite accumulation and translocation of nano-TiO_2 in cucumber cultivated on soil, their biotransformation was not reported (Servin et al. 2013). Larue et al. (2012) neither found any modifications of nano-TiO_2 accumulated in *T. aestivum* in hydroponic cultivations. On the other hand, Zhao et al. (2012) failed to confirm changes in the speciation of nano-CeO_2 in soil-cultivated corn. At the same time, insignificant biotransformations of nano-CeO_2 in *C. sativus* to cerium phosphate (in roots) and carboxylates (in shoots) (Zhang et al. 2012) were observed in hydroponic cultures.

In case of nano-ZnO, biotransformations were found in various soil-cultivated plant species. However, it must be pointed out that nano-ZnO is prone to being dissolved in soil (Dimkpa et al. 2013a), which means that it was probably Zn ions that interacted with plant metabolites. Depending on plant species, changes were found in chemical speciation of nano-ZnO, especially to Zn-citrate in soybean (Hernandez-Viezcas et al. 2013), Zn-citrate, histidine, and phytate in cowpea (Wang et al. 2013a), and Zn-phosphate in wheat (Dimkpa et al. 2012). Moreover, nano-CuO in wheat underwent partial biotransformation to Cu(I)-sulfur complexes (Dimkpa et al. 2012). Shi et al. (2014) also reported that nano-CuO may undergo modifications in *Elsholtzia splenders*, which confirmed the occurrence of nano-CuO in the form of Cu-alginate, oxalate, and cysteine. Research by Peng et al. (2015) also showed nano-CuO biotransformation in *Oryza sativa*, where it combined into cysteine citrate and phosphate ligands; furthermore, nano-CuO was reduced to Cu (I). NM biotransformations were also examined after foliar application of NMs. The study showed nano-Ag to occur mainly in the form of Ag-GSH (Larue et al. 2014a). On the other hand, carbon-based NMs (MWCNTs and fullerol) did not undergo any modifications inside tomato (Khodakovskaya et al. 2013) and bitter melon (Kole et al. 2013), respectively.

20.5 Effect of NMs on Crops

The impact of NMs on plants may be manifested by a change in mineral composition of plant, which results from changes in bioavailability of micro- and macroelements in soils or a competition of metal components of NMs and nutrients (Fig. 20.3). Research into uptake modification of nutrients was conducted using mostly inorganic NMs (nano-Au, nano-CeO_2, nano-TiO_2, nano-ZnO) on various crop plants (cucumber, corn, soybean, wheat, rice, barley) with one type of soil (Arora et al. 2012; Servin et al. 2013; Rico et al. 2013, 2014; Zhao et al. 2014b). Depending on a plant and its parts, differences in the content of some nutrients were observed, which suggests not only the influence of NMs on the uptake of macro- and microelements but also on their transport and positioning within individual plants (Rico et al. 2014).

The presence of NMs in soil and the risk of their accumulation and translocation in plants give rise to the question of their influence on the crop yield. NMs may

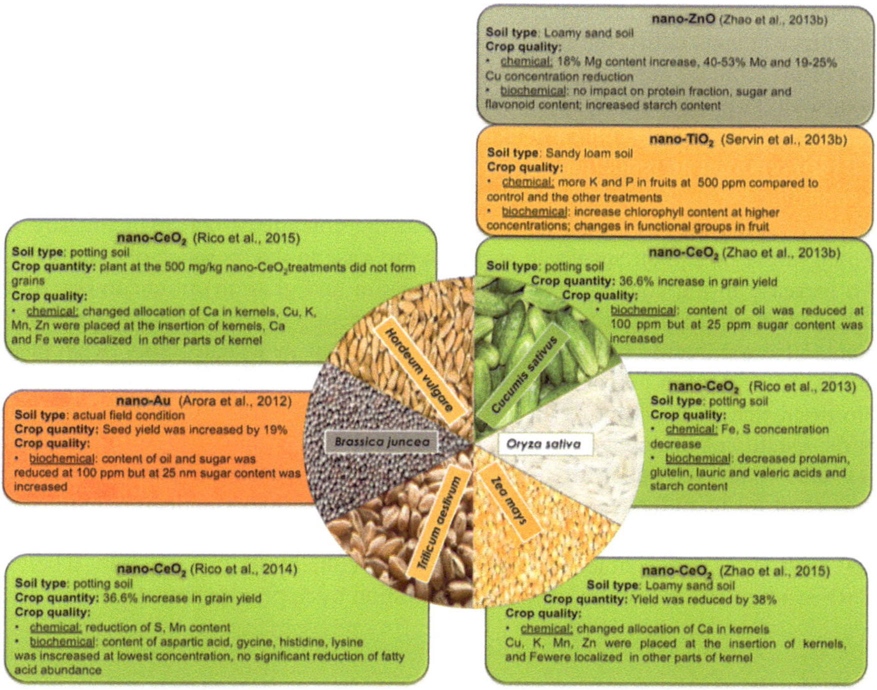

Fig. 20.3 Impact of NMs on quality and quantify crop of different plant species

affect the quality of crops through modifications in their mineral composition or biochemical properties, which determines nutritional values and taste (Fig. 20.3). These issues are investigated more and more often, and most studies concern the influence of nano-CeO_2 on plant edible parts (Rico et al. 2013, 2014, 2015; Zhao et al. 2013a;). Although no Ce presence was found in wheat grains (Rico et al. 2014), a decrease of S and Mn was observed in grains. Nano-CeO_2 was also responsible for modifications in micro- and macroelement composition in grains of rice and barley (Rico et al. 2015) and in fruits of cucumber (Zhao et al. 2013a). However, changes in the mineral composition of crops refer to various nutrients, and causes for these modifications have not yet been established. Zhao et al. (2014b) suppose that an increase in Mg content in cucumber fruits cultivated in soil with nano-CeO_2 and nano-ZnO might have been caused by inducing gene expression of aquaporins or other channels, which support Mg uptake and transport within a plant. On the other hand, Mo content was considerably reduced, which may have resulted from changes because of NMs' presence in the production of organic acids (citrate or malate) responsible for lesser Mo uptake in a plant. Servin et al. (2013) also observed an increase in the P and K content in cucumber fruits because of nano-TiO_2, and pointed out that these NMs functioned similarly to plant hormones (e.g., cytokinins), which may alter P and K uptake.

Not only does recent research focus on changes in mineral composition of grains or fruits of plants cultivated in soils with NMs but also on changes in biochemical properties (content and composition of amino acids, flavonoids, phenols, carbohydrates, or fatty acids), which determine properties of these crops. Studying the influence of nano-CeO_2 on carbohydrates' content in cucumber fruits, Zhao et al. (2014b) reported no reduction in the sugar content (glucose and fructose), which is responsible for the sweet taste of these fruits. However, with higher concentrations of nano-CeO_2, they reported an increase in the content of non-reducing sugars, which might be a plant answer to stress, in this case to the presence of nano-CeO_2. An increase in the content of proteins (globulin and glutelin) in cucumber fruits (Zhao et al. 2012a) might have been a similar reaction to stress. Research on other plants also exhibited changes in amino acid content in wheat grains (Rico et al. 2014), oil content in mustard seeds (Arora et al. 2012), and fatty acid content and profile in rice grains (Rico et al. 2013), which was triggered by various inorganic NMs. Quality assessment of *C. Sativus* fruit exposed to nano-CeO_2 and nano-CuO was also conducted under foliar application of NMs. The size of particles was less important for its influence on fruit quality after being foliage-applied than through root-based exposure. Both nano-CeO_2 and bulk-CeO_2 at the concentration of 200 mg/L were responsible for a decrease of Zn content (to 25 %) in fruit, whereas CuO resulted in a reduction of Mo concentration by approximately 50 %.

20.6 Effect of Soil Contaminants on Toxicity of NMs

In soils, NMs may interact with various components of the matrix, also with other pollutants, and in agrosystems, also with components of pesticides. NMs coexistence with other contaminants may translate into NM behavior and consequently into their toxicity as well as NM and contaminant accumulation. Few studies, mostly conducted with carbon-based NMs, exhibited a decrease in the uptake of weathered pesticides (chlordane, DDx-DDT with metabolites) by 21–80 % by zucchini, corn, tomato, and soybean in the presence of MWCNTs (De La Torre-Roche et al. 2013). An influence of NMs on the uptake of contaminants may depend on a plant species and a type of pollutant. For instance, no uptake of DDx by corn and tomato was reported in the presence of C_{60}, while the accumulation of chlordane in tomato and soybean increased by approximately 35 % (De La Torre-Roche et al. 2013). Other research also exhibited an increased accumulation of p,p′-DDE by 30–65 % in tomato, zucchini, and soybean in the presence of C_{60} (De La Torre-Roche et al. 2012). On the other hand, further studies also showed an increase in the accumulation, whereas nano-Ag decreased accumulation of p,p′-DDE in soybean and zucchini, which might have resulted from nano-Ag acting as an inhibitor of aquaporins, thus hampering pesticide transport. Moreover, Jośko and Oleszczuk (2013) studied influences of interactions between NMs on their phytotoxicity in soils. Their findings show that nano-Ni and nano-TiO_2 act antagonistically on nano-ZnO, which manifested in reducing the nano-ZnO-induced effect of root growth inhibition in cress *L. sativum*. NMs' antagonism might have resulted from nano-ZnO adsorption by other NMs, which reduced their bioavailability and phytotoxicity. The study by Dimkpa et al. (2015) showed a decrease in shoot and root growth inhibition in *Phaseolus vulgaris* when caused by nano-Cuo in the presence of nano-ZnO (1:1). These authors also highlight a potentially stimulating effect of nano-ZnO on plants, which could have decreased a toxic effect of nano-CuO.

20.7 Benefits for Plants

The effect of NMs on plants cultivated in soil is also studied in the light of their agricultural usage. NMs have become the focus of research into their potential for plant protection and fertilizing, mainly owing to their unique properties (in particular, their greater specific surface area) (Servin et al. 2015). Abovementioned investigations proved that some NMs, especially at low concentrations, stimulate plant growth and improve their biochemical parameters. There are numerous examples of beneficial (from the agricultural viewpoint) morphological and physiological NM-induced changes in plants. Positive influence of NM was typically reduced to a general increase in biomass (Jaberzadeh et al. 2013; Alidoust and Isoda 2013), growth of roots and shoots (Prasad et al. 2012; Raliya and Tarafdar 2013; Jośko and Oleszczuk 2013; Alidoust and Isoda 2013; Tarafdar et al. 2014), crop yield (Arora et al. 2012; Raliya and Tarafdar 2013; Rico et al. 2014), and biochemical parameters concerning chlorophyll content and photosynthetic rate (Linglan et al. 2008; Prasad et al. 2012; Raliya and Tarafdar 2013; Tarafdar et al. 2014). Inorganic NMs may release microelement metals (such as Zn, Cu, and Fe), which take part in plant physiological processes as cofactors. It is an increase in nutrient transport and a growth inhibition of potential plant pathogens that may be responsible for stimulating character of NMs (Gogos et al. 2012) apart from their role as microelements, which explains their usage in the production of a new generation of fertilizers and pesticides. Nonetheless, this field application of NMs has not been properly investigated yet, and many of its aspects have only been considered theoretically or with models, which makes it difficult to properly assess the usefulness of NMs for plant fertilization and protection.

20.8 Conclusion

The assessment of NMs influence on plants cultivated in soil has been conducted with a wideselection of NMs and numerous plant species. It was not until recently that studiesstarted that made use of soils with various physical and chemical properties that play animportant role in the behavior of NMs in soil and influence effects they exert on plants. Agricultural and industrial importance of plants necessitates the use of multi-parameter research into plants cultivated in soil with

NMs in order to account for complete life cycle of plants. Investigations that have been conducted so far both into phytotoxicity and potential of NMs for improving plant growth have been mostly based on individual parameters, which might fail to reflect actual benefits and losses resulting from the cultivation of plants in soils polluted or fertilized with NMs.

References

Alidoust D, Isoda A (2013) Effect of γFe_2O_3 nanoparticles on photosynthetic characteristic of soybean (Glycine max (L.) Merr.): foliar spray versus soil amendment. Acta Physiol Plant 35:3365–3375

Antisari LV, Carbone S, Gatti A et al (2014) Uptake and translocation of metals and nutrients in tomato grown in soil polluted with metal oxide (CeO_2, Fe_3O_4, SnO_2, TiO_2) or me tallic (Ag, Co, Ni) engineered nanoparticles. Environ Sci Pollut Res 22:1841–1853

Arora S, Sharma P, Kumar S et al (2012) Gold-nanoparticle induced enhancement in growth and seed yield of Brassica juncea. Plant Growth Regul 66:303–310

Carpita N, Sabularse D, Montezinos D, Delmer DP (1979) Determination of the pore size of cell walls of living plant cells. Science 205:1144–1147

Chesson A, Gardner PT, Wood TJ (1997) Cell wall porosity and available surface area of wheat straw and wheat grain fractions. J Sci Food Agric 75:289–295

Corral-Diaz B, Peralta-Videa JR, Alvarez-Parrilla E et al (2014a) Cerium oxide nanoparticles alter the antioxidant capacity but do not impact tuber ionome in Raphanussativus (L). Plant PhysiolBiochem 84:277–285

Corral-Diaz B, Peralta-Videa JR, Alvarez-Parrilla E et al (2014b) Cerium oxide nanoparticles alter the antioxidant capacity but do not impact tuber ionome in Raphanussativus (L). Plant Physiol Biochem 84:277–285

Corredor E, Testillano PS, Coronado M-J et al (2009) Nanoparticle penetration and transport in living pumpkin plants: in situ subcellular identification. BMC Plant Biol 9:45

De La Torre-Roche R, Hawthorne J, Deng Y et al (2013) Multiwalled carbon nanotubes and c60 fullerenes differentially impact the accumulation of weathered pesticides in four agricultural plants. Environ Sci Technol 47:12539–12547

De La Torre-Roche R, Hawthorne J, Deng Y et al (2012) Fullerene-enhanced accumulation of p, p′-DDE in agricultural crop species. Environ Sci Technol 46:9315–9323

Dimkpa CO, Latta DE, McLean JE et al (2013a) Fate of CuO and ZnO Nano- and microparticles in the plant environment. Environ Sci Technol 47:4734–4742

Dimkpa CO, McLean JE, Britt DW, Anderson AJ (2015) Nano-CuO and interaction with nano-ZnO or soil bacterium provide evidence for the interference of nanoparticles in metal nutrition of plants. Ecotoxicology 24:119–129

Dimkpa CO, McLean JE, Latta DE et al (2012) CuO and ZnO nanoparticles: phytotoxicity, metal speciation, and induction of oxidative stress in sand-grown wheat. J Nanoparticle Res 14:1–15

Dimkpa CO, McLean JE, Martineau N et al (2013b) Silver nanoparticles disrupt wheat (*Triticumaestivum* L.) growth in a sand matrix. Environ Sci Technol 47:1082–1090

Eichert T, Goldbach HE (2008) Equivalent pore radii of hydrophilic foliar uptake routes in stomatous and astomatous leaf surfaces—further evidence for a stomatal pathway. Physiol Plant 132:491–502

El-Temsah YS, Joner EJ (2013) Effects of nano-sized zero-valent iron (nZVI) on DDT degradation in soil and its toxicity to collembola and ostracods. Chemosphere 92:131–137

El-Temsah YS, Joner EJ (2012) Impact of Fe and Ag nanoparticles on seed germination and differences in bioavailability during exposure in aqueous suspension and soil. Environ Toxicol 27:42–49

Feng Y, Cui X, He S et al (2013) The role of metal nanoparticles in influencing arbuscular mycorrhizal fungi effects on plant growth. Environ Sci Technol 47:9496–9504

Gardea-Torresdey JL, Rico CM, White JC (2014) Trophic transfer, transformation, and impact of engineered nanomaterials in terrestrial environments. Environ Sci Technol 48:2526–2540

Gogos A, Knauer K, Bucheli TD (2012) Nanomaterials in plant protection and fertilization: current state, foreseen applications, and research priorities. J Agric Food Chem 60:9781–9792

Hawthorne J, De la Torre RR, Xing B et al (2014) Particle-size dependent accumulation and trophic transfer of cerium oxide through a terrestrial food chain. Environ Sci Technol 48:13102–13109

Hernandez-Viezcas JA, Castillo-Michel H, Andrews JC et al (2013) In situ synchrotron x-ray fluorescence mapping and speciation of CeO_2 and ZnO nanoparticles in soil cultivated soybean (*Glycine max*). ACS Nano 7:1415–1423

Hotze EM, Phenrat T, Lowry GV (2010) Nanoparticle aggregation: challenges to understanding transport and reactivity in the environment. J Environ Qual 39:1909–1924

Jaberzadeh A, Moaveni P, Moghadam HRT, Zahedi H (2013) Influence of bulk and nanoparticles titanium foliar application on some agronomic traits, seed gluten and starch contents of wheat subjected to water deficit stress. Not Bot HortiAgrobot Cluj-Napoca 41:201–207

Jośko I, Oleszczuk P (2013) Influence of soil type and environmental conditions on ZnO, TiO_2 and Ni nanoparticles phytotoxicity. Chemosphere 92:91–99

Jośko I, Oleszczuk P, Futa B (2014) The effect of inorganic nanoparticles (ZnO, Cr_2O_3, CuO and Ni) and their bulk counterparts on enzyme activities in different soils. Geoderma 232–234:528–537

Khodakovskaya MV, Kim B-S, Kim JN et al (2013) Carbon nanotubes as plant growth regulators: effects on tomato growth, reproductive system, and soil microbial community. Small 9:115–123

Kim S, Kim J, Lee I (2011) Effects of Zn and ZnO nanoparticles and Zn^{2+} on soil enzyme activity and bioaccumulation of Zn in *Cucumis sativus*. Chem Ecol 27:49–55

Kole C, Kole P, Randunu KM et al (2013) Nanobiotechnology can boost crop production and quality: first evidence from increased plant biomass, fruit yield and phytomedicine content in bitter melon (*Momordica charantia*). BMC Biotechnol 13:37

Kurepa J, Paunesku T, Vogt S et al (2010) Uptake and distribution of ultrasmallanatase TiO_2 Alizarin red S nanoconjugates in *Arabidopsis thaliana*. Nano Lett 10:2296–2302

Larue C, Castillo-Michel H, Sobanska S et al (2014a) Foliar exposure of the crop *Lactuca sativa* to silver nanoparticles: evidence for internalization and changes in Ag speciation. J Hazard Mater 264:98–106

Larue C, Castillo-Michel H, Sobanska S et al (2014b) Fate of pristine TiO_2 nanoparticles and aged paint-containing TiO_2 nanoparticles in lettuce crop after foliar exposure. J Hazard Mater 273:17–26

Larue C, Laurette J, Herlin-Boime N et al (2012) Accumulation, translocation and impact of TiO_2 nanoparticles in wheat (*Triticumaestivum* spp.): influence of diameter and crystal phase. Sci Total Environ 431:197–208

Lee S (2012) Effects of soil-plant interactive system on response to exposure to ZnO nanoparticles. J Microbiol Biotechnol 22:1264–1270

Lee S, Kim S, Kim S, Lee I (2012) Effects of soil-plant interactive system on response to exposure to ZnO nanoparticles. J Microbiol Biotechnol 22:1264–1270

Lee W-M, An Y-J, Yoon H, Kweon H-S (2008) Toxicity and bioavailability of copper nanoparticles to the terrestrial plants mung bean (*Phaseolus radiatus*) and wheat (*Triticum aestivum*): Plant agar test for water-insoluble nanoparticles. Environ Toxicol Chem 27:1915–1921

Linglan M, Chao L, Chunxiang Q et al (2008) Rubisco activase mRNA expression in spinach: modulation by nanoanatase treatment. Biol Trace Elem Res 122:168–178

Ma C, White JC, Dhankher OP, Xing B (2015) Metal-based nanotoxicity and detoxification pathways in higher plants. Environ Sci Technol 49:7109–7122

Morales MI, Rico CM, Hernandez-Viezcas JA et al (2013) Toxicity assessment of cerium oxide nanoparticles in cilantro (*Coriandrum sativu*m L.) plants grown in organic soil. J Agric Food Chem 61:6224–6230

Mukherjee A, Peralta-Videa JR, Bandyopadhyay S et al (2014a) Physiological effects of nanoparticulate ZnO in green peas (*Pisum sativum* L.) cultivated in soil. Metallomics 6:132

Mukherjee A, Peralta-Videa JR, Bandyopadhyay S et al (2013) Physiological effects of nanoparticulate ZnO in green peas (*Pisum sativum* L.) cultivated in soil. Metallomics 6:132–138

Mukherjee A, Pokhrel S, Bandyopadhyay S et al (2014b) A soil mediated phyto-toxicological study of iron doped zinc oxide nanoparticles (Fe@ZnO) in green peas (*Pisum sativum* L.). Chem Eng J 258:394–401

Navarro DA, Bisson MA, Aga DS (2012) Investigating uptake of water-dispersible CdSe/ZnS quantum dot nanoparticles by *Arabidopsis thaliana* plants. J Hazard Mater 211–212:427–435

Peng C, Duan D, Xu C et al (2015) Translocation and biotransformation of CuO nanoparticles in rice (*Oryza sativa* L.) plants. Environ Pollut Barking Essex 1987(197):99–107

Prasad TNVKV, Sudhakar P, Sreenivasulu Y et al (2012) Effect of nanoscale zinc oxide particles on the germination, growth and yield of peanut. J Plant Nutr 35:905–927

Priester JH, Ge Y, Mielke RE et al (2012) Soybean susceptibility to manufactured nanomaterials with evidence for food quality and soil fertility interruption. Proc Natl Acad Sci 109:E2451–E2456

Raliya R, Tarafdar JC (2013) ZnO nanoparticle biosynthesis and its effect on phosphorous-mobilizing enzyme secretion and gum contents in clusterbean (*Cyamopsis tetragonoloba* L.). Agric Res 2:48–57

Rico CM, Barrios AC, Tan W et al (2015) Physiological and biochemical response of soil-grown barley (*Hordeum vulgare* L.) to cerium oxide nanoparticles. Environ Sci Pollut Res 22:10551–10558

Rico CM, Lee SC, Rubenecia R et al (2014) Cerium oxide nanoparticles impact yield and modify nutritional parameters in wheat (*Triticum aestivum* L.). J Agric Food Chem 62:9669–9675

Rico CM, Majumdar S, Duarte-Gardea M et al (2011) Interaction of nanoparticles with edible plants and their possible implications in the food chain. J Agric Food Chem 59:3485–3498

Rico CM, Morales MI, Barrios AC et al (2013) Effect of cerium oxide nanoparticles on the quality of rice (*Oryza sativa* L.) grains. J Agric Food Chem 61:11278–11285. doi:10.1021/jf404046v

Schwab F, Zhai G, Kern M et al (2015) Barriers, pathways and processes for uptake, translocation and accumulation of nanomaterials in plants—critical review. Nanotoxicology 10:257–278

Schwabe F, Schulin R, Limbach LK et al (2013) Influence of two types of organic matter on interaction of CeO_2 nanoparticles with plants in hydroponic culture. Chemosphere 91:512–520

Servin A, Elmer W, Mukherjee A et al (2015) A review of the use of engineered nanomaterials to suppress plant disease and enhance crop yield. J Nanoparticle Res 17:1–21

Servin AD, Morales MI, Castillo-Michel H et al (2013) Synchrotron verification of TiO_2 accumulation in cucumber fruit: a possible pathway of TiO_2 nanoparticle transfer from soil into the food chain. Environ Sci Technol 47:11592–11598

Shams G, Ranjbar M, Amiri A (2013) Effect of silver nanoparticles on concentration of silver heavy element and growth indexes in cucumber (*Cucumis sativus* L. negeen). J Nanoparticle Res 15:1–12

Shi J, Peng C, Yang Y et al (2014) Phytotoxicity and accumulation of copper oxide nanoparticles to the Cu-tolerant plant *Elsholtzia splendens*. Nanotoxicology 8:179–188

Stewart J, Hansen T, McLean JE et al (2015) Salts affect the interaction of ZnO or CuO nanoparticles with wheat. Environ Toxicol Chem 34:2116–2125

Suriyaprabha R, Karunakaran G, Yuvakkumar R et al (2012) Growth and physiological responses of maize (*Zea mays* L.) to porous silica nanoparticles in soil. J Nanopart Res 14:1294–1307
Tan X, Lin C, Fugetsu B (2009) Studies on toxicity of multi-walled carbon nanotubes on suspension rice cells. Carbon 47:3479–3487
Tarafdar JC, Raliya R, Mahawar H, Rathore I (2014) Development of zinc nanofertilizer to enhance crop production in pearl millet (*Pennisetum americanum*). Agric Res 3:257–262
Wang P, Menzies NW, Lombi E et al (2013) Fate of ZnO nanoparticles in soils and cowpea (*Vigna unguiculata*). Environ Sci Technol 47:13822–13830
Watanabe T, Misawa S, Hiradate S, Osaki M (2008) Root mucilage enhances aluminum accumulation in *Melastoma malabathricum*, an aluminum accumulator. Plant Signal Behav 3:603–605
Watson J-L, Fang T, Dimkpa CO et al (2015) The phytotoxicity of ZnO nanoparticles on wheat varies with soil properties. Biometals 28:101–112
Watson J-L, Fang T, Dimkpa CO et al (2014) The phytotoxicity of ZnO nanoparticles on wheat varies with soil properties. Biometals 28:101–112
Whiteside MD, Treseder KK, Atsatt PR (2009) The brighter side of soils: quantum dots track organic nitrogen through fungi and plants. Ecology 90:100–108
Yoon S-J, Kwak JI, Lee W-M et al (2014) Zinc oxide nanoparticles delay soybean development: a standard soil microcosm study. Ecotoxicol Environ Saf 100:131–137
Zhang P, Ma Y, Zhang Z et al (2012) Biotransformation of Ceria nanoparticles in cucumber plants. ACS Nano 6:9943–9950
Zhao L, Peralta-Videa JR, Peng B et al (2014a) Alginate modifies the physiological impact of CeO_2 nanoparticles in corn seedlings cultivated in soil. J Environ Sci 26:382–389
Zhao L, Peralta-Videa JR, Ren M et al (2012a) Transport of Zn in a sandy loam soil treated with ZnO NPs and uptake by corn plants: electron microprobe and confocal microscopy studies. Chem Eng J 184:1–8
Zhao L, Peralta-Videa JR, Rico CM et al (2014b) CeO_2 and ZnO nanoparticles change the nutritional qualities of cucumber (*Cucumis sativus*). J Agric Food Chem 62:2752–2759
Zhao L, Peralta-Videa JR, Varela-Ramirez A et al (2012b) Effect of surface coating and organic matter on the uptake of CeO_2 NPs by corn plants grown in soil: insight into the uptake mechanism. J Hazard Mater 225–226:131–138
Zhao L, Sun Y, Hernandez-Viezcas JA et al (2013) Influence of CeO_2 and ZnO nanoparticles on cucumber physiological markers and bioaccumulation of Ce and Zn: a life cycle study. J Agric Food Chem 61:11945–11951
Zhu H, Han J, Xiao JQ, Jin Y (2008) Uptake, translocation, and accumulation of manufactured iron oxide nanoparticles by pumpkin plants. J Environ Monit JEM 10:713–717
Zhu Z-J, Wang H, Yan B et al (2012) Effect of surface charge on the uptake and distribution of gold nanoparticles in four plant species. Environ Sci Technol 46:12391–12398

Chapter 21
Toxicity of Nanoparticles and Their Impact on Environment

Pankaj goyal and Rupesh Kumar Basniwal

21.1 Introduction

As nanotechnology industry is growing fast and producing various types of nanoparticles for various applications, their toxicity concerns for human and environments have evolved (Donaldson et al. 2004). Generally dimension of nanoparticles is of nanoscale, i.e., below 100 nm (De Berardis et al. 2010). Evolution of nanotechnology has a great revolutionary impact on various sectors like electronics, biotechnology, agriculture, and aerospace engineering and special impact on medical field because of their special role in drug delivery, proteins, DNA, and monoclonal antibodies (Nowrouzi et al. 2010; De Jong and Borm 2008; Lewinski et al. 2008). The majority of nanoparticles having medical applications are liposomes, polyethylene glycol, and dendrimers. Mostly toxicological studies of these magical nanoparticles are based on nanoparticle doses and time effect on humans, animals, and environment. The impact of the dose response and their threshold levels for toxicity are the basis in deciding if they are either "safe" or "dangerous." Nanoparticle large surface areas compared to their bulk particles enhance their surface-to-volume ratio and their role in surface chemistry interaction with human or plant or animal cells, which pave the way for the toxicological studies (Jefferson 2000). Prior adverse experience with asbestos (Kane and Hurt 2008) and air pollution (Dockery et al. 1993) encourages us to study nano-toxicity. Since this experience, investigations into the toxicological potential of nanomaterials have been constantly chasing the rapid growth of nanotechnology

P. goyal
Amity Institute of Microbial Technology, Amity University Uttar Pradesh, Noida, India

R.K. Basniwal (✉)
Amity Institute of Advanced Research and Studies (M&D), Amity University Uttar Pradesh, Noida, India
e-mail: rup4kumar@gmail.comm; rkbasniwal@amity.edu

M. Ghorbanpour et al. (eds.), *Nanoscience and Plant–Soil Systems*, Soil Biology 48,
DOI 10.1007/978-3-319-46835-8_21

(Stone and Donaldson 2006). Nanoparticles can be prepared from metal, nonmetal, bioceramic, and polymeric materials. These nanoparticles present threat to human life (Oberdörster et al. 2005); because of their tiny size, they can easily enter into the human body by crossing the different biological barriers and may reach to the most sensitive tissues or cells (Pourmand and Abdollahi 2012). Deep entry of quantum dots is easily possible into human tissues which may interrupt the biochemical pathways and disturb the environment of the cell (Vishwakarma et al. 2010). Scientific studies of nanoparticles on animal has shown their presence in the liver, heart, spleen, and brain in addition to the lungs and gastrointestinal tract (Hagens et al. 2007; Nemmar et al. 2002; Takenaka et al. 2001). Although our immune system is capable to counter toxicities of compounds, nanoparticles' tiny size and long half-life period can escape them from the immune system. Half-life of nanoparticles in human lungs is about 700 days, and during the metabolism, some of the nanoparticles are gathered in the liver tissues which poses threat to the respiratory system as well as to the digestive and immune systems (Oberdörster et al. 2005; Pourmand and Abdollahi 2012; Vishwakarma et al. 2010; Hagens et al. 2007; Nemmar et al. 2002; Takenaka et al. 2001; Garnett and Kallinteri 2006). As suggested by many scientists, toxicity is inversely proportional to the size of nanoparticles (Yang and Watts 2005; Donaldson et al. 2002; Mostafalou et al. 2013). It means nanoparticles have two sides of a coin: in one side, it shows unpredictable positive health outcomes, and on the other side, it shows toxicity to living cells. Safe handling of nanoparticles requires bridging the gap of knowledge, tackling the toxicity issues by using optimum dose of nanoparticles (NPs) (i.e., below the toxic level), and defining the policies or protocols on international as well as national level for the safety usage of NPs.

21.2 Nanomaterials, Nanoparticles, Nanotechnology, and Nano-toxicity

Nanomaterials/nanoparticles are materials that have less than 100 nm structural dimension. Elements like TiO_2, ZnO, multiwalled CNTs (MWCNT), single-walled CNTs (SWCNT), etc. belong to this category. Atom arrangement of nanoparticles can be classified into two forms, i.e., amorphous and crystalline form. These forms can act as a carrier for chemicals and drugs and show great impact on the living cells compared to their bulk particles due to having distinct physicochemical properties. Nanotechnology is branch of science in which we synthesize and design the nanoparticles and exploits them for their unique properties like chemical, physical, electrical, and mechanical for the particular applications. Nano-toxicology is a latest branch of toxicology in which we study about toxic effects of nanoparticles on human, animal, and environment (Donaldson et al. 2004; Oberdörster et al. 2005).

21.3 Nanoparticle Sources

Natural nanoparticles are produced in many natural processes, including photochemical reactions, volcanic outbursts, forest fires, and erosions. Along with natural NPs, human has created various nanomaterials through the combustion of coal or fuel oil (for the power generation for vehicle/airplane engines) and through the chemical manufacturing process (Rogers et al. 2005; Linak et al. 2000). Currently many engineered NPs like carbon nanotubes (SWCNT & MWCNT), titanium oxide, zinc oxide, etc. are in the market for various applications like cosmetics, sports, tires, stain-resistant clothing, sunscreen-based creams, toothpaste, food additives, etc. Quantity of NP production can be in microgram (fluorescent quantum dots for bio-imaging) or million tons (carbon black for tier generation and other purposes) per year depending on the type of applications of particular NPs.

21.4 Classification of Nanoparticles

Classification of NPs was done by different authors in different ways; some authors classified them based on their dimensionality, morphology, composition, uniformity, and agglomeration, and others classified them into metallic and nonmetallic forms. Based on dimensionality, NPs can be one dimension like thin film, two dimensions like asbestos fibers, and three dimensions like thin film with atomic-scale porosity (Seshan 2002). Parameters like flatness, sphericity, and aspect ratio are considered for morphological characteristics of NPs. Nanoparticles can be present in the form of a single constituent, double constituent, or mixture of different constituents. These NPs exist in different forms like aerosol suspensions, colloid, or in an agglomerate state due to having different chemical and electromagnetic properties. As mentioned above, nanoparticles can be classified into metallic NPs and nonmetallic NPs. Metallic NPs are aluminum oxide, gold, copper oxide, silver, zinc oxide, iron oxide, and titanium oxide, and nonmetallic NPs are carbon-based nanomaterials, silica, and polymeric materials. The metallic aluminum nanoparticles are known to be used in fuel cells, polymers, paints, coatings, textiles, and biomaterials, and they contribute 20 % among all nano-sized chemicals. Metallic gold nanoparticles are being used as drug carriers for cancer and thermal therapy due to their easy functionalization, surface modification, and binding affinity toward amine and thiol groups (Jain et al. 2012). Applications of metallic copper oxide nanoparticles are reported in semiconductors, antimicrobial reagents, heat transfer fluids, and contraceptive devices (Aruoja et al. 2009). Metallic silver nanoparticles are well known for antibacterial activity from ancient time period. Currently it has a wide application in different fields like wound dressings, coating of surgical instruments, and prostheses (Huang et al. 2006). Applications of metallic zinc oxide are also reported for paints, wave filters, UV detectors, gas sensors, sunscreens, and many other products (Huang et al. 2006,

2010). Metallic iron oxide nanoparticles are also reported for biomedical, drug delivery, and diagnostic fields. Titanium oxides are chemically inert compound, and that's why it has been widely used in textile industries and other fields. Nonmetallic biodegradable polymeric nanomaterials show good potential for targeted drug delivery in cancer chemotherapy due their encapsulation property (Panyam and Labhasetwar 2003). Nonmetallic silica nanoparticles constitute 8 % in air among all airborne NPs in the form of silicon dioxide (Balduzzi et al. 2004). Due to easy functionalization, they can be used for drug delivery systems (Wilczewska et al. 2012). According to Huczko (2001), the most attractive and widely used nanomaterials are nonmetallic carbon-based, such as carbon nanotubes, fullerenes, and single- and multiwalled carbon nanotubes.

21.5 Nanoparticle Characterization

In order to evaluate toxicological impacts of nanoparticles on organism, they should be fully understand and characterize (Burleson et al. 2004); otherwise, pseudo-toxicity results (due to the presence of impurities and other compounds) will be produced (Sayes and Warheit 2009). It means purity of nanoparticles is also an important factor in the evaluation of toxicity. Other parameters like size, shape, and origin of nanoparticles can influence their toxicity.

21.6 Nanoparticles and Environment

The ecotoxicology and chemistry of manufactured nanoparticles have been mentioned by Handy et al. in 2008. Nanoparticles produced by either human or in natural way ultimately end up into air, water, and soil pathways. Along the cycle of NP aggregates, they can change their particle charges resulting into different surface properties of NPs. Results of these impacts on water ecosystems as well as on soil system have been mentioned by Quik et al. (2010) and Kiser et al. (2010). Popular article of Nowack about the "behavior and effects of nanoparticles in the environment" also gives important insightful details about NPs' interaction with the surrounding environment (Nowack 2009). According to Wiesner et al. (2008), nanomaterials are contaminants to environment.

21.7 In Vitro Nano-toxicology

Generally the cell is going to treat/internalize foreign particles like bacteria, viruses (natural nanoparticles), or engineered nanoparticles from their immediate environment through the two well-known basic mechanisms, i.e., phagocytosis and endocytosis. Phagocytosis requires additional recognition step compared to endocytosis. In endocytosis transmembrane transportation of liquids or molecules takes place. Endocytosis is a convenient way for spreading viruses and engineered nanoparticles from cell to cell (Garnett and Kallinteri 2006; Yacobi et al. 2010; Greulich et al. 2011), and therefore once they have entered into the body, it translocated to different parts of the body. The dose of nanoparticles is a very important factor to evaluate their in vitro toxicity. Exposure dose and the internalize quantities of NPs through the cell and their ultimate effect on the cells give the idea about their nano-toxicity. To measure the nano-toxicity, various techniques are available in the market and in scientific world (Elsaesser et al. 2010). Damage to cell can be either chemical or physical through the nanoparticles (Nel et al. 2009). Chemical damages occur through the various processes like production of free radicals or reactive oxygen species (ROS) (Nel et al. 2006; Gou et al. 2010; Li et al. 2003; Uchino et al. 2002), dissolution and release of toxic ions (Xia et al. 2008), destruction of ion exchange system of cell membrane (Auffan et al. 2008), catalyzation of oxidative damage (Foley et al. 2002), initiation of lipid peroxidation (Kamat et al. 2000), and damage to surfactant properties (Cottingham et al. 2002). Physical damage to cells through the NPs takes place with various mechanisms like disruption of membranes and membrane activity (Leroueil et al. 2008; Hussain et al. 2005; Navarro et al. 2008), generate barriers for cell transportation, initiate protein aggregation or folding, and damage to the DNA of the cell (Ovrevik et al. 2004; Hauck et al. 2008; Billsten et al. 1997; Chen and Von Mike 2005) The quantity and impact of damage to cells depends on the size and surface properties of NPs (Walczyk et al. 2010). Both chemical and physical interactions lead to internalization of NPs and show lethal effect on the living cell. So in vitro studies are mandatory to evaluate toxicities of NPs which further generate the basis for in vivo studies.

21.8 In Vivo Nano-toxicology

In vitro nano-toxicology studies are incomplete without knowing their direct or indirect effect on animal models (Rivera et al. 2010). In vivo studies help us in finding the mechanisms, pathways, and entry routes of nanoparticles in humans as well as in animals or plants (Seaton et al. 2010). Exposure of NPs is very critical to such type of workers who are regularly engaged in producing nanoparticles or working on nano-medicine industries. Industries are producing various types of NPs at a very high rate, which present high risk to their workers and stimulate us to establish some relationship between nanoparticle doses and their in vivo effect for the safe use of nanoparticles. Several nanomaterials currently fall into this category which poses threat to living organisms like titanium dioxide (Mueller and Nowack 2008), cerium dioxide (Park et al. 2008), silicon dioxide (Napierska et al. 2010), zinc oxide (Osmond and McCall 2010), silver (Ahamed et al. 2010), and carbon nanotubes (Aitken et al. 2006). Nowadays application of gold NPs has been also increased tremendously due to their outstanding bio-conjugation properties for nano-therapeutic applications (Sperling et al. 2008). According to Myllynen et al. (2008), gold nanoparticles can cross the maternofetal barrier and present possible threat to the embryo. Fullerenes are already reported for its fatal effect to mouse embryos (Tsuchiya et al. 1996). The most prominent entry routes for nanoparticles are by air or skin or via ingestion (Stern and McNeil 2008). Translocations of such NPs within the body are also reported by various authors but with limited scopes (Oberdorster 2010; Elder et al. 2009). NP uptake by the skin or ingested by the body or entry through the airway has impact on the respiratory system and on the lungs; their translocation to different body parts like the liver, spleen, heart, and other susceptible organs is also reported by various scientists (Donaldson et al. 2002; Geiser 2010; Choi et al. 2010). Protein misfolding and protein fibrillation (Linse et al. 2007) induced by nanoparticles and their entry into alveolar epithelial cells and to the olfactory bulb (Liu et al. 2009) also present potential threat to the nervous system and generate neurotoxicity (Elder et al. 2006; Oberdorster et al. 2009). Titanium dioxide nanoparticles are often used in sunscreen products and may gain access through hair follicles or wounds and lesions (Crosera et al. 2009), but their lethal impact on the skin and their translocation are little reported in the literature, for indeed conclusion on further research is required for the same. Fullerenes (Sperling et al. 2008) and quantum dots (Stern and McNeil 2008) are able to enter into the dermis of the skin (Rouse et al. 2006; Ryman et al. 2006), and their ultimate impact on the skin depends on their size and type of the surface coatings (Ryman et al. 2007). Although low concentration of NPs has been reported for various internal body parts like the liver, spleen, heart, and brain (Hillyer and Albrecht 2001; Nemmar et al. 2002; Ji et al. 2006; Oberdorster et al. 2002) after entry through the blood or other ways, their bioaccumulation in certain organs (Borm et al. 2006) present threat to the body. Nature has given us very clever immune system and body cleaning system to counter various types of NPs to reduce the toxicity of nanoparticles. Recently clearance rates of 40–50 % (He et al. 2011) of NPs have been reported but without proper sequential approach with functional deep research, and full understanding of their mechanism in vitro as well as in vivo is not enough to estimate toxicities of NPs (Longmire et al. 2008).

21.9 Application of Nanoparticles

Although we have seen many applications of nanomaterials in our daily life, comprehensive detail of those applications is still untapped. As nanotechnology field is growing very fast, we need to understand their roles in improving our quality of life through nano-medicine, electronics and microelectronics, synthetic rubber, UV absorbers for sunscreens, optical fiber cladding, cosmetics, coating, ultrafine polishing compounds, synthetic bone and adhesives, and other nano-fields like fabrics and their treatments, filtration, dental materials, surface disinfectants, fuel additives, hazardous chemical neutralizers, automotive components, electronics, scientific instruments, sports equipment, flat panel displays, drug delivery systems, and pharmaceutics. Full-fledged and safe use of nanoparticles should be based on continuous evaluation of risks and opportunities of nanotechnology for a particular application (Donaldson et al. 2004). Ecological and medical applications of NPs have been increased dramatically from the last few decades. Along with the USA, 60 countries around the world have already taken initiative for nanotechnology (Roco 2005). The following are the main applications of nanotechnology:

1. Reduced size of carbon nanotubes (CNT) has a great advantage in the field of microelectronics over other conventional-doped semiconductor crystals ranging from metallic to semiconduction (Jacoby 2002) and superconduction (Buzea and Robbie 2005).
2. Resolution of television can be enhanced by using nanostructured materials, e.g., CNT, used for low-voltage field-emission displays (Carey 2003).
3. Production of high-energy density batteries, e.g., lithium-ion-based batteries (Liu et al. 2006)
4. Nano-sized clay composite is very useful in decorating car exterior because of their light weight and resistance to scratches.
5. Single-walled carbon nanotubes (SWCNTs) have been used for molecular recognition and in atomic-force microscopy imaging of antibodies, DNA, and other biomaterials (Hafner et al. 2001).
6. Recently nanofiber scaffolds have been used to regenerate central nervous system (CNS) cells (Ellis-Behnke et al. 2006).
7. Many nanomaterials or nano-powders show antimicrobial activity (Bosi et al. 2003; Koper et al. 2002), e.g., silver and titanium dioxide are used for coatings of surgical masks (Li et al. 2006).
8. Nanoparticles can be used for bio-separation (Jirage et al. 1997; Martin and Kohli 2003), drug delivery (Uhrich et al. 1999), gene transfection (Maïté et al. 2000; Kneuer et al. 2000), medical imaging (Harisinghani et al. 2003), and, recently, nasal vaccination. The paint of titanium oxide and calcium carbonate nano-composite with silicon-based polymer (polysiloxane) can be used for absorbing the nitrogen oxide pollutant gases, emitted through vehicle exhausts (Hogan 2004).
9. Nanoparticles like iron can be used for water remediation in which they convert toxic substances into less toxic substances (He and Zhao 2005).
10. NPs are already used for textile industries because of their known antimicrobial activities.
11. Applications of NPs for biosensor, solar cells, and supercapacitors are already known (Baibarac and Gomez-Romero 2006).
12. Tungsten carbide nanoparticles are used for making cutting tools due to their built-in hardness (Stiglich et al. 1996).
13. Nanospheres of inorganic materials can be used for lubricants in many other applications (Fleischer et al. 2003).

21.10 Conclusion and Future Direction

Human exposure to nanoparticles from natural or unnatural (i.e., anthropogenic) sources has occurred since ancient times. After the discovery of combustion engines, rapid development of industries and improper burning of fuel in rural/urban areas for fulfilling energy requirement are continuously adding toxic gases and nanoparticles to our planet. Nowadays these nanoparticles are designed for a particular application due to their distinct properties and more advantages over the bulk materials. Productions of some nanoparticles are very high in concentration presenting the threat of human exposure both intentionally and unintentionally. These nanoparticles can have negative impact on public health and surrounding environments if they are not safely manufactured, handled, and disposed or recycled. Currently, the toxicity of engineered NPs is assessed with a number of approaches. Among them, the most beneficial one in terms of cost and time saving is the in vitro studies, but these studies are incomplete without in vivo studies. Consistent and reproducible results for in vitro or in vivo studies are still challenging for us, but we can achieve it by adopting multidisciplinary approaches by including different field scientists like fabrication scientists, chemists, toxicologists, epidemiologists, environmental scientists, industrialists, and even policy makers. After entering into the human body through various routes, nanoparticles translocate to distant sites within the body even at very low concentration. It means we need to develop a regulatory framework based on scientific facts which will limit human exposure to unwanted engineered nanomaterials in the environment to safe levels. If we see the other side of the coin, invented nanoparticles have a great role in the nano-medicine and in improving our standards of living. So it requires balanced framework between the therapeutic benefit and the potential risk of nanoparticles. We conclude that the development of nanotechnology and the study of nano-toxicology have increased our awareness about nanoparticle applications in different sectors and their potential threat to human as well as to environment. With tremendous applications of nanoparticles, we can make bright future of nanotechnology with better understanding about their toxicity and by covering all aspects like manufacturing, industrial, commercial, recycling, and policy framework to handle them safely and use them environmentally friendly.

References

Ahamed M, Alsalhi MS, Siddiqui MK (2010) Silver nanoparticle applications and human health. Clin Chim Acta 411:1841–1848

Aitken RJ, Chaudhry MQ, Boxall AB, Hull M (2006) Manufacture and use of nanomaterials: current status in the UK and global trends. Occup Med (Lond) 56:300–306

Aruoja V, Dubourguier HC, Kasemets K, Kahru A (2009) Toxicity of nanoparticles of CuO, ZnO and TiO2 to microalgae Pseudokirchneriella subcapitata. Sci Total Environ 407:1461–1468

Auffan M, Achouak W, Rose JR, Roncato MA, Chanéac C, Waite DT, Masion A, Woicik JC, Wiesner MR, Bottero JY (2008) Relation between the redox state of iron-based nanoparticles and their cytotoxicity toward Escherichia coli. Environ Sci Technol 42:6730–6735

Baibarac M, Gomez-Romero P (2006) Nanocomposites based on conducting polymers and carbon nanotubes: from fancy materials to functional applications. J Nanosci Nanotechnol 6:289–302

Balduzzi M, Diociaiuti M, De Berardis B, Paradisi S, Paoletti L (2004) In vitro effects on macrophages induced by noncytotoxic doses of silica particles possibly relevant to ambient exposure. Environ Res 96:62–71

Billsten P, Freskgard PO, Carlsson U, Jonsson BH, Elwing H (1997) Adsorption to silica nanoparticles of human carbonic anhydrase II and truncated forms induce a molten-globule-like structure. FEBS Lett 402:67–72

Borm PJ, Robbins D, Haubold S, Kuhlbusch T, Fissan H, Donaldson K, Schins R, Stone V, Kreyling W, Lademann J, Krutmann J, Warheit D, Oberdorster E (2006) The potential risks of nanomaterials: a review carried out for ECETOC. Part Fibre Toxicol 3:11

Bosi S, da Ros T, Spalluto G, Prato M (2003) Fullerene derivatives: an attractive tool for biological applications. Eur J Med Chem 38:913–923

Burleson DJ, Driessen MD, Penn RL (2004) On the characterization of environmental nanoparticles. J Environ Sci Health A Tox Hazard Subst Environ Eng 39:2707–2753

Buzea C, Robbie K (2005) Assembling the puzzle of superconducting elements. Supercond Sci Technol 18:R1–R8

Carey JD (2003) Engineering the next generation of large-area displays: prospects and pitfalls Philos. Transact A Math Phys Eng Sci 361:2891–2907

Chen M, Von Mikecz A (2005) Formation of nucleoplasmic protein aggregates impairs nuclear function in response to SiO2 nanoparticles. Exp Cell Res 305:51–62

Choi HS, Ashitate Y, Lee JH, Kim SH, Matsui A, Bawendi I, Semmler-Behnke M, Frangioni JV, Tsuda A (2010) Rapid translocation of nanoparticles from the lung airspaces to the body. Nat Biotechnol 28:1300–1303

Cottingham MG, Hollinshead MS, Vaux DJ (2002) Amyloid fibril formation by a synthetic peptide from a region of human acetylcholinesterase that is homologous to the Alzheimer's amyloid-beta peptide. Biochemistry 41:13539–13547

Crosera M, Bovenzi M, Maina G, Adami G, Zanette C, Florio C, Filon LF (2009) Nanoparticle dermal absorption and toxicity: a review of the literature. Int Arch Occup Environ Health 82:1043–1055

De Berardis B, Civitelli G, Condello M, Lista P, Pozzi R, Arancia G, Meschini S (2010) Exposure to ZnO nanoparticles induces oxidative stress and cytotoxicity in human colon carcinoma cells. Toxicol Appl Pharmacol 246:116–127

De Jong WH, Borm PJ (2008) Drug delivery and nanoparticles: applications and hazards. Int J Nanomedicine 3:133–149

Dockery DW, Pope CA, Xu X, Spengler JD, Ware JH, Fay ME, Ferris BG, Speizer FE (1993) An association between air pollution and mortality in six U.S. cities. N Engl J Med 329:1753–1759

Donaldson K, Stone V, Tran C, Kreyling W, Borm PJA (2004) Nanotoxicology. Occup Environ Med 61:727–728

Donaldson K, Brown D, Clouter A, Duffin R, MacNee W, Renwick L, Tran L, Stone V (2002) The pulmonary toxicology of ultrafine particles. J Aerosol Med 15:213–220

Elder A, Gelein R, Silva V, Feikert T, Opanashuk L, Carter J, Potter R, Maynard A, Ito Y, Finkelstein J, Oberdorster G (2006) Translocation of inhaled ultrafine manganese oxide particles to the central nervous system. Environ Health Perspect 114:1172–1178

Elder A, Lynch I, Grieger K, Chan-Remillard S, Gatti A, Gnewuch H, Kenawy E, Korenstein R, Kuhlbusch T, Linker F, Matias S, Monteiro-Riviere N, Pinto VRS, Rudnitsky R, Savolainen K, Shvedova A (2009) Human health risks of engineered nanomaterials. In: Linkov I, Steevens J (eds) Nanomaterials: risks and benefits, vol 3. Springer, Netherlands, p 29

Ellis-Behnke RG, Liang YX, You SW, Tay DK, Zhang S, So KF, Schneider GE (2006) Nano neuro knitting: peptide nanofibers scaffold for brain repair and axon regeneration with functional return of vision. Proc Natl Acad Sci USA 103:5054–5059
Elsaesser A, Taylor A, Yanes GS, De Mc Kerr G, Kim EM, O'Hare E, Howard CV (2010) Quantification of nanoparticle uptake by cells using microscopical and analytical techniques. Nanomedicine (Lond) 5:1447–1457
Fleischer N, Genut M, Rapoport L, Tenne R (2003) New nanotechnology solid lubricants for superior dry lubrication European Space Agency (Special Publication) ESA SP 524:65–66
Foley S, Crowley C, Smaihi M, Bonfils C, Erlanger BF, Seta P, Larroque C (2002) Cellular localisation of a water-soluble fullerene derivative. Biochem Biophys Res Commun 294:116–119
Garnett MC, Kallinteri P (2006) Nanomedicines and nanotoxicology: some physiological principles. Occup Med (Lond) 56:307–311
Geiser M (2010) Update on macrophage clearance of inhaled micro- and nanoparticles. J Aerosol Med Pulm Drug Deliv 23:207–217
Gou N, Onnis-Hayden A, Gu AZ (2010) Mechanistic toxicity assessment of nanomaterials by whole-cell-array stress genes expression analysis. Environ Sci Technol 44:5964–5970
Greulich C, Diendorf J, Simon T, Eggeler G, Epple M, Koller M (2011) Uptake and intracellular distribution of silver nanoparticles in human mesenchymal stem cells. Acta Biomater 7:347–354
Hafner JH, Cheung CL, Woolley AT, Lieber CM (2001) Structural and functional imaging with carbon nanotube AFM probes. Prog Biophys Mol Biol 77:73–110
Hagens WI, Oomen AG, de Jong WH, Casse FR, Sips AJ (2007) What do we need to know about the kinetic properties of nanoparticles in the body? Regul Toxicol Pharmacol 49:217–229
Handy RD, Von der Kammer F, Lead JR, Hassellov M, Owen R, Crane M (2008) The ecotoxicology and chemistry of manufactured nanoparticles. Ecotoxicology 17:287–314
Harisinghani MG, Barentsz J, Hahn PF, Deserno WM, Tabatabaei S, van de Kaa CH, de la Rosette J, Weissleder R (2003) Noninvasive detection of clinically occult lymph-node metastases in prostate cancer. N Engl J Med 19:2491–2499
Hauck TS, Ghazani AA, Chan WCW (2008) Assessing the effect of surface chemistry on gold nanorod uptake, toxicity, and gene expression in mammalian cells. Small 4:153–159
He F, Zhao DY (2005) Preparation and characterization of a new class of starch-stabilized bimetallic nanoparticles for degradation of chlorinated hydrocarbons in water! Environ Sci Technol 39:3314–3320
He Q, Zhang Z, Gao F, Li Y, Shi J (2011) In vivo biodistribution and urinary excretion of mesoporous silica nanoparticles: effects of particle size and PEGylation. Small 7:271–280
Hillyer JF, Albrecht RM (2001) Gastrointestinal persorption and tissue distribution of differently sized colloidal gold nanoparticles. J Pharm Sci 90:1927–1936
Hogan J (2004) Smog-busting paint soaks up noxious gases. New Scientist. http://www.newscientist.com/
Huang CC, Aronstam RS, Chen DR, Huang YW (2010) Oxidative stress, calcium homeostasis, and altered gene expression in human lung epithelial cells exposed to ZnO nanoparticles. Toxicol In Vitro 24:45–55
Huang GG, Wang CT, Tang HT, Huang YS, Yang J (2006) ZnO nanoparticle-modified infrared internal reflection elements for selective detection of volatile organic compounds. Anal Chem 78:2397–2404
Huczko A (2001) Synthesis of aligned carbon nanotubes. J Appl Phys 74:617–638
Hussain SM, Hess KL, Gearhart JM, Geiss KT, Schlager JJ (2005) In vitro toxicity of nanoparticles in BRL 3A rat liver cells. Toxicol In Vitro 19:975–983
Jacoby M (2002) Nanoscale electronics. Chem Eng News 80:38–43
Jain S, Hirst DG, Osullivan JM (2012) Gold nanoparticles as novel agents for cancer therapy. Br J Radiol 85:101–113

Jefferson DA (2000) The surface activity of ultrafine particles. Philos Trans R Soc Lond, Ser A: Math Phys Eng Sci 358:2683–2692

Ji ZQ, Sun H, Wang H, Xie Q, Liu Y, Wang Z (2006) Biodistribution and tumor uptake of C60 (OH) x in mice. J Nanopart Res 8:53–63

Jirage KB, Hulteen JC, Martin CR (1997) Nanotubule based molecular-filtration membranes. Science 278:655–658

Kamat JP, Devasagayam TP, Priyadarsini KI, Mohan H (2000) Reactive oxygen species mediated membrane damage induced by fullerene derivatives and its possible biological implications. Toxicology 155:55–61

Kane AB, Hurt RH (2008) Nanotoxicology: the asbestos analogy revisited. Nat Nanotechnol 3:378–379

Kiser MA, Ryu H, Jang H, Hristovski K, Westerhoff P (2010) Biosorption of nanoparticles to heterotrophic wastewater biomass. Water Res 44:4105–4114

Kneuer C, Sameti M, Bakowsky U, Schiestel T, Schirra H, Schmidt H, Lehr CM (2000) A nonviral DNA delivery system based on surface modified silica-nanoparticles can efficiently transfect cells in vitro. Bioconj Chem 11:926–932

Koper OB, Klabunde JS, Marchin GL, Klabunde KJ, Stoimenov P, Bohra L (2002) Nanoscale powders and formulations with biocidal activity toward spores and vegetative cells of bacillus species, viruses, and toxins. Curr Microbiol 44:49–55

Leroueil PR, Berry SA, Duthie K, Han G, Rotello VM, McNerny DQ, Baker JR, Orr BG, Banaszak Holl MM (2008) Wide varieties of cationic nanoparticles induce defects in supported lipid bilayers. Nano Lett 8:420–424

Lewinski N, Colvin V, Drezek R (2008) Cytotoxicity of nanoparticles. Small 4:26–49

Li N, Sioutas C, Cho A, Schmitz D, Misra C, Sempf J, Wang M, Oberley T, Froines J, Nel A (2003) Ultrafine particulate pollutants induce oxidative stress and mitochondrial damage. Environ Health Perspect 111:455–460

Li Y, Leung P, Yao L, Song QW, Newton E (2006) Antimicrobial effect of surgical masks coated with nanoparticles. J Hosp Infect 62:58–63

Linak WP, Miller CA, Wendt JO (2000) Comparison of particle size distribution and elemental partitioning from the combustion of pulverized coal and residual fuel oil. J Waste Manag Assoc 50:1532–1544

Linse S, Cabaleiro-Lago C, Xue WF, Lynch I, Lindman S, Thulin E, Radford SE, Dawson KA (2007) Nucleation of protein fibrillation by nanoparticles. Proc Natl Acad Sci USA 104:8691–8696

Liu HK, Wang GX, Guo Z, Wang J, Konstantinov K (2006) Nanomaterials for lithium-ion rechargeable batteries. J Nanosci Nanotechnol 6:1–15

Liu Y, Gao Y, Zhang L, Wang T, Wang J, Jiao F, Li W, Li Y, Li B, Chai Z, Wu G, Chen C (2009) Potential health impact on mice after nasal instillation of nano-sized copper particles and their translocation in mice. J Nanosci Nanotechnol 9:6335–6343

Longmire M, Choyke PL, Kobayashi H (2008) Clearance properties of nano-sized particles and molecules as imaging agents: considerations and caveats. Nanomedicine 3:703–717

Maïté L, Carlesso N, Tung CH, Tang XW, Cory D, Scadden DT, Weissleder R (2000) Tat peptide derivatized magnetic nanoparticles allow in vivo tracking and recovery of progenitor cells. Nat Biotechnol 18:410–414

Martin CR, Kohli P (2003) The emerging field of nanotube biotechnology. Nat Rev Drug Discovery 2:29–37

Mostafalou S, Mohammadi H, Ramazani A, Abdollahi M (2013) Different biokinetics of nanomedicines linking to their toxicity; an overview. Daru J Pharm Sci 21:14

Mueller NC, Nowack B (2008) Exposure modeling of engineered nanoparticles in the environment. Environ Sci Technol 42:4447–4453

Myllynen PK, Loughran MJ, Howard CV, Sormunen R, Walsh AA, Vahakangas KH (2008) Kinetics of gold nanoparticles in the human placenta. Reprod Toxicol 26:130–137

Napierska D, Thomassen LC, Lison D, Martens JA, Hoet PH (2010) The nanosilica hazard: another variable entity. Part Fibre Toxicol 7:39
Navarro E, Baun A, Behra R, Hartmann N, Filser J, Miao AJ, Quigg A, Santschi P, Sigg L (2008) Environmental behavior and ecotoxicity of engineered nanoparticles to algae, plants, and fungi. Ecotoxicology 17:372–386
Nel A, Xia T, Mädler L, Li N (2006) Toxic potential of materials at the nanolevel. Science 311:622–627
Nel AE, Madler L, Velegol D, Xia T, Hoek EM, Somasundaran P, Klaessig F, Castranova V, Thompson M (2009) Understanding biophysicochemical interactions at the nano-bio interface. Nat Mater 8:543–557
Nemmar A, Hoet PH, Vanquickenborne B, Dinsdale D, Thomeer M, Hoylaerts MF, Vanbilloen H, Mortelmans L, Nemery B (2002) Passage of inhaled particles into the blood circulation in humans. Circulation 105:411–414
Nowack B (2009) The behavior and effects of nanoparticles in the environment. Environ Pollut 157:1063–1064
Nowrouzi A, Meghrazi K, Golmohammadi T, Golestani A, Ahmadian S, Shafiezadeh M, Shajary Z, Khaghani S, Amiri AN (2010) Cytotoxicity of subtoxic AgNP in human hepatoma cell line (HepG2) after long-term exposure. Iranian Biomed J 14:23–32
Oberdörster G, Oberdörster E, Oberdörster J (2005) Nanotoxicology: An emerging discipline evolving from studies of ultrafine particles. Environ Health Perspect 113:823–839
Oberdorster G (2010) Safety assessment for nanotechnology and nanomedicine: concepts of nanotoxicology. J Intern Med 267:89–105
Oberdorster G, Elder A, Rinderknecht A (2009) Nanoparticles and the brain: cause for concern? J Nanosci Nanotechnol 9:4996–5007
Oberdorster G, Sharp Z, Atudorei V, Elder A, Gelein R, Lunts A, Kreyling W, Cox C (2002) Extrapulmonary translocation of ultrafine carbon particles following whole-body inhalation exposure of rats. J Toxicol Environ Health A 65:1531–1543
Osmond MJ, McCall MJ (2010) Zinc oxide nanoparticles in modern sunscreens: an analysis of potential exposure and hazard. Nanotoxicology 4:15–41
Ovrevik J, Låg M, Schwarze P, Refsnes M (2004) p38 and Src-ERK1/2 pathways regulate crystalline silica-induced chemokine release in pulmonary epithelial cells. Toxicol Sci 81:480–490
Panyam J, Labhasetwar V (2003) Biodegradable nano-particles for drug and gene delivery to cells and tissue. Adv Drug Del Rev 55:329–347
Park B, Donaldson K, Duffin R, Tran L, Kelly F, Mudway I, Morin JP, Guest R, Jenkinson P, Samaras Z, Giannouli M, Kouridis H, Martin P (2008) Hazard and risk assessment of a nanoparticulate cerium oxide-based diesel fuel additive—a case study. Inhal Toxicol 20:547–566
Pourmand A, Abdollahi M (2012) Current opinion on nanotoxicology. Daru J Pharm Sci 20:95
Quik JT, Lynch I, Van Hoecke K, Miermans CJ, De Schamphelaere KA, Janssen CR, Dawson KA, Stuart MA, Van De Meent D (2010) Effect of natural organic matter on cerium dioxide nanoparticles settling in model fresh water. Chemosphere 81:711–715
Rivera GP, Oberdorster G, Elder A, Puntes V, Parak WJ (2010) Correlating physicochemical with toxicological properties of nanoparticles: the present and the future. ACS Nano 4:5527–5531
Roco MC (2005) The emergence and policy implications of converging new technologies integrated from nanoscale. J Nanopart Res 7:129–143
Rogers F, Arnott P, Zielinska B, Sagebiel J, Kelly KE, Wagner D, Lighty JS, Sarofim AF (2005) Realtime measurements of jet aircraft engine exhaust. J Air Waste Manag Assoc 55:583–593
Rouse JG, Yang J, Ryman-Rasmussen JP, Barron AR, Monteiro-Riviere NA (2006) Effects of mechanical flexion on the penetration of fullerene amino acid derivatized peptide nanoparticles through skin. Nano Lett 7:155–160

Ryman-Rasmussen JP, Riviere JE, Monteiro-Riviere NA (2007) Surface coatings determine cytotoxicity and irritation potential of quantum dot nanoparticles in epidermal keratinocytes. J Invest Dermatol 127:143–153
Ryman-Rasmussen JP, Riviere JE, Monteiro-Riviere NA (2006) Penetration of intact skin by quantum dots with diverse physicochemical properties. Toxicol Sci 91:159–165
Sayes CM, Warheit DB (2009) Characterization of nanomaterials for toxicity assessment. Wiley Interdiscip Rev Nanomed Nanobiotechnol 1:660–670
Seaton A, Tran L, Aitken R, Donaldson K (2010) Nanoparticles, human health hazard and regulation. J R Soc Interface 7:S119–S129
Seshan K (ed) (2002) Handbook of thin-film deposition processes and techniques—principles, methods, equipment and applications. William Andrew, Noyes, pp 1–657
Sperling RA, Rivera GP, Zhang F, Zanella M, Parak WJ (2008) Biological applications of gold nanoparticles. Chem Soc Rev 37:1896–1908
Stern ST, McNeil SE (2008) Nanotechnology safety concerns revisited. Toxicol Sci 101:4–21
Stiglich JJ, Yu CC, Sudarshan TS (1996) Synthesis of nano WC/Co for tools and dies. Proc 1995 3 Int Conf Tungsten Refract Met. 229–236
Stone V, Donaldson K (2006) Nanotoxicology: signs of stress. Nat Nanotechnol 1:23–24
Takenaka S, Karg E, Roth C, Schulz H, Ziesenis A, Heinzmann U, Schramel P, Heyder J (2001) Pulmonary and systemic distribution of inhaled ultrafine silver particles in rats. Environ Health Perspect 109:547–551
Tsuchiya T, Oguri I, Yamakoshi YN, Miyata N (1996) Novel harmful effects of [60]fullerene on mouse embryos in vitro and in vivo. FEBS Lett 393:139–145
Uchino T, Tokunaga H, Ando M, Utsumi H (2002) Quantitative determination of OH radical generation and its cytotoxicity induced by TiO2-UVA treatment. Toxicol Vitr 16:629–635
Uhrich KE, Cannizzaro SM, Langer RS, Shakeshelf KM (1999) Polymeric systems for controlled drug release. Chem Rev 99:3181–3198
Vishwakarma V, Samal SS, Manoharan N (2010) Safety and risk associated with nanoparticles-a review. J Miner Mater Character Eng 9:455–459
Walczyk D, Bombelli FB, Monopoli MP, Lynch I, Dawson KA (2010) What the cell "sees" in bionanoscience. J Am Chem Soc 132:5761–5768
Wiesner MR, Hotze EM, Brant JA, Espinasse B (2008) Nanomaterials as possible contaminants: the fullerene example. Water Sci Technol 57:305–310
Wilczewska AZ, Niemirowicz K, Markiewicz KH, Car H (2012) Nanoparticles as drug delivery systems. Pharmacol Rep 64:1020–1037
Xia T, Kovochich M, Liong M, Mädler L, Gilbert B, Shi H, Yeh JI, Zink JI, Nel AE (2008) Comparison of the mechanism of toxicity of zinc oxide and cerium oxide nanoparticles based on dissolution and oxidative stress properties. ACS Nano 2:2121–2134
Yacobi NR, Malmstadt N, Fazlollahi F, DeMaio L, Marchelletta R, Hamm-Alvarez SF, Borok Z, Kim KJ, Crandall ED (2010) Mechanisms of alveolar epithelial translocation of a defined population of nanoparticles. Am J Respir Cell Mol Biol 42:604–614
Yang L, Watts DJ (2005) Particle surface characteristics may play an important role in phytotoxicity of alumina nanoparticles. Toxicol Lett 158:122–132

Erratum to: Nanoscience and Plant–Soil Systems

Mansour Ghorbanpour, Manika Khanuja, and Ajit Varma

M. Ghorbanpour et al. (eds.), *Nanoscience and Plant–Soil Systems*,
Soil Biology 48, DOI 10.1007/978-3-319-46835-8

In the book titled "Nanoscience and Plant–Soil Systems" the affiliation of the editor Prof. "Manika Khanuja" in the copyright page of FM is incorrect and should read as follows:

Manika Khanuja
Centre for Nanoscience and Nanotechnology
Jamia Millia Islamia
New Delhi

The updated original online version for this book can be found at
http://dx.doi.org/10.1007/978-3-319-46835-8

M. Ghorbanpour et al. (eds.), *Nanoscience and Plant–Soil Systems*, Soil Biology 48,
DOI 10.1007/978-3-319-46835-8_22

Index

M. Ghorbanpour et al. (eds.), *Nanoscience and Plant–Soil Systems*, Soil Biology 48,
DOI 10.1007/978-3-319-46835-8

MIX
Papier aus verantwortungsvollen Quellen
Paper from responsible sources
FSC® C105338

If you have any concerns about our products,
you can contact us on
ProductSafety@springernature.com

In case Publisher is established outside the EU,
the EU authorized representative is:
Springer Nature Customer Service Center GmbH
Europaplatz 3, 69115 Heidelberg, Germany

Printed by Libri Plureos GmbH
in Hamburg, Germany